International Conference on Advanced Material Research and Processing Technology (AMRPT2019)

IOP Conference Series: Materials Science and Engineering
Volume 611

Wuhan, China
19 – 21 July 2019

ISBN: 978-1-5108-9819-6
ISSN: 1757-8981

Printed from e-media with permission by:

Curran Associates, Inc.
57 Morehouse Lane
Red Hook, NY 12571

Some format issues inherent in the e-media version may also appear in this print version.

Institute of Physics
Dirac House, Temple Back
Bristol BS1 6BE UK

Phone: 44 1 17 929 7481
Fax: 44 1 17 920 0979

techtracking@iop.org

Additional copies of this publication are available from:

Curran Associates, Inc.
57 Morehouse Lane
Red Hook, NY 12571 USA
Phone: 845-758-0400
Fax: 845-758-2633
Email: curran@proceedings.com
Web: www.proceedings.com

Table of contents

Volume 611

2019 International Conference on Advanced Material Research and Processing Technology (AMRPT2019)

19–21 July 2019, Wuhan, China

Accepted papers received: 8 August 2019
Published online: 24 October 2019

Measurements and Modeling of Materials Properties and Materials Behavior

Materials for Biomedical and Chemical Technologies and Applications

Manufacturing and Equipment Technologies, Mechanical Engineering

IOP Conf. Series: Materials Science and Engineering **611** (2019) 011001 doi:10.1088/1757-899X/611/1/011001

Preface

It's our honor to announce that 2019 International Conference on Advanced Material Research and Processing Technology (AMRPT2019) will be held on July 19-21, 2019, in Wuhan, China. The goal of the conference is to provide a platform for distinguished researchers and scholars from academia and industry working in the field of material science, mechanical and manufacturing engineering to share their latest findings and discuss the faced challenges and explore future research directions. AMRPT2019 is well-prepared and welcome all contributors who share the same interest in Advanced Material Research and Processing Technology.

In order to guarantee the high quality of the whole proceedings, we invited about 80 related experts to review all the submitted papers. Each paper must be pre-screened for conformity with submission guidelines and for overall appropriateness and then peer reviewed by at least 2-3 reviewers. Papers should be originality and quality. They should not have been published elsewhere. Plagiarism in any form is not allowed. iThenticate Software was also used for plagiarism detection to ensure the originality of AMRPT2019 proceedings.

Based on the reviewing of submissions, only 90 papers have been selected as regular papers with the acceptance rate at about 35.3%. AMRPT2019 covered the topics as follows:

1. Steel, Metals and Alloys Properties and Processing Technologies
2. Measurements and Modeling of Materials Properties and Materials Behavior
3. Composites, Nanomaterial, Polymer Materials Properties and Technologies
4. Applied Mechanics and Civil Engineering
5. Materials for Biomedical and Chemical Technologies and Applications
6. Manufacturing and Equipment Technologies, Mechanical Engineering
7. Microelectronics, Power Electronics and Embedded Systems

We would like to express our entire gratitude to all the authors who presented their works as the conference AMRPT2019 and contributed in this way to the success of this event. We wish to express our hearty gratitude to IOP Conference Series: Materials Science and Engineering (MSE) for publishing these excellent papers, to the esteemed all the committee members and reviewers for their overwhelming response. We hope you enjoy the conference and contribute to future AMRPT. Finally, hope that you can find the conference both significant and enjoyable!

AMRPT2019 Organization Committees

IOP Conf. Series: Materials Science and Engineering **611** (2019) 011002 doi:10.1088/1757-899X/611/1/011002

Reviewer List

Prof. Chen Hsu, Lungwha University of Science and Technology, Taiwan
Dr. S. P. BALAKANNAN, Kalasalingam Academy of Research and Education, India
Prof. Felix Ling Ngee Leh, Universiti Tun Hussein Onn Malaysia, Malaysia
Prof. Marcello Cabibbo, Università Politecnica delle Marche, Italy
Prof. Valentin Mateev, Technical University of Sofia, Bulgaria
Dr. Ho Soon Min, Inti International University, Malaysia
Dr. Jeng Shiang-Cheng, Taipei City University of Science and Technology, Taiwan
Prof. Fateh Mebarek-Oudina, Skikda University, Algeria
Dr. Salvatore MAURO, Institute of Marine Engneering, Italy
Prof. Corneliu Burlacu, Technical University "Gh. Asachi" Iasi, Romania
Dr. Michał Banaś, Wrocław University of Science and Technology, Poland
Prof.Osman Adiguzel, Firat University,Turkey
Prof. Changpeng Chen, Wuhan University of Technology, China
Prof. Chaoqing Min, Xi'an Jiaotong University, China
Prof. Benchatti Ahmed, Université de Laghouat, Algérie
Dr. Mansoor Zoveidavianpoor, NIOC, Iran
Prof. Farid ABED-MERAIM, Arts et Métiers ParisTech, Metz, France
Dr. JAIME ANDRES PEREZ TABORDA, Universidad de los Andes, Colombia
Dr. Ming An Chung, Quanta computer Inc., Taiwan
Prof. Mohamed Salem Badawi, Alexandria University, Egypt
Prof. Deepali Kamthania, Vivekananda Institute of Professional Studies, India
Dr. Bashir Ahmad Dar, Higher education Govt. of J&K, India
Dr. Mangalraj.P, SRM University, India
Dr. Mustafa Salam Kadhm, Imam Ja'afar Al-Sadiq University, Iraq
Dr. GHULAM HASNAIN TARIQ, Khwaja Fareed University of Engineering & Information Technology, Pakistan
Dr. P. S. Samuel Ratna Kumar, Kumaraguru College of Technology, India
Dr. Rer. Nat. Syed Sajid Ali Gillani, Government College University Lahore, Pakistan
Dr. Mohsen Motahari-Nezhad, Shahid Beheshti University, Iran
Prof. Mohand DJEZIRI, Aix Marseille University, France
Prof. Vineet Jain, Amity University Haryana, India
Dr. Ahmed Kareem Abdullah, Al-Furat Al-Awsat Technical University, Iraq
Dr. Sumit Kushwaha, Kamla Nehru Institute of Technology, India
Prof. ŞTEFAN ŢĂLU, The Technical University of Cluj-Napoca, Romania
Prof. Vijayan Gurumurthy Iyer, Open Learning Programme in Entrepreneurship Cell, The EDI of India, Chennai, India
Dr. Chih-Ta Tsai, National Cheng Kung University, Taiwan
Prof. Ray-Hwa Wong, Hwa-Hsia University of Technology, Taiwan
Dr. Hedayat Omidvar, National Iranian Gas Company, Iran
Dr. Soumya Mukherjee, Amity University, INDIA
Prof. AbdelGhani AISSAOUI, University Tahri Mohamed of Bechar, Algeria
Prof. Mohammed Hadi Al-Douh, Hadhramout University, Hadhramout, Yemen
Dr. M. Dzikuc, University of Zielona Gora, Poland
Dr. Y. Polat, Nevşehir Hacı Bektaş Veli University, Turkey
Prof. Andrey V. Brazhnikov, Siberian Federal University, Russia
Prof. I. S. Sacala, University Politehnica of Bucharest, Romania
Prof. Lee-Kun-Fa, Beijing Institute of Technology, zhuhai, China
Prof. Orchidea Maria Lecian, Comenius University in Bratislava, Slovakia
Dr. M. Shokrian Zeini, Shiraz University of Technology, Iran
Prof. Ioan-Lucian Popa,"1 Decembrie 1918" University Alba Iulia, Romania

IOP Conf. Series: Materials Science and Engineering **611** (2019) 011002 doi:10.1088/1757-899X/611/1/011002

Prof. Corneliu Burlacu, Technical University "Gh. Asachi" Iasi, Romania
Dr. Bashir Ahmad Dar, Higher education Govt. of J&K India, India
Dr. Yasser GadelHak, Beni-Suef University, Egypt
Prof. Haipeng Zhang, Shangrao Normal University, China
Prof. Mohamed Ahmed El-Sadek, Exploration Division, Egypt
Dr.Farhang Jaryani, University of Applied Science and Technology, Iran
Prof. Ressol R. Shakir, University of Thi-Qar, Nasiriyah, Iraq
Dr. Mojgan Gharakhanlou, Amirkabir University of Technology, Iran
Yanurita Dwihapsari, Institut Teknologi Sepuluh Nopember Surabaya (ITS), Indonesia
Prof. ZHU Cheng, Xidian University, Xi'an, China
Prof. Nabil Safta, University of Monastir, Tunisia
Prof. Gagik Shmavonyan, National Polytechnic University of Armenia, Armenia
Dr. Liţan Daniela-Elena, Hyperion University, Romania
Prof. Khalil KASSMI, Mohamed Premier University, Morocco
Prof. Jun Chen, Northeastern University, China
Prof. P.GANESH KUMAR, K.L.N College of Engineering, India
Prof. Mezdour Douniazad, Jijel University, Algeria
Prof. Zahra Hassannejad, Tehran University of Medical Sciences, Iran
Prof. G.A. Khater, National research Centre (NRC), Egypt
Prof. Ashraf Maher Abdel Ghaffar, National Center for Radiation Research and Technology, Egypt
Prof. Jianhua Liu, The Second Hospital of Jilin University, China
Dr. K. Pratheesh, Marian Engineering College, India
Prof.Ibraheem Othman Ali Abo zeid, Al Azhar University, Egypt
Prof. Anita S. Ethiraj, VIT–AP University, India
Dr. M.Ouchabane, Advanced Technologies Development Center, Algeria
Dr. Mohamed Reda Ramadan Gomaa, Alexandria Higher Institute for Engineering & Technology, Egypt
Dr. Beya Ouertani, University of Carthage, Tunisia
Dr. Ahmad Moloodi, Materials Research Group, ACECR Mashhad branch, Iran

2019 International Conference on Advanced Material Research and Processing Technology (AMRPT2019)
July 19-21, 2019, Wuhan, China

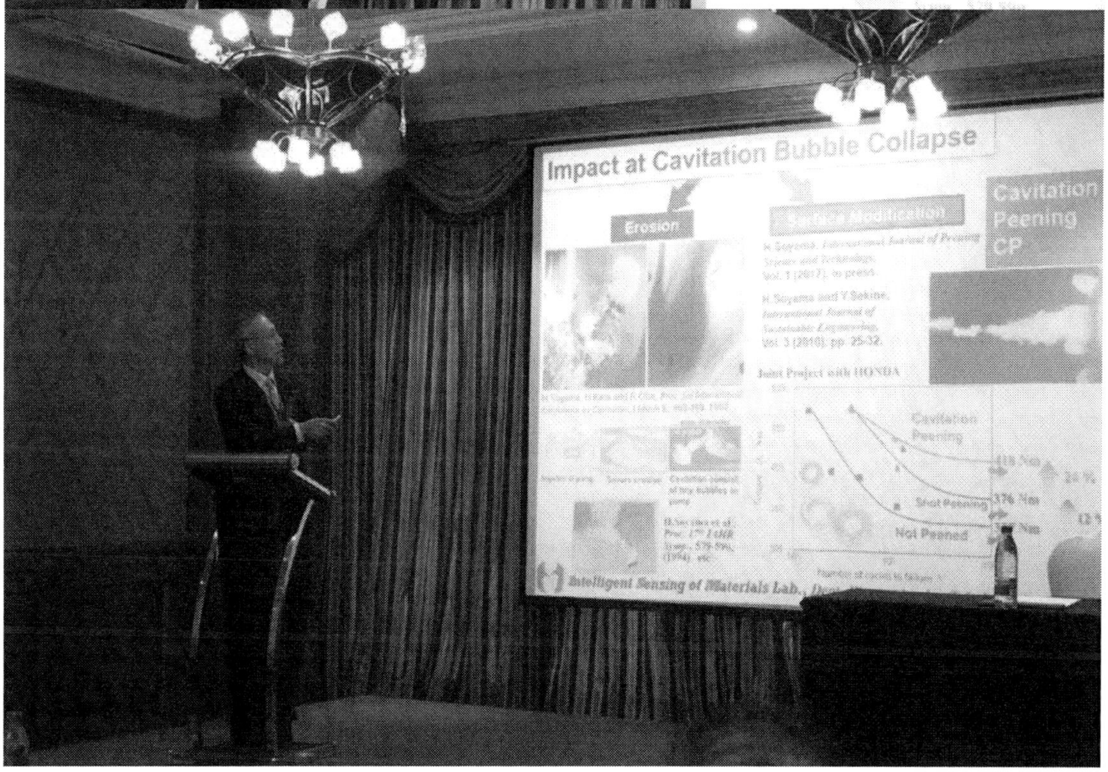
Impact at Cavitation Bubble Collapse
Erosion
Surface Modification
Cavitation Peening CP
Joint Project with HONDA
Cavitation Peening
Shot Peening
Not Peened
Intelligent Sensing of Materials Lab.

IOP Conf. Series: Materials Science and Engineering **611** (2019) 011003 doi:10.1088/1757-899X/611/1/011003

Steel for storage tanks of liquefied natural
Demand for steel for LNG storage tanks
Development trend: Nickel-saving cryogenic steel is replacin
7Ni Steel Plate
230,000 m³ Tank
Japan 7Ni Steel: in 2013, 230,000 m³ land-based
LNG storage tank was constructed
LNG-FPSO
South Korea
marine LNG stora
gineering application of international nickel-saving LNG ste

IOP Conf. Series: Materials Science and Engineering **611** (2019) 011003 doi:10.1088/1757-899X/611/1/011003

IOP Conf. Series: Materials Science and Engineering **611** (2019) 011003 doi:10.1088/1757-899X/611/1/011003

IOP Conf. Series: Materials Science and Engineering **611** (2019) 011003 doi:10.1088/1757-899X/611/1/011003

IOP Conf. Series: Materials Science and Engineering **611** (2019) 011003 doi:10.1088/1757-899X/611/1/011003

IOP Conf. Series: Materials Science and Engineering **611** (2019) 011003 doi:10.1088/1757-899X/611/1/011003

IOP Conf. Series: Materials Science and Engineering **611** (2019) 011003 doi:10.1088/1757-899X/611/1/011003

IOP Conf. Series: Materials Science and Engineering **611** (2019) 011003 doi:10.1088/1757-899X/611/1/011003

IOP Conf. Series: Materials Science and Engineering **611** (2019) 011003 doi:10.1088/1757-899X/611/1/011003

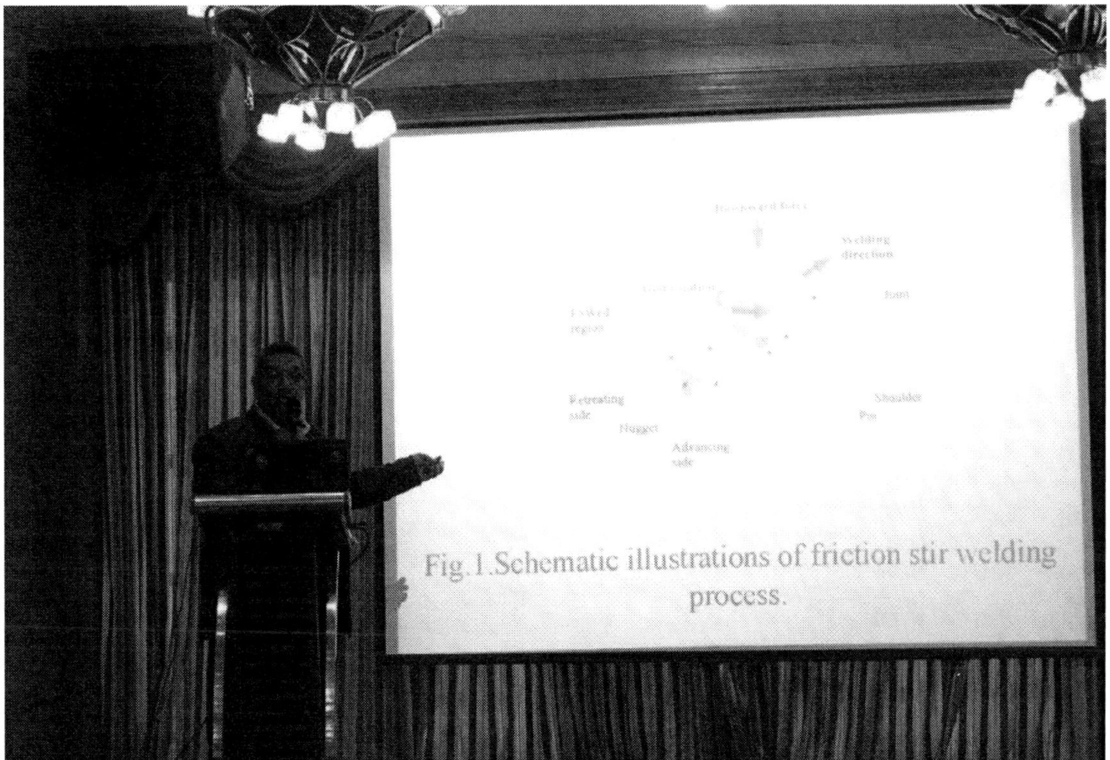

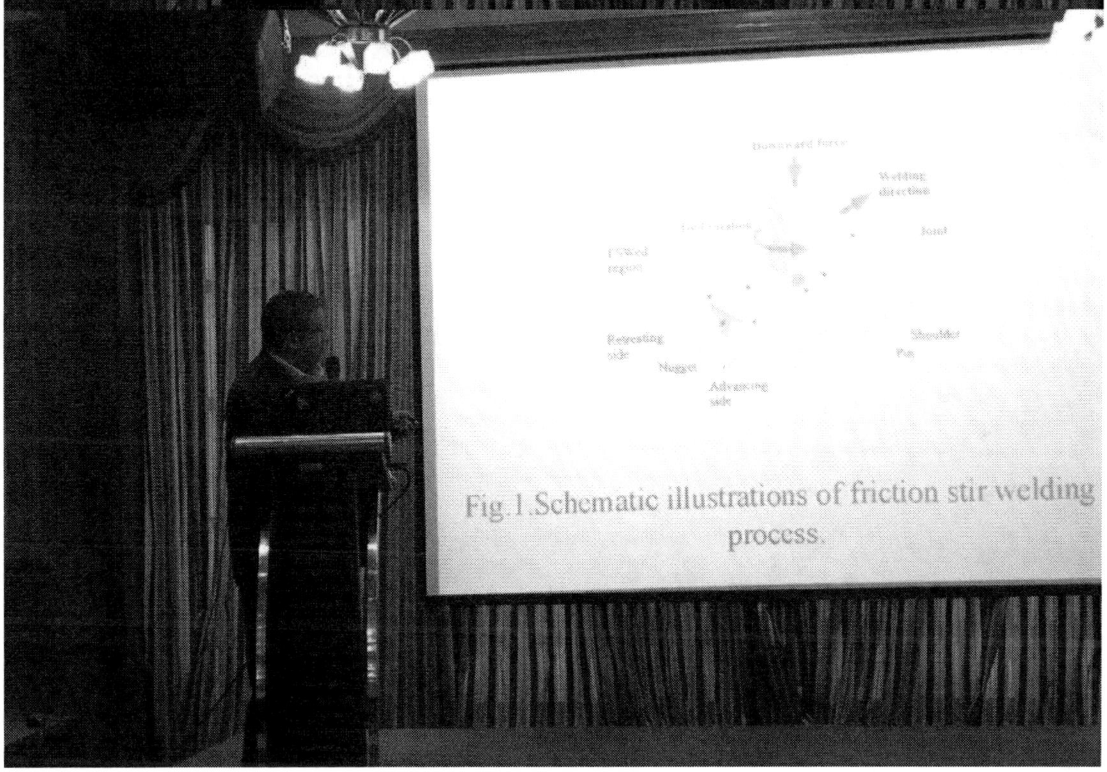

Haipeng Zhang*1,2, Xing He1, Mi Lin*2, et al.
1. Shangrao Normal University, 2. Hangzhou Dianzi University

Introduction
The truss bridge is installed on the pipe gallery to facilitate access to special sections
roads and lakes. It has large spans and is susceptible to local deformation, thinnin
settlements, which may cause safety problems. Therefore, the truss bridge needs to
safety.
The ANSYS finite element model was built to perform the thermal fluid-solid couplin

Introduction
The truss bridge is installed on the pipe gallery to facilitate access to special sections
roads and lakes. It has large spans and is susceptible to local deformation, thinnin
settlements, which may cause safety problems. Therefore, the truss bridge needs to
safety.
The ANSYS finite element model was built to perform the thermal fluid-solid couplin

简介

TCS复合试件结构组成图
加载板
Concrete
CA-Mortar
Concrete
35
30
35
75
75
100
75
75

Harmonic Propagation and Coupling Characteristics
among Wind Farm Groups
Reporter: Jiahui Yue

• Background of topic selection
• Harmonic characteristic analysis of wind farm
• Design of Wind Farm Cluster Circuits
• Simulation experiment
• Conclusion

IOP Conf. Series: Materials Science and Engineering **611** (2019) 011003 doi:10.1088/1757-899X/611/1/011003

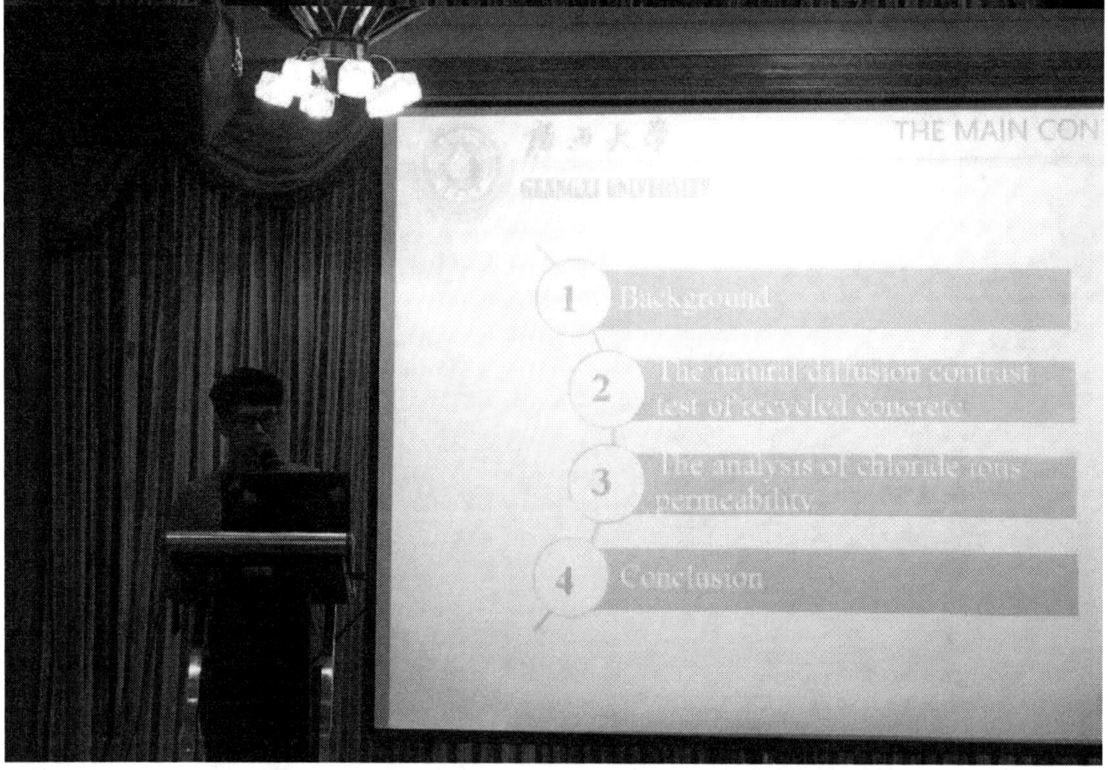

IOP Conf. Series: Materials Science and Engineering **611** (2019) 011004 doi:10.1088/1757-899X/611/1/011004

Peer review statement

All papers published in this volume of *IOP Conference Series: Materials Science and Engineering* have been peer reviewed through processes administered by the proceedings Editors. Reviews were conducted by expert referees to the professional and scientific standards expected of a proceedings journal published by IOP Publishing.

Formability and Grained Structure Refinement of Cold-Rolled Friction Stir Welded AA5754 Sheet

Marcello Cabibbo[1,*], Chiara Paoletti[1], Michela Simoncini[1,2] and Archimede Forcellese[1]

[1]Department of Mechanical Engineering and Mathematics (DIISM), Università Politecnica delle Marche, Via Brecce Bianche 12, 60131 - Ancona, Italy
[2]Università degli Studi eCampus, Via Isimabrdi 10, 22060 - Novedrate, Italy

*Corresponding author

Abstract. Among different welding techniques currently available in the market, friction stir welding (FSW) is surely considered as an effective and reliable joining technique. Friction stir welds are characterized by lack of voids, cracks and distortions, as the technique does not involve any material fusion. The grained structure of the weld joint is generally finer than the base material. In particular, the nugget zone (NZ) experiences a dynamic recrystallization process during FSW that generally guarantees a stable very-fine grained structure. In the present study, the effect of cold rolling (CR) on the formability limits, the resulted microstructure, and mechanical response of FSW but joint was investigated. To this purpose, an AA5754 aluminium alloy was used. The FSW was performed with rotational and welding speeds equal to 1200 rpm and 100 mm/min, respectively, and an initial tool sinking of 0.1 mm. Strips extracted from the FSWed sheets were CR, with the rolling direction perpendicular to the welding line. Two set-ups were used in the CR experiments. One, conventional to determine the formability limit of the FSW AA5754 sheet; a second one performed in a CR set-up designed to induce an equivalent strain $\varepsilon \sim 1$, in a single passage through the CR gage. This was aimed to make the post-FSW CR able to induce a further grained structure reduction within the NZ.
Keywords: FSW; CR; AA5754; Formability; SPD.

1. Introduction

Compared to other welding techniques, friction stir welding (FSW) has different mechanical and microstructure advantages. The key aspect refers to the welding temperature than is always within the alloy melting point. Thence, all the issues related with solidification from melting, of which the most important are porosity, possible cooling embrittlement, and formation of coarse secondary phase particles are avoided. On the other hand, the formation of residual stresses is usually minimized in FSW joints, were is typically lower than those observed in other welding techniques where alloy or filler melts are used [1]. Setting of proper welding parameters in the FSW process is mandatory to meet quality, soundness, and mechanical and microstructure properties of the joints.

FSW is particularly attractive when lightweight alloys need to be weld, such as Al and Mg alloys, since these are known to be difficult to weld, or are even unweldable through conventional fusion technologies. Typical applications include aerospace, automotive and most of the transportation industries [2, 3].

FSW sheet process is activated by the concurrent action of a rotating pin tool in contact with the sheet top surface, and by the transverse movement of the pin along the welding. The pin is usually inclined to the sheet surface with a set tilt angle. The combined effect of the plunging, the rotation, and the translation of the welding tool generates friction heating between the tool and the sheet width corresponding to the shoulder area. This pin motion induces the material plastic deformation by stirring, and promotes a complex mixing across and along the joint path. Thus, a plasticized region is created around the tool and beneath the shoulder of the pin, by stirring and shearing action, from a pin retreating side (RS), to a pin advancing side (AS) [2-6]. In addition, during FSW, the material flows around the tool while it is subjected to intense plastic deformation process and local temperature rise. Both the mechanical plastic deformation and the thermal excursion promote the formation of a fine-gained structure in the center of the welded joint, at the so-called nugget zone (NZ). The occurrence of a fine-grained structure in NZ significantly affects the mechanical properties of the welded alloy. It is well agreed that the formation of fine grains within the NZ is a dynamic recrystallization (DRX) phenomena. The NZ DRX is likely to occur continuously by the shearing of the existing grains and subgrains, or by grain formation under the pin action, which eventually are thermally induced to evolve into a high-angle boundary network, that is, into a fine-grained structure [7-9].

A key microstructure FSW issue when used to butt joints is the sheet thickness reduction that is likely to be induced as result of tool and shoulder forging-like effect [1,2,5]. In order to limit, or minimize the welded sheet thickness reduction along the weld joint (namely the NZ), the FSWed joint sheet can be cold rolled (CR) after FSW. In this case, other than a more uniform sheet thickness throughout the weld joint profile, CR generates a strain hardening coefficient rise that can possibly induce material strengthening and formability improvement [10].

Thus, the present work focuses on the effect of CR on the mechanical properties and formability of friction stir welded sheets in AA5754 aluminium alloy. Strips of the FSWed sheet were CR, with the rolling direction perpendicular to the welding line, to reach a uniform sheet thickness. Formability of the CR FSWed joints was evaluated.

Based on the observed good mechanical and formability response of the FSW and CR AA5754, a second CR set-up was design and used to obtain a very fine grained structure within the NZ, by a single passage into the CR gage. To this purpose, the CR set-up was designed and set to induce an equivalent strain of $\varepsilon \sim 1$. In this case a thickness reduction of 10% was induced, and a thickness uniformity was still maintained throughout the FSW + CR joint, from the RS to the AS.

2. Material and Experimental Procedures

2.1. Friction Stir Welding Process

AA5754 sheets were 185 mm long, 80 mm wide, and 2.5 mm thick, for the formability limit studies, and 200 mm long, 60 mm width, 1.5 mm thick, for the NZ grain refinement process obtained by post-FSW CR. The AA5754 alloy had a chemical composition (wt.%) of 3.4Mg, 0.40Mn, 0.20Si, 0.20Fe, 0.10Cr, 0.15(Ti+Zn), and Al bal.. FSW was carried out using a truncated conical-shaped pin with a base of 7.8mm in diameter, a height of 1.3mm, and a shoulder diameter of 15 mm. Pin consisted of a tool steel (H = 50HRC). During FSW, the tool was initially forced 0.1 mm down into the aluminum sheet. Welding was carried out with a pin nutting angle of 2°.

Welding parameters were set to rotational speed, ω, and welding translational speed, v, of 1200 rpm and 100 mm/min, respectively. These settings were chosen according to the weldability window of the AA5754 alloy that was determined in two previously published works by some of the present authors [11,12]. The welding line followed a path perpendicular to the alloy sheet rolling direction (RD).

2.2. Post-FSW Cold Rolling

Strips of the FSWed sheets were cut perpendicularly to the welding line, with dimensions suitable for the following CR process.

For the first study of formability, two different rolling passes were used, obtaining a final uniform thickness of the FSW-CR strip of 2.2 mm, that is, a total thickness reduction of ~10% (mode 1).

For the second study on the grain size reduction potentials at the NZ of the AA5754 sheet, by post-FSW CR, a specifically designed CR set-up able to induce an equivalent strain $\varepsilon \sim 1$ in a single passage through the CR gage was carried out (mode 2). This CR set-up is schematically reported in Fig. 1. In this case, after FSW + CR, the sheet thickness reduction on the welding NZ was of ~20%. To study the role of the severe plastic deformation induced by the single passage into the specifically misaligned CR gage (se Fig.1), the FSW strips were also annealed before CR. The annealing treatment consisted in a 415°C/3h soaking followed by a slow cooling inside the shut-down furnace. This way, the role of post-FSW CR on the $MnAl_6$ dispersoids formation and on the occurrence of Mg_2Al_3 particle precipitation in the AA5754 Al-Mg-Mn alloy was detected.

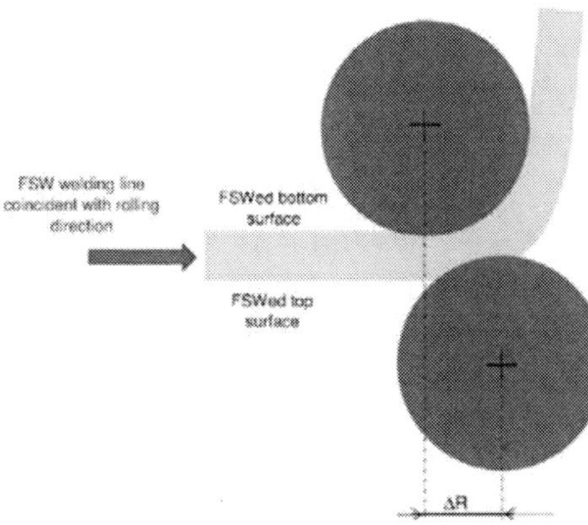

Figure 1. Cold Rolling set-up scheme designed to induce an equivalent strain of $\varepsilon \sim 1$ under a single sample passage through the gage.

2.3. Microstructural and Micromechanical Analyses
Metallographic inspections by optical microscopy (OM) were carried out on surfaces mechanically polished and electrochemically etched by a 4% HBF_4 solution at 15V for few tens of seconds. Polarized OM (POM) was used to properly highlight the grained structures.
TEM inspections were carried out by a Philips® CM20 operated at 200 keV. TEM discs were prepared by polishing, dimpling, and ion-milling using a Gatan® PIPS.
Microhardness was carried out along middle-height thickness of the FSW section. A Remet® HX-1000 tester was used. Measurements were spaced 250 μm apart.
Tensile tests were carried out on rectangular samples extracted from the base-material (BM), across the weld joint of the FSW, and the FSW + CR. Tensile tests were carried out by a conventional MTS machine equipped with extensimeter, and reported data results from 3 different tensile tests per experimental condition (i.e., BM, FSW and FSW + CR).

3. Results and Discussion

3.1. Post-FSW CR Mechanical Response
Table 1 reports the yield, ultimate strengths, and the maximum elongation obtained in the BM, FSW, and FSW + CR condition, where CR was carried out using the mode 1 (see above).
It resulted that the FSW yield strength is reduced by ~40%, compared to the base metal value. On the other hand, the FSW + CR had yield strength values quite similar to the base metals. This means that the yield strength is fully recovered to the base metal level when the FSW AA5754 sheet is CRed. Yet, the ductility response of the FSW + CR condition (5.5%) reduced 3-fold compared to the BM value of 16.5%, and halved, respect to the FSW value of 12%. The yield strength, σ_y, response corresponds, to some extent, to the ultimate strength, σ_U, among the three different experimental conditions, *i.e.* between the FSW and FSW + CR.

Table 1. Yield strength, σ_y, ultimate strength, σ_U, and maximum elongation, ε_U, of BM, FSW, and FSW + CR.

	BM	**FSW**	**FSW + CR**
σ_y, *MPa*	165 ± 5	100 ± 5	170 ± 5
σ_U, *MPa*	260 ± 5	225 ± 5	245 ± 5
ε_U, *%*	16.5 ± 0.5	12.0 ± 0.5	5.5 ± 0.3

The here reported mechanical responses are in good agreement with previous reported results on same alloy, but different welding parameters [12-15], and on different aluminium alloys [16-22].
The ductility-to ultimate strength relationship here observed, was properly addressed by applying formability tests to the BM, FSW, and FSW + CR conditions. As reported in Fig. 2, it appeared that the limiting dome height (LDH) of the FSW + CR is quite similar to that of the FSW joint. This, in turns, means that the strain hardening effect on the alloy formability (LDH) is much less pronounced than the one on ductility (ε_U). Thence, since formability decreases with the sheet thickness, knowing that the FSW + CR joint is thinner by 10% than the FSW sheet, the CR effect on the formability of FSW sheet can be considered as negligible. It is worth to highlight that the LDH obtained in the BM is only ~5% higher than that of the FSW joint, which in turns, is almost 5% higher than the one of the FSW + CR joint.

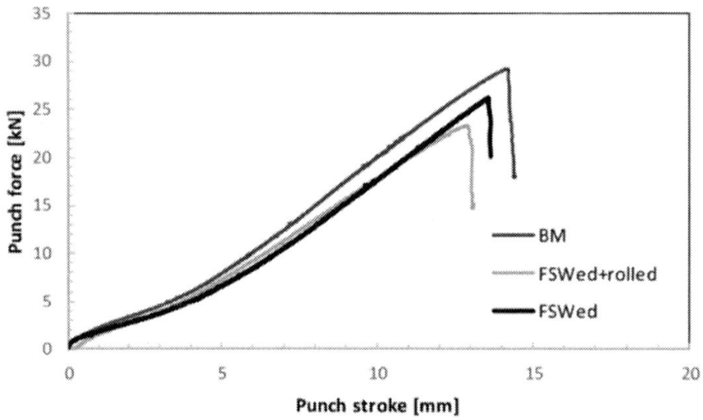

Figure 2. LDH of the BM, FSW, FSW + CR.

Given the good formability limit obtained by FSW at rotational speed, ω = 1200 rpm, and welding translational speed, v = 100 mm/min, a specifically designed shifted roll position to impose a severe shear-like plastic deformation in one single passage through the CR gage was used. The welded sheet integrity, after this specific CR process, was maintained, as shown by the OM montage of Fig. 3, referred to the joint mid-section.

Figure 3. OM montage of the FSW + A + CR section, where the annealing (A) was performed at 415°C/3h, and the CR was performed in one single passage through the CR gage to impose an equivalent $\varepsilon \sim 1$. Weldment integrity is shown from RS HAZ, TMHAZ, the NZ, throughout the AS TMHAZ, HAZ.

The joint integrity is maintained throughout the section from the left-hand side, where the retreating side (RS) is located, to the right-hand side, where is located the advancing side (AS), where no cracking formation is observed.

3.2. Grain refinement potentials of FSW + CR

Fig. 4 shows representative POM taken along the NZ section of the FSW, FSW + CR, and FSW + A + CR. It is worth to note that the coarsening of the grained structure observed in the NZ of the FSW + A + CR, was fully reduced by the dynamic recrystallization (DRX) process favored by the following CR [16,17]. No grain size coarsening occurred in the FSW + A + CR, respect to the FSW + CR condition. The mean grain size, d_g, was measured by stereological methods (EN-112) in the FSW NZ, and it resulted that d_g = 65 ± 4 μm, after FSW, d_g = 4.2 ± 0.4 μm, after FSW + CR, and d_g = 7.1 ± 0.4 μm, after FSW + A + CR. It was thus shown that the welding parameters here used to carry out the FSW, are suitable to obtain a sound weld joint also upon post-welding severe cold rolling, which greatly reduced the grained structure within the NZ by one-order of magnitude.

On the other hand, Fig. 4, also shows that no cracks were formed in the NZ, after CR. This means that the FSW + A + CR sequence is a sound process that is able to effectively refine the NZ grained structure with no excess of stresses and no cracking formation.

a) b) c)

Figure 4. POM NZ joint section of the FSW, a), FSW + CR, b), and FSW + A + CR, c).

3.3. Role of annealing on the precipitation and intermetallic particles

Fig. 5 reports representative micrographs of FSW, FSW + CR, and FSW + A + CR, where the volume fraction of the different particles and intermetallic phase are shown.

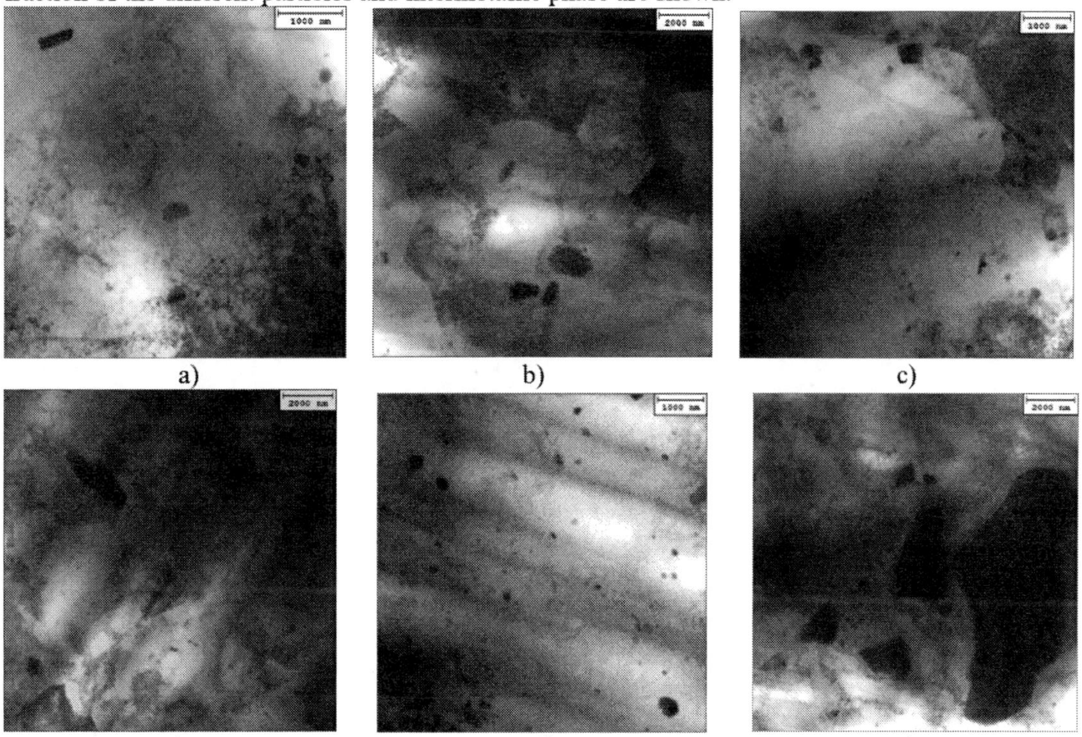

a) b) c)

c) d) e)

Figure 5. $MnAl_6$ dispersoids, Mg_2Al_3 particles, and $(Fe,Mn)_3SiAl_{12}$ intermetallic particles in the FSW, a), b), FSW + CR, c), d), and FSW + A + CR, e), f). All TEM micrographs refer to the FSW NZ.

One of the major and most interesting aspect of the deformation sequence induced by FSW and following CR, is surely the Fe-rich intermetallic agglomeration process that was promoted by the combination of plastic deformation of, first, stirring (during FSW), and, after, rolling (during CR), type (Fig. 5b, d, f).

On the other hand, the clear effect of the annealing treatment, after FSW, was observed in the mean volume fraction variation of both the $MnAl_6$ dispersoids and the Mg_2Al_3 particles. The volume fraction of both particle types was higher when the AA5754 FSW sheet is annealing prior CR. As reported in the Table 2, the post-FSW annealing induced an almost 3-fold rise of the Mg_2Al_3 particles volume fraction, respect to the FSW + CR condition, as it passed from $V_V(Mg_2Al_3)$ ~7 ‰, in FSW + CR, to 19 ‰, in FSW + A + CR. Correspondingly, also the $MnAl_6$ dispersoids increased in volume fraction from the FSW and FSW + CR, to FSW + A + CR, accounting for a 30-35% rise, i.e., $V_V(MnAl_6)$ ~40 to ~55 ‰. These particle increments were essentially due to the combined effect of post-FSW annealing thermal energy, and the shear-type plastic deformation, during CR.

Table 2. Quantitative evaluation of the $MnAl_6$ dispersoids, and Mg_2Al_3 particles.

	$V_V(MnAl_6)$, ‰	$V_V(Mg_2Al_3)$; ‰	$d_p(MnAl_6)$, nm	$d_p(Mg_2Al_3)$, nm
FSW	40 ± 10	6 ± 1	80 ± 5	150 ± 5
FSW + CR	45 ± 10	7 ± 1	75 ± 5	135 ± 5
FSW + A + CR	55 ± 10	19 ± 2	70 ± 5	160 ± 5

Another interesting result, is that the mean size of the $MnAl_6$ dispersoids, and Mg_2Al_3 particles, $d_p(MnAl_6)$ and $d_p(Mg_2Al_3)$, seemed not dependent on both CR, and annealing prior-CR.

In all the experimental conditions, the $(Fe,Mn)_3SiAl_{12}$ intermetallic particle mean size, $d_p((Fe,Mn)_3SiAl_{12})$ ranged (250±50)·(700±30)·(900±100) nm^3, the volume fraction $V_V((Fe,Mn)_3SiAl_{12})$ ranged (130 ± 30) ‰.

In order to correlate the here observed grain size reduction and particle evolution, both induced by the FSW, post-FSW annealing, and CR, microhardness profiles were measured along the FSW NZ sections, in the FSW, FSW + CR, and FSW + A + CR experimental conditions (Fig. 6).

As reported in Fig. 6, hardness profiles across the FSW joint show a certain degree of uniformity among the different characteristic FSW zones. Across the retreating thermo-mechanical heat-affected zone (R-TMHAZ), and the heat-affected zone (R-HTZ), ghe FSW-CR and FSW-A-CR showed a hardness profile that was generally lower than the advancing counterparts. The mean hardness in the FSW + CR NZ was the highest, while the two mean values of FSW and FSW + A + CR were quite similar. The higher mean harness in the FSW + CR NZ agrees with the lower mean grain side here obtained, compared to the other conditions.

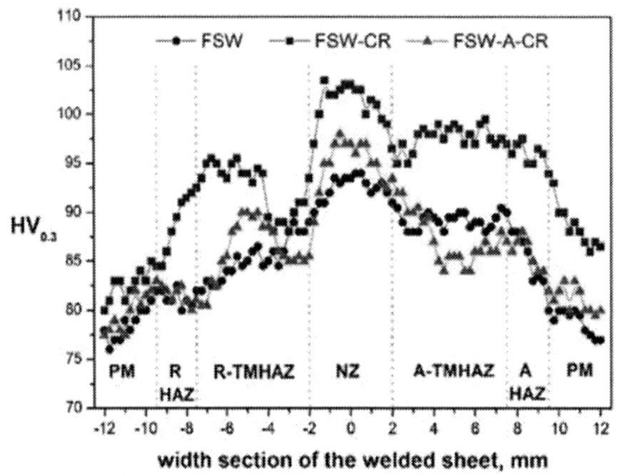

Figure 6. Microhardness profiles across all the meaningful FSW zone, i.e., the parent material (PM), the retreating heat-affected zone (R-HAZ), the thermo-mechanical heat-affected zone (R-TMHAZ), the NZ, and the advancing counterpart, A-TMHAZ, A-HAZ. The hardness were recorded at the mid-height of the FSW section, in the FSW, FSW + CR, and FSW + A + CR experimental conditions.

Some authors [13-15,19-22] also studied the role of the FSW parameters and that of the post-FSW thermal treatments on the hardness profiles across the FSW joint section. The here reported results are in good agreement with similar results obtained in terms of FSW parameters and similar thermal treatments role on the hardness profile along the different meaningful FSW zones.

4. Conclusions

The post-FSW cold rolling (CR) role in terms of mechanical properties and sheet formability of an AA5754 aluminium alloy was investigated. In addition, the soundness of a grain-refined FSW-AA5754 sheet subjected to post-FSW severe deformation by cold-rolling ($\varepsilon \sim 1$) was tested. The obtained results can be summarized as follows:

- the stress level obtained by deforming FSW sheets followed by CR, is significantly higher than the one obtained in the FSW joint;
- the FSW NZ is characterized by fine equiaxed grains, while after FSW + CR they were slightly deformed and still reasonably fine;
- cold rolling did not significantly affect formability of FSW sheets, as the obtained LDH after FSW + CR was quite similar to that of the FSW;
- By CR at $\varepsilon \sim 1$ the FSW joint in a single passage through the rolling gage, the NZ grained structure reduced by one-order of magnitude, reaching few microns in mean size;
- the annealing treatments affected the hardness profile across the FSW joint, by significantly reduce the hardness, compared to the FSW + CR hardness. On the other hand, the hardness profiles of the FSW and FSW + A + CR had a similar trend;
- the role of annealing on the $MnAl_6$ dispersoids and Mg_2Al_3 particles was determined in terms of their mean size (d_p) and volume fraction (V_V). The mean particle sizes of these two types of particles were not affected by the post-FSW annealing. Yet, their volume fraction significantly increased by the combined effect of first, post-FSW annealing, and, after, shear-like deformation on CR at $\varepsilon \sim 1$.

The here obtained results seem to confirm the possibility to scale up a CR process specifically designed to FSW joint aluminum sheets for transportation applications.

Acknowledgments
The authors acknowledge Mr. M. Pieralisi for having performed all the FSW processes.

References

[1] Gibson BT, Lammlein DH, Prater TJ, Longhurst WR, Cox CD, Ballun MC, Dharmaraj KJ, Cook GE, Strauss AM 2014 Friction stir welding: process, automation and control, J. Manuf. Process. 16 pp 56–73.
[2] Mishra RS, Ma ZY 2005 Friction stir welding and processing, Mater. Sci. Eng. R: Reports 50 pp 1-78.
[3] Geiger M, Kleiner M, Eckstein R, Tiesler N, Engel U 2001 Microforming, CIRP Annals, Manuf. Technol. 50 pp 445-462.
[4] Rodrigues DM, Loureiro A, Leitao C, Leal RM, Chaparro BM, Vilaça P 2009 Influence of friction stir welding parameters on the microstructural and mechanical properties of AA 6016-T4 thin welds, Mater. Design 30 pp 1913-1921.
[5] Simoncini M, Forcellese A 2012 Effect of the welding parameters and tool configuration on micro- and macro-mechanical properties of similar and dissimilar FSWed joints in AA5754 and AZ31 thin sheets, Mater. Design 41 pp 50-60.
[6] Cabibbo M, Forcellese A, Simoncini M 2016 New approaches to the friction stir welding of aluminum alloys, in Mahadzir Ishak (Eds.), Joining Technologies, InTech pp. 7-26.
[7] Jata KV, Semiatin SL 2000 Continuous dynamic recrystallization during friction stir welding of high strength aluminum alloys, Scripta Mater. 43 pp 743-749.
[8] Jata KV, Sankaran KK, Ruschau JJ 2000 Friction-stir welding effects on microstructure and fatigue of aluminum alloy 7050-T7451, Metall. Mater. Trans. A 31 pp 2181-2190.
[9] Su J-Q, Nelson TW, Mishra R, Mahoney M 2003 Microstructural investigation of friction stir welded 7050-T651 aluminium, Acta Mater. 51 pp 713-729.

[10] Gehring A, Saal H 2006 Mechanical properties of aluminium in structural sheeting, Thin-Walled Structures 44 pp 1231-1239.
[11] Bevilacqua M, Ciarapica FE, D'Orazio A, Forcellese A, Simoncini M 2017 Sustainability analysis of friction stir welding of AA5754 sheets, Procedia CIRP, 62 pp 529-534.
[12] Casalino G, El Mehtedi M, Forcellese A, Simoncini M 2018 Effect of Cold Rolling on the mechanical Properties and Formability of FSWed Sheets in AA5754-H114 Metals 8 pp 223-240.
[13] Cabibbo M, Forcellese A, Simoncini M, Pieralisi M, Ciccarelli D 2016 Effect of welding motion and pre-/post-annealing of friction stir welded AA5754 joints, Mater. Design 93 pp 146-159.
[14] Moshwan R, Yusof F, Hassan MA, Rahmat SM 2015 Effect of tool rotational speed on force generation, microstructure and mechanical properties of friction stir welded Al–Mg–Cr–Mn (AA 5052-O) alloy, Mater. Design 66 pp 118-128.
[15] Bisadi H, Tavakoli A, Sangsaraki MT, Sangsaraki KT 2013. The influences of rotational and welding speeds on microstructures and mechanical properties of friction stir welded Al5083 and commercially pure copper sheets lap joints, Mater. Design 43 pp 80-88.
[16] Fratini L, Buffa G 2005 CDRX modelling in friction stir welding of aluminium alloys, Int. J. Mach. Tools Manuf. 45 pp 1188-1194.
[17] Etter AL, Baudin T, Fredj N, Penelle R 2007 Recrystallization mechanisms in 5251 H14 and 5251O aluminum friction stir welds, Mater. Sci. Eng. A 445-446 pp 94-99.
[18] Sua J-Q, Nelson TW, Sterling CJ 2005 Microstructure evolution during FSW/FSP of high strength aluminum alloys, Mater. Sci. Eng. A 405 pp 277-286.
[19] Olea CAW, Roldo L, dos Santos JF, Strohaecker TR 2007 A sub-structural analysis of friction stir welded joints inan AA6056 Al-alloy in T4 and T6 temper conditions, Mater. Sci. Eng. A 454-455 pp 52-62.
[20] Khorrami MS, Kazeminezhad M, Kokabi AH 2012 Microstructure evolutions after friction stir welding of severely deformed aluminum sheets, Mater. Design 40 pp 364-372.
[21] Jayaraman M, Sivasubramanian R, Balasubramanian V 2010 Establishing relationship between the base metal properties and friction stir welding process parameters of cast aluminium alloys, mater. Design 31 pp 4567-4576.
[22] Zhang HJ, Liu HJ, Yu L 2011 Microstructure and mechanical properties as a function of rotation speed in underwater friction stir welded aluminum alloy joints, mater. Design 32 pp 4402-4407.

IOP Conf. Series: Materials Science and Engineering **611** (2019) 012002 doi:10.1088/1757-899X/611/1/012002

Improvement of Fatigue Strength of Additive Manufactured Metals by Solid-Liquid-Gas Interfacial Phenomena Induced by Pulse Laser

Hitoshi Soyama*

Department of Finemechanics, Tohoku University, Sendai, Japan

*Corresponding author

Abstract. Although additive manufactured (AM) metals are attractive materials, the fatigue strength of AM metals are considerably weak comparing with that of wrought materials. The mechanical surface treatment such as shot peening can improve the fatigue strength of metallic materials. Recently, a novel mechanical surface treatment using solid-liquid-gas interfacial phenomena induced by pulse laser has been developed. In the present paper, in order to demonstrate the improvement of fatigue strength of AM metals by solid-liquid-gas interfacial phenomena, titanium alloy Ti6Al4V manufactured by electron beam melting EBM was treated by submerged pulse laser and tested by a plane bending fatigue test. The key factors were investigated by evaluating the relation between the fatigue properties and mechanical properties of the surface treated by submerged laser peening, cavitation peening and shot peening.

1. Introduction

Metallic materials manufactured additive manufacturing (AM) processing is attractive processing for biomedical and aero-space area, as it is easy to make components directly from computer-aided design CAD with weight saving and design freedom without having to consider machining. Unfortunately, the fatigue strength of AM metals are considerably weak comparing with that of wrought materials [1, 2]. One of methods to improve the fatigue strength of AM metals is mechanical surface treatment [1, 3-5]. Thus, it is necessary to investigate the improvement of AM metals by mechanical surface treatment.

One of most popular mechanical surface treatment is shot peening. Shot peening can improve fatigue strength and elongate fatigue life. However, the increase of surface roughness by shot collisions sometimes decreases fatigue properties. In these days, novel peening methods, in which shots are not required, such as cavitation peening [6, 7] and laser peening [8-11] have been developed. It is worthwhile to investigate effect of improvement of fatigue strength by cavitation peening and laser peening.

As shown in Figure 1, in the case of mechanical surface treatments, the local plastic deformations produced by solid/liquid collision impacts or shock wave impacts were generated, and they changed mechanical properties such as residual stress, yield stress, hardness, roughness etc. However, the quantitate effects of each factor are in "black box", thus conventional way to enhance fatigue properties by mechanical surface treatments is a kind of try and error. Once inside of "the black box" was clarified, the fatigue properties can be enhanced effectively.

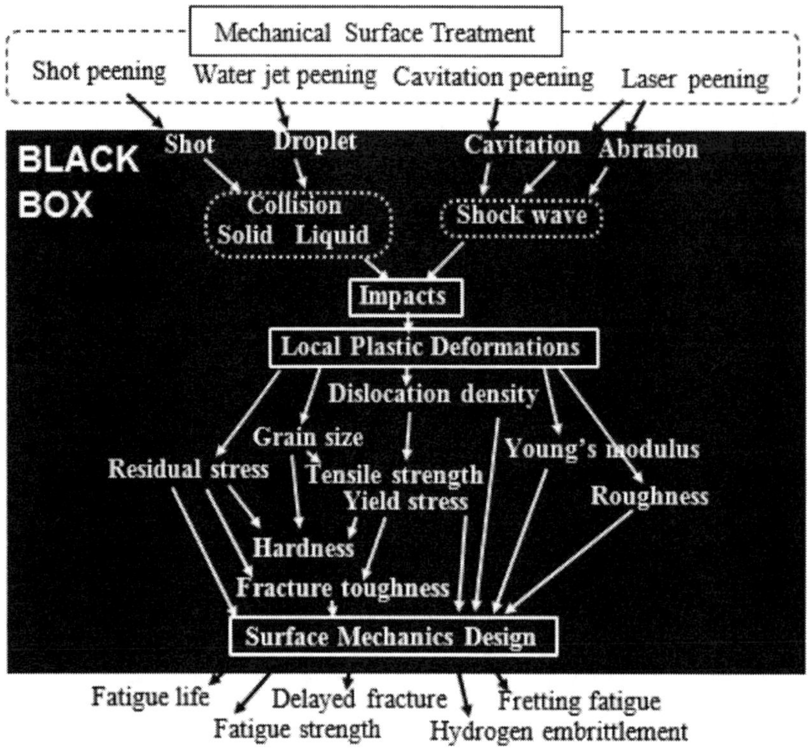

Figure 1. Surface mechanics design on improvement of fatigue properties by mechanical surface treatment considering mechanical properties.

In the case of cavitation peening and submerged laser peening, phase change phenomenon from liquid-phase to gas-phase is used to generate the local plastic deformations. Figure 2 shows (a) the typical aspect of bubble induced by a pulse laser and (b) output signal from PVDF sensor [12, 13] were shown [7]. After laser ablation, a bubble. i.e., gas-phase, was developed from water, i.e., liquid-phase, and impact was produced to the solid surface at bubble collapse. As the bubble induced by pulse laser produced the impact at bubble collapse, the bubble was named as "laser cavitation". As shown in Fig. 2 (b), the impact induced by laser cavitation was larger than that of laser ablation. Namely, submerged laser peening is a kind of mechanical surface treatment using solid-liquid-gas interfacial phenomena induced by pulse laser.

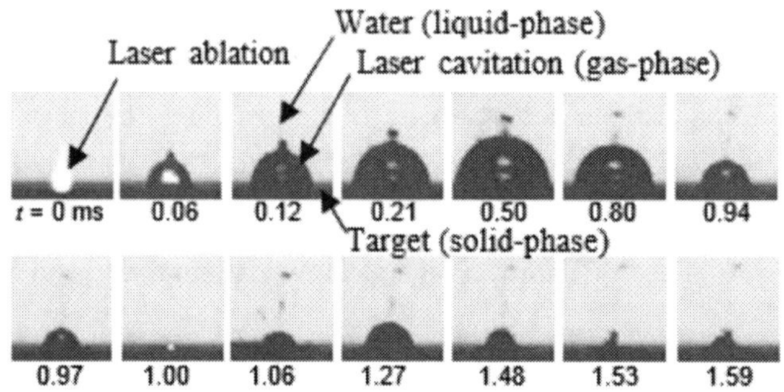

(a) Aspect of laser ablation and laser cavitation

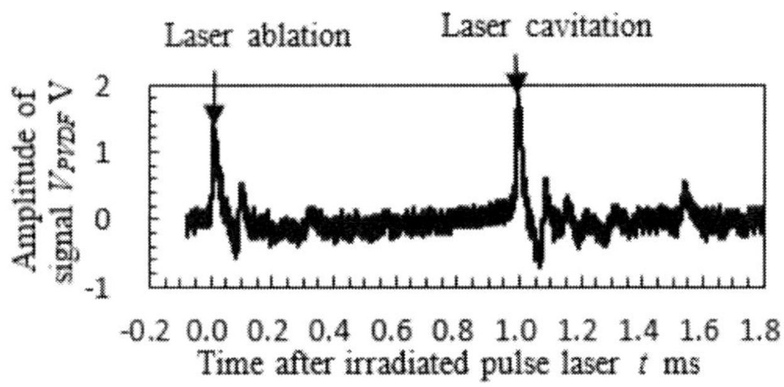

(b) Output signal from PVDF sensor

Figure 2. Solid-liquid-gas interfacial phenomena induced by pulse laser [7].

In the present paper, in order to demonstrate the improvement of AM metals by submerged laser peening and cavitation peening, the titanium alloy Ti6Al4V manufactured by electron beam melting EBM was treated by cavitation peening, submerged laser peening and shot peening, and tested by a plane bending fatigue test. And also, in order to make clear the inside of "the black box", the duralumin A2017-T3 was treated by submerged laser peening, cavitation peening and shot peening, and tested by a load controlled plane bending fatigue test. As the surface of Ti6Al4V manufactured by EBM was very rough, and it is very difficult to measure the crack propagation rate caused by the fatigue, thus the duralumin treated by the mechanical properties were examined. The mechanical properties of duralumin such as residual stress, surface roughness and crack propagation rate were evaluated.

2. Experimental Facilities and Procedures

Figure 3 illustrates a schematic diagram of submerged laser peening system in which solid-liquid-gas interfacial phenomena induced by a pulse laser was used for mechanical surface treatment. Q-switched Nd:YAG laser was used as source of the pulse laser. The fundamental harmonic and the second harmonic of a Nd:YAG laser are 1064 nm and 532 nm. The maximum energy at 1,064 nm was 0.35 J. The beam diameter, the pulse width and the repetition frequency of the laser pulse were 6 mm, 6 ns and 10 Hz, respectively. In order to avoid damage of glass chamber, the diameter of the laser beam was expanded by concave lens and then focused into the water using two convex lenses. The pulse density for the treatment of Ti6Al4V manufactured by EBM was 5 pulse/mm^2 considering previous report [7]. The standoff distances in air s_a and in water s_w were optimized by measuring arc height of metallic plate. In order to compare the other peening method, the specimens were treated by shot peening and cavitation peening. The used shot peening system was a recirculating shot peening accelerated by a water jet which injection pressure was 12 MPa [14]. Note that the water jet without shot could not introduce compressive residual stress.

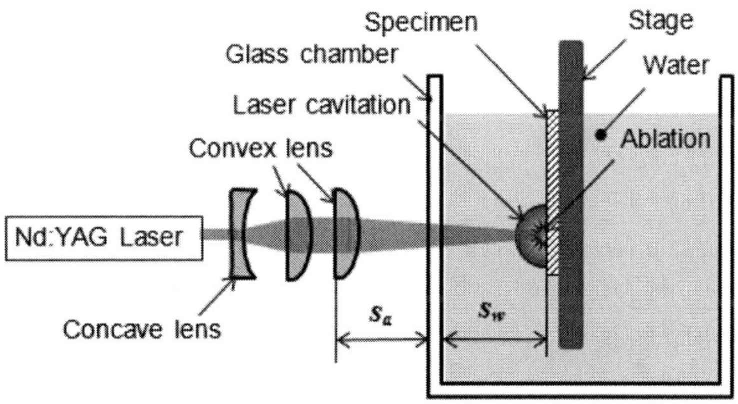

Figure 3. Schematic diagram of submerged laser peening system. The pulse laser was focused on the solid surface of specimen which was placed in water.

In the case of cavitation peening, a submerged water jet was used to generate cavitation (see Figure 4). The injection pressure of the jet was 30 MPa and the nozzle throat diameter was 2 mm. In order to enhance cavitation aggressively, the cavitator and the guide pipe were installed in the nozzle and their geometries were optimized. The cavitator was feeding the cavitation nuclei and the guide pipe enhance cavitation aggressively by controlling vertical flow around the jet. Even though the submerged water jet peening, the water column in the jet center can produce the plastic deformations. The mechanical surface treatment using water column impacts is called as "water jet peening". In the case of water jet peening, a large injection pressure is required. On the other hand, in the case of cavitation peening, the relatively low injection pressure with large nozzle is suitable. Note that the peening intensity of cavitation peening is larger than that of water jet peening. The peening condition of cavitation peening is distinguished by the standoff distance, which was defined by the distance from nozzle to the specimen surface. At the present experiment, it was 262 mm.

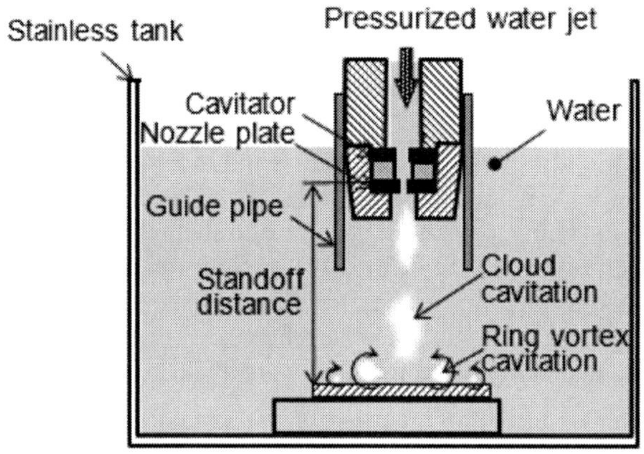

Figure 4. Schematic diagram of cavitation peening system. The submerged water jet was injected into a water filled chamber.

In order to demonstrate the improvement of fatigue strength of AM metals by mechanical surface treatment, Ti6Al4V manufactured by EBM was treated by above mentioned submerged laser peening system. The EBM conditions were follows. The averaged diameter of the powder, the spot size of electron beam for selectively melting and the stacking pitch were 75 μm, 0.2 mm and 90 μm, respectively. The solution heat treatment was carried out at 1,208 K at vacuum condition for 105 minutes then argon gas cooling. After that, aging treatment was done at 978 K at vacuum condition for 2 hours then argon gas cooling. The stress ratio at the plane fatigue test was $R = -1$. In order to investigate the comparison of peening methods on the improvement fatigue strength, the specimen was also treated by cavitation peening and shot peening [4].

In order to investigate the relation between fatigue properties and mechanical properties, specimen made of duralumin JIS A2017-T3 was treated by submerged laser peening, cavitation peening and shot peening [15], and crack propagation rate, residual stress, surface roughness and hardness were measured. The crack propagation rate was evaluated by using a load controlled plane bending fatigue tester. The applied stress σ_a was 200 MPa. Peening intensity was evaluated by curvature κ of duralumin plate JIS A2017-T3 of 3 mm thickness treated by various peening.

3. Results

Figure 5 reveals the fatigue strength of titanium alloy Ti6Al4V manufactured by EBM. The fatigue strength was obtained by Little's method [16]. The fatigue strength was 221 ± 11 MPa for the non-peened specimen, 371 ± 7 MPa after shot peening, 406 ± 11 MPa after cavitation peening and 450 ± 3 MPa after laser peening. Thus, the fatigue strength was improved by 68 % by shot peening, 84 % by cavitation peening and 104 % by submerged laser peening. Namely, mechanical surface treatments can improve fatigue strength of AM metals.

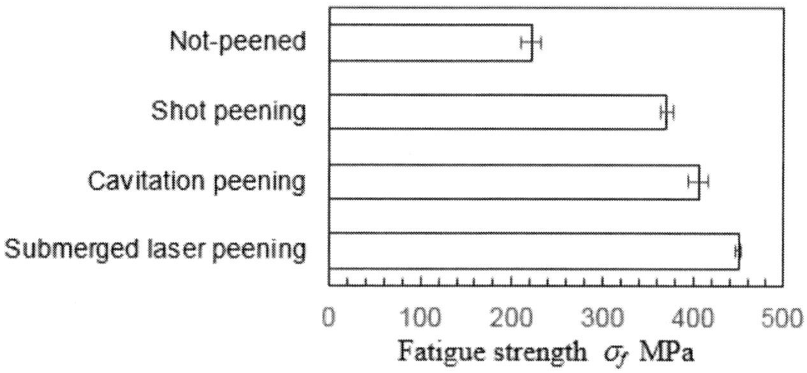

Figure 5. Improvement of fatigue strength of titanium alloy Ti6Al4V manufactured by EBM by means of shot peening, cavitation peening and submerged laser peening.

In order to investigate mechanical properties on the improvement of fatigue strength of light metallic materials by mechanical surface treatment, Figure 6 reveals the relation between the stress intensity factor range ΔK and crack growth rate *da/dn* to show the fatigue property of peened material. Tested material was duralumin JIS A2017-T3. Figure 6 shows the range of $10^{-8} < da/dn < 10^{-6}$, in which Paris law was observed. As shown in Figure 6, peened data were shifted to right hand side or downward. When the constant ΔK condition was considered, the crack propagation rate was decreased by submerged laser peening LP, cavitation peening CP and shot peening SP. If the constant crack propagation rate was assumed, ΔK was shifted to right hand side by the peening. Note that the approximation line of non-peened specimen was described by the following Eq. (1).

$$\frac{da}{dn} = 1.155 \times 10^{-11} \Delta K^{3.7622} \tag{1}$$

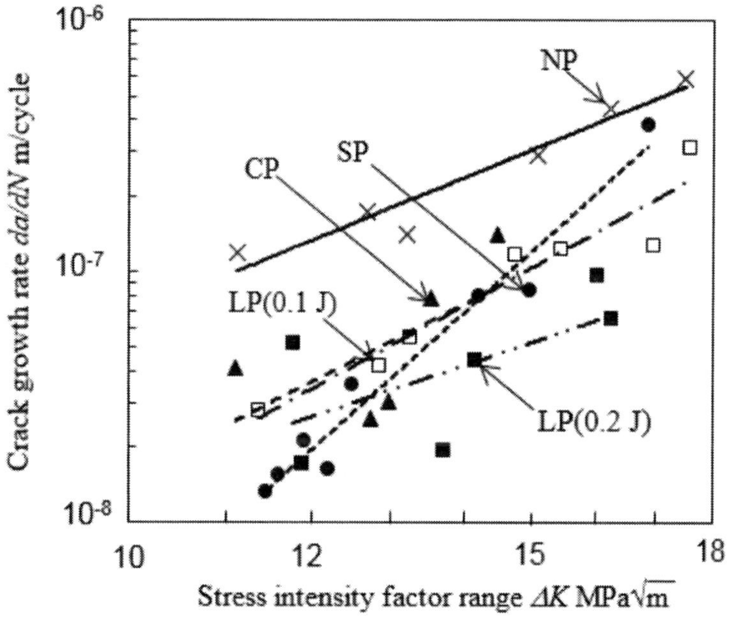

Figure 6. Improvement of fatigue property treated by submerged laser peening LP, cavitation peening CP and shot peening SP comparing with non-peened NP by describing the relation between stress intensity factor range and crack growth rate (Duralumin plate JIS A2017-T3).

Table 1 shows mechanical properties of treated surface by shot peening SP, cavitation peening CP, submerged laser peening LP and hybrid peening HP with various peening intensity κ, comparing of non-peened NP. In the present paper, the hybrid peening is combined peening with different peening methods. For example, HP of LP0.4 →CP0.4 means that 1st peening was submerged laser peening at $\kappa = 0.4$ m^{-1}

and 2nd peening was cavitation peening at $\kappa = 0.4$ m^{-1}. As shown in Table 1, compressive residual stress was introduced by peening methods and the Vickers hardness was increased. When the peening intensity was increased, the maximum height of roughness profile was increased and it seems to be saturated. As grain size D was decreased, the yield stress was increased with Hall-Petch law. And, the full width at half maximum β of X-ray diffraction pattern was increased with $1/D$. Thus, β was used as the parameter of inverse of grain size. As shown in Table 1, β of shot peening was drastically increased with the increase of peening intensity. In the case of cavitation peening, β was also increased with the increase of peening intensity. On the other hand, β of laser peening was slightly increased with the increase of peening intensity, as submerged laser peening gave the thermal effect to the surface. As shown in Table 1, the number of cycles to failure $N_{f\ exp}$ was roughly increased by peening methods. The $N_{f\ exp}$ of submerged laser peening at $\kappa = 0.6$ of 0.2 J was 4.5 times larger than that of non-peened specimen. The $N_{f\,exp}$ of hybrid peening at LP0.4→CP0.4 and CP0.4→SP0.6 are 5 and 5.2 times larger than that of non-peened specimen. Namely, $N_{f\,exp}$ of the hybrid peening at optimum condition is larger than that of single peening at optimum condition.

Table 1. Mechanical properties of peened and non-peened specimens.

Peening method	Peening intensity m^{-1}	Residual stress $_R$ MPa	Vickers hardness *Hv*	Maximum height of roughness profile *Rz* m	Full width at half maximum β deg	Number of cycles to failure $N_{f\ exp}$
NP	0	-20	130	7.55	1.80	21,423
SP	0.2	-175	150	19.5	1.89	16,629
	0.4	-212	149	35.2	1.98	22,733
	0.6	-275	167	26.7	2.08	61,634
	0.8	-248	177	29.1	2.33	69,817
CP	0.2	-204	142	17.4	1.89	14,620
	0.4	-214	154	35.9	1.91	18,848
	0.6	-288	154	38.5	1.95	65,645
	0.8	-297	160	56.7	1.98	18,149
LP	0.2	-97	185	14.8	1.90	19,468
	0.4	-222	171	13.8	1.91	76,852
	0.6 (0.1J)	-212	241	15.9	1.90	74,822
	0.6 (0.2J)	-205	220	19.6	1.91	95,883
	0.8	-211	278	19.0	1.92	39,620
HP	LP0.4→CP0.4	-238	325	21.9	1.92	107,533
	CP0.4→LP0.4	-271	207	50.2	1.96	58,208
	SP0.6→CP0.4	-241	138	44.8	2.06	44,659
	CP0.4→SP0.6	-219	147	32.1	2.11	112,331

4. Discussion

In order to key factors on the improvement of fatigue strength of light metallic materials by mechanical surface treatments, the relation between the fatigue life and the mechanical properties was investigated using the data of Figure 6 and Table 1. As Figure 6 was obtained by the test at $\sigma_a = 200$ MPa, Eq. (1) is expressed by following Eq. (2) at $\Delta K = 18$, as $\Delta K = F\ \sigma_a \sqrt{\pi\ a}$. Here, F is the constant.

$$\frac{da}{dn} = 1.155 \times 10^{-11} \times 0.09^{3.7622} \times {\sigma_a}^{3.7622} = 1.343 \times 10^{-15}\ {\sigma_a}^{3.7622} \qquad (2)$$

When the residual stress σ_R was considered, the compressive residual stress reduces the applied stress σ_a at certain ratio b, thus Eq. (2) is described as follows.

$$\left(\frac{da}{dn}\right)' = 1.343 \times 10^{-15}\ (\sigma_a - b\ \sigma_R)^{3.7622} \qquad (3)$$

As the estimated fatigue life $N_{f\,est}$ is proportional to inverse of Eq. (3), $N_{f\,est}$ can be obtained from Eq. (4), as Hv and β affective positively and Rz affects negatively. The b, c, d and e are constant, and these are obtained by a least square method. Note that the subscript $_{NP}$ means the value of non-peened value and the subscript $_P$ means the value of peened specimen.

$$N_{f\ est} = N_{f\ NP} \frac{1}{1.343 \times 10^{-15} \left(\sigma_a - b\ \sigma_{R\ P}\right)^{3.7622}} \left(\frac{Hv_P}{Hv_{NP}}\right)^c \left(\frac{Rz_{NP}}{Rz_P}\right)^d \left(\frac{\beta_P}{\beta_{NP}}\right)^e \quad (4)$$

The obtained b, c, d and e are 0.331, 1.138, 1.358 and 3.266, respectively. The calculated estimated number of cycles to failure $N_{f\,est}$ is shown in Table 2. Figure 7 reveals the relation between experimental number of cycles to failure $N_{f\,exp}$ and estimated number of cycles to failure $N_{f\,est}$. As the secondary crack was propagated in the peened specimen treated at the condition of $Rz > 30$ μm, the symbol * was marked in Table 2 and the open circles were used for the data. As shown in Fig. 7, $N_{f\,est}$ is roughly proportional to $N_{f\,exp}$. The correlation constant of 11 points except $Rz > 30$ μm was 0.654, and the probability of non-correlation is 3 %. As it was less than 5 %, the relation between $N_{f\,exp}$ and $N_{f\,est}$ is significant. Namely, the proposed equation shown in Eq. (4) can estimate the fatigue life. Thus, it can be concluded that the fatigue life treated by the mechanical surface treatment is affected by the residual stress, surface roughness, Vickers hardness and grain size.

Table 2. Estimated number of cycles to failure $N_{f\,est}$. The symbol * revealed at the condition of $Rz >$ 30 μm.

Peening method	Peening intensity κ m^{-1}	Estimated number of cycles to failure $N_{f\,est}$
NP	0	21,423
SP	0.2	24,761
	0.4	17,785*
	0.6	64,808
	0.8	67,460
CP	0.2	35,283
	0.4	16,345*
	0.6	33,467*
	0.8	24,070*
LP	0.2	25,603
	0.4	72,127
	0.6 (0.1J)	79,012
	0.6 (0.2J)	51,396
	0.8	75,187
HP	LP0.4→CP0.4	94,923
	CP0.4→LP0.4	27,700*
	SP0.6→CP0.4	17,690*
	CP0.4→SP0.6	26,213*

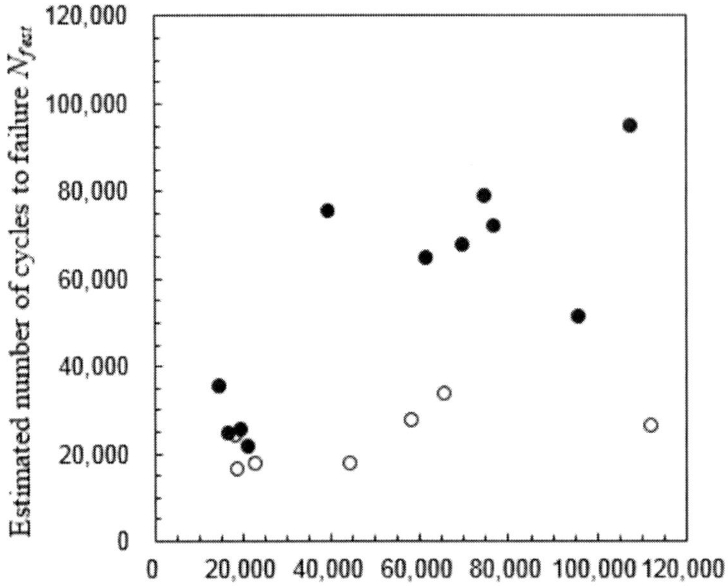

Figure 7. Relation between experimental number of cycles to failure $N_{f\,exp}$ and estimated number of cycles to failure $N_{f\,est}$. The open circle revealed the data of $Rz > 30$ μm.

5. Conclusions

In order to demonstrate the improvement of fatigue strength of additive manufactured metallic materials by mechanical surface treatment using pulse laser, titanium alloy Ti6Al4V manufactured by electron beam melting EBM was treated by submerged laser peening and tested by a plane bending fatigue test. And also the duralumin plate A2017-T3 was treated by the submerged laser peening, cavitation peening and shot peening in order to make clear the mechanical properties on the improvement of fatigue strength of light metallic materials by the mechanical surface treatment. The results obtained can be summarized as follows.

(1) The fatigue strength of Ti6Al4V manufacture by EBM at $N = 10^7$ was improved by 104 % using submerged laser peening. Namely, mechanical surface treatment using solid-liquid-gas interfacial phenomena induced by pulse laser can improve the fatigue strength of the additive manufactured metallic material.
(2) The other mechanical surface treatments such as cavitation peening using a submerged water jet and shot peening can also improve the fatigue strength of Ti6Al4V manufacture by EBM.
(3) The estimation method for the fatigue life of light metallic materials enhanced by the mechanical surface treatments was proposed. It was revealed that main factors on the improvement of the fatigue life by the mechanical surface treatment were residual stress, surface roughness and grain size.

Acknowledgements

This work was partly supported by JSPS KAKENHI Grant Number 17H03138 and 18KK0103.

References

[1] Edwards P, O'Conner A, Ramulu M 2013 Journal of Manufacturing Science and Engineering-Transactions of the ASME vol 135 pp 1-7.
[2] Edwards P, Ramulu M 2014 Materials Science and Engineering A-Structural Materials Properties Microstructure and Processing vol 598 pp 327-337.
[3] Sato M, Takakuwa O, Nakai M, Niinomi M, Takeo F, Soyama H 2016 Materials Sciences and Applications vol 7 pp 181-191.
[4] Soyama H, Okura Y 2018 AIMS Materials Science vol 5 pp 1000-1015.

[5] Sanders D, Soyama H, https://asm.confex.com/asm/aero19/webprogram/Paper47973.html, last accessed 2019/04/18.
[6] Soyama H 2017 International Journal of Peening Science and Technology vol 1 pp 3-60.
[7] Soyama H 2019 Journal of Materials Processing Technology vol 269 pp 65-78.
[8] Telang A, Gill A S, Teysseyre S, Mannava S R, Qian D, Vasudevan V K 2015 Corrosion Science vol 90 pp 434-444.
[9] Kattoura M, Mannava S R, Qian D, Vasudevan V K 2017 International Journal of Fatigue vol 104 pp 366-378.
[10] Sano T, Eimura T, Kashiwabara R, Matsuda T, Isshiki Y, Hirose A, Tsutsumi S, Arakawa K, Hashimoto T, Masaki K, Sano Y 2017 Journal of Laser Applications vol 29 pp 7.
[11] Prabhakaran S, Kulkarni A, Vasanth G, Kalainathan S, Shukla P, Vasudevan V K 2018 Applied Surface Science vol 428 pp 17-30.
[12] Soyama H, Lichtarowicz A, Momma T, Williams E J 1998 Journal of Fluids Engineering vol 120 pp 712-718.
[13] Soyama H, Sekine Y, Saito K 2011 Measurement vol 44 pp 1279-1283.
[14] Naito A, Takakuwa O, Soyama H 2012 Materials Science and Technology vol 28 pp 234-239.
[15] Kokubun T, Soyama H 2018 Transactions of the JSME vol 84 pp 18-00161-00161-00114.
[16] Little R E 1972 ASTM STP vol 511 pp 29-42.

High Temperature Deformation and Microstructure Evolution Behaviours of 7075 Cast Aluminum Alloy under Multi-pass Loading

Ranyang Zhang[1,2], Gangyao Zhao[1,*], Huan Yu[1,2], Gewang Shuai[1], Longtao Wu[1], Chenbo Jin[1], Long He[1] and Chunshuang Kang[1]

[1]School of Aeronautical Manufacturing Engineering, Nanchang Hangkong University, Nanchang 330063, China
[2]AECC Beijing Institute of Aeronautical Materials, Beijing 100095, china

*Corresponding author

Abstract. To analyse the effects of loading pass and heating temperature on the deformation behaviour and microstructure evolution of 7075 cast aluminium alloy, multi-pass high temperature loading experiments on Gleeble-3500 thermal simulation compressor were carried out. The results indicated that under each loading, all the true stress-strain curves are typical dynamic recrystallization curves. As strain increases, the stress increases sharply and then basically remains unchanged when it reaches a maximum value. As temperature rise, the peak value of flow stress slightly decreases when the temperature is less than 350°C, whereas it increases slightly when the temperature is larger than 350°C. The grain was elongated perpendicular to the axis. Only a small amount of fine recrystallized grains were found to nucleate and grow at grain boundaries at a lower temperature (250°C). As temperature increased, the recrystallized grain gradually increases, and its proportion gradually also increases until the recrystallization is basically completed. The achievements can provide a guideline to select processing parameters for the processes of metal forming of 7075 cast aluminium alloy, such as ingot blooming, spinning and extrusion.
Keywords: flow stress; hot compress deformation; 7075 cast aluminium alloy; multi-pass loading; microstructure.

1. Introduction

7075 high-strength aluminium alloy (HSA) thin-walled parts have been widely used in high-tech fields such as aviation, aerospace and weaponry because of their light weight, high strength and good corrosion resistance[1,2]. The forming process that the thin-walled parts can be directly produced by the plastic forming from a cast blank can overcome the shortcomings of the traditional manufacturing method of such parts, such as long cycle, low material utilization and high-cost [3]. However, this process of cast HSA is a complex process of multi-field and multi-factor coupling, which lead material to undergo a particularly complex process of inhomogeneous plastic deformation and microstructure evolution. Thus, various macro and micro defects, such as coarse grain, inhomogeneous structure, cracking and peeling, are easily produced during the process if the forming process is not properly controlled.

In the multi-pass loading thermal compression test process of cast HSA, hardening and softening processes, for example, dynamic recovery and dynamic recrystallization are carried out at the same time, which affect the final properties of the material [4]. As a result, how to predict the deformation and microstructural evolution behaviours of 7075 cast HSA rapidly and accurately has become urgent for the development of the process.

Researches on the prediction of deformation and microstructural evolution behaviours of 7075 cast HSA during a hot compression test process have been made for a long time. However, most of them are focused on forging HSA [5,6], and the research on cast HSA has been still scarce. For this reason, the deformation behavior and microstructural evolution of 7075 cast HSA at the high temperature are analysed by the multi-pass hot compression test. The achievements can provide a guideline to select processing parameters for the processes of metal forming of 7075 cast aluminium alloy, such as ingot blooming, spinning and extrusion.

2. Material and Method

The raw material used in the experiment is a semi-continuous cast tube of 7075 aluminium alloy. The size of the tube is ϕ170mm ×10 mm ×180 mm. The original microstructure is shown in Figure 1. And the grain size was measured by linear intercept method.

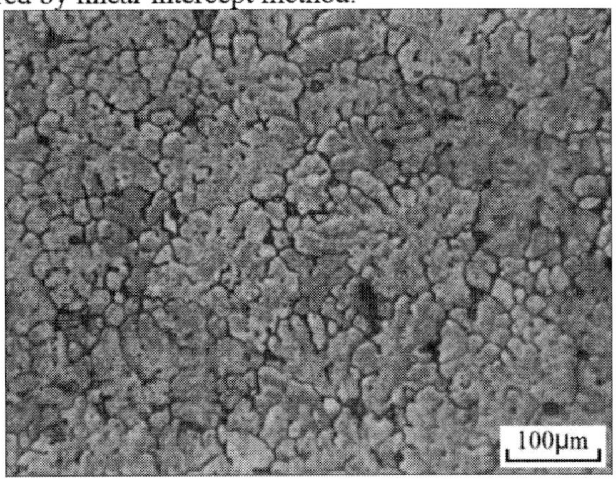

Figure 1. Original cast structure of 7075 cast aluminum alloy

The cylindrical compression specimen with the size of ϕ10 mm x 15 mm was cut out from the 7075 cast aluminium alloy tube by the ire-electrode cutting.

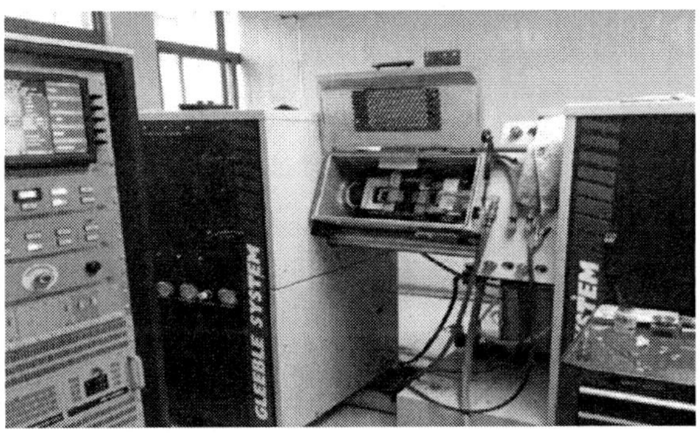

Figure 2. Gleeble-3500 thermal simulator

The quasi-static isothermal compression tests under multi-pass loading were carried out on the Gleeble-3500(see figure 2) thermal simulator at the temperatures of 250, 300, 350, 400 and 450℃ and the compression deformation 20%, 40%, 60% and 80%, respectively. And the strain rate is 0.01s^{-1}.

IOP Conf. Series: Materials Science and Engineering **611** (2019) 012003 doi:10.1088/1757-899X/611/1/012003

3. Result and Discussion

3.1. High Temperature Deformation Behavior of 7075 Cast Aluminum Alloy under Multi-pass Loading

3.1.1. Stress-strain Characteristics under Multi-pass Loading. Multi-pass hot compression experiments of 7075 cast HSA were carried out with a Gleeble 3500 thermal simulator. The multi-pass hot deformation behavior of 7075 cast aluminium alloy was obtained, as shown in Figure 3. From the figure, we can find that the stress raises fast as strain increased in the initial stage of each pass, in addition the increasing rate increases with the increase of loading passes. Moreover, when the stress reaches a maximum value, it does not change with the increase of strain. When the temperature is low (250°C) at pass 1, the true stress-strain curve does not show obvious softening characteristics. As the loading pass and temperature increased, the softening characteristics of the true stress-strain curves are becoming more and more obvious. The reason is that at higher heating temperature and loading pass, the dynamic recrystallization of the material is obvious, and the trend of dynamic recrystallization becomes more and more obvious with the increase of temperature and the deformation.

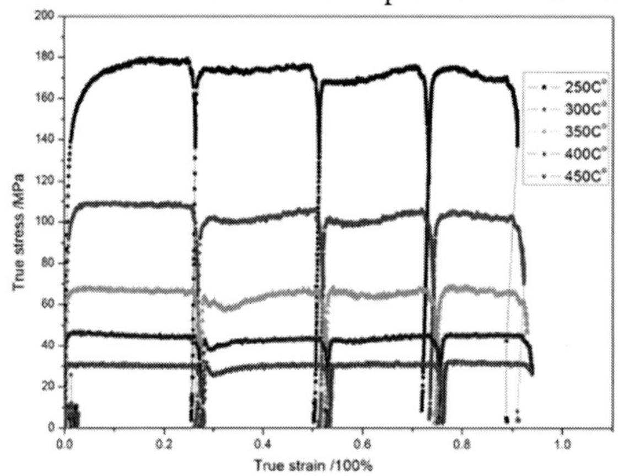

Figure 3. True stress-strain curves of 7075 cast HSA under multi-pass loading

3.1.2. Variation of Peak Stress with Loading Passes. The change regularity of peak stress with loading passes under different temperature is shown in figure 4. As can be seen from the figure, the peak stress slightly decreases as loading passes increased at relatively low temperature (<300°C), and the decreasing trend of it is gradually weakening with the increase of temperature. However, the peak stress slightly increase along with loading passes increasing at relatively high temperature (300°C<).

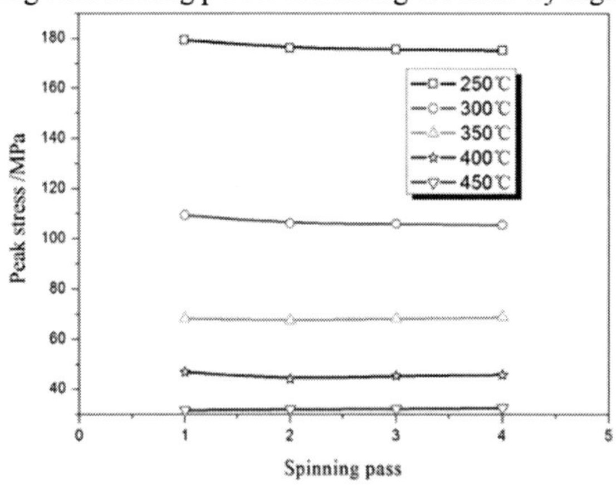

Figure 4. Variation of peak stress with loading passes

3.1.3. Variation of Peak Stress with Heating Temperature. The change law of peak stress with the heating temperature under different loading passes is shown in figure 5. From the figure, it can be seen that with deformation temperature increasing, the value of peak stress deceases gradually with heating temperature increasing, while the effect of loading passes on it can be neglected.

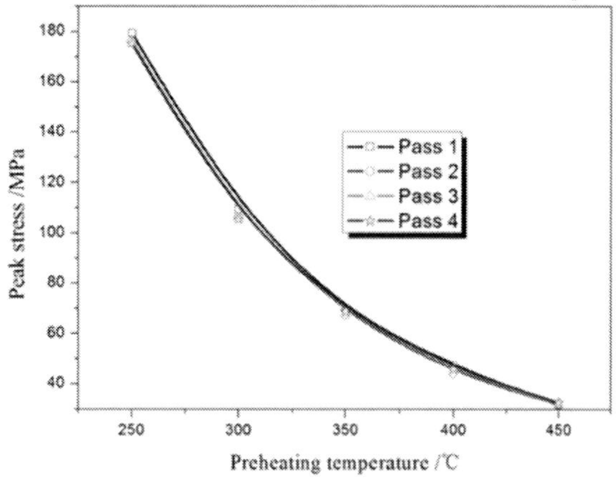

Figure 5. Curves of peak stress with preheating temperature

3.2. Microstructure Evolution Behavior of 7075 Cast HSA under Multi-pass Loading

3.2.1. Effect of Heating Temperature on Microstructure Morphology. Figure 1 shows the initial microstructures of 7075 cast HSA before uncompressing. It can be seen from the figure that the initial microstructures of 7075 cast HSA are coarse equiaxed dendrites; the average grain size measured is about 108.6μm.

Figure 6 shows the microstructures of the 7075 cast HSA at different deformation temperatures. It can be seen from the figure that the grains are elongated perpendicular to the axial direction of the compression specimen. When the heating temperature is below 250℃, the original grains are deformed and elongated, and only a small number of fine recrystallized grains nucleate and grow at grain boundaries (see Fig.6a). With temperature increasing, the elongated grains decreased, while the recrystallized grains and their proportion increased (see Figs.6 b and c). When the temperature reaches 400℃, the recrystallization has been basically completed. And the grains grow up gradually with the increase of the temperature (see Figs. 6 d and e).

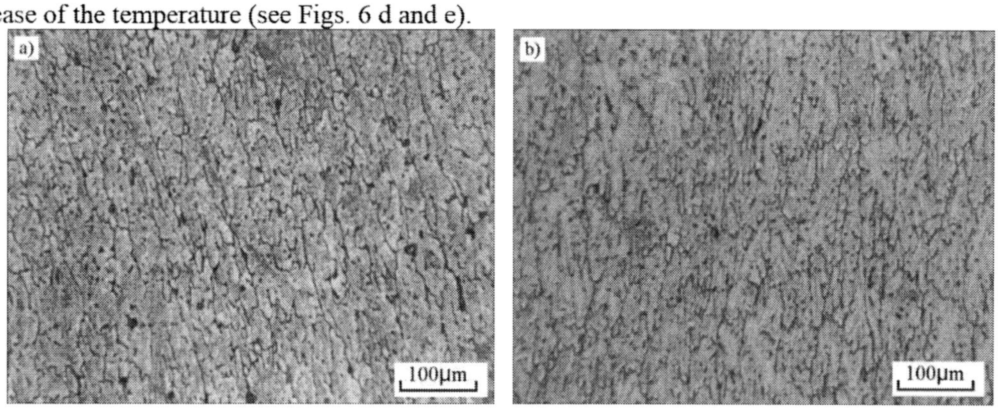

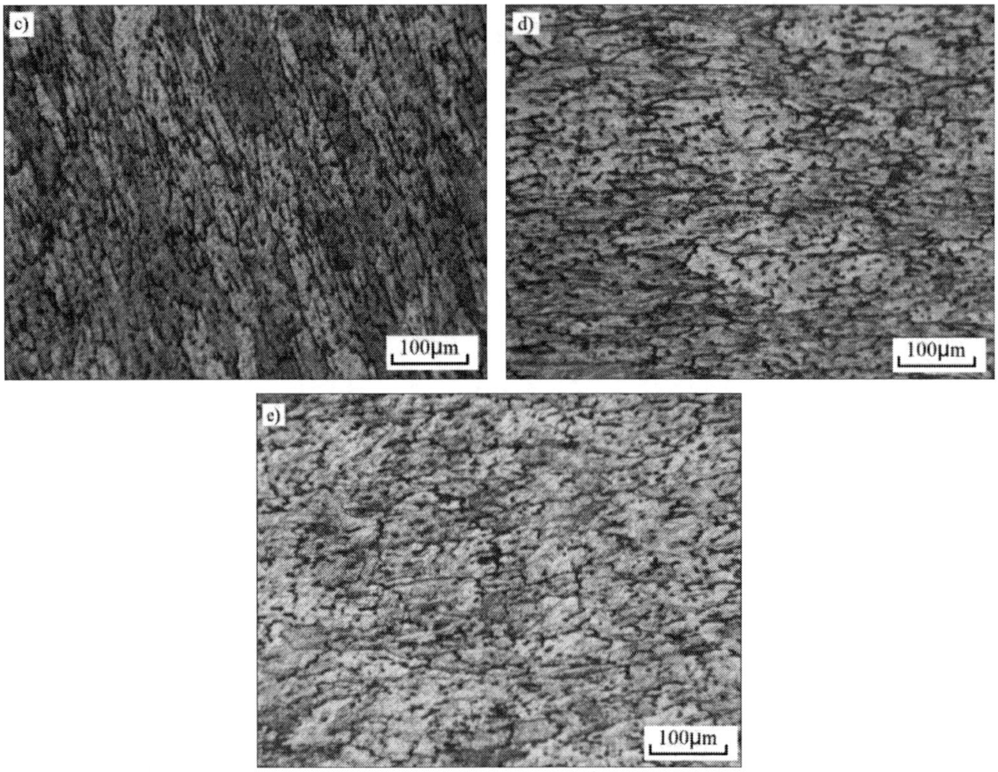

Figure 6. Microstructure of 7075 cast HSA at different deformation temperatures

3.2.2. Effect of Preheating Temperature on Grain Size. The change curve of grain size with temperature after deformation under the multi-pass loading is shown in Figure 7. From the figure, the grain size decreases firstly, and then increase with the increase of deformation temperature. The reason is that during high temperature multi-pass loading compression of 7075 cast HSA, the large plastic deformation and dynamic recovery recrystallization coexist. Large plastic deformation and appropriate dynamic recrystallization can refine grain size. However, with deformation temperature increasing, the grain size growth trend caused by recrystallization is larger than that caused by large deformation. Therefore, the grain size becomes coarser.

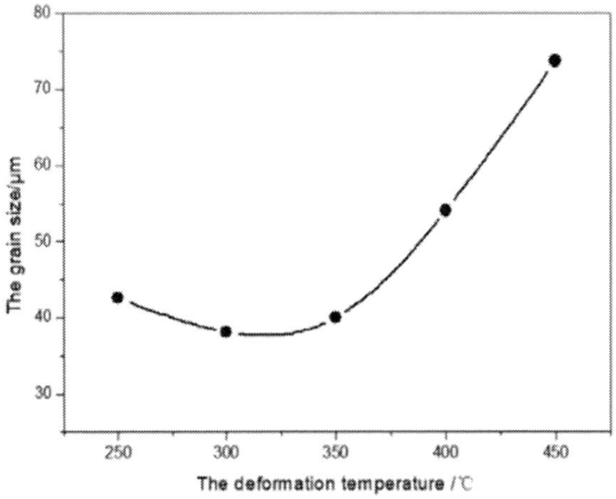

Figure 7. The variation curve of grain size with temperature after deformation under the multi-pass loading

4. Conclusions

1) The stress increases rapidly with strain increasing in the initial stage of each pass, and the increasing rate increases with the increase of loading passes. As loading pass and deformation temperature increase, the softening characteristic of the true stress-strain curve become more and more obvious.
2) As heating temperature increased, the value of peak stress deceases gradually, while the effect of loading passes on it can be neglected.
3) The grain size decreases firstly, and then increases with deformation temperature increasing.
4) During the multi-loading compression process of 7075 cast HSA, large plastic deformation and appropriate dynamic recrystallization can refine grain size.

Acknowledgments

The authors would like to thank the National Natural Science Foundation of China (Nos. 51165037 and 51665041) for the support given to this research.

References

[1] Zhao G Y, Lu C J, Zhang R Y, Guo Z H, Zhang M Y 2017 *Int. J. Adv. Manuf. Technol.* **88** 907
[2] Zhang R Y, Yu H, Zhao G Y 2019 *Int. J. Adv. Manuf. Technol.* **102** 2709
[3] Moria K I, Ishiguro M, Isomura Y 2009 *J. Mater. Process. Technol.* **209** 3621
[4] Zhao G Y, Huang F, Zhang R Y, Zhao P, Zhou Q Q *J. Plasticity Eng.* **24(2)** 64 (in Chinese)
[5] Guo H L, Sun Z C, Yang H 2013. *The Chinese J. Nonferrous Metals* **23(6)**1507 (in Chinese)
[6] Fazeli R and Ghoreishi M 2011 *Int. J. Adv. Manuf. Technol.* **52** 597

IOP Conf. Series: Materials Science and Engineering **611** (2019) 012004 doi:10.1088/1757-899X/611/1/012004

Tribology Effect of Copper Based Composites for High-Speed Train Braking Application

Xiaoming Han, Junying Yang, Rong Fu*, Linlin Su and Fei Gao

Continuous Extrusion Research Center, Dalian Jiaotong University, China

*Corresponding author

Abstract. This paper disclosed the effect of third body on tribological performance of friction pair of copper-matrix materials and H13 steel using a fixed speed test from 3.14 m/s to 31.4 m/s at 0.5MPa. Results showed that the third body increase improved and stabilized friction coefficient. When the third body decreased, the friction coefficient decreased accordingly, especially at lower speed. This was because the particle third body of low-speed was weakly bound to the matrix, so the number of particles was easy to decrease.
Keywords: the third body; copper-matrix materials; friction coefficient.

1. Introduction

As the most commonly used pad material in the braking field of high-speed train, the tribological properties of copper-matrix sintered friction materials have attracted people's attention. The performance of braking is mainly controlled by the composition of the friction material. However, in a set of publications have proved that during braking, the friction surface is worn severely and a layer organization called the third body is formed under dry conditions (1-5). A lot of research has pay attention to the effect of the third body on the friction and wear performance of friction materials (8-15). Eriksson found that the third body includes densification and granulation in organic synthetic brake pads, track running materials and carbon steel materials, which can be transformed into each other (3.6.7.16-17). Experiments showed that the existence, formation and evolution of the third body had a great influence on the product life and braking effect. The third body was thicker, which reduced the wear and friction coefficients of gray cast iron and aluminium-based materials respectively Cho(19) and Uyyuru(18). Scharf showed that the thicker third body stabilized the friction coefficient(15). Xiao pointed out that the greater the thickness of the third body, the smaller the friction coefficient of C/C-SiC material(20).These studies provided a good reference and valuable results for understanding the influence of the third bodies. It was pointed out that the influence of third bodies should be taken into account in the study of friction properties of materials.

For copper-matrix friction materials, by changing the composition, recent investigations studied the effect of surface morphology on friction and wear properties under different pressures and speeds. For example, the effect of different forms of Fe(21-23), copper(24,25), graphite(26,27) and fiber(28) and so on were investigated. At present, there is still a lack of systematic research on the role of third bodies of copper-based materials.

The current paper was based on Sheasby's(29) research method, from the view of controlling the third bodies to investigate copper-matrix friction materials at different speed.

IOP Conf. Series: Materials Science and Engineering **611** (2019) 012004 doi:10.1088/1757-899X/611/1/012004

2. Experiment

The experimental copper-matrix composites contained electrolytic copper, graphite and SiO_2 and so on. The specimens were prepared by powder metallurgy, and the sintering temperature and pressure was 820℃ and 1.77MP, respectively. The chemical composition of the specimens was listed in table 1.

Table 1. The chemical composition of the specimens.

Composition	Cu	Graphite	SiO_2	Else
Weight Percentage(%)	80	10	6	4
Particle size(μ m)	50～80	200～300	200～300	70～100

The size of specimens for tests was Φ17mm×15mm. The surface of the pin was polished with 1000 sand paper, cleaned with acetone, and then pre-polished until it was fully in contact with the pin-on-disc friction gauge. The disc (H13 steel, preferred disc material for high-speed train) with a diameter of 300mm was chosen. The hardness of specimen and disc are 60HB and 308HB, respectively.

The tribological performance of friction pair were tested on a fixed speed test bench (JF150D, jilin university of china)(figure.1). The test normal pressure applied was 0.5MPa (one of the test required pressure for pad by international Union of Railways), and the speeds were 200r/min (3.14 m/s, equivalent to 20Km/h of high-speed train), 500r/min, 1000r/min and 2000r/min.

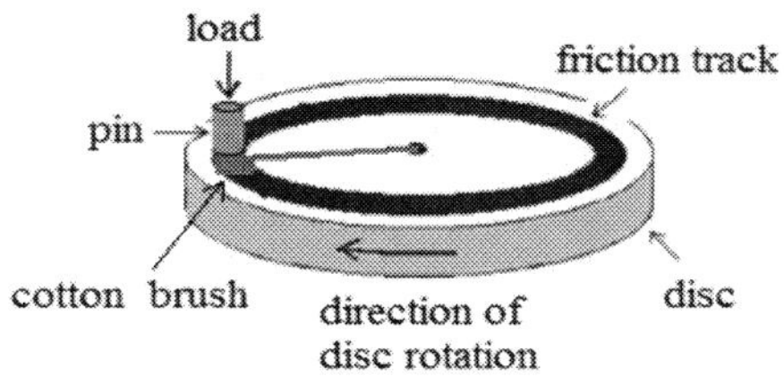

Figure 1. Schematic of the pin-on-disk tester.

Wet cotton brushes were used to reduce the third body, each brushing time was 2 to 3 seconds. In order to ensure the reliability of the results, the test repeated three times under each test condition. The pin was cut to observe the cross section (figure.2). For better understanding, the worn surface was shaped by resin to avoid the third body damage. The surface morphologies were analyzed and compared by optical microscope, confocal scanning laser microscope and SEM with energy dispersive spectroscopy.

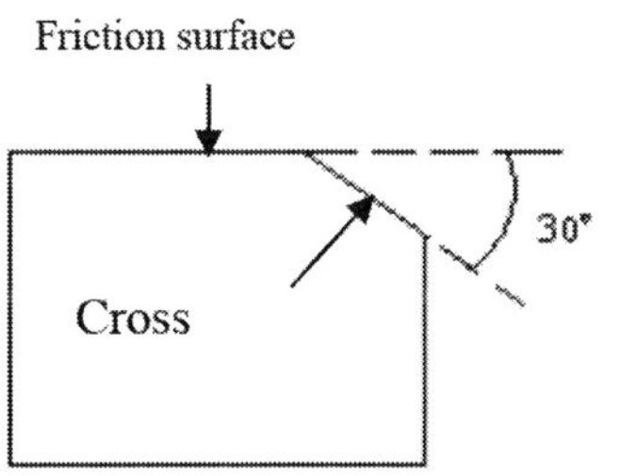

Figure 2. Schematic of pins cross section.

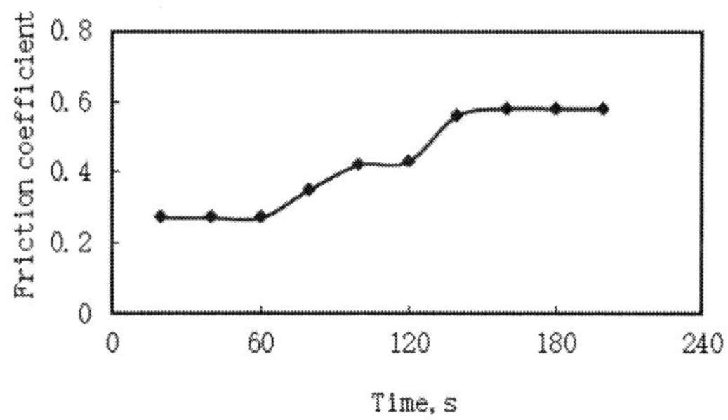

Figure 3. The frictional coefficient with time at speed 200r/min.

3. Results and Discussion

Figure 3 showed the friction coefficient of the friction process at 200r/min. In the initial stage of friction, the friction coefficient was low, then increased gradually and stabilized.

Figure 4 showed the change of surface morphology of copper matrix composites with time at 200r/min at 60s. Along the friction direction, the original surface of copper matrix composites was ploughed

(Figure.4a). There were three kinds of third body of wear surfaces: one was surface fragmentation into fine particles, the other was surface fragmentation into larger size blocks (Figure.4b), and the third was aggregation of fine particles (Figure.4c). This showed that the formation of the third body was beneficial to the increase and stability of the friction coefficient.

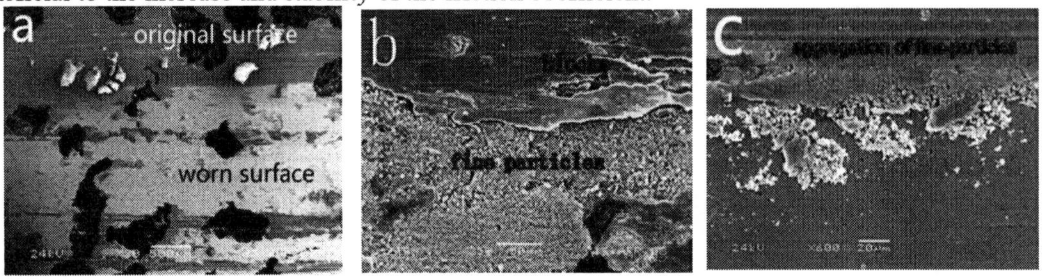

Figure 4. (a) The original surface of copper matrix composites was ploughed (b) Fine particles and blocks (c) Aggregation of fine particles.

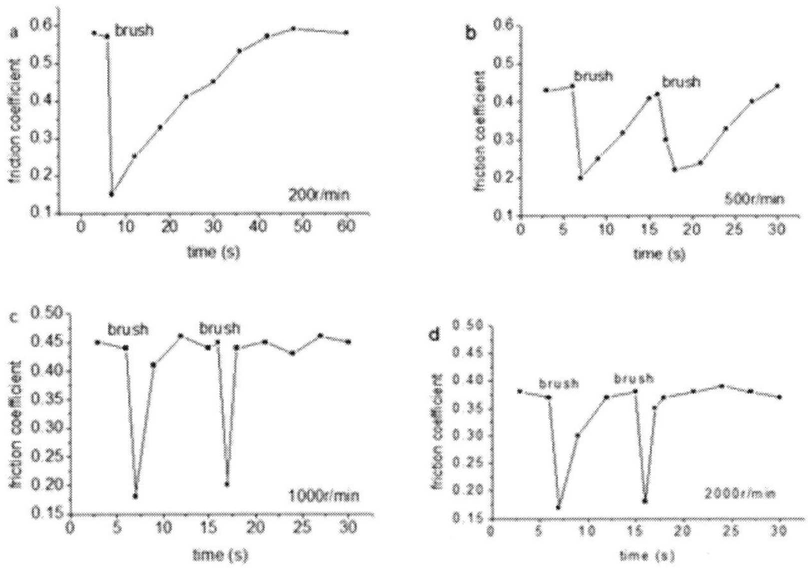

Figure 5. The friction coefficient before and after brushing.

Figure 5 showed the trends of friction coefficient before and after brushing at different speeds. At 200r/min, the friction coefficient of the brush disc immediately decreased from 0.6 to 0.15, and the decrease value was close to 0.4. When the brush stopped, the friction coefficient increased gradually and returned to its original level, which took about 36 seconds (Figure. 5a). At 500r/min, the friction coefficient of the brush disc decreased from 0.45 to 0.2, and the friction coefficient reached the original value after stopping 9-12s (Figure. 5b). When the speed raised to 1000r/min, the friction coefficient decreased 0.26 after brush the disc and then returned to 0.45 spending 3～6s (Figure.5c). When the speed at 2000 r/min, the friction coefficient difference before and after brushing was about 0.2, and the friction coefficient tended to be stable in 3-6s after brushing (Figure. 5d). This meant that, firstly, the brush disc could reduce the friction coefficient, secondly, with the increase of the speed, the decrease of the friction coefficient and the recovery time were reduced.

Figure 6 showed the surface morphology of copper matrix specimens at 200r/min. The surface before brushing was rough. After brushing, the surface was smooth and metallic. Compared with the morphology before brushing, the third-body particles on the worn surface after brushing significantly reduced.

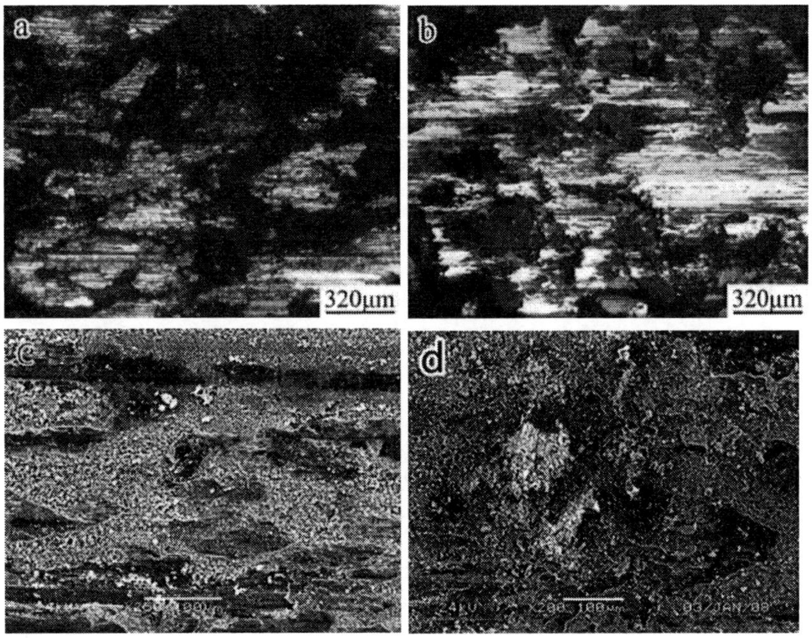

Figure 6. The micrographs of sample friction surface at speed 200r/min (a) Optical micrograph before brushing (b) Optical micrograph after brushing (c) SEM micrograph before brushing (d) SEM micrograph after brushing.

Figure 7 showed the cross-sectional morphology of copper matrix composites before brushing at 200r/min. The dark gray third body covered the worn surface, and the third body thickness was uniform, about 30-50 um. There was no obvious plastic deformation on the worn surface, only some cracks, but a clear boundary between the third body and the matrix.

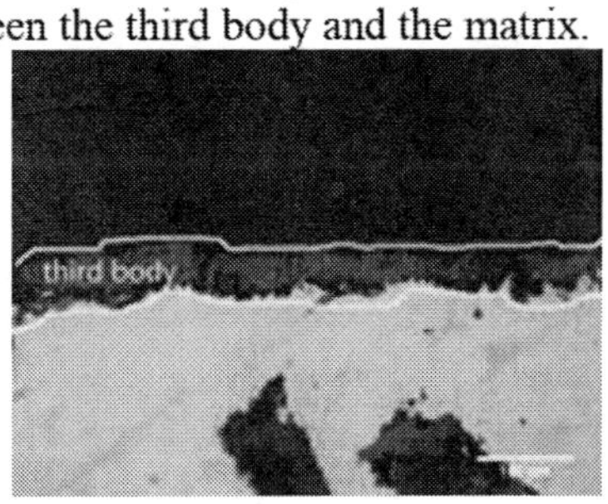

Figure 7. The cross-sectional microphotograph at 200s at speed 200r/min.

The clear boundary explained that the combination between the third body and the matrix was weak. Therefore, the fine third body particles were easy to be brushed off. Firstly, the actual contact area of the worn surface decreased with the decrease of the fine third body particles. Secondly, the number of the third body reduced, leading to the weakening or even disappearance of its function of separating the two first objects. Therefore, the friction coefficient decreased accordingly (Figure. 5a).

At 2000r/min, the worn surface before brushing was covered by the dense third body strips shown in different colors (Figure.8). The size of each strip was more than 100 μm perpendicular to the friction direction, while it was longer along the friction direction . The corresponding EDS spectra of the subsurface showed that the deep grey area was mainly iron (Figure.8c), and the gray area was mainly copper (Figure.8d).

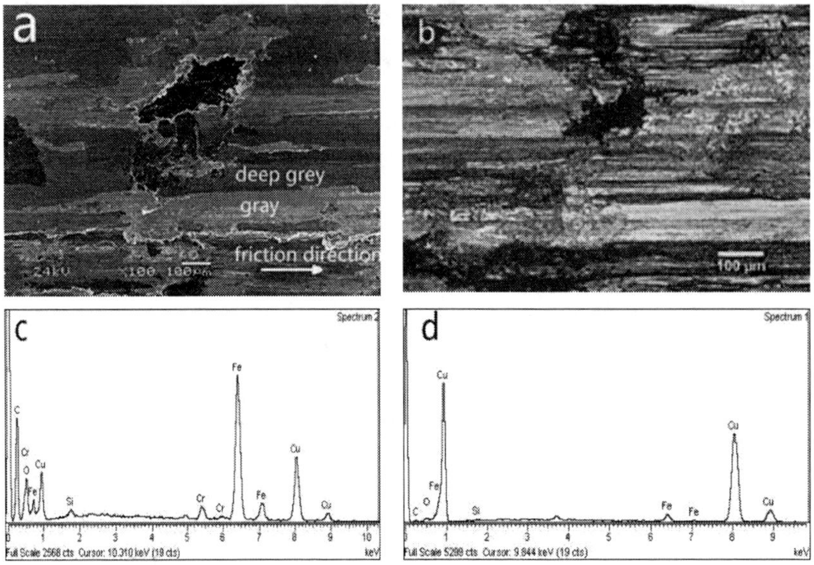

Figure 8. The worn surface of copper-matrix material before brushing at speed 2000r/min (a) SEM micrograph (b) Optical micrograph (c) The EDS spectra of deep grey (d) The EDS spectra of gray

The cross-sectional micrograph (Figure.9) showed that serious plastic deformation occurred underground, and the thickness of the third body was uneven, ranging from 0 to 100 μm. Unlike that at low speed, the third body was tightly bounded to the matrix at high speed.

Figure 9. The cross-sectional microphotograph at speed 2000r/min

Figure 10 showed the worn surface and spectrum after brushing at 2000 r/min. Compared with the surface morphology before brushing, the morphology and composition of the third body did not change significantly. It showed that the third body on the worn surface was not cut off obviously by brushing. Therefore, the friction coefficient restored to its original value quickly (Figure. 10d).

Figure 11 showed the area ratio of the blocks and aggregations. When the friction speed was 200r/min, the area of the blocks and aggregations was about 30%, the rest were fine particles. When the friction speed was 500r/min, the area ratio reached 80%. As the friction speed continued to increase, the area ratio had almost no change. This result indicated that the tribological properties were mainly controlled by the fine third body particles at low speed, while at high speed, the friction performance was mainly determined by the third body blocks and aggregations.

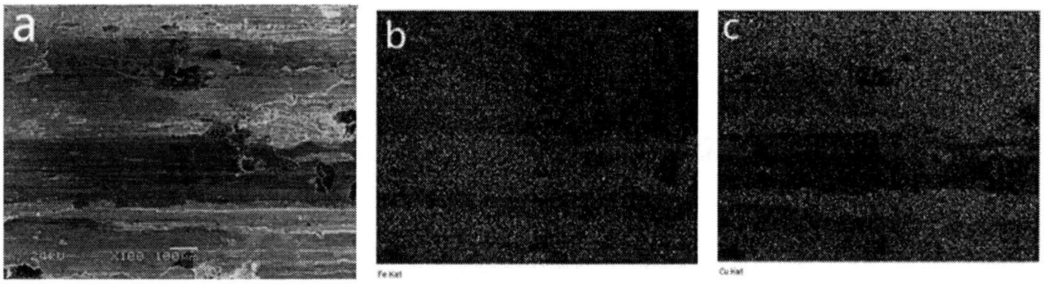

Figure 10. (a)The micrographs of friction surface at speed 2000r/min after brushing (b) The iron distribution map (c) The copper distribution map.

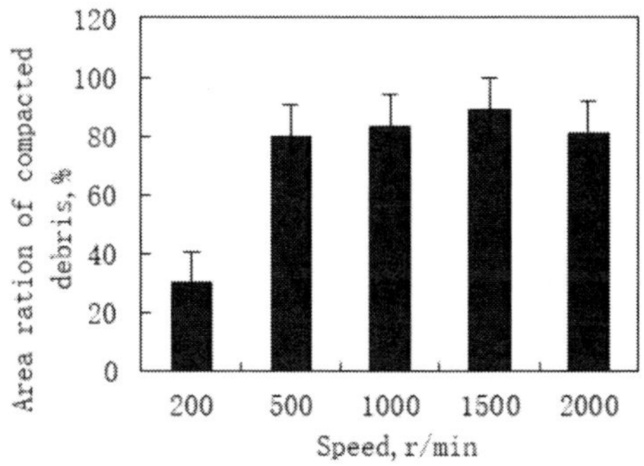

Figure 11. The effect of speed on blocks and aggregations area rations.

4. Conclusions

Based on the investigation of the effect copper based composites for high-speed train braking application, the following conclusions can be drawn:

- For copper-based friction materials, there were three kinds of third body: fine particles, larger size blocks and aggregations. The friction coefficient and its stability could be improved by the formation of third body.
- At low speed, fine particles played an important role in dominating friction and wear properties, while at high speed, the friction and wear properties were determined by the blocks and agglomerate third bodies.
- With the decrease of the number of third bodies, the friction coefficient decreased, especially at lower speeds. Compared with the high-speed accumulation and agglomeration, the low-speed particles had weak bonding with the matrix and were easy to reduce the number of particles.

Acknowledgments

This work was financially supported by the National Key Research and Development Program of China (No. 2016YFB0301403), Liaoning Natural Science Foundation (No. 20170520374), and the Doctoral research start-up fund of Liaoning Province Science and Technology Department of China (No. 20170520266).

References

[1] Osterle W, Klob H and Dmitriev A I Friction Control During Automotive Braking: Experimental Observations and Simulation at Nanometre Scale *Tribology-Materials, Surface & Interface,* 2013, p196-202

[2] Holinski R Hesse D Changes at Interfaces of Friction Components During Braking *Automobile Engineering,* 2003, vol217, p765-770
[3] Eriksson M and Jacobson S Tribological Surface of Organic Brake Pads *Tribology International,* 2000, vol33, p817-827
[4] Ozturk B, Zturk O S and Adiguzel A A Effect of Type and Relative Amount of Solid Lubricants and Abrasives on The Tribological Properties of Brake Friction Materials *Tribol. Trans.* 2013, vol 56, p428–441
[5] Godet M The Third Body Approach: A Mechanical View of Wear *Wear,* 1984, vol100, p437-452
[6] Berthier Y, Descarte S and Busquet M Fatigue Fracture Material Structure *Blackwell Publishing Ltd.* 2004,vol27, p423–436
[7] Majcherczak D, Dufrenoy P and Berthier Y Experimental Thermal Study of Contact with Third Body *Wear,* 2005, vol261, p467–476
[8] Jacko M G The Role of Friction Film in Friction *Wear,* 1991, vol146, p89-97
[9] Descartes S and Berthier Y Rheology and Flows of Third Bodies: Background and Application to An MoS1.6 Coating *Wear,* 2002, vol252, p546-556
[10] Dmitriev I A, Werner O, Heinz K and Guillermo O G A Study of Third Body Behaviour under Dry Sl iding Conditions Comparison of Nanoscale Modelling with Experiment *Estonian Joumal of Engineering.* 2012, vol18, p270-278
[11] Colas G, Saulot A, Godeau C, Michel Y and Berthier Y Decrypting Third Body Flows to Solve Dry Lubrication Issue – MoS2 Case Study under Ultrahigh Vacuum *Wear,*2013, vol305, p192–204
[12] Ghatu S, Alex D C and Maarten P B Evolution of Wear Characteristics and Frictional Behavior in MEMS Devices *Tribol. Lett.,* 2011, vol41, p177–189
[13] Uruea M J, Pitenis A A, Harris L K and Sawyer W G Evolution and Wear of Fluoropolymer Transfer Films *Tribol. Lett.* ,2015, vol57, p177–189
[14] Krick A B, Ewin J J, Blackman S G, Junk P C and Sawyer W G Environmental Dependence of Ultra-Low Wear Behavior of Polytetrafluoroethylene (PTFE) and Alumina Composites Suggests Tribochemical Mechanisms *Tribol. Int.* 2012, vol51, p42–46
[15] Scharf W T and Singer I L Monitoring Transfer Films and Friction Instabilities with In Situ Raman Tribometry *Tribol. Lett.,* 2003, vol14, p3-8
[16] Österle W, Dörfel I, Prietzel C and Rooch H A Comprehensive Microscopic Study of Third Body Formation at The Interface Between A Brake Pad and Brake Disc During The Final Stage of A Pin-on-Disc Test *Wear,*2009, vol267, p781–788
[17] Wirth A and Eggleston D A Fundanmental Tribochemical Study of The Third Body Layer Formed During Automotive Friction Braking *Wear,*1994, vol179, p75-81
[18] Uyyuru R K, Surappa M K and Brusethaug S Tribological Behavior of Al–Si–SiCp Composites/Automobile Brake Pad System Under Dry Sliding Conditions *Tribology,*2007,vol40, p365-373
[19] Cho M H, Cho K H, Kim S J, Kim D H and Jang H The Role of Transfer Layers on Friction Characteristics in The Sliding Interface Between Friction Materials Against Gray Iron Brake Disks *Tribol. Lett.,* 2005,vol20, p101-108
[20] Xiao P, Xiong X and Ren Y Y Effect of Braking Speed on Friction Properties of C/C-SiC Composites *Tribology,*2006, vol26, p12-17
[21] Xiong X, Chen J, Yao P P, Li S P andHuang B Y Friction and Wear Behaviors and Mechanisms of Fe and SiO2 in Cu-Based P/M friction Materials *Wear,*2007,vol262, p1182–1186
[22] Yao P P, Sheng H C, Xiong X and Huang B Y Worn Surface Characteristics of Cu-Based Powder Metallurgy Brake Materials for Aircraft *Transactions of Nonferrous Mzterials Society of China,*2007, vol17, p99-103
[23] Zhang P, Zhan g L, Fu K X and Cao J W Effects of Different Forms of Fe Powder Additives on The Simulated Braking Performance of Cu-Based Friction Materials for High-Speed Railway Trains *Wear,* 2018,vol414-415, p317-326
[24] Gao F, Fu R and Song B Y Effects of Speed Sequence on Friction Properties of Sintered Cu–SiO2 *Tribology-Materials,Surface&Interface,* 2013, p37-40

IOP Conf. Series: Materials Science and Engineering **611** (2019) 012004 doi:10.1088/1757-899X/611/1/012004

[25] Kovalchenko A M, Fushchich O I and Danyluk S The Tribological Properties and Mechanism of Wear of Cu-Based Sintered Powder Materials Containing Molybdenum Disulphide and Molybdenum Diselenite under Unlubricated Sliding Against Copper *Wear,*2012,vol 290, p106-123

[26] Chen B M, Bi Q L, Yang J, Xia Y Q and Hao J C Tribological Properties of Solid Lubricants(Graphite, h-BN) for Cu-Based P/M Friction Composites *Tribology international,*2008, vol41, p1145-1152

[27] Cao H Q, Qian Z Y, Zhang L, Xiao J K and Zhou K C Tribological Behavior of Cu Matrix Composites Containing Graphite and Tungsten Disulfide *Tribol. Trans.,* 2014, vol57, p1037–1043

[28] Ho S C, Lin J H and Ju C P Effect of Fiber Addition on Mechanical and Tribological Properties of A Copper/Phenolic-Based Friction Material *Wear,*2005,vol258, p861-869

[29] Sheasby J S and Vandergeest J H Debris Control in Dry Wear Testing *Wear,* 1981, vol73, p283-294

A Study on Microstructure and Mechanical Properties of AZ31 Magnesium Alloy after Constrained Groove Pressing

Pham Thi Thuy[1,2], Dang Thi Hong Hue[1], Dao Minh Ngung[1] and Pham Quang[1,*]

[1]Hanoi University of Science and Technology, No.1 Daicoviet, Hanoi, Vietnam
[2]Hanoi University of Mining and Geology, 18 Phovien, Bactuliem, Hanoi, Vietnam

*Corresponding author

Abstract. This study the AZ31 magnesium alloy trips of the dimension 60×60×3 *mm* was used for constrained groove pressing (CGP) after 4 cycles under elevated temperature condition. Microstructure and mechanical properties were analyzed for a material with ultra-fine grained structure prepared by CGP of AZ31 magnesium alloy. Analysis of the scanning electron microscope (SEM) micrographs observations showed significant drop of the grain size from (50 ÷ 60) μm to (4 ÷ 6) μm after four CGP cycles. The minimal grain size of (1÷1.5) μm was detected in the regions of the highest deformation. Moreover, the tensile test along the x and z directions showed excellent high strength (325 MPa) and also plastic elongation of 35 %. It is shown that the CGP technique is a good tool for the grain refinement and improving mechanical properties of magnesium alloys.
Keywords: severe plastic deformation; constrained groove pressing; magnesium alloy; microstructure; SEM; mechanical properties.

1. Introduction

Due to its significantly high durability, ultrafine grained (UFG) materials based on metals, light alloys have been strongly developed for applications in the automotive and aircraft industries to make finishing parts. The evolution of microstructure to improve the mechanical properties of metals such as hardness and ductility in plastic deformation was extensively studied several decades ago [1-3]. It has been confirmed that deformation at room temperature (or warm) makes the average particle size of metals drop significantly. Severe plastic deformation (SPD), developed in the 1970s by Russian scientists is an effective method for creating ultrafine grained structure materials.[4] During the process of the SPD method, large hydrostatic pressure is used to reduce lattice defects of metals. Severe plastic deformation method is defined as the process of forming metal with very high deformation can reduce particles during processing. The particle size of the material after being fabricated by this method is usually less than 1 μm, depending on which SPD technique is used. These methods include: Equal Channel Angular Pressing (ECAP, Mul-ECAP) [5-10], High Pressure Torsion (HPT) [11].

A severe plastic deformation method with great potential for the production of ultrafine materials has been developed as a CGP [12,13]. In 2001, Zhu undertook a SPD method based on the method of squeezing a flat metal plate with a cyclic groove die and then restoring the original planar shape to the flat die. Repetition of this process repeatedly causes large plastic deformation to accumulate in the sample without significantly altering the initial size (Figure 1). One of the main benefits of the CGP technique is the improvement of mechanical properties without any change in the size of the material.

In this study the AZ31 magnesium alloy was used for CGP. The investigation of the strain homogeneity along the material was done by the tensile test. The SEM were evolution observation of microstructure after CGP.

2. Experimental Process

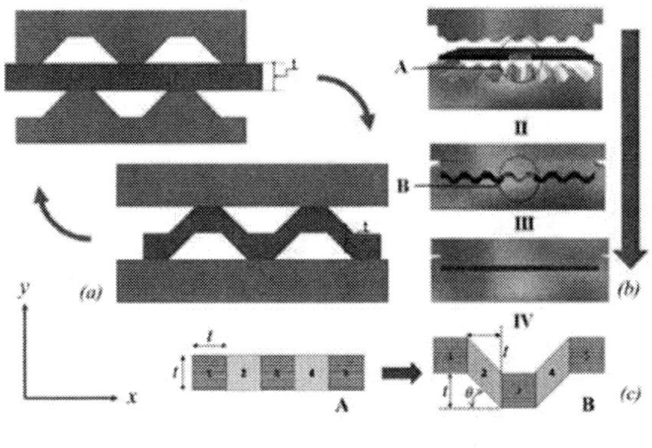

Figure 1. Schematic representation of CGP process

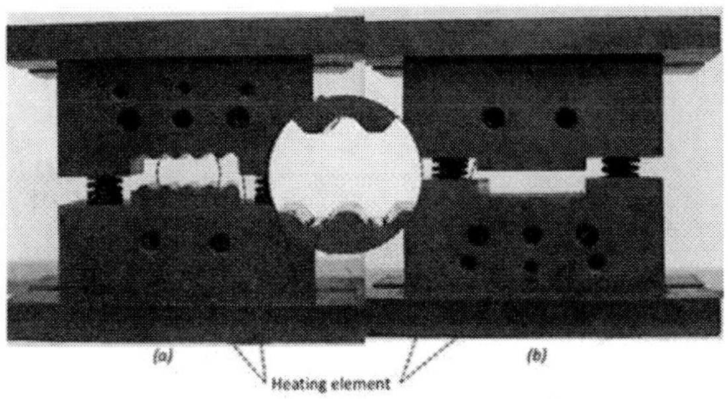

Figure 2. Designed grooved die (a), flat die (b)

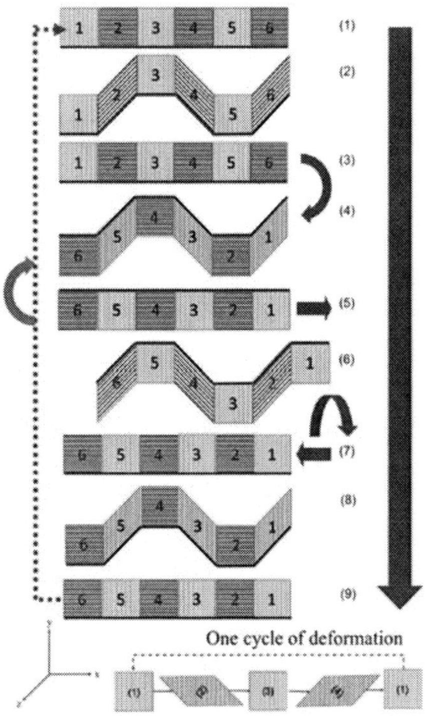

Figure 3. One CGP processing cycle

The designed grooved die (a) and flat die (b) are given in Figure 2. The sheets are subjected to total number of 16 passes (four cycles) of CGP and further processing could not be continued beyond due to cracking of plate. Every CGP cycle consists of 4 passes and 9 steps (Figure 3). In the 1st pass (1-2), a specimen is placed in the die with slight gap which is equal to the sheet thickness (1) and groove pressing begins (2). In the 2nd pass (3-4), flat pressing occurred (3); after that the specimen is rotated by 180 ° around the z direction and the groove pressing is repeated (4). In the 3rd pass (5-6), flat pressing occurred and followed by a move of t to right (5) and then a groove pressing (6). In the 4th pass (7-8), flat pressing occurred (7); after that the specimen is rotated by 180 ° around the *x* direction and moves to left and the groove pressing is repeated (8). At the end, the specimen is flattened again and rotated by 180 ° around the *z* direction (9) and the groove pressing of the next cycle is started (1). The specimen has been heated to 315 °C together with the dies during CGP. The maximum pressing load is 100 tons.

The AZ31 magnesium strips of the dimension 60×60×3 *mm* anneal at 320 °C for 2 *h* and then furnace cooled have been used as an input material. The microstructure micrograph and the chemical composition (measured on an X-ray) of AZ31 magnesium alloys are showed in Figure 4 and Table 1. The change of microstructure after deformation associated with the CGP processing at 315 °C has been investigated by the Scanning Electron Microscope (SEM). Grinding of the test specimens has been conducted with SiC papers of different grades (80, 100, 200, 400, 600, 800 and 1000) using automatic grinding machine. Then ground samples are polishedwith fine diamond paste of 1 μm. All samples have been kept in ethanol between the different polishing steps to avoid the hydrate reaction. Etching has been done at room temperature in (10 ÷ 40) *s* in a mixture of ethylene glycol, acetic acid and nitric acid. The mechanical properties in different directions (*x* and *z*) of specimens are measured using tensile test, which is performed at room temperature with universal testing machine at cross head speed of 2 *mm/min*.

The tensile specimens with the gauge dimensions of 20 *mm* in length, 5 *mm* in width, and 2.8 *mm* in thickness are prepared using wire-cutting machine.

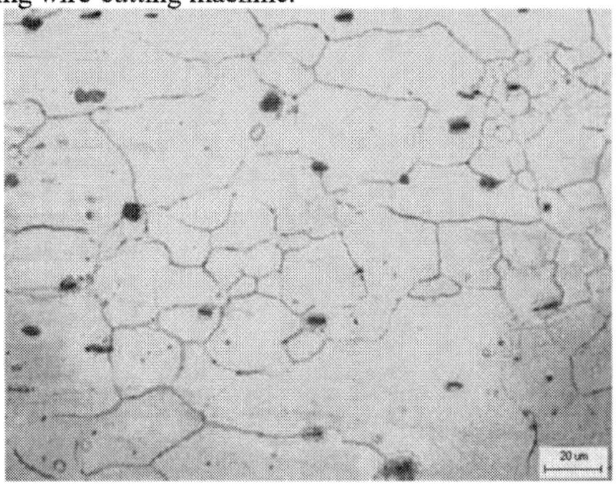

Figure 4. Initial microstructure of the AZ31 samples

Table 1. AZ31 chemical composition

ELEMENT	WEIGHT %
MG	95.55
AL	2.88
SI	0.07
CA	0.18
MN	0.35
FE	0.12
NI	0.13
CU	0.09
ZN	0.63

3. Results and Discussions

3.1. Microstructure

The SEM analysing in flat top (undeformed area 1, Figure 5a) showed the microstructure after first pressing consisted of elongated and/or equiaxed subgrains maybe segmented by dislocations. Extensive twinning found also in almost (1, 2 and 3) areas of the specimen implied that twinning is prevalent mechanism of deformation at one complete CGP cycle [12]. The presence of subgrain microstructure is an evidence that this region has also undergone quite a large amount of plastic deformation and it couldnot be denoted as an "undeformed" area, Figure 5 (a). The mechanical processing process consists of a complete CGP cycle leading to a microstructure that represents heterogeneity in the x direction. Areas with different grain sizes were observed. The grain size is reduced down to 20 μm in the deformation regions. Observation over wide microstructure areas (flat area 1 and sheared area 2) suggests a role of dislocations in formation of the substructure already observed during the first pressing. Magnesium alloy AZ31 is a polycrystalline material in which the orientation of the grains is different and between them are grain boundaries, whose the distorted atomic order obstructs the dislocation motion. Under the external forces (shear stress), because of difference in grain orientation, the slip starts only in a few well-oriented grains. The remaining grains only elastically deform and rotate to ensure the compatibility (or continuity) of the deformation. The continuous increase in external force and the relative rotation of the grains may gradually activate many other slip planes. When most of grains plastically deformed, the permanent strain of the sample will be important. Like stresses, the number of slip planes is also dependent on the deformation temperature. High temperatures enhance the dislocations motion and facilitate the rotation of the non-deformable grains, contributing to the increased number of activated slip systems.

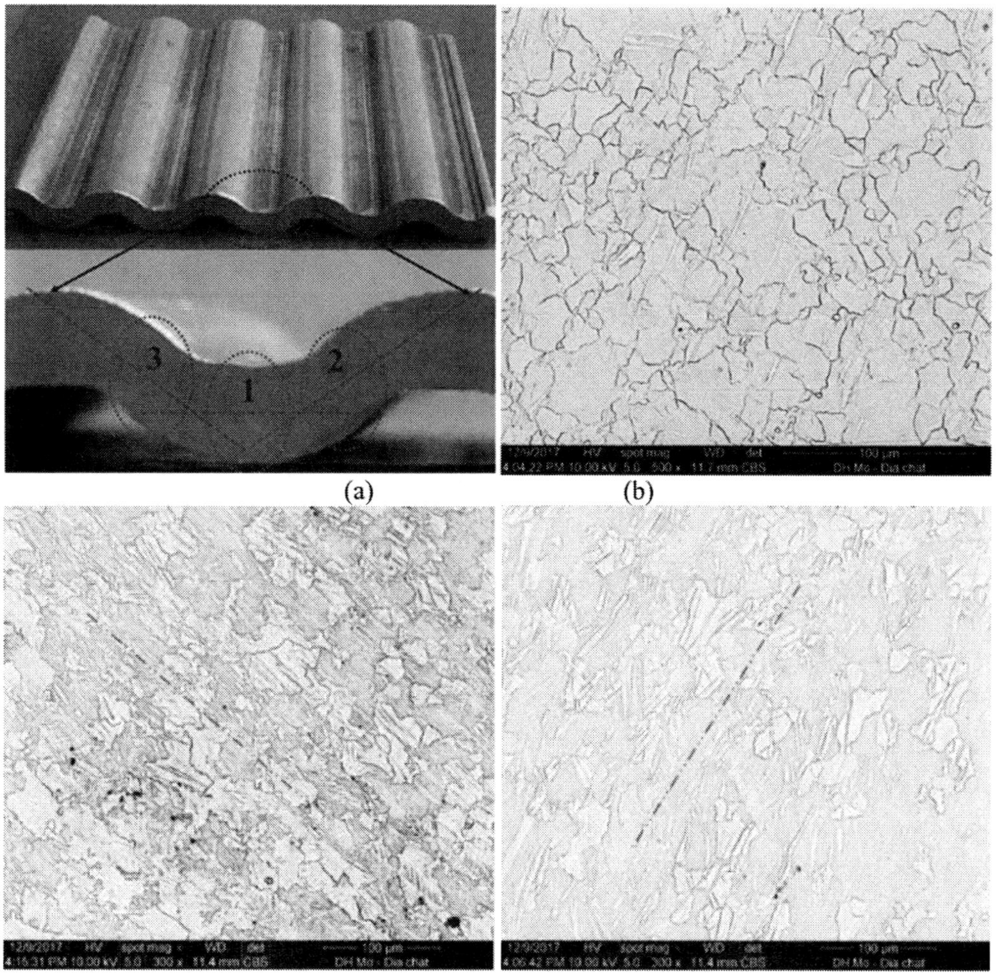

(a) (b)

(c) (d)

Figure 5. SEM micrographs of sample areas: (b) –area 1, (c) - area 3 and (d) – area 2

Dynamic recrystallization takes place in the temperature range that magnesium is usually processed by CGP (150 ÷ 327 °C) [13-16]. Refinement in magnesium alloy occurs in the way that new grains are seeded at the grain boundaries and twins of the old grains. This leads to a necklace distribution of fine grainss around coarse grains. Therefore, a multimodal distribution of grain size is expected at the early deformation stage of magnesium alloy.

This can be seen in Figure 6, where fine grained and coarse particles are interleaved. This multimodal distribution of grain sizes will be replaced by a homogeneous distribution of fine grains when the new fine grains consume the cores of the initial coarse grains. This is observed after 4 cycles of CGP in the AZ31 alloy in Figure 6d,e. Heating at 315 °C for each cycle of the CGP process dissolved the β-particles and, thus, no secondary phases inside the grains were observed. Because the material exhibits rather inhomogeneous complex microstructure containing also large elongated grains, the refinement of the microstructure by CGP technique was made.

Figure 6a,b,c,d shows the (*yz*) face of the AZ31 sample. The material exhibits inhomogeneous microstructure along the cross-section (*yz* face) with pronounced columnar structure. There are different grain structures from coarse (a) to intermediate (b), fine (c), and ultra-fine grains (d) after 1, 2, 3 and 4 cycles of CGP respectively. It also shows a possible difference in size of the newly formed grains in relation to amount of strain imposed. Namely, the Figure 6a,b corresponds to the 1, 2 cycles of CGP (low strain) while the Figure 6c,d to 3, 4 cycles (high strain). That multi-modal distribution of grain sizes are observed when comparing the large size (60 μm) of initial grains with the small size (4 ÷ 6) μm of the new grains and the minimum grain size (1÷1.5) μm. A homogeneous distribution of fine grains is

IOP Conf. Series: Materials Science and Engineering **611** (2019) 012005 doi:10.1088/1757-899X/611/1/012005

readily obtained in the early cycle of CGP if the initial grain size is sufficiently small to be totally consumed by the new grains.

In the (xz) face (Figure 6e) the larger columnar grains with $(10 \div 15)$ μm of length are observed, while smaller grains have a size of about 2 μm.

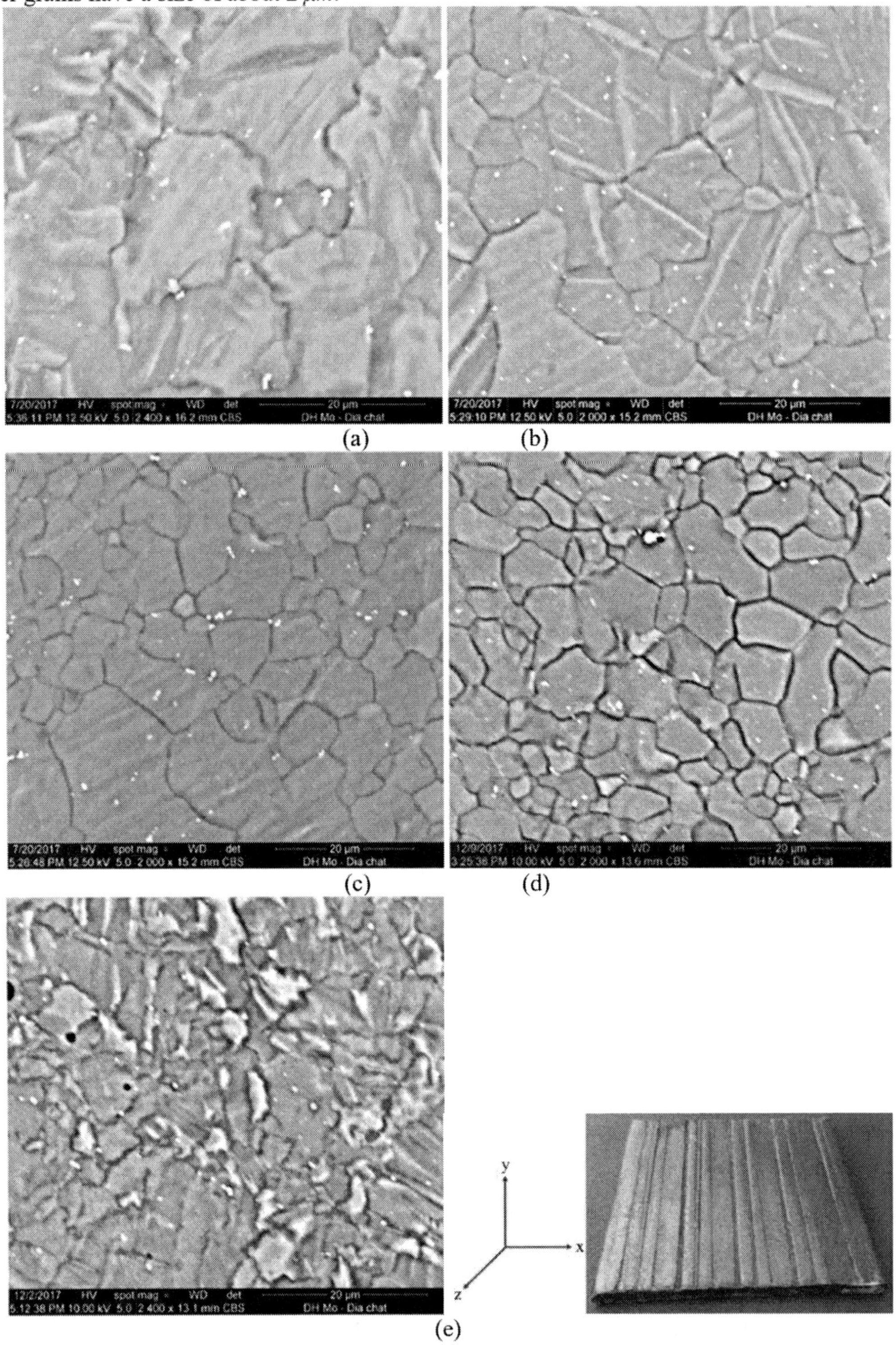

Figure 6. Microstructure evolution of AZ31 sheet after 1 (a), 2 (b), 3 (c) and 4 (d,e) cycles of CGP

3.2. Mechanical Property

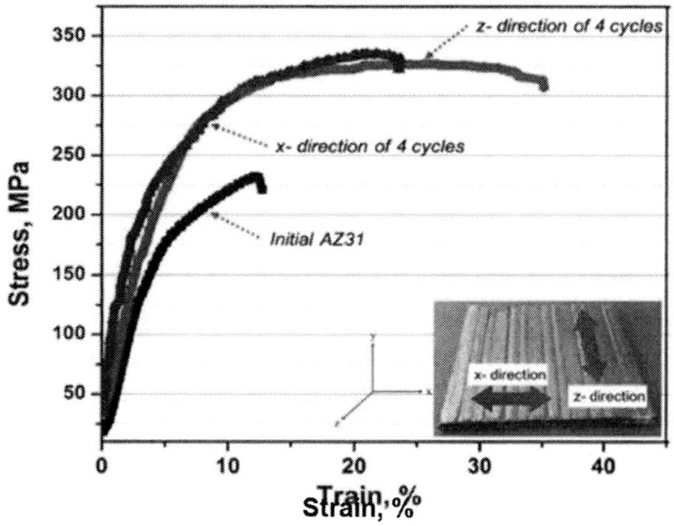

Figure 7. Tensile curves of the initial and CGPed AZ31

It's known from the Hall–Petch relation that the grain size has a direct influence on the mechanical properties of metals [17]. The analysis of the tensile test results along the *x* and *z* directions show a significant difference for the inital and CGPed samples (Figure 7). The initial alloy exhibits lower strength (235 MPa) and strain (12 %), while for the CGPed material tensile test values along the *x* and *z* directions are always higher. When the strength remains more and less 325 MPa with 24 % of strain (for *x* direction) and increases to 35 % (for *z* direction) after CGPed 4 cycles. This finding is matched with HV increase in AZ31 alloy after two cycles of CGP [18] and would be a good evidence of the microstructure evolution during CGP process resulted in grain refinement as discussed above.

4. Conclusion

The AZ31 alloy was used for the preparation of the fine-grained material with improved mechanical properties. The magnesium alloy AZ31 is the CGPed after 4 cycles under elevated temperature condition. The main results are obtained as follows.

The microstructure evolution in the CGP process and its effect on deformed behavior of AZ31 alloy has been established. Analysis of the SEM micrographs confirmed a grain refinement with distinct twinning caused by the dynamic recovery and recrystallization processes. The grain size reduced from (50 ÷ 60) μm down to (4 ÷ 6) μm. The minimal grain size of (1÷1.5) μm was detected in the regions of the highest deformation.

The mechanical properties of AZ31 magnesium alloys showed high strength (325 MPa) and elongation (35 %) after four cycles. Because the CGP groove and flat pressing have been performed within the same container combined with the multi-cycles, the CGP technique can be more suitable method for forming as well as improving the mechanical properties of CGPed AZ31 magnesium alloys.

Acknowledgments

Authors are grateful for the support of the **B2016-BKA-26** project funded by the Ministry of Education and Training of Vietnam.

References

[1] Callister W D 2001, "Fundamentals of Materials Science and Engineering", *John Wiley & Sons, Inc,* Copyright © 2001
[2] Tsuji N 2009, "in Bulk Nanostructured Materials" (Eds: Zehet-bauer M J, Zhu Y T), *Wiley -Vch, Weinheim*, 235.
[3] Valiev R Z, Islangeliev R K, Alexandrov I V 1999, *RussianFederration*

[4] Segal V M 1995 *Mater.Sci.Eng.A*, 197, pp. 157-164
[5] Quang P, Yoon S C, Hong S I and Kim H S 2006 *Materials Science Forum*, Vols. 503-504, 9pp. 931-936
[6] Yoon S C, Quang P, Hong S I, Kim H S 2007 *Journal of Materials Processing Technology*, Vols. 187-188, pp. 318-320
[7] Yoon S C, Quang P and Kim H S 2008 *Korean Journal of Metals and Materials*, Vols. 46, 7No. 3, pp. 144-149
[8] Quang P and Nghiep D M 2016 *Korean J. Met. Mater*, vol. 54-3, pp. 217-223
[9] Alexander T G L, Zhilyaev P 2008 *Progress in Materials Science*, Vol. 53, pp. 893-979
[10] Zhu Y T, Lowe T C, Jiang H and Huang J 2001 *Patent* No. 6197129, USA,
[11] Zhu Y T, Jiang H, Huang J and Lowe T C 2001 *Metallurgical and Materials Transactions A*, Vol. 32, N o. 6, pp. 1559-1562
[12] Quang P, Thuy P T, Hue D T H, Ngung D M, Schukin V Y 2019 *Acta Metallurgica Slovaca*, Vols. 25, No. 1, p. 48-54
[13] Xu C, Furukawa M, Horita Z and Langdon T G 2005 *J. Mater. Sci. Eng. A*, 398, pp. 66-76
[14] Galiyev A, Kaibyshev R and Gottstein G 2001 *Acta Mater*, 49, pp. 1199-1207
[15] Myshlyaev M M, McQueen H J, Mwembela A and Konopleva E 2002 *J. Mater. Sci. Eng. A*, 337 121
[16] Beer A G. and Barnett M R 2006 *Mater. Sci. Eng. A*, 423, pp. 292-299
[17] Callister W D 2001 "Fundamentals of Materials Science and Engineering", *John Wiley & Sons, Inc*, Copyright ©
[18] Thuy P T, Hue D T H, Ngung D M and Quang P 2018 *J. Science and Technology of Metals*, No. 78, pp.32-36

IOP Conf. Series: Materials Science and Engineering **611** (2019) 012006 doi:10.1088/1757-899X/611/1/012006

Surface Modifications of Low Carbon Steel Using Friction Stir Processing

M Newishy[1], M A Morsy[1,*], H Fujii[2], M R Elkousy[3] and N A Abd-Elraheem[3]

[1]central metallurgical research and development institute (CMRDI), Egypt
[2]Japan welding research institute (JWRI), Osaka University, Japan
[3]Mining, Petroleum and Metallurgical Department (MPM), Faculty of engineering, Cairo University, Egypt

*Corresponding author

Abstract. Surface modification was carried out on low carbon steel sheet (JISG3141 SPCC) using Friction Stir Processing (FSP). The friction stir processing slightly increased the surface hardness of the low carbon steel sheet. In order to significantly increase the surface hardness of the low carbon sheet, WC powder having a size of 5 µm was embedded into the low carbon surface layer using sandwich method where a carbon steel cover is applied on the WC powder present in the groove of 1.5 mm depth ant 3 mm width in the carbon steel sheet and the friction stir processing was applied to the sandwich using WC-Co tools. The cover was applied to prevent the loss of WC particles during processing.
SEM and TEM observation showed that the number of passes and the reversing of the direction of rotation in the next passes significantly affected the homogeneity and the distribution of the WC powder in the matrix. Moreover, crushing of WC powder to smaller ones was observed. WC fraction area analysis revealed the high density of WC distribution in the matrix. The hardness distribution showed that the surface modified hardness increased to about 700 HV at the mid depth of the surface modified sheet using WC powder compared with 400 and 500 HV at the deeper and shallower levels of depth respectively.

1. Introduction

Friction-stir welding (FSW) is a solid-state process [1–3] in which a rotating tool having a shoulder and a probe which inserted into the two intimate contacted surfaces of the two butting plates as shown in Fig. 1. Friction Stir Processing (FSP) is an adaptation of friction stir welding, was developed by Mishra et al. [4]. Compared with other metal working techniques, FSP has the advantages of low amount of heat generated, extensive plastic flow of material, very fine grain size in the stirred region, healing of flaws and casting porosity, and mechanical mixing of the surface and subsurface layers [5-7].

For many applications, the high surface wear resistance at the surface while the bulk remains tough with low wear resistance [9, 10] is required. Many techniques of surface modification such as fusion welding [11], laser beam [8–10] and other techniques [12–15] were developed which resulted in formation of unwanted phases [16–17].

S. Dodds. et al. [18] modified the surface of AISI 420 stainless steel using FSP. They compared the resulted surface properties with the surface properties of a heat treated hardening AISI 420 and with D2 tool steel. The results show that FSP of AISI 420 can produce higher hardness ~ 722 HV and better wear performance than the heat treated one. If surface modification of the very low carbon steel is required, then the only friction stir processing without the addition of reinforcement material became valueless.

FSP applied in surface modification were applied in many metals with or without using powder [19]. The aim of the work is to study the surface modification of very low carbon steel through FSP surface layer of WC composites. Surface modified layer will be studied using SEM, TEM observation. Also, micro-hardness distribution at different depth level will be determined.

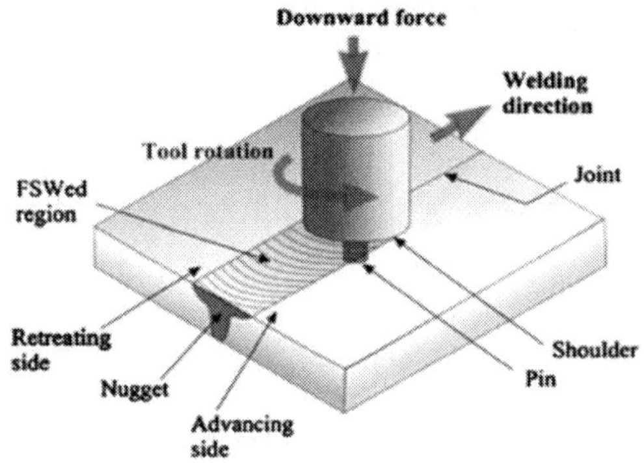

Figure 1. Friction stir welding process [20].

2. Experimental Procedures

A very low carbon steel (JIS G3141 SPCC) sheet with a length of 150 mm, width of 70 mm and a thickness of 3.2 mm was used as a base metal. The chemical composition of this low carbon steel is shown in Table 1. The surface will be modified with and without the addition of reinforcement powder. In surface modification with the addition of powder, a longitudinal groove with a depth of 1.5 mm and a width of 3 mm was machined at the mid the steel sheet. The WC powder with an average size of 5 m is added to fill this groove and a 1 mm cover sheet from the same base metal material cover the groove filled with WC powder as shown in Fig. 2. A tool of WC-Co material is used with a shoulder diameter of 15 mm with a convex probe of 6 mm diameter and a length of 2.8 mm as shown in Fig. 3. The tool holder is provided with a passing water to cool the shoulder. An argon gas is applied at the FSP location with a flow rate of 0.5 L/sec to prevent oxidation of material. The tool rotation speed is 300 rpm and the travel speed is 1.6 mm/sec with a pressure of 15 kN. The cross section of the processed passes was prepared for optical, SEM and TEM observation. The preparation includes mounting, grounding with abrasive paper ranging from 80 to 1200 mesh and polished with diamond past. The polished specimen is then etched with 3% Nital solution.

Micro-hardness distribution is evaluated at three different depth levels as illustrated in Fig. 4 using a micro-Vickers hardness tester with a load of 100 g for 15 seconds by using a micro-Vickers hardness tester with a load of 100 g for 15 seconds using a micro-Vickers hardness tester with a load of 100 g for 15 seconds. Wear Index of the WC modified surface is determined by dividing the weight loss of the WC reinforced steel by the weight loss of the base metal steel used. The wear test is done using SiC ring and applying a rotation speed of 150 rpm and a load of 0.5 bar for 900 seconds.

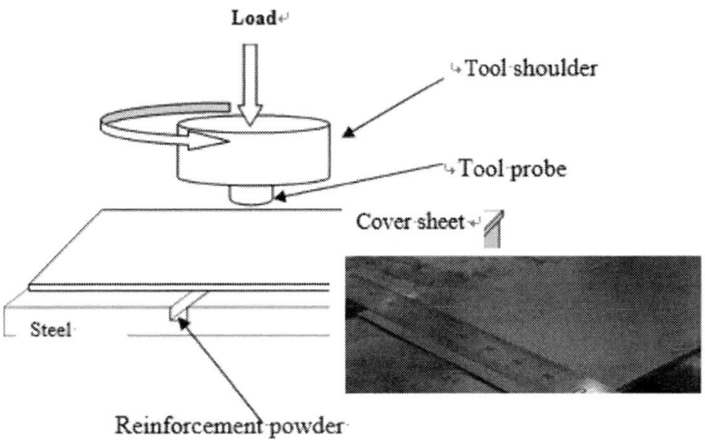

Figure 2. Schematic drawing (left) and actual specimen (right) for FSP setup

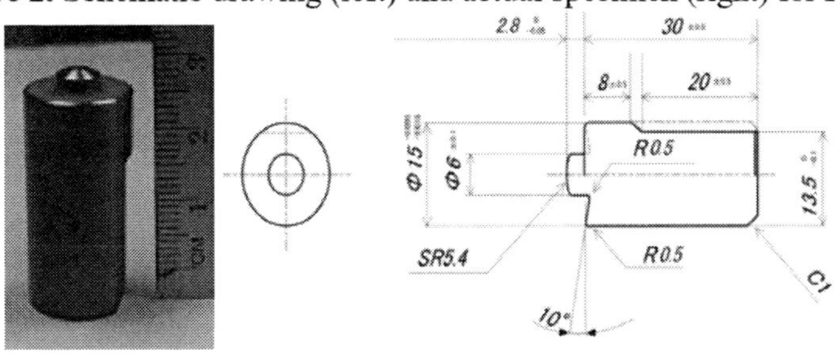

Figure 3. Tool actual and mechanical drawing view.

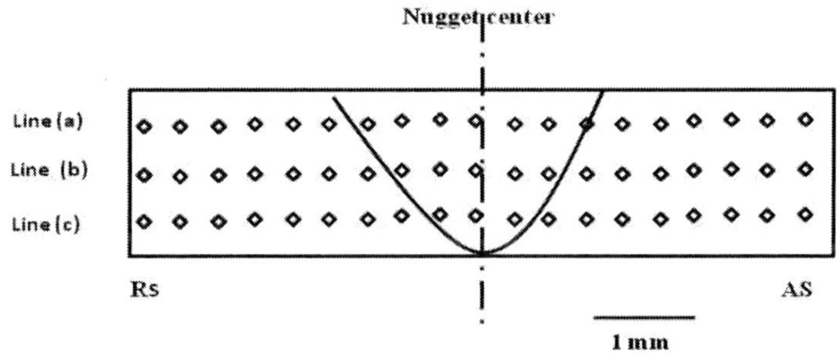

Figure 4. Schematic illustration of micro-hardness measurement

3. Results and Discussion

3.1. Friction Stir Bead without Addition of WC Reinforcement

Shape of FSP bead without addition of WC powder is shown in Fig. 5 a using the process parameters described in section 2. Bead shape shows ripples resulted from the shoulder friction with base metal which is normal with friction stir process. The cross section of the processed bead is shown in FIG. 5 b. Location a represents the base metal. However, location b and c represent the upper and lower stir zones respectively. Location d represents the thermo-mechanical affected zone (TMAZ). The heat affected zone (HAZ) is not clear. This may be attributed to the very low carbon content of the base metal and confusion with TMAZ [19]. Microstructures of these different zones are shown in Fig. 6. The base metal has coarse grains (Fig. 6 a). The upper stir zone (Fig. 6 b) and the lower stir zone (Fig. 6c) both shows grain refinement as a result of recrystallization induced by plastic deformation and the heat generated from friction [28]. The thermo-mechanically affected zone (TMAZ) is immediately adjacent to the stir zone. In this zone, the deformation and generated heat are insufficient to generate new grain formation. Deformed elongated grains are observed.

a)

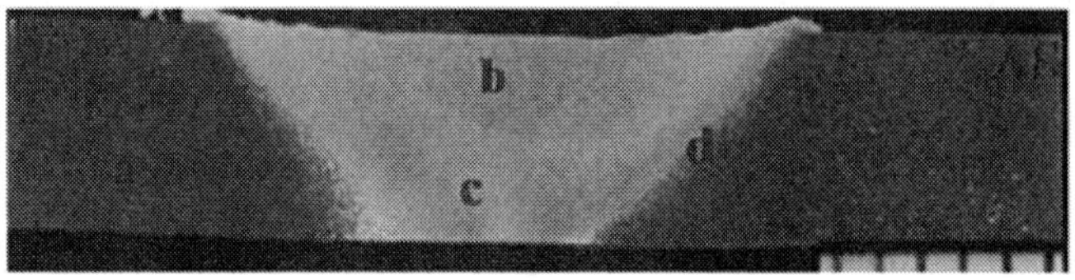

b)

Figure 5. Bead shape a) and cross section b) after FSP of low carbon sheet without addition of powder [19].

Bead shape of the friction stir processing with the addition of WC powder using sandwich method described in section 2and applying three passes with the reversing of the direction of rotation in the next passes is illustrated in Fig. 7. The bead is smooth and the figure reflects the sequence in reversing the direction of rotation. Figure 8 illustrates the macrostructure of the internal structure of the nugget at every pass. In the first pass, a defect was observed in the form of powder cluster agglomerated at the probe location in retreating side that prevents the distribution of the powder at the whole nugget as shown in Fig. 8 a. Reversing the rotation direction in the second pass resulted in diminishing the defect and significant amount of WC powder cluster are distributed in the matrix (Fig. 8 b). Reversing the rotation direction in the third pass resulted in a complete removal of defect with a significant distribution of WC powder cluster in the matrix (Fig. 8c). The WC particle volume fraction can be calculated according to ASTM [21] by point count analysis. It is about 63%. The specimen showing a removal of defect and a complete distribution of WC powder (Fig. 8 c) was observed by SEM as shown in Figs. 9a, b, c and d. The base metal shows a ferrite with 100% percentage (Fig. 9 a). This is normal with very low carbon content of base metal. Fig. 9b shows TMAZ with complete ferritic microstructure with different level grain size. At the nugget upper (upper stir zone) a homogeneous distribution of WC powder was observed after the third pass with reversing the rotation direction as shown in Fig. 9c. At the nugget center (mid stir zone) a completely homogeneous distribution of WC powder was observed as shown in Fig. 9 d.

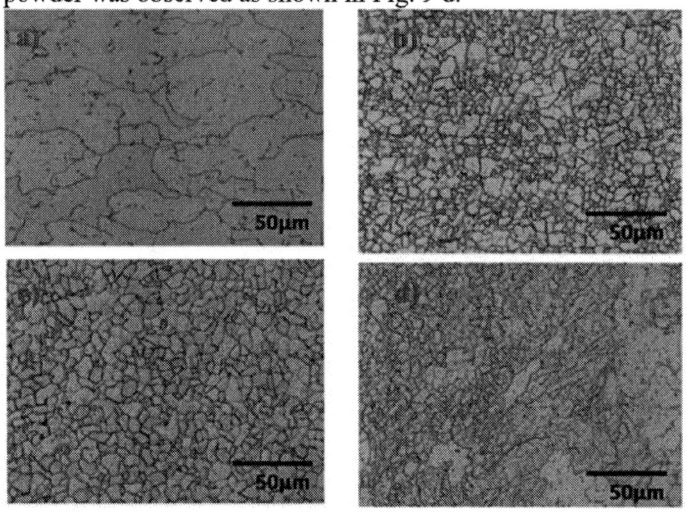

Figure 6. Microstructure of friction stir processing for low carbon steel without addition of powder a) BM, b) Lower stir zone, c) Upper stir zone and d) TMAZ [19]

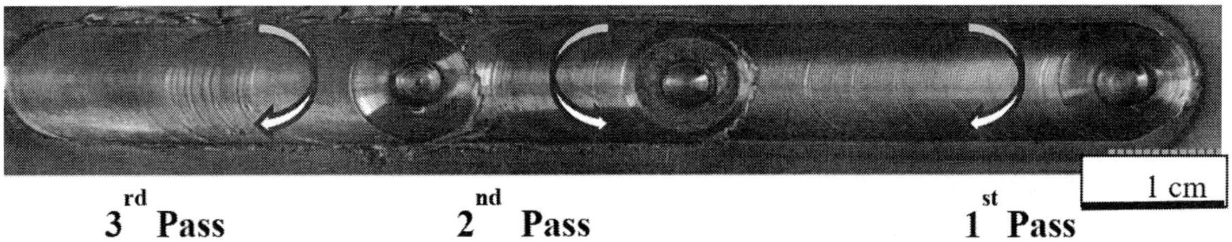

Figure 7. Bead shape after FSP of steel sheet using WC powder after three reversing passes.

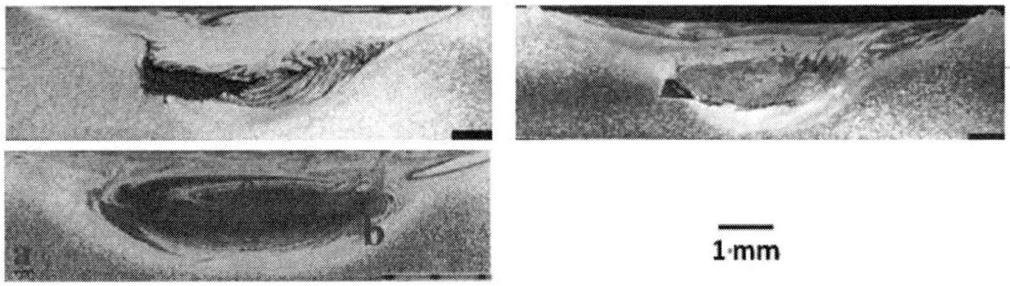

Figure 8. Macrostructure inside the nugget with different passes, 1st passes (a), 2nd pass (b) and 3rd Pass (c).

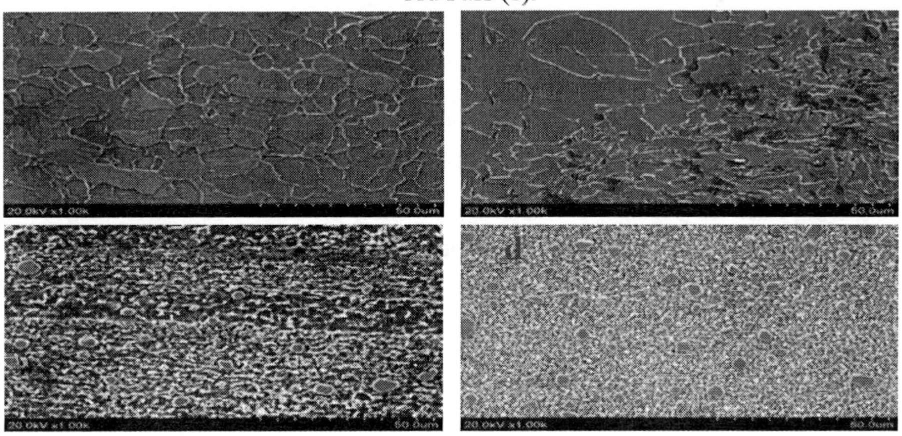

Figure 9. SEM observation of the specimen shown in Fig. 8c at different location in the macrostructure (a) BM as received, (b) TMAZ, (c) FSP with powder in the nugget upper zone and (d) FSP with powder in the nugget center.

Figure 10 (a) and (b) show the typical TEM bright field images of the matrix through the nugget center after 3rd pass, the microstructure characterization reveals that the processed zone consist of very fine deformed ferrite grains. Also WC particles sizes vary from sound particles as received to very fine resized particles in the(Nano-size order) as a result of sever plastic deformation from the pressurized tool and crushing action of the very hard WC particles.

Figure 11 shows mapping of the different elements inside the matrix of friction stir at the mid depth of low carbon steel after 3rd pass, the mapping reveals the distribution of each element, in case of W and carbon, it can be estimated that the very fine WC particle sizes was observed after FSP.

3.2. Hardness Distribution

Hardness distribution at different depth levels at the macrostructure shown in Fig. 8 c (using WC reinforced powder) is illustrated in Fig. 12. The highest hardness values (about 700 HV) located at the mid depth (line b) with a width of about 5mm. At a lower depth (line c) the hardness decreased to about 400 HV.

IOP Conf. Series: Materials Science and Engineering **611** (2019) 012006 doi:10.1088/1757-899X/611/1/012006

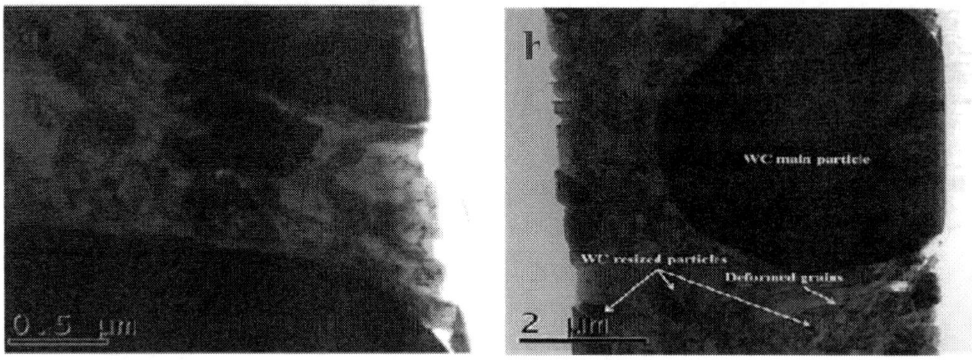

Figure 10. TEM image of WC particles distributed in the composite surface layer with different sizes and deformed grain.

However, at a higher depth level (line c), the hardness values reach to about 500 HV. The difference in these values could be attributed to the density of WC reinforced particles. The highest density is associated with the mid depth as shown in the SEM observation (Fig. 9 c and d). Hardness values were measured at the FSP of the steel sheet without powder additions. The maximum hardness values was about 150 HV compared to 100 HV at the base metal. The small increase in FSP nugget hardness values may be attributed to grain refinement observed in Fig. 6b and c.

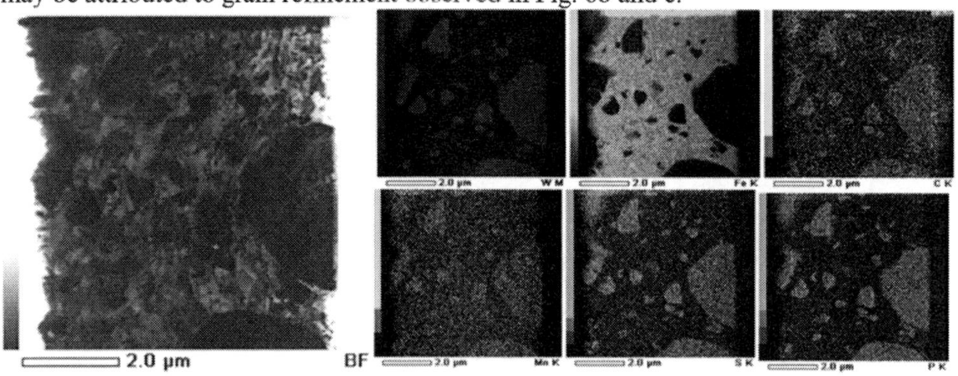

Figure 11. TEM image of WC particles distributed in the composite surface layer with different element mapping inside the matrix.

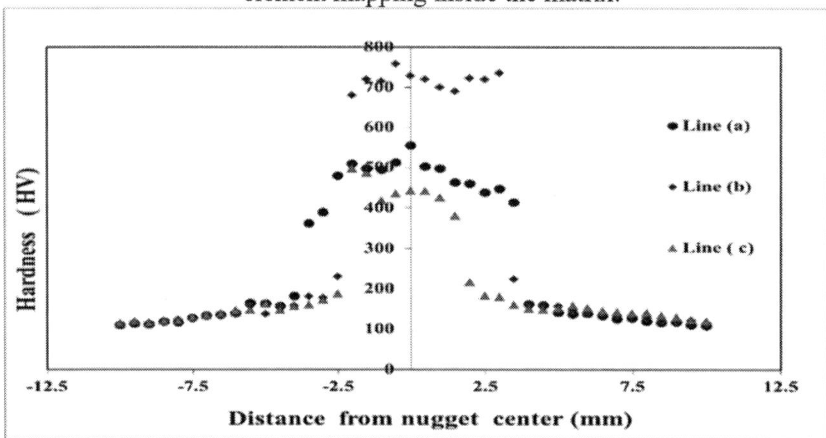

Figure 12. Micro-hardness distribution at different depth levels after the third pass shown in Fig 8c

3.3. Wear Index

The wear index of the FSP bead (steel/WC composite layer) as compared to base metal at different sliding distance is shown in Fig13. I t is clear that the wear index of the FSP with WC reinforcement powder at the 200 m sliding distance is much less than the wear index of the base metal. With the increase in sliding distance, this big difference (Fig. 13). This could be attributed to the gradual removal of WC reinforcement particles from the surface with the increase in sliding distance.

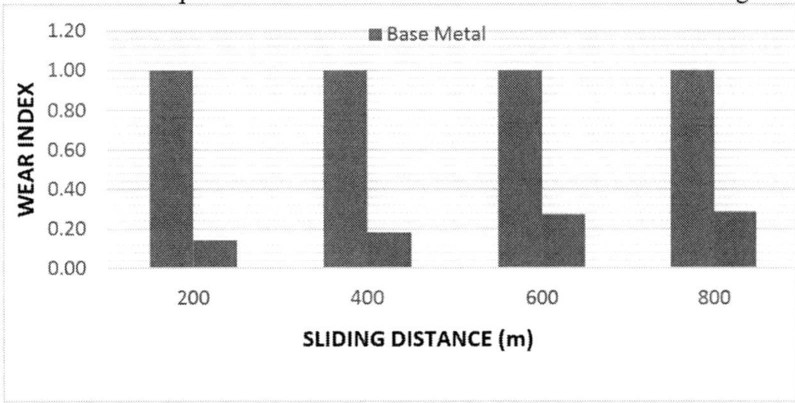

Figure 13. Wear Index of the Reinforced low carbon steel with WC powder compared with base metal

4. Conclusions

Surface modification of low carbon steel was conducted using WC powder embedded in the very low carbon steel matrix using friction stir processing. The The following results were obtained:

1. SEM and TEM observation showed that the number of passes and the reversing of the direction of rotation in the next passes significantly affected the homogeneity and the distribution of the WC powder in the matrix. Moreover, crushing of WC powder to smaller ones was observed.
2. WC fraction area analysis revealed the high density of WC distribution in the matrix.
3. The hardness distribution showed that the surface modified hardness increased to about 700 HV at the mid depth of the surface modified sheet using WC powder compared with 400 and 500 HV at the deeper and shallower levels of depth respectively.
4. The wear index of the FSP with WC reinforcement powder at the 200 m sliding distance is much less than the wear index of the base metal. With the increase in sliding distance, this big difference slightly decrease as a result of gradual removal of the WC particles from the modified surface.

References

[1] WM. Thomas, ED. Nicholas, JC. Needham, CJ. Dawes: Int. Patent App. No. PCT/GB92/02203; 1991.
[2] Dawes CJ, Thomas WM: Welding J., 75 (3):(1996) 41–45.
[3] WM. Thomas, RE. Dolby: Proceedings of the 6 int. trends in weld. res. Materials Park, OH, USA: ASM Int., (2003) 203–211.
[4] R. S. Mishra, M. W. Mahoney, S. X. McFadden, N. A. Mara and A. K. Mukherjee: Scr. Mater., 42 (2000), 163–168.
[5] R. S. Mishra and Z. Y. Ma: Mater. Sci. Eng. R, 2005, R50, 1–78.
[6] Z. Y. Ma, S. R. Sharma and R. S. Mishra: Mater. Sci. Eng. A, 2006, A433, 269–278.
[7] K. Nakata, Y. G. Kim, H. Fujii, T. Tsumura and T. Komazaki: Mater. Sci. Eng. A, A437(2006), 274–280.
[8] Y. Morisada, H. Fujii, T. Nagaoka and M. Fukusumi: Mater. Sci. Eng. A, A419 (2006), 344 - 348.
[9] K. Funatani: Surf. Coat. Technol., (2000), 133–134, 264–272.

[10] A. N. Attia: Mater. Design, 22 (2002), 451–457.
[11] M. A. Morsy and E. Elkashif Welding in the World, Le Soudage Dans Le Monde 58(4), DOI: 10.1007/s40194-014-0132-0, July 2014.
[12] C. Hu and T. N. Baker: J. Mater. Sci., 30 (1995), 891–897.
[13] A. Almeida, M. Anjios, R. Vilar, R. Li and M. G. S. Ferreira: Surf. Coat. Technol., 70 (1995), 221–229.
[14] X. B. Zhou and J. Th. M. de Hosson: Scr. Metall. Mater., 33 (8) (1995), 1345–1352.
[15] M. Yan and H. Hanqi: J. Mater. Sci., 1996, 31, 4303–4306.
[16] H. S. Li, M. X. Li, Y. Z. He, J. H. Yoon and K. Shin: Sci. Technol. Weld. Join., 11, (5) (2006), 567–571.
[17] H. C. Man, C. T. Kwok and T .M. Yue: Surf. Coat. Technol., 132 (2000), 11–20.
[18] S. Dodds, A.H. Jones, S. Cater, "Wear; Vol. 302 (2013) pp.863-877.
[19] M. Newishy, M. Elkousy, N. AbdelRahem, M. A. Morsy and S. Khodir Academic J. of Manufacturing Engineering, 11(2013), 80-85.
[20] Lee WB, Jung SB. The joint properties of copper by friction stir welding. Mater Lett. 2004, 58, 1041–6.
[21] ASTM: Standard test method for determining volume fraction by systematic manual point n count, E 562-02 (2002), USA.

Cause Analysis of Flat Concave Defects on the Inner Surface at the end of T91 Alloy-Steel Tubes

Zhenguo Sun[1,*], Yuyan Li[2], Fei Li[3] and Jun Ye[4]

[1,3,4]Jiangsu Province Special Equipment Safety Supervision Institute Wuxi Branch, Wuxi, China
[2]Jiangsu Xiya Petrifaction Equipment Ltd., Wuxi, China

*Corresponding author

Abstract. The concave defects on the inner surface at the end of T91 alloy-steel tubes were analysed by macroscopic morphology analysis, electron microscope scanning and energy spectrum analysis, metallographic analysis and anion analysis of corrosion products. The composition of water sample and oil sample of finned winding and oil immersion treatment was also analysed. It was found that this kind of defect was not caused by the preservation environment of steel tube, the treatment of finned winding and oil immersion. The defect should be related to the quality of steel tube blank or the processing technology of steel tube.

1. Introduction

A batch of T91 seamless tubes were stored for more than half a year, after finned winding, it is found that there are flat concave defects on the inner surface at some steel tube ends. The manufacturing standard of this batch of tubes is ASME SA-213M 2015[1]. During the whole cycle of the formation and existence of this flat concave defects on the inner surface, the steel tubes have gone through the processes of billet smelting, cold drawing processing, storage and preservation, finned winding and oil immersion rust prevention, etc[2]. The formation of the flat concave defects in steel tubes must be directly related to one or more of the above processes. In this paper, the effects of steel tube preservation environment, finned winding and oil immersion treatment on the formation of defects are analysed.

2. Macro Morphology Analysis of Internal Surface of Steel Tube

There is obvious corrosion phenomenon on the inner surface of the tube end, and there are a large number of flat concave defects in the uniform distribution as shown in figure 1.

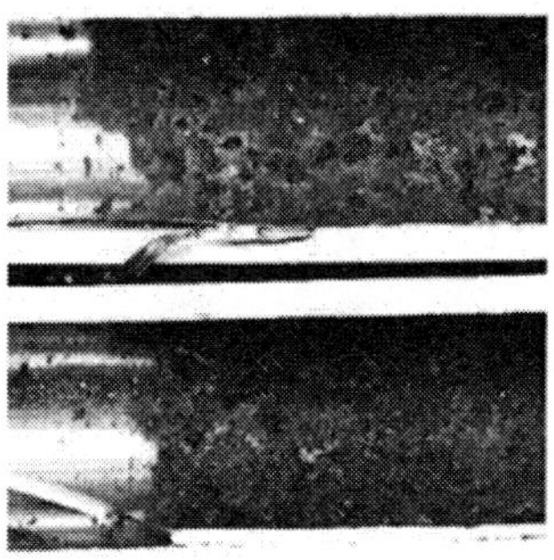

Figure 1. Macroscopic morphology of internal surface of steel tube

The corrosion products on the inner surface of the steel tube are reddish-brown, and no other forms of corrosion products are filled in the concave, and after local grinding on the inner surface, concave defects are still visible. The total length of the steel tube is 22 meters. The corrosion of the inner surface of the steel tube is slight at its 1/3 and 1/2 locations. After local grinding, the inner surface is smooth, and no concave-like defects are found.

3. Metallographic Analysis

The corrosion products on the surface of the sample were removed by pickling, and the macroscopic morphology and metallographic structure were analysed. The results show that the bottom and edge of the concave section transition smoothly, and there is no uniform expansion of the defect to the wall thickness. The microstructure is normal tempered martensite, no non-metallic inclusions are found, and the grain size is 10, as shown in figure 2 and figure 3.

Figure 2. Metallurgical structure

Figure 3. Fracture appearance

4. Electron Microscopic Scanning and Energy Spectrum Detection and Analysis of Defect Sites

The defect site of the sample was detected and analysed by electron microscope scanning and energy spectrum as shown in figure 4 and figure 5. It can be seen that the defect of the concave is approximately circular, the bottom of the concave is relatively flat, and the edge of the concave is irregular crack morphology. Under the electron microscope, Areas 4 and 5 are in the concave, areas 1, 2 and 3 are at the boundary of the concave, and area 6 is outside the concave. The results of the energy spectrum detection are similar. The black areas are mainly metal matrix and its oxides, which contains a small amount of Cl^- or does not contain Cl^-. The white area showed crystal structure under electron microscope, compared with the black area, the local concentration of Cl^- was found in the white area, but no enrichment of Cl^- was found inside and outside the concave.

Figure 4. Electron microscope scanning analysis of defect site

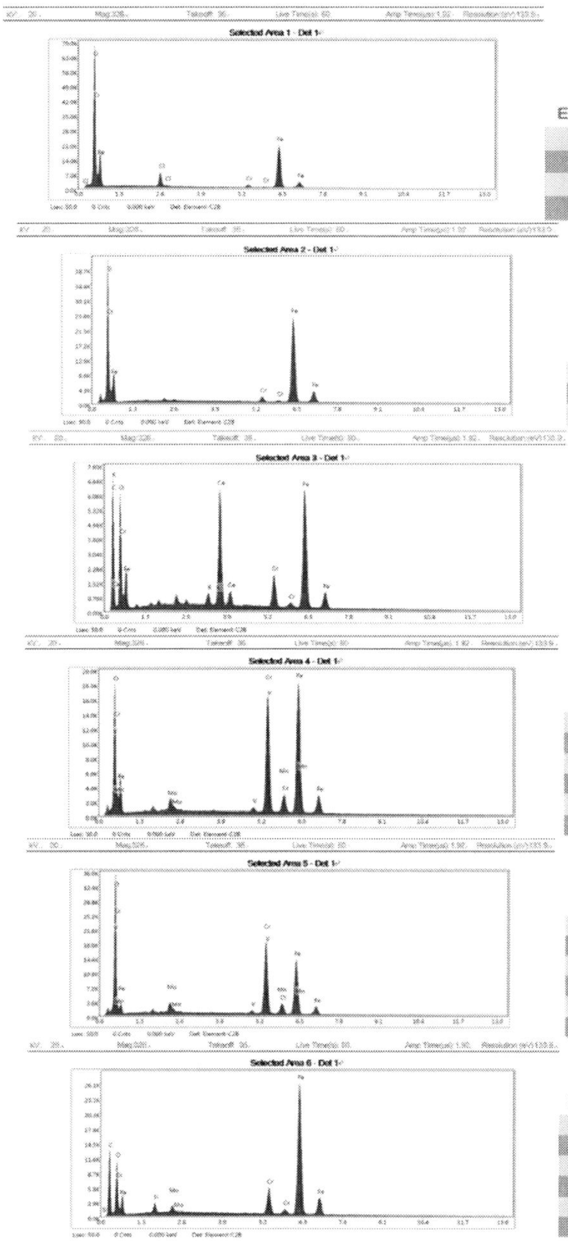

eZAF Smart Quant Results

Element	Weight %	Atomic %	Error %
O K	44.82	72.94	5.86
ClK	4.74	3.48	3.20
CrK	1.67	0.83	4.61
FeK	48.77	22.74	1.65

eZAF Smart Quant Results

Element	Weight %	Atomic %	Error %
O K	28.60	58.24	5.64
CrK	2.68	1.68	3.55
FeK	68.72	40.08	1.57

eZAF Smart Quant Results

Element	Weight %	Atomic %	Error %
C K	19.99	39.27	8.29
O K	22.94	33.84	9.49
K K	1.19	0.72	8.10
CaK	14.19	8.36	2.07
CrK	6.49	2.94	3.24
FeK	35.20	14.87	2.03

eZAF Smart Quant Results

Element	Weight %	Atomic %	Error %
O K	13.60	35.11	6.69
MoL	2.57	1.11	5.24
V K	1.29	1.05	5.95
CrK	30.26	24.04	1.68
MnK	1.60	1.20	6.58
FeK	50.69	37.50	2.16

eZAF Smart Quant Results

Element	Weight %	Atomic %	Error %
O K	24.33	52.51	6.42
MoL	3.50	1.26	3.68
V K	1.44	0.98	5.06
CrK	32.65	21.69	1.79
MnK	1.22	0.77	8.44
FeK	36.86	22.79	2.25

eZAF Smart Quant Results

Element	Weight %	Atomic %	Error %
C K	23.11	50.72	8.64
O K	10.69	17.61	8.48
SiK	0.84	0.79	7.67
MoL	1.17	0.32	7.75
CrK	7.66	3.88	2.38
FeK	56.54	26.69	1.62

Figure 5. Energy spectrum detection and analysis of defect sites

5. Analysis of Water Sample and Oil Sample of Fin Winding and Antirust Treatment by Soaking Oil

In the process of finned winding and oil immersion antirust treatment, the outer surface of the steel tube is directly in contact with cooling water and antirust oil, and the inner surface of the steel tube is likely to be in small contact with cooling water and antirust oil. The cooling water is used for recycling. The cooling water is sampled and analyzed in three periods of the day. At the same time, two parts of antirust oil are selected for composition analysis. It is found that that both water and oil samples are weakly alkaline, and the composition is not obvious abnormal as shown in table 1.

IOP Conf. Series: Materials Science and Engineering **611** (2019) 012007 doi:10.1088/1757-899X/611/1/012007

Table 1. Analysis of water sample and oil sample

	Cl^- (mg/l)	SO_4^{2-} (mg/l)	NO_3^- (mg/l)	PH	Na (mg/l)	Al (μg/l)	Fe (mg/l)	Hardness of water (mg/l)
Water sample 1	41.2	58.2	/	7.2	54.0	2.92	0.23	159
Water sample 2	61.4	87.8	/	7.8	48.0	4.37	0.21	155
Water sample 3	61.8	88.0	/	7.5	50.4	1.57	0.22	165
Oil sample 1	28	/	/	7.5	87.0	/	7.0	220
Oil sample 2	18	/	/	7.2	83.0	/	9.0	235

6. Anion Analysis of Corrosion Products

The anions of corrosion products were analysed by ion chromatography as shown in table 2. The results showed that the specific gravity of anions in corrosion products was very small and the samples were not exposed to corrosive acidic substances for a long time.

Table 2. Anion analysis of corrosion products

Wight of sample (g)	F^- (%)	Cl^- (%)	NO_3^- (%)	SO_4^{2-} (%)	Br^- (%)
0.044	0.017273	0.096477	0.022727	0.045	0

7. General Analysis

It has been confirmed that the steel tubes are superimposed and stored in the indoor standard warehouse with roof and indoor air circulation. There is no corrosive atmospheric atmosphere such as acidity in the environment of the area, and the sampling tubes are located in the middle and upper parts of the superimposed steel tubes. In the normal preservation process, the possibility of direct contact between the sample tube and the corrosive medium is relatively small. The flat circular concave defects are evenly arranged on the whole inner surface of the tube end, and there is no corrosive waterline on the inner surface, if the concave is caused by environmental oxidation corrosion. There should be multiple volume expansion corrosion products on the concave, but the results of macroscopic morphology examination of the inner surface show that there is no corrosion product filling on the concave, which does not belong to the typical environmental oxidation corrosion morphology. Therefore, the concave should not be directly caused by environmental oxidation corrosion, and the preservation environment will not directly lead to the emergence of concave defects on the inner surface of the end of the steel tube.

The concave defects on the inner surface of the steel tube only appear at the end of the steel tube, and there are no such defects on the inner surface of the other finned wound parts, such as the inner surface of the steel tube 1/3 and 1/2 locations, which indicates that the finned winding process does not directly lead to the concave defects on the inner surface.

Through metallographic analysis, it can be seen that the metallographic structure of the steel tube in the finned part is normal, which indicates that the finned processing technology has no adverse effect on the metallographic structure of the steel tube.

Through energy spectrum detection and anion analysis, it is found that the main anions in the corrosion products are Cl^- andSO_4^{2-}. Through the analysis, it can be seen that:

- The concave defects at the end of the steel tube are approximately circular, the bottom of the concave is relatively flat, and there is no enrichment of Cl^-in the inner and outer regions of the concave. At the same time, from the metallographic morphology of the concave section, it can be seen that the bottom and edge of the concave transition smoothly, there is no cracking. It was also not observed that the corrosion persistence caused by the enrichment of Cl^- was unitarily extended to the metal wall thickness.
- Both water and oil samples were weakly alkaline, and a small amount of SO_4^{2-}would only adhere to the metal surface. It does not directly lead to corrosion of metal matrix.

- The anion analysis results of corrosion products show that the specific gravity of anions in corrosion products is very small, so the samples are not exposed to corrosive acid substances for a long time. The corrosion products are mainly iron oxides. Therefore, the existence of a small amount of $Cl^-$and SO_4^{2-} is not the main factor leading to corrosion.

8. Conclusion

Through the above comprehensive analysis, it can be seen that the preservation environment of the steel tube, the processing technology of the fin winding, and the contact process between the cooling water and the antirust oil will not lead to the concave defects on the inner surface of the end of the steel tube. The occurrence of this defect should be related to the quality of steel tube blank or other links or factors such as steel tube processing technology.

References

[1] ASME SA-213M 2015 *Standard Specification for Seamless Ferritic and Austenitic Alloy-Steel Boiler, Superheater and Heat-Exchanger Tubes* (New York: The American Society of Mechanical Engineers)
[2] Caifu Wang 2014 *Research on the Corrosion Behavior and Surface Modification of T91/T92 steel* (Nan Chang: Nanchang Hangkong University)

Corrosion Study on Ni-based Alloy in Acidic Environment

Yuejuan Shi[1], Yanling Liu[1], Xuepeng Wang[1], Yongxing Zhou[1], Jun Ma[1] and Shuai Zhang[2,*]

[1]Wuzhong Instrument Co. Ltd, China
[2]School of mechanical engineering, Ningxia University, Yinchuan, China

*Corresponding author

Abstract. Ni-based alloys have excellent corrosion resistance and are widely used in the petrochemical industry. In this study, the effects of sulfuric acid concentration on the corrosion resistance of Ni-based alloys were analyzed by software simulation and electrochemical tests. The results show that the corrosion rate of the alloy is gradually increased with the increase of sulfuric acid concentration, and the influence of sulfuric acid concentration on the corrosion rate of 304 stainless steel is greater than that of nickel-based alloy. In a 10% sulfuric acid solution, the corrosion rate of the four nickel-base alloy samples is only one-twentieth of that of the 304 stainless steel substrate.
Keywords: ni-based alloy; corrosion resistance; electrochemical analysis.

1. Introduction

Ni-based alloys have excellent corrosion resistance and are widely used in the petrochemical industry, especially in the manufacture of key components of equipment [1, 2]. There are many systems for Ni-based alloys. Common systems include Ni-Mo system, Ni-Cr-Mo-Cu system, Ni-Fe-Cr-Mo-Cu system and Ni-Cr system. Different systems of Ni-based alloys have been developed for different corrosive environments[3]. Hastelloy Hastelloy B-2, Hastelloy C-2000, Hastelloy G-30 and Incoloy 925 have excellent corrosion resistance in an acidic environment. These alloy components contain a large amount of Mo, Cr and other elements that improve the corrosion resistance of the alloy, such as Cu, rare earth, etc., to ensure the excellent corrosion resistance of the alloy in an acidic environment. Many research work shows[4,5], in a corrosive environment, the Ni-based alloy indicates that a passivation film is formed to prevent the corrosive medium from entering the inside of the solution to protect the alloy, but some of the precipitated phases in the alloy also have an bad effect on the corrosion resistance of the alloy. Zhao Xuehui et al [6] studied the effect of precipitation on the corrosion resistance of nickel-based alloys, and found that $Cr_{23}C_6$ and Cr_6C and other high chromium carbides precipitated in the alloy after aging treatment significantly reduced the corrosion resistance of the alloy.

This paper mainly studies the corrosion resistance of Ni-based alloys in acidic environment, and explores the effects of Ni-based alloy composition and precipitation on the corrosion resistance of alloys.

2. Materials and Methods

Purchased 304 stainless steel and nickel-based alloy materials at Shanghai Chenghuai Special Alloy Co., Ltd. The composition of the alloy was tested in Tables 1 and 2. Sulfuric acid solutions with concentrations of 1.5%, 3%, 5%, and 10% were separately disposed, and the corrosion resistance of the alloy samples was tested using an electrochemical workstation. The sample was prepared and tested according to the requirements of GB/T24196-2009.

Table 1. Elemental composition of the 304steel (wt%)

	C	B	Cr	Si	Ni	Mn	P	Fe
304 Steel	≤ 0.08	0	18-20	≤1	8.0-10.5	≤2	≤ 0.05	Bal.

Table 2. Elemental composition of the Ni-based alloy (wt%)

Ni-based alloy	Ni	Cr	Mo	Fe	Si	Al	Cu	Co	Mn	C	Bal.
G-30	57.61	22.3	15.5	1.1	0.45	0.31	0.5	1.08	0.62	0.01	W: 0.5 La: 0.05 P: 0.01 S: 0.008
C-2000	56.46	22.56	15.8	2	0.03	0.35	1.45	1.02	0.32	0.01	S:0.001 P:0.015
Incoloy 925	43.2	21.2	2.6	24.58	0.2	0.35	2.83	2.4	0.45	0.01	S:0.01 Ti:2.18 P:0.01
B-2	65.3	0.8	28.02	4.45	0.08	0.5	0	0.62	0.5	0.01	P:0.01 S:0.01

3. Testing of Samples

Using the material thermodynamic calculation software and the corresponding Ni-based alloy database, the precipitation rule of the alloy equilibrium phase is simulated and calculated. The alloy composition (listed in Tables 1 and 2) and temperature parameters were used as the input conditions of the simulation to calculate the possible equilibrium phase (Fig. 1).

The Tafel curve of the Ni-based alloy was tested using an RST 5200 electrochemical workstation. The Tafel curve of the sample was tested by using a three-electrode potentiometric scanning method and operated in the H_2SO_4 solution as the electrolyte. The sample to be tested was used as the working electrode, the platinum wire electrode was used as the auxiliary electrode, A saturated calomel electrode (SCE) was used as a reference electrode for testing. The parameters setting are shown in Table 3.

Table 3. Electrochemical workstation related parameters

Rest time(s)	10
Scan rate(V/s)	0.0005
sampling interval(V)	0.0001
Electrode area（cm^2)	1
Curve points	2001

4. Results and Analysis.

4.1. Analysis of Software Simulation Results

The equilibrium phase diagrams of each alloys were simulated by software. The simulation result is shown as fig.1.

IOP Conf. Series: Materials Science and Engineering **611** (2019) 012008 doi:10.1088/1757-899X/611/1/012008

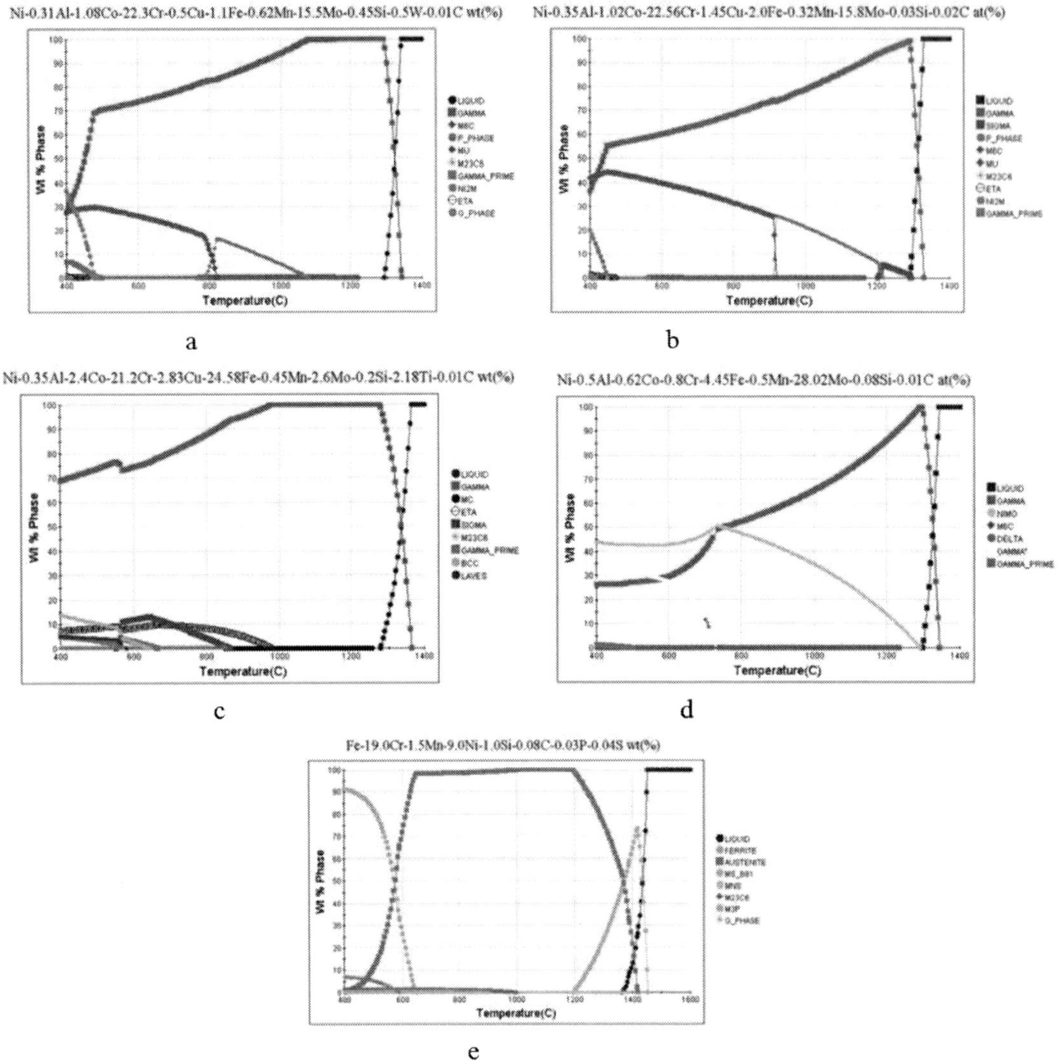

Figure 1. Thermodynamic equilibrium phase diagram

(a-Hastelloy G-30; b- Hastelloy C-2000; c-Incoloy 925; d-Hastelloy B-2; e-304 Steel)

As can be seen from Fig. 3, a large amount of γ phase was precipitated in the a, b, c and d alloy samples. The hard phase such as carbide and boride is less in the alloy, which is mainly related to the content of carbon and boron in the alloy[7]. Combined with the software simulation results and related research conclusions, On the one hand, the precipitation of carbides in the alloy acts as a dispersion strengthening, which has a positive effect on the strength and hardness of the alloy. On the other hand, the precipitation of carbides will lead to the reduce of the alloy corrosion resistance. It can be seen from the software simulation results. The content of the carbide phase in the nickel-based alloy is within 5%. During the solidification process, the carbon element preferentially reacts with Cr, Mo, etc. in the alloy, and precipitates metal carbides such as $M_{23}C_6$ and M_6C, which cause aggregation of Cr and Mo elements, causing intercrystalline chromium depletion[8]. It has a great influence on the properties of the alloy. The 304 stainless steel material has a higher carbon content than the nickel-based alloy, so there are more carbide phases.

4.2. Analysis of Electrochemical Test Results

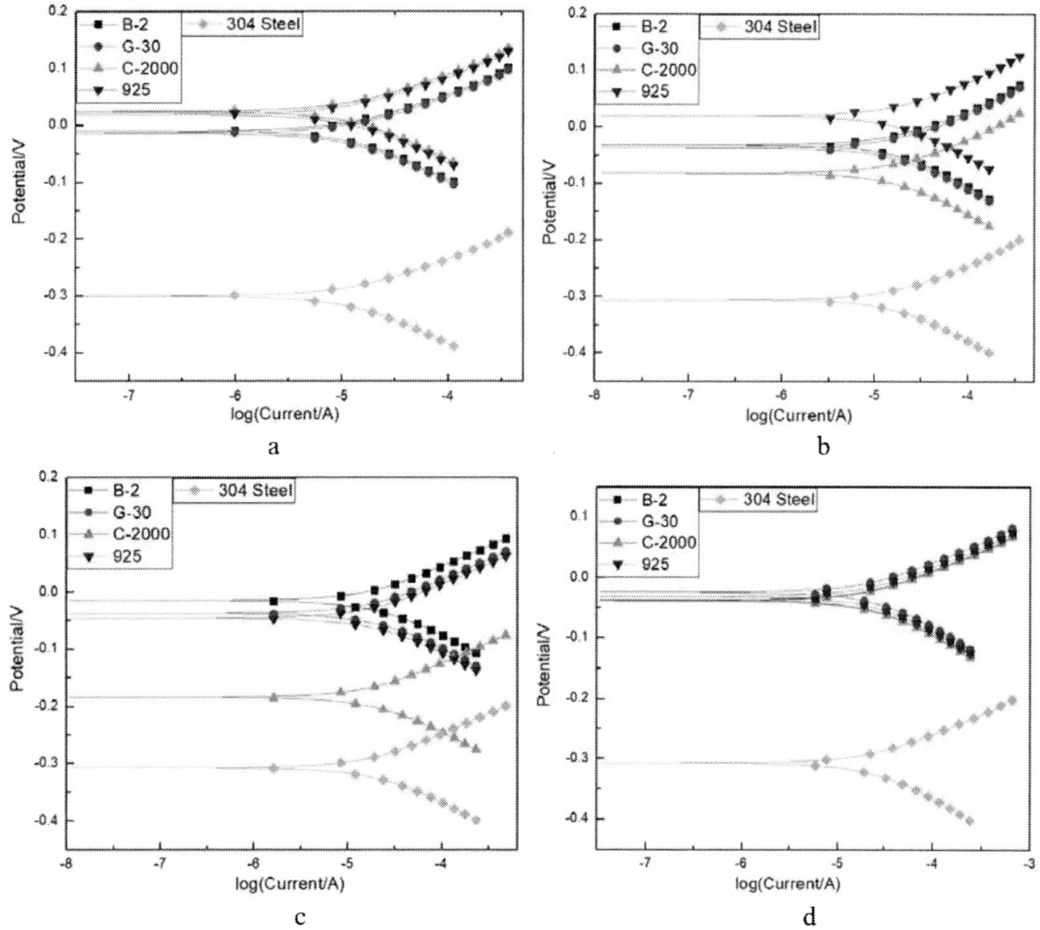

Figure 2. Polarization curves of samples in different concentrations of sulfuric acid solution (a-1.5%H_2SO_4; b-3%H_2SO; c-5%H_2SO; d-10%H_2SO)

It can be seen from Fig. 4 that the corrosion potential of the four nickel-base alloy samples is higher than the corrosion potential of 304 stainless steel in different concentrations of sulfuric acid solution. The corrosion potential can reflect the corrosion tendency of the alloy to some extent[9]. The closer the corrosion potential is to the positive direction, the less the tendency of the alloy to corrode in solution. In the test of the 304 stainless steel sample, the corrosion potential fluctuated between -0.29 V and -0.31 V, and the corrosion potentials of the other four nickel-base alloys fluctuated between -0.22 V and 0.1 V. The corrosion potential anomaly of the C-2000 sample in Figure 2(c) may be related to experimental error.

As can be seen from Fig. 3 and Table 4, with the concentration of sulfuric acid increases, the corrosion rate of the sample also added. It can be seen from the test data, that the corrosion current density of the Ni-based alloy in sulfuric acid is very small, and the corrosion rate of the material fluctuates within a certain range. The higher the concentration of the sulfuric acid solution, the faster the corrosion rate of the alloy. Moreover, in the 10% H_2SO_4 solution, the corrosion rate of theNi-base alloy sample is only one-twentieth of that of the 304 stainless steel sample.

By analysis of the composition of each alloy, it was found that the influence of Cr and Mo elements on the corrosion resistance of the alloy has a positive effect, but the content of both elements does not exceed 30%. The Cr is an element which makes nickel non-rust and has good corrosion resistance in an oxidizing medium. From an electrochemical point of view, a higher Cr content not only makes the alloy easy to passivate, but also reduces passive current density of the alloy.Thereby reducing the alloys dissolution rate and imparting passivation ability to the alloy and rapidly repairing the damaged passivation film, which is the root cause of the high corrosion resistance of the nickel-based alloy with high Cr content. Therefore, although the Mo content of Incoloy 925 alloy is only 2.6%, the higher Cr content and the addition of elements such as Cu and Fe ensure that the alloy has strong corrosion resistance in the lower concentration of sulfuric acid. The Mo element can significantly improve the corrosion tendency of nickel-based alloys in pitting and crevice corrosion environments, and improve the corrosion resistance of the alloy in reducing acid. The content of Cr in Hastelloy B-2 alloy is only 0.8% , but the extremely high Mo content in the alloy can effectively improve the corrosion resistance of the alloy. The addition of a small amount of Al, Ti, Cu, Co and rare earth elements in the alloy can also improve the corrosion potential of the alloy, reduce the corrosion tendency, promote the generation of passivation film on the surface of the alloy, refine the grain size. Further, the content of the C element in each of the Ni-based corrosion resistant alloys is extremely low. It can significantly reduce the intercrystalline chromium deficiency caused by the precipitation of carbide phases such as $M_{23}C_6$ and M_6C during solidification. This can also be verified from software simulation results, so nickel-based corrosion-resistant alloys have strong corrosion resistance[10].

Table 4. Characteristic values of Tafel curves of Ni-based alloys and stainless steels in H_2SO_4 solution

	10% H_2SO_4		5% H_2SO_4		3% H_2SO_4		1.5% H_2SO_4	
	V_{corr} /V	I_{corr}/ (μA·cm²)	V_{corr} /V	I_{corr}/ (μA·cm²)	V_{corr} /V	I_{corr}/ (μA·cm²)	V_{corr} /V	I_{corr}/ (μA·cm²)
B-2	-0.0335	0.658	-0.0084	0.424	-0.035	0.2044	-0.042	0.335
C-2000	-0.0357	0.536	-0.2252	0.44	-0.0804	0.529	0.0326	0.389
G30	-0.0306	0.273	-0.039	0.212	-0.0392	0.227	-0.0122	0.241
Incoloy 925	-0.0327	0.89	-0.0528	0.204	0.0107	0.135	0.023	0.302
304 steel	-0.3043	32	-0.2914	10.19	-0.2979	9.33	-0.2944	6.46

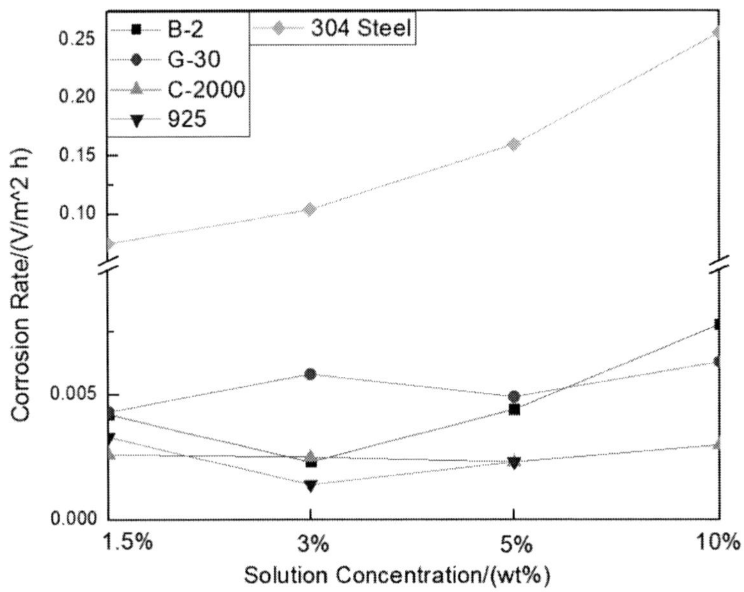

Figure 3. Corrosion rate line chart of sample in solution

5. Conclusion

1. Ni-based alloy has stronger corrosion resistance than 304 stainless steel in sulfuric acid. With the increase of sulfuric acid concentration, the corrosion rate of each alloys also increased. Among them, the corrosion rate of 304 stainless steel sample in sulfuric acid solution was about 20 times that of nickel-based corrosion resistant alloy.

2. The main precipitate in the nickel-based alloy is the γ phase, and the carbide phase in the alloy is mainly precipitated in the form of M23C6 and M6C phases, which has an effect on the corrosion resistance of the alloy.
3. The Cr and Mo elements in the nickel-based alloy have the greatest influence on the corrosion resistance of the alloy, while the addition of elements such as Cu, Al and rare earth can improve the corrosion resistance of the alloy to some extent.

Acknowledgements
This works was supported by Collaborative Innovation Center of Advanced Control Valve Project (Grant No.WZYB-XTCX-003).

References
[1] A J Sedriks. Nickel Alloys: Corrosion[J]. Encyclopedia of Materials Science & Technology, 2001:6140-6141.
[2] W Z Friend. Corrosion of nickel and nickel-base alloys[M]. 1980.
[3] X L Shang, Z J Wang, Q F Wu, et al. Effect of Mo Addition on Corrosion Behavior of High-Entropy Alloys CoCrFeNiMo_ x in Aqueous Environments[J]. Acta Metallurgica Sinica (English Letters), 2019, 32(1):41-51.
[4] Y V Deshpande, A B Andhare, P M Padole. How cryogenic techniques help in machining of nickel alloys? A review[J]. Machining Science & Technology, 2018, 22(1):1-42.
[5] D Gervasio, H Elsentriecy . Corrosion of Hastelloy in Molten Salt Used As High Temperature Heat Transfer Fluids[J]. Nature Medicine, 2015, 9(8): 1076-1080.
[6] X H Zhao, Z Q Bai, Y R Feng, et al. Effects of heart trement and precipitates phase on corrosion resistance of Ni-based alloy[J]. Transactions of Materials and Heart Treatment, 2012, 33(08):39-44.
[7] C L Hu. Corrosion Resistance Dependence of Rare Earth NiCrMoY Alloy Coatings[J]. Applied Mechanics & Materials, 2014, 665:107-110.
[8] T Hong, M Nagumo. The effect of SO 4 2− concentration in NaCl solution on the early stages of pitting corrosion of type 430 stainless steel[J]. Corrosion Science, 1997, 39(5):961-967.
[9] K. S RAJA, NAMJOSHI, et al. Improved corrosion resistance of Ni-22Cr-13Mo-4W Alloy by surface nanocrystallization[J]. Materials Letters, 2005, 59(5):570-574.
[10] J H Chang, J M Chou, R I Hsieh, et al. Corrosion behaviour of vacuum induction-melted Ni-based alloy in sulphuric acid[J]. Corrosion Science, 2010, 52(7):2323-2330.

Study of the Effect of Focusing and Oscillation of Electron Beam on the Structure and Properties of Welded Seams

I.V. Sitnikov[1,*], V.Ya. Belenkiy[2] and T.V. Olshanskaya[3]

[1,2,3]Department of Welding Production, Metrology and Materials Technology, Perm National Research Polytechnic University, Russia

*Corresponding author

Abstract. This paper presents the results of a study of the effect of focusing and oscillation of the electron beam on the structure, and the properties of welds made of heat-resistant pearlitic steel. Electron-beam welding using a static beam does not provide a uniform weld structure. According to the results of microhardness measurement, the weld metal has a high hardness with values on average 500 HV. The reason for this may be insufficient intensity of the mixing processes and insufficient residence time of the weld pool in the molten state. The use of electron beam sweeps during welding leads to a decrease in the microhardness of the weld metal to values of 450 HV. Also, the application of electron beam sweeps improves the weld geometry - a smoother transition from the upper part of the weld to the middle, more rounded shape of the root of the weld.

1. Introduction

Often, the use of electron beam welding technology with the use of a static beam results in welded joints with a number of defects inherent in this welding method, namely, the porosity of welded joints, defects in the root portion of the weld, the occurrence of cold and hot cracks, as well as uneven structure the width of the welded joint.

Recently widely used technology of electron-beam welding with various dynamic effects on the electron beam, which provide improved formation of the weld and the almost complete absence of defects. Such dynamic effects include the oscillation of an electron beam along various trajectories, the splitting of an electron beam into several thermal sources, and the modulation of current and focusing. The use of such technological methods contributes to obtaining defect-free welded joints, forming a homogeneous structure of welds when welding dissimilar materials, and the absence of porosity of welds during electron beam welding [1-6].

In this paper, studies were conducted on samples of heat-resistant steel of pearlite class 20Cr3MoVW.

Electron-beam welding was carried out using electron beam oscillations along various paths and different beam focusing, as in welding steels and alloys prone to the formation of quenching structures, to obtain a uniform structure of the welded joint, to improve the mechanical characteristics, to avoid defects, it is advisable to use electron beam oscillation along different paths.

The choice of electron beam oscillations is due to the fact that to obtain a favorable structure and properties of the compound, mixing and complete penetration of the material is required.

In order to obtain good mixing and eliminate root defects in this case, the most favorable method of oscillation are X-shaped oscillations. They allow you to capture metal in a wider range on either side of the joint, which other basic types of reamers, such as circular and longitudinal, do not allow.

IOP Conf. Series: Materials Science and Engineering **611** (2019) 012009 doi:10.1088/1757-899X/611/1/012009

2. Welding Using Various Electron Beam Focusing

The welding of samples with an electron beam with different focusing was carried out according to the modes indicated in Table 1.

Table 1. Welding modes.

Sample number	Focus type	Focusing current, mA	Beam current, mA	Accelerating voltage, kV	Welding speed, mm / s
1	under focus	790	63	60	5
2	sharp focus	800	43	60	5

As a result of welding, two welds were obtained, their macrosections are presented in Figures 1, 2.

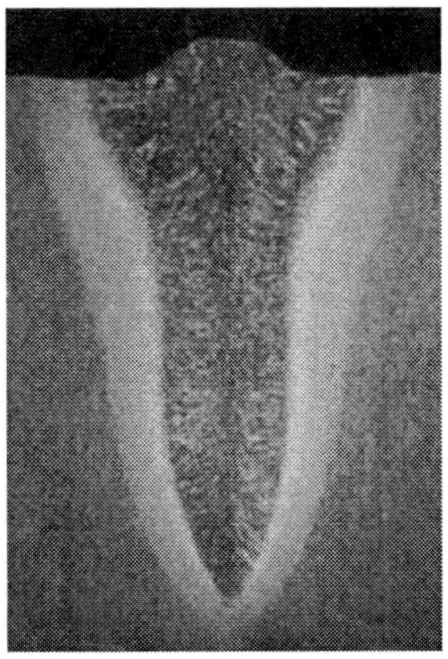

Figure 1. Sample No. 1

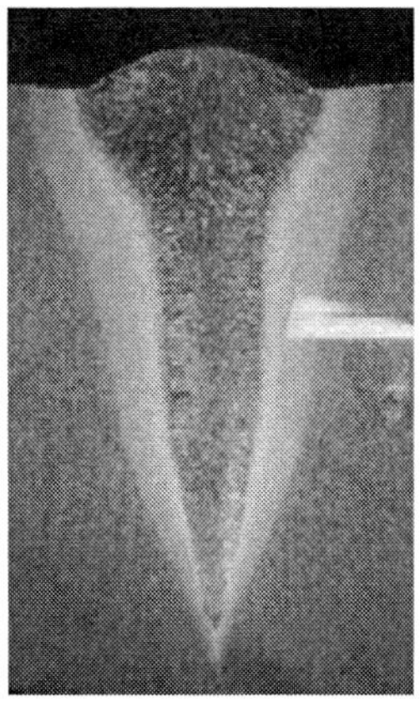

Figure 2. Sample No. 2

When using an under-focused beam, the width of the seam and the heat-affected zone increases. When using sharp focusing, the width of the seam and the heat-affected zone are reduced, but the root of the seam has a pronounced peak shape, which is an unfavorable factor.

The study of the microstructure showed that the resulting welds have a non-uniform structure in width and more uniform in depth. The microstructure of the obtained welds is shown in Figures 3, 4.

The formation of such a heterogeneous microstructure is most likely due to the different crystallization temperatures and the insufficient residence time of the weld pool in the liquid state. The resulting inhomogeneous structure of the welded joint is the cause of the non-uniformity of properties, which is clearly seen in the measurements of microhardness along the width and depth of the seam. Microhardness graphs are presented in Figures 5 and 6, the measurement step distance is 0.17 mm.

From the obtained results it is seen that when welding with an electron beam with sharp focusing, the width of the seam and the heat-affected zone are reduced. However, when welding with an unfocused beam, the microhardness values across the seam width are more uniform in comparison with the seam welded with sharp focusing of the electron beam. It is also worth noting that the microhardness values of welds are at the same level - 500 HV. This suggests that a change in the focusing of the electron beam does not have a significant effect on the microhardness of the welded joint, but only affects the final geometry of the weld.

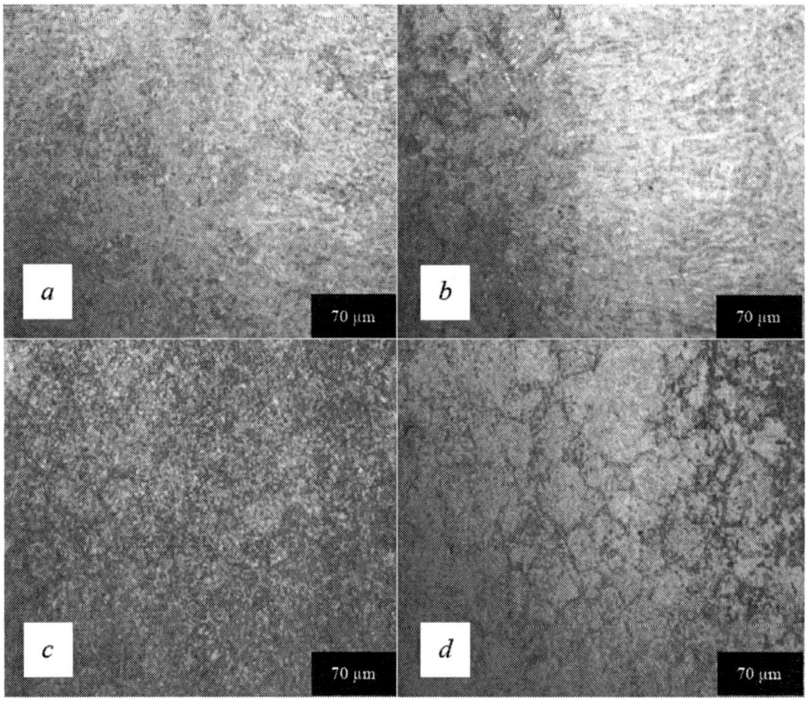

Figure 3. Microstructure of the welded joint of sample No. 1: a - seam; b - fusion line; c - HAZ; d - base + HAZ.

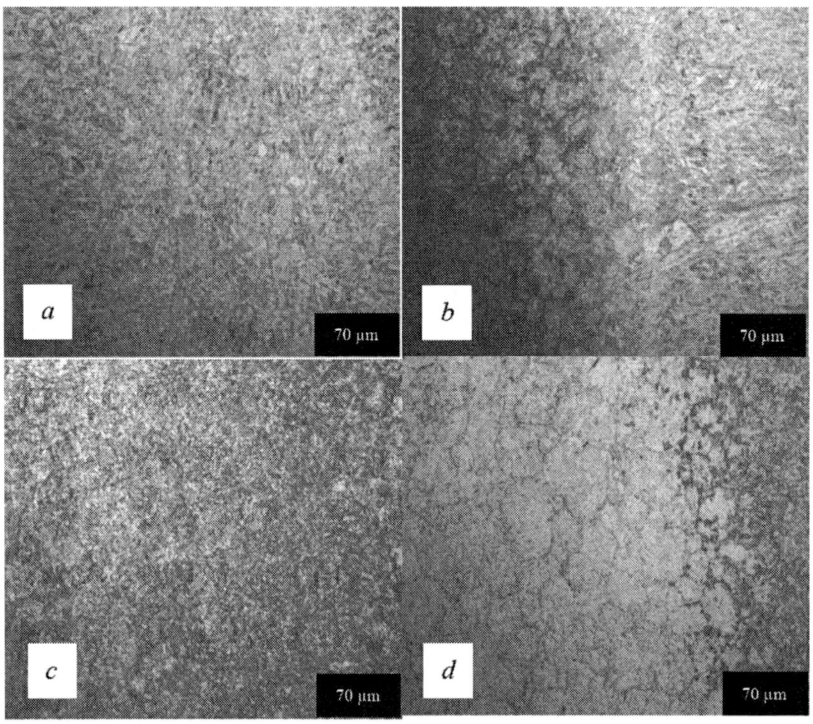

Figure 4. Microstructure of the welded joint of sample No. 2: a - seam; b - fusion line; c - HAZ; d - base + HAZ.

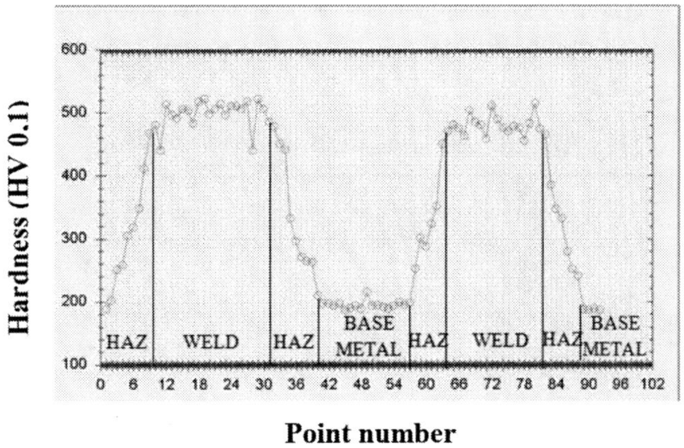

Figure 5. Microhardness graph across the width of welded joints.

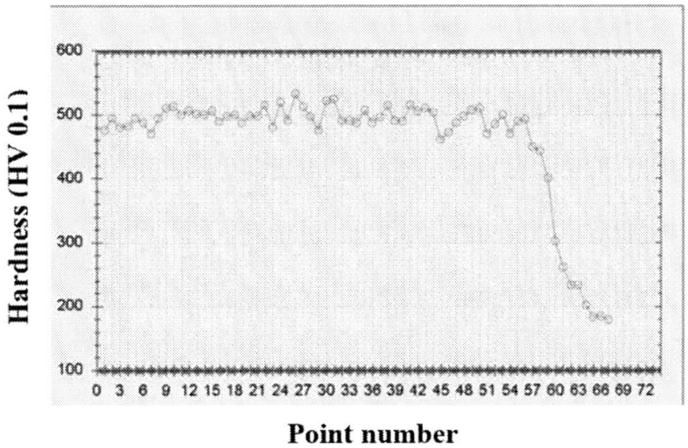

Figure 6. Microhardness graph for the depth of welded joints.

3. Welding Using Electron Beam Oscillations along Various Paths

The welding of samples with an electron beam with different types of oscillations was carried out according to the regimes indicated in Table 2.

As a result of welding, two welds were obtained, their macrosections are presented in Figures 7, 8.

Table 2. Welding modes.

Sample number	Oscillation type	Focusing current, mA	Beam current, mA	Accelerating voltage, kV	Welding speed, mm / s
3	∣ ∣	813	65	60	5
4	X-shaped	813	60	60	5

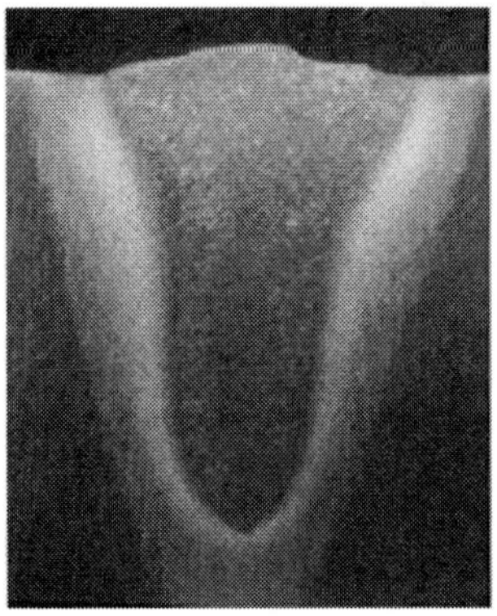

Figure 7. Sample No. 3

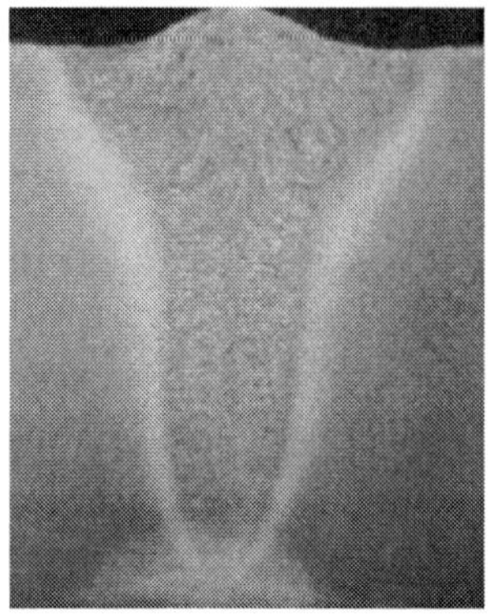

Figure 8. Sample No. 4

The macroscopes of the obtained samples show that when using the scanning of the electron beam in the form of two lines with a point, the width of the seam and the heat-affected zone increases. The root of the seam has a less peaked shape in comparison with the seam welded by a static beam. There is also a smoother transition from the top of the seam to its middle.

When using the X-shaped sweep, the width of the seam increases slightly, and the heat-affected zone decreases.

The study of the microstructure showed that the resulting welds also have a non-uniform structure in width and more uniform in depth, however, when measuring microhardness, its values in the same areas are less than the microhardness in the weld welded by a static beam. Microhardness graphs are presented in Figures 9 and 10, the measurement step distance is 0.17 mm.

The use of an electron beam scan reduces the microhardness values in a welded joint. The best sweep is the X-shaped sweep of the electron beam as the resulting geometry of the welded joint is the most optimal. However, it is worth noting that when using different types of oscillations, the level of microhardness is at the same level - 450 HV. This suggests that the sweep geometry does not have a significant effect on the microhardness of the welded joint, but only on the final geometry of the weld.

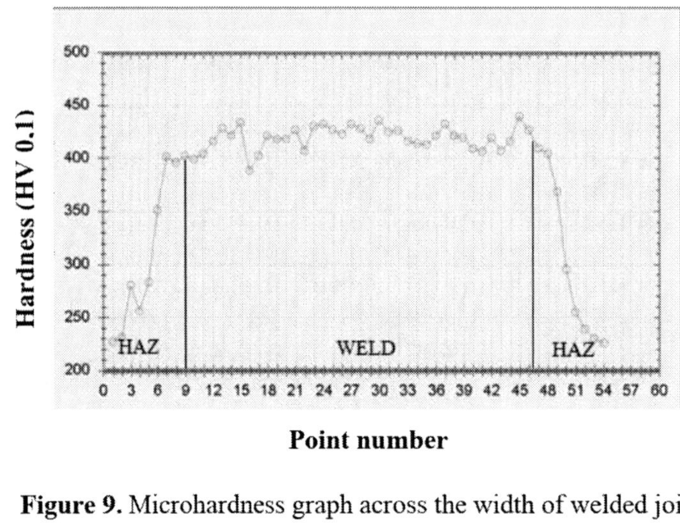

Figure 9. Microhardness graph across the width of welded joints.

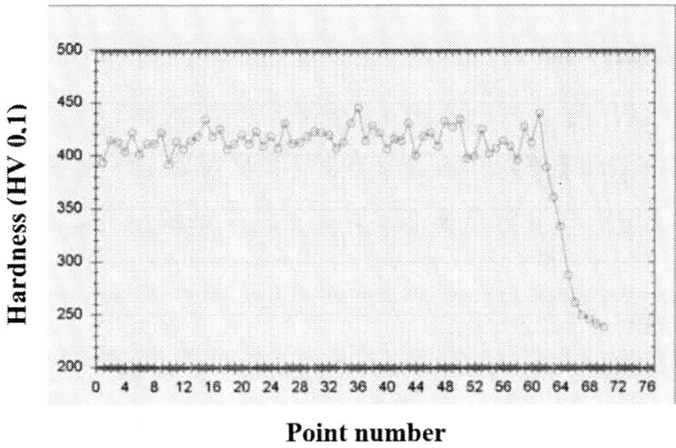

Figure 10. Microhardness graph for the depth of welded joints.

4. Conclusion

Electron-beam welding of steel 20Cr3MoVW with the use of a static beam does not provide a uniform weld structure. According to the results of microhardness measurement, the weld metal has a high hardness with values on average 500 HV. Presumably the reason for this is the insufficient intensity of the mixing processes and the insufficient residence time of the weld pool in the molten state.

The use of electron beam scans in welding steel 20Cr3MoVW leads to a decrease in the microhardness of the weld metal to values of 450 HV. Also, the application of electron beam sweeps improves the weld geometry - a smoother transition from the upper part of the weld to the middle, more rounded shape of the root of the weld. However, when using the sweep in the form of two lines with a point, the width of the seam and the heat-affected zone significantly increase. When using the X-shaped sweep, the width of the seam increases slightly, and the heat-affected zone decreases.

The results obtained are comparable with the results of previous studies in [4], [5]. The uniqueness of this work lies in obtaining the results of the regularity of the effect of focusing and oscillation of the electron beam on the structure and properties of welds when welding a specific class of steel, namely for heat-resistant steels of pearlite class.

Acknowledgments

The study was carried out with the financial support of the Government of the Perm Territory in the framework of the scientific project No. C-26/794 dated December 21, 2017.

References

[1] Olshanskaya T.V., Salomatova E.S. Review of modern methods of control of the electron beam in electron beam welding // Bulletin of the PNRPU. Engineering, materials science. - 2016. - Vol.18. №4. - p. 169-187. (in Rissian).

[2] Kaydalov A.A., Nazarenko O.K. Basic Techniques for Electron Beam Welding // Automatic Welding. –1986. - № 4.– P. 51–58. (in Russian).

[3] Masny H. Multi-beam technology in electron beam welding / Welding and Joining Institute. – 2006. – Vol. 34. – P. 1–4.

[4] Electron beam welding with X-shaped trajectory / V.M. Yazovskikh, V.Ya. Belenky, T.V. Olshanskaya // Welding production. - 1994. - № 6. - p. 5-8. (in Russian).

[5] Effect of electron beam oscillations on the formation of the structure and properties of the weld / TV Olshanskaya, D.N. Trushnikov, V.Ya. Belenkiy, G.M. Mladenov // Welding production. - 2012. - № 11. - P. 13–18. (in Russian).

[6] Sitnikov I.V. Application of electron beam oscillations in electron beam welding // Master`s Journal. - 2015. - №1. - P. 87-93. (in Russian).

IOP Conf. Series: Materials Science and Engineering **611** (2019) 012010 doi:10.1088/1757-899X/611/1/012010

Mechanical Properties and Strength Prediction of Ti Microalloyed Low Carbon Steel with Different Ti Content

Jiakuan Ren and Zhenyu Liu*

The State Key Lab of Rolling and Automation, Northeastern University, China

*Corresponding author

Abstract. The effects of Ti on microstructures, recrystallization and mechanical properties of microalloyed low carbon steels were studied. Hall-Petch relationship and Ashby-Orowan equation were applied to analyse the reason for the variation of strength. The results show that the recrystallization activation energy increases parabolically as the Ti content increases. Besides, the yield strength, tensile strength and the yield ratio increase in varying degrees with the increasing of Ti content, while the elongation decreases. The fine grain strengthening makes the greatest contribution to the strength, and the precipitation strengthening of TiC particles is the main reason why the strength of test steel increases significantly when the Ti content exceeds 0.042%. In addition, the strength prediction model was established for Ti microalloyed low carbon steel. The relative error of predicted yield strength and tensile strength of Q345B steel are within ± 8% and ± 7% respectively.
Keywords: Ti microalloyed steel; Ti content; mechanical properties; precipitation strengthening; fine grain strengthening.

1. Introduction

Since the middle of the last century, microalloy elements have been added to steels to improve the mechanical properties. Microalloying technology provides a strong support for the improvement of mechanical properties of steels. Like Nb and V, added to the steels as a microalloying element, Ti can improve the properties of the steels greatly, and Ti is a strong carbide forming element, which can play a good precipitation strengthening role[1]. At the same time, single Ti microalloying can avoid mixed crystal phenomenon which is widespread in Nb microlloying[2], higher finishing rolling temperature and coiling temperature can reduce the equipment load, which makes it easier for industrial production[3,4]. Besides, Ti has unparalleled economic advantages over Nb and V. In a word, Ti microalloying technology has great potential in improving the properties of steels and green manufacturing.

The chemical properties of Ti is lively, so it's easier to form large-volume inclusion with O, N and S, which will lead to the deteriorated performance of steels. Besides, the precipitation of TiC has a high sensitivity of temperature and cooling rate, resulting in instability of the properties of the steels. As a result, there are few applications of Ti microalloying. In recent years, with the improvement of metallurgical process and the controllability of temperature in rolling process, single Ti microalloying method has been proposed and promoted gradually. Du et al.[5] studied the nano-carbides in Ti microalloyed steels and their strengthening effects under ultra-fast cooling conditions. It was found that there were a large number of nanoscale FexC and TiC precipitates in Ti microalloyed steel. Wang et al.[6] studied the effect of Ti and Mn contents on the precipitation characteristics and strengthening mechanism in Ti microalloyed steel produced by CSP. The results showed that not only was the ratio of low-angle boundary higher, but also the number of TiC particle below 10 nm increased significantly in the steel with higher contents of Ti and Mn. Although the studies on Ti microalloyed steel are increasing, the systematic study on the effect of different Ti content on the microstructures and properties of low carbon steel of C-Mn-Ti system remains very limited so far.

According to the above reasons, five test steels with different Ti content were fabricated in this paper. The effect of different Ti content on recrystallization and mechanical properties was studied, and how the recrystallization activation energy varies with different Ti content was also investigated. In addition, the reasons for the change in mechanical properties of the test steels were calculated and analysed. Based on the traditional C-Mn steel, the strength prediction model of Ti microalloyed steel was developed, and the strength of Ti-containing steel Q345B were predicted effectively.

2. Experimental Procedures

Five 50-kg ingots of Fe-1.2Mn-0.09C-0.2Si-(0.019, 0.031, 0.042, 0.05, 0.07) Ti (wt.%) steels (hereafter, referred to as 0.02Ti, 0.03Ti, 0.04Ti, 0.05Ti, 0.07Ti steels, respectively) were melted using a high frequency vacuum induction furnace. After the ingots had been solution-treated at 1220℃ for 3 h, they were hot-rolled at temperatures ranging from ~1100 to 860℃ into a plate with 7.0 mm thickness, and ultra-fast-cooled to 640°C at a cooling rate of 35°C/s, then the plates were slow-cooled in asbestos to simulate the coiling process.

The hot deformation experiment was simulated with MMS-300 thermo mechanical simulation. The specimens (size, Φ8 × 15mm) were heated to 1220 °C at a rate of 20 °C/s and kept for 3 min. Then, cooled down to 850, 900, 950, 1000, 1050, 1100 °C at a rate of 10 °C /s and compressed with a strain of 0.6 and strain rate of 0.1/s after being held for 10 s.

Metallographic samples were taken from longitudinal section, and etched by 4% nitric acid alcohol solution before being analysed by Leica DM 2000 microscope. The thin foils were thinned by a Struers TenuPol-5 twin-jet electropolisher under 30 V at −25 °C in a 10.vol% perchloric acid alcohol electrolyte, and then examined in a FEI Tecnai G2 F20 field-emission transmission electron microscopy. The dog-bone-shaped tensile specimens were machined parallel to the rolling direction before testing in 30 t stretcher (WDW-300), The length, width and thickness of the gauge portion are 60, 12.5 and ~6 mm, respectively.

3. Results and Discussion

3.1. Recrystallization Behavior

The true stress-strain curves of test steels at $\dot{\varepsilon}$ (train rate) = 0.1 are shown in Fig. 1. When ε > ε_c, the dynamic recrystallization kinetics equation of low carbon steel can be described as [7,8,9]:

$$\ln \varepsilon_p = \ln(A \cdot d_0^{1/2}) + 0.15 \ln \dot{\varepsilon} + 0.15 \frac{Q_d}{RT} \tag{1}$$

Where, $\dot{\varepsilon}$ is train rate, ε_c is critical train, σ_p is peak stress, ε_p is peak strain, Q_d are calculated as 147, 184, 194, 198, 214 kJ/mol. As the Ti content increases, the recrystallization activation energy increases parabolically. The trend is shown in Figure 1 (f), and it can be expressed as:

$$Q_d = 71.7 + 7166.9[Ti] - 54771.2[Ti]^2, \ (\text{Ti}\% \leq 0.07) \tag{2}$$

Where $[Ti]$ is the mass percentage of Ti. The solid-solved Ti and precipitated particles could hinder the movement of dislocation and grain boundary, which inhibit the recrystallization process, so Q_d increases gradually. However, as the Ti content increases, the inhibition of the recrystallization by the solid-solved Ti and the precipitated particles is offset by the effect of refining initial austenite grain, so the rising tendency of Q_d tends to slow.

For low carbon steel and C-Mn steel, the values of α, c are 0.012, 0.67 respectively[10]. After further regression from the data in curves, the recrystallization volume fraction equation is as follows:

$$X_d = 1 - \exp\left[-0.77\left(\frac{\varepsilon}{Ad_0^{1/2}\left(\dot{\varepsilon}\exp\left(\frac{Q_d}{RT}\right)\right)^{0.15}} - 0.67\right)^{2.17}\right] \quad (3)$$

Where, A=0.034-0.176[Ti]+2.448[Ti]^2.

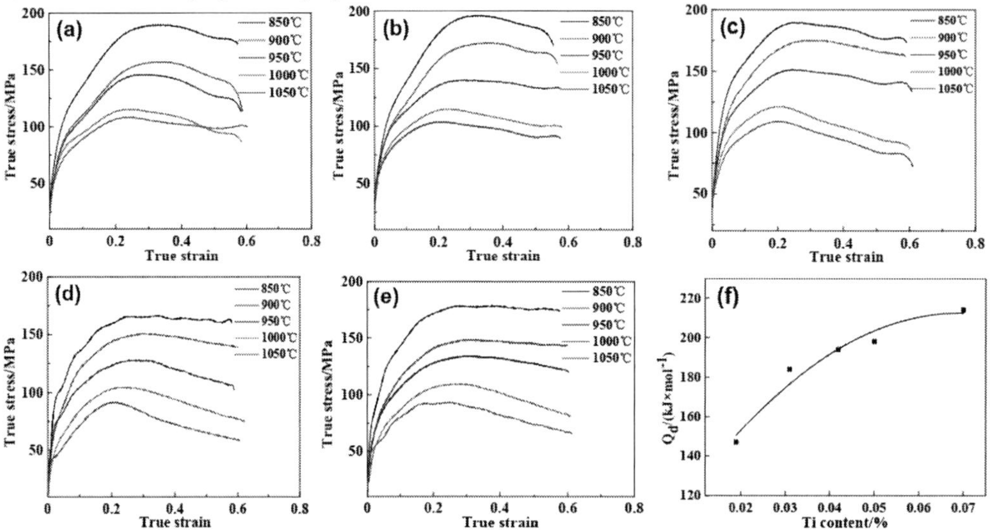

Figure 1. True stress-strain curves (a) 0.02Ti; (b) 0.03Ti; (c) 0.04Ti; (d) 0.05Ti; (e)0.07Ti, (f) the relationship between Qd and Ti content.

3.2. Microstructures and Mechanical Properties

3.2.1. Microstructure of Hot Rolled Plate. It can be seen from the optical microstructures in Fig. 2, when the Ti content is less than 0.042%, the grain size of test steel has little difference. When the Ti content exceeds 0.042%, the grain size becomes smaller. For phase proportion, when Ti content is lower than 0.05%, the microstructures are mainly composed of polygonal ferrite and a small amount of pearlite. When the Ti content exceeds 0.05%, a small amount of bainite appears.

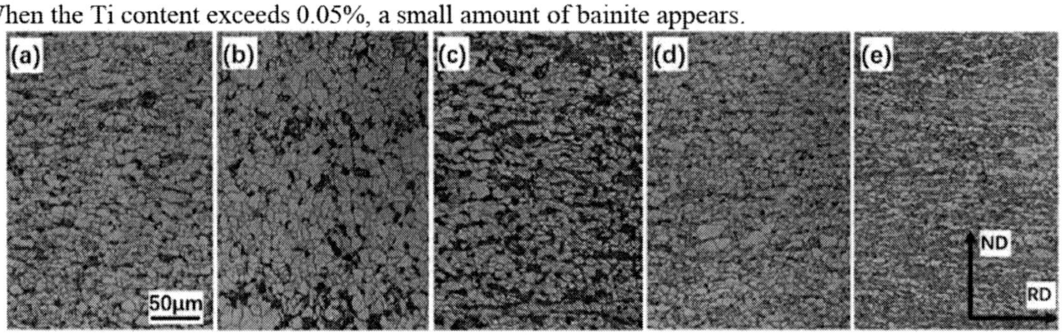

Figure 2. Optical microstructures of hot rolled plate.
(a) 0.02Ti; (b) 0.03Ti; (c) 0.04Ti; (d) 0.05Ti; (e)0.07Ti

3.2.2. Mechanical Properties. The mechanical properties of hot-rolled plate are shown in Table 1, as the Ti content increases, the yield strength, tensile strength and yield ratio increase gradually, while the elongation decreases gradually. When the Ti content exceeds 0.042%, The growing trend of strength is more significant.

3.2.3. Analysis of Strengthening Mechanism. For low-carbon steel, the main strengthening mechanisms include precipitation strengthening, solution strengthening and fine grain strengthening. The yield strength can be expressed as[11,12]:

$$\sigma_y = \sigma_0 + \sigma_s + \sigma_g + \sigma_d + \sigma_p \tag{4}$$

Where, σ_0 is the lattice frictional stress, and its value of low-carbon steel is 53MPa; σ_s, σ_g and σ_p denote solid solution strengthening, fine grain strengthening and precipitation strengthening respectively; σ_d denotes dislocation strengthening. Due to the lower dislocation density of test steel after higher temperature coiling, the dislocation strengthening is negligible in this paper.
Fine grain strengthening can be expressed by Hall-Petch formula[13,14]:

$$\sigma_g = k_y d_\alpha^{-0.5} \tag{5}$$

Where, $d_\alpha$is the average grain size of ferrite, mm; k_y is the proportionality coefficient, 17.4 MPa·mm$^{1/2}$ in low-carbon steel.
The solid solution strengthening can be expressed as[15]:

$$\sigma_s = 37[Mn]+83[Si]+59[Al]+38[Cu]+11[Mo]+33[Ni]-30[Cr]+680[P]+2918[N]+4570[C] \tag{6}$$

Where, [*M*] is the mass fraction of solid solution elements; [*C*], the content of solid-solved carbon in ferrite is taken as 0.01%; the form of elements such as manganese, copper, silicon, phosphorus, in steel are solid solution state, and their chemical composition in test steel can be directly introduced into Eq. (6).
However, Ti can easily combine with C and N to form compounds. According to the solubility product of TiN (Kunze, 1982) and TiC (Tayor, 1995) in austenite, when the coiling temperature exceeds 600 °C, the solid solubility of TiN and TiC in austenite are ~2.8×10^{-13}, ~5.4×10^{-6} respectively, this value is lower in ferrite. Besides, even in the case of low Ti content, N could combine with Al to form AlN, so the content of dissolved N in steel is very low. As a result, the solid solution strengthening effect of Ti and N are negligible.
For precipitation strengthening, Ti(C, N) belongs to the hard phase. When the precipitated phase is deformed, the dislocation almost can not cut it off. Therefore, its strengthening mechanism is Orowan strengthening theory of non-deformable particles, the strengthening mainly comes from the additional stress caused by dislocations bending around the precipitation phase and leaving Orowan dislocation loops[16]. It can be expressed by the Ashby-Orowan model[17]. In this paper, the shear modulus of matrix G = 80300MPa and the Burgers vector b = 2.5×10^{-4} μm were substituted into the model[18,19]:

$$\Delta\sigma_p = \left(\frac{10.8 f^{\frac{1}{2}}}{X}\right)\left[\ln(X/6.125\times10^{-4})\right] \tag{7}$$

Where X is the average diameter of the precipitated phase. According to TEMs statistics, X is determined as 0.01 μm; f is the volume fraction of precipitation phase.
The solubility of TiN in austenite is very low. Under soaking conditions, the solid solubility of TiN in austenite at 1220 °C is only 6.5×10^{-6}. It is considered that most of TiN particles have precipitated from the austenite. The precipitated TiN particles will coarsen during the cooling process of vacuum melting. Hence, the TiN particles is weak in precipitation strengthening, and the precipitated particles dominant in precipitation strengthening is TiC. Its volume fraction can be expressed as [12]:

$$f = (Ti-[Ti])\frac{A_{Ti}+A_C}{A_{Ti}}\cdot\frac{d_{Fe}}{100 d_{TiC}} \tag{8}$$

Where, A_{Ti}, A_c are the atomic weights of Ti, C respectively; d_{Fe}, d_{TiC} are the density of the iron matrix and the TiC phase, taken as 4.944g / cm^3 and 7.88g / cm^3, respectively; [*Ti*] is effective titanium, which

IOP Conf. Series: Materials Science and Engineering **611** (2019) 012010 doi:10.1088/1757-899X/611/1/012010

combines with carbon to form TiC, according to the stoichiometric ratio, there is following relationship[20]:

$$[Ti] = w(Ti) - 3.4w(N) - 3w(S) \quad (9)$$

Where, $w(Ti)$, $w(N)$, $w(S)$ are the mass percentage of Ti, N, S elements in steel.
Substituting the calculated results into Eq. (7) and (8), the results are shown in Table 1.
It can be seen from Fig. 3 (a) that the calculated yield strength agrees with the measured value basically, the calculated yield strength is slightly lower than the measured value due to the neglect of dislocation strengthening. In addition, the fine grain strengthening contributes most to the strength of the test steel, and the precipitation strengthening is the main reason for the difference of the strength of the test steel.

Table 1. Physical parameters of test steel.

Steel	Ferrite d/μm	σ_g/MPa	σ_s/MPa	$[Ti]$/%	f	$\Delta\sigma_p$/MPa	Calculated σ_y/MPa	Measured σ_y/MPa	R_m/MPa	TEL/%
0.02Ti	8.4	190	113	0	0	0	356	371	472	32
0.03Ti	8.9	184	112	0.0029	0.00006	23	372	387	488	29
0.04Ti	8.7	187	112	0.0079	0.00016	38	390	409	501	28
0.05Ti	7.9	196	111	0.0240	0.00048	66	426	444	525	26
0.07Ti	5.6	233	112	0.0440	0.00088	89	487	511	584	25

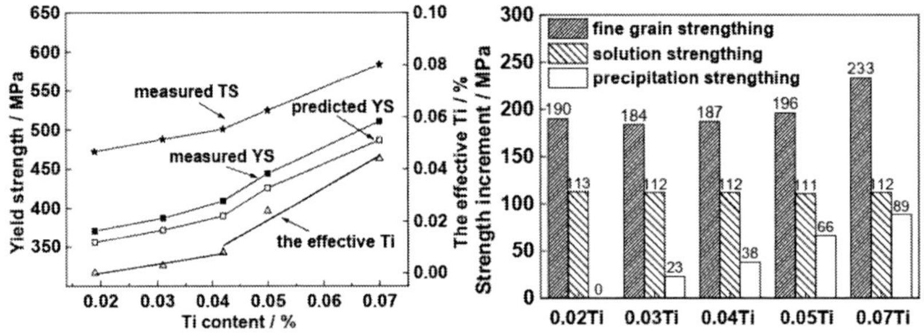

Figure 3. Strength calculation results of test steels.

According to Eq. (10)[20], the equilibrium solubility of TiN at 1220°C were calculated. As shown in Fig. 4 (a), most Ti elements are still dissolved in steel, but N elements are the opposite. In addition, the theoretical precipitation temperature of TiC was calculated as 1004, 1057, 1068, 1087, 1141°C respectively. When the temperature is below theoretical precipitation temperature, TiC will precipitate in austenite during tandem rolling, at the γ/α interface during transformation, or in supersaturated ferrite during coiling. The short gap-time during tandem rolling and rapid cooling on run-out table will inhibit the precipitation of TiC, so the most part of dissolved Ti precipitated during coiling process. As shown in fig. 4 (a), as the Ti content increases, the dissolved Ti element in matrix also increases gradually. Besides, when the Ti content exceeds 0.042%, the increasing trend becomes more obvious. Because of rapid increase of chemical driving force and dramatic decrease of diffusion velocity of elements at lower transformation temperature, a great number of nanometre-sized TiC precipitated from the ferrite during coiling process, shown in Fig. 4 (c), which play a role of pinning dislocation and improve the precipitation enhancement increment, so the strength of 0.05Ti and 0.07Ti steel increase significantly.

$$\frac{\mathrm{Ti}-[\mathrm{Ti}]}{\mathrm{N}-[N]} = 3.4 \quad (10)$$

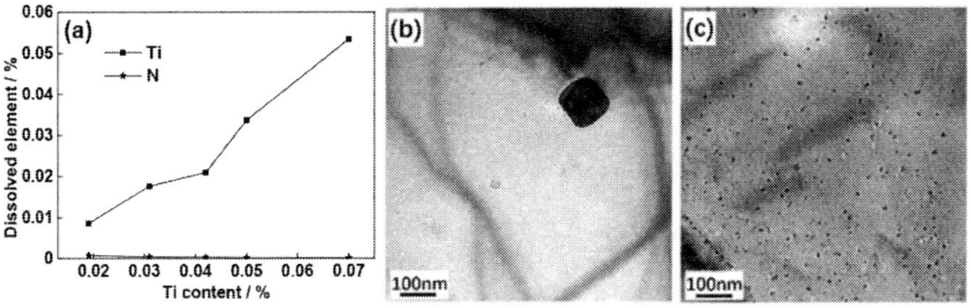

Figure 4. (a) Equilibrium solubility of TiN at 1220℃, precipitation of (b) TiN, (c) TiC.

For fine grain strengthening, 0.02Ti, 0.03Ti and 0.04Ti steel have little difference, fine grain strengthening is enhanced in 0.05Ti and 0.07Ti steel. Most of the TiN particles have precipitated during the cooling process of vacuum smelting, because of higher temperature and slow cooling speed, the TiN particles will coarsen, and its inhibition effect on recrystallization is weaker. However, the TiC nucleating in austenite during tandem rolling can inhibit the growth of recrystallized austenite grains effectively. They can also act as nucleation sites to refine grain. In addition, the TiC nucleating at the γ/α interface during transformation can pin the grain and inhibit grain growth. The dissolved Ti element and the theoretical precipitation temperature of TiC of 0.05Ti and 0.07Ti steel are significantly higher than other steels, so more precipitated TiC particles can promote grain refinement.

4. Strength Prediction of C-Mn-Ti Steel

The study on the prediction and control of microstructure-property evolution plays an important role in improving the green and intelligent development of the steel industry. Based on the research on the strengthening mechanism of Ti microalloyed low carbon steel[21], the following model to predict the strength of Ti-containing microalloyed steel was established. In this model, The effect of small amount of bainite on the strength is corrected by the coefficient of the last item.

$$YS=37[Mn]+83[Si]+59[Al]+38[Cu]+11[Mo]+33[Ni]-30[Cr]+680[P]+\sigma_p+17.4d_\alpha^{-0.5}+98.7 \quad (11)$$

$$TS=634.7[C]+53.6[Mn]+99.7[Si]+651.9[P]+472.6[Ni]+3339.4[N]+\sigma_p+11d_\alpha^{-0.5}+194.46 \quad (12)$$

Where, YS-yield strength, TS-tensile strength, MPa; d_α -ferrite grain size, mm; σ_p -precipitation strengthening.

The strength of the representative Ti microalloyed low carbon steel Q345B steel were calculated to verify the accuracy of this model. Fig. 5 shows the prediction results of Q345B steel. According to the results of 198 sets of data, it was found that the relative error of yield strength was within ± 8% and the relative error of tensile strength was within ± 7%. Besides, the average absolute error and standard deviation of the yield strength and tensile strength were small, which ensure the stability of this model.

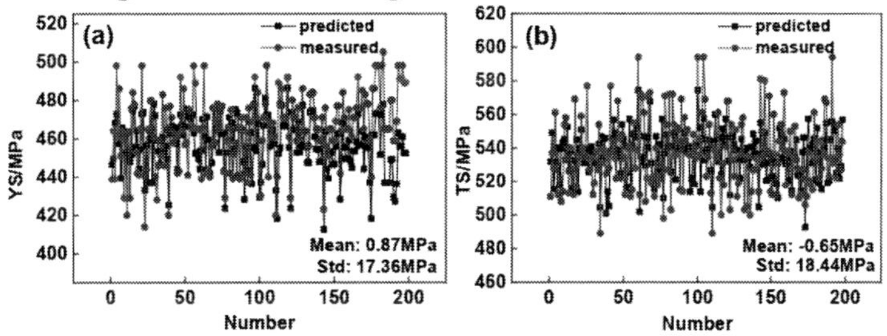

Figure 5. Strength prediction of Q345B steel.

5. Conclusions

(1) The dynamic recrystallization model of Ti microalloyed low carbon steel has been regressed. It was found that the recrystallization activation energy increases parabolically as the Ti content increases. The solid-solved Ti and precipitated particles could hinder the movement of dislocation and grain boundary,

which inhibit the recrystallization process, so the recrystallization activation energy increases gradually. As the Ti content increases, the inhibition of the recrystallization by the solid-solved Ti and the precipitated particles is offset by the effect of refining initial austenite grain, which makes the recrystallization activation energy increase slowly.
(2) As the Ti content increases, the yield strength, tensile strength and yield ratio increase, while the elongation decreases. When the Ti content exceeds 0.042%, the strength of the test steels increases significantly. Precipitation hardening of nanometre-sized TiC precipitated from the ferrite during coiling process is the main reason.
(3) The calculated yield strength agrees well with the measured value. The fine grain strengthening contributes most to the increment of strength, while the precipitation strengthening is the main reason for the difference of the strength of the test steel.
(4) On the basis of C-Mn steel, the strength prediction model of Ti microalloyed low carbon steel has been developed. In addition, the strength of the representative Ti-containing steel Q345B has been predicted, the relative error of predicted yield strength and tensile strength are within ± 8% and ± 7% respectively.

Acknowledgments
This work was financially supported by Joint Fund of Iron and Steel Key Project (UI460204), and the State Natural Sciences Foundation of China (U1660117).

References
[1] Shi Z, Chai X, Chai F, et al. The mechanism of intragranular ferrite formed on Ti-rich (Ti,V)(C,N) precipitates in the coarse heat affected zone of a V–N–Ti microalloyed steel. Materials Letters. 2016;175(14):266-270.
[2] Mao X, Chen Q, Zhu D. Recent development of microalloying technology in thin slab casting and rolling process. Iron & Steel. 2008;43(4):1-9..
[3] Mao X, Huo X, Sun X, et al. Strengthening mechanisms of a new 700MPa hot rolled Ti-microalloyed steel produced by compact strip production. Journal of Materials Processing Tech. 2010;210(12):1660-1666.
[4] Eghbali B. Microstructural development in a low carbon Ti-microalloyed steel during deformation within the ferrite region. Materials Science & Engineering A. 2008;480(1–2):84-88.
[5] Du K, Yu Y, Zhang S, et al. Nano-carbide precipitates in Ti microalloyed steel under ultra fast cooling condition and their strengthening effect. Nonferrous Metals Science & Engineering. 2016;7(4): 27-32.
[6] Wang C, Yong Q, Sun X, et al. Effects of Ti and Mn contents on the precipitate characteristics and strengthening mechanism in Ti microalloyed steels produced by CSP. Acta Metallurgica Sinica. 2011;47(12):1541-1549.
[7] Kostryzhev AG, Shahrani AA, Zhu C, et al. Effect of deformation temperature on niobium clustering, precipitation and austenite recrystallisation in a Nb–Ti microalloyed steel. Materials Science & Engineering A Structural Materials Properties Microstructure & Processing. 2013;581(581):16-25.
[8] Sellars CM, Tegart WJM. Hot Workability. Metallurgical Reviews. 1972;17(1):1-24.
[9] Sanghyun C, Kang KB, Jonas JJ. The Dynamic, Static and metadynamic recrystallization of a Nb-microalloyed steel. Transactions of the Iron & Steel Institute of Japan. 2001;41(1):63-69.
[10] Zhou, X, Liu Z, Wu D, et al. Determination of model parameters of dynamic recrystallization for Nb bearing steels during flexble thin slab rolling. Acta Metallurgica Sinica. 2008;32(4):1188-1192.
[11] Mao X, Chen Q, Sun X. Metallurgical interpretation on grain refinement and synergistic effect of Mn and Ti in Ti-microalloyed strip produced by TSCR. Journal of Iron and Steel Research, International. 2014;21(1):30-40.
[12] Xu G, Gan X, Ma G, et al. The development of Ti-alloyed high strength microalloy steel. Materials & Design. 2010;31(6):2891-2896.

[13] Li D, Ye Q, Zhou C, et al. Study on precipitation and fine grain strengthening effects of low carbon steel with Nb-Ti. Angang Technology. 2012(4):21-25.
[14] Qu Y. Principles of steelmaking. Beijing (CN): Metallurgical Industry Press; 1994.
[15] Yong Q. Microalloyed steel: physical and mechanics metallurgy. Beijing (CN): Machinery Industry Press; 1989.
[16] Yong Q. The second phase of steel. Beijing (CN): Metallurgical Industry Press; 2006.
[17] Gladman T. Precipitation hardening in metals. Metal Science Journal. 1999;15(1):30-36.
[18] Gladman T. Physical metallurgy of microalloyed steels. London (UK): Maney Publishing; 1997.
[19] Zhang G, Yang S, Jiang K. Influence of precipitation strengthening on yield strength increment $\Delta\sigma p$ and chemical composition designing of low alloy steel. Journal of Hohai University Changzhou. 2000;14(2): 16-21.
[20] Mao X. Microalloying technology of thin slab continuous casting and rolling. Beijing (CN): Metallurgical Industry Press; 2008.
[21] Hodgson PD, Gibbs RK. A mathematical model to predict the mechanical properties of hot rolled C-Mn and microalloyed steels. Transactions of the Iron & Steel Institute of Japan. 1992;32(12):1329-1338.

Effect of Ultrasonic Peening Treatment on VHCF Behavior of Friction Stir Welded Joints in Aluminum Alloys

Ruofan Zhang[1,2], Xue Li[3,4], Yongjie Liu[3,4], Chao He[2] and Qingyuan Wang[2,3,4,*]

[1]School of Architecture and Civil Engineering, Xihua University, Chengdu 610039, China
[2]School of Architecture and Civil Engineering, Chengdu University, Chengdu 610106, China
[3]MOE Key Laboratory of Deep Earth Science and Engineering, College of Architecture and Environment, Sichuan University, Chengdu 610065, China
[4]Failure Mechanics and Engineering Disaster Prevention and Mitigation Key Laboratory of Sichuan Province, Sichuan University, Chengdu 610207, China

*Corresponding author

Abstract. Studying the mechanical properties of aluminum alloy 6061 friction stir welded (FSW) joints after ultrasonic peening treatment (UPT) under very high cycle fatigue (VHCF) loading. The effect of joints ultrasonic peening treatment was explored. It can be seen from the results that the ultrasonic peening treatment can form a reinforced layer on the surface of joints, and the fatigue crack source is transferred from the surface of the specimen to the inside. However, the change of fatigue crack initiation site does not improve the fatigue strength of specimen through ultrasonic peening treatment. The fatigue strength of the aluminum alloy 6061 joint after ultrasonic peening treatment is even lower than as a received condition. The characteristics of fatigue crack initiation and the mechanism of surface strengthened layer on fatigue strength were analyzed and discussed.
Keywords: friction stir welding; S-N Curves; ultrasonic peening treatment; very high cycle fatigue; ultrasonic Fatigue; fatigue crack initiation.

1. Introduction

Ultrasonic peening treatment (UPT) was proposed by Russian scholars in the early 1970s [1, 2]. The principle of UPT was that Ultrasonic frequency vibration displacement pushed the peening head to produce pressure, which was applied to the welding seam surface. The plastic deformation layer on the specimen surface was induced and the mechanical properties of the impact site were strengthened. Ultrasonic peening produced beneficial residual compressive stress layer on the surface of the welding seam as well; the stress can significantly improve the fatigue performance of welded joints [3].
The surface condition of the friction stir welding joint had a significant influence on fatigue performance. Fatigue cracks were usually generated from the surface of the specimen, and the fatigue strength of the joint could be effectively improved by removing the flash and arc mark [4, 5]. The effect of ultrasonic peening treatment on the friction stir welding head fatigue behavior research is still not clear so far. Many vital manufacturing industries have been applied to aluminum alloys cumulatively, such as the automobile industry, aeronautics, and the military because of low density and excellent mechanical properties of aluminum alloy [6], and these applications also lead to deeper research. Rodopoulos [7] studied the influence of ultrasonic peening on the fatigue crack propagation

characteristics of FSW joint in AA6061. The resistance of this surface treatment to fatigue cracks was considered to be observed only at low-stress intensity factor amplitude, which is mainly caused by the release of residual stress during cyclic loading. However, the effect mechanism of UPT on the long life fatigue of FSW joints has not been studied and discussed by relevant experiments.

In this work, the ultrasonic fatigue experiment system is applied to the FSW joints of aluminum alloy 6061 after UPT, to study the difference of VHCF performance of joints before and after surface treatment. The fatigue crack initiation and failure mechanism were discussed based on the observation of a scanning electron microscope.

2. Experimental Methods

2.1. Material and Specimen

The present experimental material is aluminum alloy 6061 (AA 6061-T651), which is rolled as plates with a thickness of 10mm. It's composition of chemistry and properties of mechanics were showed respectively in Table 1 and Table 2. Geometric dimensions of the test piece for fatigue test are shown in figure 1. In the friction stir welding process, the diameters of stirring head at the root and top were 10mm and 7mm, respectively. The length of the stirring head was 9.2mm, while the diameter of the curved profile shoulder was 24 mm. The inclination angle of stirring head axis and plate normal was 2°, with the speed of rotation is 600r/min and the forward speed of welding is 400 mm/min. The specimen surface before and after UPT are shown in figure 2[8].

Table 1. Chemical compositions of 6061-T651 in this test (wt.%).

Cr	Cu	Fe	Si	Mn	Mg	Ti	Zn	Al
0.04-0.35	0.15-0.40	0.70	0.40-0.80	0.15	0.80-1.2	0.15	0.25	bal

Table 2. Mechanical properties of 6061-T651 in this test.

ρ (g/cm3)	E (GPa)	σb (MPa)	σs (MPa)	δ (%)	Hardness (Vickers)
2.62	68.9	310	276	12%	102

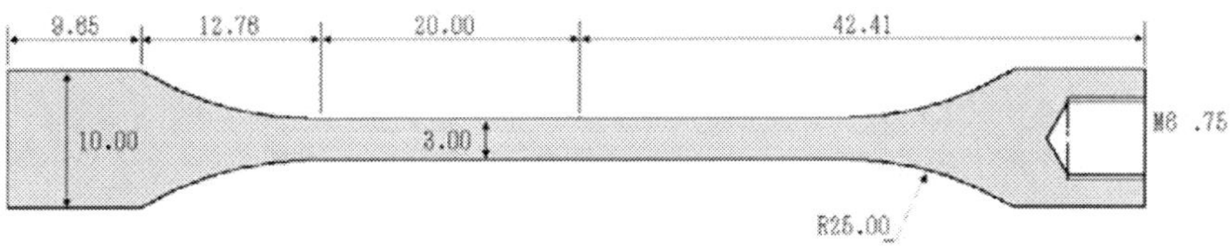

Figure 1. Geometric dimensions diagram of specimen size

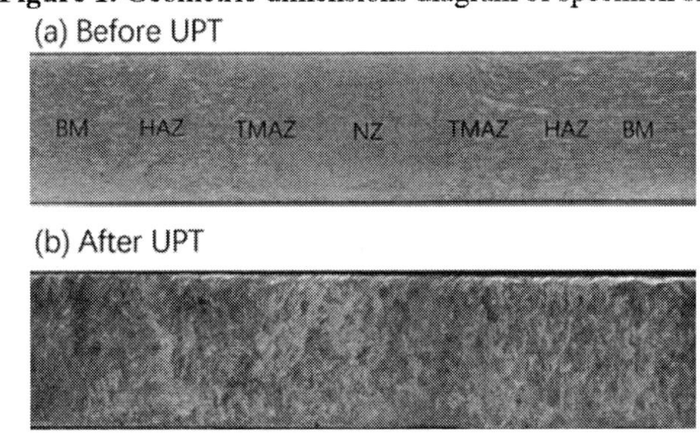

Figure 2. Microstructure of the joint (a)and the morphology of specimen surface after UPT treatment (b).

2.2. *Experimental Methods*

For surface observation, the weld cross-sections of all tested specimens were first polished using different grades of emery paper and then subjected to electrolytic etching at 30 V for 90 s using perchloric acid/ethanol solution to reveal the microstructure. The ultrasonic fatigue test system test was carried out on SHIMADZU USF-2000 at the room temperature, and with the stress ratio of R=-1. The maximum cyclic load was applied to 10^9 cycles unless the specimen failed beforehand. The specimens were sprayed with high-temperature black paint to understand the heat dissipation of the specimen under very high frequency loading, and the surface temperature field of the specimen under different loading stresses was recorded by an infrared camera. The results indicated that the temperature increase in this test was no more than 10 °C After the fatigue failure, the fracture surface was observed by scanning electron microscope (SEM, JOEL-6510LV) and optical microscope to discuss the failure mechanism of fatigue and evaluate the effect of UPT on fatigue strength.

3. Experimental Results

3.1. *S-N Curves*

Fatigue data of aluminum alloy 6061 specimens after ultrasonic impact treatment are shown in figure 3, in which the fatigue strength before UPT from He et al. is also present for comparison. The arrows which represented data showed that specimens did not cause a failure until 10^9 cycles. The fatigue life of the specimen after UPT was lower than untreated specimen at all levels of loading stress in the high-cycle region (life less than 10^7 cycles), and this phenomenon indicated that the treatment method did not enhance the fatigue strength of the joints. Instead, it decreased to some extent. In the range of VHCF, with the decrease of loading stress, the specimen before and after the UPT treatment both showed a decreasing trend, in which fatigue fail still occurred in the range of 10^7~10^9 cycles. That is to say, the traditional fatigue limit at 10^7 cycles does not exist for friction stir welded joints in AA6061.

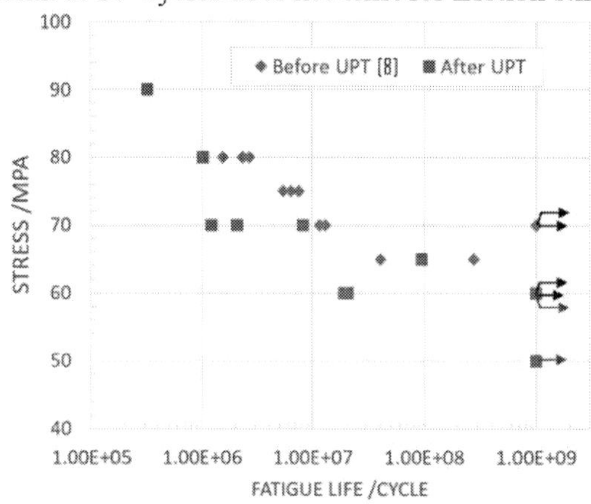

Figure 3. S-N curves for aluminum alloy 6061 ultrasonic test after UPT

In general, the study on the fatigue performance of ultrasonic peened welded joint found that the surface layer of the specimen is greatly refined, welding defects are reduced, microstructure becomes compact, and residual compressive stress could be introduced in the surface layer effectively at the same time. These factors make the specimen (after UPT) pose a greater improvement of fatigue strength, but this result in this test is opposite of above conclusions, the mechanism of friction stir welded joint in AA 6061 is probably different from the conventional fusion welded joints. The details are analyzed in the following parts via fracture analysis.

4. Fracture Analysis and Discussion

The failure specimen obtained the fracture morphology by using SEM. The mechanism information of fatigue crack initiation and propagation was investigated in studying the fracture morphology. In general, the resistances of fatigue crack initiation were deemed to play a predominant role on the mechanism of metallic fatigue failure in the VHCF regime [9]. The fatigue fracture morphology of aluminum alloy 6061 after UPT is shown in figure 4. For most metal materials, high cycle fatigue cracks usually originated from the surface of the specimen, while very high cycle fatigue cracks originating from the interior [10]. But this figure (Fig.4) showed that the fatigue crack originates from the interior of the specimen and is composed of the relatively flat plane (facet). The width of the plane zone ranges from 300μm to 450μm, which is the same as the grain size of AA 6061 in the vertical direction of the plate, this phenomenon indicates that the facets plotted by arrows should be formed along specific slip plane of a grain. After UPT, the facet region of the crack source is completely determined by the grain size. Therefore, the formation of the facet is mainly caused by the sliding cracking of grains along the slipping plane. In addition, the intracrystalline slip plane is usually formed under high stress. The width of specimen is almost occupied by limited grains, which are located at the fatigue crack source of AA6061 joint after UPT, and the two ends of the grain are directly affected by the ultrasonic peening treatment. We can infer that the formation of the intracrystalline slip plane is associated with UPT process, and this is one of the main reason to cause the decrease of fatigue strength in this test as compared with that before UPT.

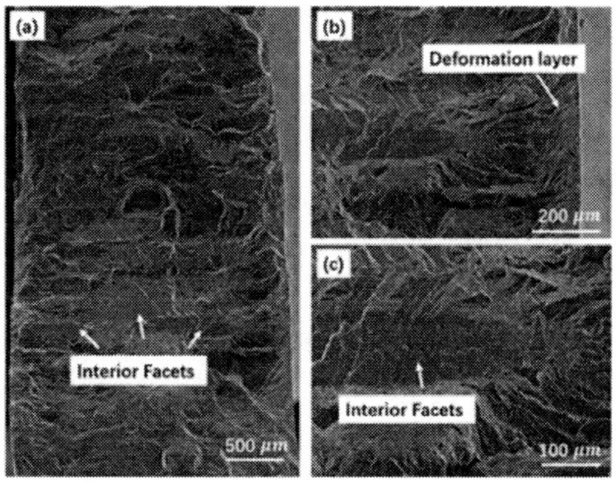

Figure 4. Fracture morphology of aluminum alloy 6061 impact specimen

In the process of ultrasonic peening, the impact head impinges on the surface of the sample at a very high speed and frequency, and strong plastic deformation is produced at specimen within a certain surface thickness range, which caused the formation of a strengthened layer at the same time. At the micro scale, dense dislocations and finer subgrains are formed in the strengthened layer [11]. The effect of UPT on the surface of the aluminum alloy 6061 joint fracture failure position in this test is shown in figure 4b. The strengthened layer is having a thickness of about 90 μm in this work. AA 6061 is presenting intracrystalline flat sliding plane in internal grain as shown in Fig. 4c, and its crack extension is mainly influenced by grain alignment and orientation. In the strengthening layer, the direction of crack growth is substantially perpendicular to the loading stress, and the two different crack propagation planes coalesced with the formation of dense river-like steps simultaneously.

For AA 6061 friction stir welded joint, fatigue crack is generated from local plastic slip band. UPT can refine the grains in the surface strengthening layer to the nanometer level. As a result, the surface slip band, grain boundary and the second phase interface increased accordingly. This phenomenon is beneficial to resist the initiation of fatigue cracks. Meanwhile, grain boundary can be used as an obstacle to the development of intra-grain slip dislocation. The reduction of grain size is conducive to the improvement of fatigue crack growth resistance at the initial stage [12]. Therefore, the UPT causes the fatigue crack initiation site of AA 6061 friction stir welded joint to transformed from the surface of the interior of specimens. The presence of the strengthened layer is beneficial to suppress the initiation and propagation of fatigue cracks. Besides, the residual compressive stress introduced at the surface can also reduce the maximum tensile stress on the surface of the specimen, thus reducing the possibility of fatigue cracks originating from the surface. In the meanwhile, due to the grain size of AA6061 is relatively large in this test, which individual length of one grain almost traverses the entire specimen, the fatigue cracks of the same grain are more likely to initiate from the sliding plane because the peening applied on both sides. Finally, the fatigue life of the specimens after UPT are reduced as compared with the untreated one.

5. Conclusion

In this paper, the ultrasonic fatigue test system is used to study the VHCF strength of AA6061 FSW joint after UPT, and the effect of surface strengthened layer on the fatigue crack initiation site and fatigue strength were analyzed and discussed. The results in this study can be summarized as follows:
(a) After UPT, the fatigue life of aluminum alloy 6061 joint is shortened than the untreated one.
(b) Fracture analysis found that the fatigue crack of specimens originated from large grains. The plane and impact treatment make the slip plane more susceptible to the occurrence of fatigue cracks, which is the main cause for the decrease of the fatigue strength of the AA6061 joint.
(c) UPT refines the grain at the surface of the specimen to form a dense strengthened layer, which can effectively improve resistance to the fatigue crack initiation and propagation. This is the main cause of the occurrence of fatigue cracks from the surface to the interior of specimen.

Acknowledgments

This work was financially supported by Chinese National Science Foundation (No.11572057) in Chengdu University.

References

[1] Mukhanov I, Golubev Y M. Strengthening Steel Components by Ultrasonically Vibrating Ball[J].Vestn. Mashin, 1966, 1152-53.
[2] Mordvintseva A. Ultrasonic treatment of welded joints[J]. Ultrasonic Applications in Welding Engineering, Moscow, TsINTIEnergomash, Proc. of NE Bauman MVTU, ed, 1959, 35.
[3] Trufiakov V. Ultrasonic impact treatment of welded joints[M]. 1995.
[4] Magnusson L, Kallman L. Mechanical properties of friction stir welds in thin sheet of aluminium 2024, 6013 and 7475[A]. In Proceedings of the Second International Symposium on Friction Stir Welding, Paper[C], 2000.
[5] Bussu G, Irving P. The role of residual stress and heat affected zone properties on fatigue crack propagation in friction stir welded 2024-T351 aluminium joints[J]. International Journal of Fatigue, 2003, 25(1): 77-88.
[6] Wang QY, Li T, Zeng XG. Gigacycle fatigue behavior of high strength aluminum alloys. Fatigue 2010;2(1):65–70.
[7] Rodopoulos C, Pantelakis S G, Papadopoulos M. The effect of ultrasonic impact treatment on the fatigue resistance of friction stir welded panels[J]. J Mater Eng Perform, 2009, 18(9): 1248-1257.
[8] Chao He, Yongjie Liu, Jiangfeng Dong, Qingyuan Wang. Through thickness property variations in friction stir welded AA6061 joint fatigued in very high cycle fatigue regime[J]. International Journal of Fatigue, 2016, 82(3), 379-386.
[9] Wang QY et al. Effect of inclusion on subsurface crack initiation and gigacycle fatigue strength. Int J Fatigue 2002;24(12):1269–74.
[10] Huang Z Y, Wagner D, Wang Q Y, etc. Effect of carburizing treatment on the "fish eye" crack growth for a low alloyed chromium steel in very high cycle fatigue[J]. Mat Sci Eng a-Struct, 2013, 559790-797.
[11] Mordyuk B N, Prokopenko G I. Ultrasonic impact peening for the surface properties' management[J]. Journal of Sound and Vibration, 2007, 308(3): 855-866.
[12] Masounave J, Baflon J-P. Effect of grain size on the threshold stress intensity factor in fatigue of a ferritic steel[J]. Scripta Metallurgica, 1976, 10(2): 165-170.

Simulation Research on Vibration Cutting Mechanism of Metal Matrix Composites

Liang Yang[1], Chunguang Xu[1,*], Bohua Fu[1] and Mingzhao Zhang[1]

[1]College of Mechanical Engineering, Dalian Jiaotong University, Dalian 116028, China

*Corresponding author

Abstract. Particle-reinforced metal composite materials have been widely used in the industry due to their excellent abrasion resistance and high strength, but they are typical hard-to-machine materials. A finite element model of the material matrix and enhanced particles have been established by the parametric modeling in this paper. Different from the previous development process, the particle is randomly distributed in this study, so the particle distribution has been better simulated than ever. The study shows that vibration-cutting particle-reinforced composites have reduced the phenomenon of particle suspension on the machined surface compared to conventional cutting, and the high-elastic-modulus reinforcing particles have been subjected to high stress, and the matrix between the particles have acted as a stress-transmitting "channel". When the particle's diameter is 20μm and the volume fraction is 30%, there is a stress transmitting channel in a certain range away from the tool tip. When the cutter contacts the particles, the vibration cutting will cause the initial crack near the particle, which cut off part of the stress transmitting channels, make the stress concentrated on the tool tip and lead to the material easier to remove.
Keywords: ABAQUS; parameter modeling; particle reinforced composites; vibration cutting.

1. Introduction

In many fields, traditional homogeneous metal materials have been unable to meet the diverse needs of human pursuit, such as light weight and high strength. Researchers have combined materials to create a wide variety of composite materials based on the desired properties. Among them, Metal Matrix Composites (MMCs) have the advantages of high strength, high modulus, high temperature resistance, non-flammability, etc., and they are widely used in military, aerospace, electronics and other industries requiring lightweight and high strength [1, 2]. The particle reinforced metal matrix composites have the characteristics of both the reinforced particles and the metal matrix. They have not only small mass and thermal expansion coefficient, but also high specific strength and stiffness [3]. The removal form of the particle reinforced composite material can be divided into four types: 1) the matrix and its inclusions together form chips; 2) the particles are pressed into the matrix material; 3) the particles are debonded due to crack propagation; 4) particles are crushed by extrusion [4]. In the four modes, there are pits on the machined surface by debonding of the particle. The broken particles plough machined surface with the cutter, and leave grooved scratches of different depths [5]. Therefore, the quality of machined surface of composite material is much lower than that of pure metal material due to the existence of reinforced particles, and there are a lot of defects (particle debonding, fracture, ploughing surface, etc.) [6]. Ultrasonic vibration cutting is based on traditional cutting, which is applied to the tool or workpiece to achieve pulsed cutting. More and more studies have shown that this cutting method can reduce the cutting force, improve the stability of the machining process, and inhibit the occurrence of machining damage [7]. Lu Shihong [8] used ABAQUS to study the formation mode of titanium alloy Ti-6Al-4V and H13 hardened steel under high strain rate and the formation mechanism of sawtooth chip under orthogonal cutting conditions. But it takes a long time to build a cutting simulation model. Jing xiu bing[9] modified J-C constitutive relation and Cockroft-Latham cutting fracture criterion. A three-dimensional ultra-precision cutting finite element model was established based on the large deformation Lagrange finite element theory. In order to study the effect of cutting conditions on the surface quality of ultra-precision cutting, a large number of mouse operations are needed in the finite element simulation process, which is particularly cumbersome.

In this study, parametric modeling is realized by means of script program, and the inefficient modeling work is avoided. The particle distribution algorithm is characterized by variable volume fraction, variable size, random radius, random particle distribution and no crossover. It lays a technical foundation for the further study of the influence of the size, the number and the distribution of the enhanced particles on the processing. By changing the amplitude curve used in velocity boundary conditions, the traditional cutting and vibration cutting are compared and analyzed by using the same model. The effects of vibration on cutting force, particle removal time and matrix damage and fracture are studied. The influence of the addition of reinforcing particles on the cutting force, surface quality, stress distribution and propagation of the metal matrix material, and the initiation and crack propagation of the metal matrix material will be studied.

2. Script-based Parameterized Finite Element Modeling of Particle Reinforced Composites

In the finite element modeling of reinforced particles, the material model of reinforced particles is set as an isotropic linear elastic model. The matrix material is the same as the pure metal material, which is simulated by the elastic-plastic model. Material parameters of matrix, reinforced particles and cutting tools are shown in Table 1. There are two simulation methods of particle distribution, one is periodic distribution, and the other is random distribution. The random distribution model of particles can fully consider the interaction between particles to a certain extent. The model can reflect the microstructure distribution of particle reinforced composites and accurately simulate the elastic modulus of composites [10]. In this study, the volume fraction of particle reinforced metal matrix composites is low volume fraction. Particle size is 20 microns.

Table 1. Material properties of metal matrix and particle and tool [11,12].

Material	**Density(kg·m^{-3})**	**Elastic Modulus(GPa)**	**Poisson's ratio**	**Thermal Conductivity(W m^{-1}·K^{-1})**	**Specific heat capacity J (Kg·°C)$^{-1}$**
AL6061	2700	68.9	0.33	167	896
SiC	3100	408	0.183	490	670
Cemented carbide tool	4250	850	0.08	2100	525

2.1. Material Constitutive Model and fracture Criterion

An elastoplastic constitutive equation is used to simulate the matrix material from the initial state until the damage in the metal matrix material model. The J-C viscous material model is used to simulate the properties of the matrix material. This model is often used to simulate the constitutive model of materials with large strain, high strain rate and high temperature. The following formula represents the basic form of the J-C model of von-Mises stress σ [13]:

$$\sigma=(A+B\varepsilon^{n})\left[1+C\ln\left(\frac{\dot{\varepsilon}}{\dot{\varepsilon}_0}\right)\right]\left[1-\left(\frac{T-T_r}{T_m-T_r}\right)^{m}\right] \tag{1}$$

Equation (1) consists of three parts. The effects of strain hardening $f_1(\varepsilon)$, strain rate strengthening $f_2(\dot{\varepsilon})$ and thermal softening $f_3(T)$ on flow stress are considered respectively. In the equation, ε is the equivalent plastic strain, $\dot{\varepsilon}$ is the plastic strain rate, and $\dot{\varepsilon}_0$ is the reference strain rate. T and T_m respectively represent the current temperature and the material melting temperature, and the T_r indicates the conversion temperature. When the temperature is below T_r, the material properties will no longer change with temperature. Five material constants A, B, C, n, and m were measured by torsion tests, static tensile tests, and Hopkinson bar dynamic tensile tests on different materials. The parameters of the J-C constitutive model are shown in Table 2 The parameters of the constitutive model of aluminum matrix can be obtained by correlative calculation.

Table 2. J-C model parameters of metal matrix materials [14].

Material	A(MPa)	B(MPa)	C	n	m	T_m(°C)	T_r(°C)
AL6061	324	114	0.002	0.42	1.34	582	20

2.2. The Establishment of Finite Element Model

Based on the commercial finite element software ABAQUS/explicit, a two-dimensional orthogonal cutting model for ultrasonic vibration cutting is established which is composed of two parts: workpiece and tool, and a micromechanical model including matrix phase and particle phase are established. The tool is set as an analytic rigid body. The cutter boundary condition assumes that the direction of speed is horizontal, and the speed is the actual linear speed of turning. The periodic amplitude curve in the finite element software ABAQUS is used to exert uniform velocity load on the tool. The amplitude curve is set for the corresponding vibration period and the vibration direction and cutting direction are in the same straight line to realize the effect of vibration cutting. Figure 1 shows the particles contacted by the tool in turn, numbered in order of contact with the tool.

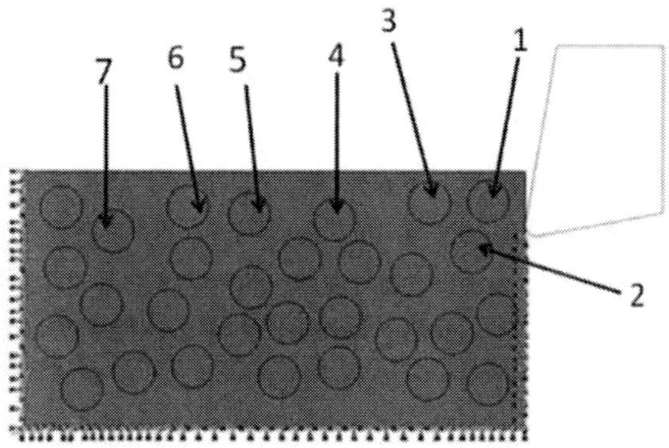

Figure 1. Two dimensional orthogonal cutting geometry model for particle reinforced composites

In the modeling process of particles reinforced composites, the geometric model is related to the random distribution of particles in metal matrix. In the two-dimensional plane, the geometric model is a solid rectangle with a certain size. In the solid rectangle, there are a certain number of solid circles. In actual conditions, the distribution of particles is basically irregular. The real particle size in composite materials is often a certain range and the radius can be changed. It also helps to study the change of particle size on the effect of processing of the same cutting parameters. The algorithm is used to realize the variable number of particles, so it is convenient to study the effect on the machining effect of the same cutting parameter when the particle volume fraction is increased or decreased. In the two-dimensional graph, there is no overlap between the particle and the workpiece boundary, and between two particles. In order to obtain a better mesh shape in the later generation of meshes, there is a certain distance between the geometric boundaries. Overall, there are the following principles: 1) the center of the circle is random; 2) the radius can be changed; 3) the quantity can be changed; 4) there is no intersection between the geometric boundaries.

3. Simulation Research on Vibration Cutting Mechanism of Particle Reinforced Metal Matrix Composites

3.1. Cutting Simulation Design

When both the particles and the matrix in the geometric model have the same properties of the metal Al6061 material, the workpiece becomes homogeneous metal Al6061 material. It can also be regarded as the aluminum reinforced aluminum matrix composite. The particle and matrix parts of the geometric model generated by the script are respectively assigned to the corresponding SiC and Al6061 material properties. The workpiece model simulates the SiC particle reinforced aluminum matrix composite material. Simulation parameters are shown in Table 3.

Table 3. Simulation vibration parameters.

Parameter	**Tool parameter**		**Vibration parameter**		**Cutting parameter**	
	Rake angle	**Relief angle**	**Amplitude**	**Frequency**	**Cutting speed**	**Back cutting depth**
Numerical	10°	10°	20μm	25kHz	$1.57m{\cdot}s^{-1}$	30μm

3.2. Study on Meso-characteristics of Surfaces of Particle Reinforced Aluminum Matrix Composites

The graphs a) and b) in figure 2 respectively reflect the machined surface of the aluminum-based material during conventional cutting and vibration cutting, and it can be seen that both are relatively flat. It is worth noting that the conventionally machined surface of aluminum-based materials has a thin stress layer, but is not uniform; and the machined surface after vibration cutting is thicker and more uniform than conventional cutting. The simulation results show that vibration cutting has the characteristics of improving the thickness and uniform distribution of the stress surface layer in the substrate metal compared with conventional cutting.

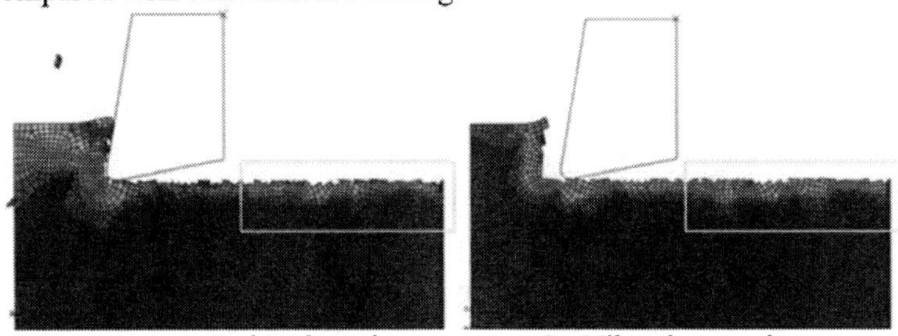

a) Conventional cutting **b)** Vibration cutting

Figure 2. Machined surfaces of aluminum matrix material

The graphs a) and b) in figure 3 show the machined surfaces of particle reinforced aluminum matrix composites in conventional cutting and vibration cutting respectively. Due to the addition of reinforcing particles, the machined surfaces of both are uneven (the pits left by the removed particles and the bulge of the particles that have not been removed.), and there are cases where the particles are suspended and the particles below the cutting path are incompletely de-bonded. The graph a) in figure 3 shows that there is a large area of stress concentration around the incompletely particles debonding below the cutting path in traditional cutting except for the obvious suspension of particles. The graph b) in figure 3 shows that the phenomenon of particle suspension on the machined surface of vibration cutting is reduced, and the area of stress concentration around incomplete debonding particles is smaller. The incomplete debonding particles and the stress concentration area under the cutting surface will affect the production and propagation of the cracks in the workpiece. The simulation results show that the vibration cutting can improve the uniform distribution of the stress on the machined surface and reduce the residual material on the machined surface compared with the common cutting for the particle reinforced metal matrix composites.

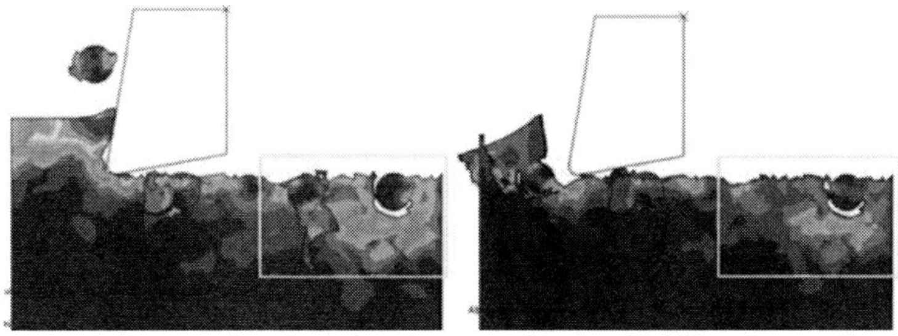

a) Conventional cutting **b)** Vibration cutting

Figure 3. Machined surfaces of aluminum matrix composites

3.3. Study on Meso-fracturing Mechanism of Cutting Zone in Granular Reinforced Aluminum Matrix Composites

It is found that the stress distribution has been changed because of the addition of reinforced particles in the metal matrix during cutting. In the cutting process of particulate reinforced metal matrix composites, "particle pairs" are easily formed between adjacent particles when the tool contacts to the reinforced particles. In addition to the large stress of the particles, the stress between the pair of particles is also large, which acts as a "channel" for stress transmission, and the stress is transferred to the adjacent particles. The graphs a) and b) in figure 4 respectively show the stress distribution and transmission of the tool contact to the No. 2 particle in conventional cutting and vibration cutting. It can be clearly seen from the figure that both traditional cutting and vibration cutting have stress transfer "channels", but the "channels" are more obvious in traditional cutting conditions. In the cutting process, due to the ultrasonic vibration applied by the tool, the stress transfering is blocked and the "channel" is cut off, which results in the stress concentration in the smaller cutting deformation zone. The matrix is more easily to failure and fracture under this stress concentration condition, and macroscopically, the material can be removed by only a small cutting force, which is beneficial to cutting.

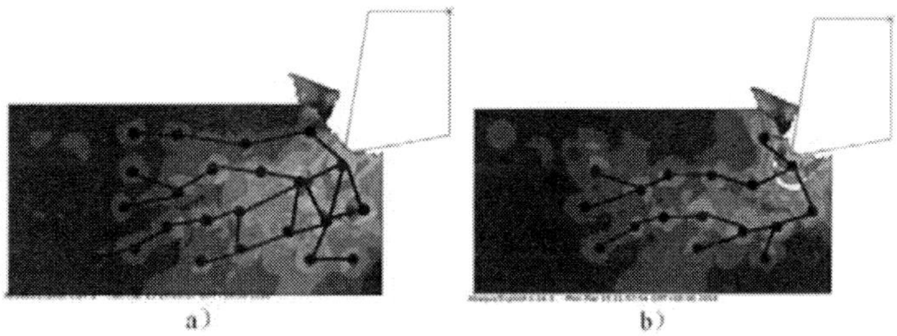

Figure 4. Stress distribution and transmission of conventional cutting and vibration cutting materials when cutting tools to particle No. 2

3.4. Study on the Cutting Force of Microscopic Model of Particle Reinforced Aluminum Matrix Composites

Figure 5 shows the tool is in contact with particle No. 5 (t=56μs) to complete particle debonding (t=58μs) during vibration cutting. At the time of 56 μs, the tool begins to contact the particle, the cutting macroscopic force reaches the maximum value, and the initial crack appears near the particle. At the time of 57 μs, the tool and the particle continue to contact, the cutting macroscopic force begins to decrease, and the stress is transmitted along the direction of crack propagation. At the time of 58 μs, the tool is still in contact with the particles, and the stress value and the macroscopic force decrease to the lowest value, which indicates that the stress has been completely released, the material has been removed and the chip is produced.

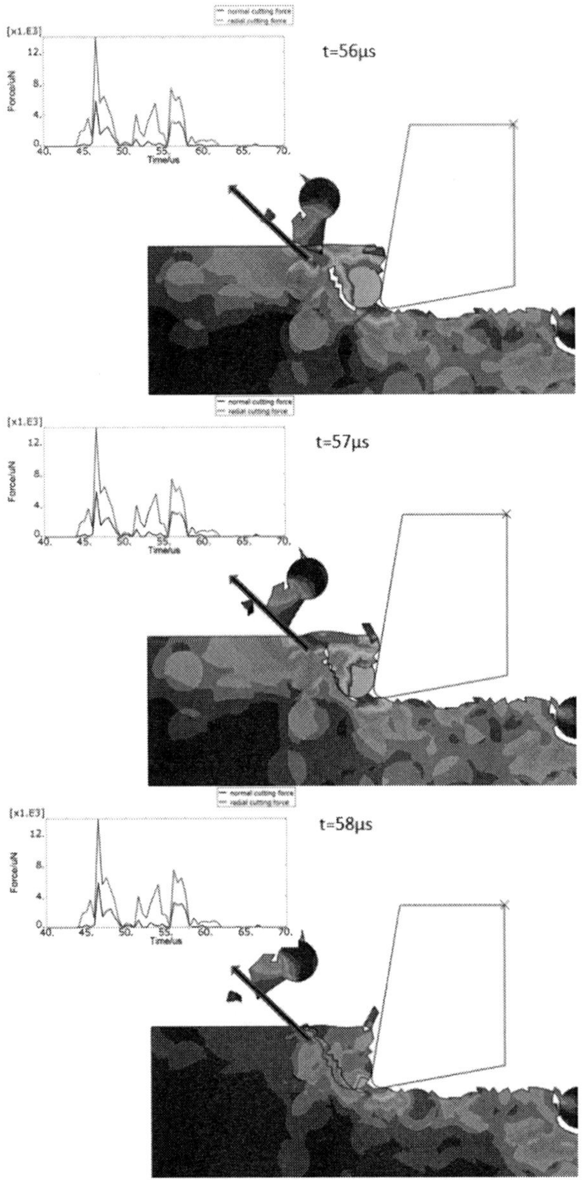

Figure 5. Stress distribution and transmission in vibration cutting

4. Conclusions

In this paper, the microscale characteristics of stress distribution and transfer on the machined surface of particle reinforced composites under vibration cutting and the micro-model cutting force are studied. The findings are as follows:

1) The particle's suspension on the machined surface of vibration cutting is reduced, and the stress concentration zone around incomplete debonding particles is smaller, the surface stress distributes uniformly and is relatively smooth;

2) When the particles reinforced composite is machined, it is easy to form "particle pair" between adjacent particles. Vibration cutting can cut off "channel", which is beneficial to cutting;

3) Under the condition that the tool contacts the same particle, the cutting force of vibration cutting is less than that of traditional cutting, which is beneficial to material removal;

4) The simulation model in this study adopts the script parametric modeling, which shortens the modeling time and improves the modeling efficiency. The random distribution of particles is helpful to

better simulate the real micro-morphology of composite materials. Ultrasound vibration cutting can cut off the "channel" of stress transfer between "particle pairs". The parameters of ultrasonic vibration machining were optimized by changing the volume fraction and particle size .It is very important for particle reinforced metal matrix composites to be used in practical machining.

Acknowledgements
The authors greatly appreciate the funding support from The Liaoning Natural Science Foundation (2015010131).

References
[1] Kang Yong, Chai Xiujuan. Progress in properties and applications of carbon/carbon composites [J]. Synthesis and Application of Composites, 2010, 39(3): 80-82.
[2] Si Chaorun. Research on Spray Forming Technology of 7055 Aluminum Alloy and Its Particle Reinforced Composites [D]. Northwestern Polytechnical University, 2015.
[3] Wei Jianning, Li Genmei. Preparation Technology and Application of Particle Reinforced Metal Matrix Composites [J]. Journal of Jiujiang University, 2008, 27(6): 33-35.
[4] He Jun. Research on ultrasonic vibration cutting technology of particle reinforced aluminum matrix composites [D]. Harbin Institute of Technology, 2008.
[5] Ge Yingfei. Basic research on ultra-precision turning of SiC particle reinforced aluminum matrix composites [D]. Nanjing University of Aeronautics and Astronautics, 2007.
[6] Quan Yanming, Zhou Zehua. Surface morphology and influencing factors of hard and brittle reinforced aluminum matrix composites [J]. China Mechanical Engineering, 1998, 9(7):10-11.
[7] Wang Minghai, Jiang Qingjie, Wang Ben, et al. Mechanism of Suppression of Damage Caused by Ultrasonic Torsional Vibration Milling of C/SiC Composites [J]. Modern Manufacturing Engineering, 2016(3): 103-109.
[8] Lu Shihong. Experimental study of high-speed cutting sawtooth chip and the construction model [D]. Nanjing University of Aeronautics and Astronautics, 2009.
[9] Jing Xiuhe, Lin Bin, Zhang Dawei. Numerical Simulation of the Influence of Cutting Conditions on Ultra-precision Cutting Process [J]. Journal of Tianjin University (Natural Science and Engineering Technology), 2011, 44(12): 1100-1105.
[10] Li Guijin, Research on Cutting Simulation of SiCp/Al Composites. Aerospace Material Technology, 2017.47 (6): 20-24.
[11] Dandekar C R, Shin Y C. Multi-step 3-D finite element modeling of subsurface damage in machining particulate reinforced metal matrix composites [J]. Composites Part A Applied Science & Manufacturing, 2009, 40(8):1231-1239.
[12] Munro M. Evaluated Material Properties for a Sintered alpha-Alumina [J]. Journal of the American Ceramic Society, 1997, 80(8):1919-1928.
[13] Johnson G R. A constitutive model and data for metals subjected to large strains, high strain rates and high temperatures[C]. Proceedings of the 7th International Symposium on Ballistics, The Hague, Netherlands, 1983.
[14] Ghandehariun A, Kishawy H A, Umer U, et al. On tool–workpiece interactions during machining metal matrix composites: investigation of the effect of cutting speed[J]. The International Journal of Advanced Manufacturing Technology, 2016, 84(9-12):2423-2435.

The Surface Chemistry of the Mild Steel and Glass Substrates for Thermal Spray

Wei Xie[1,2]

[1]College of Gongqing, Nanchang University, China
[2]Faculty of Engineering and Industrial Sciences, Swinburne University of Technology, Australia;

*Corresponding author

Abstract. The surface chemical condition of substrates plays a key role in single splats formation. Single splats are the foundation of thermal spray coatings. In this study, the surface chemistry of the mild steel and glass substrates were investigated. The result indicated that the 100 °C heat treatment released the absorbents of mild steel substrates. The adsorbed water disappears from the surface of the mild steel sample after one hour of heat treatment at 100 °C; there is no adsorbed water on the surface of the glass sample.
Keywords: surface chemistry; xps; adsorbates; thermal spray; single splat.

1. Introduction

The surface chemistry of the substrates plays a significant role in splat morphology. The volatiles of adsorbents produce bubbles to the splats, achieve splashed splats and even fragmented splats.

Many metals exhibit oxide layers of different thickness and composition as a result of their manufacture procedures. The preheating of metals gives off volatiles from the metal surface and alters the oxide thickness and composition [1]. Thermal energy can introduce surface separation of components because the oxides formation free energies are varying [2]. Hence, a different explanation for the transition temperature and surface chemicals possibly due to the volatiles of the substrates after heat treatment.

Substrate heat treatment can get rid of the adsorbates or condensates [3]. However, most substrates are impossible to be polluted with xylene, glycol or glycerol. Hence the issue is what are "adsorbates or condensates" on mild steel and glass substrates? The mild steel substrates were heated at 100 °C, hypothesized that the adsorbates were water.

Besides of remitting adsorbates or condensates, heat treatment of substrates also alternates the surface oxide thickness and composition. Chromium oxide at the inner layer and iron oxide at the outer layer was introduced by heat treatment of stainless steel, which was found by McDonald et al. [4]. The thickness of chromium oxide and iron oxide on the substrate heated at 350 °C was 124 nm and 56 nm respectively, however their thickness on the substrate heated at 650 °C was 204 nm and 84 nm respectively. But there was no difference in surface roughness between these samples [5].

Tran's [6] research identified that the input heat from particles at impact can generate gas release from the surface of stainless steel and aluminum substrates. It can be assumed that mild steel substrates should react similarly because it can be easily oxidized and absorb adsorbates. While splat morphology on gold substrates had almost no difference with and without heat treatment, founded by Fukumoto et al. [7]. The effect of surface chemicals on splat morphology was not apparent, as gold is metallic materials which is not oxidized under these circumstances. Hence it can be presumed that

IOP Conf. Series: Materials Science and Engineering **611** (2019) 012013 doi:10.1088/1757-899X/611/1/012013

glass substrates should perform similarly and that the splat morphology on glass substrates with and without heat treatment should not modify significantly.
To prove the above hypothesis, the X-ray photoelectron spectrometer (XPS) experiment was conducted on unheated mild steel, heated mild steel and glass samples. Mild steel and glass substrates were chosen because they are common materials. The surface of mild steel substrates is covered by oxides and adsorbents. The preheating amends the relative amount of these components. Such modifications can be identified by high resolution XPS.

2. Experimental Procedure

The X-ray photoelectron spectrometer (XPS) experiment was conducted with unheated mild steel, heated mild steel and glass samples. The mild steel substrates were polished. The composition of the mild steel substrates was 0.05–0.35 at%, 0.42–0.48 at%, 0.6%–0.9 at% for Si, C and Mn respectively, less than 0.06 at% for S and P totally. The glass substrates were chosen with the microscope glass slides. The composition was 72.20 at%, 14.30 at%, 1.20 at%, 6.40 at%, 4.30 at%, 1.20 at% and 0.03 at% for SiO_2 Na_2O, K_2O, CaO, MgO, Al_2O_3, Fe_2O_3 respectively. The samples were cleansed respectively in each of the following solutions: dichloromethane, hexane, acetone and methanol and sonicated for 5 minutes at the mean time; followed by rinsing with distilled water. The samples then were dehydrated with hot air and all the glass samples and half of the mild steel samples placed in a desiccator. Half of the mild steel samples were put in an oven at 100 °C for 1 hour to get rid of any adsorbed water, then left at room temperature for half an hour and then put into the desiccator. All the samples remained in the desiccator for one day. They were exposed to air for one hour prior to the XPS experiment.
The XPS instrument used in this study was Kratos AXIS UltraDLD, manufactured by Kratos Analytical Ltd, Manchester, UK. The X-ray source was monochromatic Al Kα (1486.6 eV). The analysis software was CasaXPS. High resolution scans of Fe2p, O1s and C1s were gathered to investigate oxide/oxyhydroxide components on the surface of samples. The narrow scans were curve fitted using CasaXPS software with combination of Gaussian (G) and Lorentzian (L) (GL30) for the peak profile.

3. Results and Discussion

3.1. Survey of Glass, Unheated and Heated Mild Steel Substrates

The XPS wide scan of the unheated mild steel, heated mild steel and glass samples are shown in Figure 1. The atomic compositions of the sample from this scan are calculated and presented in Table 1. The spectra from the unheated steel and the heated steel samples reveal the peaks of Fe 2p, O 1s and C 1s. The spectrum from the glass sample demonstrates the peaks of O 1s, C 1s and Si 2s. However the carbon composition analysed by XPS is 40.3 at% for unheated mild steel sample and 41.2 at% for heated mild steel sample. Interestingly the carbon composition of the glass sample is 41.1 at% and similar to the unheated and heated mild steel substrate, indicating that the samples were contaminated by adventitious carbon when they were exposed to air. The peaks of Zn 2p, N 1s, Ca 2p and Si 2s from both unheated and heated mild steel samples have lower intensities than Fe 2p, O 1s and C 1s peaks. The peaks of Na 1s, Zn 2p, N 1s, Ca 2p, Sn 3d and K 2s from glass samples also have lower intensities than O1s, C 1s and Si 2s peaks. The existence of these remnant elements is ascribed to pollution from tap water during the cleansing process.
The difference in the XPS narrow scan normalized spectra of O 1s peaks, Fe 2p peaks and C 1s peaks of unheated and heated mild steel samples is indicated in Figure 2. There was no difference

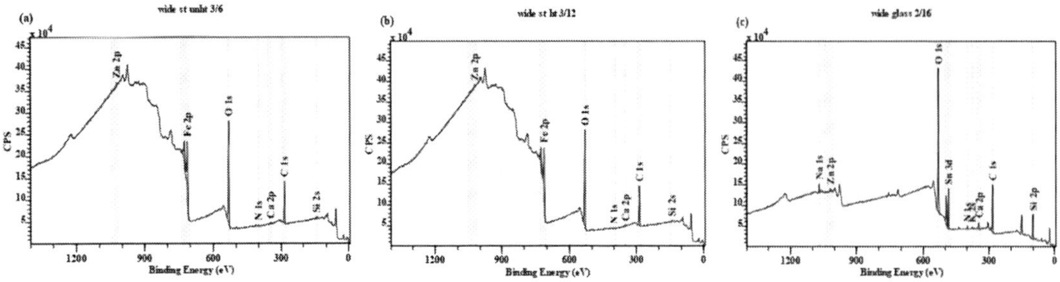

Figure 1. XPS wide scan of (a) unheated mild steel sample, (b) heated mild steel sample and (c) glass sample

between the Fe 2p peaks in the unheated and heated mild steel samples. There was a difference between the O 1s peak from unheated and heated mild steel samples; as also indicated by the C 1s peak. This indicated that the chemicals on the surface of the heated mild steel did change after the 100 °C heat treatment.

Table 1. XPS composition of unheated, heated mild steel and glass samples (at %)

Peak Area	*C 1s*	*N 1s*	*O 1s*	*Fe 2p*	*Ca 2p*	*Zn 2p*	*Si 2s*		
Untreated Steel1	39.8	0.8	42.1	15.7	0.3	0.3	1		
Untreated Steel 2	40.2	0.8	42.5	14.7	0.4	0.2	1.2		
Untreated Steel 3	41	0.5	42.2	15	0.3	0	1		
Average	40.3	0.7	42.3	15.1	0.3	0.1	1.0		
Heated Steel1	41	0.7	43.5	13.3	0.3	0.3	0.9		
Heated Steel 2	39.3	0.8	42.7	16.1	0.2	0.2	0.7		
Heated Steel 3	43.3	0.5	42.8	11.6	0.2	0.4	1.1		
Average	41.2	0.7	43	13.7	0.2	0.3	0.9		
Peak area	*O 1s*	*C 1s*	*N 1s*	*Ca 2p*	*Na 1s*	*Sn 3d*	*Si 2p*	*Zn 2p*	*K 2*
Glass1	40.8	40.6	1.6	0.9	1	1.7	12.4	0.2	0.7
Glass2	41.6	39.3	1.1	1	1.2	1.9	12.6	0.3	1
Glass3	40.9	40.5	1.4	1	1	1.7	12.2	0.3	0.9
Average	41.1	40.1	1.4	1.0	1.1	1.8	12.4	0.3	0.9

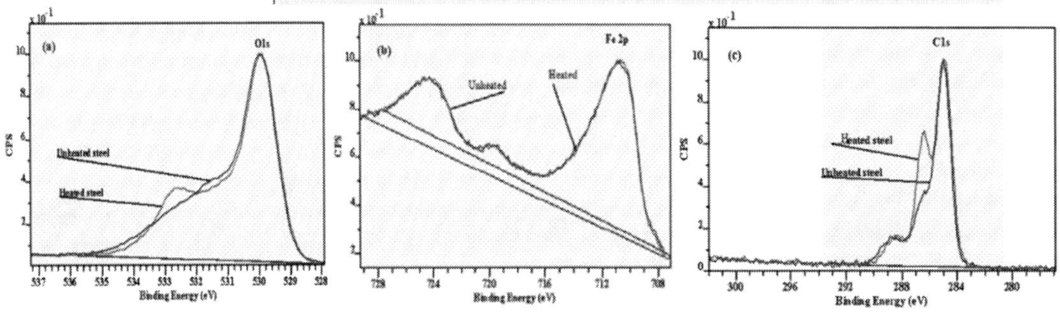

Figure 2. Differences in the XPS narrow scan normalized spectrum of (a) O 1s peak, (b) Fe 2p peak and (c) C 1s peak

3.2. Curve Fitting of Narrow Scan of O 1s and C 1s Peaks

The curve fitted narrow scans of O 1s peaks for both unheated and heated mild steel samples are depicted in Figure 3. The O 1s line shapes for oxidation were quite different from unheated and heated mild steel samples. The O 1s envelope of unheated mild steel sample was fitted with 6 components assigned as lattice O^{2-} from Fe_2O_3, Fe_3O_4 at a banding energy (BE) of 529.9 ± 0.1eV, lattice OH^- from $Fe(OH)_2$, and FeOOH at BE of 531 ± 0.1eV, adsorbed $O^{\delta-}$ (adsorbed atomic oxygen) at BE of 531.8 ±

0.1eV, adsorbed OH^- at BE of 532.8 ± 0.1eV, adsorbed OH^- at BE of 533.7 ± 0.3eV and adsorbed H_2O at BE of 534.9 ± 0.3eV [6]. The O 1s envelope of the heated mild steel sample was fitted with 5 components assigned as lattice O^{2-} at BE of 529.9 ± 0.1eV, lattice OH^- at BE of 531.1 ± 0.1eV, adsorbed $O^{\delta-}$ at BE of 531.8 ± 0.1eV, adsorbed OH^- at BE of 532.8 ± 0.1eV, adsorbed OH^- at BE of 533.9 ± 0.1eV. The adsorbed H_2O disappeared from the surface of the mild steel sample after one hour of heat treatment at 100 °C. It can be predicted that there will be less bubbles between the EMAA single splat and the mild steel substrate after one hour of heat treatment of mild steel substrates at 100 °C than substrates without heat treatment. Hence heat treatment of mild steel substrates can improve the EMAA coating properties, especially the mechanical properties.

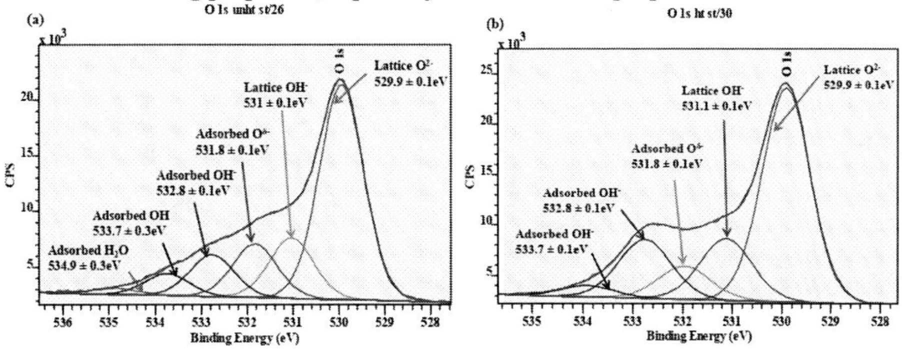

Figure 3. Curve fitted narrow scan of O 1s peak of (a) unheated mild steel, (b) heated mild steel samples

The high-resolution C1s spectra are fitted with four components for both unheated and heated mild steel samples as shown in Figure 4. The most intense peak at 285.0 eV is produced from hydrocarbon groups (C–C/C–H) such as those in the mineral oil, the propyl chain in APS and the hydrocarbon contamination [8]. The BE at 286.5, 287.9-288.2, and 289.1 eV are assigned to C–OH, C=O, and COOH or O–C=O groups (carbonyl/Ester) respectively, which are the contaminants from the cleansing process [8]. Hence the C–OH and COOH species from the C 1s envelop are from the trace of acetone during the sample cleansing process.

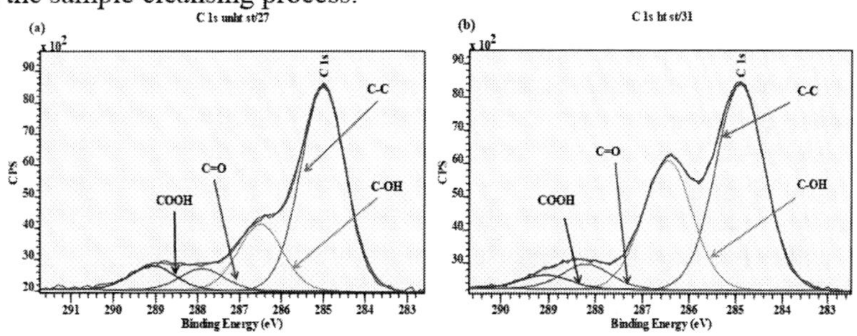

Figure 4. Curve fitted narrow scan of C 1s peak of (a) unheated mild steel and (b) heated mild steel samples

The Curve fitted narrow scan of the O 1s peak and the C 1s peak glass sample are shown in Figure 5. The O 1s envelope of heated mild steel sample was fitted with 4 components assigned as lattice OH^- at BE of 530.8 ± 0.1eV, adsorbed $O^{\delta-}$ at BE of 531.8 ± 0.1eV, adsorbed OH^- at BE of 532.6 ± 0.1eV, adsorbed OH- at BE of 533.4 ± 0.3eV [6].

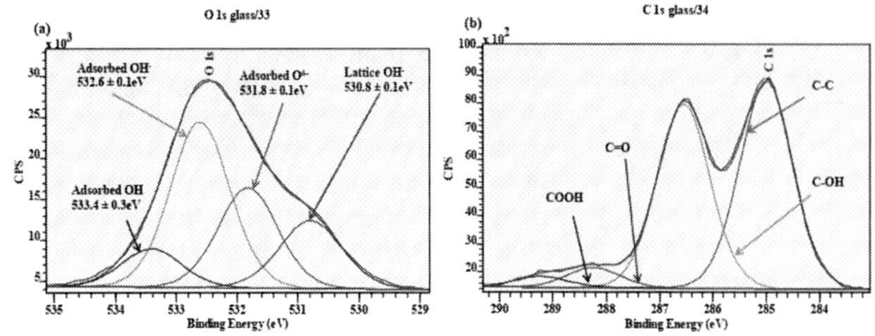

Figure 5. Curve fitted narrow scan of glass sample of (a) O 1s peak and (b) C 1s peak

There was no adsorbed water on the surface of the glass sample which was proved by lack of bubbles between the EMAA single splat and the glass substrate which has been reported by Wei et al [9]. The C 1s peak at BE of 285.0 eV was assigned as (C–C/C–H) from hydrocarbon groups such as those in the propyl chain in any hydrocarbon, the mineral oil, and APS contamination [10]. The BE at 286.6, 288.2, and 289.2 eV were assigned to C–OH, C=O, and COOH or O–C=O groups (carbonyl/Ester) respectively which were the contaminants from the cleaning regime [11]. C–OH and COOH component from the C 1s envelope was from the residue of acetone residue that evolved during sample cleaning.

4. Conclusion

The heat treatment at 100 °C has an apparent effect on the chemistry of mild steel surface. The adsorbed atomic oxygen and COOH concentration were reduced. The adsorbed H_2O was disappeared from the surface of the mild steel sample after one hour of preheating at 100 °C. There was no adsorbed water on the surface of the unheated glass sample. C–OH and COOH component from the C 1s envelope was from the residue of acetone residue. These findings are new to this area of thermal spray technology and provide a fundamental understanding to what is often assumed within common industrial practice concerning substrate heating effects.

Acknowledgments

Special thanks to Prof. Peter Kingshott for guidance in the experiment and analysing the results. Also thanks to my supervisor Prof. C.C. Berndt and vice supervisor Dr. James Wang for their guidance. Thanks Dr Robert Jones from La Trobe University allowing me to access the X-ray photoelectron spectrometer (XPS).

References

[1] Nakajima, D., et al., Mirror-finished superhydrophobic aluminum surfaces modified by anodic alumina nanofibers and self-assembled monolayers, Applied Surface Science, 2018, 440: 506-513

[2] N. Birks, G.H. Meier, and F.S. Pettit, Introduction to the high temperature oxidation of metals, 2 ed., Cambridge University Press, 2006

[3] Chen, L. and G. J. Yang, Epitaxial growth and cracking of highly tough 7YSZ splats by thermal spray technology, Journal of Advanced Ceramics, 2018, 7(1): 17-29

[4] A. McDonald, C. Moreau, and S. Chandra, Effect of substrate oxidation on spreading of plasma-sprayed nickel on stainless steel, Surface and Coatings Technology, 2007, 202(1), p. 23-33

[5] V. Pershin, M. Lufitha, S. Chandra, and J. Mostaghimi, Effect of substrate temperature on adhesion strength of plasma-sprayed nickel coatings, Journal of Thermal Spray Technology, 2003, 12(3), p. 370-376

[6] A.T.T. Tran, M.M. Hyland, T. Qiu, B. Withy, and B.J. James, Effects of surface chemistry on splat formation during plasma spraying, Journal of Thermal Spray Technology, 2008, 17(5-6), p. 637-645

[7] M. Fukumoto, H. Nagai, and T. Yasui, Influence of surface character change of substrate due to heating on flattening behavior of thermal sprayed particles, Journal of Thermal Spray Technology, 2006, 15(4), p. 759-764
[8] A.T.T. Tran, and M.M. Hyland, The role of substrate surface chemistry on splat formation during plasma spray deposition by experiments and simulations, Journal of Thermal Spray Technology, 2010, 19(1-2), p. 11-23
[9] W. Xie, J. Wang, and C.C. Berndt, Spreading behavior and morphology of ethylene methacrylic acid (EMAA) deposits via the flame spray process, Coatings, 2012, 2(2), p. 76-93
[10] A. Zafar, J. Schjødt-Thomsen, R. Sodhi, R. Goacher, and D. De Kubber, X-ray photoelectron spectroscopy and time-of-flight secondary ion mass spectrometry characterization of aging effects on the mineral fibers treated with aminopropylsilane and quaternary ammonium compounds, Surface and Interface Analysis, 2012, 44(7), p. 811-818
[11] A.T.T. Tran, "Effects of surface chemistry on splat formation during plasma spraying," Ph.D. Thesis, The University of Auckland, 2010

Investigating the Effect of TIG Welding on the Surface Roughness of the Welded Closure of Thin Stainless Steel Tubes

K. Al-Badani[1,*], E. Elmshawet[1,*], C. Tang[1], T. Kaid[1], L. Li[1], J. Liu[1], J. Wang[1] and X. J. Ren[1]

[1]Department of Maritime and Mechanical Engineering, Faculty of Engineering and Technology, Liverpool John Moores University, Liverpool L3 3AF

*Corresponding author

Abstract. Tungsten inert gas (TIG) welding of thin stainless steel tubes is an important manufacturing process with many applications, in particular for the closure of sheathed thermocouples tubes and thermocouple wells. During the welding process, the microstructure, properties as well as the surface roughness is directly influenced by the setup of the welding procedure and material properties. For the steel tubes of thermocouples, the surface roughness also influences the appearance as well as the wetting behaviour between the probe surface and liquid, which could affect the response of the probe or the maintenance when the probe is used to measure the temperature of liquid and gaseous mediums. In this work, welding tests with controlled parameters and analysis has been conducted on thin austenite stainless tubes of different sizes, the surface roughness of the welded tube is characterised for different welding settings. The data shows that the surface roughness varies between 1-3.5 microns and oxidation patch significantly reduce the smoothness of the surface. The influence of the surface roughness of the tube is investigated and their influence on the service conditions such as wetting/dripping is discussed with reference to typical results from computational fluid dynamics (CFD) simulation.
Keywords: micro-weldimg; thermocouple; TIG.

1. Introduction

Sheathed thermocouples are widely used in many industries for temperature measurement and process control. In an insulated thermocouple, the welded tip and wires are embedded within an insulation (normally MgO) covered by a sheath of protective material (normally stainless steels). Figure 1 illustrates the structure of two types of sheathed thermocouples. In an ungrounded junction, the thermocouple junction is insulated from the sheath and the ground; for a grounded structure, the thermocouple junction is an integral part of the sheath. The closure of the sheath tube directly influences the response time and protection of the sensor. The joint of the thermocouple wires is normally welded by Capacitor Discharge (CD); while the stainless tube is normally welded by Tungsten inert gas (TIG) welding process [1-3].

IOP Conf. Series: Materials Science and Engineering **611** (2019) 012014 doi:10.1088/1757-899X/611/1/012014

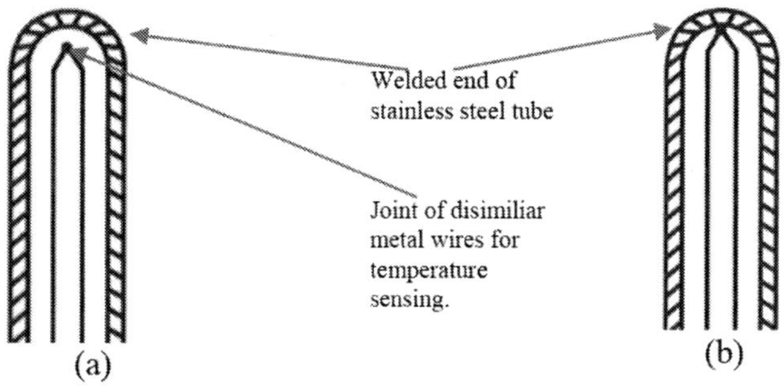

Figure 1. (a) End of a ungrounded Sheathed thermocouple (b) End of a grounded Sheathed thermocouple

Figure 1 Schematics showing the TIG welded closure on thin stainless tubes in an ungrounded Sheathed thermocouple (a) and a grounded Sheathed thermocouple. Both processes involve welding closure of thin stainless steel tubes.

The welding of the stainless tube is a complex process due to the material composition (i.e. multiple alloying elements) and properties of stainless steels. The composition needs to be balanced to maintain the corrosion resistance [4-5]. Stainless steels are known to have higher coefficient of thermal expansion compared to carbon steels, so it is prone to distortion and stress inducement if a proper welding procedure is not followed. In addition, stainless steel is a reactive metal that quickly forms an oxide layer on the surface. Therefore, welding of thin stainless steel tubes is difficult. For thermocouples, the shape of the sheath end needs to be properly controlled, as it directly influences the response time of the temperature measurements. The microstructures and the mechanical properties of stainless steels are relatively better researched [6-9], while works on the surface roughness of the welded tube are rather limited. Surface roughness is an important issues [10-12]. The surface roughness is important for thermocouples as it directly influences the colour and surface texture, it also influences the wetting between the surface and the liquid or gas being measured [10]. In addition, an improved understanding of the effect of welding parameters on the surface roughness is essential for improving the quality and properties of thermocouples. In this work, welding tests with controlled parameters and analysis have been conducted on stainless tubes of different sizes (1-3 mm in diameter), a special rig is designed to control the position of the TIG electrode tip. The effect of the electrode tip position and welding current on the surface roughness of the tube around the welded end is established; their influence on the interaction of the thermocouples and liquid is discussed with the aids of computational fluid dynamic simulations with different contact angles (wetting coefficients).

2. Testing Setup and Data Analysis

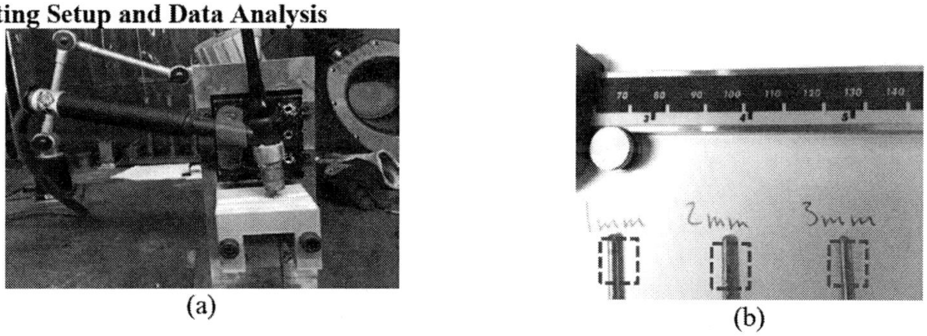

Figure 2. (a) setup of the welding tests (b) Welded end of stainless steel tubes.

Figure 2 shows the welding setup (a) and some welded samples of stainless steel tubes with different sized (b). The welding apparatus is able to control the height and position of the TIG electrode through a specifically designed frame. The TIG welding torch is fixed with a height adjustable block unit. A tungsten electrode is used and Argon gas flows through the torch; the TIG Welding machine has a Rectifier (made by FRONIUS), which covers a current range 10-180 A and voltage up to 230V, depending on the current setting that has been selected. Ar gas is supplied to the welding torch with a pre-defined flowrate. In addition, stable arc created for welding is maintained with an inert atmosphere. Figure 3 shows the ContourGT-K 3D optical microscope used to measure surface roughness of the welded tip for all specimens. The test was performed on the external wall of the tubes after the tip was closed through TIG welding as illustrated by the box (dashed lines in Figure 2(b)). This is the region in contact with the mediums being measured and the surface roughness directly influence the colour and the solid-liquid or solid-gas wetting behaviours. Roughness plays an important role in determining how a real object will interact with its environment. For example, when a sheathed thermocouple is being used to measure the temperature of a liquid, there will be residual liquid adhered onto the surface of the probe. This can result in inaccurate temperature recordings, a difficult cleaning process (when dealing with hazardous liquid) and a shorter life time span of the probe due to corrosion. Therefore, it can be understood that high surface roughness value is very undesirable. During TIG welding procedures, surface roughness is strongly influenced by complex interaction between welding parameters, such as the current, height of the torch, shielding gas flowrate and welding speed. A design of experiments (DOE) has been used to produce samples under different parameters including Electrode Workpiece distance (mm), Argon Gas Flowrate (l/min), Voltage (V) and Current (A).

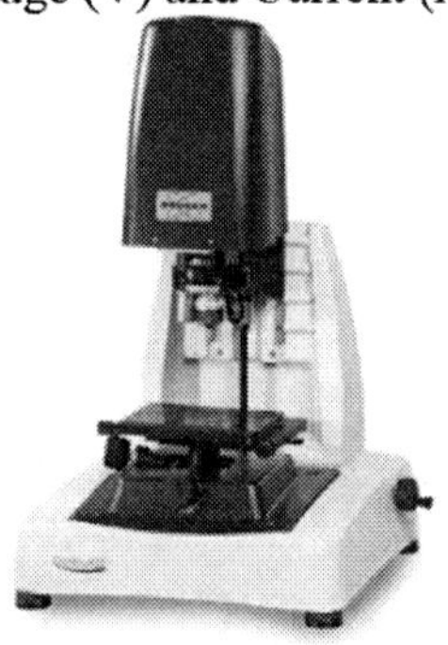

Figure 3. Brukker ContourGT-K surface texture system used to measure the surface roughness.

3. Results and Analysis

Figure 4 illustrates some typical 3D scans of the welded end of the tubes for the samples prepared using different welding heights (i.e. position of the electrode tip) and current. Samples produced by welding with a welding distance of 1mm suffered burning due to close distance of the welding operation. Detailed analysis of the cross section (a typical example is shown in Figure 4(b)) demonstrated a clear effects of the welding parameters changes, i.e. welding current and welding tip height on the surface features. The work shows that the surface texture system is able to highlight the welding penetration changes with the adjustment of the welding tip height and current settings as comparatively shown in Figure (a) and Figure (c). In addition, the surface roughness tests can help identify the oxide zone on the tube as illustrated in Figure (d). These oxide layer may cause uneven corrosion to the stainless steel tube surface. Surface roughness of the weld zone for all the tested samples were measured and the average surface roughness value was calculated from three readings. Some typical data is shown in Figure 5, roughness value found in the range of 1.1 to 3.5 micron is quite low for a welded specimen. Therefore it can be said that using automated system is possible to produce good surface quality of welding which may not require any further finishing operation. These roughness values also show that the applied current has no consistent effect on the surface roughness value but the position of the electrode tip could be a crucial issue.

IOP Conf. Series: Materials Science and Engineering **611** (2019) 012014 doi:10.1088/1757-899X/611/1/012014

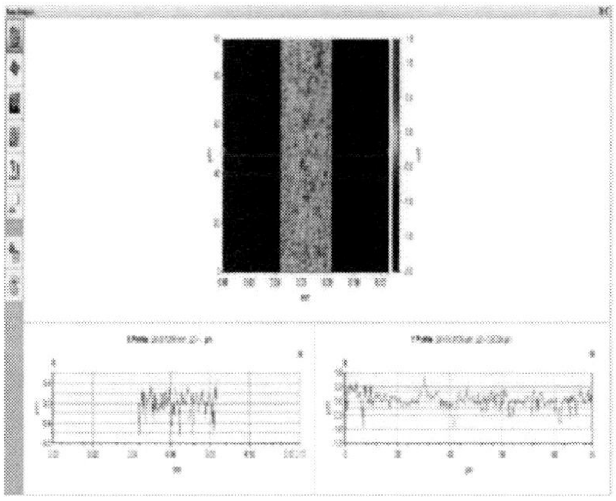

(a) Electrode tip height = 2mm, Current = 30AMPs

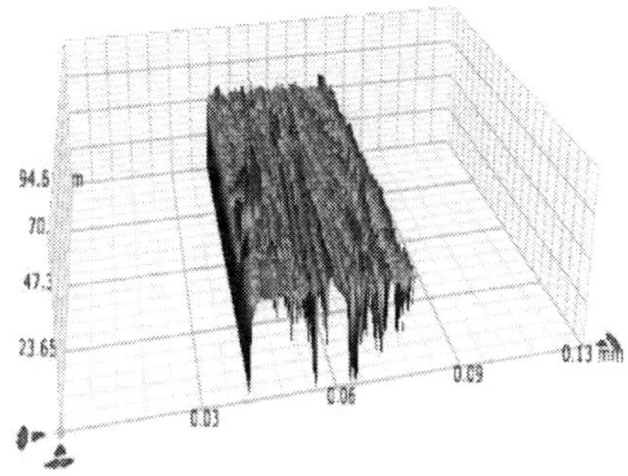

(b) Enlarged view of the (a) showing the cross sectional view of the surface

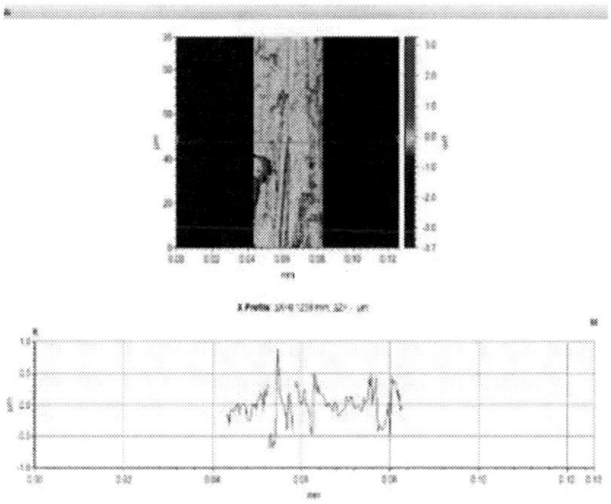

(c) Electrode tip height = 2mm, Current = 50AMPs

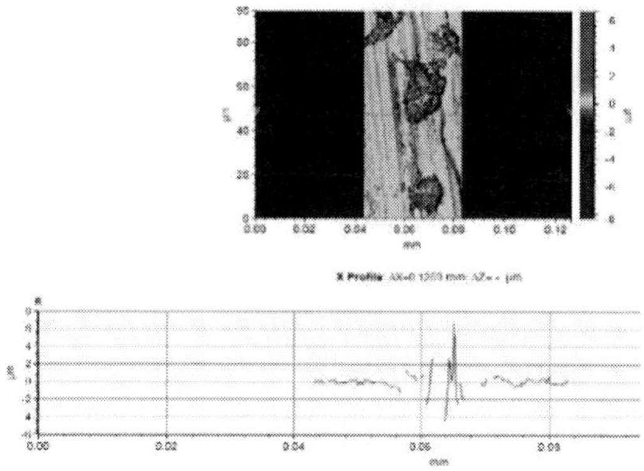

(d) Typical example showing the defects on the surface identified through Surface Roughness Scan.

Figure 4. Typical example of surface roughness characterisation of the welded tubes. (a) Welding tip height=2mm, Current =30AMPs; (b) enlarged 3D sectional view of the surface roughness, (c) welding current=50AMP; (d) Surface roughness and surface defects when the electrode is close to the end of the tube.

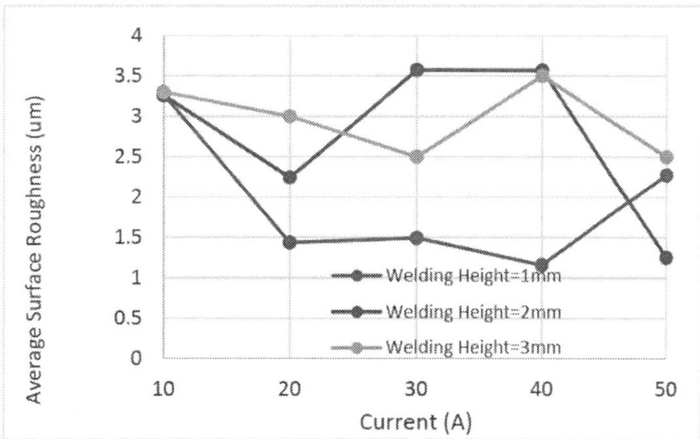

Figure 5. Average surface roughness of the samples with different welding current and welding tip height.

As an extension of this project, a computation fluid dynamics model has been developed to study the link between surface roughness and surface wetting (represented by the contact angle between the solid surface and the liquid) on the tip of the thermocouple sensor [10]. A wetting process is the ability of the solid surface to maintain contact with the liquid. This phenomena often occurs due to intermolecular interactions when the liquid and solid are brought together. The hypothesis is that surface roughness can affect the adhesive forces between the solid and the liquid, which controls the wetting process and dripping residuals, which is relevant to the thermocouples performances when measuring temperature of a liquid. Less residual will be a better option, in particular when dealing with acidic liquid for safe operation or a liquid system for which maintaining the composition balance is critical (e.g. pharmaceutical system). Establishing a quantitative link directly between surface roughness to wetting is challenging, one approach is to use numeric modelling to map the effective zones (regime of contact angles/wetting coefficients) within which the adhesion between the solid surface and a liquid is less severe.

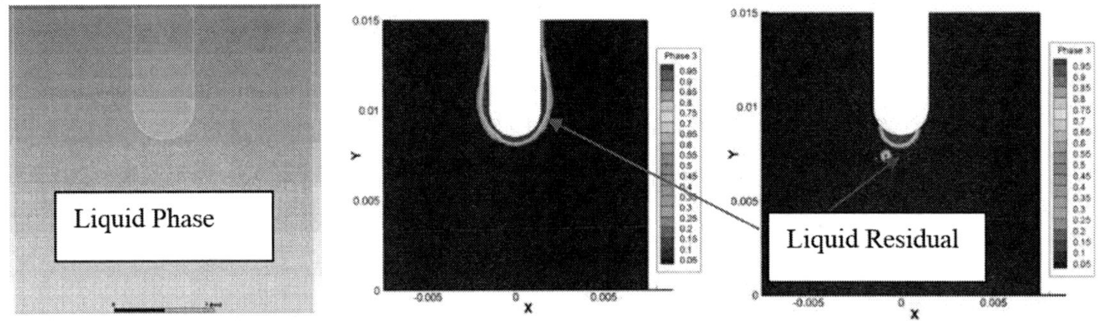

(a) Computational Fluid Dynamic (CFD) model.

(b) Dripping of liquid from the tip of tube closure, at contact angle of 0 (High surface roughness)

(c) Dripping of liquid from the tip of tube closure, at contact angle of 60 (High surface roughness)

Figure 6. Example numerical model developed to assess the effect of contact angle (wetting coefficients) on the residual of liquid on the tube when a tube is pulled from a liquid. The contact angle is affected by the surface roughness.

Two typical CFD examples are shown in Figure 6, which illustrate the dripping of a liquid from the thermocouple stainless steel tube when it is taken out of the liquid. This model is able to assess the effect of contact angle and wetting characteristics on the liquid-stainless steel tube interaction. Initial simulation results showed that when the contact angle is low (representing strong wetting) fully wetting is achieved and a residual on both sides of the tip can be observed (Figure 6(b)). This can cause the probe to produce inaccurate temperature readings due to the wetting affect and difficult to clean. On the other hand, at a contact angle of 60, a mall volume of residual liquid is adhered to the end of the tube. Future works will focus on developing systematic studies on the contact angle of flat and curved face with different surface roughness.

4. Conclusions

This paper reported a systematic work of controlled TIG welding tests and analysis on welding closure of stainless tubes with different sizes, the quality of the weld and surface roughness is investigated and the key factors controlling the welding process is established. Surface roughness of welded tube is found to change with the distance from the welding zone, which highlight the influence of the thermal process or potential phase changes. The result shows that microstructure burns formed in some welding conditions, which is quantitatively identifiable through surface roughness testing, this may help to manage the potential welding defects on the corrosion resistance of the material. The data shows that the surface roughness varies between 1-3.5 microns and oxidation significantly reduce the smoothness of the surface. CFD data shows that wetting coefficient could significantly influence the dripping mode of the liquid from the tube. Future work will combine the experimental results with modelling and detailed materials characterisation to develop a welding map for guiding the optimisation of the welding process directly linking surface roughness to the wetting behaviour of the welded zones.

Acknowledgments
This project has received funding from the European Union's Horizon 2020 research and innovation programme under the Marie Skłodowska-Curie grant agreement "No 689857" and "No 823786".

References

[1] Loto R., 2017, Study of the corrosion behaviour of s32101 duplex and 410 martensitic stainless steel for application in oil refinery distillation systems. *Journal of Materials Research and Technology*, 6(3), 203-212.

[2] Raveendra, A., & Kumar, B. R., 2013, Experimental study on pulsed and nonPulsed current TIG welding of stainless steel sheet (SS304). *International Journal of Innovative Research in Science, Engineering and Technology*, 2(6), 5-15.

[3] Zong, R., Chen, J., Wu, C., 2019, A comparison of TIG-MIG hybrid welding with conventional MIG welding in the behaviors of arc, droplet and weld pool, Journal of Materials Processing Technology, 270, 345-355.

[4] Li, D., Lu, S., Dong, W., Li, D., & Li, Y., 2012, Study of the law between the weld pool shape variations with the welding parameters under two TIG processes. *Journal of Materials Processing Technology*, 212(1), 128-136.

[5] Brytan Z., Niagaj J., Reiman Ł., 2016, Corrosion studies using potentiodynamic and EIS electrochemical techniques of welded lean duplex stainless steel UNS S82441, *Applied Surface Science*, 388 A, 160-167.

[6] Hsieh C. and Wu W., 2012, Overview of intermetallic sigma phase precipitation in stainless steels. ISRN Metallurgy, 1, 1-16.

[7] Ghosh, N., Pal, P.K., Nandi, G., 2016, Parametric optimization of MIG welding on 316L austenitic stainless steel by Taguchi method, Archives of Materials Science and Engineering 79(1):27-36.

[8] Satija, A.P., Archit, P. Patel, A.P., 2017, Effect of different fluxes on DOP and bead width in TIG welding: a literature review; IOSR Journal of Mechanical and Civil Engineering (IOSR-JMCE) 14(4), 1-6.

[9] Sharma, P., Dwivedi, D.K., 2019, TIG welding of dissimilar P92 steel and 304H austenitic stainless steel: mechanisms, microstructure and mechanical properties, Journal of Manufacturing Processes, 44, 166-178

[10] Starvin, S. Ganesh, K.C., Vasudevan, M., 2018, Numerical simulation of creep behaviour of 316LN stainless steel weld joint, Materials Today: Proceedings, 5 (2), 8193-8198.

[11] K. Al-Badani' Numerical and Experimental Investigation of Liquid-Solid Interaction, *PhD Thsis,* Liverpool John Moores University, 2019.

[12] Manaf, N., Fukuda, K., Subhi, Z., Radzi, M.F., 2019, Influences of surface roughness on the water adsorption on austenitic stainless steel, Tribology International, 136, 75-81

[13] Li, J., Pan, L., Fu, Q., Zhou, Y., Guo, N., 2019, Wettability and corrosion behavior of a Ni coating on 304 stainless steel surface, Surface and Coatings Technology, 357, 740-747.

Elastic-plastic Sliding Asperity Interaction of Machined Multiscale Metal Surface

Guoxi Li[1], Kai Zhang[1,*] and Jingzhong Gong[2]

[1]College of Mechatronics Engineering and Automation, National University of Defense Technology, No.109 Deya Street, Changsha, China
[2]School of Mechanical Engineering, Hunan International Economics University, No.822 Fenglin Street, Changsha, China

*Corresponding author

Abstract. Sliding process between asperities of machined metal surface is important to analyse wear mechanisms. Elastic-plastic sliding asperity model based on fractal geometry is proposed in this paper to extend the sliding process of single micro-asperity to a macro-asperity. The contact force, the critical interference and residual interference are calculated with fractal dimension *D* and fractal roughness *G* in the multiscale views. The influence of fractal parameters on normal and tangential contact force are studied. The results of the simulation show that the plastic deformation and residual interference are the main factors affecting the contacting force and asperity radius. For smaller *G* and bigger *D* surfaces, the sliding interaction of asperities is less remarkable and the plastic force accounts for less in the contact force.
Keywords: sliding process; elastic-plastic interaction; fractal geometry; multiscale model.

1. Introduction

Sliding wear is indispensable and considerable for metal surfaces. Most works focused on special metals based on experiments. It is essential to study the mechanism of the sliding process. Asperity-based theory[1], fractal theory[2] and Persson's theory[3] are classical methods to analyse the contact mechanics of rough metal surfaces. Simplifications and approximations have to be made because different surface machining processes (grinding, milling, turning, etc.), multiple physical characteristics (wear, friction, lubrication, thermal, etc.) and a wide range of factors often affect the contact and sliding between rough surfaces[4,5]. Thus, most previous models are derived from specific condition.

Asperity-based semi-analytical model and finite element model are commonly analysed for sliding asperity interaction. Wang and Schipper[6] extended the Boucly-Nelias-Green (BNG) model[7] of layered semispherical asperity. The factors affecting friction are studied by discrete fast Fourier transform (DC-FFT) technique. Zhao et al.[8] described a frictionless sliding process of parabolic asperity by considering the effect of strain hardening exponents based on the elastic-plastic model proposed by Jackson et al[9]. Gao et al.[10] presented lateral contact and interaction between shoulder-shoulder asperities and uses the distribution of the contact azimuthal angle to discriminate the contact performances. In addition to asperity-based theory, Persson's theory is an advanced mechanics way to extrapolate the lateral interaction, adhesion and coalescence between asperities[11]. Mishra et al.[12] studied friction during the Loading and sliding process based on the Material Point Method (MPM). They[13] also analysed the influence of asperity geometric size on sliding forces of elliptical asperities.

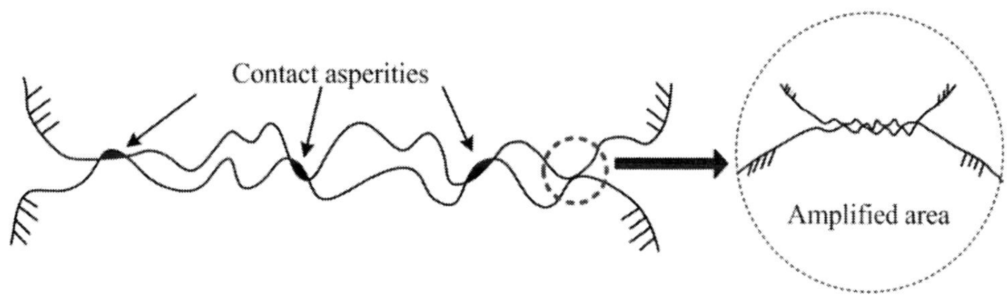

Figure 1. Simplified contact between rough surfaces with oblique contact asperities.

However, the contact between real machined metallic surfaces is self-affine as shown in Figure 1. Fractal theory can describe the multi-scale structures of the asperities which are independent of the resolution of the measurement devices[14]. And sliding process of single micro-asperity can be extended to a macro-asperity by fractal geometry. Meng et al.[15] identified Interfacial defect and adhesion wear of α -iron asperity-asperity and described the multiscale sliding interaction. Zhang et al.[16] presented an elastic-plastic model of fractal surfaces. Some relevant parameters about oblique asperity contact such as contact angle are introduced to improve the accuracy of load and contact area. Based on this paper, sliding asperity interaction considering the loading and unloading process is depicted in this paper. The surface is assumed to be isotropic and the material is homogeneity. The elastic and plastic contact force is calculated by fractal model. And numerical simulations are applied to acquire the influence of fractal parameters on normal and tangential contact force. With the qualitative analysis of contact force, this work can obtain a more accurate sliding model and might give some helpful results on the sliding asperity interaction in fractal surfaces.

2. Theory and Methodology

Microscopically, the contact points of single asperity contact pairs are not just located at the summits of the asperities according to the real surface topography as shown in Figure 1. And the azimuthal angle along the normal of the contact point is not perpendicular to the substrate of the rough surface generally. Figure 2 describes the asperity sliding interaction process from initial state until removal from contact. The contact angle α is defined as the angle between the tangent plane and the substrate plane and $\tan\alpha = h^{-1}r_{a}$. Friction, adhesion, coalescence and temperature effects are ignored during the sliding process. And the shape of the asperity is assumed to be sphere. r_a is the tangential deviation of two summits. R_{upper} and R_{lower} are the radii of the upper asperity and lower asperity, respectively. The equivalent radius can be calculated by $R^{-1} = R_{upper}{}^{-1} + R_{lower}{}^{-1}$. It can be seen that during the sliding process, the contact angle has persistent changes because of the movement of the contact region and contact point. And permanent deformation is caused by residual interference in the contact area[9].

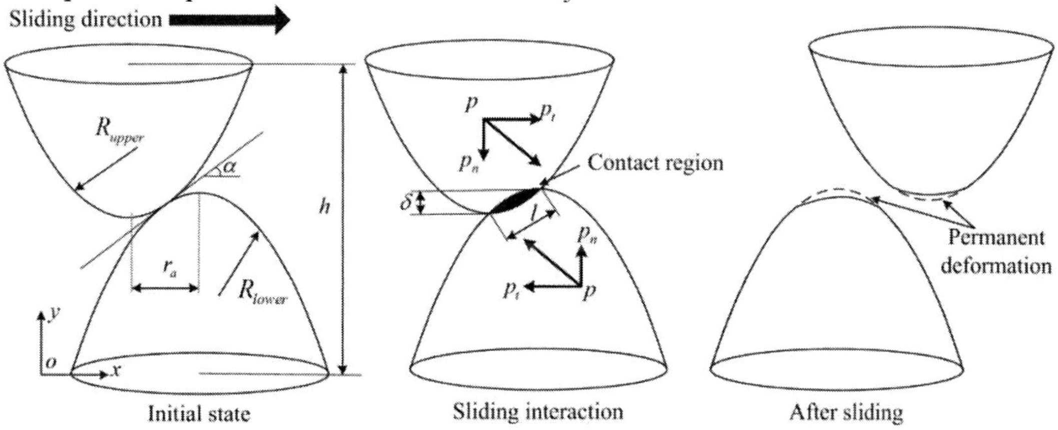

Figure 2. Schematic of spherical asperity sliding process.

Generally, the mating interfaces of an oblique asperity contact pair is treated as asperities condensed by a rigid plane in fractal theory. To exhibit the interaction of the upper and lower asperities, the contact region is extended and exhibited as Figure 2. δ is the interferences perpendicular to the surface. p is the Hertz force and p_n, p_t are its normal and tangential components. l is the length scale. The relationship of these parameters can be obtained as[16]:

$$p=\begin{cases} p_e = 3^{-1}4(2\pi)^{0.5} E'G^{D-1}(a')^{1.5-0.5D} & ,a' > a'_c \\ p_p = Ha' & ,a' < a'_c \end{cases}, \tag{1}$$

$$R=(a')^{0.5D}(2\pi)^{-1}G^{1-D}, \tag{2}$$

$$\delta = G^{D-1}(a')^{1-0.5D}, \tag{3}$$

$$l=(0.5a')^{0.5}, \tag{4}$$

$$\delta_c = \left((2E')^{-1}\pi H\right)^2 R, \tag{5}$$

$$a'_c = G^2 H^{\frac{2}{1-D}}(2E')^{\frac{1-D}{2}}(0.5\pi)^{\frac{1}{1-D}}, \tag{6}$$

where p_e and p_p are the elastic and plastic components. E' is the equivalent elastic modulus, defined as $\frac{1}{E'}=\frac{1-\nu_1^2}{E_1}+\frac{1-\nu_2^2}{E_2}$, E_1, ν_1 and E_2, ν_2 are the elastic moduli and Poisson ratios of two contact surfaces, respectively. δ_c is the critical interference. a' is the truncated area and a'_c is the critical area. σ_y and H are yield strength and hardness of the softer material. D is the fractal dimension and G is the fractal roughness parameter of the surface profile. Due to the instant value of contact force of each sphere asperity pair[9], critical interference δ_c, contact force p_c and its tangential and normal components, p_t and p_n can be written by equations (1)-(6) as:

$$\delta_c = \frac{2^{0.5D-3}\pi H^2 l^D}{E'^2 G^{D-1}}, \tag{7}$$

$$p_c = 3^{-1}2^{4-0.5D}\pi^{0.5}E'G^{D-1}l^{3-D}, \tag{8}$$

$$\begin{cases} p_t = p_c \sin\alpha \\ p_n = p_c \cos\alpha \end{cases}. \tag{9}$$

According to method proposed by Zhao et al.[8], the continuous sliding process can be approximated as a quasi-static process. In this way, the loading and unloading process can be divided into many static steps and the contact force can be calculated by iterative method. i is represented as the index of an arbitrary step.

The value of contact angle α changes with the decrease of r_a and equal to zero when the contact point moves to the asperity summit. The elastic-plastic contact force during this loading process is continuously increasing. And the asperities begin to rebound elastically during the unloading process. The residual interference δ_{res} which would remain indented can be givens as[17]:

$$\frac{\delta_{res}}{\delta} = 0.18\ln\frac{\delta}{\delta_c} - 0.06. \tag{10}$$

And the changes on the contact radii in each step can be expressed as:

$$R_{i+1} = R_i - B\max\left[\left(\delta_{\mathrm{res}}\right)\right]_{0\to i}. \tag{11}$$

The elastic-plastic deformation is simulated after δ_{res} is calculated. B is the factor of influence of the residual deformation and plastic force.

3. Simulations and Discussions

3.1. Influence of Fractal Dimension and Fractal Roughness on Normal Contact Force

Figure 3 presents the normal contact sliding process in the sliding forces with different fractal parameters. The radii and material of the computed asperities are the same and C45E4 is used to simulate whose material properties are listed in Table 1. The range of asperity overlaps δ is selected as $0.01R$. The contact force and tangential deviation are effectively normalized by Hertz force and initial tangential position of the contact point.

Table 1. Material Properties[16].

Elastic modulus (Gpa)	Poisson ratio ν	Hardness H (HV)
180	0.3	255

It can be seen that fractal dimension D and fractal roughness G have great influences on normal contact force. Because of the existence of residual interference and plastic deformation, the spheres are no longer in contact before the tangential deviation arrive the other side. When xr_a^{-1} is smaller than zero, the sliding interaction is loading process. And there is no obvious difference in the initial loading state of different fractal parameters. After $xr_a^{-1} = -0.6$, the difference becomes evident and bigger fractal parameters has bigger normal contact force until the force is decrease to zero in the unloading process. Furthermore, the plastic deformation is larger for smaller G and bigger D.

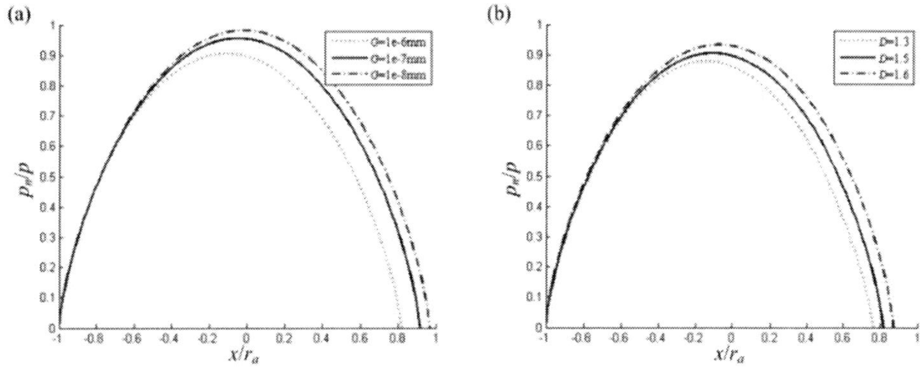

Figure 3. The normal contact force in the sliding process, (a) $D=1.5$, (b) $G=1e-6mm$.

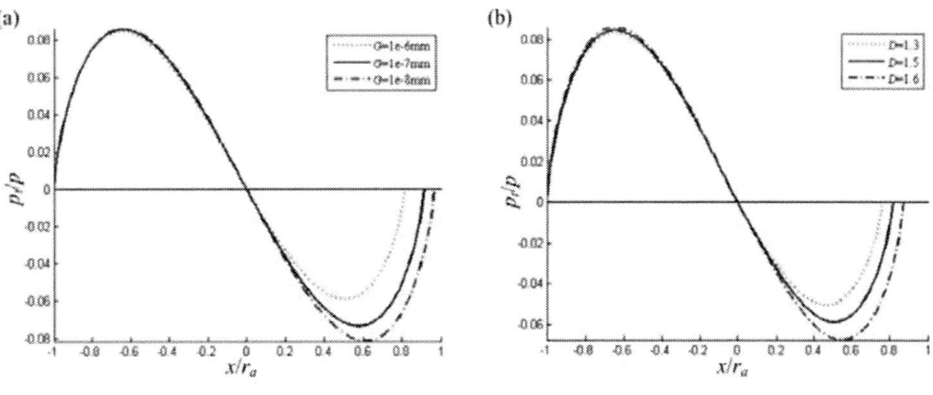

Figure 4. The tangential contact force in the sliding process, (a) $D=1.5$, (b) $G=1e-6mm$.

3.2. Influence of Fractal Dimension and Fractal Roughness on Tangential Contact Force

The tangential contact force resembles a sine function and behaves like the dragging effect in Figure 4. However, the negative force in the unloading process is not the same as the loading part. The maximum absolute value of $p_t p^{-1}$ is 0.06 while it is 0.08 in the loading state. The reason is that the plastic deformations produced in the loading process have reduced the radii of the asperities and decrease the contact force in the next iteration. This distinction is also caused by plastic residual interference. And compared to the normal contact force, the tangential contact force begins to different after xr_a^{-1} is equal to zero. It is because sliding occurs at small scales with small the contact angle and asperity overlap. Thus, the value of $p_t p^{-1}$ is much smaller than $p_t p^{-1}$ and it needs a longer loading process to accumulate distinct plastic deformation. The relative tendency of the terminal force is almost the same as the normal contact force. And the tendency of the influence of the fractal parameters is also similar. The increments of the contact force and the decrements of the contact angle lead to this comprehensive curve.

The simulation results described in Figure 3 and Figure 3 basically reflected the qualitative influence of fractal dimension D and fractal roughness G on contact forces. In the multi-scale condition, the sliding interaction is calculated with the different asperities when the length scale l keeps changing with the tangential displacement according to equation (7). And the surface is smoother when the D is bigger and G is smaller. This means more micro-asperities within a macro-asperity region are bearing the load. Hence, the plastic deformation and residual interference play a smaller role. Furthermore, it is effective to select rough surfaces with bigger D and smaller G to decrease elastic-plastic sliding asperity interaction.

4. Conclusion

This paper presented a sliding asperity model based on fractal theory. The contact force and the critical interference are calculated considering the elastic and plastic deformation. The numerical simulation is developed to study the influence of fractal dimension and fractal roughness on the contact normal and tangential force. The results are compared and the effect of plastic deformation and residual interference are analysed. Some conclusions can be drawn as followed.

Fractal dimension D and fractal roughness G have great influences on contact force. For smaller G and bigger D surface, the sliding interaction of asperities is less remarkable and the plastic force account for less in the contact force. The plastic deformation and residual interference are important factors for sliding interaction. In the initial state of loading process, the normal component of the contact force is more obvious than the tangential component. It is because sliding occurs at small scales with small the contact angle and asperity overlap which makes the tangential component much smaller relatively in frictionless surface. It can be seen that sliding interaction model based on fractal theory is an effective method to study the multi-scale characteristics of rough machined surfaces. And sliding interaction with friction, energy loss, material hardening and adhesive should be taken into consideration in further research.

Acknowledgments

The authors acknowledge the National Natural Science Foundation of China (Grant: 51775552) and Key Research Project of The Education Department of Hunan Province (Grant: 15A107).

References

[1] Greenwood J A 2006 *A simplified elliptic model of rough surface contact* (Wear) pp 191–200
[2] Yuan Y, Gan L, Liu K and Yang X 2017(30) *Elastoplastic Contact Mechanics Model of Rough Surface Based on Fractal Theory* (Chin J Mech ENG) pp 207–215
[3] Persson B N J and Scaraggi M 2014(141) *Theory of adhesion: Role of surface roughness* (J Chem Phys) p 124701.
[4] Huang J, Gao C, Chen J, Lin X and Ren Z 2015(7) *Thermomechanical analysis of an elastoplastic rough body in sliding contact with flat surface and the effect of adjacent contact asperity* (Adv Mech Eng) pp 1–10

[5] Xu H and Komvopoulos K 2019(171) *A fracture mechanics analysis of asperity cracking due to sliding contact* (Int J Solids Struct) pp 1–9
[6] Wang C and Schipper D J 2018(6) *On an elastoplastic sliding model for a coated single asperity* (Lubricants) https://doi.org/10.3390/lubricants6040096.
[7] Boucly V, Nelias D and Green I 2007(129) *Modeling of the rolling and sliding contact between two asperities* (J Tribol) pp 235–245
[8] Zhao B, Zhang S and Keer L M 2016(364-365) *Semi-analytical and numerical analysis of sliding asperity interaction for power-law hardening materials* (Wear) pp 184–192
[9] Jackson R L, Duvvuru R S, Meghani H and Mahajan M 2007(262) *An analysis of elasto-plastic sliding spherical asperity interaction* (Wear) pp 210–219
[10] Gao Z, Fu W, Wang W, Kang W and Liu Y 2018(126) *The study of anisotropic rough surfaces contact considering lateral contact and interaction between asperities* (Tribol Int) pp 270–282
[11] Aferrante L, Bottiglione F, Putignano C, Persson B N J and Carbone G 2018(66:75) *Elastic contact mechanics of randomly rough surfaces: an assessment of advanced asperity models and Persson's theory* (Tribol Lett) https://doiorg/101007/s11249-018-1026-x
[12] Mishra T, Ganzenmüller C G, Rooij M D, Shisode M, Hazrati J and Schipper D J 2019(418–419) *Modelling of ploughing in a single-asperity sliding contact using material point method* (Wear) pp 180–190
[13] Mishra T, Rooij M D, Shisode M, Hazrati J and Schipper D J 2019(137) *An analytical model to study the effect of asperity geometry on forces in ploughing by an elliptical asperity* (Tribol Int) pp 405–419
[14] Zuo R and Wang J 2016(164) *Fractal/multifractal modeling of geochemical data: A review* (J Geochem Explor) pp 33–41
[15] Meng X, Fang C and Niu K 2018 Tribological behavior anisotropy in sliding interaction of asperities on single-crystal α -iron: A quasi-continuum study (Tribol Int) pp 347–359
[16] Zhang K, Li G, Gong JZ and Zhang M 2019(11) *Normal contact stiffness of rough surfaces considering oblique asperity contact* (Adv Mech Eng) pp 1–14
[17] Quicksall J J, Jackson R L and Green I 2004(218) *Elasto-plastic hemispherical contact models for various mechanical properties* (J Eng Tribol J) pp 313–322

Advanced Micromorphology Analysis of Cu/Fe NPs Thin Films

Ş Ţălu[1,*], A Ghaderi[2], K Stępień[3] and F M Mwema[4]

[1]The Directorate of Research, Development and Innovation Management (DMCDI), The Technical University of Cluj-Napoca, Constantin Daicoviciu Street, no. 15, Cluj-Napoca, 400020, Cluj county, Romania
[2]Young Researchers and Elite Club, West Tehran Branch, Islamic Azad University, Tehran, Iran
[3]Faculty of Mechatronics and Mechanical Engineering, Kielce University of Technology, Al. 1000-lecia P. P. 7, 25-314 Kielce, Poland
[4]Department of Mechanical Engineering Science, University of Johannesburg, Auckland Park Campus, South Africa

*Corresponding author

Abstract. In this work, an advanced analysis of Cu/Fe NPs thin films using atomic force microscopy (AFM) has been discussed to characterize at nanoscale 3-D surface microtexture. Samples of Cu/Fe thin films were fabricated by Direct Current-Magnetron Sputtering technique with two controlled thicknesses (group I: Cu 55 nm/Fe 55 nm and group II: Cu 55 nm/Fe 70 nm) in specific conditions of pressure and power. The results obtained from experimental measurements suggested that the surface of group I has the lowest values for fractal dimension (D = 2.28 ± 0.01) and root mean square height (Sq = 4.40 ± 0.1 nm); while the highest values for fractal dimension (D = 2.31 ± 0.01) and root mean square height (Sq = 4.67 ± 0.1 nm) were found in group II. Stereometric and fractal analyses applied for thin films are modern tools for accurate quantitative morphometric characterisation.

1. Introduction

Nanoparticles (NPs) are defined as a various class of materials including a variety of particulate substances that have in at one external dimension approximately 1 to 100 nm [1, 2].

Metal nanoparticles can be considered a research field in full expansion with many industrial applications (automotive, electronic, petrochemical, chemical, military, aerospace, power generation, solar electric power, energy) due to their wide range of tunable physicochemical characteristics [3-5].

In recent decades, different types of Metal NPs have been reported and classified in various research reports depending on their chemical composition and crystalline forms [1]. In different studies about thin films prepared by various techniques of DC magnetron sputtering were applied different mathematical methods for evaluation of micro-roughness parameters [6-10].

Scientific reports with computerized procedures and interpretation in detail regarding the stereometric analysis of Cu/Fe NPs thin films and optimal research techniques for description of 3-D surface micromorphology are recorded in the scientific literature [3-5]. Furthermore, recent studies confirmed the 3-D (three dimensional) surface microtexture of thin films could be estimated using fractal [11-14] and multifractal [9, 15-18] analyses, as well as by power spectral density (PSD) functions [6, 19-21]. Also, it can be noted that microroughness of a surface texture obtained by AFM is usually considered in analyses only along z-axis (vertical axis) in correlation with measurement factors [22-25], while PSD analysis give information about lateral and vertical signals obtained from AFM micrographs [21,

26-32]. In this work, the stereometric and fractal analysis, as well as PSD functions are applied to characterize different surface microtextures of samples (Cu/Fe NPs thin films) prepared by DC Magnetron sputtering technique considering AFM micrographs and the corresponding parameters, computed according with the standard ISO 25178-2:2012.

2. Materials and Methods

The experimental procedure to obtain samples with two controlled thicknesses (group I: Cu 55 nm/Fe 55 nm and group II: Cu 55 nm/Fe 70 nm) by DC Magnetron sputtering technique, in specific conditions of pressure (basic pressure, 10.7×10^{-3} Pa; work pressure 8 Pa) and power (20 Watt) has been published in details in our previous works [3-5].

Atomic force microscopy imaging was performed using a commercial Nanoscope Multimode AFM system (Digital Instruments, Santa Barbara, CA). The surfaces were scanned in non contact mode at a scan rate of 10–20 μm/s to record AFM images of 256 x 256 pixels for areas of 1 μm x 1 μm.

3. Results and Discussion

The stereometric, fractal and PSD analyses applied for AFM data were described in chosen terms of stereometric analysis using the MountainsMap Premium software version 7.4.8872 [33] and were calculated based on the definitions given in the standard ISO 25178-2:2012 [34].

The relevant AFM micrographs of the samples, for 1 μm x 1 μm scanning square areas, are illustrated in fig. 1.

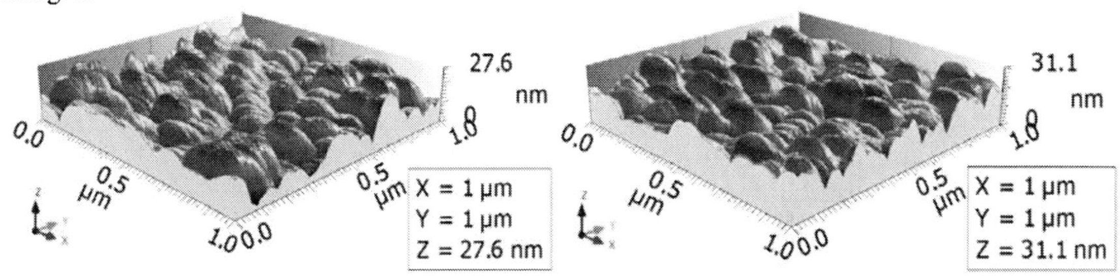

a) b)

Figure 1. 3-D AFM micrographs of samples: a) group I; b) group II.

The root mean square height for group I is Sq = 4.40 [nm] and for group II is Sq = 4.67 [nm], whilst arithmetical mean height for group I is Sa = 3.58 [nm] and for group II is Sa = 3.79 [nm]. Also, the maximum height for group I is Sz = 27.6 [nm] and for group II is Sz = 31.1 [nm].

The representation of surface texture directions of analyzed samples using Cartesian graphs are shown in figure 2 and the corresponding values are given in Table 1.

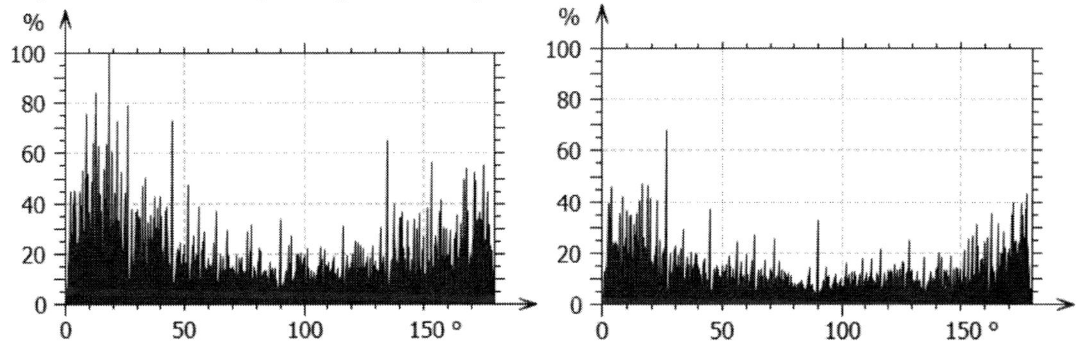

a) b)

Figure 2. The representation of surface texture directions of analyzed samples using Cartesian graphs for: a) group I; b) group II.

Table 1. Surface texture directions of analyzed samples using Cartesian graphs.

Samples	Isotropy [%]	First direction [°]	Second direction [°]	Third direction [°]
Group I	60.8	18.5	13.0	26.5
Group II	49.5	0.10	26.5	16.0

The graphical representation of surface texture isotropy of analyzed samples using the autocorrelation functions are presented in figure 3.

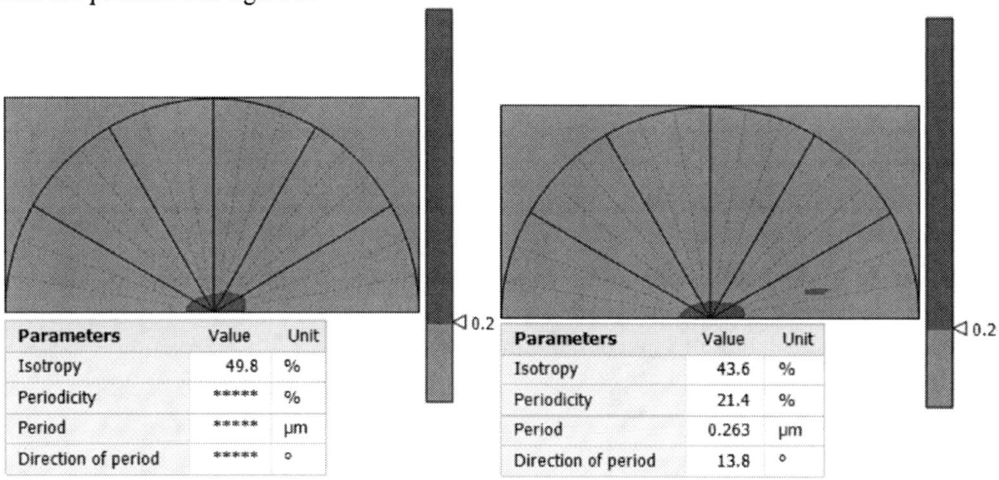

a) b)

Figure 3. The representation of surface texture isotropy of analyzed samples for: a) group I; b) group II.

Graphical representations of the furrows of the analyzed samples are illustrated in figure 4.

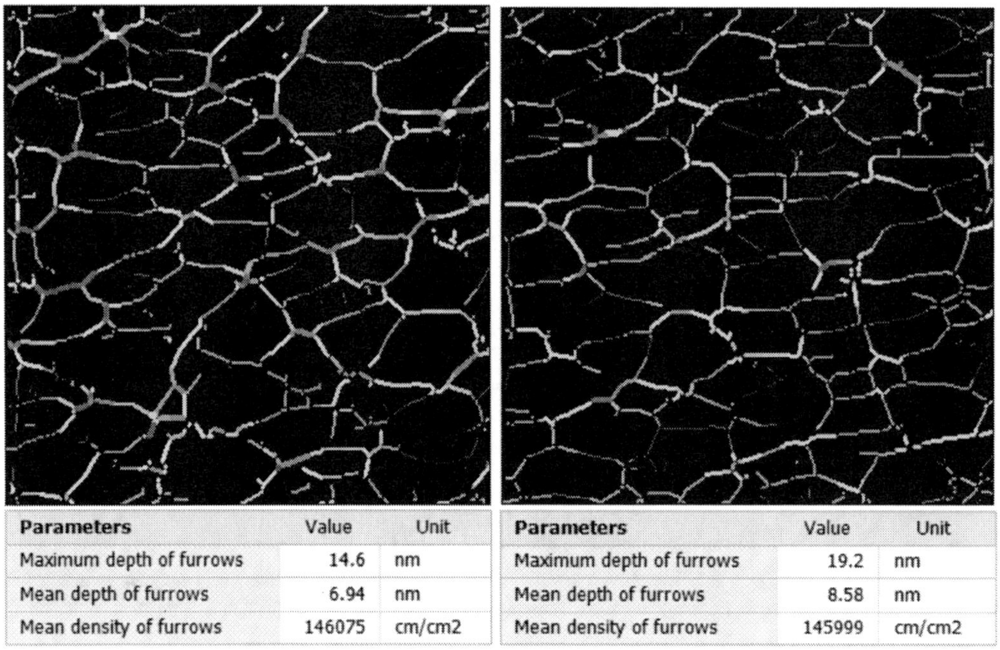

a) b)

Figure 4. The representation of the furrows for: a) group I; b) group II.

The values of the maximum depth of furrows and mean depth of furrows parameters are greater for group II, whereas the greater value of mean density of furrows is associated with group I.

Computation of the fractal dimension (Df) based on the enclosing boxes method [33] with determination of coefficients of correlation (R^2), are detailed shown in fig. 5.

IOP Conf. Series: Materials Science and Engineering **611** (2019) 012016 doi:10.1088/1757-899X/611/1/012016

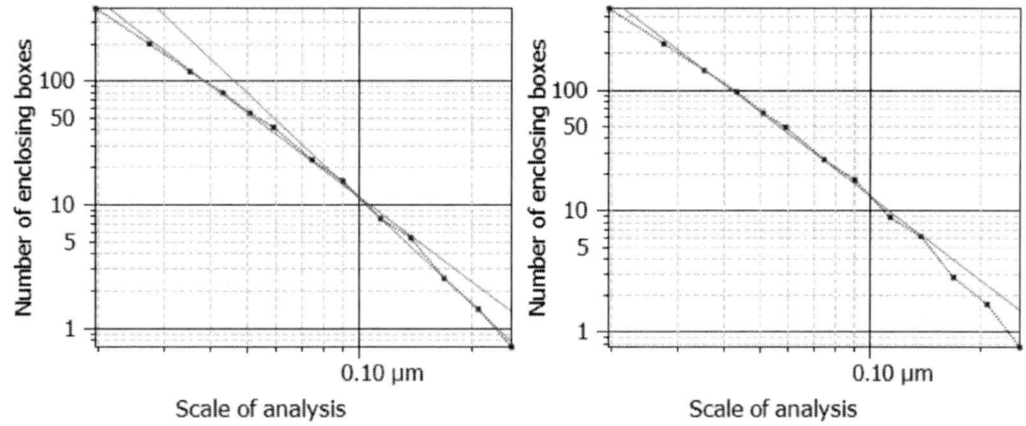

Information	
Method	Enclosing boxes

Parameters	Value
Fractal dimension	2.28
Slope(1)	-2.28
R²(1)	0.997
Slope(2)	-2.84
R²(2)	0.994

a)

Information	
Method	Enclosing boxe

Parameters	Value
Fractal dimension	2.31
Slope(1)	-2.31
R²(1)	0.997
Slope(2)	-2.61
R²(2)	0.993

b)

Figure 5. Graphs of the fractal dimension with coefficients of correlation (R^2), computed by the enclosing boxes method for: a) group I; b) group II.

The representation of averaged power spectrum density (PSD) of surface textures is shown in figure 6 with the corresponding parameter values.

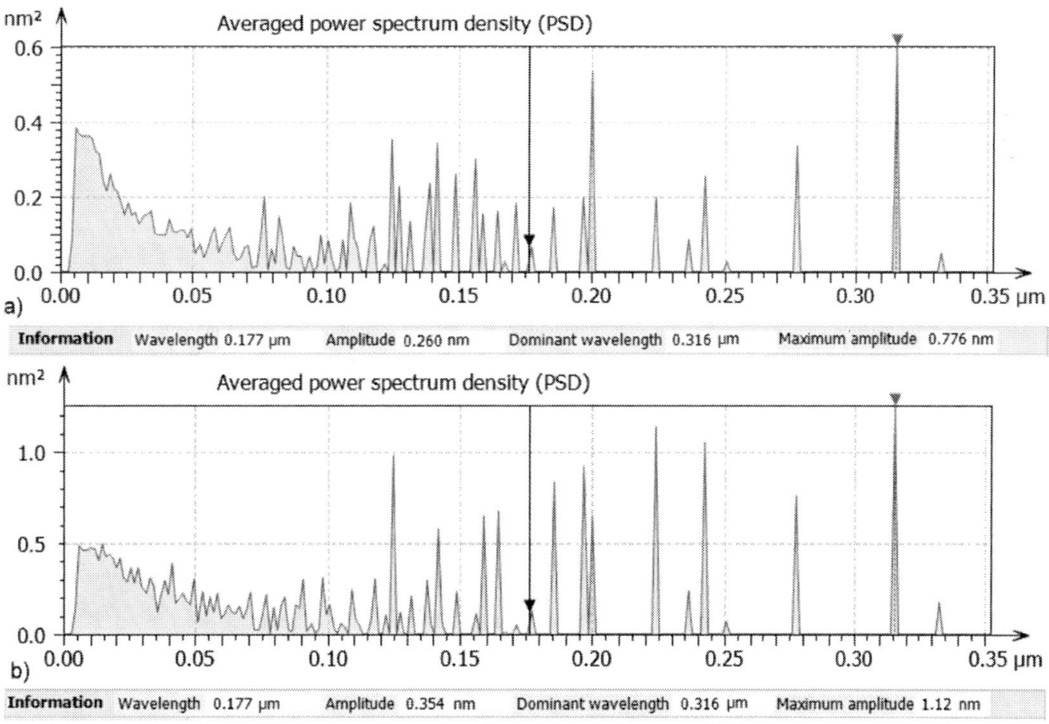

Figure 6. The representation of averaged PSD of surface textures for: a) group I; b) group II.

For both samples, wavelength and dominant wavelength have the same values, whilst amplitude and maximum amplitude is greater for group II.
Graphical representations of the slices of the analyzed samples are presented in figure 7.

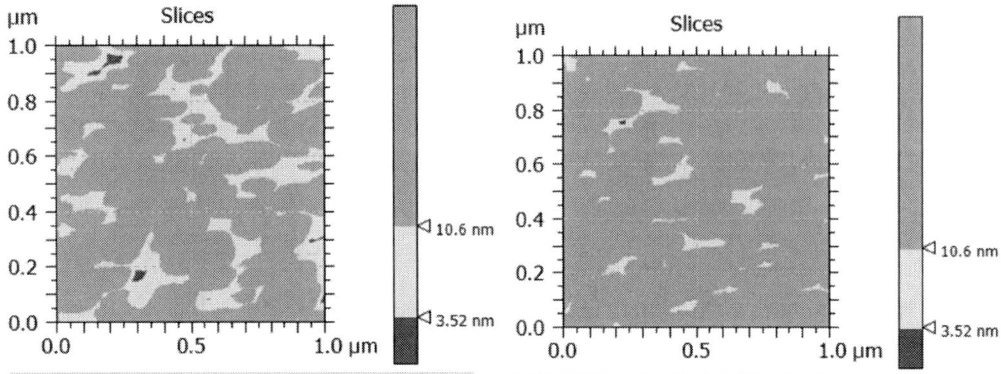

Parameters	Unit	■	□
Projected area	%	0.494	19.4
Volume of void	%	0.152	5.94
Volume of material	%	99.8	94.1
Volume of void	nm³/µm²	5352	420088
Volume of material	nm³/µm²	3510760	6651245
Mean thickness of void	nm	0.00535	0.420
Mean thickness of material	nm	3.51	6.65

a)

Parameters	Unit	■	□
Projected area	%	0.0305	5.25
Volume of void	%	0.0119	1.18
Volume of material	%	100	98.8
Volume of void	nm³/µm²	420	83519
Volume of material	nm³/µm²	3515726	6987786
Mean thickness of void	nm	0.00042	0.0835
Mean thickness of material	nm	3.52	6.99

b)

Figure 7. The slices for: a) group I; b) group II.

It can be noted that there are significant differences between the values of the two slices expressed in all parameters: volume of void, volume of material, mean thickness of void and mean thickness of material.
The Sk parameters of the analyzed samples are computed in Table 2.

Table 2. Sk parameters of analyzed samples.

Samples	Sk [nm]	Spk [nm]	Svk [nm]	Smr1 [%]	Smr2 [%]	Sa1 [$nm^3/\mu m^2$]	Sa2 [$nm^3/\mu m^2$]
Group I	12	3.74	3.57	10.9	92.9	204393	127503
Group II	12.7	4.27	3.71	9.93	92.2	211901	145184

The Sk values of following parameters of group II (Sk, Spk, Svk, Sa1 and Sa2) are greater than the corresponding values of group I, whilst parameters Smr1 and Smr2 are greater for group I than group II.
Graphs of the contour lines (imaginary lines with points with the same elevation) and photo simulation of the surface microtexture are presented in figure 8.

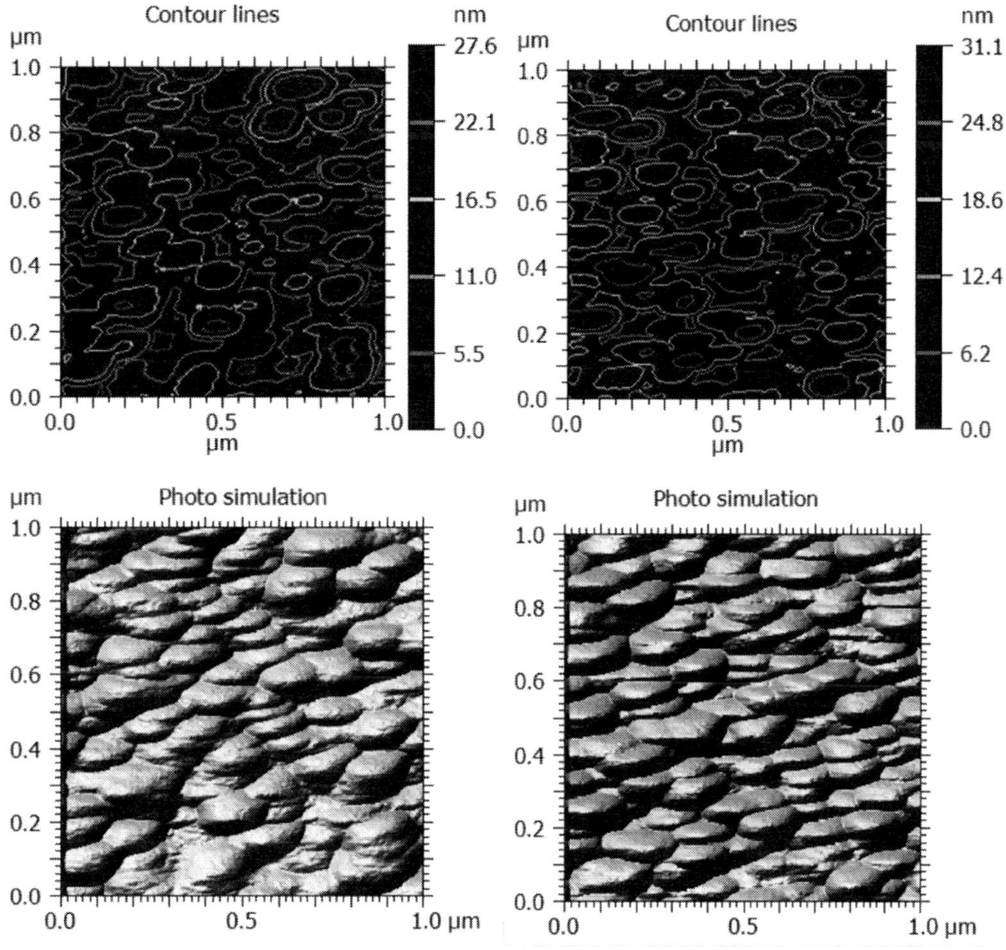

Figure 8. The graphical representation of the contour lines and photo simulation of the surface for: a) group I; b) group II.

These contours appear on associated topographic micro scale maps as colour lines and give sufficient and accurate details to facilitate reading them. Also, this comprehensive information highlights individual surface features.

These results demonstrate the applicability of the AFM system and are essential for deciphering the complexity of micromorphology of samples and for mapping their topography.

The results of the numerical calculations suggested that the surface of group I has the lowest values for fractal dimension (D = 2.28 ± 0.01) and root mean square height (Sq = 4.40 ± 0.1 nm); while the highest values for fractal dimension (D = 2.31 ± 0.01) and root mean square height (Sq = 4.67 ± 0.1 nm) were found in group II. On the other hand, the PSD highlights new information about the surface micromorphology including fractal and superstructure contributions.

4. Conclusion

In our work it is shown an advanced analysis of Cu/Fe NPs thin films fabricated by a DC magnetron sputtering technique with two controlled thicknesses 55 nm and 70 nm of Fe over layer.

The micromorphology patterns of the samples were analyzed using AFM data with a high-resolution using MountainsMap Premium software according with the standard ISO 25178-2:2012. Also, based on the quantitative tests the morphological changes of samples were examined, and we concluded that in the engineering design of Cu/Fe NPs thin films, synthesized by DC magnetron technique, this approach has several advantages over other methods to identify sets of topography parameters.

AFM platform and MountainsMap Premium software applied highlighted the multiparametric characterization capabilities and provided new perspectives in the detection of surface 3-D morphology of Cu/Fe NPs thin films.

References

[1] Laurent S, Forge D, Port M, Roch A, Robic C, Vander Elst L and Muller R N, Magnetic iron oxide nanoparticles: synthesis, stabilization, vectorization, physicochemical characterizations, and biological applications, Chem. Rev., 110, 2010. DOI: 10.1021/cr900197g.

[2] Jeevanandam J, Barhoum A, Chan Y S, Dufresne A and Danquah M K, Review on nanoparticles and nanostructured materials: history, sources, toxicity and regulations, Beilstein J Nanotechnol., 9: 1050-1074, 2018. DOI: 10.3762/bjnano.9.98.

[3] Moradian R, Ghaderi A and Elahi S M, Synthesis of Cu/Fe bilayers: micro structure and optical properties, J Mater Sci: Mater Electron, 2016. DOI 10.1007/s10854-016-4793-x.

[4] Stach S, Garczyk Z, Ţălu Ş, Solaymani S, Ghaderi A, Moradian R, Nezafat N, Elahi S M and Gholamali H, Stereometric parameters of the Cu/Fe NPs thin films, J. Phys. Chem. C 119 (31): 17887e17898, 2015. DOI: 10.1021/acs.jpcc.5b04676.

[5] Ţălu Ş, Stach S, Solaymani S, Moradian R, Ghaderi A, Hantehzadeh M R, Elahi S M, Garczyk Ż and Izadyar S, Multifractal spectra of atomic force microscope images of Cu/Fe Nanoparticles based films thickness. J Electroanal. Chem., 749: 31-41, 2015. DOI: 10.1016/j.jelechem.2015.04.009.

[6] Ţălu Ş 2015 *Micro and nanoscale characterization of three dimensional surfaces. Basics and applications* (Cluj-Napoca, Romania: Napoca Star Publishing House).

[7] Ţălu Ş, Stach S, Valedbagi S, Elahi S M and Bavadi R, Surface morphology of titanium nitride thin films synthesised by DC reactive magnetron sputtering, Mater Sci-Poland, 33(1): 137-143, 2015. DOI: 10.1515/msp-2015-0010.

[8] Stach S, Dallaeva D, Ţălu Ş, Kaspar P, Tománek P, Giovanzana S and Grmela L, Morphological features in aluminum nitride epilayers prepared by magnetron sputtering, Mater Sci-Poland, 33(1): 175-184, 2015. DOI: 10.1515/msp-2015-0036.

[9] Ţălu Ş, Stach S, Valedbagi S, Bavadi R, Elahi S M and Ţălu M, Multifractal characteristics of titanium nitride thin films, Mater Sci-Poland, 33(1): 541-548, 2015. DOI: 10.1515/msp-2015-0086.

[10] Arman A, Ţălu Ş, Luna C, Ahmadpourian A, Naseri M and Molamohammadi M, Micromorphology characterization of copper thin films by AFM and fractal analysis, J. Mater. Sci. Mater. Electron., 26: 9630–9639, 2015. DOI: 10.1007/s10854-015-3628-5.

[11] Shikhgasan R, Ţălu Ş, Dinara S, Sebastian S and Guseyn R, Epitaxy of silicon carbide on silicon: Micromorphological analysis of growth surface evolution, Superlattices Microstruct. 86: 395-402, 2015. DOI: 10.1016/j.spmi.2015.08.007.

[12] Ţălu Ş, Solaymani S, Bramowicz M, Kulesza S, Ghaderi A, Shahpouri S and Elahi S M, Effect of electric field direction and substrate roughness on three-dimensional self-assembly growth of copper oxide nanowires, J. Mater. Sci. Mater. Electron., 27(9): 9272-9277, 2016. DOI: 10.1007/s10854-016-4965-8.

[13] Sobola D, Ţălu Ş, Solaymani S and Grmela L, Influence of scanning rate on quality of AFM image: Study of surface statistical metrics. Microsc. Res. Tech., 80(12): 1328-1336, 2017. DOI: 10.1002/jemt.22945.

[14] Ţălu Ş, Bramowicz M, Kulesza S, Dalouji V, Solaymani S and Valedbagi S, Fractal features of carbon–nickel composite thin films, Microsc. Res. Tech., 79(12): 1208-1213, 2016. DOI: 10.1002/jemt.22779.

[15] Ţălu Ş, Stach S, Mendez A, Trejo G and Ţălu M., Multifractal characterization of nanostructure surfaces of electrodeposited Ni-P Coatings, J. Electrochem. Soc., 161: D44–D47, 2013. DOI:10.1149/2.039401jes.

[16] Yadav R P, Kumar M, Mittal A K and Pandey A C, Fractal and multifractal characteristics of swift heavy ion induced self-affine nanostructured BaF_2 thin film surfaces, Chaos, 25(8): 083115, 2015.

[17] Ţălu Ş, Marković Z, Stach S, Marković B T and Ţălu M, Multifractal characterization of single wall carbon nanotube thin films surface upon exposure to optical parametric oscillator laser irradiation, Appl Surf Sci, 289: 97-106, 2014. DOI: 10.1016/j.apsusc.2013.10.114.
[18] Ţălu Ş, Morozov I A and Yadav R P, Multifractal analysis of sputtered indium tin oxide thin film surfaces, Appl Surf Sci., 484: 892-898, 2019. DOI: 10.1016/j.apsusc.2019.04.170.
[19] Senthilkumar M, Sahoo N K, Thakur S and Tokas R, Characterization of microroughness parameters in gadolinium oxide thin films: A study based on extended power spectral density analyses, Appl Surf Sci., 81(5), 2005. DOI: 10.1016/j.apsusc.2005.02.122.
[20] Mwema F M, Akinlabi E and Oladijo P 2019 The use of Power Spectrum Density for surface characterization of thin films, *Photoenergy and Thin Film Materials.* (John Wiley & Sons). DOI: 10.1002/9781119580546.ch9.
[21] Gong Y, Misture S T, Gao P and Mellott N P, Surface roughness measurements using power spectrum density analysis with enhanced spatial correlation length, J. Phys. Chem. C, 120(39): 22358-22364, 2016. DOI: 10.1021/acs.jpcc.6b06635.
[22] Méndez A, Reyes Y, Trejo G, Stępień K and Ţălu Ş, Micromorphological characterization of zinc/silver particle composite coatings. Microsc. Res. Tech., 78(12): 1082-1089, 2015. DOI: 10.1002/jemt.22588.
[23] Zare M, Solaymani S, Shafiekhani A, Kulesza S, Ţălu Ş and Bramowicz M, Evolution of rough-surface geometry and crystalline structures of aligned TiO_2 nanotubes for photoelectrochemical water splitting. Sci Rep., 8: 1-11, article number: 10870, 2018. DOI: 10.1038/s41598-018-29247-3.
[24] Stach S, Sapota W, Ţălu Ş, Ahmadpourian A, Luna C, Ghobadi N, Arman A and Ganji M, 3D surface stereometry studies of sputtered TiN thin films obtained at different substrate temperatures. J. Mater. Sci. Mater. Electron., 28(2): 2113-2122, 2017. DOI: 10.1007/s10854-016-5774-9.
[25] Naseri N, Solaymani S, Ghaderi A, Bramowicz M, Kulesza S, Ţălu Ş, Pourreza M and Ghasemi S, Microstructure, morphology and electrochemical properties of Co nanoflake water oxidation electrocatalyst at micro- and nanoscale. RSC Advances, 7(21): 12923-12930, 2017. DOI: 10.1039/c6ra28795f.
[26] Jacobs TDB, Junge T and Pastewka L, Quantitative characterization of surface topography using spectral analysis. Surf. Topogr.: Metrol. Prop., 5(1): 013001, 2017. DOI: 10.1088/2051-672X/aa51f8.
[27] Martínez JFG, Nieto-Carvajal I, Abad J and Colchero J, Nanoscale measurement of the power spectral density of surface roughness: how to solve a difficult experimental challenge. Nanoscale Res Lett., 7(1): 174, 2012. DOI: 10.1186/1556-276X-7-174.
[28] Gavrilă R, Dinescu A and Mardare D, A power spectral density study of thin films morphology based on AFM profiling, Rom J Inf Sci Tech., 10(3): 291-300, 2007.
[29] Karana S and Mallik B, Power spectral density analysis and photoconducting behavior in copper(II) phthalocyanine nanostructured thin films. Phys. Chem. Chem. Phys., 10: 6751-6761, 2008. DOI: 10.1039/B809648A.
[30] Itoh T and Yamauchi N, Surface morphology characterization of pentacene thin film and its substrate with under-layers by power spectral density using fast Fourier transform algorithms. Appl Surf Sci., 253(14): 6196-6202, 2007. DOI: 10.1016/j.apsusc.2007.01.056.
[31] Dash P, Mallick P, Rath H, Tripathi A, Prakash J, Avasthi DK, Mazumder S, Varma S, Satyam PV and Mishraa NC, Surface roughness and power spectral density study of SHI irradiated ultra-thin gold films. Appl Surf Sci., 256(2): 558-561, 2009. DOI: 10.1016/j.apsusc.2009.08.046.
[32] Yu L, Xu J, Dong S and Kojima I, Surface morphology and growth mechanisms for sputtered amorphous silicon nitride thin films, Thin Solid Films, 516(8): 1781-1787, 2008. DOI: j.tsf.2007.07.158.
[33] MountainsMap® Premium 7.4.8872 software (Digital Surf, Besançon, France).
[34] ISO 25178-2: 2012, Geometrical product specifications (GPS) - Surface texture: Areal - Part 2: Terms, definitions and surface texture parameters.

Effect of Bismuth on Sulfide in High Sulfur Free-cutting Steel

Shaohui Hu[1], Zhiwei Li[2], Tian Fan[2] and Jianxun Fu[1,2,*]

[1]Materials Genome Institute of Shanghai University, Shanghai University, Shanghai 200072, China
[2]Center for Advanced Solidification Technology, School of Material Science and Engineering, Shanghai University, Shanghai 200072, China

*Corresponding author

Abstract. In order to investigate the effect of bismuth content on the morphology of sulfide in free-cutting steel, different pure bismuth particles were added to SAE1215 high sulfur free-cutting steel to modify the sulfide. The samples were observed and analyzed by metallographic microscope and scanning electron microscope. It was found that the size of sulfide decreased with the increase of bismuth content. Bismuth exists in steel in three forms: 1) Bismuth exists alone in steel matrix; 2) Bismuth adheres to sulfide; 3) Bismuth encapsulates sulfide. Sulfide exists in the form of MnS in steel with two morphologies: spherical MnS and rod/dendritic MnS. The sulfide morphology changes from rod or dendritic to spherical or ellipsoid with the increase of bismuth content.
Keywords: bismuth; sulfide; morphology.

1. Introduction

With the development of intellectualization, automation and digitalization of mechanical processing [1], more and more attention has been paid to free-cutting steel. Lead free-cutting steel has excellent cutting performance[2], but it is toxic to producers and the environment, and it is difficult to recycle lead in steel. The addition of lead substitutes has become an important direction in the study of free-cutting steel [3].
Bismuth and lead have similar physical properties, and bismuth is non-toxic. Bismuth does not dissolve in steel and has less macrosegregation [4]. Bismuth has better wettability than lead. Adding bismuth with half lead content in free-cutting steel can obtain the same cutting performance as lead-based free-cutting steel [5]. Bismuth is therefore a reasonable substitute for lead[6].

2. Experiment

In order to adjust the content of bismuth in SAE1215 free-cutting steel, three groups of bismuth particles with 99.99% purity were added in tube high temperature resistance furnace to melt SAE1215 free-cutting steel(diameter - 20mm, high - 120mm,rolled product). The composition of SAE1215 free-cutting steel is shown in Table 1.

Table 1. the composition of SAE1215 free-cutting steel (wt.%)

Element	C	Si	Mn	P	S
Content	0.05~0.07	0.01~0.03	1.20~1.30	0.046~0.055	0.34~0.36

The experimental steps are as follows: 900g free-cutting steel is placed in alumina crucible (inner diameter - 53mm, outer diameter - 61mm, high - 120mm), and slag-free smelting is carried out in high temperature resistance furnace. Then, under the protection of high purity argon (1L/min), the sample is heated to 1620 C. After melting, the steel is kept for 30 minutes to achieve enough homogenization. Then turn off the gas, add pure bismuth particles (calculated), and then open the gas. After 5 minutes, mix with quartz tube (quartz tube to insert the bottom, touch the bottom wall) for 20 to 30 seconds. After stirring, heat preservation for 5 minutes, sampling process with pipette; after another five minutes, activity oxygen in steel was detected by oxygen determinator. At this time, activity oxygen was measured as terminal activity oxygen. After holding for 10 minutes, the sample is cooled with the furnace.

3. Results and Discussion

After the experiment, about three grams of debris were drilled in the centre of the ingot and sent to the National Iron and Steel Material Testing Centre to detect the bismuth content. The results of the test and the active oxygen content in the steel are shown in Table 2.

Table 2. test results (wt.%)

Furnace number	Content of bismuth	Activity oxygen /ppm
1	0.021	10
2	0.054	12
3	0.071	21

Zeiss metallographic microscopy was used to analyze sulphides in ingots with three stoves. It can be seen from Fig. 1 that sulphides precipitate along grain boundaries and distribute in clusters. When bismuth content is low, most sulphides are long strips. With the increase of bismuth content, the size of sulphides decreases. The shape of sulphides changes from long strips to round or oval ones, sulfides become smaller. In order to further illustrate the inhibitory effect of bismuth on the growth of sulfides, the average equivalent diameter of sulfides was counted. When the content of bismuth increased from 21ppm to 71ppm, the average equivalent diameter of sulfides decreased from 4.40μm to 3.49μm. It shows that bismuth plays an active role in improving the morphology of sulfides.

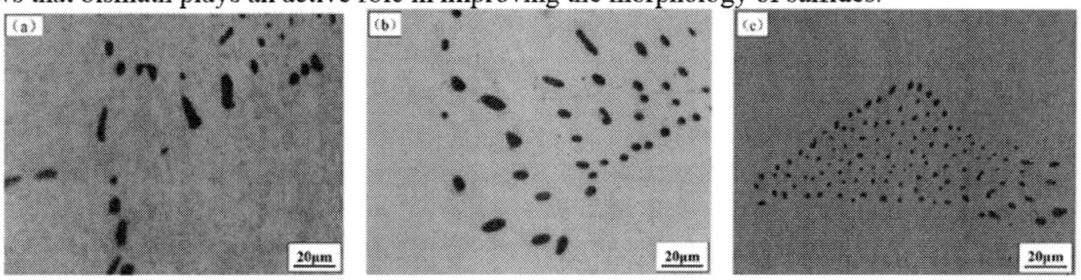

Figure 1. OM image of sulphide (a) furnace number 1; (b) furnace number 2; (c) furnace number 3

In order to further analyse the effect of bismuth on sulfide composition, the sulfide was examined by Phenom-World desktop scanning electron microscopy. The results show that bismuth is distributed in steel in three forms: 1) Bismuth is separately distributed in steel matrix; 2) Bismuth is adhering to sulfide; 3) Bismuth is encapsulated in sulfide, as shown in Fig. 2. Energy dispersive spectrometer shows that the white part is bismuth phase and the dark part is sulfide. It can be seen that when bismuth content is low, Bismuth adheres to the surface of sulfide, and there is hardly any single bismuth phase. With the increase of bismuth content, bismuth begins to encapsulate sulfide and exists in the steel matrix alone. The more bismuth content, the more Bismuth-MnS composite inclusions.

IOP Conf. Series: Materials Science and Engineering **611** (2019) 012017 doi:10.1088/1757-899X/611/1/012017

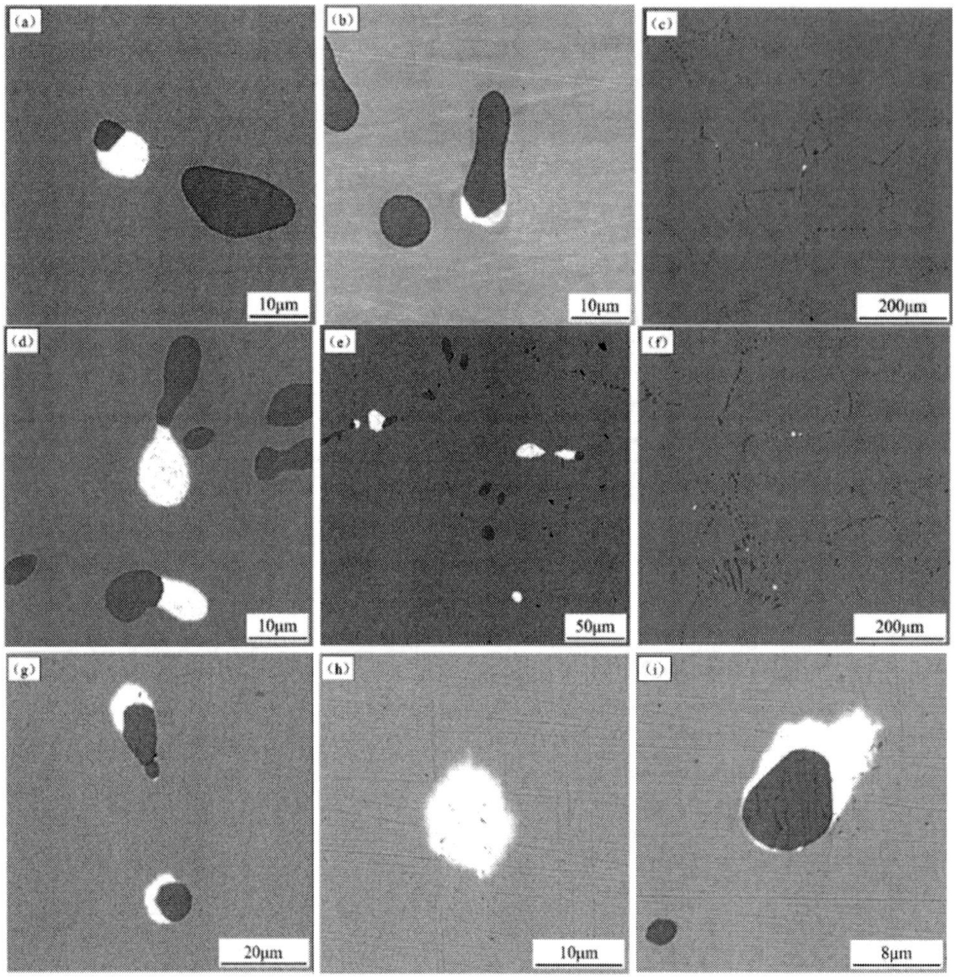

Figure 2. SEM image of sulphides:(a~c) furnace number 1; (d~f) furnace number 2; (g~i) furnace number 3

In order to better observe the three-dimensional morphology of sulfides in steel, electrolysis experiments were carried out on steel samples for 20 minutes. As can be seen from Fig. 3, there are three shapes of bismuth in steel: 1) spindle shape; 2) irregular shape; 3) approximate spherical shape. When bismuth content is low, bismuth mostly adheres to the surface of sulfide irregularly, and the sulfides are rod-like or dendritic. With the increase of bismuth content, bismuth adhering to sulfide increases. Bismuth phase occurs as spindle or nearly spherical and encapsulates sulfides. The sulfides are transformed from dendritic or rod-like to spherical or ellipsoid.

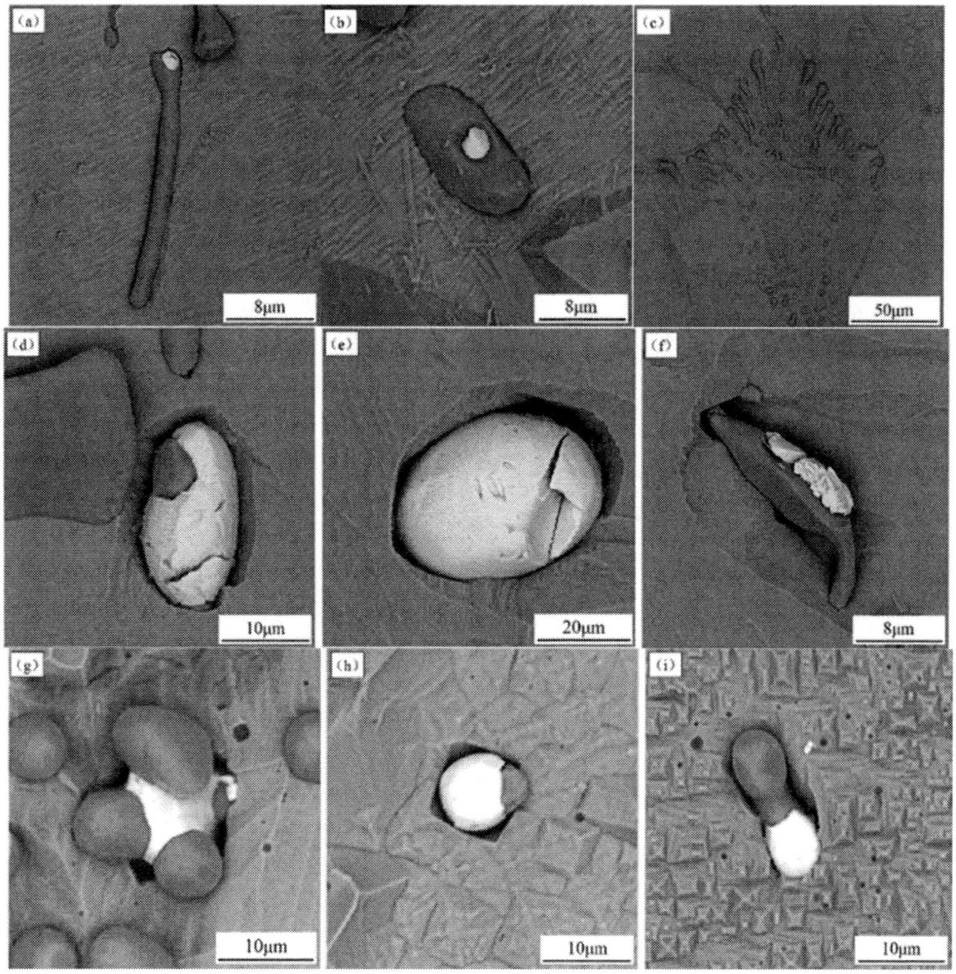

Figure 3. Three-dimensional morphology of sulfides: (a~c) furnace number 1; (d~f) furnace number 2; (g~i) furnace number 3

4. Conclusion

With the increase of bismuth content, the bismuth adhering to or encapsulating sulfides increases, the average equivalent diameter of sulfides in steel decreases, and the size of sulfides tends to be smaller.
Scanning electron microscopy (SEM) shows that bismuth exists mainly in three forms in free-cutting steel: 1) bismuth exists alone in the steel matrix; 2) bismuth coated sulphide (MnS+Bi composite phase); 3) bismuth adheres to sulphide; in addition, with the increase of bismuth content, the number of composite inclusions increases.
Electrolytic corrosion shows that there are three main shapes of bismuth in steel: 1) spindle shape; 2) irregular shape; 3) approximate spherical shape. There are two main forms of sulfides: type I (spherical) and type II (rod or dendritic).

References

[1] Ran Sun and Qiming Sun 2018 *Characteristics and Development Evaluation of Automation Technology in Machinery Design and Manufacturing* Practical Electronics p 42-43
[2] Xiaobing Zhang 2007 *Development of a New Generation of Self-lubricating Lead-free Free Cutting Steel* Papers Collection of 2007 China Iron and Steel Annual Conference
[3] Chuntang Liu and Daowen Tang 2012 *Production Practice of Low Carbon Sulfur-Bismuth Free Cutting Steel* Steelmaking p 22-24
[4] Yongjun Zhang and Chen Zhu 2005 *Development of Lead-free Free Cutting Steel for Mechanical*

Structures Materials Review p 68-71
[5] Yuxuan Chen and Chunjian Cao 2016 *Research Progress and Prospect of Free Cutting Steel at Home and Abroad* Hot Working Technology p 25-28
[6] Jianlun Xu and Senlong Yang 1993 *Study on Bi-Containing Free-Cutting Stainless Steel* Journal of Shanghai University of Technology p 257-263.

Analysis of Sulfide Morphology in Non-quenched and Tempered Steel

Menghao Zhao, Kenan Ai, Zhiwei Li, Dong Zhang, Ping Shen and Jianxun Fu*

Center for Advanced Solidification Technology, material science and Engineering School, Shanghai University, shanghai 200072, China

*Corresponding author

Abstract. Non-tempered steel without quenching and tempering heat treatment, saving energy and protecting the environment. Sulfide form is an important indicator of its production level. To explore the gap between the levels of sulphide control in 3 steel mills, metallographic microscope, scanning electron microscope and small sample electrolysis technology were used to analysis and comparison of sulfides in non-quenched and tempered steel produced by plant A (Germany), plant B (domestic), and plant C (domestic). The ratings of Class A inclusions in the three plants are: plant A is coarse grade 2.5 and fine grade 2.0, the sulfide are short rod shape; plant B is coarse grade 2.5 and fine grade 2.5, most of the inclusions are short rods, a few are long strips, and there are inclusions with MnTe; plant C is coarse grade 2.5 and fine grade 3.0, inclusions are generally long strips. The non-quenched and tempered steel inclusions produced by plant A are evenly distributed, and the sulfide form is well controlled.
Keyword: sulfide; morphology; non-quenched and tempered.

1. Introduction

Non-quenched and tempered steel has the advantages of process simplification, energy saving, short production cycle, etc., in line with the development trend of the current era[1-3]. The steel can be used to make the crankshaft of an automobile engine [4-6]. Crankshaft is one of the most important and most complicated transmission parts in the engine. Its reliability directly determines the reliability of the whole engine[7-8]. Therefore, the selection of engine crankshaft is particularly important. At present, there is a certain gap between non-tempered steel produced in China and international advanced levels such as Japan and Germany. Sulfide control level difference is an important reason for poor product reliability. It is important to analyze the difference between China and other advanced countries, and find out the key factors for improving quality. Inclusions morphology is an important factor in evaluating the quality of the steel. The form of sulfide has a significant impact on the quality of the product and the pass rate of magnetic flaw detection[9-10]. By comparing and analyzing non-quenched and tempered steel inclusions, it is helpful to analyze the production level of different steel, and establish sulfide control targets for the development of high-quality non-tempered steel.
In order to find the control level of sulfide form in three enterprises, metallographic microscope, scanning electron microscope and sample electrolysis technology are used in non-quenched and tempered steel produced by plant A (Germany), plant B (domestic) and plant C (domestic). The inclusions were analyzed and compared to find the gap between the control levels of sulfides in different technical routes, laying the foundation for the development of domestic high-quality non-quenched and tempered steel.

IOP Conf. Series: Materials Science and Engineering **611** (2019) 012018 doi:10.1088/1757-899X/611/1/012018

2. Experiment

The steel production process used in the experiment is converter-LF-RH-CC-heat transfer rolling. Sample 1 is a rolled material of plant A, sample 2 is a strip sample taken from a steel bar of B steel mill, and sample 3 is a sample of non-quenched and tempered steel rolled material of C steel mill, respectively. Sampling from the 1/4 position of the samples. The steel produced by the three steel mills belongs to the same grade, and their composition is shown in Table 1.

After the samples were pre-ground, it was polished with a polishing paste of W2.5, W1.5, W0.5, and microscopically observed using a Zeiss metallographic microscope (AVxo) and a scanning electron microscope with energy dispersive X-ray spectrometer(SEM-EDS), the sample was photographed 15 times in a random and random different parts. Metallographic photographs were selected (100 times) as inclusion statistics, and the type and composition of inclusions were observed by scanning electron microscopy with energy spectrum. The polished sample was electrolytically etched with a sample of 10% AA non-aqueous solution (10% tetramethylammonium chloride-1% acetylacetone-89% methanol) [11], and the electrolysis current was 0.2-0.4 A. The shape of the inclusions was obtained by scanning electron microscopy at 25-40 min.

Table 1. Chemical composition of the experimental steel (wt.%).

Element	C	Si	Mn	P	S	Cr
Content	0.36~0.40	0.50~0.65	1.40~1.55	<0.025	0.010~0.040	0.10~0.25

3. Results and Discussion

The morphology of inclusions in plant A, B and C are shown in Figure 1. According to Figure a, the inclusions are gray, distributed in chains and distributed evenly in the steel. The number of rod-shaped sulfide is the least. Figure b shows the inclusions of plant B are distributed in a chain and the number of long inclusions increases. The inclusions of plant C are unevenly distributed and basically elongated.

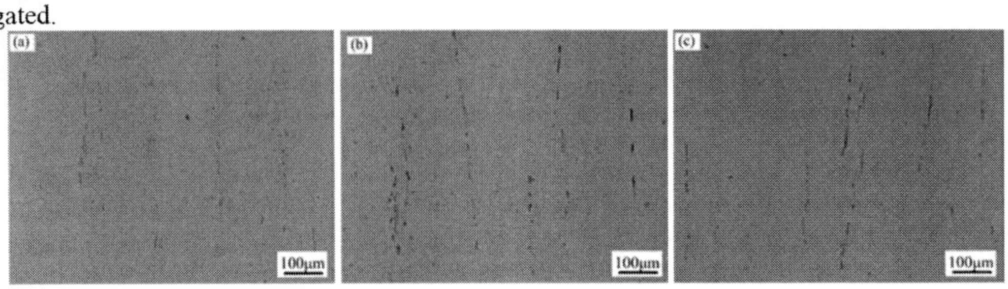

Figure 1. Morphology of MnS inclusions. (a) Plant A; (b) Plant B; (c) Plant C.

In order to further analyze the form of MnS, the inclusions produced by three steel mills were further analyzed by SEM-EDS, as shown in Figure 2. According to Figure a, MnS inclusions are generally short rods, but there are also long strips of MnS of plant A. From Figure b, it can be found that most of the MnS of plant B are nearly spherical, and there are more long rod-shaped inclusions than plant A. According to Figure c, the MnS inclusions are almost all distributed in a long rod shape, and the morphology of the inclusions is extremely poor in non-quenched and tempered steel produced by plant C.

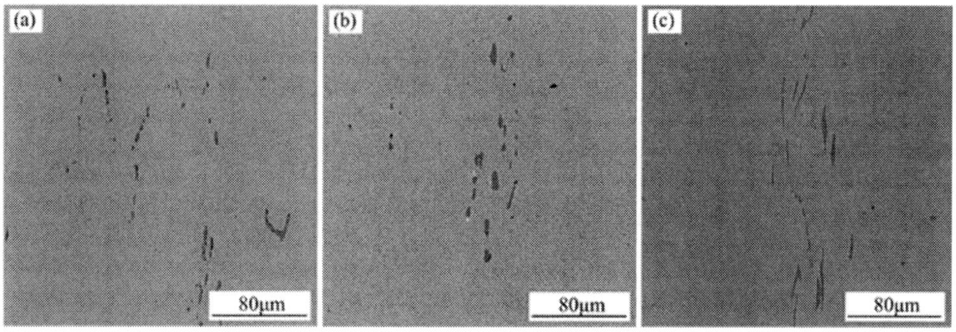

Figure 2. SEM image of MnS inclusions (a) Plant A; (b) Plant B; (c) Plant C.

In the SEM, white matter was found and attached to manganese sulfide in non-quenched and tempered steel produced by plant B, typical MnS inclusions with white matter are shown in Fig. 3. After EDS analysis, the analysis results are shown in Table 2.

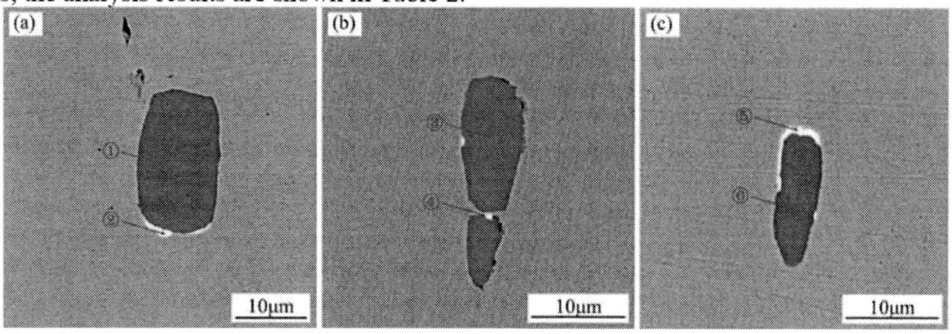

Figure 3. SEM image of typical MnS inclusions with white matter.

Table 2. The atomic percentage of MnS inclusions with white matter(%).

Point	Mn	S	Te	Fe
1	46.07	42.08	-	11.85
2	8.60	-	11.07	80.34
3	40.23	37.11	-	22.67
4	33.45	22.68	12.82	31.05
5	40.58	11.12	29.29	19.01
6	45.27	34.62	-	20.12

According to Table 2, the atom number of Mn element and the S element are approximately 1:1 at point 1, the gray-black substance is MnS. The atom number of Te element and the Mn element are approximately 1:1 at point 2, the white matter is TeMn. Points 2~6 are consistent with the results of point 1 and 2. As can be seen from Table 1, inclusions of plant A are mainly MnS and MnTe.

National standard GB/T 10561-2005/ISO 4967:1998 (E) for the steels produced by plant A, plant B and plant C in Class A inclusions national standard rating. The worst rating results of inclusions are shown in Table 3.

Table 3. The worst rating results of inclusions.

Steel mill	Plant A	Plant B	Plant C
Fine length (μm)	713.9	758.9	845.3
Fine rating	2.5	2.5	2.5
Coarse length (μm)	504.7	669.7	1277.3
Coarse rating	2.0	2.5	3.5

A subsection comparing the rating results of three steel mills, plant A has the best rating for Class A inclusions: fine grade 2.5, coarse grade 2.0, and the inclusions are substantially spherical. The rating of Class A inclusions in plant B is fine grade 2.5, coarse grade 2.5 and the total length of inclusions has increased. The rating of Class A inclusions of plant C is fine grade 2.5, coarse grade 3.5 and the level of inclusions control is poor.

The equivalent diameter and area distribution of the inclusions of the three steel mills are shown in Figure 4 and Figure 5, respectively. The average equivalent diameter of the inclusions of the three steel mills are 3.71μm (plant A), 5.04μm (plant B) and 3.93μm (plant C). The average area of the inclusions of the three steel mills are 17.58μm (plant A), 24.98μm (plant B) and 17.51μm (plant C). The average equivalent diameter and average area of inclusions produced by plant A and C are not much different, but the number of small size inclusions in the factory A is significantly higher than that in the factory C.. The equivalent diameter and average area of plant B are larger than plant A and C.

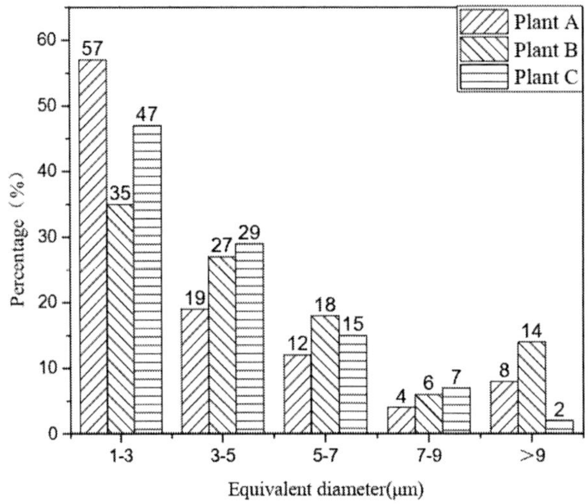

Figure 4. Map of inclusions equivalent diameter distribution.

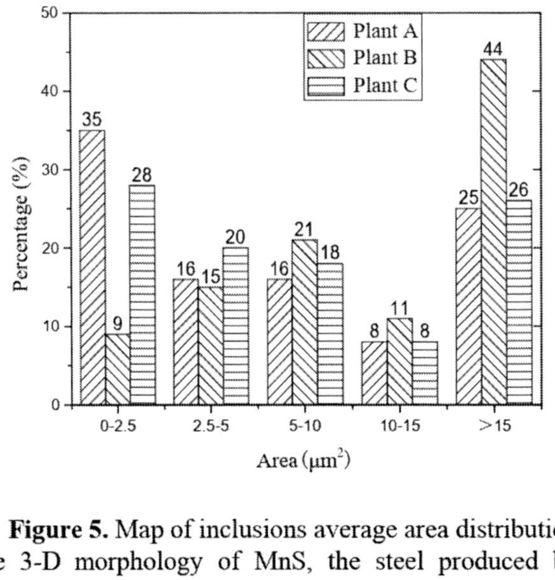

Figure 5. Map of inclusions average area distribution.

To further analyze the 3-D morphology of MnS, the steel produced by three steel mills were electrolytically etched, and the steel matrix was dissolved instead of non-metallic inclusions to preserve the real form of the inclusions. The three-dimensional morphology of MnS inclusions are shown in Figure 6. The MnS of the steel produced by plant A are mainly spherical and short rods. The

MnS inclusions surrounded by white MnTe are mainly short rods and long strip in plant B. The MnS inclusions in plant C are mainly strips, fans and irregulars.

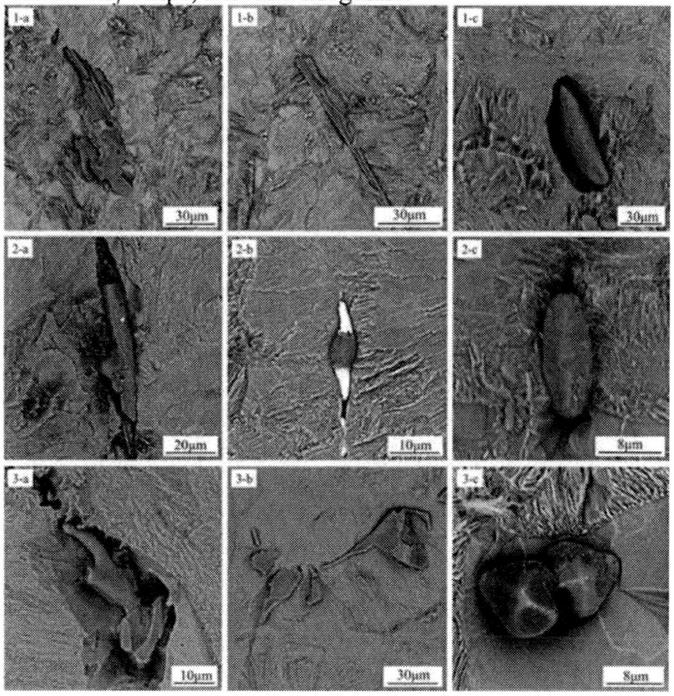

Figure 6. 3-D morphology of MnS inclusions: 1-a, 1-b, 1-c plant A; 2-a, 2-b, 2-c plant B; 3-a, 3-b, 3-c plant C.

4. Conclusion

(1) The steel produced by plant A has the best rating for Class A inclusions: fine grade 2.5, coarse grade 2.0. The rating of Class A inclusions in plant B is fine grade 2.5, coarse grade 2.5. The rating of Class A inclusions in plant C is fine grade 2.5 and coarse grade 3.0, the level of inclusions control is poor.

(2)The MnS inclusions of the steel by plant A are evenly distributed and generally short rods and nearly spherical, the morphology control of MnS inclusions is good. The MnS inclusions of plant B surrounded by white MnTe are mainly short rods and long strip. The MnS inclusions of plant C are mainly strips, fans and irregulars.

Acknowledgements

The authors would like to thanks the State Natural Science Fund Projects of China (51474142 & 51671124) for supporting this work.

References

[1] Zeng Z Q, Yang Q K and Ai K N 2019 *Analysis on nozzle clogging of 49MnVS steel during production* Iron and Steel Vol 54 p 43-48.

[2] Chang K D, Wang P and Liu W P 2011 *Development status and application prospect of non-quenched tempered steel* Heat Treatment of Metals Vol 36 p 80-85.

[3] Pi J, Bai Y and Zhen R 2017 *A Study on the Effect of Aging on Mechanical Properties of Cold-Formed Non-quenched Steel via Nanoindentation* Journal of Materials Engineering and Performance Vol 26 p 28-33.

[4] Liu D L, Tan L and Yang H X 2013 *Microalloy of ferrite-pearlite non-quenched and tempered steels and their forging technology* Journal of Iron and Steel Research Vol 25 p 41-46.

[5] Yang X W, Tu Q and Liu J 2017 *Study on process of eliminating banded microstructure of non-quenched and tempered steel C38N2 for crankshaft* Heat Treatment of Metals Vol 46 p 234-237.

[6] Yang Y, Zhou L Y and Jiang P 2018 *Influence of die-forging deformation on microstructure of 1538MV non-quenched and tempered steel for crankshaft* Chinese Journal of Engineering Vol 40 p 579-586.

[7] Aliakbari K 2019 *Failure analysis of four-cylinder diesel engine crankshaft* Journal of the Brazilian Society of Mechanical Sciences and Engineering Vol 41 p 1-8.

[8] You D L, Cui H and Wu H 2013 *Research on Inclusions in Non-quenched and tempered Sulfur-bearing Steel for Autos* Iron Steel Vanadium Titanium Vol 34 p 73-78.

[9] Shao X, Lv L G and Du Q 2018 *Research on inclusions in non-quenched and tempered sulfur-bearing steel for autos* Heat Treatment of Metals Vol 43 p 144-147.

[10] Shao X J, Wang X H and Ji C X 2015 *Morphology, size and distribution of MnS inclusions in non-quenched and tempered steel during heat treatment* International Journal of Minerals Vol 22 p 483-491.

[11] Zhang D, Shen P and Xie J B 2018 *A method for observing tridimensional morphology of sulfide inclusions by non-aqueous solution electrolytic etching* Journal of Iron and Steel Research Vol 26 p 275-284.

Deep Analysis of Free-cutting Phase and Its Distribution in Japanese SF20T Pen Tip Steel

Panpan Zhang, Zhiqi Zeng, Tian Fan, Ping Shen and Jianxun Fu*

Center for Advanced Solidification Technology (CAST), School of Materials Science and Engineering, Shanghai University, Shanghai 200072, China.

*Corresponding author

Abstract. This paper aims at investigating the existence state of free-cutting phase in Japan SF20T free-cutting stainless steel (one of the most widely used pen tip steel). By using the devices of an optical microscope (OM), a scanning electron microscope (SEM), electrolytic etching, and Vickers hardness meter, the quantity, composition, shape, size, distribution of the free-cutting phase in the steel and the change of hardness of steel were analysed and characterized, strive to find out the reason of the performance of SF20T steel is far superior to that of other countries'pen tip steel. The results show that the morphologies of inclusions in the steel were mostly in spindle shape, and their distributions were dispersed and quite uniform. The free-cutting phases, MnS in spindle, Pb, and MnTe, were beneficial to enhancing the cutting performance of the steel. In contrast to domestic materials, the distribution of inclusions in the Japanese steel was more uniform, thereby providing key significance for production of pen-making and steelmaking in domestic plants.
Keywords: pen tip steel; free-cutting phase; morphology; hardness.

1. Introduction

As the "crown" of manufacturing industry for production of the pen tip steel, the ballpoint pen steel in recent years has been widely concerned. In early 2016, China Premier Li Ke-qiang even issued a sigh of difficult ball-making of the pen for pen-making industries. More than 3000 enterprises annually produce more than 40 billion ballpoint pens, nearly occupying 80% of the global market share. But the materials of pen tip (pen holder)-super free-cutting stainless steel are totally derived from the imports all the time. In 2017, Tai-steel Group[1], a Chinese enterprise, first announced a successful research and developed the pen tip steel. But the actual effect was not satisfactory due to poor machinability and great costs. Therefore, full scales of mastering the advanced technologies for the production of pen tip steel are of great importance for Chinese steelmaking plants.

In the early days, the materials for pen tip steel were lead-containing brass and nickel-white copper. Due to the poor corrosion resistance, both materials were replaced by free-cutting stainless steel as the most widely used material for ballpoint pen tip steel[2]. At present, most of it is ferritic stainless steel. The hardness of ferrite is low and its plasticity is good. So the ferritic stainless steel is easy to deform and produces a sticky knife phenomenon during cutting, affecting the cutting performance of steel. The usual solution to improve the cutting performance of the material is to add the free-cutting elements to steel, with easier cutting of ferrite stainless steel[3].

From many studies on pen tip steels, it can be found that the distribution, size and shape of free-cutting phases of domestic pen-point steels are significantly different from those of foreign high-end pen tip steels, which results in great differences in steel properties. Nowadays, because of excellent properties, the Japanese pen tip steel has still been dominating the worldwide pen-making industries, occupying the first place. Thus, this paper analysed the best SF20T pen tip steel of Japan currently to detect the free-cutting phase and its distribution. Most importantly, it provides a comprehensive understanding of the production of Japanese pen tip steel for development of new materials, and gives more useful references for domestic related pen-making and steel enterprises.

2. Analysis of Steel Composition

The SF20T steel is an S-Pb-Te ferritic free-cutting stainless steel, and its chemical composition was obtained in this study by an inductively coupled plasma spectrometer determination and carbon sulfur analysis. The chemical composition of steel is shown in **Table 1**.

The carbon content of SF20T ballpoint tip steel was 0.010%, which was a ferrite stainless steel. The sulfur content was 0.290%, classified into high sulfur steel. The Mn content was 1.070%, and the Mn/S was over 3.5 to form a large number of MnS, an easy-cutting phase.

By generating stress concentration at the cutting contact, the cutting performance of steel is improved due to the production of melting brittle and lubrication effects. The Te content was 0.028% and was easy to form MnTe, attached to the outer surface of MnS to prevent its deformation. MnTe plays a role in controlling the form of MnS, thereby further improving the cutting performance of steel[4].

Table 1. The chemical composition of pen tip steels (wt. %)

C	Si	Mn	Cr	Mo	S	Pb	Te
0.010	0.330	1.070	19.650	1.780	0.290	0.180	0.028

3. Analysis of Free-cutting Phases

3.1. Methodology

Firstly, two detection planes of metallographic samples from the rolled rod were that one was parallel to the rolled direction (longitudinal direction) and the other was perpendicular to the rolled direction (transverse direction). The shape and distribution of inclusions were observed with the OM after polishing. The statistics for each sample with 20 photos×100 magnification were carried out with Image-Pro Plus software to quantitatively count the number, size, and distribution of the inclusions. The non-aqueous solution was used to observe the three-dimensional morphologies of the inclusions. The electrolyte solution consists of 1% ammonium tetramethyl chloride-10% triethanolamine-methanol, located in a low temperature thermostatic tank to maintain the electrolysis temperature of the sample in the experiment steady. The electrolytic parameters were that electricity density of 0.2 A, time~25 min. The two-dimensional and three-dimensional morphologies of the inclusions in polished and electrolytic samples were observed using the OM and SEM with spectrum analyser. The inclusion composition was detected to explore the free-cutting phases in the tip steel. Also, the Vickers hardness meter was used to test the hardness of the steel matrix and inclusions.

3.2. OM Analysis

The horizontal and vertical polished planes of the samples were observed by OM. **Fig. 1** shows the morphology and distribution of inclusions under×200 viewing field. A large number of inclusions were observed. **Fig.** 1(a) shows that majority of inclusions were in round or oval shape and distributed uniformly with sizes of 10 μm. **Fig.** 1(b) shows that most of inclusions were in oval shape in the longitudinal direction, illustrating that the inclusions in SF20T steel were in spherical or spindle-shaped. In order to analyse the variation of the inclusion size along the rolling direction, a result of the finished pen tip in the previous study by our team was chosen for comparison with SF20T steel, as shown in **Fig. 1**(c). Compared to the transverse direction, was significantly reduced to 5~8 μm. And after the drawing of the steel, the inclusions were in fractures, and their sizes were also significantly reduced by 3~6 μm, Thus, the size of the inclusion was reduced to avoid the formation of continuous long-bar inclusions caused by the anisotropy of steel.

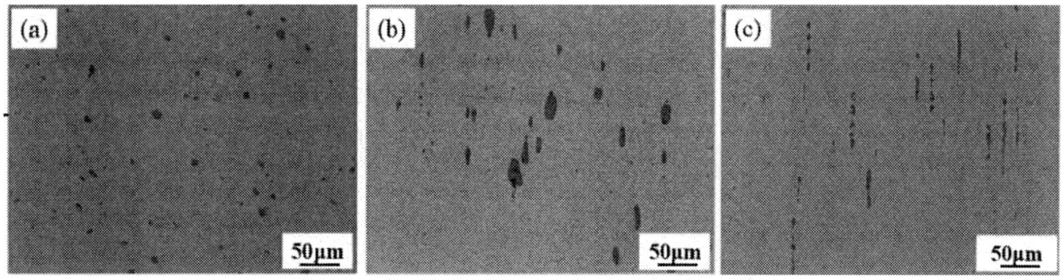

((a): horizontal; (b): vertical; (c): drawing)

Figure 1. Metallographic morphology

According to the statistics for the inclusions, the total number of inclusions was 3199 at viewing field area of 5814600 μm^2, and the density of inclusion was 550/mm^2. The average area of an inclusion was 34.8 μm^2 with an equivalent diameter of 4.87 μm^2. **Fig. 2** shows detailed statistical results of the inclusion equivalent diameter and aspect ratio. **Fig. 2**(a) shows that the distribution state of inclusion equivalent diameter was that 1~2.5 μm occupied 39.95%, the 2.5~5μm occupied 19.193%, and > 5 μm occupied about 40%. **Fig. 2**(b) shows that the aspect ratio of 1~3 occupied 83.964%. Combined with the inclusion pattern in **Fig. 1**, it is confirmed that most of the inclusions in the steel are spindle-shaped, spherical and spheroidal, and 15.067% of the inclusions had a aspect ratio of 3~6, and few inclusions showed a aspect ratio of more than 15, accounting for only 0.96% of the total. Most inclusions in steel were spherical or spindle-shaped, and the average aspect ratio of the inclusions was 2~3.

Usually the smaller the length and width value of the inclusion in steel is, closer to the ball, the more favorable the cutting performance of the steel. The small length and width of the inclusions in steel can relatively avoid the anisotropy caused by softness phases in the cutting process. The breaking chips are usually in short and small curls to ensure that the steel has a good cutting performance[5].

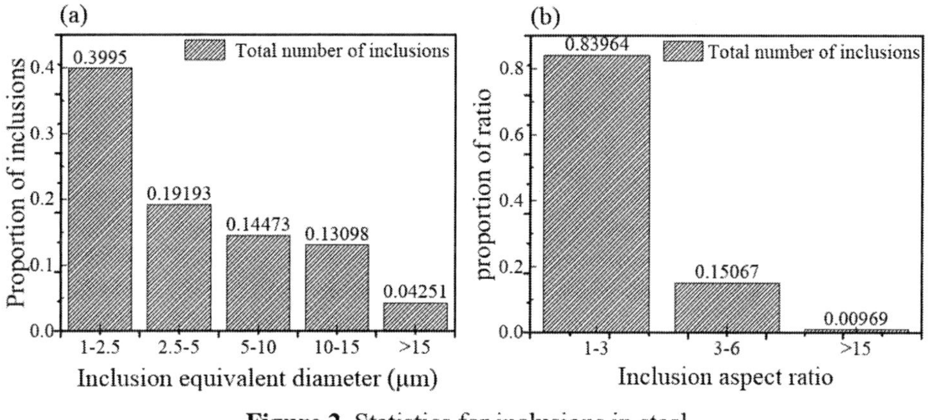

Figure 2. Statistics for inclusions in steel

3.3. Analysis of Inclusion Composition

The use of energy spectrum was adopted for further observation of the inclusions. The point 2 in **Fig. 3**(a) shows that the composition of light grey inclusions was a free-cutting phase, MnS. Because of a good plasticity, MnS in the rolling process is easily elongated and sticky with the tool, so it is usually in long strip, with a strip or curling chip, and not easy to be segregated, with poor surface finish. The point 2 in **Fig. 3**(b) shows that MnS inclusions in SF20T steel were in spheroid shape. At same time, the chip was in C-type, and the debris was easy to be segregated and to improve the finish of the steel surface in the cutting process [6]. Meanwhile, the white bright color area of point 4(a) after detection was MnTe, MnTe-MnS. Both MnTe and MnTe-MnS tended to be in spherical or spheroidal shape, playing key roles in controlling the form, beneficial to improving the cutting of steel. In addition, MnTe is also an important free-cutting phase, which plays the role of stress concentration source, lubrication, etc. during cutting [7]. The compositions of point 1 in **Fig. 3**(a)-(c) were detected a certain amount of lead. Pb is difficult to be dissolved into iron matrix. Pb is rarely combined with alloy and non-metallic elements to form compounds, often attached to ends of MnS, or iron matrix in a free state, thus forming a morphological distribution in **Fig. 3**. In the process of cutting, strong friction between cutting tools and chips, so the lead particles are melted, resulting in lubrication effect, beneficial to chip-breaking and brittle effect, further improving the cutting performance of steel and tool life[8-10].

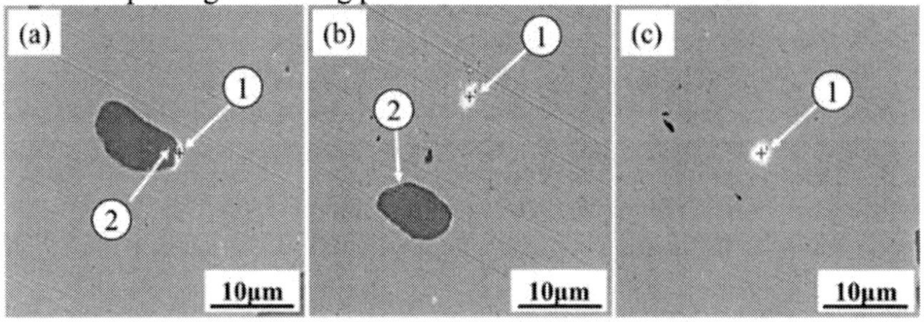

Figure 3. Schematics of two dimensional SEM morphologies

3.4. Analysis of 3D Morphology

After the electrolysis etching of the sample with non-aqueous solution, steel matrix was dissolved; whereas the inclusions inlayed in the steel were not dissolved exposed to the surface of the steel in a three-dimensional form. The tri-dimensional morphologies of inclusions were observed by SEM, which were divided into three types according to the form categories of sulphide [11]: I-type, which is mainly spheroidal, spherical. The transverse and longitudinal tri-dimensional morphologies in inclusions were respectively shown in **Fig. 4**(a) and (b). As evident in the figure, the MnS in SF20T tip steel is a typical Class I and III inclusions.

Further analysis of the inclusions using the OM, is shown in **Fig. 4**(c). The inclusions were mainly free-cutting phase MnS, and there was a small amount of white bright Pb (point 3) at ends. Attached to the package, the spherical inclusions are not easy to deform during the rolling. Presence of free-cutting phases can effectively improve the cutting performance of steel.

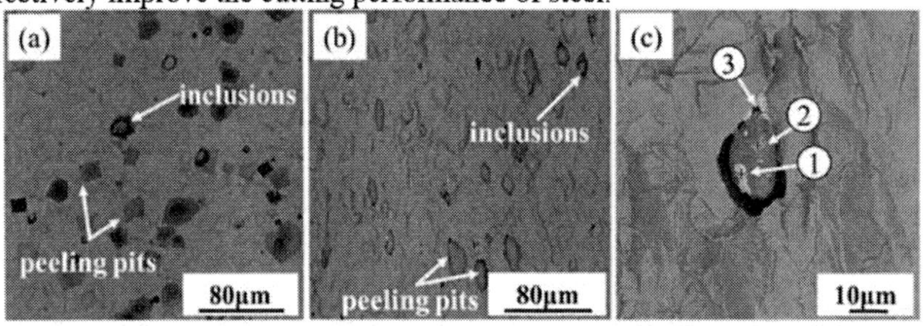

Figure 4. Tri-dimensional morphologies of inclusions after etching

3.5. Micro-hardness Analysis of Steel and Inclusions

The micro-hardness values of the steel and inclusion were detected, with 10 points for steel and 21 points for inclusions. A load of 10 kgf and a loading and unloading time of 5 s were performed. The hardness in **Fig. 5** shows that the hardness of steel matrix was larger, and varied little with the average value of 269.9 kgf/mm^2. In contrast, the micro-hardness of the inclusions was relevantly smaller, with an average of 109.0 kgf/mm^2. The hardness of the inclusion was more unstable, and a small number of the points owned larger hardness values but far less than steel.

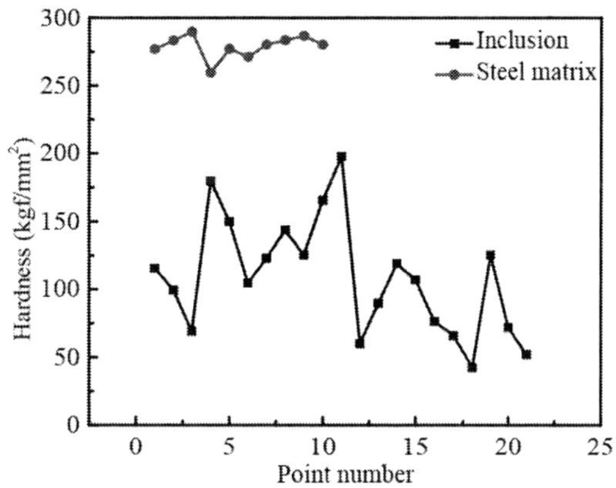

Figure 5. Comparison of micro-hardness of inclusions and steel substrates

In general, the hardness of steel was in 170.0~230.0 kgf/mm^2 and had enough embrittlement[12], with better processing performance. The great hardness harms the knife life, and accelerates the tool wear. However, the smaller hardness causes the sticky knife, deteriorating the process under the cutting force. All of the cutting performance has a negative impact.

4. Conclusions

(1) The inclusions in SF20T tip steel are mostly in spindle and spheroid shape (Class I, Class III). The equivalent diameter of inclusions in the cross-section was about 2.5, and the average aspect ratio was 2~3, with the uniform and dispersed distribution in steel.

(2) The free-cutting phases in steel were MnS, MnTe, Pb; where MnS was in light gray, spheroidal. MnTe and part of Pb in white color mainly wrapped the MnS.

(3) The micro-hardness value of steel matrix in pen tip steel was 269.9 kgf/mm^2; while the hardness of the inclusions was in range of 42.3~179.8 kgf/mm^2, with average value of 109.0 kgf/mm^2, beneficial to reducing the cutting force of the steel.

Acknowledgments

The authors would like to thank the State Natural Science Fund Projects of China (51474142, 51671124) for supporting this work.

References

[1] Wang F, Hu L N and Huang C B 2017 *Tens of billions of pens will be equipped by Chinese tip* Pioneers p 42-43

[2] Li G Z, Hui R and Li J H 2002 *Development of 30MnVS for free cutting and quenching and tempering steel for passenger cars* Iron & Steel p 42-45

[3] Feng M Y, Yan W B and Zhang Y M 2014 *Quality analysis and improvement of X1215 free cutting steel* Hebei Metallurgy p 46-48

[4] Wang Y J 2016 *Numerical investigation of processing properties of pen tip drilling free-cutting stainless steel* Xi'an University of Architecture and Technology

[5] Ma Y J, Li L L, Xin W and Li J S 2018 *Study on micro-inclusions behavior of 1215 free-cutting steel* Value Engineering p 207-208

[6] Watson J D 1986 *Microscopy and the development of free-machining steels*. Applied Metallography Springer Boston MA

[7] Shen P, Yang Q K and Zhang D 2018 *Application of tellurium in free-cutting steels* Journal of Iron and Steel Research, International p 787-795

[8] Luo G and Wang H M 2015 *Analysis of microstructure and property of a free-cutting stainless steel* Shanghai Metals p 10-14

[9] Mao Z Q 2002 *Development of domestic and overseas free-cutting steels* Automobile Technology & Material p 1-4

[10] Liu H L 1998 *The function mechanism of MnS in YF45V steel* Machinery Development p 25-27
[11] Sims C E and Dahle F B 1938 *Effects of various deoxiders on the structure of sulfide inclusions* Transactions Amwerican Foundry Society p 65
[12] Information on https://zhidao.baidu.com/question/262644469790890045.html

Simulation of the Thermal Contact Resistance Characteristics on Rod Fastening Rotor

Yiping Li

School of Chemical and Environmental Engineering, Shanghai Institute of Technology, Shanghai 201418, China

Abstract. The thermal contact resistance model of the rod fastening rotor is improved, which can be more suitable for the engineering condition. By means of numeric simulation, the relationships between characteristics of the heat transfer between contact interfaces and the pre-tightening force of the tie rod, the surface roughness and the temperature of the contact interface are revealed intuitively. The simulation results show that the coefficient of contact heat transfer between the contact interfaces will increases with the pre-tightening pretension of the tie rod, will decreases with the increase of the surface roughness, and will increases with the increase of the surface temperature. It is also shown that the thermal contact resistance between the contact interfaces will have an important influence on the overall temperature distribution and heat transfer of the rod fasten rotor, and the thermal contact resistance between the interfaces must be considered when the temperature field distribution of the rod fastening rotor are accurately researched.
Keywords: rod fastening rotor; thermal contact resistance; numeric simulation; temperature field analysis.

1. Introduction

Because of the advantages of higher rigidity, higher strength, lighter weight, easier assembly and disassembly, and realization of cooling channels, rod fastening rotor is widely used in heavy duty gas turbines[1]. Structurally, it is no longer a continuous whole. The structural non-continuity makes its thermodynamic analysis more complex than conventional rotors. The connecting surface of the rod fastening rotor discs is complete state in contact with the two surfaces in the macroscopic view. However, because of the microscopic unevenness of the surface, the two contact surfaces only contact with each other at some micro-convex peaks, and the actual contact area of the bonding surface less than nominal contact area[2].

Research shows that thermal contact resistance is a strongly nonlinear problem influenced by many factors such as surface topography, material thermal properties and load, temperature, and medium[3]-[6]. The traditional model uses statistical parameters such as root mean square gradient and instrument resolution. The measurement results of different instruments are not unique. Fractal theory uses fractal dimension-independent parameters such as fractal dimension when describing rough surfaces to provide surface roughness information that exists on fractal surfaces. For this reason, literature[7] considered the shrinkage thermal resistance at the contact point based on the M-T contact thermal model and corrected the M-T model. Literature[8] and[9] used the fractal theory to perform heat transfer analysis on elastic contact interfaces. Literature[10] applied fractal theory and traditional heat conduction theory to establish a fractal model of thermal contact resistance, which makes it more suitable to study the contact thermal resistance of the connecting surfaces.

2. Model of Contact Surface Thermal Resistance

Figure 1 is a simple model to study the thermal resistance characteristics of the connecting interfaces of the rod fastening rotor. The rod fastening rotor is consist of five discs pressed by eight circumferential rods. The outer diameter of the disc *D*=140mm, inner diameter *d*=80mm, and thickness of each disk were same, and uniformly distributed 8 holes in the circumferential direction.

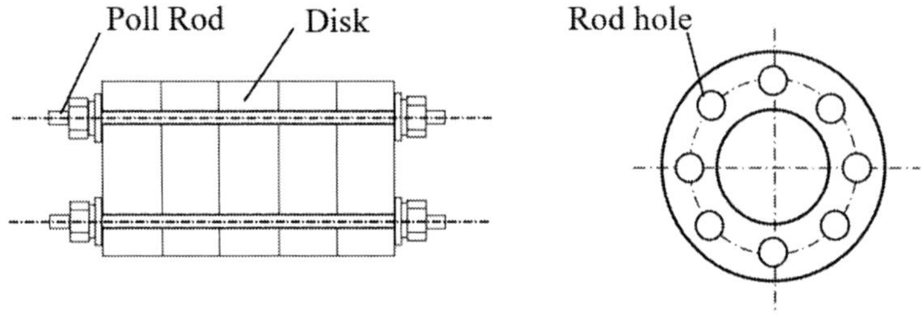

Figure 1. Model of the thermal contact resistance of the connecting interfaces.

There are three ways to transfer heat between the coupling surfaces of the rod fastening rotor. It has shown that for metal contact problems, when the temperature is lower than 900K, the radiation heat transfer is less than 2% in the total contact heat transfer [11]. Therefore, the radiative heat transfer between the gaps can be ignored. The medium in the connecting interfaces is lubricating oil or air, and its thermal conductivity is much smaller than that of metal materials. Therefore, the medium heat conduction can be neglected, and the heat transfer coefficient of the contact surface h_c can be expressed as:

$$h_c = \frac{1}{L_g}\frac{A_c}{A}\frac{2\lambda_A\lambda_B}{\lambda_A + \lambda_B} \tag{1}$$

$$R_c = \frac{1}{h_c A} \tag{2}$$

Among them, h_c is the heat transfer coefficient of the contact surface, W/(m^2.K); L_g is the thickness of the connecting interfaces gap to participate in heat transfer, m; A_c is the actual contact area of the connecting interfaces, m^2; A is the nominal contact area of the connecting interfaces, m^2; λ_A and λ_B are the thermal coefficient of the disk, W/(m.K). The thermal contact resistance R_c is in inverse proportion to the contact heat transfer coefficient h_c.

When the surface roughness of the two disks is the same, the thickness L_g of the connecting interfaces gap can be expressed by the following equation:

$$L_g = 2(z - \delta_L) \tag{3}$$

δ_L is the maximum deformation of the micro-convex peak of the connecting interfaces, m; z can be taken as the height of the roughness R_z, m. According to the fractal theory [12]:

$$\delta_L = G^{D-1}(2a_L)^{(2-D)/2} \tag{4}$$

a_L is the largest contact area between the connecting interfaces, m^2. Bringing equation (4) into equation (3), it can be obtained:

$$L_g = 2[z - G^{D-1}(2a_L)^{(2-D)/2}] \tag{5}$$

Based on the fractal theory, the actual contact area A_c of the connecting interfaces is obtained [8]:

$$A_c = \frac{D}{2-D}\psi^{(2-D)/2}a_L \tag{6}$$

In the formula, D is the fractal dimension, which is related to the surface roughness; ψ is the domain expansion factor of the micro-contact size distribution (ψ>1), which can be found in the literature [8]. The relationship between the rod preload and the actual contact area of the connecting interfaces is as follows [13]:

When 1 < D < 2 and D ≠ 1.5

$$P=\frac{2^{\frac{12-3D}{2}}DG^{D-1}E^{*}(\ln\gamma)^{\frac{1}{2}}\psi^{\frac{2-D}{2}}}{3\pi^{\frac{3-D}{2}}(3-2D)}a_L^{\frac{D}{2}}\left(a_L^{\frac{3-2D}{2}}-a_c^{\frac{3-2D}{2}}\right)+\frac{2KYD\psi^{\frac{2-D}{2}}}{2-D}a_L^{\frac{D}{2}}a_c^{\frac{2-D}{2}} \tag{7}$$

When D =1.5

$$P=2^{\frac{7}{4}}\pi^{-\frac{3}{4}}G^{\frac{1}{2}}E^{*}\psi^{\frac{1}{4}}(\ln\gamma)^{\frac{1}{2}}a_L^{\frac{3}{4}}\ln\frac{a_L}{a_c}+6KY\psi^{\frac{1}{4}}a_L^{\frac{3}{4}}a_c^{\frac{1}{4}} \tag{8}$$

$$\frac{1}{E^{*}}=\frac{1-\upsilon_1^{2}}{E_1}+\frac{1-\upsilon_2^{2}}{E_2} \tag{9}$$

E^* is the composite elastic modulus, MPa. E_1, E_2, υ_1 and υ_2 are the elastic modulus and Poisson's ratio of the materials respectively; G is the fractal roughness parameter, m, which is related to the surface roughness; γ is the spatial frequency of the random profile, γ=1.5 is more suitable for high spectral density; a_c is the critical contact area of the connecting interfaces, m^2; K is the ratio of the hardness H of the softer material to its yield strength Y. According to literature [13], σ_b is the tensile intensity, MP$_a$, a_c and H are as follows:

$$a_c=\frac{G^2}{\left(H/2E^{*}\right)^{2/D-1}} \tag{10}$$

$$H=\frac{\sigma_b}{3.4} \tag{11}$$

The relationship of the fractal parameters D, G and the surface roughness parameter R_a is as follows [14]:

$$D=\frac{1.515}{Ra^{0.088}} \tag{12}$$

$$G=10^{\frac{-8.259}{Ra^{0.088}}} \tag{13}$$

The traditional thermal contact resistance model only considers the variation of individual or partial parameters with temperature, which does not completely calculate the actual thermal contact resistance. When the rod fastening rotor actually operates in the gas turbine, the physical parameters of each component will change with the temperature. The rod fastening rotor model materials in figure 1 are all No. 45 steels. The physical properties of the rods can be obtained from the manual of the metal materials as shown in table 1. The parameters are brought into equations (7) ~ (13) to find a_L, and bring a_L to equations (5) and (6) to find L_g, A_c; then bring L_g, A_c into equation (1). The surface contact heat transfer coefficient h_c can be obtained.

In summary, the contact surface heat transfer coefficient and the rod pre-tightening force are both functions of the actual contact area of the connecting interfaces, and the nonlinear relationship between them is established by the above equations. The above solution step can obtain h_c at different conventional temperatures. From equation (2), the thermal contact resistance can be obtained. h_c is often used in finite element analysis. Therefore, only h_c will be analyzed and solved in the following.

Table 1. Physical parameters of rod fastening rotor

Temperature (°C)	Elastic Modulus ($\times 10^5 MP_a$)	Tensile Strength (MP_a)	Poisson's ratio υ	Thermal Conductivity (w/m.k)	Thermal diffusivity ($\times 10^{-6} m^2/s$)	Specific heat capacity (J/kg.℃)
20	2.09	639	0.269	48.2	11.7	--
100	2.07	605	0.270	48.1	11.2	--
200	2.02	702	0.290	46.5	10.3	578
300	1.96	728	0.312	44.0	9.2	624
400	1.86	573	0.309	41.4	8.3	649
500	1.74	383	0.308	38.1	7.0	716

3. Analysis of Heat Transfer Characteristics on the Connecting Interfaces

Based on the above improved model, the effects of different roughness, different rod pre-tightening forces, and different temperature on the thermal contact resistance of the connecting interfaces were calculated. Three kinds of discs with roughness R_a = 0.2, 0.4, and 0.8μm are set, pre-tightening forces P = 30, 40, 50, 75, 100, 125kN, temperature T = 20, 100, 200, 300, 400, 500℃. By analyzing and comparing the calculated data, the change rule of h_c with R_a, P and T can be drawn.

Figure 2(a) to (f) respectively shows the relationship between the contact heat transfer coefficient h_c of the connecting interfaces and the surface roughness R_a of the disc and the pre-tightening force P at different temperatures T. h_c increases as the P increases. This is because the increase of the pre-tightening force makes the micro-convex peaks in the connecting interfaces deformed. As the pre-tightening force increases, the contenting interfaces undergo elastic deformation, which results in contact between the bonding surfaces. The number of micro-convex peaks and the contact area increase, which directly results in an increase in the area where the heat transfer between the bonding surfaces actually participates, resulting in an increase of h_c. h_c decreases as the roughness R_a increases. When R_a is the smaller, the surface of the wheel is the smoother. Under the ideal conditions, there is no thermal contact resistance for the smooth contact surfaces.

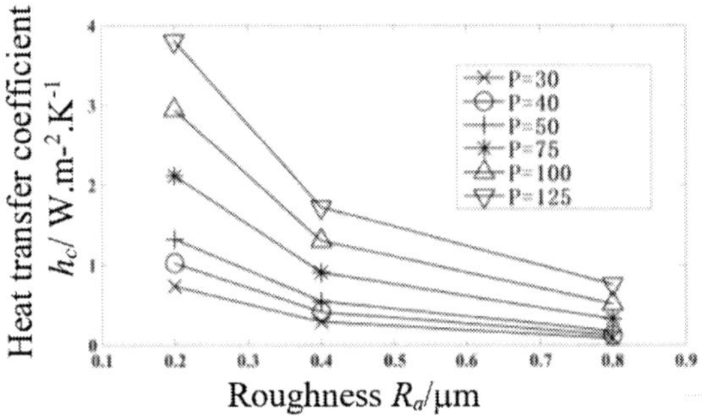

(a) T=20℃

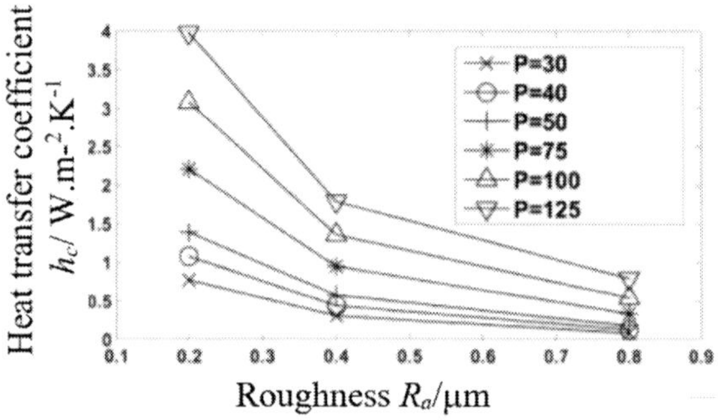

(b) T=100℃

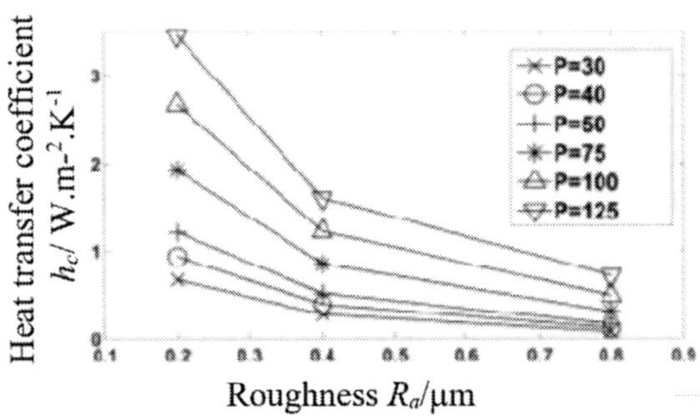

(c) T=200℃

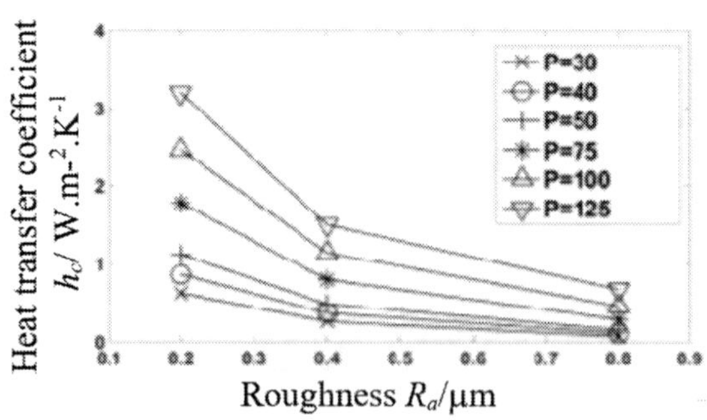

(d) T=300℃

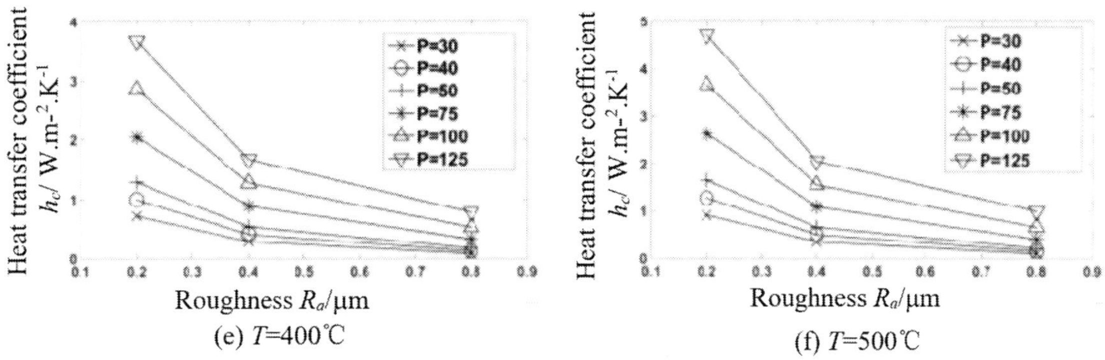

Figure 2. The influence of R_a, P on h_c.

The surface contact heat transfer coefficient h_c first increases, then decreases, and then increases again as the joint surface temperature T rises, and h_c increases as a whole. The main reason is that the temperature will affect the physical properties and surface contact conditions of the disk material. For steel 45, as T increases, its elastic modulus E and thermal conductivity λ decrease, while the hardness H changes first, then increases, then decreases.

4. Temperature Field Analysis

The hindrance to the heat flow is mainly caused by the thermal contact resistance between the connecting interfaces of disks. Each disk is simplified to have the thickness, the same inner and outer diameters. Since the rod fastening rotor is an axisymmetric structure, its eighth model is used to simplify the analysis. The heat transfer coefficient between the connecting interfaces solved by the above equations is used in ANSYS.

Figure 3 shows cloud diagram of the temperature field distribution of one-eighth rod fastening rotor with one-hour heating under conditions of R_a=0.4μm and P=30kN. The highest temperature occurs on the first disk. It can be found that the temperature distribution is not continuous at the first disk. The temperature distribution at the connecting interfaces of the second, third, and fourth disks is actually discontinuous, since ANSYS only uses one colour to represent the temperature difference section, the display is not obvious.

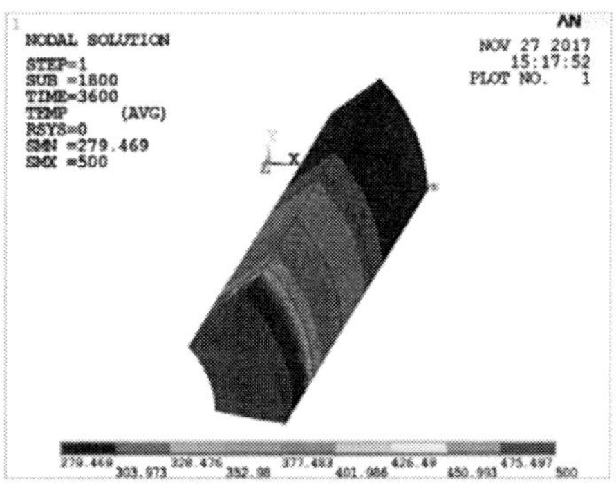

Figure 3. Rod fastening rotor temperature field.

Figure 4 shows the variation of the axial temperature of the outer surface of the rod fastening rotor. The horizontal ordinate indicates the axial position of the rod fastening rotor, and the longitudinal coordinates indicates the temperature. Due to the fact that the variations of the roughness are consistent, the results of the roughness R_a=0.4μm are only shown. In figure 5(a), it is found that there are jumps in the temperature distribution at the axial positions of 37mm, 74mm, 111mm, and 148mm. The temperature difference occurs at the positions. Comparing figure 5 (a) to (f), it can be found that, when the pre-tightening force P increases, the temperature of the rotor end also increases after heating for 1 hour. The rotor end temperature with different pre-tightening forces was 279.469℃, 290.734℃, 297.34℃, 305.746℃, 309.733℃, 312.006℃. It is clear that at P=125kN, the end temperature is higher than that at P=30kN. In addition, with the increase of the pre-tightening force P, the temperature jump at the connecting interfaces between the same discs becomes less and less obvious. That is, the temperature difference at the connecting interface becomes smaller and smaller, and the continuity of the temperature field becomes better and better.

Further research results show that, when the surface roughness R_a increases, the temperature of the rotor end decreases. When R_a=0.2μm, the rotor end temperature is higher than when R_a=0.8μm. The pre-tightening force P, the surface roughness R_a of the disc, and the temperature T of the joint surface have a great influence on the contact heat transfer coefficient h_c of the connecting surface. Heat flow has an important impact. Therefore, in order to obtain a precise temperature field, the contact thermal resistance of the wheel and the disk must be taken into consideration.

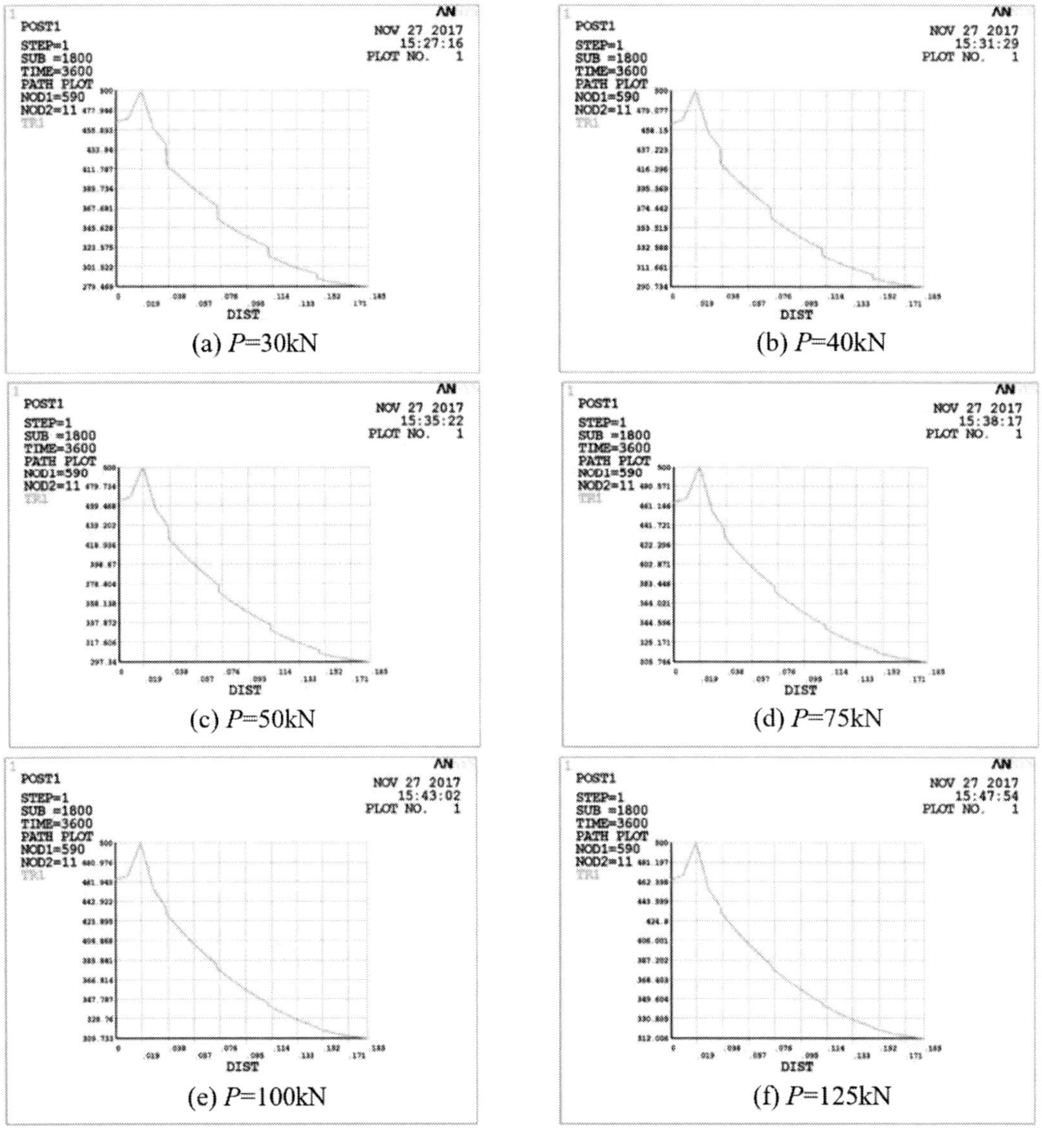

(a) P=30kN (b) P=40kN

(c) P=50kN (d) P=75kN

(e) P=100kN (f) P=125kN

Figure 4. The axial temperature variation of the rotor with pre-tightening force P

5. Conclusion

The improved fractal model makes it more suitable to study the thermal contact resistance of the connecting interface of the rod fastening rotor, which reveals the non-linear implicit function

relationship between the contact heat transfer coefficient of the connecting interface and the pre-tightening force of the rods, and the surface roughness of the discs.
The heat transfer coefficient of the connecting interface would increase when the pre-tightening force of the rod increases, and surface roughness of the disks decreases, the temperature of the disk surface increases.
The thermal contact resistance of the connecting interface have an important influence on the overall temperature field distribution and heat transfer of the rod fastening rotor. Therefore, in order to obtain a precise temperature field of the rod fastening rotor, the thermal contact resistance of the connecting interface must be taken into consideration.
In the paper, when the fractal model of contact thermal resistance of the joint surface is established, the relationship between fractal parameters D, G and roughness R_a is established according to the reference [14], which may have some influence on the research results. The future research would be to establish the relationship between the parting parameters and the roughness of the joint surface, which could further improve the model accuracy.

Acknowledgments
This project is supported by National Natural Science Foundation of China (Grant No. 51475311).

References
[1] Peng HE, Zhan-shen LIU, Guang-hui ZHANG and Yong-zhi FENG 2010 Dynamic Modeling and Analysis of Distributed Rod Fastening Rotor *Turbine Technology*, vol 52(1) pp 4–8
[2] Jian-feng ZHAO, An-liang WANG and Chun-xin YANG 2003 The Statisical Characterisitic of Roughness Profiles on Thermal Contact Conductance Surface *Cryogenics*, (4) pp 49–56
[3] An-liang WANG and Jian-feng ZHAO 2011 Review of prediction for thermal contact resistance *Sci China Tech Sci*, (5) pp 545–556
[4] Wen-zhen FANG, Xing-jie REN and Wen-quan TAO 2017 Numerical Predictions of Thermal Contact Resistance Based on The Multi-Block Lattice Boltzmann Method *Journal of Engineering Thermophysics*, vol 38(3) pp595–599
[5] Zhi-feng LIU, Cheng-yu MA, Yong-sheng ZHAO and Cui FANG 2016 Analysis of High Speed Motorized Spindle Thermal Characteristics Based on Thermal Contact Resistance *Journal of Beijing University of Technology*, vol 42(1) pp 17–23
[6] Yang WU, Xue-liang ZHANG, Shu-hua WEN and Yong-hui CHEN 2015 Three-dimensional Fractal Model Combined with Surface Thermal Contact Conductance of Machine Tool *Journal of Taiyuan University of Science and Technology*, vol 36(5) pp 368–374
[7] Lan-ping ZHAO and Lie XU 2003 Fractal Model Including Constriction Resistance for Thermal Contact Conductance *Journal of Tongji University* (*Natural Science*), vol 31(3) pp 296–299
[8] Wang S and Komvopoulos K A 1994 A Fractal Theory of the Interfacial Temperature Distribution in the Slow Sliding Regime: Part I—Elastic Contact and Heat Transfer Analysis *Journal of Tribology, Transactions of the ASM*, vol 116(4) pp 812–822
[9] Wang SHAO 2004 Real contact area of fractal-regular surfaces and its implications in the law of friction *Journal of Tribology, Transactions of the ASME*, vol 126(1) pp 1–8
[10] Xu M, Jiang S and Cai Y 2007 An improved thermal model for machine tool bearings *International Journal of Machine Tools & Manufacture*, vol 47(1) pp 53–62
[11] Fenech H and Rohsenow W M 1963 Prediction of Thermal Conductance of Metallic Surfaces in Contact *Journal of Heat Transfer*, vol 85(1) pp 15
[12] Shu-hua WEN 2012 *Theoretical modeling and simulation of surface contact characteristics* (Beijing: National Defense Industry Press)
[13] Guo-sheng LAN, Xue-liang ZHANG and Hong-qin DING 2011 Modified Contact Model of Joint Interfaces Based on Fractal Theory *Transactions of the Chinese Society for Agricultural Machinery*, vol 42(10) pp 217–223
[14] Bing FANG, Lei ZHANG and Ji ZHAO 2012 Dynamic Parameter Identification and Modeling for Bearing Joint Interfaces *Journal of Xi'an Jiaotong University*, vol 46(11) pp 69–74

FEA of BTA Deep Hole Drilling Based on ANSYS

Xin Liu* **and Yazhou Feng**

Xi'an Shiyou University, Xi'an, China

*Corresponding author

Abstract. This paper takes BTA deep hole drilling as the research object. ANSYS software is used in the static and dynamic analysis of the force of BTA deep hole drilling process from theory and in modeling and simulation. Firstly, According to the finite element analysis, the most concentrated stress and the largest deformation of the cutting tool are the cutting edge and the cutting surface of the tool block. The main vibration of drill stem is in the form of twist and bending. Sceondly, it is concluded that the main vibration of drill stem is in the form of twist and bending through the modal analysis. Finally, the natural frequency of this type of BTA drill stem is obtained by analyzing the amplitude response curve, and it is suggested to avoid the rotational speed near the dangerous speed during machining to avoid resonance of the drill stem.

Keywords: BTA deep hole drilling; ANSYS; FEA; inner-chip removal

1. Introduction

Deep hole drilling processes, which differ significantly from conventional drilling processes, are relevant for a lot of different applications where holes with high length-to-diameter-ratios and very good qualities are necessary[1]. Deep hole drilling usually adopts two processing methods: inner-chip removal processing and outside-chip removal processing. The main form of inner-chip removal is BTA deep hole drilling[2]. The working principle is that during deep hole drilling, the cutting fluid enters the cutting zone from the outside of the drill stem and the annular space formed by the processed hole, and the chips are removaled from the inside of the drill stem, as shown in Fig. 1.

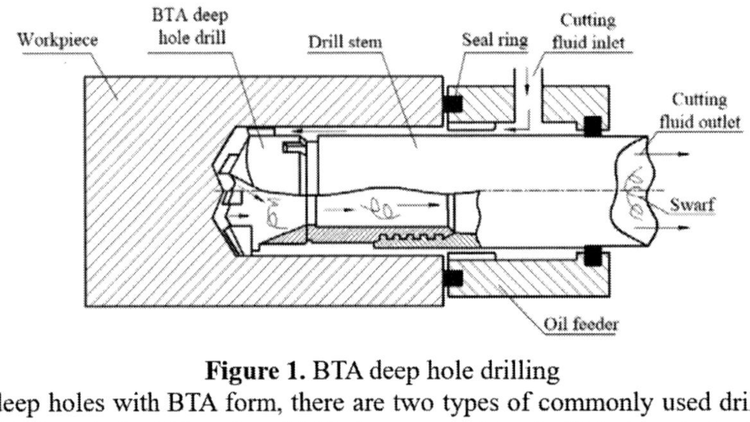

Figure 1. BTA deep hole drilling

When drilling deep holes with BTA form, there are two types of commonly used drill tool structures: single-blade BTA deep hole drill and multi-blade staggered teeth BTA deep hole drill[3]. However, deep hole drilling technology currently presents certain difficulties at the technical level. For example, the overall rigidity of the drilling system is relatively poor, the chip exit is blocked, the drilling is unstable, vibration occurs, the tool wear and breaking[4]. By the use of the finite element method, the analytical modeling method and the experimental method[5-7]. The numerical approaches used finite element analysis (FEA) to simulate the machining process. Li made use of Deform-3D finite element analysis software to simulate the single-ended near-dry BTA drilling, and analyzed the distributed situation of cutting force, cutting temperature and tool wear in different cutting parameters[8]. Wei analyzed the thermal stress and transient dynamics of common bit, solved the optimization problem of cutting parameters before machining, improved the machining efficiency and saved the cost[9]. Dong Built a three-dimensional model of BTA deep hole drilling, simulation the static characteristic of the three-dimensional model using Ansys simulation software, observed the wear condition of the BTA drilling teeth, middle teeth and side teeth in the Experiment, verify the results of simulation, providing a basis for the BTA drilling structural optimization[10]. At present, the existing research focuses on the simulation of turning and milling tools. There are few researches on cutting tools of deep hole machining, especially the inner chip removal cutting tools of BTA system.

In this paper, the single-edged BTA deep hole drill is taken as the research object. The cutting part of the drill bit is mainly composed of an inner edge, an outer edge, a drill tip, a guide block and a clearance hole. Two guide blocks are arranged on the circumference of the drill bit, and the cutting edge is generally ground with two or more chip steps, as shown in Fig. 2. Generally only suitable for deep hole drilling of small and medium diameter. It is suitable for a range of 6 to 65 mm, the aspect ratio can be up to 100, and maximum up to 250.

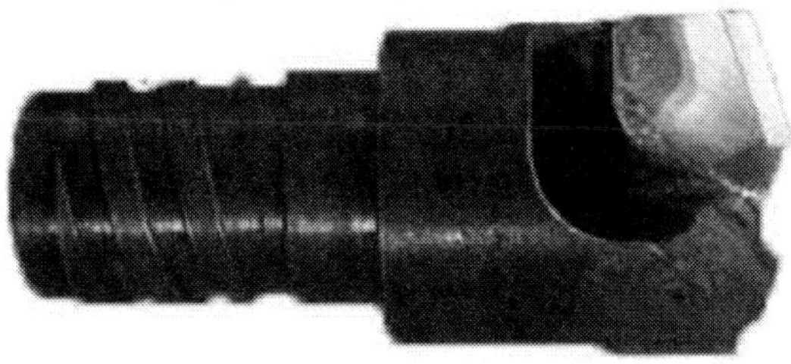

Figure 2. Structure of single-edged BTA deep hole drill tool

2. Analysis and Simulation of Static for BTA Deep Hole Drilling

Static analysis refers to the change of structural stress and strain of the BTA tool structure under the condition of stable load, without considering the tool is subject to varying loads or changes over time. In the analysis, it is assumed that the material, structure and state of the BTA tool are linear, and the entire model load and deformation are linear variation.

2.1. Establishment and Import of Finite Element Model for BTA Deep Hole Drilling

Modeling was constructed using UG software based on the boundary dimension of the BTA deep hole drill. Create a new static structure static analysis project in ANSYS software, import the created 3D model into ANSYS software, and generate geometry.

2.2. Enter Material Properties in the Material Library

After importing the model of the BTA deep hole drill tool, the basic characteristics of the material of the tool structure are defined, including the unit system, the unit type, the unit constant, the elastic modulus of material, the Poisson's ratio, the density and the use of material library files, etc. In this paper, the Linear Elastic Isotropic model is used to define material properties as isotropic. In order to simplify the mesh size in the finite element meshing and reduce the amount of computation in the solution, the model is simplified, and the key dimensions are retained. Meanwhile, the connection of the thread and the welding of the blade are neglected. The new material (40Cr) of the cutter body is added to the material library, where the material density is $7.9\times103Kg/cm^3$, the elastic modulus is 210 GPa and the Poisson's ratio is 0.3. The material for the blade and the guide block that added is YG8, the density is $14.5\times103Kg/cm^3$, the elastic modulus is 540 GPa and the Poisson's ratio is 0.3, etc. Workpiece material is TC4 titanium alloy, its diameter is 56mm, the processing aperture is 34mm, and

IOP Conf. Series: Materials Science and Engineering **611** (2019) 012021 doi:10.1088/1757-899X/611/1/012021

the length is 1995mm.

2.3. Mesh Generation

Mesh Generation is an important step in finite element analysis. The finer the mesh is, the more accurate the simulation results. Based on the automatic division of ANSYS software, the mesh density is increased for the parts where the force is concentrated, such as the tip of the knife, the corner, and the guide block, so as to increase the accuracy of the simulation. For other parts with relatively complete structure and large size, a sparse mesh division is adopted to reduce the amount of calculation, as shown in Fig. 3. The size of the important structural part is 0.3 mm, the other part is 1 mm.

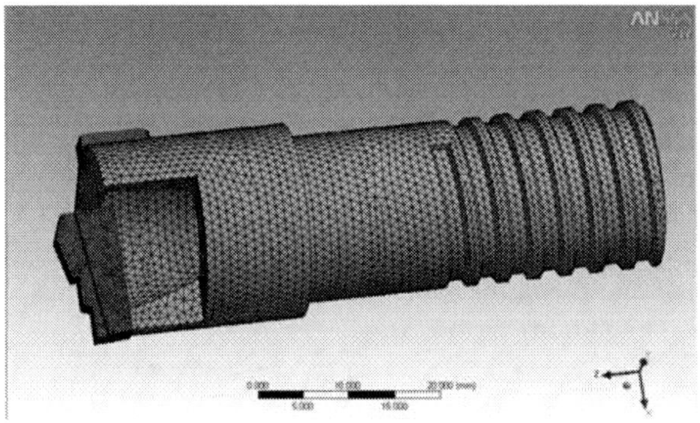

Figure 3. Mesh generation and optimization of BTA drill tool

2.4. Applying Loads and Defining Constraints

According to the above mechanics model analysis of BTA drill tool, the model in ANSYS is loaded. The fixed constraint conditions are cutter body and guide block. The position where apply the force load is the blade , as shown in Fig. 4.

(a) Axial load Diagram of BTA tool

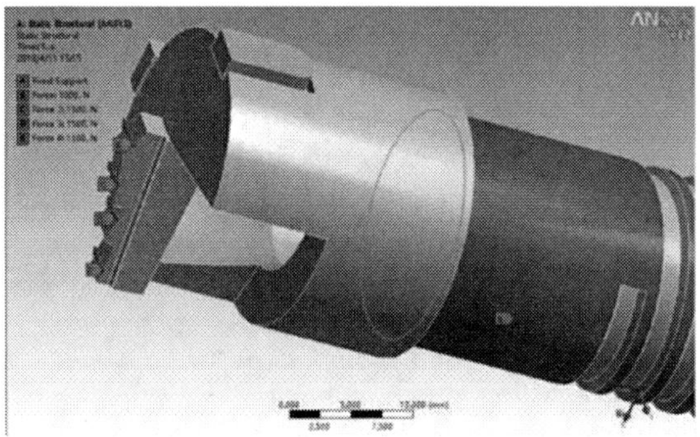

(b) Radial load Diagram of BTA tool

Figure 4. Axial load and radial load Diagram of BTA tool

Add Stress and Strain in the Solution solver toolbar. And then, solve the Solve Command. The stress and strain calculated by ANSYS12.0 software are expressed in the form of cloud picture respectively. The position of stress concentration, the maximum stress point and the structure with the largest deformation can be seen intuitively. Fig. 5 shows that the stress and strain cloud diagram is divided into two parts: the cutting edge and the cutting surface of the tool block.

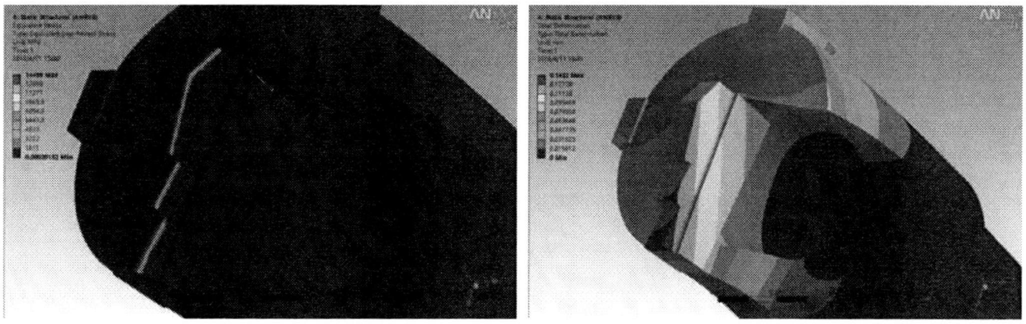

(a) Stress nephogram of BTA drill tool (b) Strain nephogram of BTA drill tool

Figure 5. Stress and strain nephogram of BTA bit

3. Analysis and Simulation of Dynamic Characteristics for BTA Deep Hole Drill

3.1. Modal Analysis

It is possible to analyze the main modal characteristics of the drill stem when it is subjected to a certain range of frequencies more directly by modal analysis, and then the actual vibration response of the drill stem under the action of the vibration source can be obtained. The dynamic characteristics of the drilling system can be evaluated, the structural dynamic characteristics of the newly designed members can be estimated and optimized, the failure of the system can be diagnosed and predicted, and the load of the drill stem system can be identified.

From the vibration dynamics, it can be seen that most of the energy of vibration is concentrated in the first ten-orders. Therefore, it is generally only necessary to analyze the modal deformation nephograms of the first order to the tenth order. As shown in Fig. 6, the first ten modal shapes of drill stem are calculated through the modal analysis, it is concluded that the main vibration of drill stem is in the form of twist and bending

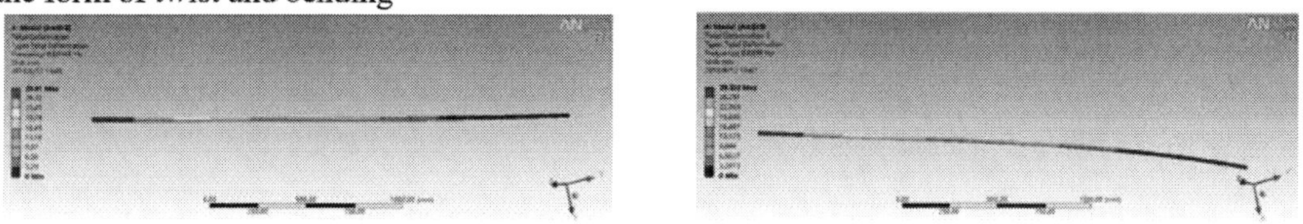

(a)First-order modal deformation nephogram (b)Second-order modal deformation nephogram

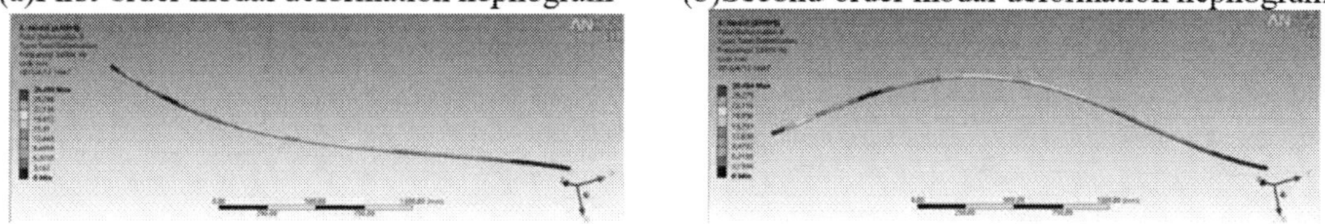

(c)Third-order modal deformation nephogram (d)Fourth-order modal deformation nephogram

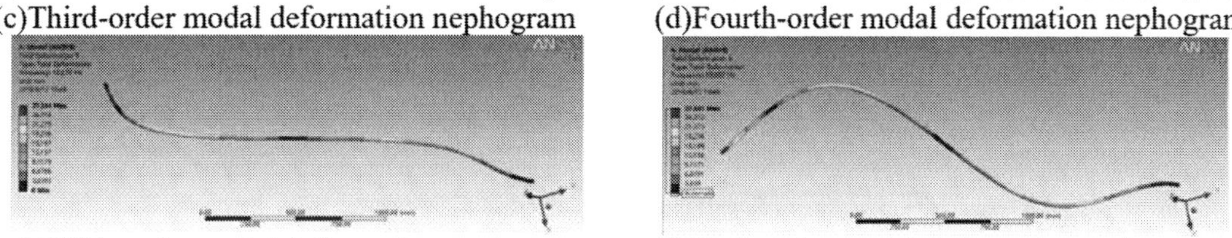

(e)Fifth-order modal deformation nephogram (f)Sixth-order modal deformation nephogram

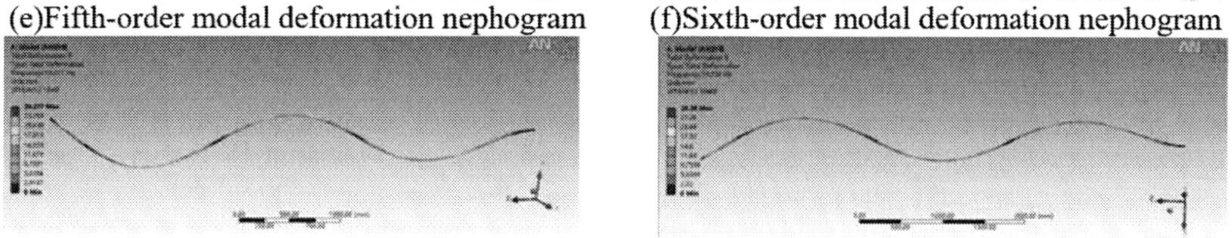

(g)Seventh-order modal deformation nephogram (h)Eighth-order modal deformation nephogram

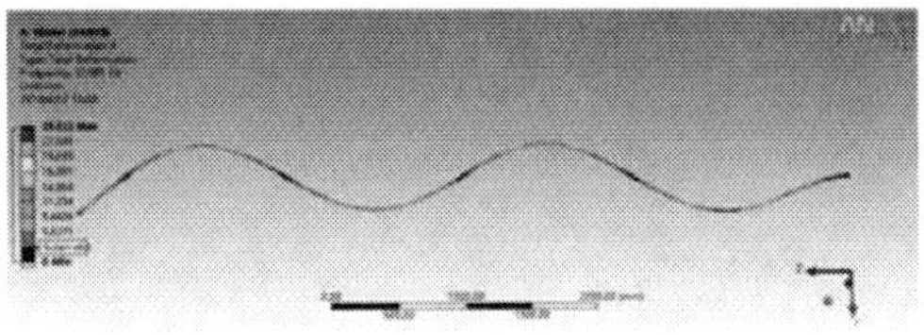

(i)Nineth-order modal deformation nephogra

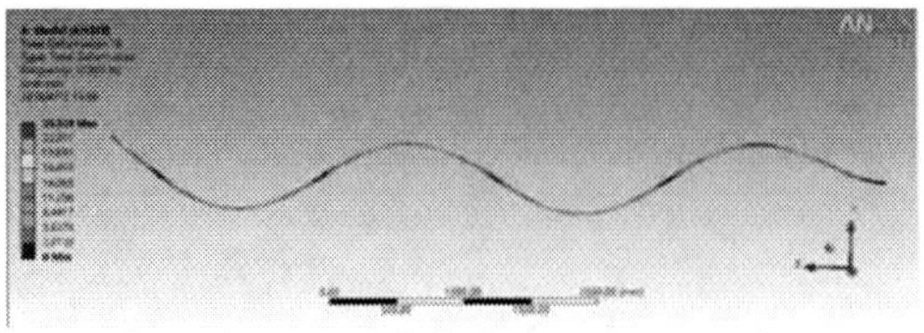

(j)Tenth-order modal deformation nephogram

Figure 6. The first ten-order modal deformation nephogram of BTA drill stem

3.2. Harmonic Response Analysis

Harmonic response is used to analyze the cyclical response of the drill stem under cyclic loading. During the analysis, only the steady-state forced vibration of the drill stem is calculated. The purpose of the analysis is to obtain the curve of the displacement and frequency of the drill stem at different frequencies, so as to determine the sustained dynamic characteristics of the drill stem, thereby avoiding resonance, fatigue and other results from forced vibration. Fig. 7 shows that the results of the harmonic response analysis. It can be seen that when the frequency is 300HZ, the amplitude in the X direction reaches the maximum, and the drill stem may produce resonance at this frequency.

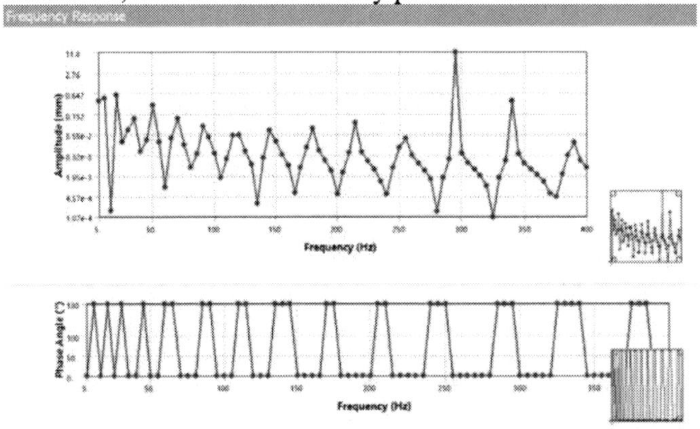

Figure 7. Harmonic response and the natural frequency of BTA drill stem

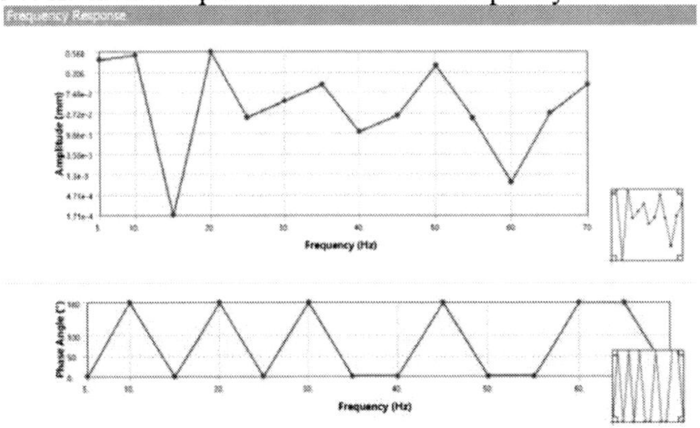

Figure 8. Harmonic response analysis of the first ten-order vibration

Due to the harmonic response analysis of the drill stem based on the modal analysis, the frequency range is between 0-70 Hz and and the load is applied to the part of drill tool, as shown in Fig. 8. The frequency at which the amplitude reaches the peak is consistent with the modal frequencies from the first to tenth order of the excitation frequency and it is also the natural frequency of the drill stem. In the process of deep hole drilling, the external excitation frequency should be controlled to avoid resonance, which can reduce the influence of vibration on the processing quality.

4. Conclusion

By using ANSYS finite element analysis software, the statics and dynamics of BTA deep hole drilling process are analyzed theoretically and on the level of modeling and simulation, and the following conclusions can be drawn:

(1) The most concentrated stress and the largest deformation are the blade parts. Since the cross section of the BTA bit is a complete annular structure, the structural integrity of the drill tool is greatly enhanced. The torque is larger and the deformation is smaller than the gun drill.

(2) Through the ANSYS simulation analysis, the stress cloud diagram of the bit is solved, which provides the basis and data support for improving the size structure of the bit.

(3) If the external excitation frequency is close to a certain order value, the drill stem will resonate and the amplitude will increase significantly, which will affect the processing quality. In order to reduce the influence of vibration on the processing quality, the external excitation frequency should be controlled to avoid resonance. Or by optimizing the design of drill stem, the amplitude can be reduced by changing the frequency range of drill stem and staggering the excitation frequency. According to the results of the modal analysis above, torsion and bending are the main vibration forms of drill stem.

References

[1] Smith, David J. . Deep Hole Drilling. Practical Residual Stress Measurement Methods. John Wiley & Sons, Ltd, 2013.

[2] Liu, Zhanfeng, et al. "Experimental study on deep hole drilling of 17-4PH material." Materials Science & Engineering Conference Series Materials Science and Engineering Conference Series, 2018.

[3] Liu, Zhan Feng, and R. L. Li. "Experimental Research on Superlong Deep-Hole Drilling Processing Technology for Beryllium Bronze." Applied Mechanics and Materials 130-134(2011):1634-1637.

[4] Liu, Zhanfeng, et al. "Study on super-long deep-hole drilling of titanium alloy." Journal of Applied Biomaterials & Functional Materials 16.1_suppl(2018):150-156.

[5] Biermann, Dirk, A. Sacharow, and K. Wohlgemuth. "Simulation of the BTA deep-hole drilling process." Production Engineering 3.4-5(2009):339-346.

[6] Wang, Tianqi, et al. "ANSYS is Used on Drilling Inner Blind Hole on Deep Hole." Wireless Personal Communications 10(2018):1-12.

[7] Tiffe, Marcel, and D. Biermann. "Modelling of Tool Engagement and FEM-Simulation of Chip Formation for Drilling Processes." Advanced Materials Research 1018(2014):183-188.

[8] Daobo, Li, et al. "FEA of the near-dry deep hole drilling based on Deform-3D." Modern Manufacturing Engineering (2013).

[9] Wei, Ping, J. Han, and H. L. Geng. "Based on the ANSYS of Small and Deep Hole a Bit Cutting Transient Dynamic Study." Advanced Materials Research 503-504(2012):802-804.

[10] Dong, Zhen, X. Q. Shen, and H. Zhang. "The Analysis and Experimental Verification on Static Characteristics of BTA Deep Hole Drilling Based on Ansys Software." Applied Mechanics and Materials 556-562(2014):1178-1182.

Finite Element Numerical Simulation Research on Fractured Horizontal Well in Stress-dependent Tight Reservoirs Based on Heat Transfer Theory

Pengfei Zhang and Juhua Li*

Institute of Petroleum Engineering, Yangtze University, Wuhan, Hubei, China

*Corresponding author

Abstract. We present an efficient numerical method to handle stress-dependent behavior for fractured wells in tight reservoirs using the large-scale general finite element software ANSYS as the platform. A new similarity criterion is deduced to describe the seepage flow based on the similarity principle. The similarity criterion is a key parameter to transfer unstable heat problems to well performances. Then a comparison between the numerical and history-matched results is made to validate the accuracy of the results, by which effective stress-dependent coefficient calculated by numerical simulation is in agreement with that from experiment. The result shows that it is feasible to apply the ANSYS function of temperature field analysis in the computation of the unstable seepage flow considering permeability change in stress-dependent tight reservoirs. Finally using a certain field data and history-matched reservoir model, the influence of the fractures on horizontal well is investigated, including fracture spacing, fracture length and parameters of fracture with included angle on the basis of the new method. The results from the model considering the stress-dependent factor are significantly different from that without considering the factor. This work provides reservoir engineers with a simple and feasible method for optimizing the fracture parameters of tight oil development.
Keywords: fractured well, productivity, similarity principle, stress-dependent, tight reservoir, finite element numerical simulation method.

1. Introduction

Because of the poor connectivity of low permeability reservoirs and the small pore throat passage, the traditional method is no longer applicable. Therefore, in order to improve production, the method of fracturing multiple fractures in horizontal wells is usually adopted. With the application of fracturing, the design of fracture length and spacing is very important. Due to the existence of fracturing fractures, the flow and pressure distribution near the fractures are also complex. So it is necessary to carry out numerical simulation of reservoir.

In the paper, we will present a new efficient numerical method to handle stress-dependent behavior for fractured wells in tight reservoirs using the large-scale general finite element software ANSYS as the platform. Then a new similarity criterion is deduced to convert transient thermal simulation results into unstable seepage phenomena, which is proved to be feasible in the productivity calculation of fracturing wells in tight reservoirs. Finally case studies are presented. Results show that the new simulation method is appropriate for the optimization of fractured wells in tight reservoirs. Based on the similarity between temperature field and seepage field, the temperature field is used to simulate the seepage problem of horizontal fractured wells, and a new idea is proposed. The temperature field replaces the seepage field, and the temperature represents the fluid seepage velocity. It provides the simulation study productivity. A new method that is closer to the actual value. Therefore, it is very necessary to perform finite element simulation on it.

2. Methodology

There are similarity mathematical models between heat flow field and seepage field (Arunn, 2012). The similarity of temperature field and seepage field is shown from three aspects: the basic theory, differential equations, initial and boundary conditions. Table 1 shows the physical quantities corresponding to the temperature field and seepage field.

Table 1. The comparison of physical quantities of temperature field with seepage fields

The seepage field		The temperature field	
Differential equation	$\nabla^2 P = \mu C_t \frac{\partial P}{\partial t}$	Differential equation	$\nabla^2 T = C \frac{\partial T}{\partial t}$
Darcy's law	$v = -\frac{K}{\mu}\frac{dP}{dL}$	Fourier law	$q = -K_\gamma \frac{dT}{dn}$
The fluid flow	Q	Heat	Q
Mobility	$\frac{K}{\mu}$	Thermal conductivity	K_γ
Seepage velocity	v	Heat flow intensity	q
Total compressibility	C_t	Specific heat	C

C_t— total compressibility, the pore compressibility coefficient and the coefficient of compressibility of reservoir rocks,1/MPa.

C —The amount of heat absorbed or emitted by a substance of unit mass that rises or falls at a unit temperature. $J/(kg \cdot °C)$.

Table 2. The comparison of boundary conditions of temperature field with seepage fields

	Seepage field	Temperature field
Initial condition		$T\|_{t=0} = T_0(x,y,z,0)$
The first boundary condition	$P\|r_1 = P(x,z,y,t)$	$T\|r_1 = T(x,z,y,t)$
The second boundary condition	$\frac{K}{\mu}\frac{dP}{dL}\|\Gamma_2 = -v(x,z,y,t)$	$K_r \frac{dP}{dL}\|\Gamma_2 = -q(x,z,y,t)$

The differential equation of seepage flow is:

$$\frac{\partial}{\partial x}\left(K_x \frac{\partial P}{\partial x}\right) + \frac{\partial}{\partial y}\left(K_y \frac{\partial P}{\partial y}\right) + \frac{\partial}{\partial z}\left(K_z \frac{\partial P}{\partial z}\right) = 0 \tag{1}$$

The differential equation of heat conduction is:

$$\frac{\partial}{\partial x}\left(K_{rx} \frac{\partial T}{\partial x}\right) + \frac{\partial}{\partial y}\left(K_{ry} \frac{\partial T}{\partial y}\right) + \frac{\partial}{\partial z}\left(K_{rz} \frac{\partial T}{\partial z}\right) = 0 \tag{2}$$

3. The Application of New Simulation Method in ANSYS

Because reservoir permeability is low, fracture length is long and reservoir flow time is long, the transient effect can not be ignored, but the transient effect is more complex in well entry and around the fracture. In order to take into account the effect of stress sensitivity, we propose to use the transient heat transfer module in ANSYS to simulate the incoming wells and fractures.

According to the given field oil well data and similarity index, ANSYS thermal module is used for numerical simulation. The change of effective fracturing will change the fluidity. The fluidity corresponds to the thermal conductivity in the temperature field, which will also change.

Fig. 1 shows the distribution of temperature field simulated by ANSYS software, which can be compared intuitively. Fig.1 (a) and (b) show the temperature distribution considering the influence of thermal conductivity or not respectively. The area within the isotherm line of 1.84 ℃ in Fig.1 (b) is significantly larger than that in Fig.1 (a). The isotherm line in temperature field corresponds to the isopiestic line in seepage field, the area within the isotherm of 1.84 ℃ in temperature field corresponds to the area within the isopiestic line of 18.4 MPa in seepage field. The area within isopiestic line of 18.4 MPa without considering stress-dependent effect is significantly greater than that considering stress dependent effect. In other word, the drainage area around the wellbore in no stress-dependent reservoir is greater than that in stress-dependent reservoir.

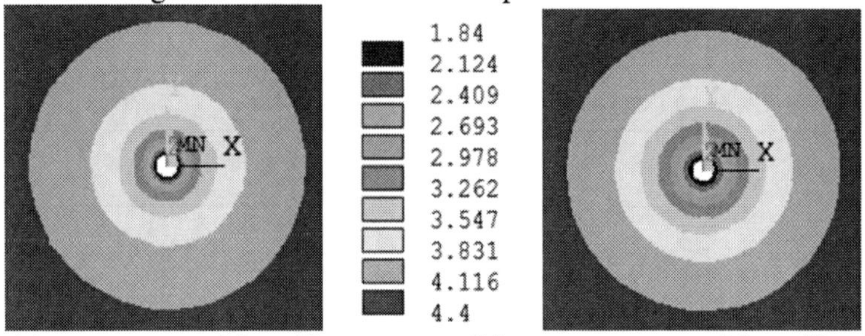

(a) pressure-sensitive eff (b) no pressure-sensitive effect

Figure 1. Temperature distribution around the wellbore

The transient thermal simulation results are transformed to equivalent seepage parameters by the new similarity criterion for specific heat capacity and the comprehensive coefficient in stress-dependent media. After the production performance history curves of the fractured vertical wells are matched, the stress-dependent coefficient is determined as 0.055MPa^{-1} in Fig. 2. The results show that production performance considering reservoir stress-dependent properties can be simulated well using the ANSYS finite element method.

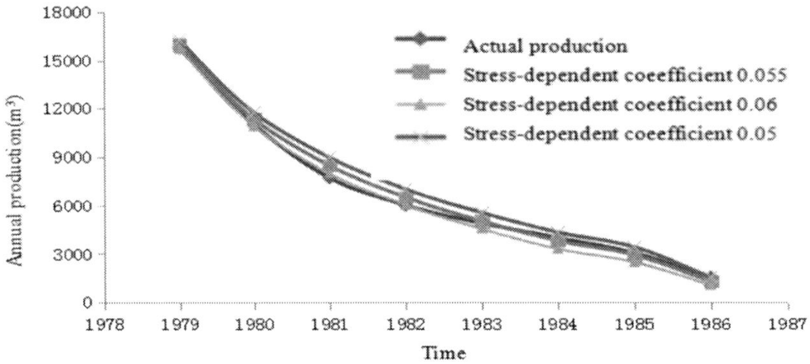

Figure 2. Production Fitting Chart of Vertical Well Considering Pressure Sensitivity

4. Optimization of Fractured Horizontal Well in Tight Reservoirs with Finite Element Numerical Simulation

4.1. Free Surface Adapting Grid of Fractured Horizontal Well

For the half of the velocity vector diagram near wellbore area in Fig.3 (b), the isopiestic line is denser at periphery fracture shown in Fig.3. The part of the eservoir fluid directly flows into the wellbore, but most of the fluid flows into the wellbore through the fracture, and the seepage velocity is also different at different endpoints of fracture.

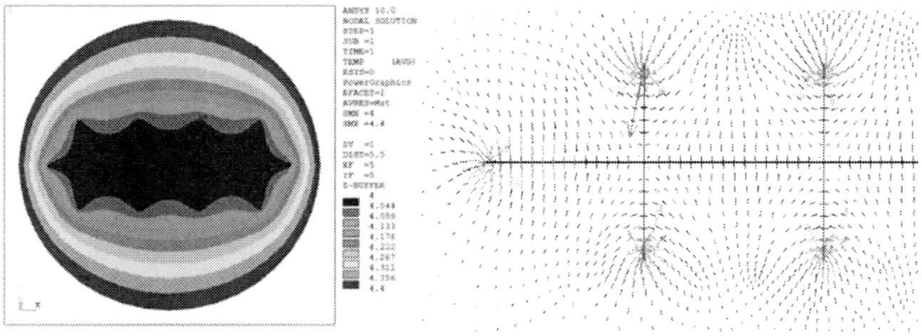

4.1. Seepage field (b) vector diagram

Figure 3. Seepage chart of near wellbore in fractured horizontal well

4.2. The Sensitivity Analysis of Fracture Parameters in Stress-dependent Tight Reservoirs

4.2.1. The Fracture Number. Assumed that stress-dependent coefficient is 0, 0.03 and 0.08, the fracture half-length is 150 m, the horizontal well length is 1000 m. The finite element software ANSYS is applied to simulate oil production rate with different fractures number of 2, 3, 4, 5, and 6. As shown in Fig.4, the recovery ratio of fractured horizontal well in stress-dependent reservoirs generally increases with the increase of the number of fractures when the fracture length is constant. The higher the stress correlation coefficient, the higher the oil content. Meanwhile with the gradual increase of the fractures number, oil production rate increases first and then gradually decreases since the fracture interference results in. To efficiently develop fractured horizontal well, we optimize the number of fracture to reduce fracture interaction.

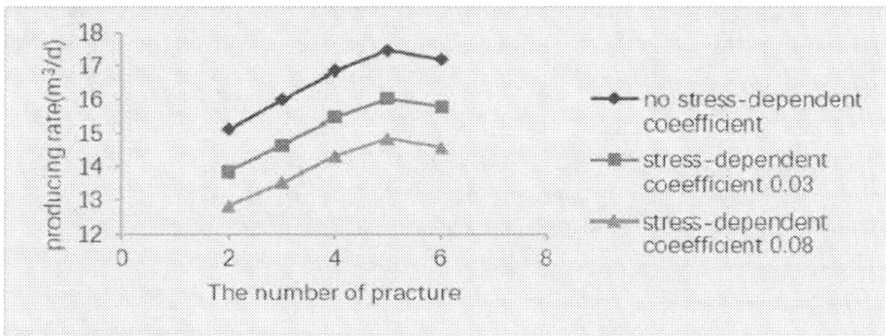

Figure 4. Productivity of fractured horizontal well effect with different fracture number

4.2.2. The Fracture Half-length. In this section, we assumed that the horizontal well has three hydraulic fractures with fractures spacing of 270 m. The finite element numerical simulation software ANSYS is applied to simulate the oil production with fracture half-length of 125 m, 150 m, 175 m, 200 m, 225 m, 250 m, respectively. As shown in Fig.5, the recovery ratio of fractured horizontal well in stress-dependent reservoir generally increases with the increase of fracture half-length. When the crack length is constant, the smaller the stress coefficient is, the larger the recovery is.

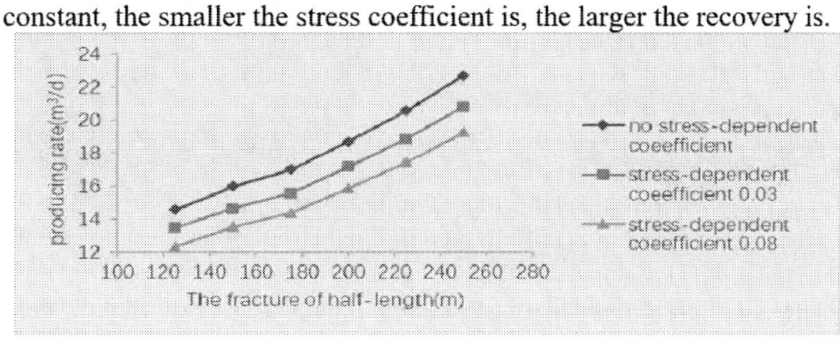

Figure 5. Productivity of fractured horizontal well effect with different fracture half-length

IOP Conf. Series: Materials Science and Engineering **611** (2019) 012022 doi:10.1088/1757-899X/611/1/012022

4.2.3. The Fracture Spacing. As shown in Fig.6, if other fracture parameters remain unchanged, the production of fractured horizontal wells in formation generally decreases first with the increase of fracture spacing, and decreases with the increase of stress correlation coefficient. This is mainly due to the short distance between cracks and the serious interference of cracks. Therefore, in fracturing design of horizontal wells, fracture spacing should be optimized.

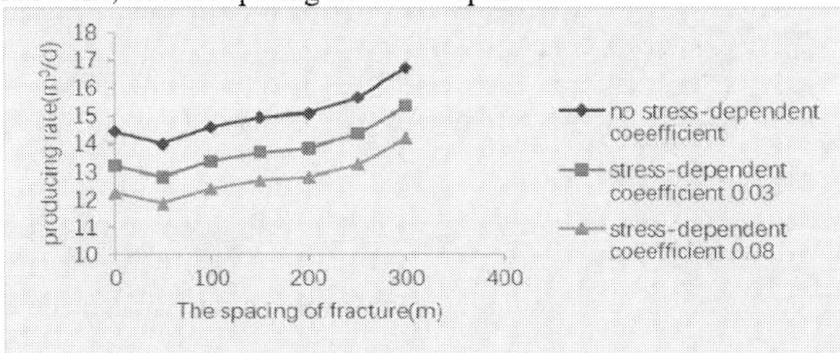

Figure 6. Productivity of fractured horizontal well effect with different fracture spacing

4.2.4. The Fracture Angle. In Fig. 7, It can be seen that, for the same fracture half length, fracture number and fracture spacing, the higher fracture angle yields the higher oil production rate. In addition, the larger the stress-dependent coefficient is, the lower the oil production rate is. For example, in Figure 8, the fracture angle of 60°, The larger the heat capacity is, the smaller the low-temperature region area forms. In other word, converted into seepage field, the larger reservoir stress-dependent coefficient results in the smaller flowing area controlled by the fractures and wellbore and the lower production of the fractured horizontal well.When the crack Angle 90°, the highest yield, the effect is best. Reason is that as the Angle between the fracture plane and the horizontal wellbore is more and more big, the greater the vertical distance between cracks, the less the cause mutual interference between cracks, increasing the effective drainage area, so as to increase the oil production.

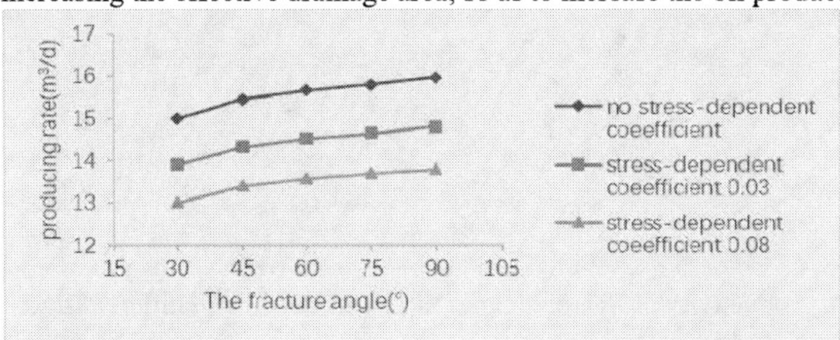

Figure 7. Productivity of fractured horizontal well effect with different fracture angle

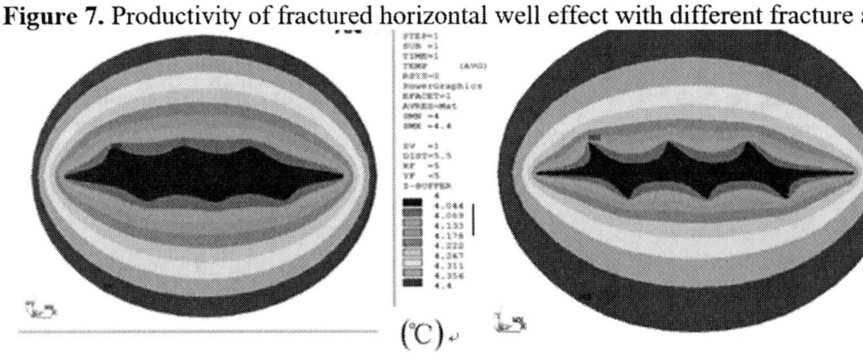

(a) Pressure sensitive coefficient 0.03 (b) Pressure sensitive coefficient 0.08

Figure 8. The temperature distribution of the same fracture angle of different pressure sensitive coefficient on fractured horizontal well

From Fig. 4-7, the greater the pressure sensitivity coefficient of the reservoir, the greater the rate of decline in horizontal fracturing production, and the lower the production of fracturing horizontal wells. The pressure sensitivity of the reservoir has a considerable effect on the productivity of the fractured horizontal well. Due to the rapid change of the near well zone pressure in the fracture and the horizontal well, the pressure loss mainly occurs in the small area near the wellbore, and the average pressure is getting closer to the initial pressure of the reservoir.

5. Conclusions

The paper presents a new efficient numerical method to handle stress-dependent behavior for fractured wells in tight reservoirs using the large-scale general finite element software ANSYS as the platform. It is feasible to apply the ANSYS function of temperature field analysis in the computation of the unstable seepage flow considering permeability change in stress-dependent tight reservoirs. It is very convenient to use ANSYS transient thermal model in reservoir with single phase slightly compressible fluid and deformed porous media.

Considering the economic benefit, there should be an optimal crack length. As the supply radius increases, the daily output decreases gradually. Daily production increases with the increase of fracture spacing. Oil production is greatest when fractures are distributed in a fishbone - like manner. The greater the Angle between the fracture plane and the horizontal wellbore, the greater the production of fracturing the horizontal well. The higher the pressure sensitivity, the lower the oil production.

Acknowledgments

We gratefully acknowledge the supported by the National Key Technologies R & D Program of China during the 13th Five-Year Plan Period (Grant No. 2016ZX05060).

References

[1] Arunn N (2012) Essentials of heat and fluid flow in porous media. CRC Press, Taylor & Francis/Ane Books Pvt. Ltd., pp 92-95
[2] Norbeck J, Huang H, Podgorney R, et al. (2014) An integrated discrete fracture model for description of dynamic behavior in fractured reservoirs SPEJ 253:183-194
[3] Chen JH, Kang YL, You LJ. et al (2011) Review and prospect about study on stress- sensitivity of low - permeability reservoir. Natural Gas Geoscience, 22(1): 31-34.
[4] Burgoyne MW, Little AL (2012) From high perm oil to tight gas - a practical approach to model hydraulically fractured well performance in coarse grid reservoir simulators. SPE#156610 presented in the SPE Asia Pacific Oil and Gas Conference and Exhibition, Perth, Australia
[5] Geiger S, Matthai S, Niessner J, et al (2009) Black-Oil simulations for three-component, three-phase flow in fractured porous media. SPE#107485 prepared for presentation at the SPE Europec/EAGE Annual Conference and Exhibition, London, United Kingdom
[6] Hoteit H, Firoozabadi A (2006) Compositional modeling by the combined discontinuous Galerkin and mixed methods. SPE#90276 prepared for presentation at the SPE Annual Technical Conference and Exhibition, Houston, Texas, USA
[7] Ibrahim M (2013) Development of new well index equation for fracture wells. SPE#164017 prepared for presentation at the SPE Middle East Unconventional Gas Conference and Exhibition
[8] Joun Gyun Kim, Milind D. Deo (2000) Finite element, discrete-fracture model for multiphase flow in porous media AICHE Journal, 46(6):1120~1130
[9] Lei ZhD, Tian ChB, Wang F, et al (2014) A dynamic discrete fracture model for simulation of tight low-permeability reservoirs IPTC-17992 prepared for presentation at the international Petroleum Technology Conference, Kuala Lumpur, Malaysia
[10] Lin M, Chen S, Xu J (2014) Fractured reservoir modeling: effects of hydraulic fracture geometries in tight oil reservoirs. SPE#167761 presentation at the SPE/E AGE European Unconventional Conference and Exhibition, Vienna, Austria

The Generalized Inverse Computing Method of a Class of Block Matrix from Models for Flow in Porous Materials

Tangwei Liu[1,*], Jingying Zhou[2], Kangxiu Hu[3], Shuo Chen[4] and Jinfeng Lai[5]

[1,2,3,4,5]School of Science, East China University of Technology,418 Guanglan Road, Nanchang, 330013, China

*Corresponding author

Abstract. The generalized inverse method of block matrix plays an effective and important role during the computing procedure of certain mathematical models. Some numerical methods for flow models in porous materials are discussed in this paper. Firstly, some partial differential equation models of flow in porous materials are given. Then a class of block matrix from the discrete systems of mathematic models by the multiscale finite element methods are analyzed. Finally, the generalized inverse of the two-by-two block matrix are obtained and the algorithm for computing the minus inverse is presented. The presented methods have an advantage of concise computing procedure for the simulation of models.
Keywords: porous materials; partial difference equation model; inverse computing method; block matrix.

1. Introduction

Flow in porous materials can be described by some partial differential equation (PDE) mathematic models[1,2,4]. Recently, there have been some advances in numerical methods of these models[3,5,6]. In general speaking, multi-scale finite element methods(MsFEM) are regarded as the general numerical methods to solve the PDE models with special boundary conditions[1,2,4]. In previous studies, some scholars have given many expressions for some generalized inverse of partitioned matrix [7,9,10]. However, it is still difficult to calculate the generalized inverse of some special block matrix[8].In this paper, the generalized inverse of two by two block matrices is used to solve a special class of linear systems derived from the numerical solution of PDEs for flow in porous materials. The brief and efficient computing methods is presented for the procedure simulation of models.

2. Two by Two Block Matrix from the Flow Models

In this section, we first consider the following coupled PDEs model for porous materials consisting of two equations for pressure P and total Darcy velocity U in a spatial domain [4].

$$\nabla \cdot U(x,t) + q(x) = (\phi\mu C)\frac{\partial P(x,t)}{\partial t}, x \in \Omega, t > 0, \tag{1}$$

$$U(x,t) = -\lambda K(x)\nabla P(x,t), \tag{2}$$

$$P(x,0)=P_0, \tag{3}$$

where Ω is a domain in R^d $(d=2)$, ϕ , μ , c , q and $k(x)$ denote the porosity viscosity, the compressibility, source term and the permeability tensor, respectively. We assume that $U=U(x,t)$,

$P = P(x,t)$ are unknown functions. And here we assume that the equations (1)-(3) are equipped with the following boundary conditions

$$\vec{U} \cdot \vec{n} = 0 \; on \; \partial\Omega , \tag{4}$$

where $\vec{n}$ denotes the outward normal of $\partial\Omega$.

The Laplace transform of the function p and its inversion formulas are given by

$$\overline{P}(x,s) = L[P(x,t)] = \int_0^\infty e^{-st} P(x,t)dt ,$$

$$P(x,t) = L^{-1}[\overline{p}(x,s)] = \frac{1}{2\pi i}\int_{c-i\infty}^{c+i\infty} e^{st}\,\overline{p}(x,s)ds ,$$

where $s = c + i\omega; c, \omega \in R$.

By using the mixed MsFEM, we obtain the approximations of the Laplace transform of the pressure P and total Darcy velocity U:

$$\overline{u}_H = \sum_{i=1}^{N} \tilde{U}_i(s)\psi_i(x) \; and \; \overline{p}_H = \sum_{j=1}^{N} \tilde{P}_j(s)\cdot\varphi_j(x)$$

where the coefficients $\tilde{p}$ and $\tilde{U}$ are obtained by solving the following linear equations [4]:

$$\begin{bmatrix} H & C^T \\ C & L \end{bmatrix}\begin{bmatrix} \tilde{U} \\ -\tilde{P} \end{bmatrix} = \begin{bmatrix} \mathbf{0} \\ F \end{bmatrix} . \tag{5}$$

We note that

$$\mathrm{H} = [\int_\Omega \psi_i \cdot (\lambda K)^{-1} \psi_j dx], \tag{6.1}$$

$$C = [\int_\Omega \varphi_j div(\psi_i) dx], \tag{6.2}$$

$$L = [\varphi s \int_\Omega \varphi_i \cdot \varphi_j dx], \tag{6.3}$$

$$F = [\int_\Omega h \cdot \varphi_j dx], \tag{6.4}$$

where ψ_i, φ_j represent velocity and pressure basis function ($i, j = 1, 2, \cdots N.$), respectively.

3. The Generalized Inverse of a Class of Two-by-Two Block Matrix

3.1. The Basic Concept of Inverse Matrices

There are many kinds of inverse matrices for block matrices such as minus inverse, plus inverse, least square generalized inverse, reflexive generalized inverse and minimum norm generalized inverse. In this section, we consider only the minus inverse of four-block matrices.

For the matrix $A \in F^{m \times n}$, we consider the following formulas:

$$(i)\, AXA = A, \forall A \in F^{m \times n}$$

For an $m \times n$ matrix A, if there exists a matrix X satisfy (i), then matrix X is called a minus inverse of A, and denoted by A^-. And for any $A \in F^{m \times n}$, minus inverse A^- exists.

3.2. Minus Inverse of a Two-by-Two Block Matrix

In this section, we consider the generalized inverse of the following two-by-two block matrix

$$A=\begin{pmatrix} H & C^T \\ C & L \end{pmatrix}. \tag{7}$$

We discuss the minus inverse A^- of matrix A when $C = C^T = \mathbf{0}$.
We partition A^- into appropriate blocks as follows

$$A^- = \begin{pmatrix} B_{11} & B_{12} \\ B_{21} & B_{22} \end{pmatrix}. \tag{8}$$

Since $AA^-A = A$, we have

$$AA^-A=\begin{pmatrix} H & \mathbf{0} \\ \mathbf{0} & L \end{pmatrix}\begin{pmatrix} B_{11} & B_{12} \\ B_{21} & B_{22} \end{pmatrix}\begin{pmatrix} H & \mathbf{0} \\ \mathbf{0} & L \end{pmatrix}=\begin{pmatrix} HB_{11}H & HB_{12}L \\ LB_{21}H & LB_{22}L \end{pmatrix}=A=\begin{pmatrix} H & \mathbf{0} \\ \mathbf{0} & L \end{pmatrix}.$$

From $HB_{11}H = H$, $LB_{22}L = L$, we obtain $B_{11} = H^-$, $B_{22} = L^-$.

The general solution of the matrix equation $AXB = \mathbf{0}$ is $X = Y - AA^-YBB^-$,where Y,B are some arbitrary matrices of appropriate order.

So the general solutions of the matrix equation $HB_{12}L = \mathbf{0}$ and $LB_{21}H = \mathbf{0}$ are

$$B_{12} = Y - H^-HYYLL^-, B_{21} = Z - L^-LZHH^-,$$

where Y and Z are arbitrary matrices of appropriate order.
We have

$$A^- = \begin{pmatrix} H & \mathbf{0} \\ \mathbf{0} & H \end{pmatrix}^- = \begin{pmatrix} H^- & Y-H^-HYYLL^- \\ Z-L^-LZHH^- & L^- \end{pmatrix}. \tag{9}$$

As A_{12} and A_{21} are nonzero, we discuss some singular and nonsingular cases about matrix A .

I) When matrix A is singular, assume that $\det(A_{11}) \neq 0$,by the elementary transformation of the partitioned matrix

$$\begin{pmatrix} E & \mathbf{0} \\ -CH^- & E \end{pmatrix}\begin{pmatrix} H & C^T \\ C & L \end{pmatrix}\begin{pmatrix} E & -H^-C^T \\ \mathbf{0} & E \end{pmatrix}=\begin{pmatrix} H & \mathbf{0} \\ \mathbf{0} & L-CH^-C^T \end{pmatrix}.$$

Notes

$$P=\begin{pmatrix} E & \mathbf{0} \\ -CH^- & E \end{pmatrix}, Q=\begin{pmatrix} E & -H^-C^T \\ \mathbf{0} & E \end{pmatrix}, D=\begin{pmatrix} H & \mathbf{0} \\ \mathbf{0} & L-CH^-C^T \end{pmatrix},$$

such that $PAQ = D$.

According to A singular, $\det(A_{22} - A_{21}A_{11}^{-1}A_{12}) \neq 0$,the following formula can be obtained.

$$A^- = \begin{pmatrix} H & C^T \\ C & L \end{pmatrix}^- = QD^-P = \begin{pmatrix} E & -C^T H^- \\ \mathbf{0} & E \end{pmatrix} \begin{pmatrix} H & \mathbf{0} \\ \mathbf{0} & L - CH^-C^T \end{pmatrix}^- \begin{pmatrix} E & \mathbf{0} \\ -C^T H^- & E \end{pmatrix}$$
$$= \begin{pmatrix} E & -H^-C^T \\ \mathbf{0} & E \end{pmatrix} \begin{pmatrix} H^- & \mathbf{0} \\ \mathbf{0} & (L - CH^-C^T)^- \end{pmatrix} \begin{pmatrix} E & \mathbf{0} \\ -CH^- & E \end{pmatrix} \tag{10}$$
$$= \begin{pmatrix} H^- + H^-C^T(L - CH^-C^T)^- CH^- & -H^-C^T(L - CH^-C^T)^- \\ -(L - CH^-C^T)^- CH^- & (L - CH^-C^T)^- \end{pmatrix}.$$

If $\det(L) \neq 0$,then $\det(H - C^T L^- C) \neq 0$. We have

$$A^- = \begin{pmatrix} (H - C^T L^- C)^- & -(H - C^T L^- C)^- C^T L^- \\ -L^- C(H - C^T L^- C)^- & L^- + L^- C(H - C^T L^- C)^- C^T L^- \end{pmatrix}. \tag{11}$$

II) When A is nonsingular, if $\det(H) \neq \mathbf{0}$,then $PAQ = D \Rightarrow A = P^- DQ^-$.
Since

$$(P^-DQ^-)(P^-DQ^-)^-(P^-DQ^-) = P^-DQ^-QD^-PP^-DQ^- = P^-DD^-DQ^- = P^-DQ^-,$$

D is a quasi-diagonal matrix, the following formula can be obtained

$$D^- = \begin{pmatrix} H & \mathbf{0} \\ \mathbf{0} & L - CH^-C^T \end{pmatrix}^- = \begin{pmatrix} H^- & \mathbf{0} \\ \mathbf{0} & (L - CH^-C^T)^- \end{pmatrix},$$

$$A^- = \begin{pmatrix} H^- + H^- C^T(L - CH^- C^T)^- CH^- & -H^- C^T(L - CH^- C^T)^- \\ -(L - CH^- C^T)^- CH^- & (L - CH^- C^T)^- \end{pmatrix}. \tag{12}$$

Similarly,when $\det(A_{22}) \neq 0$, we can get

$$A^- = \begin{pmatrix} (H - C^T L^- C)^- & -(H - C^T L^- C)^- C^T L^- \\ -L^- C(H - C^T L^- C)^- & L^- + L^- C(H - C^T L^- C)^- C^T L^- \end{pmatrix}. \tag{13}$$

If A is semi-positive definite Hermite matrix, similar conclusions will be drawn regardless of whether H or L is invertible or not.
A is semi-positive definite Hermite matrix, there exists matrix $\overline{B}$, such that $A = \overline{B}^{\mathrm{H}} \overline{B}$.
Note that

$$\overline{B} = (B_1 \quad B_2), A = \begin{pmatrix} H & C^T \\ C & L \end{pmatrix},$$

one has the following formula

$$\begin{pmatrix} H & C^T \\ C & L \end{pmatrix} = \begin{pmatrix} B_1^{\mathrm{H}} B_1 & B_1^{\mathrm{H}} B_2 \\ B_2^{\mathrm{H}} B_1 & B_2^{\mathrm{H}} B_2 \end{pmatrix}.$$

The equations $A^{\mathrm{H}} Ax = \mathbf{0}$ and $Ax = \mathbf{0}$ have the same solution.
Since $A^{\mathrm{H}} A(A^{\mathrm{H}} A)^- A^{\mathrm{H}} A = A^{\mathrm{H}} A$, the following formulas be obtained.

$$A^{\mathrm{H}}(A(A^{\mathrm{H}} A)^- A^{\mathrm{H}} A - A) = \mathbf{0}, A(A^{\mathrm{H}} A)^- A^{\mathrm{H}} A - A = \mathbf{0}, A = A(A^{\mathrm{H}} A)^- A^{\mathrm{H}} A,$$

$$C^T H^- H = B_2^{\mathrm{H}} B_1 (B_1^{\mathrm{H}} B_1)^- B_1^{\mathrm{H}} B_1 = B_2^{\mathrm{H}} B_1 = C^T,$$

$$HH^- C^T = B_1^{\mathrm{H}} B_1 (B_1^{\mathrm{H}} B_1)^- B_1^{\mathrm{H}} B_2 = B_1^{\mathrm{H}} B_2 = C^T.$$

Notes

$$P = \begin{pmatrix} E & \mathbf{0} \\ -C^T H^- & E \end{pmatrix}, Q = \begin{pmatrix} E & -H^- C^T \\ \mathbf{0} & E \end{pmatrix}, D = \begin{pmatrix} H & \mathbf{0} \\ \mathbf{0} & L - CH^- C^T \end{pmatrix},$$

we obtain the following formulas

$$D = PAQ = \begin{pmatrix} E & \mathbf{0} \\ -CH^- & E \end{pmatrix} \begin{pmatrix} H & C^T \\ C & L \end{pmatrix} \begin{pmatrix} E & -H^- C^T \\ \mathbf{0} & E \end{pmatrix}$$
$$= \begin{pmatrix} E & \mathbf{0} \\ -CH^- & E \end{pmatrix} \begin{pmatrix} B_1^{\mathrm{H}} B_1 & B_1^{\mathrm{H}} B_2 \\ B_2^{\mathrm{H}} B_1 & B_2^{\mathrm{H}} B_2 \end{pmatrix} \begin{pmatrix} E & -H^- C^T \\ \mathbf{0} & E \end{pmatrix} = \begin{pmatrix} H & \mathbf{0} \\ \mathbf{0} & L - CH^- C^T \end{pmatrix}.$$

Obviously, P and Q are invertible, we have

$$A^- = QD^- P = Q \begin{pmatrix} H & \mathbf{0} \\ \mathbf{0} & L - CH^- C^T \end{pmatrix}^- P = Q \begin{pmatrix} H^- & \mathbf{0} \\ \mathbf{0} & (L - CH^- C^T)^- \end{pmatrix} P$$
$$= \begin{pmatrix} E & -H^- C^T \\ \mathbf{0} & E \end{pmatrix} \begin{pmatrix} H^- & \mathbf{0} \\ \mathbf{0} & (L - CH^- C^T)^- \end{pmatrix} \begin{pmatrix} E & \mathbf{0} \\ -CH^- & E \end{pmatrix} \tag{14}$$
$$= \begin{pmatrix} H^- + H^- C^T (L - CH^- C^T)^- CH^- & -H^- C^T (L - CH^- C^T)^- \\ -(L - CH^- C^T)^- CH^- & (L - CH^- C^T)^- \end{pmatrix}.$$

Using the above results, an algorithm for computing matrix A^- is given as following.

Algorithm1 Implementing process of computing matrix $\begin{bmatrix} H & C^T \\ C & L \end{bmatrix}^-$
1. Compute H^-;
2. Compute $(L - CH^- C^T)^-$;
3. Compute A^- according to (14).

4. Generalized Inverse Solutions of a Class of Linear System

For the linear system(5), we obtain the inverse of the block matrix as following by the algorithm 1.

$$\begin{bmatrix} H & C^{\mathrm{T}} \\ C & L \end{bmatrix}^- = \begin{pmatrix} H^- + H^- C^T (L - CH^- C^T)^- CH^- & -H^- C^T (L - CH^- C^T)^- \\ -(L - CH^- C^T)^- CH^- & (L - CH^- C^T)^- \end{pmatrix}. \tag{15}$$

We have

$$\begin{bmatrix} \tilde{U} \\ -\tilde{P} \end{bmatrix} = \begin{pmatrix} H^- + H^- C^T (L - CH^- C^T)^- CH^- & -H^- C^T (L - CH^- C^T)^- \\ -(L - CH^- C^T)^- CH^- & (L - CH^- C^T)^- \end{pmatrix} \begin{bmatrix} \mathbf{0} \\ F \end{bmatrix}, \tag{16}$$

$$\tilde{U} = -H^- C^T (L - CH^- C^T)^- F, \quad \tilde{P} = -(L - CH^- C^T)^- F. \tag{17}$$

At last we can compute the numerical value of $U(x,t)$, $P(x,t)$ by $\tilde{U}(x,s)$, $\tilde{P}(x,s)$ using the numerical inversion formula of Laplace transform.

5. Conclusion
In this paper, we derive a new brief computing procedure by the generalized inverse of the block two-by-two linear systems, which arise from partial differential equation models of flow in porous materials. Numerical solutions of pressure and Darcy velocity in the mathematic models can be obtained. The results is meaningful for the numerical simulation of flow in porous materials.

Acknowledgments
The authors are grateful to the referees and Dr. Jianhua Zhang for their valuable comments and suggestions which helped to improve the presentation of the paper.
Funding: This research was funded by the National Natural Science Foundation of China under Grant No. 41766001.

References
[1] Kippe V Aarnes J E and Lie K A 2008 *A comparison of multiscale methods for elliptic problems in porous media flow* vol 12 Comput Geosci pp 377-398.
[2] Aarnesand EfendievY 2008 *Mixed multiscale finite element methods forstochastic porous media flows* vol 30 SIAM Journal on Scientific Computing pp2319-2339.
[3] Bultreys T De Boever W & Cnudde V 2016 *Imaging and image-based fluid transport modeling at the pore scale in geological materials: A practical introduction to the current state-of-the-ar.* Earth-Science Reviews155 pp93-128.
[4] Liu T W Qiu X L et al 2013 *A Hybrid Laplace Transform Mixed Multiscale Finite-Element Method for Flow in Porous Media* vol 10Journal of Information and Computational Science pp 4773-4781.
[5] BearJ 2018 *Modeling phenomena of flow and transport in porous media (Vol. 31)* Springer.
[6] Anbari A Chien H T Datta S S Deng W Weitz D A & Fan J 2018 *Microfluidic model porous media: fabrication and applications* Small 14(18) 1703575.
[7] Tian Y and Takane Y 2009 *More on Generalized Inverse of Partitioned Matrices with Banachicwicz-Schur forms* Linear Algebra and its Applications pp 1641-1655.
[8] Yan Z Z 2012 *New representations of the Moore-Penrose inverse of 2ʹ 2 block matrices* Linear Algebra and its Applications pp 35-37.
[9] Zhao Y and Jia S H 2009 *Solution of generalized inverse matrix and linear equations* Mathematical practice and cognition pp 239-244.
[10] Zhang L P and Yu H B 2008 *Expression and calculation of generalized inverse matrix* Journal of Xichang university pp 39-42.

PSO-SVM Model for Classifying Spontaneous Combustion Tendency Grade of Sulfide Ores

Nianping Liu*, **Xiaojun Xie and Siqi Huang**

School of Environmental and Resources, Southwest University of Science and Technology, No. 59, Middle Section of Qinglong Avenue, Fucheng District, Mianyang, Sichuan Province, China. 621010

*Corresponding author

Abstract. Three main factor indexes that most reflect spontaneous combustion tendency of sulfide ores were selected such as oxidation increment of ore sample under low temperature, self-heating point temperature and self-burning point temperature as the basic discriminate factors of the classification model. Based on the theory of cross validation and particle swarm optimization(PSO) to optimize the parameters of support vector machine (SVM),the PSO-SVM model for classifying spontaneous combustion tendency grade of sulfide ores was built by using training samples from measured data, and was verified by engineering data. It shows that the model has high predication accuracy, which also is scientific and feasible, it provides a new approach to evaluate the spontaneous combustion tendency grade of ores in any sulfide mines.
Keywords: sulfide ores; spontaneous combustion tendency; support vector machine (SVM); particle swarm optimization.

1. Introduction

Because of the autothermal and the spontaneous combustion of sulfide ores (SCSO) in air, it is easy to cause mine fire, so SCSO is one of the major hazards in the mining of high sulfur deposit. in the mining of sulfide deposit, accurately determining the level tendency of SCSO is an important prerequisite for protection methods to determine the design and safety deposit mining. There are two main types to determine the tendency of SCSO at home and abroad. One is determined by testing some oxidation performance indicators of ore, such as the constants of oxygen and the electrochemical performance are used as judgment indicators in former USSR, the oxidized ore and the rate of temperature that are determined by H2O2 are used as judgment indicators in Changsha Research Institute of Mining And Metallurgy[1]. The other is determined by the method of data mining based on the typical sample tendency of SCSO, such as the multiple discriminate analysis[2], the extension method[3],the neural network method[4] and the fuzzy clustering analysis method[5]. The method has some advantages in the determination of the SCSO tendency, but there are some limitations, such as discriminate analysis needs certain typical samples, neural network has the problems of low convergence speed and poor fault tolerance etc. Based on the above research results, the tendency of SCSO is discriminated by taking advantage of SVM for small sample classification. Genetic algorithms and parallel search using global optimization features for SVM parameters were optimized with the actual engineering and experimental results, through the establishment of a representative sample of index system was proposed based on particle swarm discriminate model of SVMs, and cross verify the validity of the model for spontaneous combustion tendency of sulfide ore grade

IOP Conf. Series: Materials Science and Engineering **611** (2019) 012024 doi:10.1088/1757-899X/611/1/012024

determination provides a new way of thinking. And by using genetic algorithm parallel optimization of SVM parameter was optimized, combined with the practical engineering and experimental results, through the establishment of index system of a typical sample, a discriminative model PSO based on SVM was presented, and the validity of the model was cross verified, a new ideas for the spontaneous combustion tendency of sulfide ores grade determination was provided.

2. The Support Vector Machine and the Particle Swarm Algorithm

SVM is a machine learning method based on statistical learning theory [6]. Its basic idea is that input vectors are mapped to a high dimensional feature space through a pre –selected nonlinear mapping. Then optimal classification hyperplane is constructed in this space. The optimal classification hyperplane makes the maximum distance from the nearest sample of each class to hyperplane[7-8]. In which the nearest sample to the optimal hyperplane is called support vector. Discriminate function optimal classification is marked

$$y=\mathrm{sgn}[(w\cdot x)+b] \tag{1}$$

In the first formula, sgn() is sign function ,w is weight vector ,x is input vector, B is constant.
The data are non-linear transformed. The data are mapped into a high dimensional feature space by a nonlinear mapping function $\varphi(x)$, discriminate function becomes

$$y=\mathrm{sgn}\{[w\cdot\varphi(x)]+b\} \tag{2}$$

Constraints of no error classification in feature space become

$$yi[w\cdot\varphi(xi)+b]-1\geq 0 \tag{3}$$

In this formula, xi is the i-th training data, yi equals ±1.
According to the VC (Vapnik-Chervonenkis Dimension) theory, The empirical risk minimization and an upper bound model of complex degree can be expressed as a quadratic programming problem under conditions of constraints:

$$\min\varphi(w)=(w\cdot w)/2 \tag{4}$$

Introduce slack variables ζi ≥ 0, It is the index for characterization of samples can be correctly identified, if it can be correctly classified ζi=0, Otherwise ζi>0, the formula (4) can be rewritten as

$$yi[w\cdot\varphi(xi)+b]-1+\zeta i\geq 0 \tag{5}$$

Meanwhile, penalty was introduced in the optimization targets, we will get:

$$\min\phi(w,\xi)=(w\cdot w)/2+c\sum_{i=1}^{n}\xi_i \tag{6}$$

In which, c is the penalty factor, it indicates the degree of misclassification of punishment. According to Kuhn-Tucker conditions, using Lagrange optimization method, the optimal hyperplane problem can be changed into an optimization objective function:

$$\max W(\alpha)=\sum_{i=1}^{n}\alpha_i-\frac{1}{2}\sum_{i,j=1}^{n}\alpha_i\alpha_j y_i y_j(x_i\cdot x_j) \tag{7}$$

a_i is Lagrange Multiplier, Constraint condition is

$$\sum_{i=1}^{n}a_i y_i=0, 0\leq a_i\leq c(i=1,2,...,n) \tag{8}$$

Answering the above problem of convex quadratic programming, gets the Optimal classification function:

$$f(x) = sgn\left[\sum_{i=1}^{I} y_i \cdot a_i(x \cdot x_i) + b\right] \tag{9}$$

In which, I is the number of support vectors. For the nonlinear problem, only introducing kernel function can satisfy the Mercer condition can be solved. Let kernel function $K(x_i \cdot x_j)$, Classification and recognition function become

$$f(x) = sgn\left[\sum_{i=1}^{I} y_i \cdot a_i \cdot K(x \cdot x_i) + b\right] \tag{10}$$

Common kernel functions have polynomial functions, radial basis function, the sigmoid function and so on. In this paper, the radial basis function is used,

$$K(x, x_i) = \exp\left\{-\frac{\|x - x_i\|^2}{\sigma^2}\right\} \tag{11}$$

Where σ is parameter.

PSO is widely used in recent years based on iterative optimization algorithm, its basic idea comes from the study of predation on birds [9]. In the PSO algorithm, the solution to optimization problem is initialized to a group of random particles, each particle has a corresponding potential optimal solution, position, velocity and fitness are used to characterize the particle, in each optimization iteration, particles through dynamic tracking individual optimal value pbest and group gbest to update their optimal values. Assuming in the D-dimensional space, Group X= composed of N particles (X_1, X_2,... , X_n),In which, the i-th particle is represented as a D-dimensional vector $Xi=(x_{i1},x_{i2},\ldots,x_{id})T$ It represents the position of the i-th particle in D-dimensional space, velocity of the i-th particle is $Vi=(V_{i1},V_{i2},\ldots,V_{iD})T$, its individual optimal value $pbest=(P_{i1},P_{i2},\ldots,P_{iD})T$, group optimal value $gbest=(Pg1,Pg2,\ldots,PgD)T$. when the optimal value is found, Particles will update their transferring speed and position by the following formula:

$$V_{id}^{k+1} = wV_{id}^{k} + c_1r_1(P_{id}^{k} - X_{id}^{k}) + c_2r_2(P_{gd}^{k} - X_{id}^{k}) \tag{12}$$

$$X_{id}^{k+1} = X_{id}^{k} + V_{id}^{k+1} \tag{13}$$

In the formula, V_{id} is the particle velocity; K is the current number of iterations; w is the inertia weight; c_1 and c_2 are non-negative learning factors; r_1 and r_2 are the random numbers between [0,1] .

3. Judgment Application tendency of SCSO Based on the PSO-SVM Model

Spontaneous combustion tendency of sulfide ores has many kinds of index, On the basis of relevant research achievements, the weight gain rate of sulfide ores (X_1)under the low-temperature oxidation, the hot spot temperature sulfide ores (X_2) and spontaneous ignition temperature (X_3) of sulfide ores are selected to represent the index of discriminating features of spontaneous combustion tendency grading factor. According to the literature[2], The sulfide ore spontaneous combustion tendency level is divided into three categories: easy to spontaneous combustion (I), easy to self-heating but not easy to spontaneous combustion (II), and not easy to spontaneous combustion (III).In order to verify the applicability of the PSO-SVM model, fourteen samples of typical ore spontaneous combustion tendency index value are collected as training samples 3. Training sample data for the forecast model shown in table 1.

Table 1. Training sample data for the forecast model

sample number	Discriminate factors			Discrimination results	Actual results
	Oxidation weight gain (X_1)%	Self-heating temperatures (X_2)/℃	Self-burning temperatures (X_3)/℃		
1	9.50	60	296	I	I
2	1.20	127	453	II	II
3	2.10	270	422	II	II
4	14.70	140	270	I	I
5	2.10	122	436	II	II
6	0.90	243	301	II	II
7	1.00	297	385	II	II
8	3.20	400	414	III	III
9	2.20	400	430	III	III
10	9.30	56	233	I	I
11	0.45	400	435	III	III
12	1.56	220	408	II	II
13	2.70	220	366	II	II
14	2.11	243	385	II	II

Table 2. The result of the forecast model

sample number	Discriminate factors			Discrimination results	Actual results
	Oxidation weight gain (X_1)%	Self-heating temperatures (X_2)/℃	Self-burning temperatures (X_3)/℃		
15	10.80	146	239	I	I
16	2.20	230	437	II	II
17	2.54	220	374	II	II
18	1.60	400	425	III	III

PSO-SVM is programmed by using MataLab with Libsvm Toolbox[10]. Through the determination of the PSO parameters are as follows: the number of group particle is 20, The maximum number of evolution is 200, the elastic coefficient of speed in the formula of velocity updating and in the formula of group updating are both one. Cross-validation parameters is three, Learning factor $c1$=1.49, $c2$=1.69。 Parameter optimization of fitness curve as shown in Figure 1:

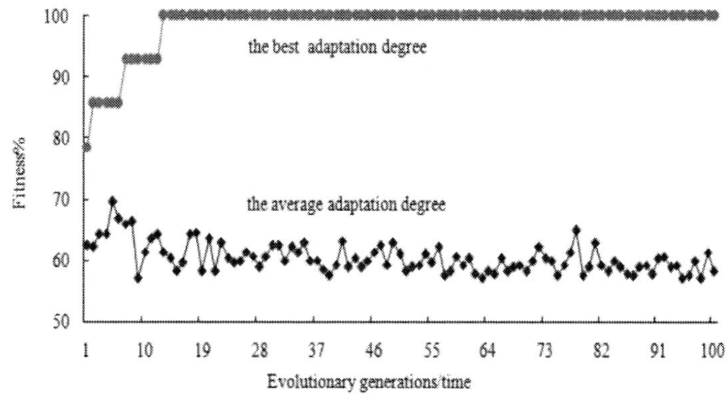

Figure 1. Fitness curves

It can be seen from Figure 1, In the evolution to thirteenth generations, The best fitness value converges to 100%,After 100 iterations the penalty parameter is 3.438, The kernel function parameters is 4.4574. Based on the PSO-SVM model study of inversion analysis of training samples, The discriminant accurate rate was 100%, it shows that the model has high accuracy.

4. Engineering Application

In order to verify the accuracy of the model further, four samples which are selected from Xinqiao Pyrite in Tongling Nonferrous, Yinjiagou pyrite in Jinyuan Mining Co will be identified, the PSO-SVM model is used to identify results, if discrimination results is entirely consistent with the actual situation, it further illustrates the usefulness and accuracy of the engineering model. The result of the forecast model see table 2.

5. Conclusions

(1)Through the optimal method of particle swarm optimization combined with cross validation, the best penalty parameter and kernel parameter to make the training samples with the highest classification accuracy rate are obtained, the limitations of any given or testing experience given by the penalty parameter and kernel function parameter are avoided, the accurate discrimination ability of this model is improved.

(2)The accuracy of PSO-SVM model to the training samples of validation is 100%.The trained PSO-SVM model is applied to the discriminate analysis on the different kinds of spontaneous combustion tendency of the actual supplied ore mines, the Discrimination results completely match with the actual situation of ore samples, it further proves the effectiveness of the proposed method, therefore, the model can be judged an auxiliary discrimination method of SCSO tendency.

(3)The same way as all the data mining models, the accuracy and reliability of the PSO-SVM model largely depends on the choice of typical and representative training samples. Therefore, It is necessary to collect a large number of actual data of sulfide ores, establish a sound database of SCSO, and further improve the accuracy and applicability of the model.

Acknowledgments

This research is a result of partial support under Sichuan Science and Technology Program(Project ID:19GJHZ0103) and the grant of the China Scholarship Council.

References

[1] Changnian Xia, Chao Wu. Study of risk assessment technique for evaluating the spontaneous combustion of sulfide ore inmining stope[J]. Fire Safety Science, 2005, 15(2): 106-110.

[2] Fuqiang Yang, Chao Wu, Zijun Li, et al. Grade classification of spontaneous combustion tendency for sulfide ores based on distanced discriminate analysis theory[J].Journal of China coal society,2010,35(12):2111-2115.

[3] Hanhua Hu, Zheng Liu, Zijun Li, et al. Fisher discriminate analysis to the classification of spontaneous combustion tendency grade of sulphide ores[J], Journal of China coal society,2010,35(10):1675-1678 .

[4] Ming Li, Chao Wu, Zijun Li. Research on ANN dynamic prediction model for spontaneous combustion of sulfide ores with spontaneous combustion of sulfide ores with multi-factors coupling[J]. China safety science journal, 2007,17(8):32-36.

[5] Fuqiang Yang, Chao Wu, Zijun Li. Matter-element model and its application to comprehensive determination on spontaneous combustion tendency of sulfide ores[J]. Journal of central south university(science and technology),2011,42(11):3459-3464.

[6] Vapnik V N. Statistical Learning Theory[M]. New York： John and Wiley,1998.

[7] Vladimir N Vapnik. An overview of statistical learning theory [J].IEEE Transactions on Neural Networks (S1045-9227),2002, 10(5):988-999.

[8] Ortes C, Vapnik V. Support-vector networks[J]. Machine Learning, 1995,5(20): 273-297.

[9] KENNKEDY J, EBERHART R C. A discrete binary version of the particle swarm algorithm[c]//Pro. 1997 Conf. on System, Man, and Cybernetics. Piscata way, NJ: IEEE Service Center 4104-4109.

[10] R.E. Fan, P.H. Chen, and C.J. Lin, Working Set Selection using Second Order Information for Training Support Vector Machine, Journal of Machine Learning Research,2005, 6:1889-1918.

Numerical Model and Simulation Analysis of Composite Stringer Docking Structures

Jingchao Wei*, Gang Wang, Su Cao and Xiangming Chen

Aeronautics Science and Technology Key Laboratory of Full Scale Aircraft Structure and Fatigue, Aircraft Strength Research Institute of China, Xi'an 710065, China

*Corresponding author

Abstract. Due to the manufacturing restrictions or design requirements, composite structures are usually connected in the form of butt joint near the wing rib or fuselage frame to solve the problem of discontinuity of load transfer. In this paper, the finite element model of butt joint structure is established by simulating the engineering model of metal fasteners and common joint connected by beam element under the action of in-plane load, and the mechanical simulation analysis of typical butt joint configuration is carried out. In the engineering model, the allowable stress value is used to predict the failure of composite material and the classical strength theory is used to predict the fracture failure of metal fasteners. The result show that the strength prediction obtained by the engineering model is in good agreement with the test results, and the application of this method can improve the calculation efficiency of connection strength with complicated multi-fastened composite structures.
Keywords: numerical model; simulation analysis; strength prediction; composite structure.

1. Introduction

Composite structure has been widely used in a variety of main/secondary load bearing structures of aircraft, and composite stringer due to manufacturing, transportation, cost and other considerations or restrictions, usually adopts the segmented design and docking mode. The connection of multiple parts is mainly realized through mechanical fasteners, whose connection strength directly affects flight safety and aircraft service life[1, 2]. Due to the importance of connection design, a large number of scholars have conducted extensive research on the connection strength of composite laminates. The three-dimensional cumulative damage calculation can vividly and objectively reflect the damage generation and evolution process at the connection, which is the current mainstream computational simulation method[3, 4, 5]. Xiao et al. established a two-dimensional damage accumulation model based on the nonlinear shear theory of composite materials and Hashin, yamada-sun failure criterion. The nonlinear material term and the degradation mode of the failure layering were included in the stress analysis of the model, and the finite deformation and the contact boundary conditions of the nail holes were considered[6]. Wei et al. used three-dimensional solid element model to simulate the single bolt loading and structural progressive failure process through friction contact between bolt and hole[7].
Due to the structural characteristics of composite laminates, it is necessary to use nonlinear finite element method to calculate the damage and extension repeatedly, which makes the calculation of the connection strength large scale and long time. For large-scale composite structures, a simple and fast finite element analysis method is needed. Based on the beam element joint method, this paper establishes an engineering model of the connection structure, which can quickly judge the weak links and failure locations of the structure and better predict the failure strength of the structure.

2. Composite Structure Design and Finite Element Model

2.1. Configuration of Composite Structure

The composite stringer docking structure refers to the butt-joint area of stringer near the wing root of the aircraft structure. The test specimens are designed symmetrically without considering the influence of the back Angle on the outer wing. In order to avoid the early failure of the test pieces in the loading area, the stringer flange is widened in the clamping area. In the test pieces, the actual structure of the cross joint is simplified into a t-joint, and the web of the t-joint is perpendicular to the edge strip.

Both sides of the stringer are connected by two diagonal boxes and horizontal cross joint, and the web of the stringer is obliquely cutting in the root; The corner boxes is connected with two tensile bolts, and the corner box is not connected with the web. There are four rows of nails along the span. The first three rows of nails connect the corner box, reinforced siding and horizontal cross joint, while the fourth row of nails only connect the reinforced siding and horizontal cross joint.

2.2. Material Property

The test pieces are made of carbon fiber reinforced epoxy resin based prepreg (as shown in Table 1), and the stringer and the skin are made of co-bonded technology. Corner box and cross joint material is titanium alloy Ti-6Al-4V, after forging and forming and rib material is aluminum alloy 7050.

Table 1. Formatting sections, subsections and subsubsections.

Longitudinal modulus E_{11}/ MPa	Transverse modulus E_{11}/ MPa	Shear modulus G_{12} / MPa	Poisson's ratio ν_{12}
178000	8260	5200	0.34

2.3. FE Model

Shell element is used to simulate the structure of skin, stringer, butt joint, corner box, cross joint and rib. For the girders and webs, the shell element nodes are not biased and located on the middle plane of the structure. For the flat plate connected with the stringer such as corner box, cross joint and trident joint, the shell element node is positioned on the outer surface close to the skin and the long truss, and the element is set with corresponding bias. For the vertical plates of rib webs, cross joints and trident joints, the shell element nodes are located on the middle plane of the structure, and the element is not biased. All fasteners are simulated with Beam element. In order to ensure equal displacement loading, generate a node outside the structure for loading, apply a 10kN tensile load to the point, and couple the node with the end node of the test piece with constraints of Coupling.

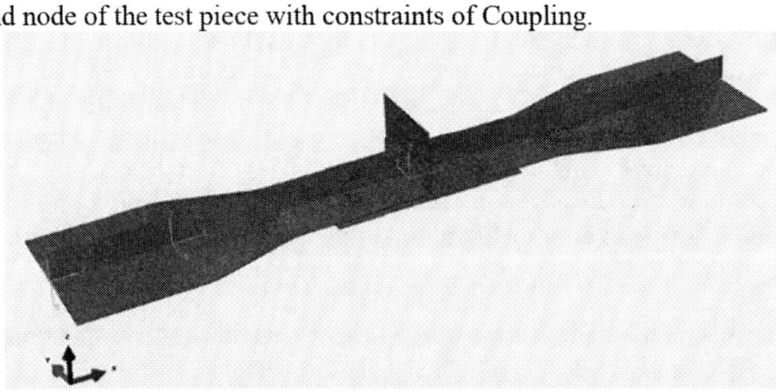

Figure 1. Finite element model of composite stringer butt-joint structure.

2.4. Results of Calculation

2.4.1. Distribution of Fastener Load. The fastener load is basically measured in the rib position. In order to describe the fastener Numbers on the side of the test piece, fasteners no. 1 to no. 16 endure cutting load on the long girder strip, and fasteners no. 20 to no. 23 endure pulling load on the corner box joint, as shown in Figure 2.

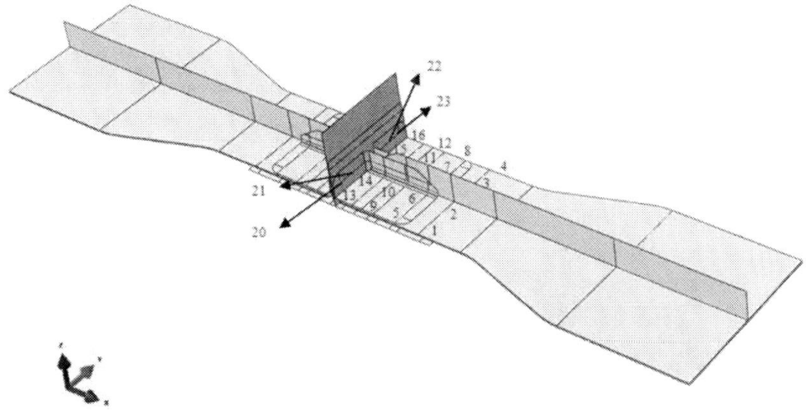

Figure 2. Schematic diagram of fastener number in the finite element model.

It can be seen from the table 1 that the shear load distributed by fasteners no.6, no.7, no.14 and no.15 is relatively large, accounting for 9.21% and 6.92% of the total shear load respectively. The maximum single shear force is 525N, and the maximum double shear force is 921N. The ratio of tensile load transmitted through the cross joint and corner box is about 2.3 : 1.

Table 2. Shear load on fasteners.

No.	Shear load between skin and cross joint/ N	Shear load between stringer and angle box / N	Total shear load / N	Percentage
1	413	/	413	4.13%
2	428	/	428	4.28%
3	428	/	428	4.28%
4	413	/	413	4.13%
5	416	361	777	7.77%
6	453	468	921	9.21%
7	453	468	921	9.21%
8	416	360	776	7.76%
9	369	165	534	5.34%
10	408	252	660	6.60%
11	408	252	660	6.60%
12	369	164	533	5.33%
13	487	89	576	5.76%
14	525	167	692	6.92%
15	524	168	692	6.92%
16	486	87	573	5.73%
Total	6996	3001	9997	100.00%

2.4.2. Composite Strain. The x-direction strain of the composite material is mainly tensile strain, which is prone to stress concentration due to extrusion along the edge of hole. Therefore, when the strength are checked, the edge of the composite material hole adopt the design allowable values method of single/double shear stress, and the design allowable values of strain are checked at the location away from the hole. The maximum stress value of the double shear hole edge is located near the hole of fastener No.6 and 7 on the edge strip, and the maximum stress value of the single shear hole edge is located near the hole of fastener No. 2 and no. 3(as shown in Figure 3). The maximum strain value away from the edge of the hole is located at the straight section of the tensile necking root with a strain value of 47με. The maximum stress value of the double-shear hole edge of the skin is located near the holes of fasteners No. 5 to No. 8 on the edge strip, and is 5.7MPa. The maximum stress value of the single-shear hole edge is located near the holes of fasteners no. 1 to no. 4, and is 5.7MPa. The maximum strain value away from the edge of the hole is located at the straight section of the tensile necking root with a strain value of 49με.

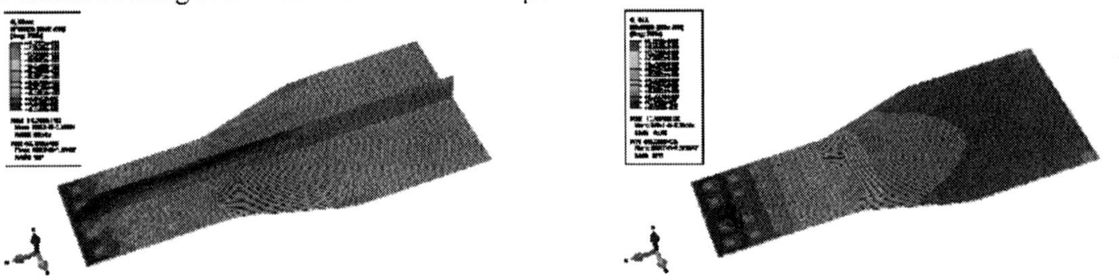

(a) Longitudinal C:\Users\asus\AppData\Local\Youdao\DictBeta\Application\7.1.0.0421\resultui\dict\stress of stringer

(b) Longitudinal C:\Users\asus\AppData\Local\Youdao\DictBeta\Application\7.1.0.0421\resultui\dict\stress of skin

Figure 3. Longitudinal C:\Users\asus\AppData\Local\Youdao\DictBeta\Application\7.1.0.0421\resultui\dict\stress C:\Users\asus\AppData\Local\Youdao\DictBeta\Application\7.1.0.0421\resultui\dict\nephogram of composite stringer and skin.

3. Strength Prediction

3.1. Strength of Composite Structures

The material strength of composite structure is checked by the structural main strain. The maximum tensile strain, compressive strain and shear strain of FE model are compared with tensile strain allowable value, compressive strain allowable value as well as shear strain allowable value of composite material, and the strength on the side of composite hole are checked by bearing stress and single/double shear strength. Strength margins are calculated by equations (1) - (5).

Tensile strength margin of composite materials:

$$M.S.=\frac{[\varepsilon_t]}{\varepsilon}-1 \tag{1}$$

Compression strength margin of composite materials:

$$M.S.=\frac{[\gamma]}{\gamma_{xy}}-1 \tag{2}$$

Shear strength margin of composite materials:

$$M.S.=\frac{[\sigma^d_{bru}]}{\sigma}-1 \tag{3}$$

Single-lap bearing strength margin of composite materials:

$$M.S.=\frac{[\sigma^s_{bru}]}{\sigma}-1 \tag{4}$$

3.2. Strength of Composite Structures

The tensile strength of fasteners is checked by the tensile load of bolts. The maximum bolt tension in the structure is compared with the tensile stress allowable value of bolts, and the strength margin is calculated according to equation (5). The shear strength of fasteners was checked by the shear load of bolts. The maximum single shear force of bolts in the structure was compared with the single shear allowable value of bolts, and the strength margin was calculated according to equation (6).

$$M.S. = \frac{[F]}{F} - 1 \quad (5)$$

$$M.S. = \frac{[Q]}{Q} - 1 \quad (6)$$

3.3. Structural Failure Mode

The strength margins of composite structures are obtained through finite element analysis and strength check, and the failure mode of the structure is estimated. Among them, the composites mainly consider the tensile/compression strength margin and pin-hole bearing strength margin, while the bolts mainly consider the tensile strength margin and shear strength margin, which serve as a reference for the prediction of structural failure mode. By comparing the above strength margin values, the tensile strength margin of bolts is the minimum (M.S. = 2.05).

It can be seen that the possible failure mode is tensile failure of No.21 and No.22 bolts, with the initial failure load of about 726kN, and the failure load of about 1088kN if the failure is estimated to be 1.5 times the allowable value of the design.

4. Conclusion

In this paper, the finite element analysis and strength prediction of composite stringer docking structure are carried out. According to the structural characteristics, loading and clamping mode, an engineering finite element model is established, and the stress, strain and bolt load distribution of the structure are obtained through linear static analysis. The margin analysis of the material strength, bolt tensile strength and bolt shear strength of the composite stringer butt-joint test pieces is carried out, and the strength margin of each part of the structure is obtained, and the failure load and possible failure mode of the structure are estimated by linear extrapolation. Although this article is based on linear analysis, the material design allowable values as a benchmark are adopted in the analysis, rather than material real strength value, structure progressive failure process cannot be simulated. However, the application of engineering experience and simplified model can predict the failure position and failure load of structure well and quickly, and greatly improve the efficiency of connection calculation in engineering applications.

References

[1] Aviation research institute of China 2004 Handbook of Composite Structure Design (Beijing: Aviation industry)

[2] Xie M J 2011 Composite material connection (Shanghai: Shanghai jiao tong university)

[3] Sun HT, Qing X, Chang F-K. The response of composite joints with bolt-clamping loads, part II: model verification. J Composite Material 2002; 36(1): 69-91.

[4] Binnur Goren Kiral. Effect of the clearance and interference fit on failure of the pin-loaded composites. Materials and Design 2010; 31: 85-93.

[5] Irisarri F -X, Laurin F, Carrere N, Marie J -F. Progressive damage and failure of mechanically fastened joints in CFRP laminates - Part I: Refined Finite Element modeling of single-fastener joints. Composite Structures 2012; 94: 2269-2277.

[6] Yi Xiao, Takashi Ishikawa. Bearing strength and failure behavior of bolted composite joints (Part II: modeling and simulation). Composites Science and Technology 2005; 65: 1032-1043.

[7] Wei JC, Jiao GQ, Liu FL, Yan ZM. Bearing strength of composite joints interference-fitted with blind bolts. Acta Aeronautica et Astronautica Sinica 2013; 34: 1627-1635.

IOP Conf. Series: Materials Science and Engineering **611** (2019) 012026 doi:10.1088/1757-899X/611/1/012026

Hydrothermal Liquefaction of Model Compounds Protein and Glucose: Effect of Maillard Reaction on Low Lipid Microalgae

Xiaohan Tang, Chao Zhang and Xiaoyi Yang*

School of Energy and Power Engineering, Beihang University, 37 Xueyuan Road, Haidian District, Beijing, P. R. China

*Corresponding Author

Abstract. Reaction between protein and carbohydrate is important for the hydrothermal liquefaction (HTL) of low lipid microalgae. Model compounds glucose and soy protein were used to simulate the HTL of low lipid microalgae under different conditions. Transfer of element and distributions of compounds were quantified to study the effect of Maillard reaction during HTL. Solid residue was mainly caused by dehydration of glucose. Deamination was the main reaction for protein, although most of the nitrogen was presented in aqueous or gas products, high protein content had bad effects on the quality of bio-crude. Interaction between protein and glucose at optimal ratio could prevent the production of solid residue from glucose and counteract bad effects of protein during HTL, thus promote the quality and yield of bio-crude.

1. Introduction

Microalgae have emerged as feedstocks for bio-energy production due to their high photosyntheic efficiency and fast growth rate [1]. Hydrothermal liquefaction (HTL) is believed as a cost-effective way because it avoids the drying process [2]. Chemical properties of the hydrothermal products mostly depend on the microalgal composition [3]. Current microalgae-to-biofuel technology focus on utilization of lipids rich strains because high lipids content can improve the yield and quality of bio-crude [4,5]. While lipids accumulation for most microalgae requires severe cultivation condition, such as phosphate and nitrogen limitation [6]. Potential conversion methods should prove the ability to convert cheap, low quality feedstock. Thus, the conversion research about low lipid microalgae is required.

Low lipid microalgae are rich in protein and carbohydrate. When protein and carbohydrate processes HTL reaction individually, either of them generated different products; but it may interplay each other and the products thus become complex when they process together [7,8]. The complex process of polymerization during HTL of cabohydrate and protein, named Maillard reaction, generate some polymeric products with different molecular weights, structures and element compositions [9]. Since the Maillard reaction is complex [10,11], it is simple to evaluate the performace of hydrothermal liquefaction by the element distrubution and understand the reaction mechanism by the element transfer route instead of characterizing every single compounds of product.

Up to date, Maillard reactions between carbohydrate and protein were mostly investigated in the field of food chemistry, and most of the research was conducted at relative lower temperature of 60~120 °C, which is different from the high react temperature above 200 °C during HTL process. In addition, most of the HTL research focus on the quality and yield of bio-crude. The shortage of information of aqueous and solid products after HTL makes it difficult to understand the reaction mechanism. In this study, glucose and soy protein were chosen as the model component to simulate low lipid microalgae, the mixture of them at different ratios were run HTL tests from 160 to 280 °C. Effects of reaction conditions on distribution of element and compounds in the bio-crude, aqueous fraction and solid residue were analyzed, respectively.

The element and compounds distributions under different reaction conditions may be helped to explore the HTL reaction mechanism and optimizing the reaction process. With better knowledge of the reaction mechanism, the properties of HTL products can be improved by adjusting operating conditions or biomass substrate design.

2. Material and Methods

2.1. Materials

In order to reflect the reaction between carbohydrate and protein in actual microalgae, glucose and soy protein were chosen as the model compounds. Glucose and soy protein were purchased from Sinopharm Chemical Reagent Co., Ltd (Shanghai, China). The feedstocks were used as received.

2.2. Hydrothermal Liquefaction Experiment

HTL experiments were carried out in a 500 mL stainless-steel autoclave batch reactor with a magnetic stirring. The reaction procedure can be found in previous published work [12]. In a typical experiment run, 100 g of feedstock (10 wt.% solid) were added into the reactor. The residue air was replaced by nitrogen gas after reactor was sealed. The reaction time for all the HTL experiments was 60 min. The reaction time didn't include the heat-up times which took 45-60 min. for the reactor to increase to 280 °C.

2.3. Products Separation and Yields Calculation

After the reaction was finished, the reactor was cooled down to room temperature by compressed air. The reactor was opened and the products mixture was transferred to a flask. 100 mL of dichloromethane (DCM) was poured to the reactor to wash the reactor wall and stirrers and the washed fraction was also transferred to the flask. The reaction mixture was separated by filtration. The filter residue was defined as solid residue, the filtrate can be separated to DCM phase and aqueous phase. The DCM phase after solvent removal was defined as bio-crude, and the aqueous phase got after HTL reaction was dried at 70 °C, the left material was defined as involatiles. The yields of bio-crude, involatiles and solid residue were calculated as the mass of bio-crude, involatiles and solid residue dividing by the mass of dry substrate, respectively.

2.4. Element and Compounds Analysis of Each Fraction

Bio-crude was identified by GC-MS (Agilent 7890A-5975C, USA) equipped with a HP-5 capillary column (30 m×0.25 mm i.d. ×0.25 μm). The inlet temperature was 250 °C, the split ratio was 10:1 and the flow rate was 1 mL/min. Oven temperature was initially maintained at 50 °C for 5 min. followed by heating to 260 °C (5 °C /min.) and maintained at this temperature for 5 min. Carbon, hydrogen and nitrogen content HTL products were analyzed by an element analyzer (Euro Vector EA3000, Italy), and the oxygen content was calculated by difference. All the analysis was conducted twice and the standard error was less than 0.5 wt.%.

3. Results and Discussion

3.1. HTL Products yields

Effects of temperature and the ratio between protein and glucose on the yields of each HTL products are presented in figure 1. As presented in figure 1 (a), the total recovery of bio-crude, involatiles and solid residue decreased with temperature increasing, which could be caused by the formation of more volatile or gas products at higher temperature. When temperature was increased from 160 °C to 280 °C, the yield of bio-crude gradually increased from 1.38% to 9.06%, while the yields of involatiles and solid residue decreased from 38.16% and 31.68% to 23.86% and 13.19%, respectively. There appeared to be some inter-conversion of material in the solid residue, involatiles and bio-crude during the reaction processes, the aqueous phase and solid residue were converted into bio-crude phase at higher temperature. It can be deduced that high temperature could elevate products yields of Maillard reaction and promote the transformation of solid residue and aqueous products into bio-crude, while cause more mass loss due to more gas production.

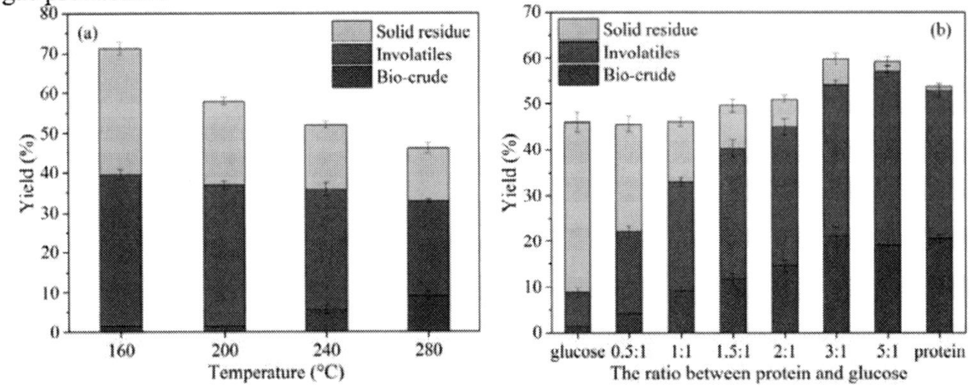

Figure 1. Effect of (a) temperature and (b) ratio between protein and glucose on the yields of HTL products.

Figure 1 (b) depicts the change in yields of HTL products with respect to variant ratios of glucose and protein. The bio-crude and involatiles yields of glucose were much lower than those of protein. The solid residue yield of glucose was 37.2%, while trace of solid residue left for protein (0.97%). Bio-crude and involatiles products were the main products of protein. Over 60% of carbon from glucose of glucose was generated as Hydroxymethylfurfural (HMF) ring structure [13]. Considering the high solid residue of glucose, it can be deduced that dehydration and condensation was the main reaction for glucose during HTL process. The pH of the aqueous, which was measured right after HTL test, was 1.5 (table 1) for pure glucose. That means some acids were produced during the HTL of glucose and thus accelerated the polymerization of dehydration products of glucose. Therefore, we hypothesized that the formation of solid residue during HTL of glucose was caused by the polymerization of furan material, such as 5-HMF, and the formation of solid residue inhibited the transfer of glucose to bio-crude.

Table 1. The pH of post-HTL aqueous at different ratios between glucose and protein.

Ratio	glucose	0.5:1	1:01	1.5:1	2:01	3:01	4:01	5:01	protein
pH	1.5	6.4	7.5	7.4	7.3	8.2	8.2	8.3	9.5

For the mixture of glucose and protein, the yield of bio-crude increased with the ratio between protein and glucose and achieved maximum of 21.23% at the ratio of 3:1, then decreased slightly. Peterson et al. reported that the reaction between glycine and glucose was correlated with the glucose concentration [14]. This result may suggest that there was an optimum ratio between glucose and protein to achieve the highest yield of bio-crude. The yield of involatiles followed almost the same trend with bio-crude. The involatiles yield, 7.61% for individual glucose, increased with the ratio between protein and glucose and achieved 37.96% at the ratio of 5:1. It seemed like the reaction between glucose and protein can promote the generation of aqueous products compared with individual glucose. The solid residue significantly decreased from 37.15% to 2.30% when the ratio between protein and glucose increased from 0 to 5:1. The presence of protein in HTL process made more solid residue converted to aqueous phase and bio-crude. The results indicated that the reaction between protein and glucose can reduce the reaction rate of polymerization, resulted in decrease of solid residue yield. In addition, the total recovery of bio-crude, solid residue and involatiles increased gradually with the ratio, suggested the decrease of gas yield. Each fraction yields of Maillard reaction could be controlled by adjusting the reaction temperature and the ratio between carbohydrate and protein.

3.2. Characterization of Bio-crude

Bio-crude of HTL was a complicated mixture and various compounds were detected by GC-MS.

Temperature had a significant influence on compounds distribution of Maillard reaction products. At 160 °C, the molecular weight of the products was small and peaks were shown before 10 min, most of which was heterocyclic oxygen, such as furan. With temperature increasing, more N-heterocyclic compounds with heavy molecular weight or complicated structure were generated.

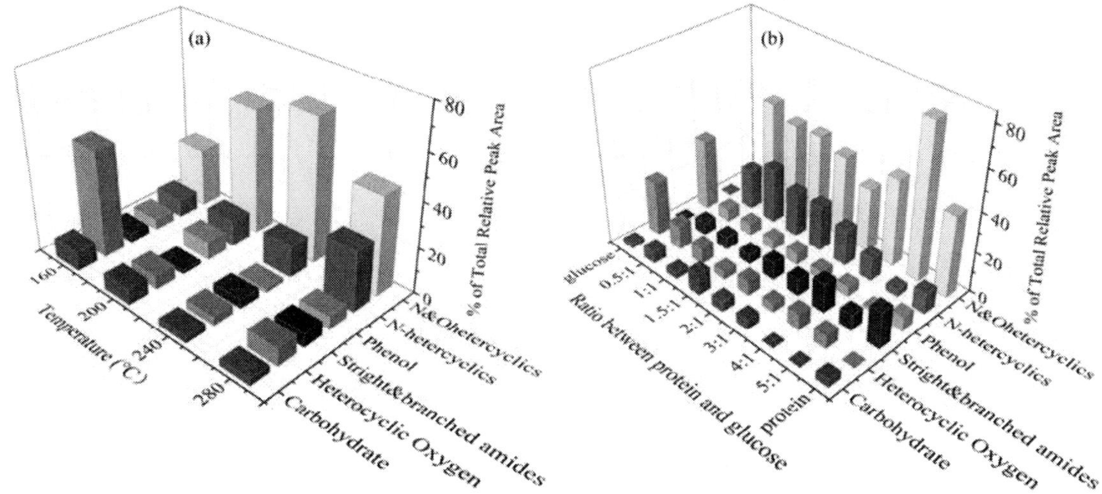

Figure 2. Composition of bio-crude got at different (a) temperatures and (b) ratio between protein and glucose.

Chemical compounds characterized by GC-MS were categorized according to the functionalities. All the experiments presented in figure 2(a) were conducted for the mixture of soy protein and glucose at the ratio of 1:1 and the results indicated that reaction temperature significantly affected the chemical composition of bio-crude. About half of the peak area was taken by the heterocyclic oxygen content at 160 °C. With temperature increasing to 200 °C, heterocyclic oxygen content sharply decreased to 6.03% and kept small variation as temperature continuing rising. While the N-heterocyclic content consecutively increased as temperature rising. The N&O heterocyclic content increased from 25.83% to 64.06% as temperature rising from 160 °C to 240 °C, and decreased to 43.62% at 280 °C. These results indicated that most bio-crude was derived from glucose at 160 °C and less protein was decomposed to ammonia acid that can react with glucose at this temperature. With temperature increasing, protein start to decompose, resulted in the higher content of N-heterocyclic and N&O heterocyclic in bio-crude.

Heterocyclic oxygen and phenol were the main products in bio-crude from glucose, while straight & branched amides and N&O heterocyclic were the main component in bio-crude from protein HTL due to the generation of amines and aldehydes through decarboxylation and deamination reactions of amino acids. The mixture of glucose and protein generated more N-heterocyclic and N&O heterocyclic but less heterocyclic oxygen than the individual feedstock. The maillard reaction between protein and glucose lead to nitrogen containing cyclic compounds [15]. Most part of the heterocyclic oxygen was furan, such as 5-HMF. Under sub-critical condition, 5-HMF processed via two routes simultaneously: decomposing to gas or polymerizing to char [16]. The results presented in figure 2 (b) suggested that the reaction between glucose and protein at an appropriate ratio could inhibit the polymerization of 5-HMF and thus decreased the solid residue yield. In addition, it also made more protein transferred into bio-crude but not aqueous or gas products during HTL process, which could be deduced from the increase in N-heterocyclic and N&O heterocyclic content of the mixture feedstock.

3.3. Element Recovery

The element recoveries of each fraction for different substrate after HTL (280 °C for 1 hour) are presented in figure 3 and they were calculated as below:

Element recovery of each faction= (element content of each fraction × yield of each fraction)/the element content of raw material

The element recoveries of glucose and protein were totally different. For glucose, most part of the carbon was left in solid residue, thus the carbon recovery of bio-crude and involatiles was just 2.25% and 9.51%, respectively. The carbon of protein was mainly recovered in bio-crude (31.49%) and involatiles (32.65%). Little solid residue was generated from HTL of protein, resulted in just 0.75% of carbon recovery. The carbon recovery in bio-crude gradually increased with the ratio between protein and glucose and reached the maximum of 34.57% at the ratio of 3:1, then decreased to 30.68% at the ratio of 5:1. With the ratio between protein and glucose increasing, the carbon recovery of solid residue kept decreasing and the involatiles performed inversely. When glucose mixed with protein, the condensation of glucose reduced, therefore, the solid residue rapidly decreased (figure 1) and the pH of aqueous phase sharply increased. It can be concluded that the reaction between glucose and protein favored the transfer of carbon from the solid residue to gas, bio-crude or aqueous products.

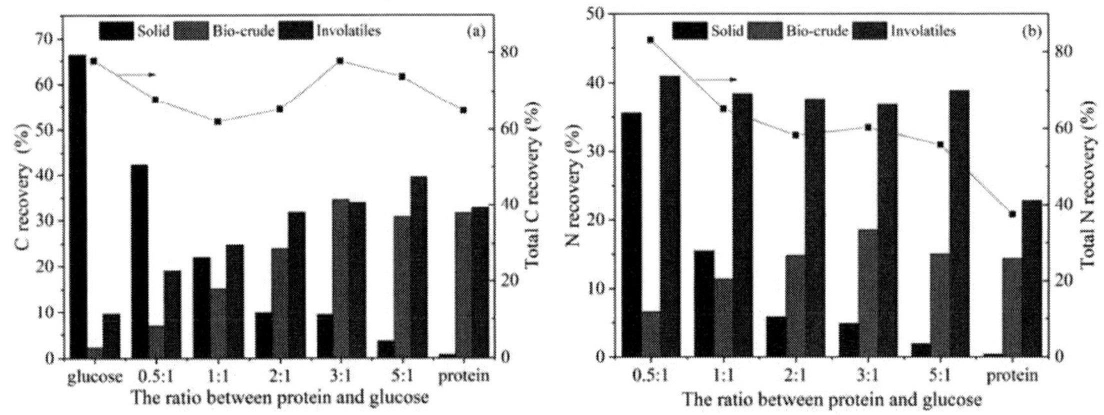

Figure 3. C and N recovery of HTL products.

The study of nitrogen recovery is important for the optimization of HTL process, because nitrogen has become the biggest obstacle during the bio-crude upgrading process. After mixed with glucose, the nitrogen recovery of involatiles was around 40% for all the samples at different ratio of protein and glucose, much higher than that of individual protein (22.80%). The nitrogen recovery of solid residue at the ratio of 0.5:1 (protein: glucose, *wt./wt.*) was as high as 35.62%. The presence of glucose may inhibit the nitrogen transferred to bio-crude and aqueous. With the ratio between protein and glucose increasing, the nitrogen recovery of solid residue decreased to 1.89% at the ratio of 5:1. Nitrogen recovery of bio-crude performed the same trend with carbon and hydrogen recoveries. The maximum value of nitrogen recovery in bio-crude was 18.53%, gained at the ratio of 3:1.

Besides the element recovery of each product fraction, the total recovery contained bio-crude, solid residue and involatiles are also presented in figure 3. The gap between 100% and the total recovery was caused by the gas and volatile products. The gas products were not considered because carbon recovery of gas was less than 10% and over 90% of carbon in HTL gas was CO_2 [17].

The total carbon recovery decreased from 77.99% to 62.09% when the ratio between protein and glucose increased from 0 to 1:1 and then increased with the ratio. Pure glucose had the highest total carbon recovery but the lowest hydrogen recovery. Reactions such as decarboxylation of protein resulted in about 35% loss of carbon.

The total nitrogen recovery of protein was the least (37.50%) compared with other substrates. The exist of glucose increased the total nitrogen recovery. For the mixture feedstock, the highest total nitrogen recovery achieved at the ratio of 0.5 and then decreased with the ratio increasing. The pH of the HTL aqueous for pure protein was 9.5. Yang et al. [18] reported that half of the total nitrogen in the aqueous was determined to be NH_3-N, little NO_3-N or NO_2-N. The NH_3-N caused the high pH of the aqueous and the loss of nitrogen because NH_3 is alkaline and easily soluble in water.

4. Conclusions

Maillard reaction had a predominant role in HTL process of low lipid microalgae. Temperature can affect the severity of Maillard reaction during HTL, and thus determines the products yield. The ratio of glucose and protein determined the distribution of compounds and element among different product fractions. Glucose and protein at an appropriate ratio during HTL can inhibit the polymerization of glucose and thus decreased the solid residue yield. The reaction between protein and glucose can promote more C content but less N content transfer from involatiles into bio-crude.

Acknowledgments
This project was supported by National key research and development program (2018YFB1501505).

References

[1] Chisti Y, 2007, *Biotechnol. Adv.*, 25, 294-306.
[2] Peterson A A, Vogel F, Lachance R P, Fröling M, Antal J M J and Tester J W, 2008, *Energ. Environ. Sci.*, 1, 32-65.
[3] Biller P and Ross A B, 2011, *Bioresource Technol.*, 102, 215-225.
[4] Li H, Liu Z, Zhang Y, Li B, Lu H, Duan N, Liu M, Zhu Z and Si B, 2014, *Bioresource Technol.*, 154, 322-329.
[5] Yoo G, Min S P, Yang J W and Choi M, 2015, *Appl. Energ.*, 156, 354-361.
[6] Chen W, Zhang Y, Zhang J, Yu G, Schideman L C, Zhang P and Minarick M, 2014, *Bioresource Technol.*, 152, 130-139.
[7] Teri G, Luo L and Savage P E, 2014, *Energ. Fuel*, 28, 7501-7509.
[8] Zhang C, Tang X, Sheng L and Yang X, 2016, *Green Chem.*, 18, 2542-2553.
[9] Yaylayan V A and Kaminsky E, 1998, *Food Chem.*, 63, 25-31.
[10] Valdez P J and Savage P E, 2013, *Algal Research*, 2, 416-425.
[11] Torri C, Garcia L, Samorì C, Fabbri D and Brilman D W F W, 2012, *Energ. Fuel.*, 26, 658-671.
[12] Tang X, Zhang C, Li Z, Yang X, 2016, *Bioresource Technol.*, 2016, 202: 8-14
[13] Baccile N, Laurent G, Babonneau F, Fayon F, Titirici M and Antonietti M, 2009, *J. Phys. Chem. C.*, 113, 9644-9654.
[14] Peterson A A, Lachance R P, Tester J W, 2010, *Ind. Eng. Chem. Res*, 49(5) 2107-2117
[15] Kruse A, Maniam P, Spieler F, 2007, *Ind. Eng. Chem. Res.*, 46(1) 87-96
[16] Chuntanapum A and Matsumura Y, 2009, *Ind. Eng. Chem. Res.*, 48, 9837-9846.
[17] Yu G, Zhang Y, Schideman L, Funk T and Wang Z, 2011, *Energ. Environ. Sci.*, 4, 4587.
[18] Yang Y F, Feng C P, Inamori Y and Maekawa T, 2004, *Res. Cons. Recyc.*, 43, 21-33.

Global Stability of a Predator-Prey Model with the Logistic Term

Yali Yang[1], Zhiwei Li[2,*] and Fei Wang[1]

[1]Department of Basic Sciences, Air Force Engineering University, Xi'an, China
[2]Graduate College, Air Force Engineering University, Xi'an, China

*Corresponding author

Abstract. This paper is mainly to analyze stability of equilibria in a new predator-prey model, Logistic-Volterra model. It exists three different equilibria for the model. The stability of these equilibria for the model are discussed by Liapunov function or other principle. Through theoretical analysis, (0,0) is unstable and existent in any situations. (N_1 ,0) is existent and global asymptotic stability if $\delta_2 < k_2 / N_1$, and (x_1^*, x_2^*) is existent and uniformly and globally asymptotic stability if $\delta_2 > k_2 / N_1$. Then, some simulation are given and verify the proof of the result. At last, some conclusions about biology were raised.
Keywords: Predator-Prey Model; Stability; Liapunov Function; Logistic-Volterra Model

1. Introduction

In the last century, Volterra proposed the following model (1) which describes the changes in the number of two species as prey and predator [5].

$$\begin{cases} \dfrac{dx_1}{dt} = k_1 x_1 - \delta_1 x_1 x_2 \\ \dfrac{dx_2}{dt} = -k_2 x_2 + \delta_2 x_1 x_2 \end{cases} \tag{1}$$

In model (1), x_1 and x_2 are non-negative variables which are the quantities of prey and predator respectively. k_1, k_2, δ_1 and δ_2 are positive constants. k_1 and k_2 are the natural change rates of prey and predator, respectively. δ_1 and δ_2 are coefficients meaning what is the influence they make to another species. In an ideal environment, the quantity of a species will change with a rate which is in direct proportion to its total. Meanwhile, it also is affected by another. The predator curbs the growth rate of prey, but the prey promotes its growth under the predation relation.

In figure 1, it gives the numerical simulation results of model (1), where k_1=1.2, k_2=0.6, δ_1 =0.1 and δ_2 =0.02. It finds that the quantities of prey and predator change periodically, and the change of predator always lag behind another. In practice, the number of predator and prey may fluctuate in a certain period. This model correctly reflects the quantitative relationship between predator and prey in the case of sufficient resources.

However, resources are often inadequate in practice. There will be a new competitive relationship between different individuals within the same species. Model (1) is no longer applicable in this situation. Then there is a new predator-prey model with the Logistic term to describe this kind of new condition.

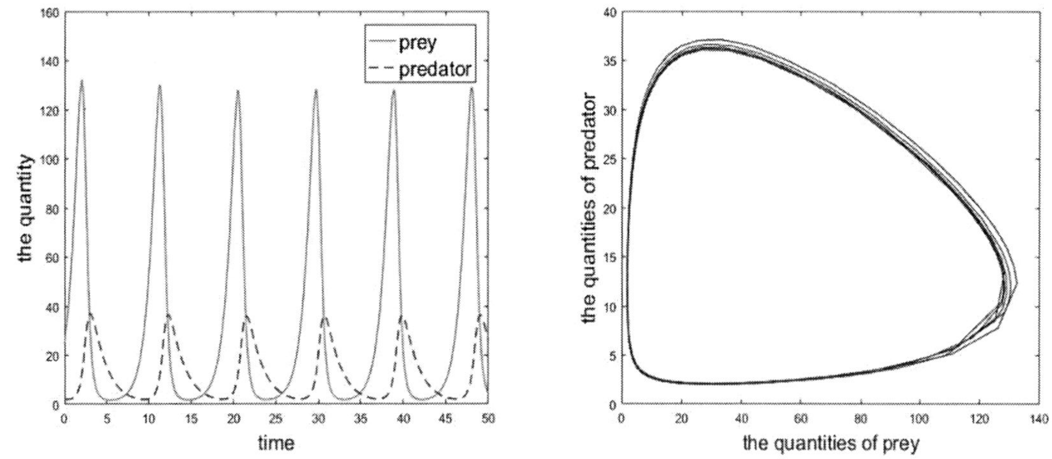

Figure 1. Quantity changes of prey and predator for the model (1)

The main purpose of this paper is to analyze the existence and the global stability of equilibrium for the new model.

In order to correctly describe the quantitative relationship between predator and prey with limited resources, it is necessary to add the Logistic term into model (1). Equation set (2) is the new model, and it is the research object of this paper [6].

$$\begin{cases} \dfrac{dx_1}{dt} = k_1x_1\left(1-\dfrac{x_1}{N_1}\right)-\delta_1x_1x_2 \\ \dfrac{dx_2}{dt} = -k_2x_2\left(1+\dfrac{x_2}{N_2}\right)+\delta_2x_1x_2 \end{cases} \tag{2}$$

In the model (2), N_1 and N_2 are maximum environmental capacities of prey and predator, respectively. k_1 and k_2 are internal competition coefficients of those two species. N_1, N_2, k_1 and k_2 are positive constants. $-k_ix_i^2/N_i(i=1,2)$ represents the competition within the same species respectively.

Paper [1] is discussed with Holling Ⅳ type functional response and Leslie - Gower numerical response of fractional order prey-predator model. Paper [2] proves the existence and uniqueness of global positive solution of a stochastic time-delay predator-prey model in polluted environment. In paper [3], a global state analysis was performed for a predator-prey model that predators have a stage - structured. Paper [4] analysis the SEIR model of predator disease in the predator system and obtained the necessary and sufficient conditions for predator extinction and disease becoming endemic.

In this paper, the second part discusses the existence of the equilibrium, the third part proves the stability of the equilibrium, the fourth part verifies the proof of the third part through numerical simulation, and the fifth part puts forward the conclusion.

2. Existence of the Equilibrium

In order to get the equilibrium of model (2), it needs to obtain the solution of the equations (3).

$$\begin{cases} f(x_1,x_2) \triangleq k_1x_1\left(1-\dfrac{x_1}{N_1}\right)-\delta_1x_1x_2 = 0 \\ g(x_1,x_2) \triangleq -k_2x_2\left(1+\dfrac{x_2}{N_2}\right)+\delta_2x_1x_2 = 0 \\ x_1,x_2 \ge 0 \end{cases} \tag{3}$$

Then, there are three equilibria: (0, 0) which always exists, (N_1,0) which is existent in any situations for the model (2) , and (x_1^*, x_2^*) is existent if $\delta_2 > k_2 / N_1$, where

$$x_1^* = \frac{k_2 N_1 (k_1 + \delta_1 N_2)}{N_1 N_2 \delta_1 \delta_2 + k_1 k_2}, \quad x_2^* = \frac{k_1 N_2 (\delta_2 N_1 - k_2)}{N_1 N_2 \delta_1 \delta_2 + k_1 k_2} \tag{4}$$

3. Stability of Equilibrium

Lemma 1 [7] [8]. For this equation set,

$$\frac{dx}{dt} = Ax, \quad A = (a_{i,j})_{n \times n}, \quad a_{i,j} \in R, \quad x = (x_1, x_2, \cdots, x_n)^T \in R^n \tag{5}$$

If all eigenvalues of matrix A have negative real parts, then the equilibrium of the equation set (5) is asymptotically stable.

According to lemma 1, it is feasible to judge the local stability of the equilibrium based on Jacobi matrix [9].

$$J = \begin{bmatrix} f_{x_1} & f_{x_2} \\ g_{x_1} & g_{x_2} \end{bmatrix} = \begin{bmatrix} k_1\left(1 - \frac{2x_1}{N_1}\right) - \delta_1 x_2 & -\delta_1 x_1 \\ \delta_2 x_2 & -k_2\left(1 + \frac{2x_2}{N_2}\right) + \delta_2 x_1 \end{bmatrix} \tag{6}$$

Let $p = tr\, J = f_{x_1} + g_{x_2}$, $q = \det J = f_{x_1} g_{x_2} - f_{x_2} g_{x_1}$, then the stability condition for those the equilibrium are $p < 0$ and $q > 0$.

For the equilibrium (0, 0),

$$p = k_1 - k_2, \quad q = -k_1 k_2 < 0 \tag{7}$$

Therefore, (0, 0) is unstable and existent in any situations.

For the equilibrium (N_1, 0),

$$p = -k_1 - k_2 + \delta_2 N_1, \quad q = k_1 k_2 - k_1 \delta_2 N_1 \tag{8}$$

If $\delta_2 \le k_2 / N_1$, then $p < 0$ and $q > 0$. That is, (N_1, 0) is an equilibrium with local stability if $\delta_2 \le k_2 / N_1$.

For the equilibrium (x_1^*, x_2^*)

$$p = -\frac{k_1 x_1^*}{N_1} - \frac{k_2 x_2^*}{N_2}, \quad q = \frac{k_1 k_2 x_1^* x_2^*}{N_1 N_2} + \delta_1 \delta_2 x_1^* x_2^* \tag{9}$$

If $\delta_2 > k_2 / N_1$, then $x_1^* > 0, x_2^* > 0$, $p < 0$ and $q > 0$. That is, (x_1^*, x_2^*) is an equilibrium with local stability if $\delta_2 > k_2 / N_1$.

In sum, we have the following theorem.

Theorem 1. The equilibrium (0, 0) is unstable always; if $\delta_2 \le k_2 / N_1$, the equilibrium (N_1,0) is locally stable; if $\delta_2 > k_2 / N_1$, the equilibrium (x_1^*, x_2^*) is locally stable.

Furthermore, we discuss the global asymptotic stability of equilibrium (N_1, 0).

Theorem 2. If $\delta_2 \le k_2 / N_1$, then (N_1,0) is an equilibrium with global asymptotic stability;

Proof. According to the first equation of (2), it gives

$$\frac{dx_1}{dt} < k_1x_1 - \frac{k_1x_1^2}{N_1} \Rightarrow x_1 < \frac{N_1}{ce^{-k_1t}N_1+1} \Rightarrow \limsup_{\delta t\to\infty} x_1 \le N_1 \tag{10}$$

According to the second equation of (2), it gives

$$\frac{dx_2}{dt} < -k_2x_2 - \frac{k_2x_2^2}{N_2} + \delta_2N_1x_2 \Rightarrow x_2 < \frac{1}{ce^{-(\delta_2N_1-k_2)t} + \frac{k_2}{N_2(\delta_2N_1-k_2)}}$$

$$\Rightarrow \limsup_{\delta t\to\infty} x_2 \le \frac{N_2(\delta_2N_1-k_2)}{k_2} \tag{11}$$

And there is a known condition, $\delta_2 \le k_2/N_1$ and $x_2 > 0$. Then $\lim_{\delta t\to\infty} x_2 = 0$. That is, $\forall\varepsilon\in(0,1)$, $\exists$T, if $t \ge T$ then $0 < x_2 < \varepsilon$.

According to the first equation of (2), it gives

$$\frac{dx_1}{dt} > k_1x_1\left(1-\frac{x_1}{N_1}\right) - \delta_1x_1\varepsilon \Rightarrow x_1 > \frac{1}{ce^{-(k_1-\delta_1\varepsilon)t} + \frac{k_1}{N_1(k_1-\delta_1\varepsilon)}}$$

$$\Rightarrow \liminf_{\delta t\to\infty} x_1 \ge N_1 \tag{12}$$

Therefore, $\lim_{\delta t\to\infty} x_1 = N_1$.

In a word, if $\delta_2 \le k_2/N_1$, then (N_1,0) is an equilibrium with global asymptotic stability.

For the equilibrium (x_1^*, x_2^*), we have the following theorem for its global asymptotic stability.

Lemma 2 [3]. For the following equations (13),

$$\frac{dx}{dt} = f(t,x),\ f(t,0)=0,\ x\in R^n \tag{13}$$

If $V(t,x)$ is a positive definite function in the field $I\times R^2$ with infinitesimal upper bound and infinite lower bound, and $\dot{V}(t,x)$ is a negative definite function, then the equilibrium of equations (13) has uniformly and globally asymptotic stability.

Theorem 3. If $\delta_2 > k_2/N_1$, then (x_1^*, x_2^*) is an equilibrium with uniformly and globally asymptotic stability.

Proof. If $\delta_2 > k_2/N_1$, (x_1^*, x_2^*) is existent.

Analyze the expression (4), then reason out those expressions,

$$k_1 = \frac{k_1}{N_1}x_1^* + \delta_1x_2^*,\ k_2 = -\frac{k_2}{N_2}x_2^* + \delta_2x_1^* \tag{14}$$

According to (13), convert the form of model (2) into (15):

$$\begin{cases} \frac{dx_1}{dt} = x_1\left[\frac{k_1}{N_1}(x_1^*-x_1) + \delta_1(x_2^*-x_2)\right] \\ \frac{dx_2}{dt} = x_2\left[\frac{k_2}{N_2}(x_2^*-x_2) - \delta_2(x_1^*-x_1)\right] \end{cases} \tag{15}$$

Let $V(t,x_1,x_2)=\sum_{k=2}^{2}C_k\left(x_k-x_k^*-x_k^*\ln\frac{x_k}{x_k^*}\right)$, where C_1 and C_2 are positive constants and $C_1\delta_1-C_2\delta_2=0$.

1) This part is to prove that $V(t,x_1,x_2)$ is a positive definite function in the field $I\times R_+^2$.

$$V(t,x_1^*,x_2^*)=0\,,\quad \frac{\partial V(t,x_1,x_2)}{\partial x_k}=\frac{C_k\left(x_k-x_k^*\right)}{x_k}\,,$$

$$\frac{\partial^2 V(t,x_1,x_2)}{\partial x_k^{\ 2}}=\frac{C_k x_k^*}{x_k^{\ 2}}\,,\quad \frac{\partial^2 V(t,x_1,x_2)}{\partial x_1\partial x_2}=0 \tag{16}$$

Therefore, $V(t,x_1,x_2)$ is a positive definite function in the field $I\times R_+^2$.

2) This part is to prove that $V(t,x_1,x_2)$ is a function with infinitesimally upper bound.

Let $W_1(t,x_1,x_2)=\sum_{k=2}^{2}C_k\left(x_k-x_k^*\ln\frac{x_k}{x_k^*}\right)$, it gives

$$W_1(t,x_1^*,x_2^*)=C_1x_1^*+C_2x_2^*,\quad \frac{\partial W_1(t,x_1,x_2)}{\partial x_k}=\frac{C_k(x_k-x_k^*)}{x_k}\,,$$

$$\frac{\partial^2 W_1(t,x_1,x_2)}{\partial x_k^{\ 2}}=\frac{C_k x_k^*}{x_k^{\ 2}}\,,\quad \frac{\partial^2 W_1(t,x_1,x_2)}{\partial x_1\partial x_2}=0 \tag{17}$$

Therefore, $\forall V(t,x_1,x_2)\in I\times R_+^2$, $W_1(t,x_1,x_2)\ge C_1x_1^*+C_2x_2^*$, and $W_1(t,x_1,x_2)>V(t,x_1,x_2)>0$. So $W_1(t,x_1,x_2)$ is a positive definite function in the field $I\times R_+^2$.

It gives $V(t,x_1,x_2)$ is a function with infinitesimal upper bound.

3) This part is to prove that $V(t,x_1,x_2)$ is a function with infinitely lower bound.

Let $W_2(t,x_1,x_2)=V(t,x_1,x_2)/2$. According to that $V(t,x_1,x_2)$ is a positive definite function in the field $I\times R_+^2$, it gives that $W_2(t,x_1,x_2)<V(t,x_1,x_2)$, and $W_2(t,x_1,x_2)$ is also a positive definite function. And $\lim_{x_1,x_2\to\infty}W_2(t,x_1,x_2)=\infty$. Therefore, $W_2(t,x_1,x_2)$ is an infinite positive definite function.

It gives $V(t,x_1,x_2)$ is a function with infinite lower bound.

4) This part is to prove that $\dot{V}(t,x_1,x_2)$ is a negative definite function in the field $I\times R_+^2$.

$$\dot{V}(t,x_1,x_2)=\frac{\partial V(t,x_1,x_2)}{\partial x_1}\frac{dx_1}{dt}+\frac{\partial V(t,x_1,x_2)}{\partial x_2}\frac{dx_2}{dt}=-\frac{C_1k_1}{N_1}\left(x_1-x_1^*\right)^2-\frac{C_2k_2}{N_2}\left(x_2-x_2^*\right)^2 \tag{18}$$

Therefore, $\dot{V}(t,x_1,x_2)$ is a negative definite function in the field $I\times R_+^2$.

According to the Lemma 2 and above analysis, if $\delta_2>k_2/N_1$, $\left(x_1^*,x_2^*\right)$ is an equilibrium with uniformly and globally asymptotic stability.

4. Numerical Simulation

For the Logistic-Volterra model (2), let $k_1=1$, $k_2=0.8$, $N_1=100$, $N_2=25$, $\delta_1=0.2$, $\delta_2=0.004$, then $\delta_2\le k_2/N_1=0.008$. The simulation results of this case are shown in figure 2.

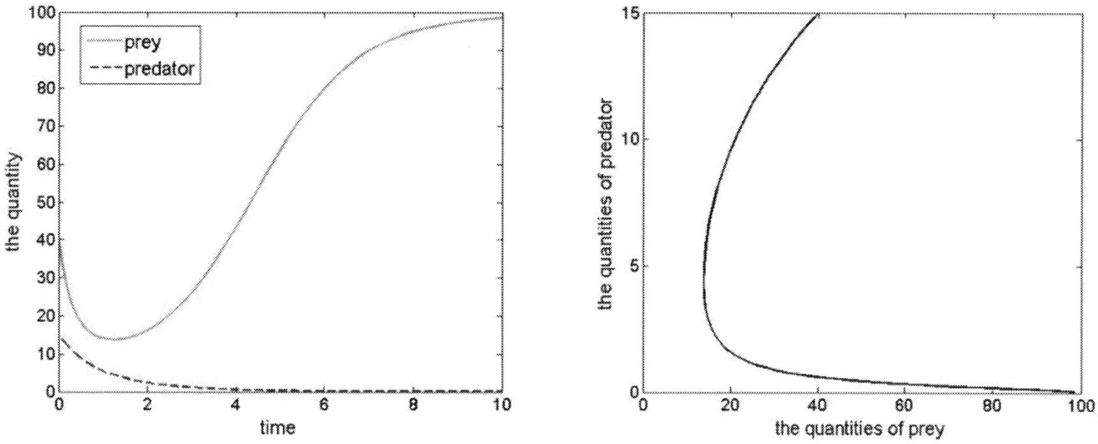

Figure 2. Quantity changes of prey and predator for the model (2) if $\delta_2 \leq k_2/N_1$.

In figure 2, it gives that if $\delta_2 \leq k_2 / N_1$, (N_1,0) is an equilibrium with global stability, and the quantities of prey and predator will change towards it without fluctuations. This means if $\delta_2 \leq k_2 / N_1$, prey and predator can't exist at the same time after a period, and predator will gradually disappear. In this case, although food is plentiful for the predator, it is mainly the survival pressure within the population that affects the development of the predator population and ultimately leads to its extinction.

For the model (2), let $k_1 = 1$, $k_2 = 0.8$, $N_1 = 100$, $N_2 = 25$, $\delta_1 = 0.2$, $\delta_2 = 0.04$, then $\delta_2 > k_2 / N_1 = 0.008$. The simulation results of this case are shown in figure 3.

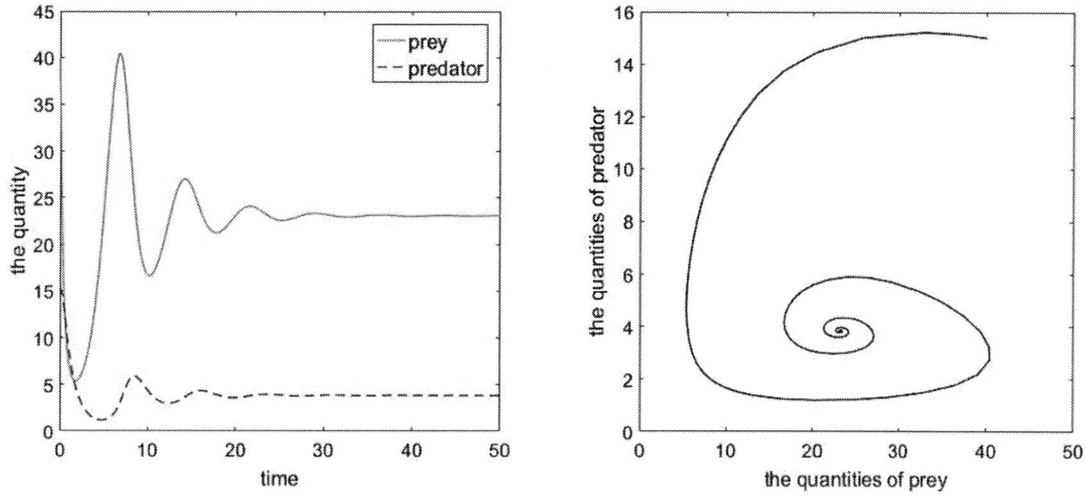

Figure 3. Quantity changes of prey and predator for the model (2) if $\delta_2 > k_2/N_1$.

In figure 3, it gives that if $\delta_2 > k_2 / N_1$, $\left(x_1^*, x_2^*\right)$ is an equilibrium with global stability, and the quantities of prey and predator will fluctuate towards it. And the fluctuating time for the situation $\delta_2 > k_2 / N_1$ is longer than the time for $\delta_2 \leq k_2 / N_1$. In the situation $\delta_2 > k_2 / N_1$, prey and predator will exist stably in a certain proportion which depends on k_1, k_2, δ_1, δ_2, N_1 and N_2 after a period. In this case, for predators, food and internal survival stress alternate as dominant factors and finally tends to balance.

5. Conclusions

According to the analysis of this paper by Liapunov function or other principle, for the predator-prey model (2), (N_1,0) is an equilibrium with global asymptotic stability and predator will gradually disappear after a period if $\delta_2 \le k_2 / N_1$. (x_1^*, x_2^*) is an equilibrium with uniformly and globally asymptotic stability and prey and predator can exist stably in a certain proportion after a period if $\delta_2 > k_2 / N_1$. And this proportion depends on k_1, k_2, δ_1, δ_2, N_1 and N_2, not on the initial quantities of prey and predator. That is, for the same predator-prey system, no matter how much the initial quantities of prey and predator are, the equilibrium is same.

δ_1 represents the ability of the predator to cope with reproductive pressure. If $\delta_2 > k_2 / N_1$, this ability is greater than the total reproductive pressure on the species, and the predator will not become extinct. In industries such as agriculture, forestry, animal husbandry and fishery, it is necessary to reduce the total reproductive pressure on economically harmful species such as pests to let it, and increase the total reproductive pressure on economically harmful species such as fish, cattle and sheep to get more revenue.

References

[1] Z. G. Du. Has the Holling IV type functional response of fractional order kinetics analysis of prey-predator model [J]. Journal of Jinggangshan university (natural science edition), 2019, 40(03):9-13+23.
[2] W. Q. Yi, G. R. Liu. Qualitative analysis of a class of random time-delay predator-prey models [J]. Journal of Chongqing university of technology (natural science), 2019, 33(03):202-208.
[3] C. Wang, F.Q Zhang, H.W Liu. Global state analysis of a predator-prey model with phase structure [J]. Practice and understanding of mathematics, 2018, 48(03):301-307.
[4] C. R. Xue. Global stability of the SEIR model of predator disease [J]. Computer engineering and applications, 2015,51(23): 74-77.
[5] Y. N. Xiao, Y. C. Zhou, S.Y. Tang. Principles of mathematical biology, Xi 'an: Xi'an Jiaotong University Press. 2012, pp.42-45.
[6] Z. G. Li, S. K. Kang. The new model of population stable development, Journal of Shandong Normal University (Natural Science) 27, pp. 25-27 (2012).
[7] Z. E. Ma, Y. C. Zhou, C. Z. Li. Qualitative and stability methods of ordinary differential equations, Beijing: Science Press, 2015, pp.70-77.
[8] X. X. Liao. Mathematical theory and application of stability, Wuhan: Central China Normal University Press, 2001, pp.58
[9] L. L. Yu. Global asymptotic stability of a predation model, Journal of Huaihai Institute of Technology (Natural Science) 46, pp. 4-6 (2008).

Failure Analysis of the Semi-permanent Coupler Used in Metro Rail

M A Morsy[1,*], S Khafagy[2] and ES Mosa[3]

[1]Welding and NDT Dept., CMRDI, Cairo, Egypt
[2]TIMS, Cairo, Egypt
[3]Al-Azhar University, Faculty of Engineering, Mining and Petroleum Engineering Dept., Egypt

*Corresponding author

Abstract. Failure analysis of the semi-permanent coupler which failed in service of Metro rail was conducted in order to know the reason for failure. The examinations include visual, photo detection, chemical analysis, hardness measurements, tensile and v-notch impact test, microstructure and SEM observation of fracture surface. Chemical analysis and tensile test shows that the coupler material can be classified as grade B cast steel in AAR M201 steel casting specification. Although the microstructure of the coupler material shows incomplete homogenization indicating the improper conduction of heat treatment, the v-notch impact results is accepted according to this specification. The fracture surface shows the existence of much porosity at the internal surface of the hollow like pipe- coupler and the fracture started at these sites showing beach marks. SEM observation of the fracture surface shows fatigue striations indicating the occurrence of fatigue started at the porosity defects. The cast defects act as a stress concentration site for fatigue crack initiation. Crack occurred, propagated, and connected the defects to each other. The occurrence of collision resulted in bending of the female part of the coupler (plastic deformation occurs) at the lower thin wall (~ 5.41 mm thickness). Moreover, opening of the previously formed fatigue crack and accelerating the failure were occurred. Recommendations were explained to prevent the occurrence of this failure.
Keywords: failure analysis; coupler; chemical analysis; mechanical testing; microstructure; fracture surface; fatigue; porosity; crack

1. Introduction

The automatic coupling Scharfenberg coupler was designed in 1903 [1].Firstly, it was used in transit trains and later on, it was utilized in passenger service trains. The main advantage of this coupler is the applying of electrical and pneumatic engagement and automatic disengagement. Otherwise, the major disadvantage is its low maximum tonnage, which is unsuitable for cargo operations.

This coupler connects two cars together in order to transfer the axial compressive and tensile loads and to damp the received impact by each car through shock absorption that increases the crash worthiness. Any increase in train's speed causes the acting loads to be augmented. This will lead to reduce the lifetime of the coupler and increase the damage intensity of the coupler [2]. Also, semi permanent coupler now has anti climbing function that reduces the risk of car damage and increase the survival of passengers [3].

Seshu et al [4] changed the design of the rail way coupler of AAR type E in order to avoid the high stress concentration in the knuckle of this coupler. Huang et al [5] discovered that aside from the coarse dendritic cast structure containing porosity, the wrong repair welding procedure may affect on the formation of micro-cracks that propagated during service and connected to the cast defects and as a result the knuckle failed.

(a) Male coupler **(b) Female coupler**

(c) Connection of coupler **(d) Female part after fracture**

(e) Deformation of female part **(f) fracture surface at female part**

Figure 1. Semi-permanent coupler before and after fracture.

In Egypt metro train, a fracture was occurred at the train coupler, which resulted in a complete stop of the metro train. Central Metallurgical research and Development Institute was asked to investigate the cause of failure of the coupler.

The received coupler had failed at the female coupler shank (without shock absorption) as shown in Figs.1a to 1f. The aim of this study is to determine the failure causes of the coupler with regarding to the metallurgical and mechanical point of view.

2. Experimental Procedure

The fracture parts of the screw coupler were first subjected to visual examination, photographed by camera, fracture surface was macro-photographed to determine the origins of fracture and the direction of crack growth. Ultrasonic cleaning was conducted to the fracture surface and then observed by Scanning Electron Microscope (SEM). The material was analyzed using optical emission spectrometry. The material was grinded and polished for observation of microstructure by optical microscope after etching with Nital. Tensile testing was conducted using universal testing machine according to ASTM -E8 M. and V-notch impact test was conducted according to ASTM E23-12.

3. Results

3.1. Chemical and Mechanical Testing of Fractured Coupler Material

Table1 explains the chemical analysis of the coupler material. Chemical composition shows that the coupler material meet the requirement of AAR-M201 steel casting specification Grade B cast steel [12].

Table1. Chemical analysis of the coupler material, wt%.

C	Si	Mn	P	S	Cr	Mo	Ni	Al	Bal.
0.26	0.48	1.29	0.019	0.02	0.11	0.017	0.06	0.016	Fe

3.2. Microstructure of the Coupler Material

Figure 2 shows the microstructure of the coupler material. It is ferrite and perlite structure in the cast form. The formation of pearlite at the austenite grainboundary, Fig. 2 indicating that the normalizing process does not conduct properly.It is decided that the microstructure of tempered martensite has a high fatigue resistance than normal ferrite-pearlite structure [8, 9, 11]. Also, the grain refinement structure has a high resistance to fatigue faluire[8,9].

3.3 Mechanical Properties of the Coupler Material

Mechanical properties were determined from the reminder part of the coupler with a thickness of 9.5 mm and the test result is shown in Table 2 and 3. The mechanical properties fulfill the requirement of AAR-M201 steel casting grade B.

Table 2. Tensile test properties and hardness of the coupler material.

Average ultimate tensile strength, Mpa	Average Yield Strength, Mpa	Average Elongation percentage, %	Average hardness, HV
537	382.7	28	170

Table 3. V-notch impact test results of coupler material

Specimen number	A, J	B, J	C, J	Average, J
	36	38	34	36

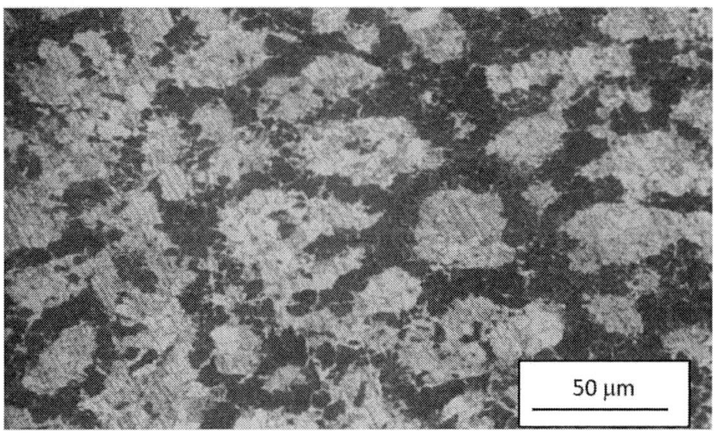

Figure 2. Microstructure of the link material (ferrite-pearlite structure).

3.4. Fracture Surface of Coupler

Figure 3 shows the fracture surface of the coupler (the coupler like a tube and this figure represent the part that has the origin of cracks). The existence of surface porosity which is a cast defect may acts as a stress concentration sits for initiation of fatigue cracks [3-8]. Beach marks were observed at the initiation sites as shown in Fig. 3.

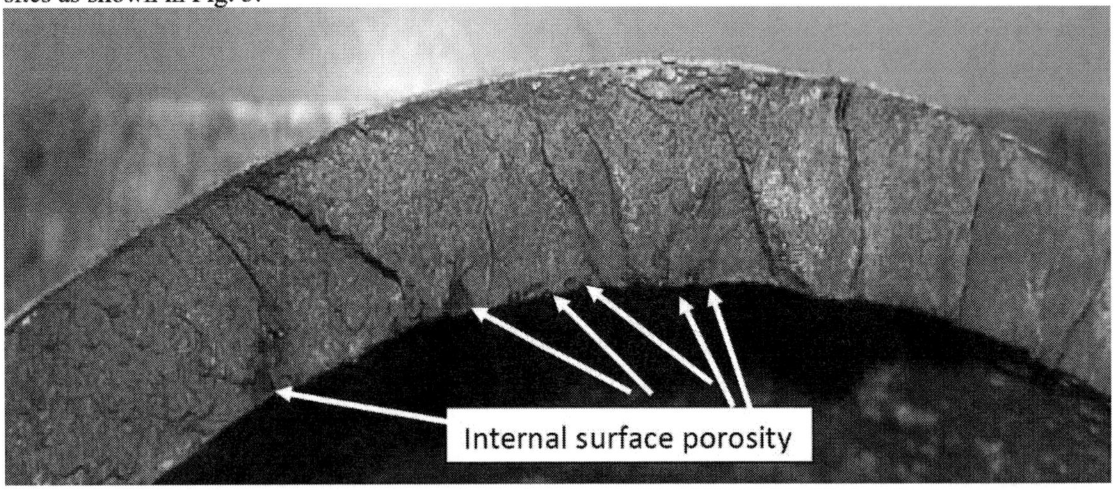

Figure 3. Crack initiation site showing the existence of internal surface porosities at the same line.

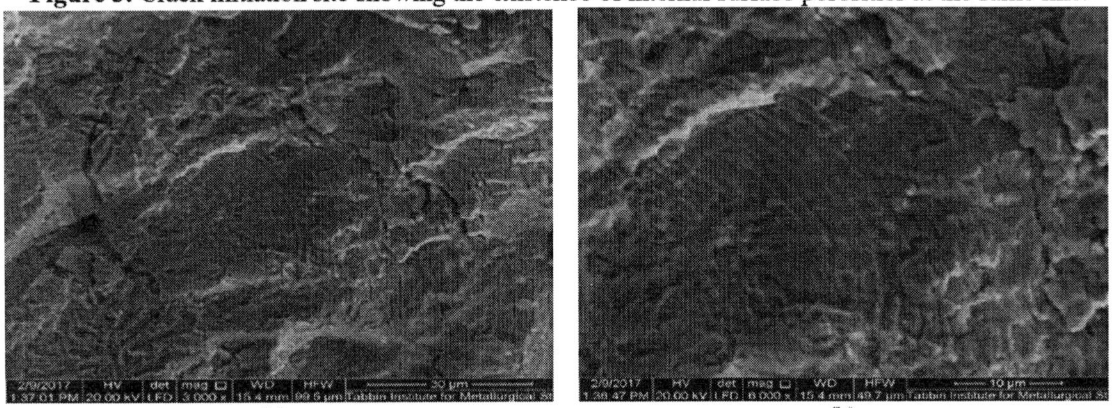

(a) (b)

Figure 4. SEM observation showing the fatigue striations of the fracture surface at the starting point of crack, a) low magnification and b) High magnification.

Figure 4 shows the SEM observation at the initiation site of fracture surface shown in Figs. 1f and 3. Fatigue striations was observed that indicating the fatigue failure mechanism (see Figs. 4 a & b).

Figure 5. SEM observation of the crack propagation zone.

Figure 5 shows the SEM observation of crack propagation zone where the secondary cracks were observed. The cast defects act as a stress concentration site for fatigue crack initiation [4-7]. The fatigue cracks start at the cast defects as shown in Fig. 3. Crack occurs and propagates, and connects the defect to other defects [5, 6]. The occurrence of collision resulted in bending of the female part (plastic deformation occurs) of the coupler at the lower thin wall (~ 5.41 mm thickness) as shown in Fig. 1e. Moreover, this also causes opening of the previously formed fatigue crack and accelerating the propagation and termination of the fatigue failure as shown in Figs. 1 and 3.

Checking for the existence of internal porosity in the other new couplers resulted in the observation of much internal porosity in many couplers that may affect on the occurrence of fatigue and a decrease in the fatigue resistance of the coupler.

4. Concluding Remark and Recommendation

From the foregoing results of internal surface finish of the coupler (very rough surface finish with the existence of many surface porosities), the improper homogenization of microstructure as a result improper heat treatment and the abrupt change in the velocity of metro train; all contribute to the occurrence of fatigue failure of coupler. The following recommendation should be applied:

1. The coupler standard concerning the material specification, heat treatment processes, mechanical properties, dimension, surface finish, and inspection criteria should be applied before, during and after manufacture of the coupler.
2. Regular check for the existence of fatigue cracks is important to prevent the catastrophic failure of the coupler.
3. The existence of internal or external surface porosity has a great adversely affect on the fatigue resistance.

References

[1] Arthurton R..I.D., "Automatic couplers for railway Rolling Stock and Continuous Braking of Freight Trains", Proceeding of the Institution on Mechanical Engineers, Vol. 156,1: pp74-78, Frist published Jun 1, 1947.

[2] http://voith.com/corp-en/connection-components-couplings/railway-couplers.html

[3] Daunys P., Putnaite D. Stress strain analysis for railway carriage automatic coupler SA3. Mechanika 2004; 2:46.

[4] Chunduru S. P., Kim M. J., Mirman C., failure analysis of rail road couplers of AAR type E, "Engineering failure analysis journal, 18 (2011) 374-385.

[5] Huanhang J., Xia L., Zhang Y. and Li S., "Investigation on brittle fracture mechanism of a grade E cast steel knuckle" Case studies in Engineering failure analysis journal 2(2014)15- 24.

[6] Hertzberg RW. Deformation and fracture mechanics of engineering materials. John Wiley & Sons; 1996.

[7] Brooks CR., Choudhury A. Metallurgical failure analysis. New York: McGraw Hill; 1993.
[8] Fatigue and microstructure, in proceeding of the ASM Materials Science Seminar, American Society of Metals, 1979.
[9] ASM Handbook Volume 1 Properties and selection: Iron, steels, and High performance alloys ASM Handbook Committee, Fatigue Resistance of Steels, 1990 P673-688.
[10] Mousa S.M., Noughabib V., Dehghania K., Pouranvaric M.. "Failure analysis of automatic coupler SA-3 in railway carriages". Engineering Failure Analysis Journal 14 (2007) 903–912
[11] ASM Handbook Volume 11 Failure Analysis and Prevention ASM Handbook Committee, Failure related to casting, (1990) 242-255.
[12] AAR_M201 Steel Casting Specification Grade B cast steel for coupler and freight car draft components (2012).

IOP Conf. Series: Materials Science and Engineering **611** (2019) 012029 doi:10.1088/1757-899X/611/1/012029

Assessing the Dispersion Quality of Carbon Nanotubes Suspensions by Low Field Nuclear Magnetic Resonance

Yongping Huang, Shuxian Hong, Weiwen Li* and Shijie Guo

Guangdong Provincial Key Laboratory of Durability for Marine Civil Engineering, College of Civil Engineering, Shenzhen University, Shenzhen, China

*Corresponding author

Abstract. In this study, a novel method based on low field nuclear magnetic resonance measurements were used to characterize the dispersion quality of carbon nanotube suspensions. The liquid specific surface area, which calculated with the relaxation time, were used as an indicator of dispersion quality of the carbon nanotubes suspensions. The experimental results were compared with those measured by the conventional method methods, including Zeta potentiometer method and ultraviolet visible spectrophotometer method. The results show that low field nuclear magnetic resonance has the advantages in the characterizing the dispersion quality of high concentration carbon nanotube suspensions and tracking its re-agglomeration process non-destructively.

1. Introduction

Previous studies have shown [1-3] that carbon nanotubes (CNTs) are nanomaterials with low density, high strength, high hardness, high elastic modulus and excellent electrical properties. Carbon nanotubes, which are widely used as new nanomaterials, are widely used to reinforce matrix materials. And many scholars in the field of building materials have applied them to cement-based materials. Many studies have shown that [4-6], with the addition of carbon nanotubes, the flexural strength, compressive strength, freeze-thaw resistance, damping and microscopic properties of cement-based materials have improved. However, CNTs are hardly soluble in water and organic solvents. And it has a high van der Waals force and aspect ratio, so the phenomenon of intertwining and agglomeration between carbon nanotubes is very easy [7]. The uniform dispersion of CNTs is a prerequisite for the better use of CNTs.

Preparation of homogenous dispersed carbon nanotubes suspension is the premise of homogenous dispersion of carbon nanotubes in cement-based composites. It is critical to find a means to quickly and accurately assess the dispersion quality of carbon nanotubes. The Ultraviolet visible (UV -Vis) spectrophotometer method [8] and the Zeta potentiometer method [9] are currently used to characterize the dispersion effect of carbon nanotube suspensions. But both methods of characterization have limitations. The Zeta potentiometer method has requirements on the conductivity and potential of the dispersion itself, which greatly limits the scope of application. The UV-Vis spectrophotometer method needs to dilute the stock solution, which may destroy the original dispersion state of the carbon nanotubes in the stock solution. Due to the deficiencies of the above methods, it is necessary to find a method which can quickly and accurately determine the dispersion stability of carbon nanotubes in solution.

Baranowska-Korczyc et al. [10] observed that the improvement in dispersibility of PEG-MWNCTs / Fe was associated with an increase in the relaxation rate of water. Since CNTs have a strong influence on the molecular structure of water in solution, Davide Porrelli et al. [11] used low field nuclear magnetic resonance (LF-NMR) to characterize aqueous dispersions of different covalently functionalized multi-walled carbon nanotubes. It has been found that the transverse relaxation rate r_2 of water molecules can be used to assess the concentration and dispersion of CNTs. Studies have shown that the relaxation rate and transverse relaxation time measured by LF-NMR techniques can characterize the specific surface area of suspended particles [12]. The surface of the imaginary suspended particles is adsorbed with a layer of water molecules having a thickness of L, which is adsorbed water. The water molecules are free water outside the layer. According to the relaxation time of water molecules in two different states, the average relaxation rate ($R_2 = \frac{1}{T_2}$) of the suspension can be obtained as:

$$R_2 = P_b R_{2b} + P_f R_{2f} = P_b (R_{2b} - R_{2f}) + R_{2f} \tag{1}$$

where P_b is the ratio of the bound solvent; R_{2b} is the relaxation rate of the bound solvent; P_f is the ratio of the free solvent; and R_{2f} is the relaxation rate of the solvent.

Therefore, the specific surface area of the suspended particles depends on the specific surface area of the surface adsorbed water thickness L. The relaxation time measured by LF-NMR technology is related to the specific surface area of the material [13]. Therefore, the specific surface area (S_a) of the suspended particles can be calculated using the transverse relaxation time (T_2) measured by LF-NMR. Its calculation equation (2) is:

$$S_a = \frac{R_2 - R_{2f}}{K_p \cdot \varphi} \tag{2}$$

Where K_p is the specific surface area relative coefficient and φ is the volume ratio of suspended particles to solvent.

These findings have focused our attention on whether the specific surface area of suspended particles measured by LF-NMR techniques can be used to characterize the dispersion stability of carbon nanotube suspensions.

This study used the specific surface area measured by LF-NMR to characterize the dispersion quality of carbon nanotube suspensions. Then, the dispersive quality of the carbon nanotube suspension was characterized by Zeta potential method and UV-Vis spectrophotometer to verify the accuracy of the LF-NMR test results. Then, the advantages and disadvantages of these characterization methods are summarized.

2. Experimental Materials and Methods

2.1. Materials

Carbon nanotubes are multi-walled carbon nanotubes and produced by the Chinese Academy of Sciences Chengdu Organic Chemistry Co., Ltd. by chemical vapor deposition (CVD). The basic parameters of multi-walled carbon nanotubes are shown in Table 1. The experimental water uses deionized water produced by Huanan High-Tech Water Treatment Equipment Co., Ltd. Surfactant (SAA): Polyvinylpyrrolidone (PVP), produced by Shanghai Jingchun Biochemical Technology Co., Ltd.; Arabic gum powder (GA), produced by Sinopharm Chemical Reagent Co., Ltd.

Table 1. Basic parameters of carbon nanotubes

Outer diameter (nm)	Inner Diameter (nm)	Length (μm)	Special Surface Area (m^2/g)	Tap density (g/m^3)	True density (g/m^3)	Purity (%)	Electric Conductivity (S/cm)
20~30	5~10	10~30	>110	0.28	~2.1	>98	>100

2.2. Preparation of Carbon Nanotube Solution

Three kinds of carbon nanotube suspensions were prepared in this experiment, named C1, C2, and C3. The preparation ratio is shown in Table 2. First, 0.6 g of each surfactant was added to a beaker containing 200 ml of deionized water, and stirred with a glass rod for 5 minutes. After the surfactant was completely dissolved, the carbon nanotubes were directly added, and a magnetic stirrer (model RH digital, Stirred by Ningbo Xinzhi Technology Co., Ltd.) at a rate of 600 rpm for 15 minutes. Finally, it was placed in an ultrasonic cell pulverizer (model JY92-IIN, produced by Ningbo Xinzhi Technology Co., Ltd.) with a power adjustment of 60%. The ultrasonic wave was treated for 3 seconds at intervals of 3 seconds for 30 minutes.

Table 2. Formulation of Carbon Nanotubes Dispersions

Sample	CNTs(g)	GA(g)	PVP(g)	water(ml)	Sonication time(min)
C1	0.2	-	-	200	30
C2	0.2	0.6	-	200	30
C3	0.2	-	0.6	200	30

Note: GA: gum arabic powder; PVP: polyvinylpyrrolidone;

20 ml of the dispersed carbon nanotube suspension was poured into a 30 ml centrifuge tube. Then, the centrifuge tube was placed in a centrifuge (the centrifuge model was TG16-WS, produced by Xiangyi Centrifuge Instrument Co., Ltd.). It was centrifuged at 8000 rpm for 10 min, 30 min and 60 min, named C1-10, C1-30, C1-60, C2-10, C2-30, C2-60, C3-10, C3-30 and C3-60.

2.3. Characterization of Dispersion Quality of Carbon Nanotube Suspensions

The dispersion quality of the above carbon nanotube suspensions was characterized by using New Size's particle size specific surface area analyzer model PQ001. 1.5 ml of the dispersed carbon nanotube suspension was placed in a glass sample tube, and then the glass sample tube containing the sample was inserted into the instrument probe coil to start the test. The specific surface area (Sa) measurement of the suspended particles was carried out at 32 ° C using a CPMG (Carr-Purcell-Meiboom-Gill) sequence.

The dispersed carbon nanotube suspensions C1, C2 and C3 were diluted 100-fold. Immediately measure a certain volume of suspension and characterize the dispersion quality of the carbon nanotube suspension by Zeta potential meter (Model Zetasizer Nano-ZS90, produced by Malvern Instruments, UK)and UV-vis spectrophotometer(Model Lambda 750 is produced by American PerkinElmer company).

3. Results and Discussion

3.1. Dispersion Quality of Carbon Nanotube Suspensions with Different Dispersants

Figure 1 shows the three dispersed carbon nanotube suspensions doped with different dispersants. Since the carbon nanotubes were well dispersed in the suspension just after sonication, and no precipitation occurred, the dispersion quality of the three sets of carbon nanotube suspensions could not be directly distinguished by the naked eye. Therefore, it is necessary to use a relatively sophisticated instrument to characterize its dispersion quality.

Figure 2 shows the dispersion quality of carbon nanotube suspensions doped with different dispersants was characterized by LF-NMR. The transverse relaxation time of the carbon nanotube suspension measured by LF-NMR. The relaxation time was calculated according to the equation (2) to obtain the liquid specific surface area of each suspension, and Figure 3 shows the results. The dispersed carbon nanotubes have a larger specific surface area than the carbon nanotubes which are agglomerated together. Therefore, the dispersion quality can be judged according to the specific surface area of the carbon nanotube suspension. According to this principle, it can be concluded from figure 3 that the dispersion quality of the three groups of dispersions is C2, C3, and C1 from high to low. That is, the carbon nanotube suspension with the addition of gum arabic powder has the best dispersibility, followed by the addition of polyvinylpyrrolidone. Carbon nanotube suspensions without the addition of a dispersant have the worst dispersibility.

Figure 1. Dispersed carbon nanotube suspension mixed with different dispersants.

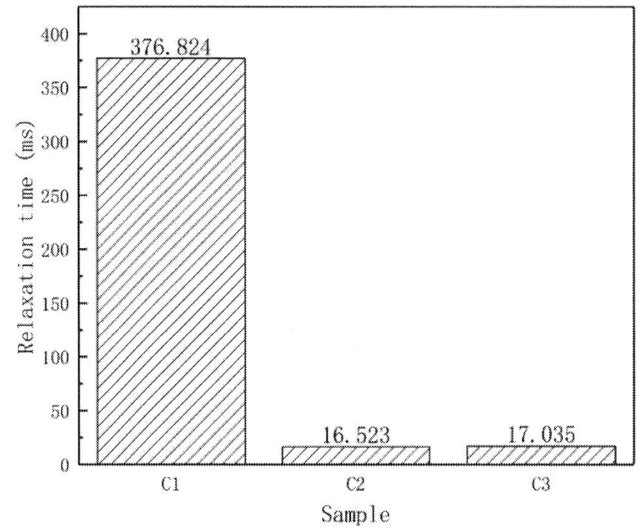

Figure 2. Relaxation time of carbon nanotube suspensions with different dispersants.

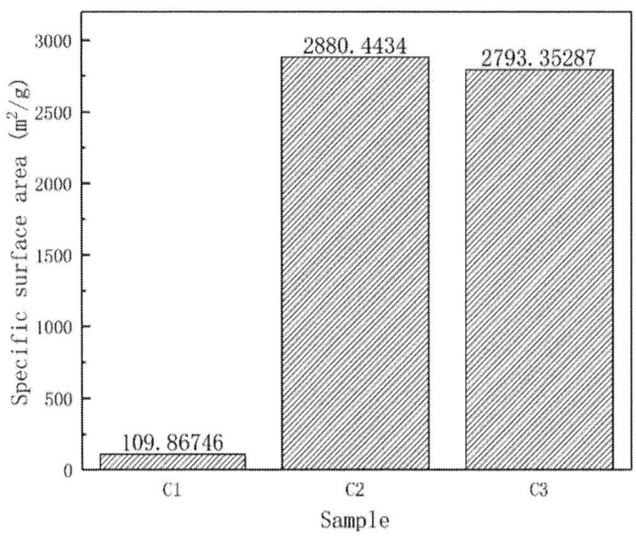

Figure 3. Specific surface area of carbon nanotube suspensions with different dispersants added.

UV-Vis spectrophotometer and Zeta potential meter are currently the most common methods for characterizing the dispersion quality of carbon nanotube suspensions. This study uses both characterization methods to verify the correctness of the LF-NMR method.

The Zeta potential value can reflect the structure and characteristics of the diffused electric double layer in the system. It is an important indicator of the electrokinetic properties of colloids. The higher the electron density of the surface of the nanoparticles, the greater the electrostatic force generated, the more effectively the agglomeration between the particles is prevented from dispersing well [14]. Figure 4 shows the results. The potentials measured by the carbon nanotube suspensions were all negative, the potential of C2 was the lowest, and the potential of C1 was the highest. It can be seen that the dispersion quality of each dispersion is C2, C3, and C1 from high to low. High concentrations of carbon nanotube suspensions also need to be diluted to lower concentrations for testing. However, the dispersion state of the diluted carbon nanotube solution may change to affect the accuracy of the detection result.

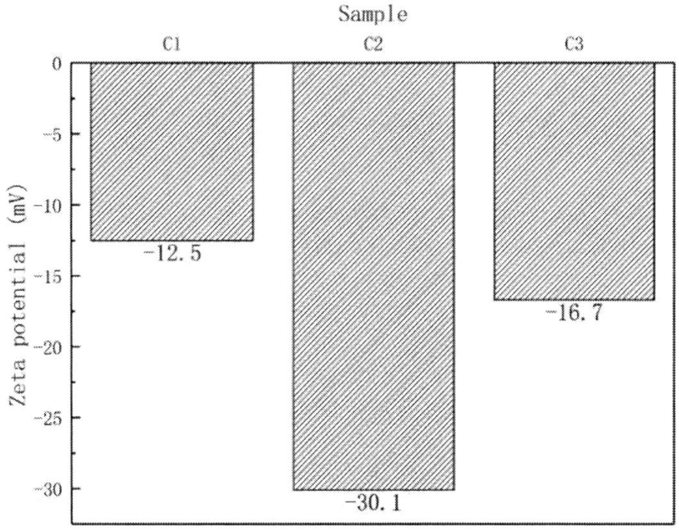

Figure 4. Zeta potentiometer test results.

The UV-Vis spectrophotometer method determines the quality of the dispersion by the absorption of the specific peaks of ultraviolet light and visible light by the carbon nanotubes in the suspension, and the more the specific peak absorption amount, the better the dispersibility.

According to the existing research, the absorption peak of carbon nanotubes is between 250 nm and 260 nm [15]. Due to the high concentration of the three samples, the dispersion quality of the three samples could not be distinguished directly by the UV-Vis spectrophotometer. So the three samples were diluted a hundred times. And figure 5 shows the analysis results of the UV-Vis spectrophotometer. Each suspension had the highest absorbance at a wavelength of 260 nm. The suspension containing the surfactant has a higher absorbance than the suspension without the surfactant, and the absorbance of the suspension C2 is the highest. Therefore, the degree of dispersion of the dispersion measured by UV-Vis spectrophotometer is C2, C3, and C1 in descending order.

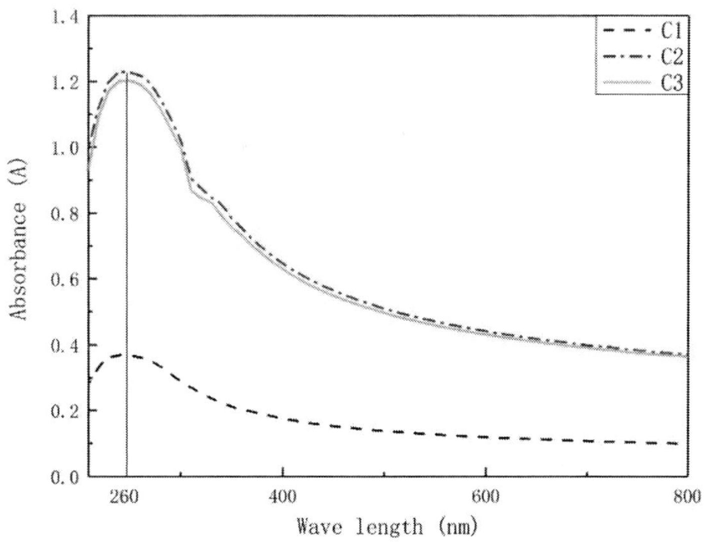

Figure 5. UV-Vis spectrophotometer test results.

The test results of Zeta potentiometer and UV-Vis spectrophotometer are consistent with the results obtained by LF-NMR in this experiment, and the accuracy of the results of LF-NMR test is verified. However, the test principle of the Zeta potentiometer is to judge the dispersibility according to the conductivity of the solution. The solution added with the surfactant has a charge affecting the conductivity of the carbon nanotube solution, which leads us to misjudge the dispersion quality of the carbon nanotube solution. Therefore, the Zeta potentiometer is not suitable for the characterization of the dispersion quality of carbon nanotube suspensions with strongly charged surfactants. The UV-Vis spectrophotometer method is not affected by the conductivity of the solution. However, the concentration of the carbon nanotube suspension has an effect on the UV-Vis spectrophotometer. Therefore, two different concentrations of carbon nanotube suspensions need to be diluted to the same concentration to compare their dispersibility. Therefore, the use of UV-Vis spectrophotometer to measure the dispersion of carbon nanotube solution has certain limitations. The characterization method of LF-NMR used in this study is not affected by solution concentration and conductivity. It can take the solution directly for testing, which is convenient and quick, and the test time is short. In summary, the use of LF-NMR to characterize the dispersion quality of carbon nanotube suspensions is accurate and has few influencing factors, which is a accurate characterization method.

3.2. LF-NMR Characterization of Suspension Dispersion Quality with Increasing Centrifugation Time

The LF-NMR method is further applied to investigate the re-agglomeration process of three groups of carbon nanotube suspensions. C1, C2 and C3 were centrifuged for 0 min, 10 min, 30 min and 60 min, respectively, to accelerate the re-agglomeration procession of the carbon nanotubes. The LF-NMR method was used to characterize the dispersion quality of the suspensions at different centrifugation times, and the dispersion quality of the suspension was observed with time. Figure 6 shows the suspension of carbon nanotubes at different centrifugation times. It can be clearly seen that the suspension of C1 group decreases with centrifugation time, as the carbon nanotubes precipitated. The re-agglomeration of C2 and C3 cannot be identified visually, which will be further investigated with LF-NMR and UV-Vis spectrophotometer.

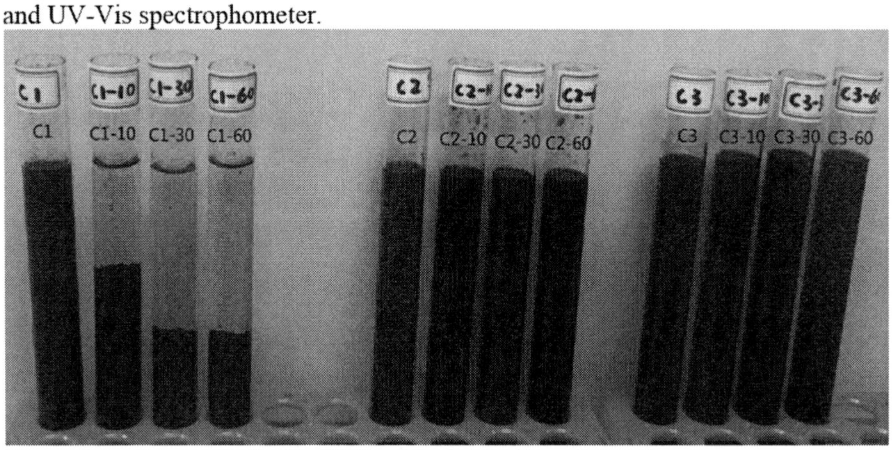

Figure 6. Comparison of dispersion quality of carbon nanotube suspensions under different centrifugation times.

The solution needs to be diluted when characterizing the dispersion quality of the carbon nanotube suspension using an UV-Vis spectrophotometer and a Zata potentiometer. However, the dilution process changes the precipitation state of the sample. The the re-agglomeration process can only be tested by LF-NMR. Figure 7 shows the specific surface area of a carbon nanotube suspension at different centrifugation times as measured by nuclear magnetic resonance. Because of the severe sedimentation of C1, the specific surface area value is much smaller than C2 and C3. The agglomeration of carbon nanotubes increased under centrifugation, so the specific surface area of the three samples decreased with time. The specific surface area of C2 decreased more slowly than C3, indicating that the C2 reagglomeration degree is lower than C3. So the LF-NMR can be also used to determine the rate of re-agglomeration of the carbon nanotube suspension. And the results again confirmed that C2 has the best dispersion quality, followed by C3 and finally C1. It further demonstrates the accuracy of characterizing the dispersion quality of carbon nanotube suspensions using LF-NMR.

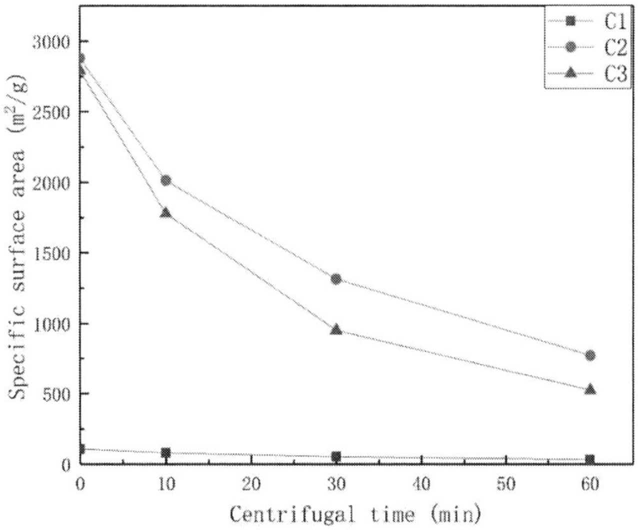

Figure 7. The specific surface area of the CNTs suspension as a function of centrifugation time.

4. Conclusion

In this study, a novel method based on LF-NMR were used to characterize the dispersion quality of carbon nanotube suspensions. The following conclusions were obtained:

- LF-NMR can be used to characterize the dispersion quality of carbon nanotube suspensions, and the results are in agreement with the UV-Vis spectrophotometer and Zeta potential.
- The Zeta potentiometer is not suitable for the determination of carbon nanotube suspensions as its test results are greatly affected by the conductivity of the suspension. The UV-Vis spectrophotometer needs to dilute the solution when testing the high concentration of carbon nanotube suspension. Therefore, the LF-NMR method has obvious advantages for the characterization of the dispersion quality of the high-concentration carbon nanotube dispersion.
- As the LF-NMR method does not require dilution of the sample, it can be used to determine the rate of re-agglomeration of the carbon nanotube suspension.

Acknowledgments

The authors would like to gratefully acknowledge Shenzhen Science and Technology Plan Projects (Project No.JCYJ20170818100641730) for financial support on this study.

References

[1] Salvetat J P, Bonard J M, Thomson N H, Kulik A J, Forr'o L, Benoit W and Zuppiroli L 1999 *Mechanical Properties of Carbon Nanotubes (*Applied Physics A) vol 69 (Holland: Materials Science Processing) pp 255–260

[2] *Yu M F,* Lourie O, Dyer M J, Moloni K, Kelly T F and Ruoff R S 2000 *Strength and Breaking Mechanism of Multiwalled Carbon Nanotubes Under Tensile Load* (Science) vol 287 (New York : Amer Assoc Advancement Science) pp 637–640

[3] Treacy M M J, Ebbesen T W and Gibson J M, 1996 *Exceptionally High Young's Modulus Observed for Individual Carbon Nanotubes* (Nature) vol 381 (London: Macmillan Magazines Ltd) pp 678-680

[4] Li W W, Ji W M, Wang Y C, Liu Yi, Shen R X and Xing F 2015 *Investigation on the Mechanical Properties of a Cement-Based Material Containing Carbon Nanotube under Drying and Freeze-Thaw Conditions (Materials) vol 8* (Switzerland: MDPI) *pp 8780-8792.*

[5] Li W W, Ji W M, Liu Y, Xing F and Liu Y K 2015 *Damping Property of a Cement-Based Material Containing Carbon Nanotube* (Journal of Nanomaterials) vol 2015 (England: Hindawi Ltd) pp 1-7

[6] Alafogianni P, Dassios K, Tsakiroglou C D, Matikas T E and Barkoula N M 2019 *Effect of CNT Addition and Dispersive Agents on the Transport Properties and Microstructure of Cement Mortars* (Construction and Building Materials) vol 197 (England: Elsevier Sci Ltd) pp 251-261.

[7] Harris P J F 2004 *Carbon Nanotube Composites* (International Materials Reviews) vol 49 pp 31-43

[8] Alafogianni P, Dassios K, Farmaki S, Antioho S K, Matikas T E and Barkoula N M 2016 *On the Efficiency of UV–vis Spectroscopy in Assessing the Dispersion Quality in Sonicated Aqueous Suspensions of Carbon Nanotubes* (Colloids and Surfaces A: Physicochemical and Engineering Aspects) vol 495 (Netherlands: Elsevier Science Bv) pp 118-124

[9] Hu H, Yu A P, Kim E, Zhao B, Itkis M E, Bekyarova E and Haddon R C 2005 *Influence of the Zeta Potential on the Dispersability and Purification of Single-Walled Carbon Nanotubes* (The Journal of Physical Chemistry B) vol 109 (Washington: Amer Chemical Soc) pp 11520-11524.

[10] Anna B K, Małgorzata J D, Barbara M M, Alicja W, L Emerson C, Tomasz Z, Krzysztof K K and Stefan J 2016 *PEG–MWCNT/Fe Hybrids as Multi-modal Contrast Agents for MRI and Optical Imaging* (RSC Advances) vol 6 (England: Royal Soc Chemistry) pp 49891-49902.

[11] Porrelli D, Cok M, Abrami M, Bosi S, Prato M, Grassi M, Paoletti S and Donati I 2017 *Evaluation of Concentration and Dispersion of Functionalized Carbon Nanotubes in Aqueous Media by Means of Low Field* nuclear magnetic resonance(Carbon) vol 113(England: Pergamon-Elsevier Science Ltd) pp 387-394.

[12] Yang P Q, Gao Y W, Yang Y, Hou H Y and Chen L 2016 *Test Method for Specific Surface Area of Particulate Matter in Suspension System* (China: patent)

[13] She A M and Yao W 2013 *Characterization of Microstructure and Specific Surface Area of Pores in Cement Paste by Low Field* nuclear magnetic resonance *Technique* (China: Journal of Wuhan University of Technology) vol 35 pp 11-15

[14] Wu J, Li T T, Qi T, Qin Q W, Li G Q, Zhu B L, Wu R and Xie C S 2012 *Influence of Dopants on Electrical Properties of ZnO-V2O5 Varistors Deduced from AC Impedance and Variable-Temperature Dielectric Spectroscopy* (Journal of Electronic Materials) vol 41(New York: Springer) pp 1970-1977

[15] Walters D A, Ericson L M, Casavant M J, Liu J, Colbert D T, Smith K Aand Smalley R E 1999 *Elastic Strain of Freely Suspended Single-wall Carbon Nanotube Ropes*(Applied Physics Letters)vol 74 (New York: Amer Inst Physics) pp 3803-3805.

Finite Element Analysis of Overhead Pipeline Truss Bridge under Thermal Fluid Loads

Guantang Lin[1], Si Huang[2,*], Qifeng Niu[3], Zhengnan Xu[4], Cong Zhang[5] and Yinghong Chen[6]

[1,6]Guangdong Institute of Special Equipment Inspection and Research Zhuhai Branch, Zhuhai, China, 519002
[2,3,4,5]School of Mechanical and Automotive Engineering, South China University of Technology, Guangzhou, China, 510641

*Corresponding author

Abstract: The subject used in this study is an overhead pipeline truss bridge which sits in a chemical industrial park. The overhead pipeline truss bridge is a part of a pipeline system consisting of 10 sets of columns, 4 layers of pipe racks and 27 pipelines that transports flowing media under various temperature and pressure conditions. An integrated ANSYS finite element model was built considering the factors such as gravity, medium temperature, and flow load, etc. The results of the distributions of displacement, axial force, shear force, bending moment and stress of the overhead pipeline truss bridge were calculated. The primary stress and secondary stress were checked to judge the reliability of the overhead pipeline truss bridge. It concluded that the computation model and calculation used in this study is feasible and effective. The feature in this paper is to explore a practical method and to provide a solid technological foundation for the safety operation and assessment of overhead pipeline truss bridge.
Keywords: overhead pipeline truss bridge; thermal fluid loads; finite element analysis.

1. Introduction

The overhead pipelines in the chemical industry park are generally several or even dozens of concentratedly laid on the pipe gallery to transport chemical raw materials and finished products under various temperature, pressure and flow conditions, and the span of the pipe gallery is large, and it will encounter deep trenches, roads, rivers and other areas with complex engineering geology. In order to avoid the influence of complex terrain on the pipeline, the pipe gallery is generally provided with a spanning device such as a pipeline truss bridge. In recent years, relevant investigations have carried out research work on pipeline truss bridges. Isting and Thier [1] performed a survey of criteria in the design of pipeline systems for the viewpoint of safety. They concluded that static and dynamic loads acting on the pipeline system owing to thermal stresses called for an analysis of the forces and moments as well as for possibilities of compensation. Ralph Alan Dusseau [2] studied the wind load response of two suspension pipeline bridges. The suspension pipe profiles and the pipeline stiffness were improved, then reducing the risk of vertical wind oscillation of the pipeline suspension bridge. Yanqiang Liu et al [3] analyzed effects of possible loads such as earthquake, temperature change, wind-force, weight of truss and pipe, pressure test and cross-sectional type and size of the truss on structural safety. Urdea M [4] studied about truss deformations to find the best form for trusses using Solidworks software. However, the above mentioned investigations for overhead pipeline truss bridge were mainly based on some commonly design principles, and lacked sophisticated and integrated computational methods and assessments.

In this work, an overhead pipeline truss bridge in a chemical industrial park in South China was selected as the research object. By taking into account the factors such as gravity, medium temperature, flow forces on the pipeline, the ANSYS finite element model was established systematically to perform the thermal fluid loads calculation. The article aims to explore a practical method for the safety assessment of a complex overhead pipeline truss bridge, and to provide a solid technological foundation for the safety operation of the overhead pipeline truss bridge.

2. Finite Element Model

2.1. Geometric Model

By simplifying the actual parameters, the geometric model of the overhead pipeline truss bridge is shown in Figures 1 and 2, for longitudinal and transverse directions respectively. The horizontal layout of the steel frame is shown in Figure 3. The pipe gallery had 10 sets of columns and 4 layers of pipe racks, the first layer of pipe rack had 5 pipes, the second layer had 3, the third layer had 3, the fourth layer had 16 and a total of 27 pipes. The pipe number is shown in Figure 2.

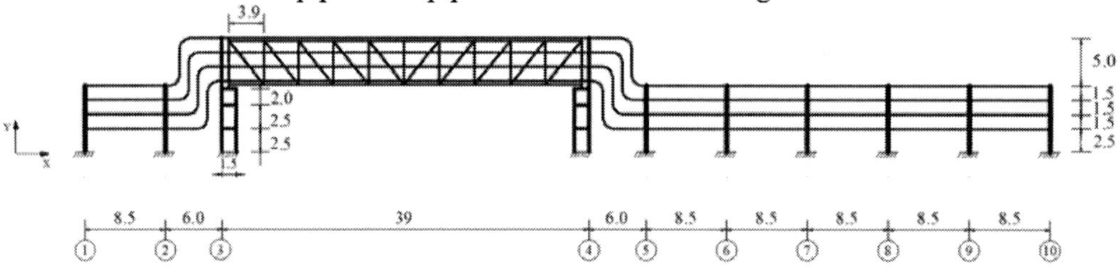

Figure 1. Longitudinal schematic of the model.

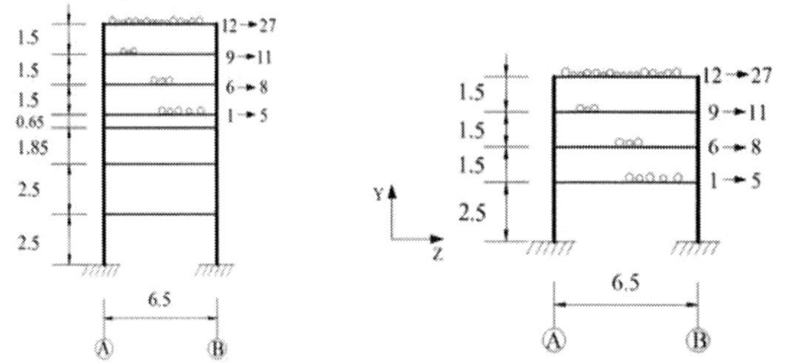

(a) Pipe racks 3 and 4 (b) The rest pipe racks

Figure 2. Longitudinal schematic of the model.

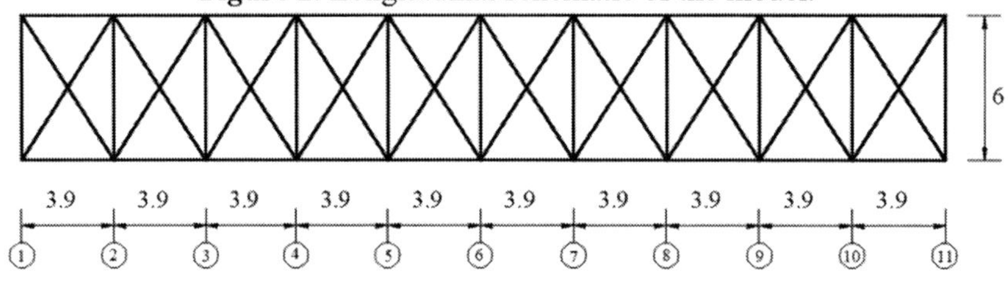

Figure 3. Horizontal layout of steel frame.

2.2. Element Type in Finite Element Method (FEM)

According to the layout of the overhead pipeline truss bridge shown in Figure 1, 2 and 3, the ANSYS Parametric Design Language (APDL) finite element model of ANSYS was established. Pipe racks and steel frame were modelled by using beam elements (Beam 189) and pipes modelled by using pipe elements (Pipe 289) [5-8]. Table 1 shows the specific data of pipe rack, steel frame and pipe elements, where B is the sectional length of the beam, H is the sectional height, D is the outer diameter of the pipe, δ is the thickness of the pipe, T_1 is the thickness of the steel frame web, and T_2 is the thickness of the steel frame wing.

2.3. Material Properties
The material constants used in the finite element model are shown in Table 2. The integrated overhead pipeline FEM model of ANSYS was built by the material of the structure, the element type and the element section size, which included 29780 elements and 58834 nodes. Figure 4 shows the FEM grids of the overhead pipeline truss bridge.

Table 1. Finite element parameters.

Structural part		Element type	Section size (mm)	Material type
Pipe rack	Pipe column	Beam189	B =600, H =500	Concrete K475
			B =500, H =300	
			B =950, H =500	
	Beam of Pipe rack		B =700, H =500	
Pipeline (Number)	15	Pipe289	D=89, δ =3.5	1020 steel
	9, 13, 20, 21, 22		D=108, δ =3.5	
	10, 23, 24, 25, 27		D=159, δ =3.5	
	6, 7, 11		D=219, δ =4	
	3, 16, 18, 19		D=273, δ =4.5	
	2, 8, 14, 17, 26		D=325, δ =5	
	4, 5		D=377, δ =5	
	1		D=426, δ =5	
	12		D=529, δ =6	
Steel frame	String beam and vertical rod	Beam189	B =350, H =350, T_1=12, T_2 =19	Grade 50
	Steel frame Abdominal rod		B =300, H =300, T_1=10, T_2 =15	
	Horizontal support		B =70, H =5	
	Steel frame beam		B =294, H =200	

Table 2. Material parameters.

Material	Elastic modulus E (MPa)	Poisson's ratio μ	Density ρ (kg/m^3)	Thermal expansion coefficient α (℃$^{-1}$)	Part
Concrete K475	3.15×10^4	0.2	2450	1.0×10^{-5}	Pipe rack
1020 steel	2.1×10^5	0.3	7800	1.2×10^{-5}	Pipeline

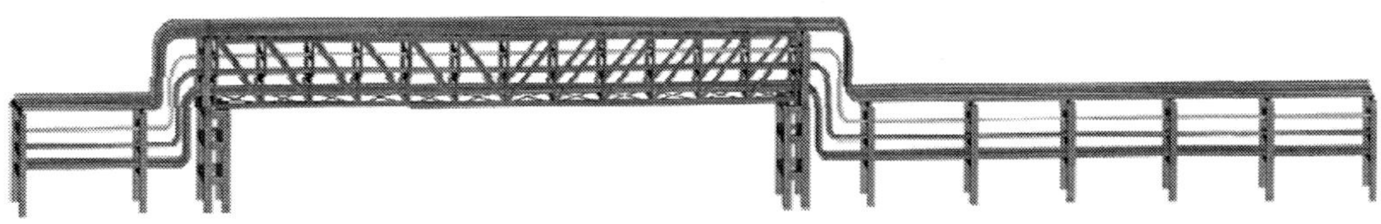

Figure 4. Finite element mesh of overhead pipeline truss bridge.

2.4. Transport Media and Operational Parameters

According to the actual situation on site, the transported media and their physical properties and operating conditions are shown in Table 3.

Table 3. Medium parameters in pipeline.

Media	Pipeline (Number)	Density ρ (kg/m³)	Viscosity υ(10^{-6}m²/s)	Flow Q (m³/h)	Pressure p (MPa)	Temperature T (℃)	Thermal Conductivity k (W/m·K)
Superheated steam	13	15.4	14.8	721	1.25	194	0.07
Oily sewage	15	850	1	56	0.8	25	0.5
Naphtha	3, 11, 23	653, 760, 760	8	591, 378, 196		40	0.11,0.14,0.11
Aviation kerosene	24	780	3.5	196		40	0.12
Condensate	8	800	8	842		40	0.14
Hydrocracking oil	27	832	13.2	196		40	0.12
Diesel oil	14, 26	855, 955	8	842,842		40	0.12, 0.12
Wax oil	2	877	8	842	0.8	40	0.12
Fuel oil	7, 19	900, 994	8	378,591		80	0.11, 0.12
Crude oil	1, 12	959, 950	8	1468,2267		40	0.14, 0.14
Solvent oil	22	970	2.74	87		40	0.12
Distillate	6	1000	8	378		80	0.11
92#gasoline	17	746	8	842		40	0.15
95# gasoline	18	746	8	591		40	0.15
Light aromatic hydrocarbon	4	730	8	1142			0.14
Aromatic hydrocarbon	5	890	8	1142	0.8	80	0.14
Heavy aromatic hydrocarbon	16	1000	30	591			0.12
Xylem mixture	25	860		196			0.13
Toluene	20	866	8	86	0.8	80	0.13
Benzene	21	879		86			0.16
Aromatic products	10	680	8	196	0.8	80	0.14
Non-aromatic products	9	800		56		40	0.12

3. Load and Constraints

The load in FEM computation mainly included the gravity, medium temperature, flowing forces acted on the pipeline. Fixed constraints were imposed on both ends of the pipes and the bottoms of the columns in the calculation.

3.1. Thermal Load

The pipeline is expanded and compressed due to the difference between the flowing medium temperature and the local temperature while installation. Therefore, the thermal and structure options were chosen together in FEM calculation [9-11]. Tables 2 and 3 show the material expansion coefficient, media operating temperature and thermal conductivity for FEM thermal load.

3.2. Flow Load

The force acted by the flowing medium on the pipeline mainly includes the resistance along the path caused by the viscous flow and the force generated by the 90° elbow [12]. The resistance along the pipe path is written as:

$$F_X = (p_1 - p_2)A = \lambda \frac{l}{d}\frac{\rho v^2}{2}A \tag{1}$$

Where λ is resistance coefficient dependent on Reynolds number. l is the length of the pipe, d is the inner diameter of the pipe, v is the flow velocity, A is the cross-sectional area of the pipe, P_1 is

IOP Conf. Series: Materials Science and Engineering **611** (2019) 012030 doi:10.1088/1757-899X/611/1/012030

pipe inlet pressure, P_2 is pipe outlet pressure, as shown in Figure 5. The force on the 90° elbow is expressed in equation (2). The direction and the position of force are shown in Figure 6.

$$F_X = (p_1 + \rho v^2)A, \quad F_Y = (p_2 + \rho v^2)A \tag{2}$$

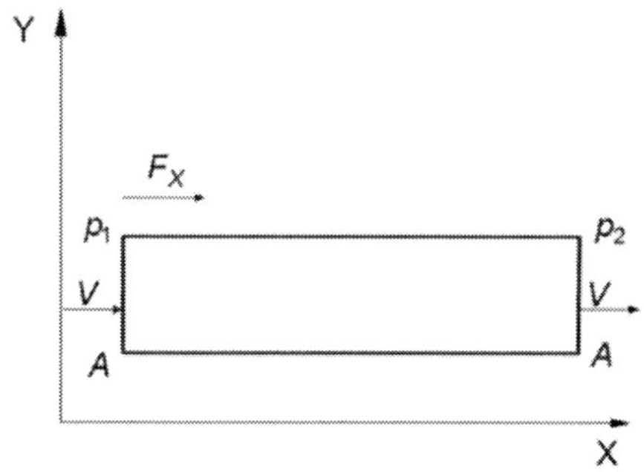

Figure 5. Resistance along the pipe path.

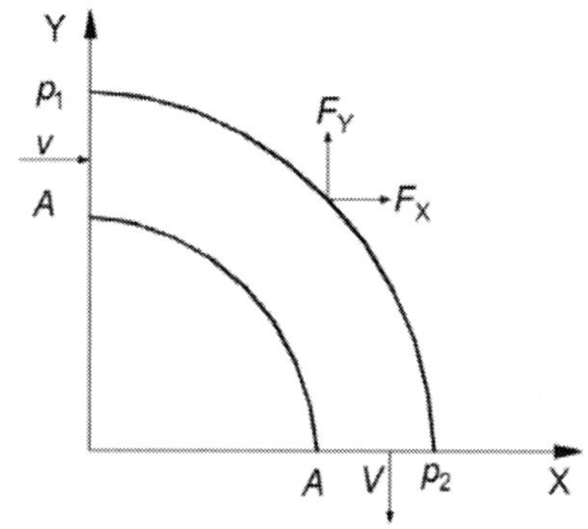

Figure 6. Force on the 90° elbow.

4. Calculation Results and Analysis

The analysis and judge for calculation results mainly focused on the overall displacement and stress distribution of the overhead pipeline truss bridge. The horizontal support structure focuses on the maximum value of the bending moment and shear force, while the column focuses on the maximum axial force and its corresponding position.

4.1. Displacement Distribution

Figure 7 and 8 are respectively X-direction and Y-direction calculated displacements of the overhead pipeline truss bridge. From the figures, the maximum X-direction displacement (13.48 mm) appeared in the pipe 15 conveying oily water between the rack 2 and the rack 3. The maximum Y-displacement (14.01 mm) occurred between 5 and 6 horizontally arranged in the steel frame shown in Figure 3. In addition, the general structure under static action needs to meet the $\delta / L \le 1/900 = 1.11\times10^{-3}$, δ represents the maximum deflection, L represents the length of the truss bridge, that is, $\delta / L = 0.01401/39 = 3.59\times10^{-4} \le 1.11\times10^{-3}$, it is known that the truss bridge satisfies the stiffness requirement under the combined force.

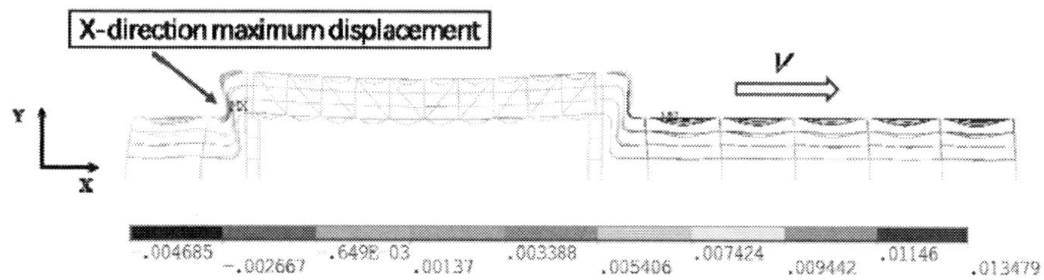

Figure 7. X-direction displacement distribution.

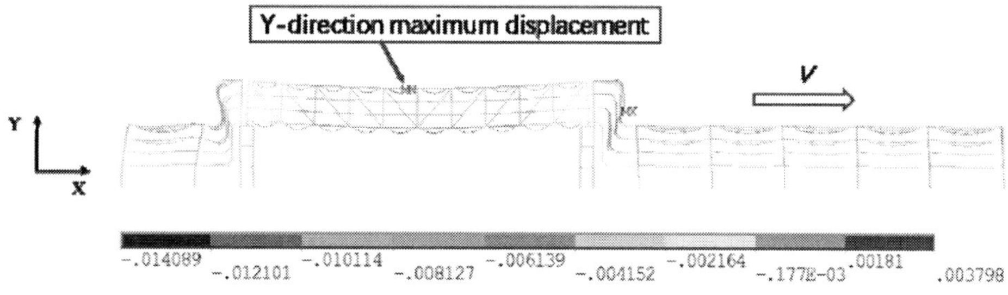

Figure 8. Y-direction displacement distribution.

IOP Conf. Series: Materials Science and Engineering **611** (2019) 012030 doi:10.1088/1757-899X/611/1/012030

4.2. Von Mises Stress Distribution

Figure 9 shows the calculated Von Mises stress distribution of the overhead pipeline truss bridge. From the figure, the maximum stress (149 MPa) appeared in the pipe 15 conveying oily water in contact with the fourth-layer beam of the rack 6. Since the inner diameter of the pipe was small, the medium temperature and the flow velocity were high, the stress in the pipe was higher than others.

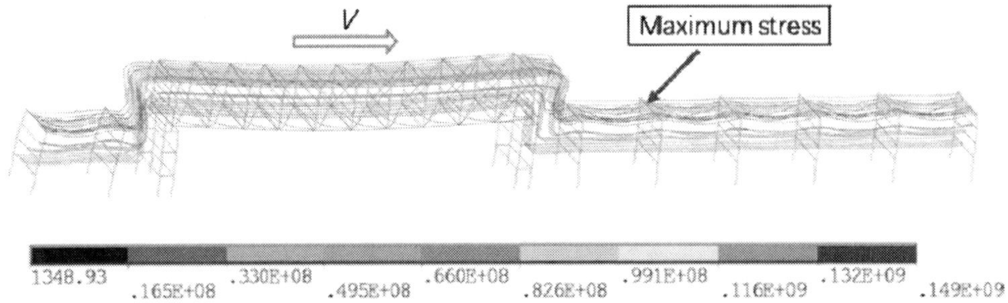

Figure 9. Von Mises Stress distribution.

4.3. Axial Force Distribution

Figure 10 shows the column axial force distributions of the overhead pipeline truss bridge. From the figure, the bottoms of the two highest columns in the overhead pipeline truss bridge, the racks 3 and 4, the largest axial force is 2230 KN, cross section is 600×600 mm, and the compressive strength value is 6.2 N·mm^{-2}, which is much less than the design value of Concrete K475 concrete compressive strength 16.3 N·mm^{-2} [13].

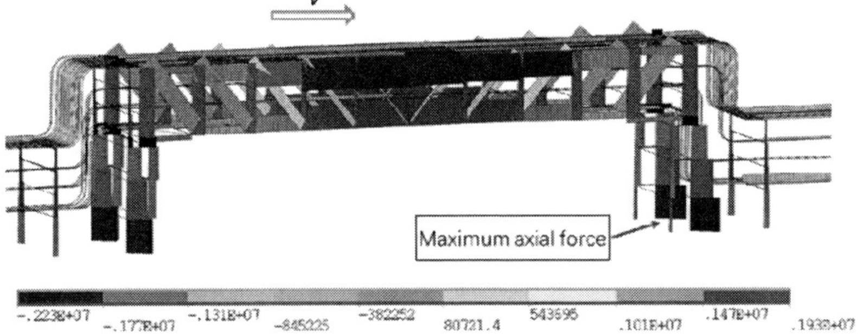

Figure 10. Column axial force distribution.

4.4. Shear Force Distribution

Figure 11 is a shear force distribution of the overhead pipeline truss bridge. It can be seen from the figure that the maximum shear force on the beam was 315.78 kN, which appeared at the fourth-layer beam of the rack 4 connecting with the pipeline 12. The maximum shear force on the pipe was 41.91 kN, which occurred on the pipe 12 conveying crude oil. This means that the maximum shear force of the beam and the pipe occurs at the higher rack on the transition path between high and low in the truss bridge.

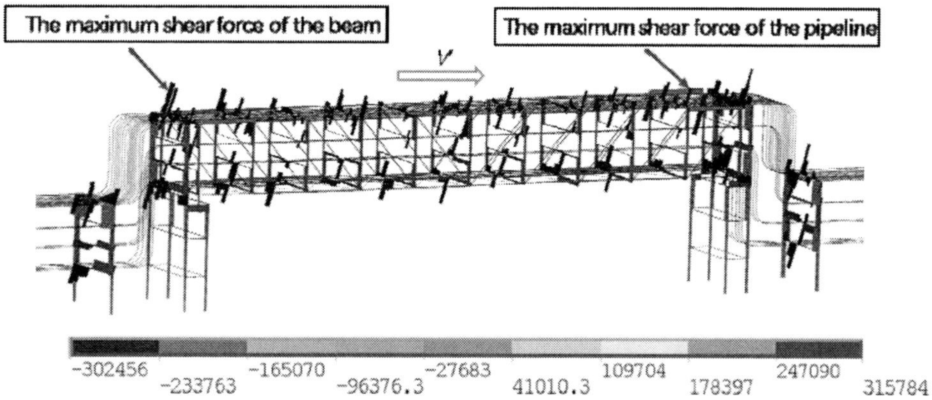

Figure 11. Shear force distribution.

4.5. Bending Moment Distribution

Figure 12 shows the calculated bending moment distributions of the overhead pipeline truss bridge. It can be seen from the figure that the maximum bending moment of the rack beam was 238.89 kN·m, which appeared at the junction of the pipeline 8 and the second-layer beam of the rack 5 , The maximum bending moment of the pipeline was 129.63 kN·m, which appeared at the pipeline 12 conveying crude oil. This means that the maximum bending moment of the beam and the pipeline also appears on the lower rack on the transition path from high to low in the compensator.

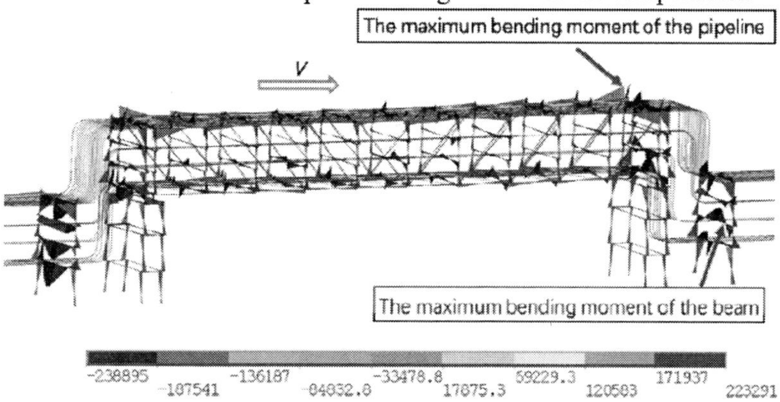

Figure 12. Bending moment distribution.

5. Pipe Strength Check

According to the technical standards of industry [14-15], the primary stress σ_1, the secondary stress σ_2 and total stress σ_t of the pipeline are required to satisfy the following conditions.

$$\sigma_1 \leq 0.75\sigma_s,\ \ \sigma_2 \leq 0.72\sigma_s,\ \ \sigma_t \leq \sigma_s \tag{3}$$

Where σ_s is the yield strength of the pipe. σ_1 is caused by gravity and other loads. σ_2 is caused by thermal deformation only. σ_t is caused by all loads and thermal deformation. It can be seen in Table 4 that the calculated stress of the overhead pipeline truss bridge satisfies the above condition (3). This indicates that the overhead pipeline truss bridge is reliable in service.

Table 4. Strength check.

Stress type	Allowable stress(MPa)	Max calculated value(MPa)	Whether the strength met requirement
σ_1	221.3	125	Yes
σ_2	212.4	122	Yes
σ_t	433.7	247	Yes

6. Conclusion

(1) The maximum displacement of the overhead pipeline truss bridge is in the middle of the steel frame. Through the δ / L stiffness check, the stiffness of the pipeline truss bridge meets the design requirements.

(2) The bottom of the highest column in the overhead pipeline truss bridge bears the maximum axial force. The maximum shear and bending moments of the rack beam occur in the rack beam from the high to low transition of the rack. The maximum shear force and bending moment of the pipeline appear at the junction of the higher rack between the high and low rack.

(3) In the entire overhead pipeline truss bridge, the maximum stress occurs at the contacting position of the pipe with a smaller inner diameter, higher medium temperature and flow velocity with the beam. By the strength check, the present studied model meets the design requirements.

References

[1] Isting C and Their B 1977. Chemical-Engineering and Safety in Design of Pipeline Systems in Chemical-Plant. Chemie Ingenieur Technik, vol 7, p 528-535.

[2] Dusseau R A 1990. Wind analysis of pipeline suspension bridges. Journal of Wind Engineering and Industrial Aerodynamics, p 927-936.

[3] Liu Y Q and Rui S H I 2017. Analysis of Environmental Effect and Optimization on Truss Overhead Crossing Structure Design of Oil and Gas Transmission Pipeline. DEStech Transactions on Engineering and Technology Research 3rd/amma.

[4] Urdea M 2018. Static linear analysis for trusses structure for supporting pipes. IOP Conference Series: Materials Science and Engineering, Vol 399, p 012051.IOP Publishing.

[5] Zhang X and Du Q 2011. Research on heat-solid coupled numerical simulation of engine exhaust pipe based on ANSYS. 2011 International Conference on Electric Information and Control Engineering. p 6070-6073. IEEE.

[6] Deng C, Wan X, Zhang Z, et al 2012. Numerical Modelling and Analysis of Infrared Testing in Delaminations of Pressure pipes Using ANSYS. 2012 Symposium on Photonics and Optoelectronics. p 1-4. IEEE.

[7] Jinmei T. 2008. Comparison of Beam Element and Shell Element in Analyzing Inherent Vibration of Structures. Nuclear Power Engineering, vol 1, p 50-52.

[8] Feng L, Deguo W, Lu L, et al 2010. Simulation of APDL-Based Super-High Pressure Compressor Pipeline System Vibration Control. 2010 International Conference on Digital Manufacturing & Automation,vol 2, p 366-369. IEEE.

[9] Bellet M, Jaouen O and Poitrault I 2005. An ALE-FEM approach to the thermomechanics of solidification processes with application to the prediction of pipe shrinkage. International Journal of Numerical Methods for Heat & Fluid Flow, vol 2, p 120-142.

[10] Krushinski V I 1974. Calculation of L-Shaped Pipelines to Provide for Compensation of Thermal Elongations, Taking into Account Friction in Sliding Supports. Energomashinostroenie, p 20-22.

[11] Shevchenko Y U N, Babeshko M E and Gololobov V I 1993. Calculation Procedure of Axisymmetrical Thermoelastoplastic Stress Strained State of Compensating Elements of Thermoinsulated Pipelines. Problemy Prochnosti (Russia), vol 8, p 87-93.

[12] Yamaguchi H 2008. Engineering fluid mechanics Vol. 85. Springer Science & Business Media.

[13] ACI Committee 2005. Building code requirements for structural concrete (ACI 318-05) and commentary (ACI 318R-05). American Concrete Institute.
[14] American Society of Mechanical Engineers 2009. Pipeline Transportation System for Liquid Hydrocarbons and Other Liquids: ASME Code for Pressure Piping, B31.4 American Society of Mechanical Engineers.
[15] American Society of Mechanical Engineers 2018. Gas Transmission and Distribution Piping Systems: ASME Code for Pressure Piping, B31.8. American Society of Mechanical Engineers.

Novel Teaching-learning-based Optimization Algorithm for Design of Digital Filters

Xiangyong Kong[1,2,*], Zhinan Li[3], Dexuan Zou[1] and Gai Pan[1]

[1]School of Electrical Engineering & Automation, Jiangsu Normal University, Xuzhou, Jiangsu 221116, P.R. China
[2]Jiangsu Normal University Kewen College, Xuzhou, Jiangsu 221116, P.R. China
[3]Zaozhuang Vocational College, Zaozhuang, Shandong 277800, P.R. China

*Corresponding author email: kxy2006@126.com

Abstract. An improved teaching-learning-based optimization algorithm named NTLBO is proposed for IIR digital design in this paper. Conventional mathematical methods have failed when reduced order adaptive models were used for the purposes of identification of problems. NTLBO utilizes a multi-learning strategy and opposition learning to overcome this disadvantage of the basic TLBO. The quasi-opposition-based learning has been applied to increase the diversity of solutions and broaden the search space to improve the global search ability. The multi-learning strategy makes local search more effective so as to speed up the convergence. To make a tradeoff between exploration and exploitation properly, the teaching factor is redesigned to increase the likelihood of solutions jumping out of local optima. Experiments are carried out on the classical examples and comparisons are made as well. The results indicate that the NTLBO algorithm achieved preferable performance in both reduced and same order models of IIR digital filters.
Keywords: digital filters; Teaching-Learning-Based Optimization algorithm; random multi-learning strategy; quasi-opposition learning.

1. Introduction

Design of digital filter has become a hot issue for many researchers due to the popularization of digital system and as well as due to the demand for date processing. The digital filter consists of two basic filters, namely Finite Impulse Response (FIR) filters and Infinite Impulse Response (IIR) filters. IIR filter has higher precision, lower system delay, fewer number of calculations per time step etc. It has been used to solve several problems including communication systems, image processing, signal processing, biomedical imaging, system identification and modeling, and seismic survey[1] etc.

These methods can be divided into two classes: the conventional design methods and heuristic optimization methods. One design analog filters first and then utilizing the bilinear transformation method to transform them into digital filter, which is very simple, but sometimes the result is not ideal and usually trapped into the local optima. In order to offset the defect, several meta-heuristic optimization algorithms are introduced to design the IIR filter[2-7].

However, these algorithms are easily getting trapped into local optima and have slow convergence and stagnation. Teaching-Learning-Based Optimization algorithm (TLBO)[8-9] with different strategies is introduced to overcome all these problems. TLBO mimics the teaching-learning process and through the interaction among the students in one class to improve the performance of all, which is not only has characteristics of parameter-free, fast convergence and simple computation, but also its capacity to global search is better than other algorithms. However, this algorithm was easily trapped into minimum or backwater status. A novel TLBO called NTLBO is proposed to identify the IIR filter. Multi-learning strategy and quasi-opposition-based learning are utilized for adaptive IIR filtering applications.

Section 2 expounds the identification problem clearly. A brief overview of TLBO is presented in Section 3. The novel TLBO (NTLBO) algorithm is given in Section 4. Section 5 provides the experimental settings and analyses the results in detail. Finally, in Section 6, the conclusions and the future direction of research are offered.

2. Overview of Filter Model

The input-output relationship of IIR digital filters is given as follows:

$$a_0 y(p) + \sum_{k=1}^{v} a_k y(p-k) = \sum_{k=0}^{u} b_k x(p-k) \tag{1}$$

where $x(p)$ is the input of the filters and $y(p)$ is the output of the filters; v is the order of the filter. a_0=1, the transfer function is expressed as:

$$H(z) = \frac{\sum_{k=0}^{u} b_k z^{-k}}{1 + \sum_{k=1}^{v} a_k z^{-k}} \tag{2}$$

In fact, the design of IIR filter can be viewed as an identification problem of system coefficients. Figure 1 shows the design of IIR digital filter.

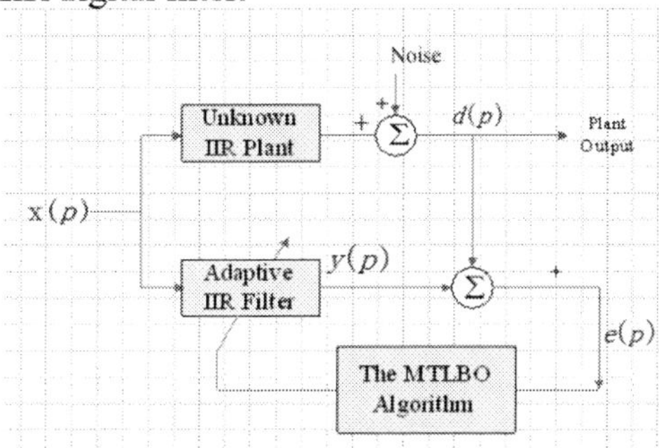

Figure 1. Principle diagram of IIR digital filter design

$$MSE = J(\omega) = E[e^2(p)] \approx \frac{1}{N}\sum_{p=1}^{N} e^2(p) \tag{3}$$

$MSE(dB)=10log_{10}(J)$ in dB. The error signal can be calculated as $e(p)=d(p)-y(p)$. $d(p)$ is the responses of the unknown plant and $y(p)$ for IIR filter. N samples are used for the calculation of the objective function. Here, w is the filter coefficient vector: $w=[a_0, a_1, \cdots a_v, b_0, b_1, \cdots b_u]$.

3. The Basic TLBO Algorithm

TLBO[8] was proposed in 2011 by the Indian scholars Rao et al. It simulates the process of the teaching and learning in daily life and has teaching phase and learning phase. During teaching, class students learn from the teacher and students with the average values to improve their performance; in the learning stage, students communicate with other more knowledgeable students, which improve their performance accordingly.

3.1. Teaching Phase

During teaching, a teacher is selected from the best learners in the class to improve the average score from the initial level to his own level depending on their capabilities, which effectively incorporates the influence of the population mean. The teacher tries his or her best to share his or her knowledge with the students, which will not only increase the knowledge level, but also enhance the students' results. However, average result of the students is usually not up to the level of the teacher and only reaches some extent in terms of the capability of the class. The mathematic expression of the teaching process as below.

$$X^{new} = X^{old} + rand \cdot (X^{teacher} - T_F \cdot Mean) \quad (4)$$

Where, $X^{teacher}$ represents the best learner in current population. *rand* is a uniform random number between 0 and 1. X^{new} is updated value of X^{old}. The teaching factor T_F is gotten as $T_F = round(1 + rand)$, *Mean* is the average result of the class.

3.2. Learning Phase

Learning large amounts of knowledge from the teacher is not only one way of learning knowledge. In order to gain more knowledge, the students learn something new from their classmates who have more profound knowledge than through interaction among themselves. The following expressions are for minimization problem. The arithmetic expression of the phase is

$$X^{new} = \begin{cases} X^{old} + rand \cdot (X^{r1} - X^{old}), & f(X^{r1}) < f(X^{r2}) \\ X^{old} + rand \cdot (X^{r2} - X^{old}), & otherwise \end{cases} \quad (5)$$

Where X^{r1} and X^{r2} are two different learners, X^{new} is replaced by X^{old} if X^{new} gives a better fitness values, otherwise, X^{old} is retained to be a parent vector in the next generation.

4. The Proposed NTLBO Algorithm

Multi-learning is utilized to overcome the disadvantage of the basic TLBO that the local search capacity is poor and the solutions are intend to be trapped around the local optima. At the same time, quasi-opposition based learning introduced can broaden the search space and increase the diversity of the individuals to raise the probability of jumping out of local optima. Besides, considers the fact in our teaching-learning process, a new teaching factor is designed, which can balance the aggregation and divergence of the algorithm.

4.1. Analysis of Teaching Factor and Improvement

The teaching factor in TLBO is randomly selected and can be either 1 or 2. At the early iteration, the learners improve their results through learn something from their teacher. The individuals move towards the best individual in current population so as to accelerate the convergence and the results obtained have better solutions. When one is better than the mean level of the class, it can get close to the best one in current population. However, assumed that a learner is superior to the mean level of the class, the learner will get close to the best one among current population in the opposite direction with a certain probability which can lead to low optimization efficiency and longer computation time.
According to the above shortcomings of the TLBO algorithm, the teaching factor is redesigned using the form of the piecewise function. The modified teaching factor is given by

$$\begin{aligned} & if\ iIter \le \frac{nIter}{2} \\ & \qquad TF = \begin{cases} round(1 + rand(0,1)), & if\ fitness_{mean} \ge fitness_i \\ round(1.05 + \frac{rand(0,1)}{2}), & otherwise \end{cases} \\ & else \\ & \qquad TF = round(1 + rand(0,1)) \\ & end \end{aligned} \quad (6)$$

where *iIter* is current iteration and *nIter* is the maximum iteration. $fitness_{mean}$ is mean of all the fitness values. $fitness_i$ is fitness value of current individual.

4.2. *Random Multi-learning*

The learning way of the learners in basic TLBO algorithm is simple. It will slow the convergence and may lead to be trapped into the local optima. In view of these shortcomings, random multi-learning strategy is employed. Through the random multi-learning strategy, the learners gain knowledge not only through interaction among them, but also learn new things through the difference between two difference learners. The expression of random multi-learning strategy is given by:

$$\begin{aligned}&\textit{if}\ \ rand < 0.5\\ &\qquad X_i^{new} = \begin{cases} X_i^{old} + rand\cdot(X_i^{old} - X_i^{r1}) & f(X_{old}) < f(X_{r1}) \\ X_i^{old} + rand\cdot(X_i^{r1} - X_i^{old}) & otherwise \end{cases}\\ &else\\ &\qquad X_i^{new} = \begin{cases} X_i^{old} + rand\cdot(X_i^{r2} - X_i^{r1}) & f(X_{r2}) < f(X_{r1}) \\ X_i^{old} + rand\cdot(X_i^{r1} - X_i^{r2}) & otherwise \end{cases}\\ &end\end{aligned} \tag{7}$$

where X_i^{r1} and X_i^{r2} are random select two different individuals. X_i^{new} is the updated value of X_i^{old}.

4.3. *Quasi-opposition Learning*

In 2007, Rahnamayan[10] raised a quasi-opposition-based learning technique. However, compared with the opposite point, a quasi-opposite point gets closer to the solution[11-12]. To take advantage of different strategies, the local search and quasi-opposition learning are combined to broaden the searching space and enhance the individual diversity to avert premature convergence. Thus, we use the quasi-opposite point instead of opposite point to generate an opposite population when meet a certain probability J_r. The arithmetic expression is given by:

$$x_i^{new} = \begin{cases} rand(\frac{x_i^l + x_i^u}{2}, x_i) & \textit{if } rand < J_r \\ x_i^{old} & otherwise \end{cases} \tag{8}$$

5. Simulation Results and Discussions

Two classical IIR plants are carried out with MATLAB (2009b) to examine the performance of NTLBO in comparison with PSO-w, DE and HS. The input signal of system is a uniform white sequence, in order to compare with CSO, a Gaussian white signal with variance 10^{-3} is set as the additive noise. In each case, the experimental results are obtained on account of 50 independent runs. Parameters of NTLBO is set as follows: population size=30, iteration cycle=150; the parameters setting of the WPSO, DE and HS algorithms are given. For the PSO-w, population size=30, iteration cycle=300, w_{max}=0.9, w_{min}=0.1, c_1=c_2=2.05. For the DE, the population size is 30; iteration cycles is 300; Cr=0.8; F=0.6. For the CSO, the parameters setting and simulation results are taken from [2].

5.1. *Example 1*

Take the unknown third order plant as the second example, which is considered from [2, 6] and the transfer function is

$$H_s(z) = \frac{0.05 - 0.4z^{-1}}{1 - 1.131z^{-1} + 0.25z^{-2}} \tag{9}$$

5.1.1. Case 1. A second order model is employed and its transfer function is given as

$$H_{af}(z) = \frac{a_0 + a_1 z^{-1}}{1 + b_1 z^{-1} + b_2 z^{-2}} \tag{10}$$

Table 1. Results of 2 order filter model

MSE(dB)	WPSO	DE	NTLBO
Best	-24.6883	-89.8605	-120.5037
Worst	-15.7917	-74.7761	-100.4938
Mean	-18.926	2.2363	1.0207
Std	2.3743	1.618	3.9586

5.1.2. Case 2. A first order (reduced order) model is employed and its transfer function is given as

$$H_{af}(z) = \frac{a}{1 + bz^{-1}} \tag{11}$$

Table 2. Results of 1 order filter model

MSE(dB)	WPSO	DE	NTLBO
Best	-16.7911	-17.8605	-19.5037
Worst	-15.1355	-15.7761	-17.4938
Mean	-16.0649	-16.2363	18.0207
Std	0.7315	1.618	0.9586

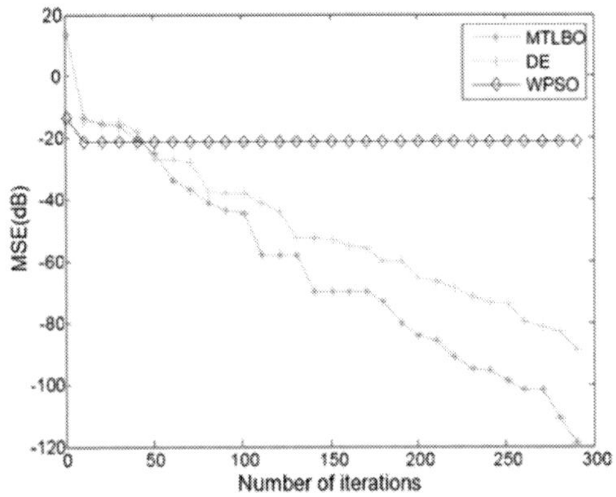

Figure 2. Convergence curve for 2 order filter model

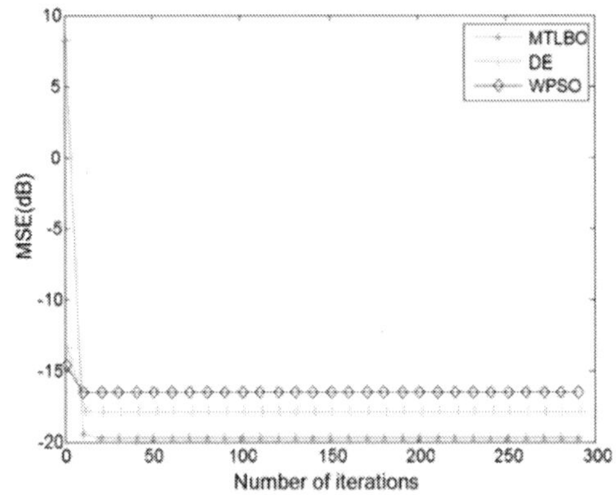

Figure 3. Convergence curve for 1 order filter model

Figure 2 and figure 3 depict the convergence characteristic of this test system. Table 1 and table 2 summarize the best and worst variances obtained from NTLBO and other methods. The standard deviation values are also included. For this example, NTLBO has faster convergence and finds greater solutions.

5.2. Example 2

Take the unknown fourth order plant as the second example, which is taken from [2, 6] and the transfer function is.

$$H_s(z) = \frac{1 - 0.9z^{-1} + 0.81z^{-2} - 0.729z^{-3}}{1 + 0.04z^{-1} + 0.2775z^{-2} - 0.2101z^{-3} + 0.14z^{-4}} \tag{12}$$

5.2.1 Case 1. A fourth order model is employed and its transfer function is given as

$$H_{af}(z) = \frac{a_0 + a_1 z^{-1} + a_2 z^{-2} + a_3 z^{-3}}{1 - b_1 z^{-1} - b_2 z^{-2} - b_3 z^{-3} - b_4 z^{-4}} \tag{13}$$

Table 3. Results of 4 order filter model

MSE(dB)	WPSO	DE	NTLBO
Best	-22.184	-27.7508	-89.3209
Worst	-18.1651	-21.4617	-79.5307
Mean	-19.6603	-23.8232	-82.0988
Std	1.8853	1.9354	4.415

5.2.2 Case 2. A third order (reduced order) model is employed and its transfer function is given as

$$H_{af}(z)=\frac{a_0+a_1z^{-1}+a_2z^{-2}}{1-b_1z^{-1}-b_2z^{-2}-b_3z^{-3}} \tag{14}$$

Table 4. Results of 3 order filter model

MSE(dB)	WPSO	DE	NTLBO
Best	-10.7256	-16.4028	-23.7584
Worst	-8.1116	-15.7179	-21.6125
Mean	-8.7883	-16.2088	-22.3448
Std	0.8591	0.4679	0.5733

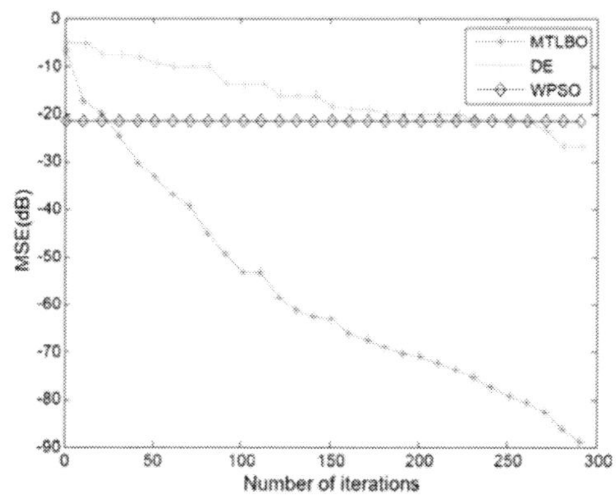

Figure 4. Convergence curve for 4 order filter model

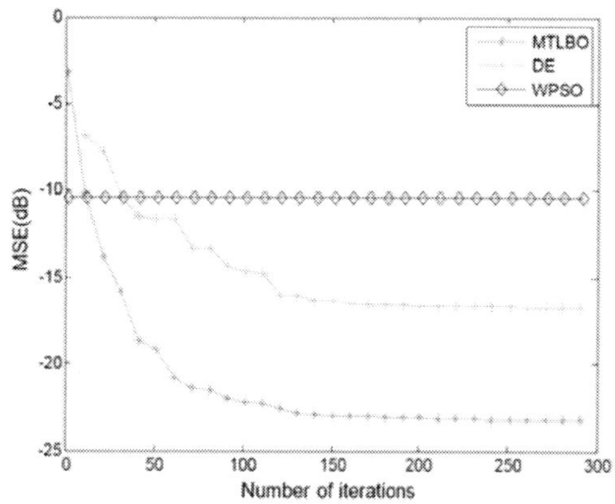

Figure 5. Convergence curve for 3 order filter model

From table 3 and figure 4, we can know that the proposed NTLBO algorithm has better convergence than others and found best solution among of all. Besides, the standard deviation of the results by NTLBO is very small. Seen from table 4 and figure 5, with the reduced order models, the proposed method could find a better solution around the optimal solution in contrast to other approaches. In summary, the NTLBO algorithm can achieve better performance for two IIR digital filter models. From the results, the proposed approach has higher convergence precision and better robustness.

6. Conclusions

A new algorithm named NTLBO is raised to solve the IIR digital filter design problem. The opposition learning techniques is introduced in NTLBO to enhance the diversity and broaden the searching space. In an attempt to improve the ability of the local search, random multi-learning strategy is proposed which can avert local optima trappings. The teaching factor was redesigned to make a compromise between exploring and exploiting properly. The simulation results on two IIR filters models verified the superiority of the novel NTLBO algorithm. The comparative results illustrate that NTLBO has better convergence and search precision than some other algorithms. In order to better balance aggregation and divergence, further research could focus on adaptively adjust the teaching factor.

Acknowledgements
This work was supported by the Natural Science Foundation of Jiangsu Province (No. BK20160220) and Natural Science Foundation of the Jiangsu Higher Education Institutions of China (Grant No. 18KJB413002).

References

[1] Liu Q, Lim Y C, Lin Z. Design of Pipelined IIR Filters Using Two-Stage Frequency-Response Masking Technique[J]. IEEE Transactions on Circuits and Systems II: Express Briefs, 2019, 66(5): 873-877.
[2] Panda G, Pradhan P M, Majhi B. IIR system identification using cat swarm optimization[J]. Expert Systems with Applications, 2011, 38(10): 12671-12683.
[3] Chang W D. A modified PSO algorithm for IIR digital filter modeling[J]. Journal of Circuits, Systems and Computers, 2018, 27(05): 1850073.
[4] Karaboga N, Latifoglu F. Elimination of noise on transcranial Doppler signal using IIR filters designed with artificial bee colony–ABC-algorithm[J]. Digital Signal Processing, 2013, 23(3): 1051-1058.
[5] Shafaati M, Mojallali H. IIR filter optimization using improved chaotic harmony search algorithm[J]. Automatika, 2018, 59(3-4): 331-339.
[6] Karaboga N. Digital IIR filter design using differential evolution algorithm[J]. EURASIP Journal on Applied Signal Processing, 2005, 2005: 1269-1276.
[7] Chen S, Istepanian R, Luk B L. Digital IIR filter design using adaptive simulated annealing[J]. Digital Signal Processing, 2001, 11(3): 241-251.
[8] Rao R V, Savsani V J, Vakharia D P. Teaching–learning-based optimization: a novel method for constrained mechanical design optimization problems[J]. Computer-Aided Design, 2011, 43(3): 303-315.
[9] Rao R V, Savsani V J, Vakharia D P. Teaching–learning-based optimization: an optimization method for continuous non-linear large scale problems[J]. Information sciences, 2012, 183(1): 1-15.
[10] Saha S, Mukherjee V. A novel quasi-oppositional chaotic antlion optimizer for global optimization[J]. Applied Intelligence, 2018, 48(9): 2628-2660.
[11] Rahnamayan S, Tizhoosh H R, Salama M M A. Quasi-oppositional differential evolution[C]// 2007 IEEE Congress on Evolutionary Computation. IEEE, 2007: 2229-2236.
[12] Warid W, Hizam H, Mariun N, et al. A novel quasi-oppositional modified Jaya algorithm for multi-objective optimal power flow solution[J]. Applied Soft Computing, 2018, 65: 360-373.

Study on the Design of Cutting Disc in Ultrasonic-assisted Machining of Honeycomb Composites

Yaozhi Xia[1], Jianfu Zhang[1,*], Zhijun Wu[1], Pingfa Feng[1,2] and Dingwen Yu[1]

[1]Department of Mechanical Engineering, Tsinghua University, Beijing, China
[2]Graduate School at ShenZhen, Tsinghua University, ShenZhen, China

*Corresponding author

Abstract. To study the influences of geometrical parameters on the design of ultrasonic-assisted cutting disc tool for honeycomb composites, a finite element model integrating the cutting disk and ultrasonic horn was established. The performance of the output amplitude of the cutting disc was analyzed under different geometrical parameters including cutter radius, cutter thickness and anterior angle. Then the influences of the geometrical parameters on the energy density and bending rigidity of the cutting disc were discussed. The simulation results shown that with the increase of cutter radius, both the energy density and rigidity decreased, and the amplitude at the cutting edge increased firstly and then decreased under the interaction of the energy density and rigidity. There was an ideal interval of radius within which the amplitude reached the maximum. With the increase of cutter thickness and anterior angle, the energy density decreased while the rigidity increased, the amplitude at the cutting edge presented a declining trend.

1. Introduction

Honeycomb composite has been more and more widely applied in such fields as the aerospace, auto, ship, etc. for its excellent physical properties and chemical properties [1,2]. It is mainly processed in two methods, traditional high-speed milling and ultrasonic-aided processing. Specifically, the latter applies ultrasonic vibration in traditional tool motion to effectively reduce the cutting force and improve the processing quality and processing efficiency [3,4].

Special cutters applied in the ultrasonic-assisted processing of honeycomb composites mainly include triangular blade knife and cutting disc. Cutting disc is mainly used for plane and surface machining, as a tool for finish machining. As an actuating element in the cutting process, cutting disc's geometrical shape and dimension parameters would impact the vibration performance, machining accuracy and processing efficiency of the ultrasonic system greatly. Therefore, the selection of optimal parameters in the design of cutter is of great importance.

Regarding the technical research on the ultrasonic-assisted processing of honeycomb composites in foreign countries, Austrian GFM Company took the lead in producing a special ultrasonic machine by aiming at honeycomb composites; French CRENO also made a planer-type ultrasonic machine and developed a mechanical-arm machine, which could finish the processing of honeycomb materials of complicated shape. But since related applications involve the military industry, there were few research papers. In terms of the design of ultrasonic vibration system, Lucas [5] et al. established a finite element model of a single-blade ultrasonic-assisted cutting device by simulating the cutting of cheese and two other kinds of materials that could be used for simulating food and biological tissue; Nick [6] et al. built a mathematical model of the design process and came up with the optimization method. With the maximum output amplitude in resonant mode as the objective, an objective function was defined to

optimize the ultrasonic vibration system; BJORNSSON [7] et al. studied the ultrasonic system of triangular blade applied on the mechanical arm, achieved the maximum amplitude 29um, and carried out the machining experiment of carbon fiber reinforced materials.
As for the shape design of special cutters used for ultrasonic-assisted cutting of honeycomb composites in China, finite element simulation is mainly applied in the design of cutter and ultrasonic horn. For instance, Shen Y B [8] applied the finite element analysis method to study the dynamic characteristics of sharp cutter, designed the sharp cutter for cutting honeycomb composites, and decided the geometric dimensioning of cutter in the best mode of vibration. Wu X [9] put forward the concept of substitution method and finite element analysis method, and designed ultrasonic straight-blade cutter based on the ultrasonic-assisted processing principle. Huang S [10] designed a disc cutter with sawtooth, built a cutting force model, and tested that sawtooth disc cutter could reduce the cutting force. Fang L [11] optimized the cutter by combining the statics and modal analysis results of disc cutter, made the structure of cutter compact, and reduced the motor power. Zhang Y D [12] et al. designed a broad-blade ultrasonic cutter, and concluded that slotting at a specific position of the cutter would realize the consistent lateral oscillation. Some scholars also optimized the design of cutter through the research on the processing technology. For instance, Hu X P [13] et al. built a theoretical model arguing that the cutting force of straight-blade cutter was related to such parameters as the cutting angle, depth, anterior angle of cutter, etc. which was of instructive to the design and optimization of acoustic system in ultrasonic-assisted cutting. Zhang S F [14,15] analyzed the impact of geometrical parameters of the disc cutter on the cutting temperature based on ABAQUS software, and built the ultrasonic vibration frequency model with the numerical fitting method. Zhuang Y [16] et al. argued that the key to the effective cutting of AFRP lay in the sharpness of cutting blade, designed the ultrasonic cutter and explored cutting parameters through the cutting experiment. Chen J Q [17] analyzed the wear pattern of the cutter during the machining process, conducted the software development based on the computer digital image, and implemented the real-time detection of the wear pattern of disc cutter.
In conclusion, domestic and foreign studies on the special disc cutter for the ultrasonic-assisted cutting of honeycomb composites mainly focus on the modal analysis of cutter and matching with the ultrasonic vibration system. Besides, technological tests mainly focus on the impact of machining parameters on the machining quality. But there are few studies on the influence law of geometrical parameters on the output amplitude and cutting effect. This paper carries out the modelling and simulation analysis of the harmonic response of ultrasonic disc cutter based on ANSYS Workbench software, concludes the laws of influence of geometrical parameters on the amplitude output of the ultrasonic vibration system, and analyzes the results from the energy density and structural rigidity.

2. Simulation Model of the Cutting Disc and Ultrasonic Horn

Ultrasonic horn is the core of the ultrasonic vibration system, which mainly magnifies the supersonic microvibration caused by the transducer till the amplitude meets the requirement of machining. Generally, when the shape of ultrasonic cutter is similar to the end of ultrasonic horn or the cutter is light in weight, the impact of cutter on the resonant frequency of ultrasonic vibration system can be ignored, and the ultrasonic horn can be optimized directly. But when the structure of the disc cutter used for the ultrasonic cutting of honeycomb composites differs greatly from the ultrasonic horn and it is heavy in weight, with a great impact on the structure of the ultrasonic vibration system, integrated design of the disc cutter and ultrasonic horn is necessary.
The disc cutter is mainly applied for facing or surface machining, a type of finish machining in the machining process of paper honeycomb. Specifically, it is in threaded connection with the ultrasonic horn, as shown in Figure 1.

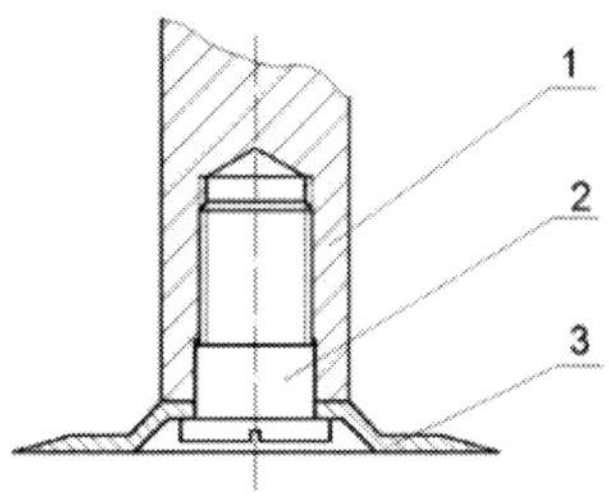

1 ultrasonic horn 2 screw 3 cutting disc

Figure 1. Assembly diagram of the disc cutter

Figure 2. Simulation model of the tool

Traditional ultrasonic vibration system is mainly designed by setting up a dynamic differential equation and solving the equation according to the boundary conditions to obtain all parameters. But once the disc cutter is integrated with the ultrasonic horn, it is complicated in structure, and it is also difficult to solve the equation. This paper mainly conducts the finite element analysis of the integrated structure of disc cutter and ultrasonic horn with ANSYS Workbench to study its dynamic characteristics, so that it shows better dynamic acoustic features at the resonant frequency. Besides, the geometric dimensioning of the cutter at the optimal inherent frequency and vibration mode is selected.

2.1. Simulation Model

According to the geometrical shape of existing ultrasonic horn and cutter, a finite element simulation model is set up, as shown in Figure 2.

The key to the ultrasonic-assisted cutting of honeycomb composites lies in the smooth cutting of honeycomb fiber, and the rigidity of material in the cutting direction is small, so the blade of cutter shall be sharp, and the material of cutter shall show good machined characteristics. Moreover, the cutter is sheet-like, it is required to have good abrasive resistance. But since honeycomb material, in hallow cell-size structure, displays low harness, and the area of cut is small, it sets a low requirement on the hardness and heat-conducting property of the cutter. The principle of ultrasonic-assisted machining is to exert ultrasonic vibration on the cutter, so the material is required to display strong toughness, and endure long-term high-frequency vibration. At present, common cutter materials include high-speed steel, alloyed tool steel, cemented carbide, tool carbon steel, ceramics, etc. Since alloyed tool steel is more suitable for making low-speed cutting tool, while ceramic material is poor in toughness, and tool carbon steel is poor in abrasive resistance, and cemented carbide is fragile, featuring poor shock resistance and anti-vibration performance, high-speed steel is relatively suitable, and its material performance is shown in table 1.

Table 1. Properties of the cutting disc

Material	Elastic modulus	Density	Poisson's ratio
W18Cr4V	2e11Pa	7800kg/m^3	0.3

2.2. Analysis of Vibration Characteristics

Related studies have shown that during the cutting of honeycomb composites with special ultrasonic disc cutter, the greater the amplitude of ultrasonic vibration at the blade is, the greater the role of ultrasonic-assisted cutting will be, namely the smaller the cutting force will be, and the better the cutting effect will be. For this reason, the mode of vibration of the greatest vibration amplitude at the blade should be selected as the target mode.

Integrated modal analysis of the cutter and ultrasonic horn is carried out with the modal modular in ANSYS Workbench. Different modes of vibration of the cutter at different resonant frequencies can be obtained. In general, the mode of vibration of the cutter consists of three types, namely the wave type, longitudinal vibration type and twist vibration, as shown in Figure 3.

(a) wave type (b) longitudinal vibration type (c) torsional vibration type

Figure 3. Integrated simulation mode diagram

Harmonic response simulation analysis is carried out for the cutter within 18 kHz~22 kHz by making use of the harmonic response module of ANSYS Workbench, and the curve of the mean value of longitudinal vibration amplitude and excitation frequency at the blade is obtained, as shown in Figure 4. It can be seen from the frequency-amplitude curve of the harmonic response analysis that the system reaches the maximum amplitude at point A near 20 kHz, and the mode of vibration corresponding to the modal analysis is shown in Figure 5.

According to the simulation result, in case of the same input, the vibration amplitude of resonance at the blade of longitudinal vibration is the greatest, and it is a mode of vibration featuring the highest energy conversion efficiency, as the target mode of vibration for subsequent designs.

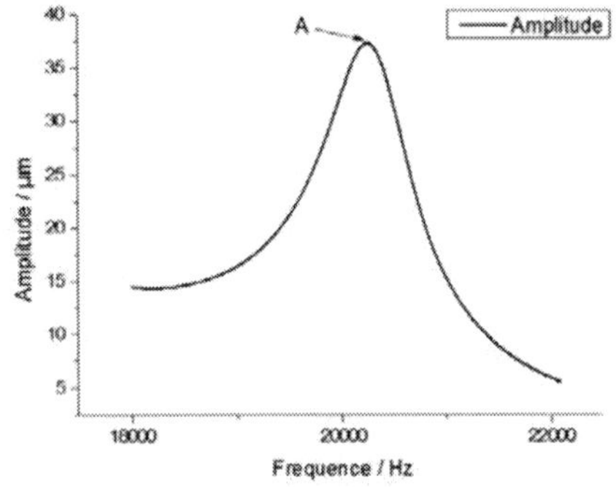

Figure 4. Frequency-Amplitude curve of harmonic response analysis

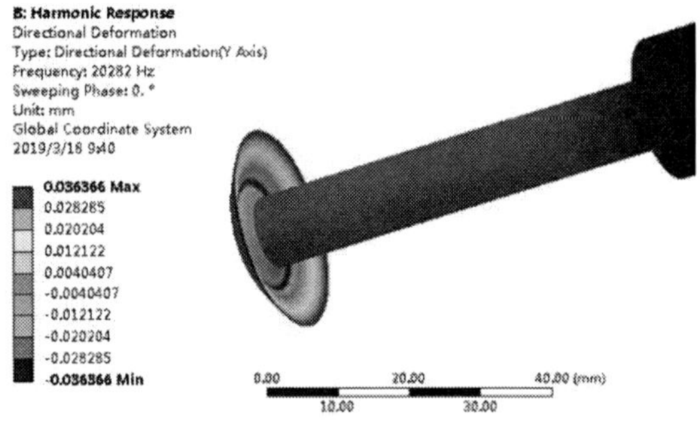

Figure 5. Mode diagram at maximum amplitude

As stated previously, under the ultrasonic assistance, the greater the vibration amplitude of the disc cutter at the blade is, the greater the role of ultrasonic-assisted cutting will be. This paper takes different geometrical parameters for the harmonic response analysis to get the mean vibration amplitude of the cutter at the blade in longitudinal resonance, and it also analyzes the laws of influence of different geometrical parameters on the vibration amplitude of the cutter.

3. Influences of Geometrical Parameters on the Output

This paper studies the laws of influence of the cutter thickness, anterior angle and blade radius on the vibration amplitude in longitudinal vibration mode with the single factor method. At first, the resonant frequency H of longitudinal vibration mode is obtained through the modal analysis. According to the shell size and material characteristics of the matched transducer of ultrasonic vibration system, 2 MPa of pressure is input, and the scope of harmonic response frequency is set to be (H-100, H+100) Hz, for solving the output amplitude of the ultrasonic vibration system at the blade in case of longitudinal resonance, and analyzing the change rules of output vibration amplitude in case of the same input.

3.1. Effect of Tool Radius R on the Amplitude

According to the requirement on the inner arc surface machining size and machining frequency of work piece, the radius of small disc cutter is decided to be about 15mm. When the cutter radius *R* ranges between 12 mm and 17 mm, it can be learnt from the modal analysis results that when the thickness b ranges between 0.6 mm and 1.3 mm, and the anterior angle γ_0 between 11°and 17°, the resonant frequency of the vibration system is about 20 kHz, and it is matched with the transducer of the system.

This paper conducts the harmonic response analysis of the output vibration amplitude of the system within the choice range of parameters.

In Figure 6, it is the curve of influence of value on the vibration amplitude when the blade thickness b is 0.9 mm, the anterior angle γ_0 is 13°, 15° and 17°, and radius R ranges between 12 mm and 17mm.

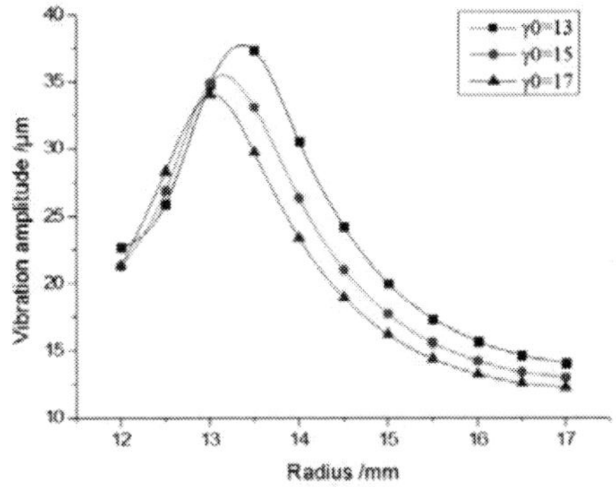

Figure 6. Curve chart of amplitude change with radius

It can be seen from the three curves in the figure that when the radius ranges between 12 mm and 13.5 mm, with the increase of the radius, the output vibration amplitude increases from 25 μm to 35μm. When the radius is over 13.5, the output vibration amplitude decreases gradually from 35 μm to 15μm at a slower speed with the increase of the radius, and the curve tends to be flat. Generally, the maximum vibration amplitude shows when the radius ranges between 13 mm and 14 mm, within which the optimal radius of the vibration system occurs.

3.2. Effect of tool Thickness b on the Amplitude

When the tool radius R is 13 mm and the thickness b of the tool is in the range of 0.6 mm to 1.2 mm, the overall longitudinal vibration resonance frequency of the vibration system is around 20 kHz. However, it is difficult to machine a tool with a thin thickness, and the rigidity of the tool will be too weak to guarantee stability when cutting honeycomb materials, the value of the tool thickness b is set from 0.8 mm to 1.2 mm

In Figure 7, it is the curve of influence of tool thickness b on the vibration amplitude when the cutter radius R is 13 mm, and anterior angle γ_0 is between 11° and 17°.

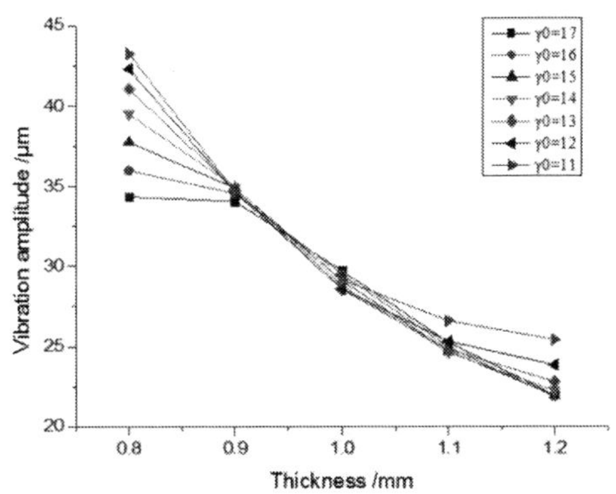

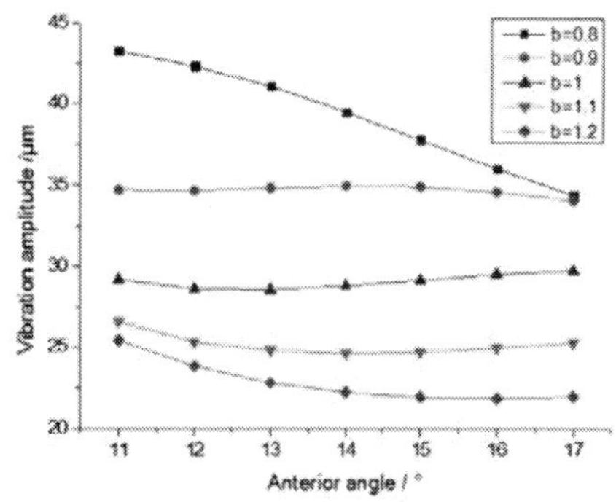

Figure 7. Curve chart of amplitude change with thickness

Figure 8. Curve chart of amplitude change with anterior angle

It can be seen from the curves in the figure that when the tool thickness ranges between 0.8 mm and 1.2 mm, with the increase of the thickness, the output vibration amplitude significant decreases from 40 μm to 25 μm. Therefore, the thickness of the tool should be taken as small as possible during design.

Considering the tool machining process, tool stiffness and vibration effect, the tool thickness should be around 0.9 mm.

3.3. Effect of Anterior Angle γ_0 on the Amplitude

According to the section 2.2, the value of the tool thickness should not be too small. Therefore, in this section, the tool thickness is in the range of 0.8mm to 1.2mm and the tool radius R is 13 mm. The influence of the anterior angle on the amplitude is simulated. The result is shown in Figure8.

It can be seen from the curves that when the anterior angle ranges between 11° to 17°, with the increase of the anterior angle, the output vibration amplitude significant decreases. When the tool is thin, the change of the anterior angle will reduce the output amplitude value from 43 μm to 35 μm, which is a big change. When the tool is thick, the total variation of the output amplitude is about 2 μm as the angle of the rake angle changes. In this case, the change in the rake angle has little effect on the output amplitude.

Since the smaller the tool's anterior angle is, the sharper the tool is, which means more conducive to cutting the fiber. Therefore, when the tool thickness is taken to a larger value, the tool rake angle should be taken as small as possible while preserving the tool life.

4. Analysis of Simulation Results

The influences of the geometrical parameters on the vibration characteristics were analyzed from the energy density and bending rigidity

4.1. Analysis of Energy Density

The energy accumulation principle of ultrasonic horn is: on the premise of ignoring the energy loss during the transmission, the energy remains the same when going through any cross section of the ultrasonic horn. Therefore, the expression of the energy density ρ_0 on the cross section of the sectional area S_0 is:

$$\rho_0 = E/S_0 \tag{1}$$

In which: E is the sectional energy, which is a constant value.

It is clear in equation (1) that the smaller the sectional area is, the greater the energy density will be. The relational expression between the vibration amplitude A and the energy density is[18]:

$$\rho_0 = \frac{1}{2} K_e A^2 \tag{2}$$

Namely

$$A = \sqrt{\frac{2\rho_0}{K_e}} \tag{3}$$

In which: K_e is a coefficient, related to the system structure.

According to equation (3), the greater the energy density is, the greater the vibration amplitude will be. To make the vibration amplitude transfer of the ultrasonic vibration system steady, the diameter at the joint of disc cutter is set as 10 mm according to the end size of the ultrasonic horn, while the total height of the cutter is set to be 2.7 mm according to the experience. A rectangular plane coordinate system is set for the disc cutter, as shown in Figure 9, in which, b is the thickness of blade, γ_0 is the anterior angle, R is the radius of cutter. The expression of two straight lines in the figure is:

Line 1:

$$y_1 = tan\alpha(R - x_1) \tag{4}$$

Line 2:

$$y_2 = 7.7 - \sqrt{2}b - x_2 \tag{5}$$

By ignoring the stepped surface, and taking the sectional area of the disc cutter within y (0, b) into account, the sectional area equation of the cutter when $y = y_0$ is:

$$S_0 = \pi(x_1^2 - x_2^2) = \pi[(R - y_0/tan\alpha)^2 - (7.7 - \sqrt{2}b - y_0)^2] \tag{6}$$

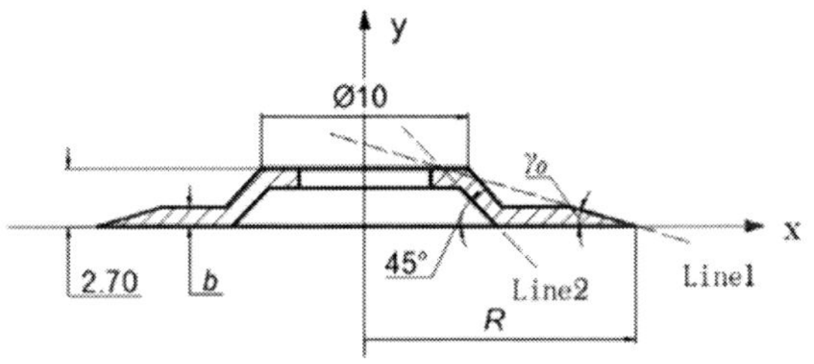

Figure 9. Coordinate system diagram of the disc cutter

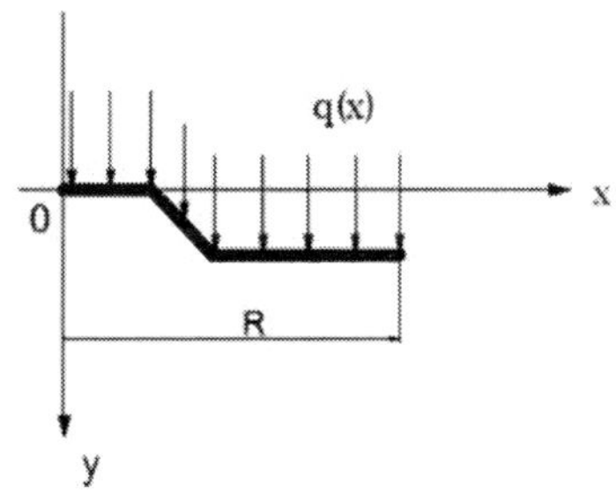

Figure 10. Structure diagram of the cantilever beam

The change relation between geometrical parameters R, b, γ_0 and sectional area can be obtained according to equation (6), and the relation between geometrical parameters and the energy density and vibration amplitude of the cross section can be obtained according to equation (1) and (3), as shown in table 2. It is clear in table 2 that when only the energy density is taken into account and other factors are ignored, the energy density of the cross section would decrease with the increase of the radius R, thickness b and anterior angle γ_0.

Table 2. Change relation between geometric parameters and energy density

Geometric parameters	Increase or decrease	Sectional area	Energy density	Vibration amplitude
Radius R	↑	↑	↓	↓
Thickness b	↑	↑	↓	↓
Anterior angle γ_0	↑	↑	↓	↓

4.2. Analysis of Bending Rigidity

When the energy density is the same, the greater the bending rigidity is, the smaller the vibration amplitude will be. Therefore, bending rigidity of the cutter is also one of the influence factors. This paper mainly measures the structural rigidity by calculating the deflection at the cross section of the disc cutter.

The cross section of the disc cutter is abstracted to be the cantilever beam in Fig 10. *q(x)* is related to the energy density. During the calculation, *q(x)* is regarded as a constant value. The elasticity modulus of the beam *E* is the elasticity modulus of the cutter material; the inertia moment *I* of the beam is not a constant value in practical conditions, but it is regarded as a constant value for simplifying the calculation.

The equation of bending moment is:

$$\mathrm{M(x)} = \int q(x)dx = \int q \cdot xdx = \frac{1}{2}qx^2 \ (0 \le x \le R) \tag{7}$$

The differential equation obtained is:

$$\frac{d^2\omega}{dx^2} = \frac{M(x)}{EI} \Rightarrow EI\omega'' = \frac{1}{2}qx^2 \tag{8}$$

According to the first integration:

$$\mathrm{EI}\omega' = EI\theta = qx^3/6 + C \tag{9}$$

According to the double integration:

$$\mathrm{EI}\omega = qx^4/24 + Cx + D \tag{10}$$

According to the boundary conditions, x=0, θ=0 are replaced in equation (8), and it can be worked out that C=0; x=0, ω=0 are replaced in equation (9), and D=0. C and D are replaced in equation (10), the deflection curve equation can be obtained:

$$\omega = qx^4/24\mathrm{EI} \tag{11}$$

The maximum deflection at the blade is:

$$\omega_{max} = qR^4/24\mathrm{EI} \tag{12}$$

It can be seen from equation (12) that the greater the radius R is, the greater the deflection at the blade will be, namely the greater the vibration amplitude will be. When the thickness b increases, the the equivalent inertia moment of the structure I increases, and the deflection at the blade and the vibration amplitude decrease. When the anterior angle increases, the equivalent inertia moment increases, while the deflection decreases. Therefore, the relation between geometrical parameters of the cutter and the inertia moment, and deflection at the blade can be obtained, as shown in table 3.

Table 3. Change relation between geometrical parameters and deflection

Geometric parameters	Increase or decrease	Inertia moment	Deflection	Vibration amplitude
Radius R	↑	→	↑	↑
Thickness b	↑	↑	↓	↓
Anterior angle γ_0	↑	↑	↓	↓

According to the equations in table 2 and table 3, it can be seen that with the increase of radius, the energy density decreases, but the rigidity becomes weaker. Therefore, there would be the choice range of optimal radius. With the increase of cutter thickness, the energy density decreases, and the rigidity becomes stronger, which would decrease the vibration amplitude at the blade. With the increase of the anterior angle, the energy density decreases, and the rigidity becomes stronger, so it would also decrease the vibration amplitude at the blade.

4.3. Analysis of Simulation Results

By analyzing the simulation results in Section 3 from two aspects of energy density and bending rigidity, it can be seen that when the radius is less than 13.5 mm, although the energy density at is high, the overall rigidity of the tool is strong and the vibration effect is suppressed, resulting in a small amplitude value. Conversely, when the tool radius is larger than 13.5 mm, the energy density reduced too much, also resulting in a smaller amplitude value. This means the vibration system has an ideal radius interval to maximize the amplitude value, and the parameter value of this interval should be preferentially selected in the design parameter selection. For thickness and anterior angle, as the value of the parameter increases, the energy density of the tool decreases while the bending rigidity increases, which causes the amplitude to decrease. These two parameters should take as small a value as possible within the allowable range.

To design the geometric parameters of the cutting disc, the influences of the geometric parameters of the tool on the energy density and the tool stiffness should be considered comprehensively to maximize the conversion efficiency of the tool.

5. Conclusion

1) A finite element model of the disc cutter and amplitude transformer was established, and modal analysis and harmonic response analysis were carried out. Characteristics of vibration amplitude at the blade of the integrated structure with the frequency were obtained. The simulation results shown that the maximum mean longitudinal vibration amplitude could be achieved for the integrated structure, and longitudinal vibration mode can be used in the ultrasonic-assisted cutting.

2) Harmonic analysis results shown the vibration amplitude increased firstly and decreased later with the increase of radius R, and the ideal values of the vibration amplitude was within 13mm and 14mm, the vibration amplitude decreased with the increase of cutter thickness, and within 0.8mm and 1mm, the cutter shown better stability in case of great vibration amplitude. The vibration amplitude decreased with the increase of anterior angle, the smaller the anterior angle the better within enough bending rigidity.

3) From the aspects of energy density and bending rigidity, the increase of the tool radius would reduce the energy density and decrease the bending rigidity, so there existed an optimized value range for the tool radius. The increase of the tool thickness and the anterior angle would reduce the energy density and increase the bending rigidity, which could increase the output amplitude.

Acknowledgements
The authors gratefully acknowledged the financial support for this research provided by National Natural Science Foundation of China (Grant No.51761145103 and 5187050582).

References
[1] Jin C Z 2006 *Study on High Speed Machining Process and Reliability of Fixture Method for NOMEX Honeycomb* (Zhejiang:Zhejiang University) (in Chinese)
[2] Roy R, Park SJ and Kweon JH 2014 *Characterization of Nomex honeycomb core constituent material mechanical properties* (Composite Structures) 117,255-266.
[3] Li X L, Xue W F, et al 1985 *Vibration Cutting and Its Application in Maching* (Beijing:Beijing Science & Technology Press). (in Chinese)
[4] Zha H T, Zhang J F,et al 2017 *An Experimental Study on Rotary Ultrasonic Drilling Small-diameter Holes of High Volume Fraction Silicon Carbide-reinforced Aluminum Matrix Composites (SiCp/Al)* (Proceedings of 2017 14th International Bhurban Conference on Applied Sciences and Technology) pp 34-39.
[5] Lucas M, MacBeath A, McCulloch E, at al 2006 *A finite element model for ultrasonic cutting* (Ultrasonics) 5-6, 519-526
[6] Nick MG, Akbari J,et al 2011 *Modeling and Optimization of an Ultrasonic Step Based on Combination of Finite Element Method and Mathematical Full Factorial Design* (Advanced Materials Research) 320, 553-558.
[7] Björnsson A, Johansen K and Alexandersson D 2013 *Three-dimensional ultrasonic cutting of RTM-preforms-a part of a high volume production system* (19th International Conference on Composite Materials(ICCM19)) pp 8960-8969
[8] Shen Y B 2014 *Study of An Ultrasonic Acoustic System of The Pointed Cutter* (Hangzhou: Hangzhou Dianzi University) (in Chinese)
[9] Wu X , Hu X P and Yu B H 2015 *Design of Ultrasonic Cutting Tool for an Ultrasonic Assisted Cutting Process of Nomex Honey-comb Materials Based on Substitution Method* (China Mechanical Engineering) 26(6), 809-813 (in Chinese)
[10] He W, Huang S, et al 2017 *Dynamics Analysis of Ultrasonic Milling Serrated Disc Cutter of NOMEX Honeycomb Composite Material* (Tool Engineering) 51(05), 37-40. (in Chinese)
[11] Fang L 2013 *Development of ultrasonic milling circular tool made of NOMEX cellular materials* (Hangzhou: Hangzhou Dianzi University) (in Chinese)
[12] Zhang Y D, Chen J, et al 2017 *Structural design of wide edge ultrasonic cutter* (Journal of Mechanical & Electrical Engineering) 34(12), 1382-1386. (in Chinese)
[13] Hu X P, Yu B H, Li X Y, et al 2017 Research on Cutting Force Model of Triangular Blade for Ultrasonic Assisted Cutting Honeycomb Composites (1st CIRP Conference on Composite Materials Parts Manufacturing (CIRP CCMPM 2017)) pp 159-163
[14] Zhang S F,Pang H W， et al 2018 *Study on Numerical Fitting Method of Vibration Frequency of Ultrasonic Cutting Vibration System* (Tool Engineering) pp 108-113 (in Chinese)
[15] Zhang S F, Wang J F, et al 2017 *Simulation Analysis of Tool Parameters Influence on Feeding Force and Cutting Temperature of Ultrasonic Cutting for Honeycomb* (Dalian: Journal of Dalian Jiaotong University) (in Chinese)
[16] Zhuang Y 2013 *Study on the Combined Machining Technology of Sawing and Grinding for Aramid Composites* (Dalian:Dalian University of Technology) (in Chinese)
[17] Chen J Q 2014 *Research on Wear Detection of NOMEX Honeycomb Core Ultrasonic Finishing Round Cutter* (Hangzhou: Hangzhou Dianzi University) (in Chinese)
[18] Cao G F 2004 Ultrasound Machining Technology (Chemical Industry Press). ISBN 7-5025-6102-1.(in Chinese)

A Study on Conductive Polymer in PCB Manufacturing

Binyun Liu[1], Liang Xiao[1], Yaqi Zhang[2], Meng Wang[2], Xuefeng He[2], Wang Xia[2], Zihong Zhou[2] and Liqin Qiu[2,*]

[1]Guangdong Toneset Science & Technology Co., Ltd, Guangdong Printed Circuit Chemicals Engineering Technology Center, Guangzhou 510288, China.
[2]School of Chemistry, The Key Laboratory of Low-Carbon Chemistry & Energy Conservation of Guangdong Province, Sun Yat-Sen University, Guangzhou 510275, China.

*Corresponding author

Abstract. Making Hole Conductive is one of the key technologies in Printed Circuit Board manufacture. For a long time, electroless copper was the unique technology of hole metallization process. However, it's long process, high water consumption and the resulting huge amount of waste water. A large number of toxic and harmful chemicals are also discharged into the environment, causing great harm to the health of workers and the environment. In this paper, the environmentally materialization technology-selective organic conductive coating (SOC) technology was studied. The resource consumption, product quality, and environmental and economic benefits of the traditional PTH and new SOC technology applied in the making hole conductive were compared and analyzed in detail through the investigation of long-term operation examples. The necessity, advantages and practicality of SOC technology were revealed.

1. Introduction

With the increasing demand for electronic products, the structure of PCB is becoming complex, resulting in high-density interconnect technology, commonly known as HDI (High Density Interconnect). HDI embodies the most advanced technology of contemporary PCB manufacturing, bringing the fine wire of PCB and the miniaturization of the aperture. With the high density and high performance of electronic equipment, the through holes and blind vias on the PCB are widely used, and the requirements for hole metallization are getting higher and higher. The hole metallization technology determines the reliability of the metallization performance between the layers of electronic products. It is one of the key process of PCB manufacturing. However, the traditional plating through hole (PTH) generates hydrogen gas during the process of reducing copper[1], which is not conducive to the manufacture of micro-blind holes, and it is necessary to find a new and reliable alternative process. On the other hand, the traditional hole metallization process is complicated. Whether it is from the manufacturing technology or from the cost, Making Hole Conductive (MHC) is one of the key process in the PCB manufacturing, and its technological progress is very important to improve PCB quality and reduce production costs. In addition, serious negative environmental impacts are also the main problems that traditional PTH urgently needs to solve[2]. At present, the hole metallization process is transitioning from the traditional electroless copper plating to the new technology direction, especially the selective organic conductive coating (SOC). SOC is an ideal hole metallization process[3]. In this paper, through theoretical analysis and specific examples, the technical, economic and environmental benefits of traditional methods and new conductive polymers used in the conductance of holes are studied in detail, and the necessity, advancement and feasibility of traditional electroless copper plating techniques are replaced by selective organic conductive polymers.

2. Research Methods

On the basis of analyzing the development of hole metallization technology, this paper firstly defines the traditional PTH and SOC process design parameters, and compares the material consumption, energy consumption, emissions and cost on a case-by-case basis to explore their respective characteristics. Then, through the company's five years operation data of these two types of large-scale production, long-running actual data statistics, through case comparison, analysis and discussion of the advanced performance and feasibility of SOC process.

3. Results and Discussion

3.1. Hole Metallization Principle, Application and Technical Analysis

(1) The core technology of circuit board production

No matter what kinds of printed circuit boards, the formation of line graphics is the first priority, so far, the mainstream method of forming line graphics is the graphics transfer method. For traditional two-layer or more boards, the line between the layers is conducted through the metal through holes, and the HDI board adds the micro-blind hole metallization technology. All the via holes must be metallized to achieve electrical conductivity. In order to metallize the interlayer insulating material, it is necessary to first conduct the conductors for subsequent copper plating deposition. Currently, the industrial metallization process is still based on traditional electroless copper.

(2) Analysis of the Principles and Technical Features of Traditional PTH and SOC Process

The traditional PTH process is to form a layer of conductive copper on the wall of the insulating hole [4]. The principle is to first adsorb a layer of palladium on the hole wall, and then under strong alkaline conditions, Using the palladium as catalyst, the Cu^{2+} is reduced to Cu by formaldehyde, and deposited on the wall of the insulating hole to conduct a metallization function for the subsequent thick copper plating. The strong complexing agent EDTA is used to stabilize the copper-containing material ion plating solution. The reaction mechanism is as follows:

$$Cu^{2+} + 2HCHO + 4OH^- \xrightarrow{Pd} Cu + 2HCOO^- + 2H_2O + H_2\uparrow \quad (1)$$

The process is long and generally requires 6 steps, as follows: Conditioner→Microetch→Predip→ Activator→Acceleration / Reduction→Electroless copper. The water and electricity resources and energy used in the PTH process are numerous, and the process control is complicated, which also brings many uncertainties to the stability of the process production. In order to solve this problem, the most effective method is to simplify the process and not use or use less pollutants that affect the environment. Through the unremitting efforts of the industry, different alternative technologies have been introduced, ATOTECH has successfully developed the use of sodium potassium tartrate [5] to replace EDTA complexing agent used in PTH process. Waste water treatment is relatively easy, and it has been widely used in PCB manufacturing; Shipley integrated sensitization and activation for the first time in one step [6]; later, RohmHaas developed a low palladium, low formaldehyde and non-accelerated PTH process, which can also save energy. Improvements in consumption and wastewater treatment [7]; Macdermid invented the Phoenix process, using sodium hypophosphite as a reducing agent to replace formaldehyde[8]; Guangdong Guanghua has studied glyoxylate as a reducing agent to replace formaldehyde[9]. However, these technologies have been improved in the original process, the process has not been shortened, the process is still complicated, and the problem cannot be solved fundamentally. In latest and simpler energy-saving and environmentally friendly processes, such as palladium[10], Carbon[11], Graphite[12], Graphene/Graphene Oxide[13], Pyrrole[14]/ Thiophene[15] conductive polymer, etc. are used, and a series of direct metallization processes have been developed, and the development prospects of conductive polymer processes are promising.

In the 1970s, Japanese scientist Hideki Shirakawa, American scientist Heeger AJ and Macdiarmid AG jointly invented conductive polymers and won the 2000 Nobel Prize in Chemistry. The invention of conductive polymers laid a new process for PCB hole metallization. The theoretical basis. In the 1990s,

Blasberg Oberflachentechnik GmbH introduced a direct metallization process using conductive polymers as conductive layer[16] and named DMS-E. Initially they chose polypyrrole as conductive polymer[17], but polypyrrole monomer has a very high vapor pressure at room temperature, affecting the health of operators; Zhang Peng et al. studied the technology of pyrrole film formation after secondary oxidation and find the process is complicated[18], while Poly(3, 4-ethylenedioxythiophene) (PEDOT) has better conductivity. Under the same conditions, the monomer vapor pressure is small, and the process only requires three steps [19]. Polythiophene replaces polypyrrole as conductive polymer for new generation of printed circuit boards since then.
The selective organic conductive coating(SOC) [20] process deposition of initiator on resin and glass fiber. The 3,4-ethylenedioxythiophene (EDOT) monomer is polymerized into poly 3,4-Ethylene dioxythiophene (PEDOT), which forms a layer of conductive polymer on an insulating substrate, it's a simple MHC process. The reaction mechanism of this process as follows:

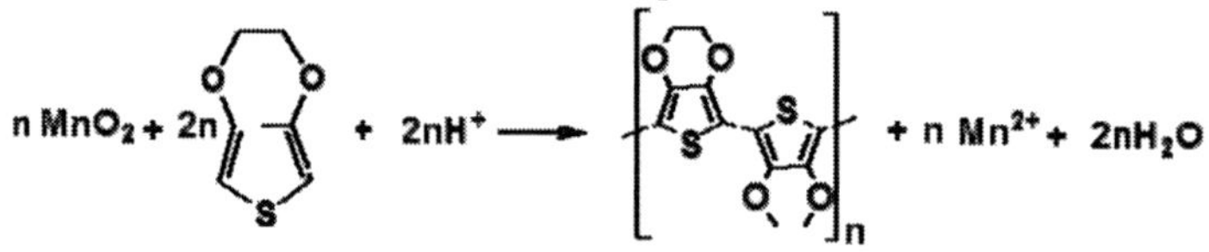

3.2. Analysis of Material Consumption, Energy Consumption and Cost of PTH and SOC Technology

Table 1 and Table 2 respectively show the control parameters of PTH and SOC process design of Guangdong Toneset Sci-Tech Co., Ltd.

Table 1. PTH process control parameters and main pollutants

Process	Material	Cont.	Consumption*	Time/min	Temp./℃	Main Pollutant
Conditioner	Cleaner 2613A	45 g/L	0.3 kg	5-8	45~55	Alkaline
	Cleaner 2613B	2.5%	0.2 kg			Surfactant
Micro-etch	$Na_2S_2O_8$	45 g/L	3.0 kg	1-2	20~30	Oxidant
	H_2SO_4 (98%)	2% (v/v)	2.7 kg			Acid
Pre-dip	Pre-dip 2624	200 g/L	2.0 kg	1-3	R.T.	Inorganic Salt
	HCl (37%)	1.0%	0.1 kg			Acid
Activator	Pre-dip 2624	200 g/L	0.1 kg	5-8	35~45	Inorganic Salt
	Activator 2616B	1.0%	0.5 kg			Pd^{2+}、Sn^{2+}
	HCl (37%)	1.0%	/			Acid
Acceleration	Accelerator 2627	8%	2.0 kg	2-3	25~35	Reducing Agent
Electroless Copper	PTH 2638A	5.2%	8.0 kg	18-25	28~34	Cu^{2+}
	PTH 2638B	6%	3.5 kg			Chelator
	PTH 2638E	0.2%	0.1 kg			Chelator
	HCHO (37%)	1.3%	4.0 kg			HCHO
	PTH 2638C	-----	14.0 kg			Alkaline
	PTH 2638D	3.0%	0.1 kg			Alkaline

*Consumption indicates the needed amount of the materials for the production of 100 m^2 printed circuit boards.

Table 2. SOC process control parameters and main pollutants

Process	Material	Cont.	Consumption*	Time/min	Temp./℃	Main Pollutant
Conditioner	Conditioner 2301A	5%	0.5 kg	0.9-1.2	48~52	Surfactant
	Conditioner 2301B	2%	0.2 kg			Surfactant
Initiator	Initiator 2302	120 mL/L	1.2 kg	1.0-1.5	86~90	$NaMnO_4$
	Boric Acid	10 g/L	0.1 kg			Boric Acid
Polymerization	Polymerizer 2303A	20 mL/L	0.5 kg	1.5-2.0	16~22	EDOT
	Polymerizer 2303B	40 mL/L	0.5 kg			Polysulfonate
	Regulator 2303C	pH1.9~2.1	0.1 kg			Inorganic Salt
	Phosphoric Acid	4 g/L	0.1 kg			Phosphoric Acid

*Consumption indicates the needed amount of the materials for the production of 100 m^2 printed circuit boards

It can be found that: (1) PTH chemical treatment process requires 6 steps and 14 materials, and the production of 100m^2 printed circuit board requires about 42.65 kg of chemicals; water consumption is 50 L/min, and treatment time is at least 1h. (2) The SOC process requires only 3 steps and 8 materials. The production of 100 m^2 printed circuit board requires about 3.2 kg of chemicals; the water consumption is 24 L/min, and the processing time is within 10 min. Therefore, the traditional PTH has a long process, requires a lot of chemical, consumes a large amount of water, and has a long time. The SOC process has made great progress in reducing material consumption and reducing resource use, and the production cycle is significantly shortened, product delivery is accelerated and efficiency is improved.

3.3. Analysis of the Types and Quantities of PTH and SOC Process Waste

According to the monthly production capacity of 30,000 square meters per line, based on the amount of chemical used in Table 1, Table 2 and the water consumption of the process, it is assumed that all chemical will not be affected by other factors such as bringing in, bringing out, evaporating, etc. The amount of waste bath and waste water discharged can be calculated as in Table 3.

Table 3. Theoretical Waste Liquid and Wastewater Discharge*

Process	Waste Solution (kg)	Waste Water (L)
PTH	12,795 = 42.65/100*30000	1,716,000 = 10*5*34320
SOC	9600 = 3.2/100*30000	823,680 = 8*3*34320

*Calculated according to effective working hours per month (34320 min = 26 days/month*22 h/day*60 min/h)

It can be seen that the SOC emissions are greatly reduced compared to the traditional PTH process, in which the waste liquid discharge is only 3/4 of the PTH, and the wastewater discharge is reduced to half. On the point of the main material components, PTH needs to use copper, palladium, tin and other metal materials, as well as formaldehyde, complexing agents and other materials, so there are corresponding emissions of these substances, copper bath waste also needs to be dealt with, the process is complicated and processing cost is high; while the SOC not only has a small amount of waste liquid discharge, but the processing is relatively simple, the corresponding processing cost is low, and the amount of harmful substances is greatly reduced.

Correspondingly, cost comparison accounting for two processes shows that PTH requires 10 kinds of special materials, while SOC only needs 6 kinds. From the cost of raw materials and usage, the cost of SOC special materials is only half of PTH. Other acid and alkali and other auxiliary materials are also used less frequently; in terms of labor, PTH requires manual upload and download plates, while SOC can use automatic discharge and closing machine, all the chemical can be automatically added which greatly saves labor costs; In terms of cost of water, electricity and wastewater treatment, the cost of SOC also has obvious advantages. Overall, the cost of SOC is about 44% lower than that of traditional PTH.

3.4. Actual Operation of Two Processes

Based on above PTH and SOC process design and analysis, we take the SOC of the PCB manufacturer Guangzhou EE factory and the PTH line of Huizhou GCS factory as an example (the average production capacity is 30,000 square meters / month) and further conduct the actual operation inspection. To produce the same amount of PCB, statistical analysis of its water and electricity consumption, waste water discharge, waste disposal, labor costs, etc., as well as efficiency, yield power and other quality management. It can be seen that the SOC process exhibits significant advantages in practical applications.

The SOC process has been applied in the PCB manufacturing plant for more than five years. Compared with the traditional PTH process, it has obvious quality, environmental protection and cost advantages. The following results are obtained through statistics: Firstly, from the perspective of quality control, the traditional Vertical electroless copper plating line needs to check the back-light every 4 hours. When the quality control is fine, the average back-light check is at least 2 times per month. 0.1% to 0.3% of the PTH process rejection is caused by hole skipping plating. The SOC process only needs to check backlight per shift, basically no backlight problems, and the proportion of the SOC process affected by the finished product scrap is below 50 ppm, which not only improves the yield power, but directly reduce the loss caused by the quality scrap. It ensures on-time delivery and produces significant economic benefits.

Secondly, from the perspective of environmental protection, for production scale of 30,000 ㎡ of circuit boards per month, the amount of waste solution and waste water produced by statistics are shown respectively in Table 5 and Table 6. It can be seen that the SOC waste discharge types and emissions quantity are significantly reduced. Among them, the total amount of SOC waste liquid is less than 1/3 of PTH and the discharge of waste water is only 1/2 of PTH.

Table 4. The use of chemicals and pollutant discharge in GCS PTH process*

Process	Chemical	Consumption	Daily waste solution (T)	Daily waste water (T)	Main Pollutant
Conditioner	2613A/B	0.8 kg//100m^2	0.01	9.6	COD
Micro-etch	$Na_2S_2O_8$	6.0 kg//100m^2	0.075	9.6	Oxidant, Cu^{2+}
Pre-dip	2624	2.2 kg//100m^2	0.0275	9.6	Inorganic
Activator	2616B	1.0 L//100m^2	0.0125	9.6	Pd^{2+}
Acceleration	2627	2.0 L//100m^2	0.025	9.6	Reducing agent
Electroless Copper	2638A/B/C	21.57 L//100m^2	0.2695	9.6	Cu^{2+}, EDTA HCHO, COD
	Total (T/day)		0.4195	57.6	All above
	Total (T/month)		10.907	1497.6	All above

*The data were acquired at the production scale of 30,000 m^2 PCB per month

Table 5. The use of chemicals and pollutant discharge in EE SOC process*

Process	Chemical	Consumption	Daily waste solution (T)	Daily waste water (T)	Main Pollutant
Micro-etch	$Na_2S_2O_8$	7.7 kg//100m^2	0.096	7.2	Oxidant, Cu^{2+}
Conditioner	2301A/B	0.74L//100m^2	0.01	7.2	COD
Initiator	2302	1.1 L/ /100m^2	0.014	7.2	Oxidant
Polymerization	2303A/B/C	0.5 L//100m^2	0.006	7.2	COD
	Total (T/day)		0.126	28.8	
	Total (T/month)		3.276	748.8	

*(a) A micro-etching process was added in the front of SOC process for the production of multi-layer boards. (b) The data were acquired at the production scale of 30,000 m^2 PCB per month.

4. Conclusion

Through research and running case analysis, it can be found that under the premise of ensuring product quality, the SOC process is convenient to operate, the process is short, and the production process is easy to be automatically controlled, which can greatly reduce the production cycle of the product, reduce the production cost, and improve the production efficiency. At the same time, the production process has low emission of toxic waste and is environmentally friendly, which greatly reduces the wastewater discharge and environmental pressure during the PCB production process. Therefore, whether it is from the perspective of technology, economy or environmental protection, SOC technology has obvious advantages, it is general trend of SOC TO replacing the traditional PTH process.

Acknowledgements

We would like to thank Science and Technology Program of Guangzhou (201604016037) for financial support.

References:

[1] Radovsky D. A., Ronkese B. J., US Pat. 3099608(A), 1963.
[2] Wang G., Han J., Li N., Li D., Plating and Finishing, 2005, 27(2), 21-23
[3] Liu B., Xiao L., Su C., 2017 Spring International PCB Technology / Information Forum, 218-223
[4] Stafstrom E., Circuit Tree, 2004, 10, 12-14
[5] Liu S., Yang H., Long C., J. Kaili Uni., 2007, 25(6), 37-40
[6] Song B. Z., Zhao Y. P., Cai Z. S., Papers Collection of Shanghai Symposium on New Dyeing and Finishing Technologies, Energy Conservation and Envir. Prot., 2013, 197-200
[7] Shipley Jr. C. R., US Pat. 3011920, 1961
[8] Xie Y., Zhang X. S., Chem. Eng. Equip., 2011(1), 16-18
[9] Li W., Liu B., Wang H., Xue H., Printed Circuit Information, 2006, (8), 24-27
[10] Vaskelis A., Norkus H. J., Rozovskis G. & Vinkevicius H. J., Transaction of Institute of Metal Finishing 1997, 75(1), 1-3
[11] Lee H., US Pat. 2005199504 (A1), 2005
[12] Carano M., Circuit World, 1999, 25(3), 18-22
[13] Dow W.-P., Chang S.-C., US Pat. 9105696(B1), 2015
[14] Ragge A., Meyer H., Urrutia D. G., US Pat. 5597471(A), 1997
[15] Hupe J., Kronenberg W., US Pat. 5194313(A), 1993
[16] Kronenberg W., Hupe J., DE Pat. 3806884C1, 1989
[17] Cai J. Q., Electroplating & Pollution Control, 1999, 19(2), 10-12
[18] Zhang P., He D., Journal of Sichuan Institute of Light Industry and Chemical Technology, 2000, 13(3), 63-67
[19] Liu B., Xiao L., He C., Eletroplating & Finishing, 2018, 37(8), 350-355
[20] Liu B., Su C., Gao J., Xiao L., Printed Circuit Information, 2013, (Z1), 131-136

Development of CPP Based Low Temperature Heat Sealing Film Material

Ligui Xiong[1], Zhigang Tu[2,*], Shangxian Zhang[3] and Liqiong Zhang[4]

[1,2,3]Guangdong Polytechnic Institute, Zhongshan, China
[4]Zhongshan torch polytechnic, Zhongshan, China

*Corresponding author

Abstract. The casting polypropylene (cpp) heat sealing film was prepared by adding the homemade smooth masterbatch, in the storage period of 30 days measured the surface tension is greater than 38 mN/m, the non-corona surface friction coefficient between 0.1 ~ 0.2, corona surface is 0.5, the longitudinal tensile strength reached 60 MPa, transverse tensile strength reached 27 MPa, the films have high surface tension, low friction coefficient, high mechanical strength properties.Measured at the same time its initial heat sealing temperature, heat sealing strength and optical properties are better than the national standard, to maintain film comprehensive properties are stable during the storage period.

1. Introduction

CPP (casting polypropylene) heat sealing film has the characteristics of high transparency, good mechanical properties and high heat sealing strength, and is widely used in the packaging industry. However, it tends to be brittle at low temperatures, and the toughness is deteriorated, which limits its use under low temperature conditions. Domestic low temperature modification of CPP heat sealing film has been a research hotspot. For example, the use of polyolefin elastomer to blend and modify it can reduce the film haze and improve the transparency. It is expected to replace the low transparency used in low temperature refrigeration. PE (polyethylene) film. However, the elastomer has problems of stickiness and blemish, which causes the mechanical strength of the film to decrease and the friction coefficient to increase. [1-6]

In this paper, the self-made smooth masterbatch is added in the preparation process of CPP heat sealing film. The produced film has the characteristics of high surface tension, low friction coefficient and large mechanical strength, and can maintain the performance of the film stable within 30 days of storage period.

2. Experimental Part

2.1. Main Raw Materials

Polypropylene homopolymer (FC-801, Sinopec Shanghai Petrochemical Co., Ltd.), polypropylene copolymer (H531PL, Singapore Polyolefin Corporation) and other commercially available. Sleek masterbatch homemade.

The self-made smooth masterbatch carrier is a polypropylene homopolymer which is granulated by a twin-screw extruder and contains 10% high-purity non-toxic erucamide, water ≤0.06%, and melt flow rate of 5～7g/10 min (temperature 230°C, load 2.16 kg), added to the core layer of the film.

2.2. Basic Formula and Sample Preparation

Basic formula (parts by mass): heat seal layer (layer A): copolymerized polypropylene resin 95.0% to 97.0%, auxiliary agent (3.0 to 5.0) g/kg; core layer (layer B): homopolymer polypropylene resin 97.0%~97.5%, auxiliary agent (2.5-3.0) g/kg; corona layer (C layer): homopolymer polypropylene resin 96.0% to 96.5%, auxiliary agent (2.0-3.5) g/kg.

Sample preparation: CPP heat-sealing film was prepared by three-layer co-extrusion casting method of A, B and C. The process flow was: CPP resin, other additives→metering→extrusion plasticization→shunt→die→casting cooling Molding→Thickness measurement→Corona treatment→Trimming→Winding→Finished product after 48h normal temperature aging treatment →slitting→inspection→CPP heat sealing film.

2.3. Performance Test Method

The surface tension is tested according to GB/T 14216—2008 "Determination of Wetting Tension of Plastic Film and Sheet"; the friction coefficient is tested according to GB 10006-1988 "Method for Measuring Friction Coefficient of Plastic Film and Sheet"; Light transmittance/haze reference GB/ T 2410-2008 "Determination of transparent plastic transmittance and haze" test; mechanical properties refer to GB/T 13022-1991 "plastic film tensile properties test method" test; gloss according to ASTM D523-1989 (1999) material gloss Degree test; heat shrinkage rate is tested according to GB/T 12027-2004 "Test method for dimensional change rate of plastic film and sheet heating"; heat seal strength is tested according to QB/T 2358-1998 "Test method for heat seal strength of plastic film packaging bags". [7-11]

3. Results and Discussion

3.1. The Effect of Storage Time on the Surface Tension of the Film

A self-developed smooth masterbatch is added to the formulation to produce a 3-layer structure of 25 μm thick CPP heat seal film to track the surface of the CPP film within 30 days (1, 2, 3, 14, 21, 30). The tension changes and the result is shown in Figure 1.

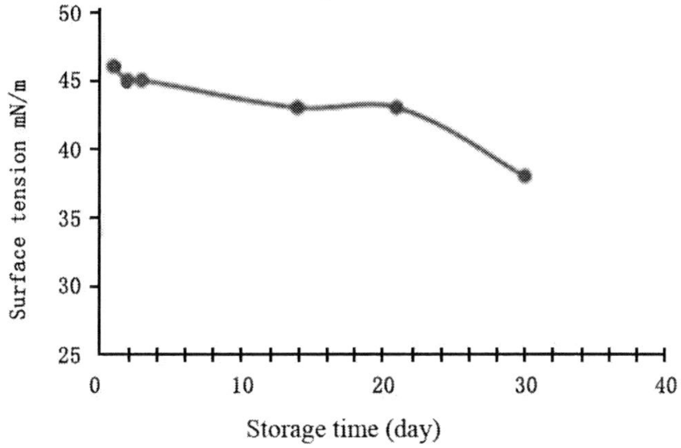

Figure 1. The surface tension change of CPP heat sealing film within 30 days.

When the CPP heat-sealing film was just off the line, the surface tension value reached a peak value of 46 mN/m, and began to decrease with time in the 30 days after tracking. At the 30th day, it was 38 mN/m. It can be seen that the CPP heat seal within 30 days of the storage period. The film can meet the requirements of surface tension of more than 38 mN/m for secondary processing such as printing and coating.

During the processing of CPP heat-sealing film, the external tensile force causes the mechanical properties of the orientation direction to increase significantly. Under the condition of rapid quenching, the viscosity of the CPP melt is high, and the internal stress cannot be relaxed quickly, usually after 3 days of the offline. At the beginning of the relaxation, the properties of the film changed. It can be seen from Fig. 1 that the surface tension of the film decreased significantly after the third day. In the study, the self-made smooth slip masterbatch contains 10% high-purity non-toxic erucamide, which has smooth and anti-blocking effect, and can migrate from the film core layer to the film surface during storage, which relieves to some extent. The decrease in surface tension value is due to the fact that the lubricity of erucamide increases the mobility of movement between the macromolecular chains or segments of the polypropylene, and the surface of the film maintains a high surface tension for a long period of time.

3.2. The Effect of Storage Time on the Friction Coefficient of the Film

A self-developed smooth masterbatch is added to the formulation to produce a 3-layer structure of 25 μm thick CPP heat seal film, and the corona film is tracked within 30 days (1, 2, 3, 14, 21, 30) after the CPP film is taken offline. The coefficient of friction between the surface and the non-corona surface is shown in Figure 2.

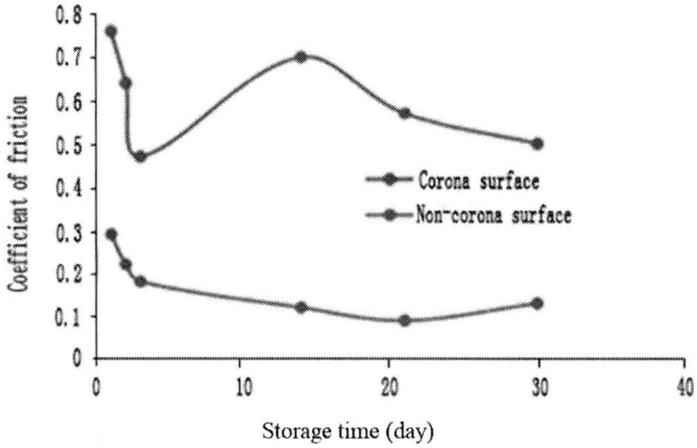

Figure 2. The friction coefficient change of CPP heat sealing film within 30 days.

Under normal circumstances, the migration of the smoothing aid can rapidly reduce the friction coefficient of the film, but the loss of the smoothing aid in the later migration also causes the surface friction coefficient of the film to gradually increase. It can be seen from Fig. 2 that the friction coefficient of the non-corona surface when the film is just off the line is 0.3, and then starts to decrease, and is stable between 0.1 and 0.2 during the tracking period. The corona surface is forced to break due to electron breakdown, air oxidation, etc. on the surface of the film to form a partial region. The unevenness of the film surface causes the various additives in the inner layer of the film to migrate to the film. After the surface is easy to disperse unevenly, the friction coefficient of the corona surface is obviously increased after being placed for a period of time, and the stability of the friction coefficient can be impaired. After adding the self-made slippery masterbatch, the friction coefficient of the corona surface of the CPP heat seal film changed greatly after the line was removed, but the loss increase after aging treatment was stable below 0.6 after 21 days of tracking, and decreased to 0.5 after 30 days. The self-made smooth masterbatch can effectively ensure the stability of the friction coefficient of the CPP heat seal film.

3.3. The Effect of Storage Time on the Tensile Strength of the Film

A self-developed smooth masterbatch was added to the formulation to produce a 25-μm thick CPP heat seal film with a 3-layer structure, and the pull of the CPP film within 30 days (1, 2, 3, 14, 21, 30) was traced. The tensile strength is shown in Figure 3.

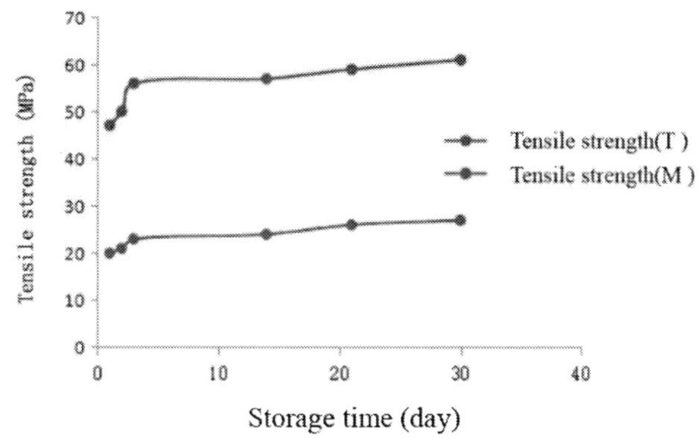

Figure 3. The tensile strength change of CPP heat sealing film within 30 days.

As the storage time increases, the tensile strength of the film continues to increase, with the largest change in the first 3 days of the off-line. After 30 days, the film has a tensile strength of 60 MPa and a transverse tensile strength of 27 MPa. It is because when the film is just off the wire, the polymer chain and the segment are oriented in the direction of the tensile force in the film, and the internal stress is generated. After 3 days, the internal stress begins to relax, and the polymer chain and the segment occur. Retracting or entanglement, the straightening state is destroyed, and the variation range of the transverse and longitudinal tensile strength of the film fluctuates obviously; with the increase of the standing time, some polymer chains and segments are under the action of the environment. The change of stress returns to equilibrium, so the tensile strength increases gently in the first 3 days before the line, and the growth rate is basically small in the 21- to 30-day interval, which tends to be stable.

3.4. Comprehensive Performance of CPP Heat Sealing Film

When the CPP heat seal film with a total thickness of 25 μm was placed for 30 days, it was compared with the national standard GB/T 27740-2011 "cast polypropylene (CPP) film". The results are shown in Table 1. It can be seen that the film produced by the CPP heat sealing film prepared by using the self-made smooth masterbatch can maintain the high surface tension, low friction coefficient, high mechanical strength and meet the national standard during the storage period, and the initial heat sealing temperature and heat sealing. Both strength and optical performance are superior to national standards, and the comprehensive performance of the product is stable.

Table 1. Effect of slipping masterbatch on other Porperties of CPP heat sealing film.

Test items	GB/T 27740-2011	Measured
Thickness,μm	-	25
Tensile Strength MPa,MD	≥35	60
Tensile Strength MPa,TD	≥25	27
Elongation at break %,MD	≥280	570
Elongation at break %,TD	≥380	650
Gloss %	-	82.9
Haze %	≤5.0	2.5
Initial heat sealing temperature℃	<145	115
Heat seal strength N/15mm	≥8	10.14
Friction coefficient (hot cover)	≤0.5	0.5
Surface tension (treatment surface)mN/m	≥38	38

4. Conclusion

The surface tension of CPP heat-sealing film prepared by using self-made smooth masterbatch is more than 38 mN/m when stored for 30 days; the friction coefficient of non-corona surface is stable between 0.1 and 0.2 during the tracking period, and the corona surface is stable at 0.5.

With the increase of storage time, the tensile strength of the film continued to increase. After 30 days of standing, the longitudinal tensile strength of the film reached 60 MPa, and the transverse tensile strength reached 27 MPa, which was higher than the national standard.

The initial heat-sealing temperature, heat-sealing strength and optical performance of the CPP heat-sealable film prepared by using the self-made smooth masterbatch are better than the national standard at the storage period of 30 days, and the comprehensive performance is stable.

References

[1] D.N.LIAO, N.N.DUAN and Y.Z.LIN, *"Study on properties of CPP heat sealable film low temperature heat seal layer"*.China Plastics,vol.29,no.8,pp.77-81,2015.

[2] S.L.TENG. *"Selection and optimization of CPP cast film "*.New Chemical Materials, vol.41,no.8,pp.190-191,2013.

[3] F.Y.SHI. *"Study on Corona Treatment of Low Temperature Cast Polypropylene Packaging Film in Keli Printing"*. China Plastics, vol.26,no.5,pp.74-77,2012.

[4] Y.Z.LIN and W.Z.Chen. *" Development of CPP heat sealing film for high performance modified atmosphere packaging "*.China Plastics, vol.25,no.10,pp.69-71,2011.

[5] F.Y.SHI. *" Research on Formulation Process of High Transparent and Low Temperature Casting Polypropylene Packaging Film "*.China Plastics, vol.26,no.2,pp.59-62,2012.

[6] L.ZHANG,*"Structure and properties of polypropylene terpolymer for CPP film"*.China Synthetic Resin and Plastic, vol.33,no.5,pp.49-52,2016.

[7] S.JIANG,P.SHEN and X.XIE,*"The platelet structure of polypropylene cast base film with different chain structure and its tensile pore formation"*.Polymer Materials Science and Engineering, vol.30,no.11,pp.98-101,2014.

[8] X.XIE,XC.YANG and Q.YANG,*"Hard elastic structure and tensile pore formation of polypropylene cast base film"*. Polymer Materials Science and Engineering, vol.29,no.5,pp.109-116,2013.

[9] H.W.SHEN,X.J.LI and Y.LI, *"Extrusion casting process and tensile pore formation of polypropylene hard elastic film"*.Engineering Plastics Application, vol.44,no.8,pp.54-59,2016.

[10] G.WANG, *"Structure and properties of polypropylene resin for cast film"*. China Synthetic Resin and Plastics, vol.29,no.6,pp.36-39,2012.

[11] K.F.YU, *"Effects of Annealing Conditions on Oriented Lamellar Structures of PP Cast Precursor and Microporous Membrane Formation by Uniaxilly Stretching"*. China Synthetic Resin and Plastics, vol.29,no.2,pp.84-89,2015.

Facile Preparation of Ag-TiO₂ Nanocomposite with Enhanced Photocatalytic Properties

Tao Tian[1] and Chuangye Yao[2,3,*]

[1]School of Metallurgy and Environment, Central South University, Changsha 410083, China
[2]School of Physical Science and Technology, Lanzhou University, Lanzhou 730000, China
[3]School of Materials Science and Engineering, Sun Yat-Sen University, Guangzhou 510275, China

*Corresponding author

Abstract. For decades, photocatalytic semiconductor materials were regretfully limited to low-yield preparation method and poor catalytic efficiency although it has dramatically captured attention in sewage treatment applications. To expand this slim range of semiconductor catalysts, the nanocrystalline silver embedded in titanium dioxide (Ag-TiO_2) heterojunction was prepared by means of facile mechanical milling technique capable of large-scale production, in terms of optimum Ag content. The as-obtained Ag-TiO_2 nanocomposite particles were systematically characterized by means of X-ray diffraction technique, Raman spectra, X-ray photoelectron spectroscopy, transmission electron microscopy analysis, and UV-vis spectrophotometry. Results suggest that in comparison with pure TiO_2 catalyst, Ag-TiO_2 nanocomposite catalyst significantly enhanced photocatalytic activity for the degradation of methylene blue (MB), whereas its band gap decreased in view of the red-shift phenomenon. In terms of the experimental results, a mechanism was proposed to explain the enhancement of photocatalytic activity. Our findings offer a promising route towards the large-scale preparation of high-efficiency photocatalysts for commercial applications.

1. Introduction

Semiconductor photocatalysis is one of the most straightforward and efficient approaches for treating environmental water pollution. Semiconductor materials in photocatalytic technology have the advantages of high efficiency, stable chemical properties, and processing pollutant degradation at room temperature in sewage rehabilitation engineering.[1] As early as in 1972, Fujishima and Honda presented a hydrogen generation effect of TiO_2 based on electrochemical decomposition of water, which motivated intensive efforts devoting to the TiO_2 based catalyst study in the past decades.[2] TiO_2 is a versatile and technologically important material with abundant photocatalytic activity, which has been intensively investigated in the field of decomposition of organic pollution thanks to their low-cost and good environmental stability in nature.

However, the available experimental results about this TiO_2 based photocatalysis are unfortunately limited to two key points in preparing suitable samples for studying the photocatalytic effect as follows: (i) wind band gap and high electron-hole recombination rate restrict the photocatalytic efficiency,[3] and (ii) current reports on TiO_2 photocatalysis are limited to small-scale laboratory production due to the confinement of preparation method of photocatalyst, preventing the practical commercial applications. For improving the photocatalytic efficiency of TiO_2, many approaches have been carried out to overcome this scientific issues through several technique such as introducing metal nanoparticles, coupling of semiconductor, and synthesis of titania with high surface area (thin films, nanosheets, etc.).[1,4,5] Among them, introducing metal nanoparticles is the most extensive and feasible method, which is highly effective for enhancing photocatalytic activity. The role of metal nanoparticles can not only promote the separation of photoinduced charge carriers but also provided an alternative reaction pathway to the reaction products with a lower activation energy, subsequently enhancing the photocatalytic activity. Also, the sterilization effect for metal nanoparticles (i.e. Ag) is beneficial so that the metal-semiconductor heterojunction structure can be a promising application for biological wastewater treatment of enriched bacteria.[6] To expand the slim range of small-scale preparation is also the key issue for realizing degradation of organic pollution in large-area commercial applications. Previous methods preparing photocatalysts are mainly to focus on chemical solution method.[5] This method is considerably limited as compared to mass production and industrialization in treating large-area sewage owing to that it needs strict reaction conditions, complex steps, and after-treatment. Recently, mechanical milling for preparation of photocatalysts has become an interesting research hot due to its characteristics via a simple and feasible technique for processing advanced materials.[7] In particular, this method allows us to develop a large-scale preparation of semiconductor photocatalysts for wastewater treatment.

Herein, we prepared the Ag-TiO_2 nanocomposite photocatalyst via a facile mechanical milling, and then characterized their crystal structure, particle size, surface morphology, band gap, and chemical bonding states. We also investigated the photocatalytic properties of the Ag-TiO_2 nanocomposite. In addition, the photocatalytic mechanism of the Ag-TiO_2 nanocomposite has been discussed in detail.

2. Experiment Section

2.1. Synthesis

Commercial silver (purity 99.95%) and titanium oxide (purity 99.99%) powders were used as raw materials, the nanosized Ag-TiO_2 heterojunction catalyst was prepared by milling small-amount Ag and high-amount TiO_2 particles in terms of the optimum ratio (i.e. 2 wt.%), which occurred in a WC vial with WC balls under air atmosphere by means of a Fritsch P4 planetary ball mill. The mixed raw powders (Ag and TiO_2 particles) were milled at 250 rpm for 4 h complying with ball-to-powder weight ration of 50:1, then the ball-milled samples were thermally treated in air ambient at 400 °C for 1 h. Two types of catalyst, pure TiO_2 (control group) and 2% Ag-TiO_2 nanocomposite were synthesized for comparative study.

2.2. Characterization and Measurements

The crystal structure of the catalysts was detected by means of a X-ray diffraction technique(XRD: Rigaku D/Max-2400 X-ray diffractometer) measurements with K_α radiation of copper target. The average grain sizes of silver and titanium oxide nanoparticles were estimated by using the Scherrer equation according to the line broadening of the XRD diffraction peaks. The Raman spectra (Horiba Jobin Yvon LABRAM H800) was recorded with a laser wavelength of 532 nm. The chemical states of the catalysts were examined by a X-ray photoelectron spectroscopy analysis (XPS: PHI-5702) with Al Kα radiation. Observation based on transmission electron microscopy (TEM: FEI Tecnai G^2 F30) characterization was carried out with a beam energy of 300 KeV. The morphology of the catalysts was observed. Light absorption measurements were implemented by using UV-Vis spectrophotometer (CSHI-MADZU, UV-3101) to analyze the absorption range and band gap of the catalysts. Photocatalytic performance was tested on a UV-2401 spectrophotometer. Lamp source of 500 W monitored ultraviolet by using a mercury lamp with a wavelength focused on 365 nm. In the experiment, the as-prepared photocatalyst samples of 20 mg were added into the MB solution of starting concentration 20 mg/L about 10 mL. Before the irradiation treatment, the suspension was magnetically stirred under dark conditions for 0.5 h in order to obtain the establishment of the adsorption/desorption equilibrium of the MB solution onto the surface of photocatalysts. Then, light irradiation was added into the mixture of MB solution and photocatalysts.

3. Results and Discussion

Figure 1a illustrates the typical XRD pattern of Ag-TiO_2 nanocomposite particles. Partially indexed diffraction peaks can be assigned to those of the tetragonal TiO_2 phase (JCPDS No. 21-1272), being anatase structure. In addition to TiO_2 matrix diffraction peaks, four obvious diffraction peaks were indexed to the (111), (200), (220), and (311) planes in the standard spectrum of metallic silver (JCPDS No. 04-0783).[8] No impurity phase such as Ag_2O was examined in the XRD patterns. According to the Scherrer equation, the average grain sizes of metallic Ag and TiO_2 phase were estimated to be ~25 nm and ~40 nm, respectively, which confirmed nanocrystalline in nature.

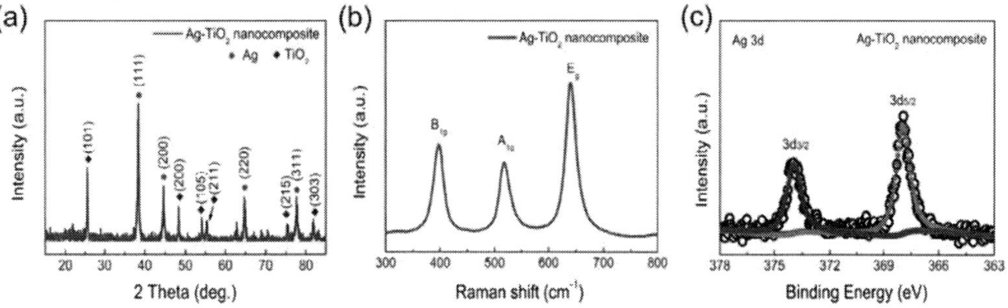

Figure 1. (a) XRD pattern and (b) Raman spectra of Ag-TiO_2 nanocomposite photocatalyst. (c) XPS spectra of Ag-TiO_2 nanocomposite photocatalyst for Ag 3d region.

Raman spectroscopy is one of the powerful methods to further investigate the structural characteristics of Ag-TiO_2 nanocomposite photocatalyst (figure 1b). The Raman vibration peaks of 400, 518.2, and 640.9 cm^{-1} can be assigned to the vibration modes of B_{1g}, B_{2g}, and E_g of anatase TiO_2 phase, respectively,[9] which is in accordance with the Raman analysis of ~30 nm anatase TiO_2 reported by Choi et al. and further confirmed the nanocrystalline in nature.[9]

The chemical state of embedded Ag was further studied by XPS technique, as displayed in figure 1c. It was observed that two obvious energy binding peaks appear at 367.9 eV and 373.9 eV, which corresponds to Ag $3d_{5/2}$ and Ag $3d_{3/2}$ peaks in Ag-TiO_2 nanocomposite, respectively.[8] This result suggested that Ag exists in the metallic state (Ag^0), agreed well with the XRD analysis.

Figure 2 illustrates TEM observations of the synthesized Ag-TiO_2 nanocomposite particles. These particles can be clearly distinguished in the bright field image (figure 2a) though agglomerated particles occurred. The crystal structure of Ag phase was investigated by means of the selected area electron diffraction (SAED) pattern which could be marked by the red circles, as shown in figure 2b. The diffraction spots of the SAED image were successfully indexed by comparing XRD JCPDS cards (No. 04-0783),[8] which further confirmed the XRD and XPS results. Figure 2c depicts the high-angle annular dark field (HAADF) image recorded with a scanning transmission electron microscope (STEM) for Ag-TiO_2 nanocomposite particles, where the bright field contrast is from embedded Ag nanoparticles due to that image formation in HAADF-STEM is sensitive to the difference of elemental atomic number. A magnified STEM image for further analysis, as shown in figure 2d, in which the diameter of the Ag nanoparticles was estimated to be about 18-27 nm, the same as the particle size found from XRD measurements.

The light absorption characteristics *versus* wavelength range curves of pure TiO_2 sample and Ag-TiO_2 nanocomposite sample were investigated and are displayed in figure 3. The band gap energy, calculated from the intercept of the tangents ($Eg= 1240/\lambda$) of the plots,[4] is 3.25 eV and 3.13 eV for the pure TiO_2 and Ag-TiO_2 nanocomposite particles, respectively. The UV-Vis spectra of the prepared Ag-TiO_2 nanocomposite indicated an obvious red-shift phenomenon in comparison with the pure TiO_2 particles, which should be ascribed to the size effect of the band gap energy. For the milled Ag-TiO_2 nanocomposite, the Femi level between the conduction band (CB) and valence band (VB) shifts down, accompanying with that the electrons in VB is easy to be excited by lower energy light and then migrate to CB. Nanocrystalline Ag-TiO_2 heterojunction structure materials exhibited a larger absorption range in the 200-400 nm region, which is effective to enhance the photo-quantum efficiency and photocatalytic activity.

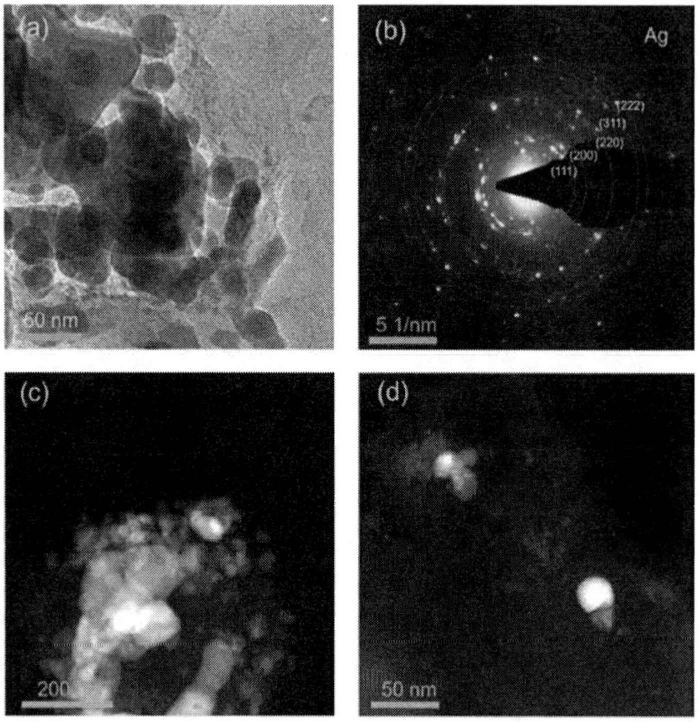

Figure 2. TEM images of Ag-TiO_2 nanocomposite particles: (a) bright field, (b) SAED pattern, (c) STEM image, and (d) a magnified STEM image.

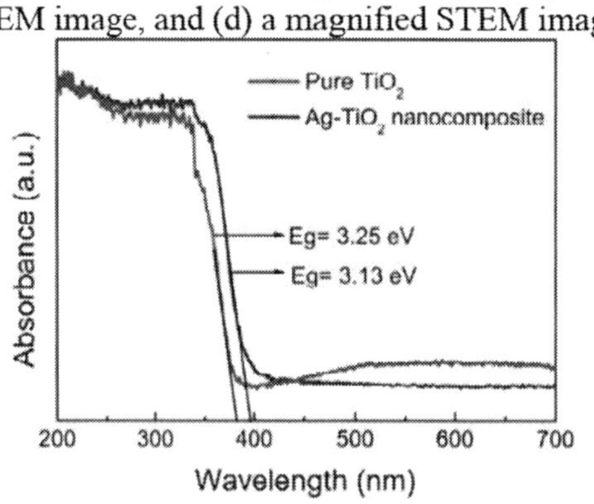

Figure 3. UV-visible absorption spectra of pure TiO_2 and Ag-TiO_2 nanocomposite particles.

Figure 4a illustrates the photo-degradation rate of MB in different photocatalysts under UV irradiation (wavelength peak at 365 nm). Prior to photocatalytic treatment, MB dye solutions with catalysts were continuously stirred under dark environment for 30 min in order to achieve a stable equilibrium condition. Experimental results indicated that the adsorption of MB molecules on catalysts is almost negligible due to the nearly unchanged C/C_0 value. After light irradiation for 60 min, the maximum of MB degradation efficiency has achieved to be 85% in Ag-TiO_2 nanocomposite catalyst, whereas only 41% of MB dye molecules can be decomposed by the pure TiO_2 catalyst. As shown in figure 4b, the resultant degradation process of MB is in well agreement with the first-order kinetics model according to Langmuir-Hinshelwood expression, and their kinetics can be expressed as follows:

$$\text{Ln}\,(C_0/C) = kt \qquad (1)$$

Where C_0 denotes the initial concentration of aqueous MB, C denotes the concentration of aqueous MB at the react time t, and k is the apparent reaction rate. The determined k value was shown as linear slope, and the determined reaction rate constants (k) were $8.21\text{x}10^{-3}$ min^{-1} and $32.01\text{x}10^{-3}$ min^{-1} for the pure TiO_2 and Ag-TiO_2 nanocomposite catalysts, respectively.

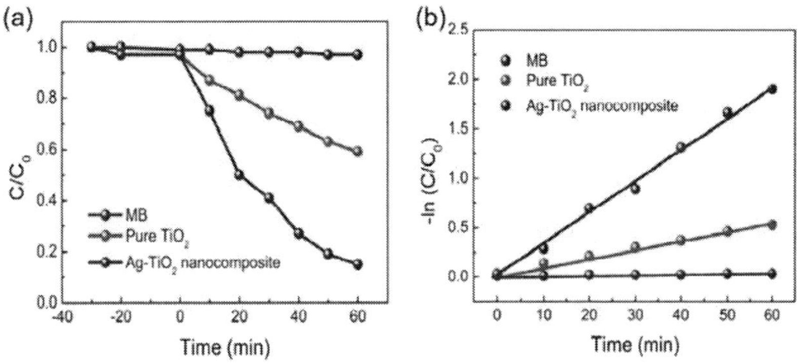

Figure 4. (a) Photo-degradation rate of MB dye by different photocatalysts, (b) The fitting linearly of -ln (C/C$_0$) versus the irradiation time.

On the basis of the experimental results and analysis, the probably photocatalytic reaction mechanism was proposed to explain the enhancement of photocatalytic activity, as shown in figure 5. A tentative pathway of charge transfer in the photocatalytic process was revealed. We know that light irradiation is helpful to generate photoinduced electrons (e^-) and hole (h^+) (Eqs. 2). The electrons of the TiO_2 conduction band is easy to access reducing ability, then reacting with O_2 that is dissolved in the aqueous solution to create superoxide radical ($\cdot O_2^-$) (Eqs. 3).[1] Whereas, hydroxy (OH^-) in the adsorbed water can be oxidized to hydroxyl radical ($\cdot OH$) by photo-induced holes (Eqs. 4-5).[10] Finally, these resultant free radicals will act as the reactive species to generate oxidation of organic MB dyes (Eqs. 6-7).[1,5] The possible photocatalytic reactions as follows: [1,5,10]

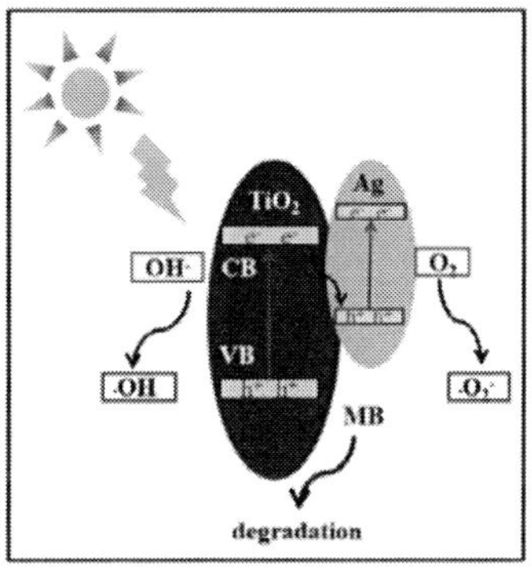

Figure 5. The schematic photocatalytic mechanism.

$$Ag\text{-}TiO_2 + hv \rightarrow Ag\text{-}TiO_2\ (e^- + h^+) \quad (2)$$

$$O_2 + e- \rightarrow \cdot O_2^- \quad (3)$$

$$H_2O \leftrightarrow H^+ + OH^- \quad (4)$$

$$OH^- + h^+ \rightarrow \cdot OH \quad (5)$$

$$\cdot O_2^- + MB \rightarrow CO_2 + H_2O \quad (6)$$

$$\cdot OH + MB \rightarrow CO_2 + H_2O \quad (7)$$

In terms of the Ag-TiO_2 nanocomposite based photocatalytic effect, the charge transfer from the excited conduction band of TiO_2 to Ag is feasible.[5] The tight attachment of Ag and TiO_2 is conducive to the enhancement of photocatalytic performance due to preventing the recombination effect of the electron-hole pairs.[5] On the other side, MB molecules, adsorbed in the surface of Ag nanoparticles, can be directly oxidized by means of the positive charges of Ag nanoparticles, which also provides an approach to degrade the MB molecules.[5]

4. Conclusion

In summary, silver embedded in titania nanocomposite photocatalyst was prepared by means of a facile mechanical milling. The Ag-TiO_2 nanocomposite catalyst exhibited enhancement of photocatalytic activity such as a photocatalytic efficiency as high as 85% for the degradation of MB after 1 h irradiation, which is superior to that of the pure TiO_2 catalyst (only ~41% with photocatalytic efficiency). Additionally, a narrowed band gap (3.13 eV) in Ag-TiO_2 nanocomposite was inspected by UV-visible absorption spectra as compared to pure TiO_2 catalyst (3.25 eV). The photocatalytic results are in accordance with the first-order kinetics equation. This enhancement with photocatalytic activity can be ascribed to that the Ag nanoparticles embedded in the TiO_2 surfaces can efficiently collect the free electrons. This study serves as a feasible, promising, and high-yield approach for the large-scale catalysis of waste water in practical applications.

Acknowledgments

This work was supported by the Open Sharing Fund for the Large-scale Instruments, Equipment of Central South University (No. CSUZC201904).

References

[1] Li HC, Ming T, Wang JC, Wang JF, Yu JC, Yu SH. 2014, Ultrasonic aerosol spray-assisted preparation of TiO_2/In_2O_3 composite for visible-light-drive photocatalysis, *J. Catal.* 310, 84-90.
[2] Fujishima A, Honda K. 1972, Electrochemical photolysis of water at a semiconductor electrode, *Nature* 238, 37-38.
[3] Tian J, Sang YH, Yu GW, Jiang HD, Mu XN, Liu H. 2013, A Bi_2WO_6-based hybrid photocatalyst with broad spectrum photocatalytic properties under UV, visible, and near-infrared irradiation, *Adv. Mater.* 25, 5075-5080.
[4] Yang J, Jiang YL, Li LJ, Muhire E, Gao MZ. 2016, High-performance photodectectors and enhanced photocatalysts of two-dimensional TiO_2 nanosheets under UV light excitation, *Nanoscale* 8, 8170-8177.
[5] Xie W, Qin N, Li BJ, Bao DH. 2015, Enhanced visible-light catalytic activity of Au nanoparticles loaded c-axis oriented $Bi_2VO_{5.5}$ porous thin films, *Ceram. Int.* 41, 8433-8443.
[6] Xia ZK, Ma QH, Li SY, Zhang DQ, Cong L, Tian YL, Yang RY. 2016, The antifungal effect of silver nanoparticles on Trichosporon asahii, *J. Microbiol. Immunol. Infect.* 49, 182-188.
[7] Hao L, Yan JC, Guan SJ, Cheng LJ, Zhao Q, Zhu Z, Wang Y, Lu Y, Liu JZ. 2019, Oxygen vacancies in TiO_2/SnO_x coatings prepared by ball milling followed by calcination and their influence on the photocatalytic activity, *Appl. Surf. Sci.* 466, 490-497.
[8] Hao AZ, Ismail M, He S, Qin N, Huang WH, Wu J, Bao DH. 2017, Ag-NPs doping enhanced resistive switching performance and induced changes in magnetic properties of $NiFe_2O_4$ thin films, *RSC Adv.* 7, 46665-46677.
[9] Choi HC, Jung YM, Kim SB. 2005, Size effect in the Raman spectra of TiO_2 nanoparticles, *Vib. Spectrosc.* 37, 33-38.
[10] Zhou X, Wu J, Zhang J, He P, Ren JX, Zhang JJ, Lu J, Liang PK, Xu K, Shui F. 2017, The effect of surface heterojunction between (001) and (101) facets on photocatalytic performance of anatase TiO_2, *Mater. Lett.* 205, 173-177.

IOP Conf. Series: Materials Science and Engineering **611** (2019) 012036 doi:10.1088/1757-899X/611/1/012036

Synthesis and Dispersion of Organic Pigments by Amphiphilic Hyperbranched Polyesteramides Dispersant

Qing Xu[1], Shangjun Long[2], Guohua Liu[1], Xuelin Zhang[3] and Yuanjing Zhou[2,*]

[1]Institute of Biology, Guizhou Academy of Sciences, Guiyang 550009, China
[2]Institute of Analysis and Testing, Guizhou Academy of Sciences, Guiyang 550002, China
[3]Center for Research and Development of Fine Chemicals of Sun Yat-sen University

*Corresponding author

Abstract. The amphiphilic hyperbranched polyesteramides (AHPAs) consisting of hydrophobic hyperbranched polyether cores and hydrophilic poly (ethyleneglycol) arms with urethane bonds in the molecular structure were synthesized by solution polymerization in chloroform using hexamethylene diisocyanate (HMDI) as a coupling agent. And their chemical structures were confirmed by FT-IR and and ^{13}C NMR. Further, the AHPAs were applied to disperse organic pigments and their dispersibilities were investigated. The results showed that AHPAs had good dispersing effect on organic pigments. When the dosage of AHPA was 12% (on the weight of pigment), the dispersing effect of AHPA on phthalocyanine green was the best with the lowest relative settlement rate, viscosity and particle size. The dispersibilities of AHPA on organic pigments were phthalocyanine green > benzidine yellow G > pigment red F5R. AHPAs have the characteristics of high solubility, low viscosity, good rheological properties and no pollution and would be the ideal materials for using as the pigment dispersants.

1. Introduction

Hyperbranched polymers with a three-dimensional dendritic architecture exhibiting special properties have aroused extensive attention. Especially, there are a large number of terminal functional groups in their molecular structures.By modification of these end-groups, the properties of hyperbranched polymers such as glass transition temperature, solubility, melt viscosity and surface properties are significantly modulated to define mainly possible applications [1-5]. In recent year, the emergent applications of hyperbranched polymers including biomedicine, electrical conductivity, chemical sensing, and catalyst carrier have been explored successfully [6-9].

Organic pigments have the characteristics of small particle size, large specific surface area and low polarity. In order to give full play to the coloring effect of organic pigments, measures must be taken to make them uniformly and stably disperse in the medium. Traditional organic small molecules or inorganic dispersants show effective dispersing effect in the medium, but they do not adsorb firmly on the surface of organic pigment particles, and the long-term stability of the system is not good. They also have adverse effects on the appearance, hardness and aging resistance of the paint film [10, 11]. A series of amphiphilic hyperbranched polyethers (AHPs) with different branching degrees were synthesized in our previous work and applied as pigment dispersant on pigment yellow with good dispersing effect [12]. Here, we prepared their derivatives (AHPAs) through modification of end groups and appliedthem to organic pigments. Their dispersibility to organic pigments was investigated.

2. Experimental

2.1. Reagents and Instruments
All chemical reagents used in this study were bought from Sinopharm Chemical Reagent Co., Ltd and were of the highest purity available. Methoxy polyethylene glycol(mPEG, Mn=600, 1000, 2000) was dried by azeotropic distillation with benzene.$CH_2Cl_2$and $CHCl_3$ were dried by distillation in presence of CaH_2 under reduced pressure.
Fourier transform infrared spectroscopy (FT-IR) tests were conducted on a Spectrum 100 FT-IR spectrometer (Perkin-Elmer, US) using KBr discs. ^{13}C NMR spectra were recorded on a Varian MERCURY plus 400 NMR spectrometer operating at 100 MHz.

2.2. Synthesis of Amphiphilic Hyperbranched Polyesteramides (AHPAs)
By two-step reactions, AHPAs were synthesized. First, the hydroxyl end-group of mono methoxy polyethylene glycol (mPEG) reacted with one NCO group ofhexamethylene diisocyanate (HMDI) to obtain the isocyanate-terminatedintermediate (mPEG-NCO). And then, the unreacted other NCO group in the mPEG-NCO reacted withthe hydroxyls of HP to obtain the amphiphilic hyperbranched polyesteramide. The detailed synthesis process was described as below.

2.2.1 Synthesis of mPEG-NCO. Methoxy polyethylene glycol (mPEG1000, 27 g) was dissolved in dry CH_2Cl_2 (80 mL). Under nitrogen, the reactor was heated to 40 °C in an oil bath, and then hexamethylene diisocyanate (HMDI) (48 mL) was slowly added by a drip funnel. After 8 h, the reaction mixture was precipitated in petroleum ether and filtered to obtain the light yellow intermediate mPEG-NCO. FT-IR (KBr, cm^{-1}): 3440, 2890, 2270, 1715, 1540, 1110, 848, 755.

2.2.2 Synthesis of hyperbranchedpolyesteramides (AHPAs). HP was synthesized by cation ring-opening polymerization according to the previous report [12]. Under nitrogen, the HP was dissolved in dry $CHCl_3$. After the reactor was heated to 50 °C, mPEG-NCO in dry $CHCl_3$ was slowly added into the reactor by a drip funnel. When the reaction went on for 12 h, the reaction product was precipitated in alcohol and filtered to obtain the yellow waxy solid. FT-IR (KBr, cm^{-1}): 3438, 2885, 1715, 1540, 1110, 848, 752. ^{13}C NMR (100 MHz): δ (ppm) 7.9 ($-CH_3$), 22.8($-CH_2-$), 43.5($-C$), 63.0 ($-CH_2OH$), 67.0– 72.0 ($-CH_2-O-CH_2-$).

2.3. Preparation of Pigment Dispersion System
A certain amount of pigment AHPA and deionized water were mixed and stirred evenly. The pigment dispersed system was obtained by using JKQ2200DE numerical control ultrasonic cleaner to disperse the pigments with 800 W powers for a certain time.
The blank control system without hyperbranched polyether dispersant was prepared under the same conditions.

2.4. Determination of Dispersion Performance
The dispersion stability, viscosity and particle size of dispersion system were determined.

*2.4.1 Dispersion stability.*Place the dispersed pigment system on the test tube rack, observe the settlement, record the location of the suspension settlement at intervals, and calculate the relative settlement rate (R) as following (1)[13]:

$$R = (H - h) / H \times 100\% \tag{1}$$

In the formula (1): h means settlement height (mm), H means total height (mm).

2.4.2. Viscosity. The viscosity of dispersed pigment systemwas measured by Brookfield digital viscometer at room temperature.

2.4.3 Particle size. The particle size of dispersion systemwas measured by ZS-90 Nano particle analyzer at room temperature.

3. Result and Discussion

3.1. Synthesis and Characterizationof Amphiphilic HyperbranchedPolyesteramides (AHPAs)

The synthetic route of AHPAs was shown in Scheme 1. By controlling the reaction temperature and the amount of reactants, one NCO group of HMDI was first introduced ontothe hydroxyl end-group of mono methoxy polyethylene glycol (mPEG) to obtain the isocyanate-terminatedintermediate (mPEG-NCO). And then, the other unreacted NCO group of mPEG-NCO was grafted onto the hydroxyls of HP.

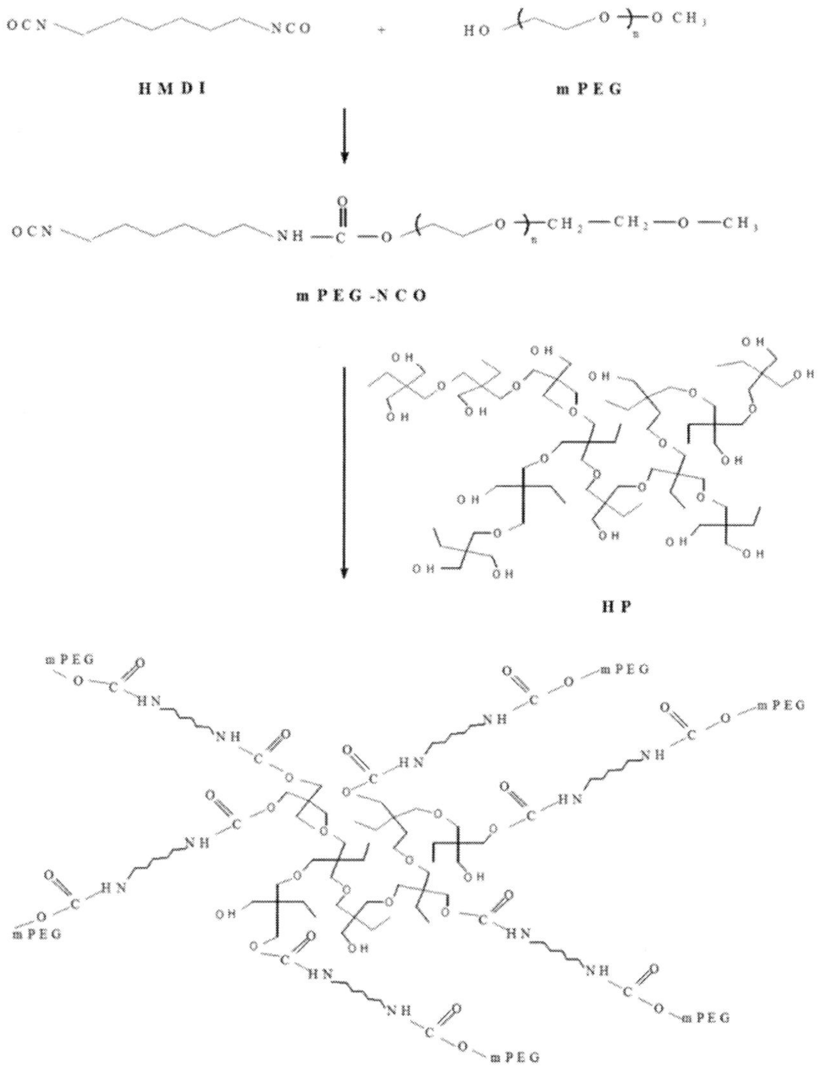

Scheme 1. Synthetic routes of amphiphilichyperbranched polyesteramides (AHPAs).

Figure1 gave the FT-IR spectra of mPEG-NCO and AHPA. The FT-IR spectrum of mPEG-NCO (Figure 1(a)) showed a strong hydroxyl peak at 3440 cm^{-1} for hydroxyl, a strong absorption peak at around 2890 cm^{-1} for methyl and methylene groups, a sharp peak at about 2270 cm^{-1} for NCO groups, a sharp peak at 1715 cm^{-1} for C=O groups, at 1540 cm^{-1} for NH groups, at 1110 cm^{-1} for ether bonds.

By grafting mPEG-NCO onto HP, the NCO peak at 2270 cm^{-1} disappeared in the FT-IR spectrum of AHPA (Figure 1(b)) [11], which meant that the NCO group had been changed into the carbonate group, and the absorption of amide NH groups at 1540 cm^{-1} and ether bonds at 1100 cm^{-1} were retained. These results suggested that graft hyperbranched polyesteramides, AHPAs, were formed via the reactions of the two NCO groups of HMDI with hydroxyls of mPEG and HP, respectively.

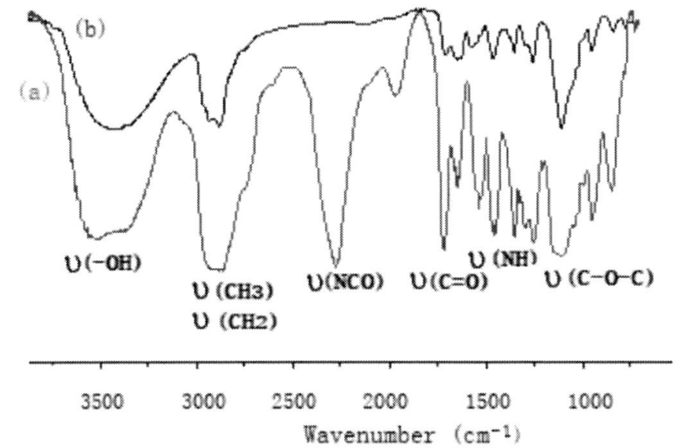

Figure 1. FT-IR spectra of mPEG-NCO (a), AHPA (b).

The chemistry structure of AHPAs was further confirmly by the ^{13}C NMR measurement. As shown in Figure 2, the triple peaks at around 43.7 ppmrepresented linear unit (L), dendritic unit (D) and terminal unit (T) of hyperbranched polymer, respectively. The characteristic mPEG carbons from the polymerized methylenes were seen at about 71.2 ppm. These results indicated the formation of the graft hyperbranched polyesteramides (AHPAs) [12].

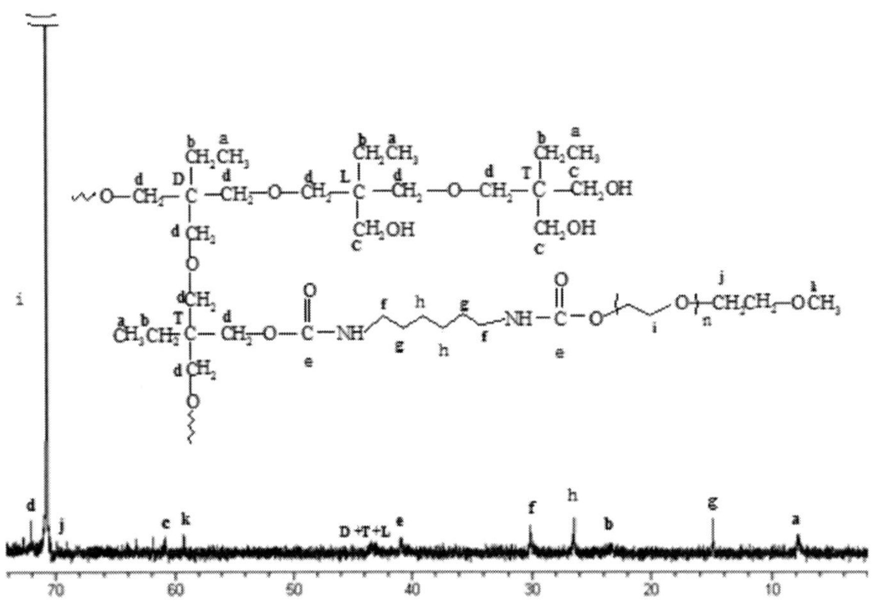

Figure 2. ^{13}C NMR spectrum of AHPA in $CDCl_3$.

3.2. Dispersion Stability

Settlement method is the most direct and simple method to observe the stability of dispersed systems [13]. Figure 3 showed the change of relative settlement rate (R) with natural settlement time. The smaller the relative settlement rate, the better the dispersion. The relative sedimentation rate of the sample with dispersant (S1, S2) was obviously lower than that of the sample without dispersant (S0). Figure 4 showed the change of relative settlement rate with addition of dispersant. When the dosage of dispersant increased, the relative settlement rate of the system decreased first, and then increased. This might be due to the fact that when the dosage of dispersant was small, the dispersant could not completely cover the surface of the pigments, and could not form a complete adsorption layer on the surface of the pigments. The uncovered part of the surface of the pigments aggregated in order to reduce the surface energy, so that the pigments aggregated and dispersed unsteadily. When the dosage of dispersant was further increased, enough dispersants were adsorbed on the surface of the pigments in order to prevent particle agglomeration. When the dosage of dispersant increased to a certain extent, polymer molecular chains entangled with each other, resulting in the re-aggregation of pigment particles, but the dispersion stability of pigments decreased. Only when the dispersant formed a dense monomolecular adsorption layer on the surface of the pigment particles, the dosage of the dispersant would be optimum.

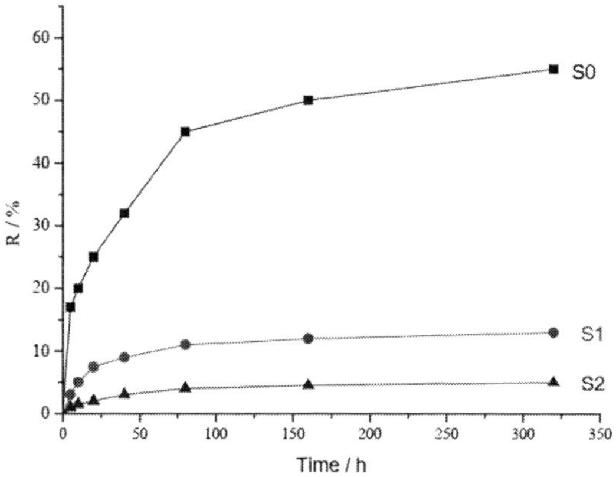

Figure 3. The change of relative settlement rate with natural settlement time.

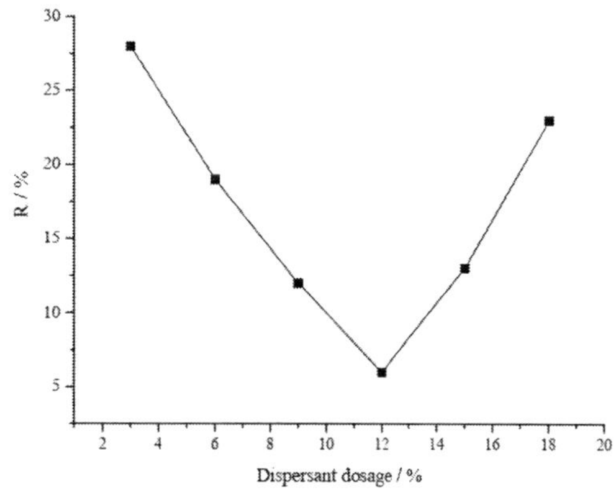

Figure 4. The effect of dispersant dosage on relative settlement rate.

3.3. Viscosity

Dispersants can not only maintain the dispersion stability of the pigment system, but also reduce the viscosity of the system. As can be seen from Figure 5, the addition of dispersant significantly reduced the viscosity of pigment system. However, with the increase of dispersant dosage, the viscosity of the system increased inversely. Therefore, the dosage of dispersant in the dispersion system was not the higher the better, which was consistent with the experimental results of sedimentation (Figure 3).

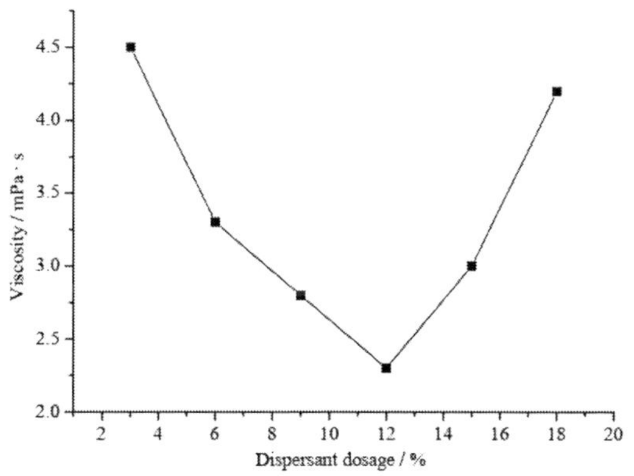

Figure 5. The effect of dispersant dosage on the viscosity of pigment system.

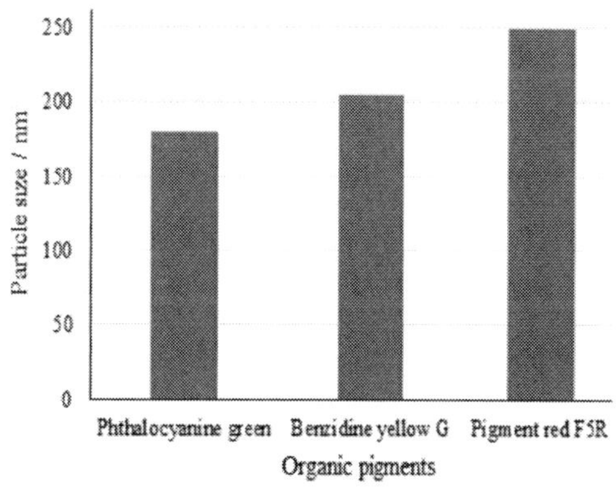

Figure 6. The effect of AHPA on the particle size of different pigment system.

3.4. Particle Size

Figure 6 showed the effect of AHPA on the particle size of different pigment system. The particle size of phthalocyanine green, benzidine yellow G, and pigment red F5R was around 180, 205, and 250 nm, respectively. This is related to the molecular structure of pigments themselves, the molecular structure of dispersants, and the interaction between pigments and dispersants.

4. Conclusion

By solution polymerizationusing hexamethylene diisocyanateas a coupling agent, the amphiphilic hyperbranched polyesteramides (AHPAs) consisting of hydrophobic hyperbranched polyethers cores and hydrophilic poly (ethyleneglycol) arms with urethane bonds in the molecular structure were synthesized. Being using as the pigment dispersant, AHPAs had good dispersing effect on organic pigments, and had great potential in the application of pigment dispersants. Further research will continue.

Acknowledgments

This work was financially supported by the Science and Technology Project of Guizhou Province, China ([2016]1139, JZ [2015]2006, Z [2015]6013), and [2017]01).

References

[1] Huang Y, Wang ST and Zhu XY 2017*Acta Polym. Sin.***2**245
[2] Mahmoodi Y, Dodel T, Moieni M, Hasanzadeh M, Horak D, Haghi A and Zaikov G 2014*Key Engineering Materials***1**133
[3] Zhang J, Xie S, Zhang X, Lu Z, Xiao H, Li C, Li G, Xu X, Chen X and Bo Z 2017*Chem. Commun.***53**537
[4] Mu B, Liu T T AND Tian W *2018 Macromol. Rapid Comm.* e1800471
[5] Qi M, Huang W, Xiao G, Zhu X, Gao C and Zhou Y 2017 *ActaPolym.Sin.***2** 214
[6] Gurunathan T, Mohanty S and Nayak, S K 2016 *Polym-Plast Technol.* **55** 92
[7] Tirthankar J 2016, *Paintindia***66** 59
[8] Nabae Y and Kakimoto M A 2018 *Polymers***12** 10
[9] Han J R, Liu T, Zhang S, Hao C, Xin J N, Guo B H and Zhang J W 2019 *Ind. Eng. Chem. Res.* **16** 6466
[10] Rojin B and William P 2018 *Paint & Coatings Industry***4** 86
[11] Nagose S, Rose E and Joshi A 2019 *Prog. Org. Coat***133** 55
[12] XuQ,Zhou YJ, Long S J, Liu Y G and Li J H2017*MaterialsScienceandEngineering***292**012109.
[13] Liu A H, Wu B Y and Jiang Z D 1996 *Coatings industry***4** 7

Preparation of Nanocomposites of Polypyrrole-metal (or Metal Oxide) on the Surface of V_2O_5 Coated Cathode through a One-step Method

Ziyang Li[1], Jiawei Li[1], Yuan He[1], Kun Zhou[1], Huaran Ji[1], Wei Shi[1,*], Yanan Sun[1], Ting Wu[2,*] and Dongtao Ge[1,*]

[1]Key Laboratory of Biomedical Engineering of Fujian Province University, Research Center of Biomedical Engineering of Xiamen, Department of Biomaterials, College of Materials, Xiamen University, Xiamen 361005, China
[2]School of Medicine, Xiamen University, Xiamen 361005, China

*Corresponding author

Abstract. Simultaneous formation of polypyrrole (PPy) and metal (or metal oxide) on a single electrode is particularly challenging because the electrodeposition of PPy and metal (or metal oxide) usually requires opposite potentials. Namely, the preparation of PPy needs an oxidation potential while the metal (or metal oxide) needs a reduction potential. In the present work, nanocomposites of PPy and metal (or metal oxide) were prepared successfully on the cathode by a one-step process, in which V_2O_5 coated titanium sheet was applied as cathode. In fact, two chemical reactions happened here at the same time. The first reaction was the chemical preparation of PPy on the cathode by the use of V_2O_5 as oxidant and the second reaction was the electrochemical formation of metal (or metal oxide) on the surface of the same cathode. As a result, nanocomposites of PPy and metal (or metal oxide) were one-step prepared on a single electrode. The morphology, composition and structure of nanocomposites were investigated via field emission scanning electron microscopy (FESEM), energy dispersive X-ray (EDX) and X-ray diffraction (XRD). Furthermore, PPy-Ag nanocomposite was prepared and used as a sensor to evaluate the electrocatalytic performance of H_2O_2. It was found that the sensor displayed excellent H_2O_2-sensing properties and a feasibility for practical applications. These results imply that the present one-step process not only presents a simple, facial and efficient way for the formation of PPy-based nanocomposites but also exhibits great potential to improve the application performance and broaden the applied fields of the nanocomposites.
Keywords: polypyrrole; V_2O_5; cathode; electrodeposition; nanocomposite; sensor.

1. Introduction

During the past decades, conducting polymer polypyrrole (PPy)-based nanocomposites have addressed considerable attention due to their novel properties and potential applications in various fields [1]. Among numerous PPy based nanocomposites, PPy-metal and PPy-metal oxide nanocomposites have raised particularly interest because the incorporation of PPy with metal or metal oxide may produce nanocomposites with novel properties that derived from effectively incorporation of the feature of parent components [2]. To date, a large number of PPy-metal (or metal oxide) nanocomposites have been prepared and found diverse applications in the fields of actuators [3], sensors [4], fuel cells [5], supercapacitors [6], corrosion protection [7], and drug release [8].

Generally, PPy-metal (metal oxide) nanocomposites were able to be prepared by chemical oxidation or electrochemical oxidation method [9, 10]. Compared with chemical oxidation method, electrochemical polymerization could obtain functional films with different structure and properties by changing the electrode, electrolyte or electropolymerization methods [11, 12]. For the electrochemical preparation of PPy-metal (metal oxide) composites, it was usually carried out by two steps. PPy was electropolymerized on one anode in the first step. In the second step, metal or metal oxide was electrodeposited on surface of PPy-coated cathode which was used as the anode in the first step. It is obvious that the two-step process for the preparation of PPy-metal (metal oxide) nanocomposites usually requires opposite potentials [13-16]. Therefore, it is always a challenge to achieve pyrrole polymerization and metal nanoparticle formation simultaneously on a single electrode [17].

Recently, a novel method was developed to synthesize PPy-metal nanocomposites on the cathode using a one-step method [18-20]. NO^+ generated from cathode in nitric acid electrolyte was used as an oxidizing agent of pyrrole monomers, and then the direct preparation of PPy-metal composites was accomplished on the surface of cathode. However, the polymerization condition of strong acid electrolyte is too hard and dangerous, which limits its wider applicability. Given this point, it is urgent to develop a facial and efficient method to achieve the synthesis of PPy-metal (or metal oxide) nanocomposites by one-step process.

In the present work, the synthesis of PPy-metal (or metal oxide) nanocomposites by one-step process has been successfully accomplished on the surface of V_2O_5 film coated cathode. As shown in Figure 1, the titanium electrode coated with V_2O_5 film was used as cathode and immersed in electrolyte containing pyrrole monomers and metal salts. Here, V_2O_5 acted as an oxidant agent for pyrrole polymerization. After immersion of the cathode, a negative potential was applied. Then the oxidation of pyrrole and the reduction of the metal (or metal oxide) were conducted simultaneously on the V_2O_5 film. As a result, PPy-metal or PPy-metal oxide nanocomposite could be obtained in one-step process. Energy dispersive X-ray (EDX), field emission scanning electron microscopy (FESEM) and X-ray diffraction (XRD) were conducted to verify the obtained nanocomposites. The feasibility and versatility of this method were demonstrated by successful preparation of several kinds of PPy-metal (metal oxide) nanocomposites, such as PPy-Sn, PPy-Ag, PPy-ZnO, and PPy-NiO/$Ni(OH)_2$ nanocomposites. Furthermore, in order to confirm the applicability of this strategy, PPy-Ag nanocomposite was applied to detect current response of H_2O_2, and it accessed a satisfactory result. These traits suggest that the one-step process for the preparation of PPy-metal (or metal oxide) may shed a new light for relevant fields in the future.

2. Materials and Methods

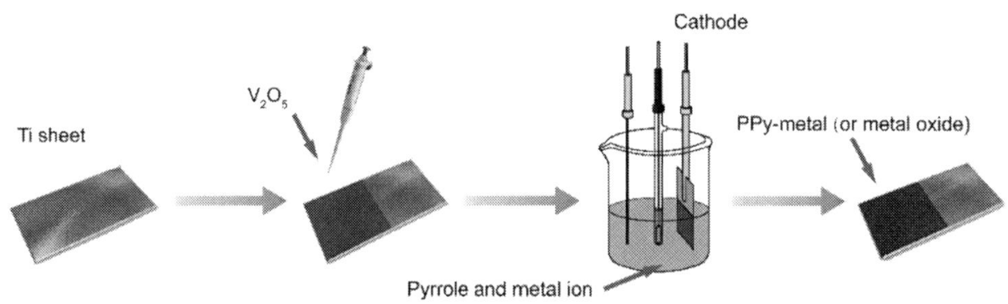

Figure 1. Schematic illustration of PPy-based composites by one-step process on cathode.

2.1. Materials

The titanium plate was cut into rectangular pieces of 2×1 cm, cleaned with 1500-grit sandpaper firstly, cleaned ultrasonically three times in deionized water, and then polished to a smooth mirror surface on 2000-grit sandpaper, ultrasonically washed three times with distilled water, ethyl alcohol and acetone, and dried in nitrogen ambience. Pyrrole (Py) (98%) was taken from Sigma Aldrich and further purified by twice distillation before use. The V_2O_5 was prepared as the previous report [21]. Hydrogen peroxide (30 wt% solution) (H_2O_2) and all other common solvents and reagents were analytical grade and purchased from Sinopharm Group.

2.2. Characterization
Electrochemical deposition and measurements were accomplished with an electrochemical workstation (CHI660D). All electrochemical experiments were employed with a conventional three electrodes system, with a reference electrode (RE) of saturated calomel electrode (SCE), a counter electrode (CE) of Pt wire, and a modified Ti sheet or 3 mm Au as working electrode (WE). The surface topography of all samples was observed by field emission scanning electron microscopy (FESEM, SU-70, Japan). The analysis of the elemental composition was performed with energy dispersive X-ray detector (EDX, SU-70, Japan). Attenuated total reflectance-Fourier transform infrared reflection (ATR-FTIR, is10, USA) spectroscopy was employed to characterize surface features of the modified electrode, by using an ATR accessory which background spectra were obtained by measuring the air directly. The structural analysis of the films was investigated by X-ray diffraction (XRD, Bruker-axs, USA) on a diffractometer with Cu Kα radiation.

2.3. The Preparation of PPy-metal (or Metal Oxide) Nanocomposites
To prepare the PPy-Sn nanocomposites, under the three-electrode system, a 0.1 M HCl solution with 0.01 M $SnCl_2$, 0.1 M NaCl and 0.1 M Py was employed as the electrolyte solution. The pretreated titanium sheet with about 5 μL V_2O_5 on its surface was used as the working electrode (WE). The reaction was lasted for 120 s with the constant potential of -0.8 V, and the WE was taken out after the electrodeposition was completed. The electrolyte solution used for the preparation of the PPy-Ag nanocomposites included 0.1 M $NaNO_3$, 0.01 M $AgNO_3$, 0.01 M *p*-toluenesulfonic acid (TsOH) and 0.1 M Py. The electrodeposition was the same as above method at room temperature except the two-electrode system was employed. The co-deposition reaction was performed with constant current of -2.5 $mA \cdot cm^{-2}$ for 150 s. PPy-ZnO nanocomposites was prepared in the electrolyte containing 0.05 M $Zn(NO_3)_2$, 0.1 M Py, 0.01 M TsOH, 0.1 M KNO_3 at 70 ℃ of water bath temperature, with -0.9 V constant potential for 150 s. The electrolyte including 0.01 M $NiSO_4$, 0.1 M $NaNO_3$ and 0.01 M sodium dodecyl sulfonate (SDS) was used to prepare PPy-NiO/$Ni(OH)_2$ electrode and activated with -0.9 V for 300 s under constant potential mode and constant temperature of 30 ℃. All samples were cleaned three times in deionized water, dried with nitrogen and stored at 4 ℃ for later assessment.

2.4. Construction of Sensors
Under the same system to the preparation of PPy-Ag modified electrode, a 3 mm Au electrode was employed as a cathode (WE). Firstly, about 2 μL of V_2O_5 was coated evenly on the surface of Au electrode. The electrodeposition reaction was conducted at -2.5 $mA \cdot cm^{-2}$ constant current for 120 s in the above electrolyte, and then it was washed cleaned with deionized water, subsequently dried in the nitrogen and then stored at 4 ℃. All electrochemical detections were conducted in the electrolyte of 0.1 M (pH=6.5) phosphate buffer solution (PBS). The electrolyte were saturated with N_2 before electrochemical detection.

3. Results and Discussion

3.1. Synthesis of PPy Using V_2O_5 as Oxidant
In order to prove that pyrrole can be oxidized by V_2O_5, V_2O_5 coated titanium sheet was immersed into a solution of 0.1 M pyrrole for 60 s. Then the sheet was taken out for characterization. It can be seen from insert of Figure 2c, the color of V_2O_5 coated titanium translated from red (before immersion) to black (after immersion). Figure 2a and Figure 2b displays the microscopic image of the sheet surfaces before and after immersion, respectively. It was found that before immersion V_2O_5 exhibits wire-like morphology with diameter of about 40 nm. After immersion, the diameter of the nanowire was increased to about 60 nm. Meanwhile, the peaks of C, N element were also found in the EDX spectrum and the molar ratio of C, N is 1.43: 0.38 (Figure 2c), indicating the formation of PPy. Figure 2d displays the ATR-FTIR spectroscopy of the reaction products. The characteristic peaks of PPy at 1169 cm^{-1} and 1092 cm^{-1} were corresponded to C–N in-plane deformations and N–H bonds. The characteristic peaks at 1309

and 1053 cm^{-1} were attributed to the in-plane C−H deformations. The bands at 1535 cm^{-1} usually reflects the C=C stretching vibration in the pyrrole ring [22-24]. A weak peak at 1690 cm^{-1} is attributed to stretching mode of V–O–C in PPy-V_2O_5 composite. The peaks of 1169 and 911 cm^{-1} also imply the doping state of PPy composites [23]. And before the oxidation of PPy, the spectrum of the V_2O_5 is verified by few absorption bands located at 989, 843, and 579 cm^{-1}, which corresponding to the stretching vibration of V=O band (vanadyl oxygen), deformation modes of V–O–V band [25, 26]. These traits demonstrated the formation of the PPy and laid the foundation for the following work.

3.2. Synthesis of PPy-metal Nanocomposite

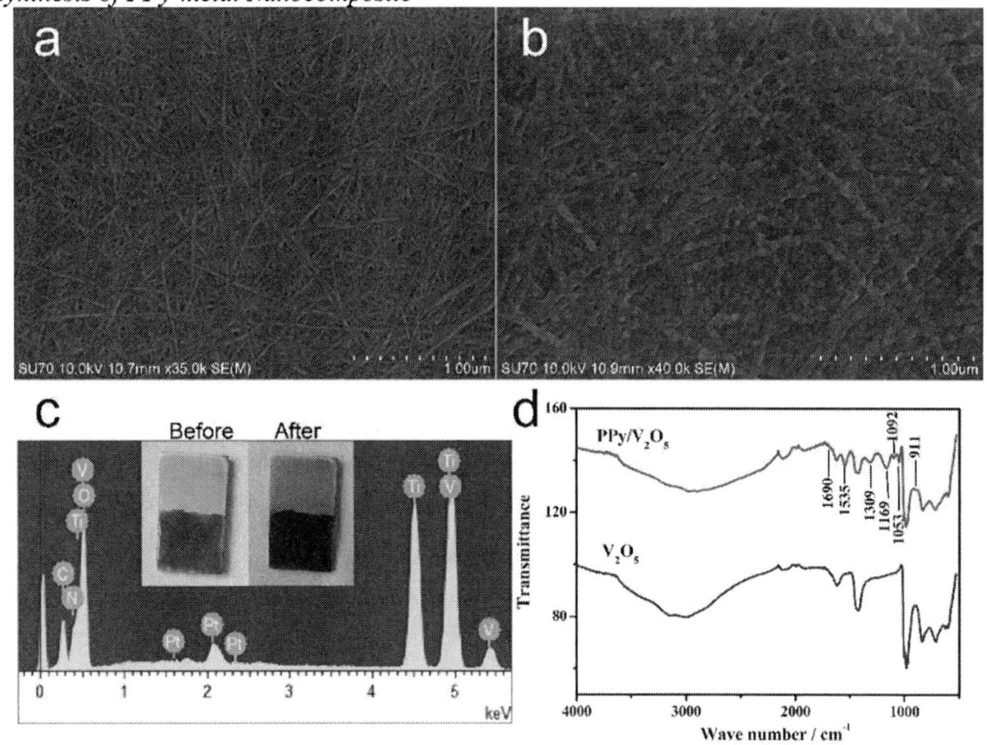

Figure 2. The FESEM images of a V2O5 and b PPy oxidized by V2O5. c The EDX spectrum of Ppy-V2O5, and insert image is V2O5 coated titanium before and after immersed in the solution of 0.1 M pyrrole. d The FTIR spectra of V2O5 and PPy oxidized by V2O5.

Based on the above results, PPy-Sn nanocomposite was synthesized. The synthesis process was illustrated in Figure 1. Figure 3a showed the FESEM image of the PPy-Sn nanocomposite. It can be observed numerous aggregated particles with uneven size on the PPy film. The chemical constitution of the nanocomposite was carried out by EDX. As displayed in the Figure 4a, EDX spectrum reveals a strong Sn peak, suggesting the existence of Sn element. The peaks of S, Na, Cl, K, presented in EDX spectrum, were caused by unavoidable trace impurities in electrolyte. The crystal analysis of the nanocomposite was made use of XRD analysis and the result showed that the diffraction peaks of 31.2°, 32.5°, 45.4° and 64.2° at 2θ values could be assigned to the (200), (101), (121) and (301) tetragonal crystalline planes of Sn, respectively (Figure 5a) [27, 28]. In addition, a broad peak at about 10° could be observed, which reflects the amorphous form of PPy [29, 30].

Figure 3. The FESEM images of a PPy-Sn, b PPy-Ag, c PPy-ZnO and d PPy-NiO/Ni(OH)2 nanocomposite.

Then PPy-Ag nanocomposite was synthesized following the similar process in order to prove the generality of the present one-step method. The FESEM image of the PPy-Ag nanocomposite was shown in Figure 3b. The typically dendritic crystals are uniformly spread on the surface of the PPy film. The chemical constitution of the PPy-Ag nanocomposite was carried out by EDX. The peaks of Ag are observed, confirming the existence of Ag elements. The peaks of C, N represented the characteristic elements of PPy (Figure 4b). In the previous study, it had been reported that dendritic crystals of silver could be deposited using electrochemical method [31, 32]. XRD was conducted to analyze the crystal structure of the PPy-Ag nanocomposite (Figure 5b). Diffraction peaks at 37.98°, 43.98°, 64.32° and 77.42° were corresponding to (111), (200), (220), (311) and (222) in face-centered cubic crystal of Ag, and it further indicated that the dendritic structures belong to silver crystal [33, 34]. We can infer from above results that a large amount of Ag nanoparticles were one-step co-deposited with PPy film on the V_2O_5 coated titanium electrode. Compared with the method of Ag could be reduced with PPy directly, the process of this strategy is more fast and convenient [35].

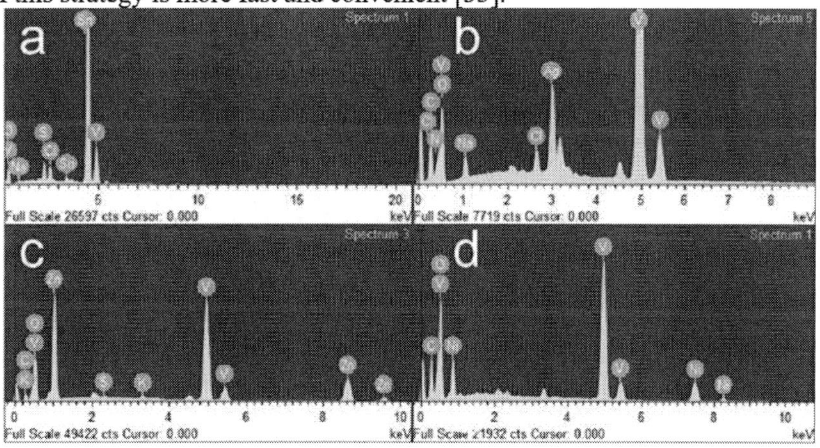

Figure 4. The EDX spectra of a PPy-Sn, b PPy-Ag, c PPy-ZnO and d PPy-NiO/$Ni(OH)_2$ nanocomposites.

3.3. Synthesis of PPy-metal Oxide Nanocomposite

To further demonstrate the versatility of this one-step method, co-deposition of metallic oxide and PPy on V_2O_5 coated titanium cathode was performed. In this regards, PPy-ZnO nanocomposite was prepared first. Figure 3c represents the FESEM image of the obtained PPy-ZnO nanocomposite. It clearly shows that the crystal exhibited a three-dimensional petaliform structure on the surface of rough PPy film. XRD peaks of 31.77°, 34.42° and 36.25° corresponded to (100), (002), (101) of hexagonal phase ZnO (Figure 5c) [36, 37]. Moreover, the characterization of EDX also shows the intensive peak of Zn, implying the Zn was existed in petaliform crystal (Figure 4c).

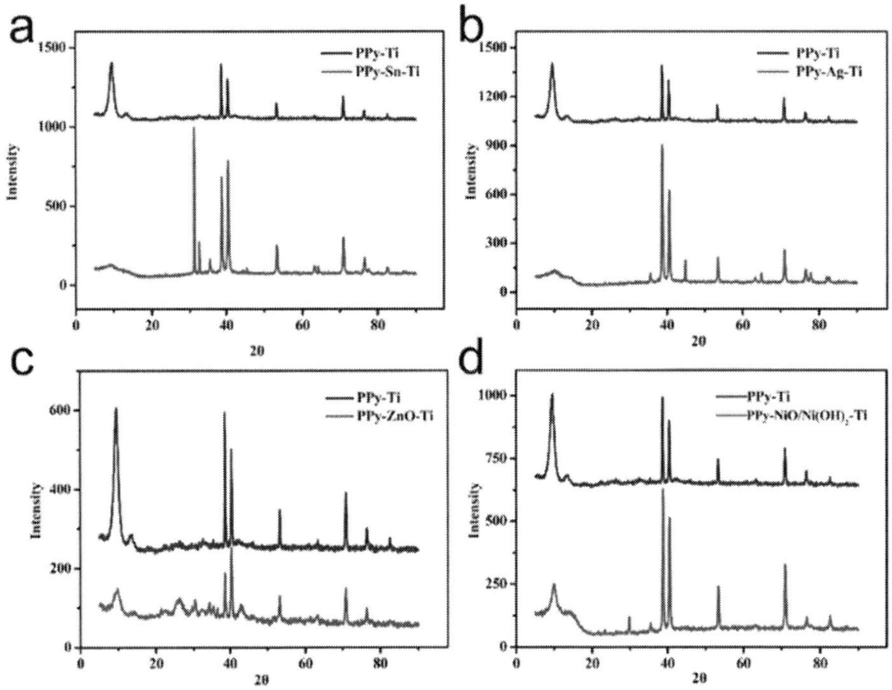

Figure 5. XRD patterns of a PPy-Sn, b PPy-Ag, c PPy-ZnO and d PPy-NiO/Ni(OH)$_2$, respectively.

In addition, another nanocomposite of PPy-metal oxide, PPy-NiO/Ni(OH)$_2$ nanocomposite, was also prepared. Figure 3d shows FESEM image of the nanocomposite. Similar to the previous report [38, 39], NiO/Ni(OH)$_2$ prepared with electrodeposition presented the morphology of compact blocky grains. The PPy substrate was completely covered by lots of NiO/Ni(OH)$_2$ nanoparticles. Moreover, two peaks of Ni are found from EDX spectrum (Figure 4d), suggesting the existence of Ni element. On the other hand, it could not find out any relative characteristic peak from the result of XRD analysis (Figure 5d), maybe the NiO/Ni(OH)$_2$ was essentially poor crystallization or amorphous structure under this preparation system [40, 41]. All of the above results confirmed that various PPy-metal (or metal oxide) nanocomposites could be synthesized by the present one-step method and thus consolidated its versatility for preparing PPy-metal or PPy-metal oxide nanocomposites in a fast, simple, and efficient way.

3.4. The Sensor Application of PPy-Ag Composite

Given the good catalytic properties of Ag nanoparticles for H_2O_2 [34], PPy-Ag nanocomposites was designed to be a non-enzyme sensor for H_2O_2 detection. From the Figure 6a, the response current of parallel electrodes was shown when 0.1 mM H_2O_2 was added. The response current of PPy-Ag modified electrode exhibited several times higher than the Ag modified electrode after 0.1 mM of H_2O_2 was added, while it has hardly any changes on bare Au electrode. Intrinsically, the linear PPy structure provided a larger specific surface area and reaction sites for Ag particles, and the overpotential characteristics of PPy would also contribute to the catalytic reaction of H_2O_2 [29]. The large area of current response also reflected the excellence pseudocapacitance properties of PPy [42]. The PPy-Ag modified electrode exhibited an increasing current response with increased concentrations of H_2O_2 to the electrolyte, suggesting that H_2O_2 could be easily reduced on PPy-Ag modified electrode in a wide concentration range (Figure 6b). The peak value of response current also increased with the square root of scan rates linearly, implying the response process to H_2O_2 was diffusion controlled (Figure 6c). In comparison, we investigated the H_2O_2 sensing ability of bare Au electrode, Ag, and PPy-Ag by chronoamperometry (Figure 6d). The bare Au electrode showed a negligible response to the added H_2O_2. The Ag modified electrode exhibited an observed increase with the continuous addition of H_2O_2. However, the response current increment of the PPy-Ag electrode was enhanced significantly and the response time was only 3 s inferred from the insert image (Figure 6d), indicating an excellent detection performance. From Figure 7a, the response current was linearly increased with the concentration of H_2O_2 ranging from 0.01 mM to 43.55 mM and the correlation coefficient is calculated as 0.99951 (Figure 7b). The linear regression equation for the H_2O_2 detection is presented as I (mA) = -0.0145-0.0435×C (mM). The sensitivity of the modified electrode is 621.43 $\mu A \cdot mM^{-1} \cdot cm^{-2}$, and the detection limit is 4.52 μM (S/N=3). Based on the previous study, the PPy-Ag modified electrode is comparative or superior to most of Ag-based H_2O_2 sensors (Table 1).

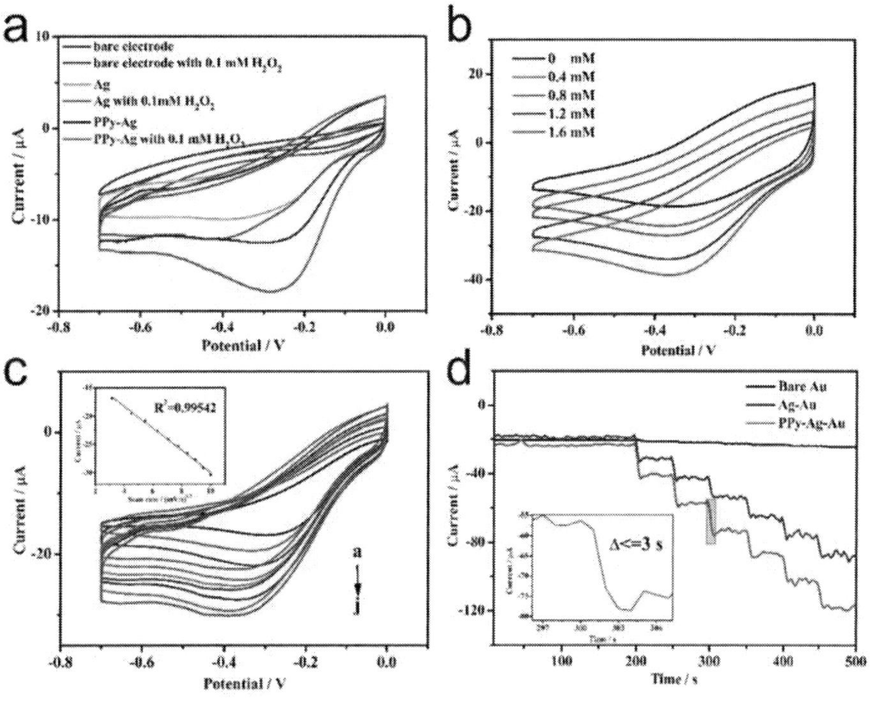

Figure 6. a CVs of bare electrode, Ag and PPy-Ag modified electrode with and without 0.1 mM H_2O_2 at a scan rate of 50 mV·s-1. b CVs of PPy-Ag modified electrode with different concentrations of H_2O_2 at a scan rate of 50 mV·s-1. c CVs of PPy-Ag modified electrode at different scan rates (from a to j: 10, 20, 30, 40, 50, 60, 70, 80, 90, 100 mV·s-1), and inset image displays the plot of the peak current vs. square root of scan rates. d I-t curves depending on the response of bare electrode, Ag and PPy-Ag modified electrode to H_2O_2 in 0.1 M PBS (pH=6.5) at -0.3 V upon successive addition of 0.5 mM H_2O_2 under continuous stirred conditions.

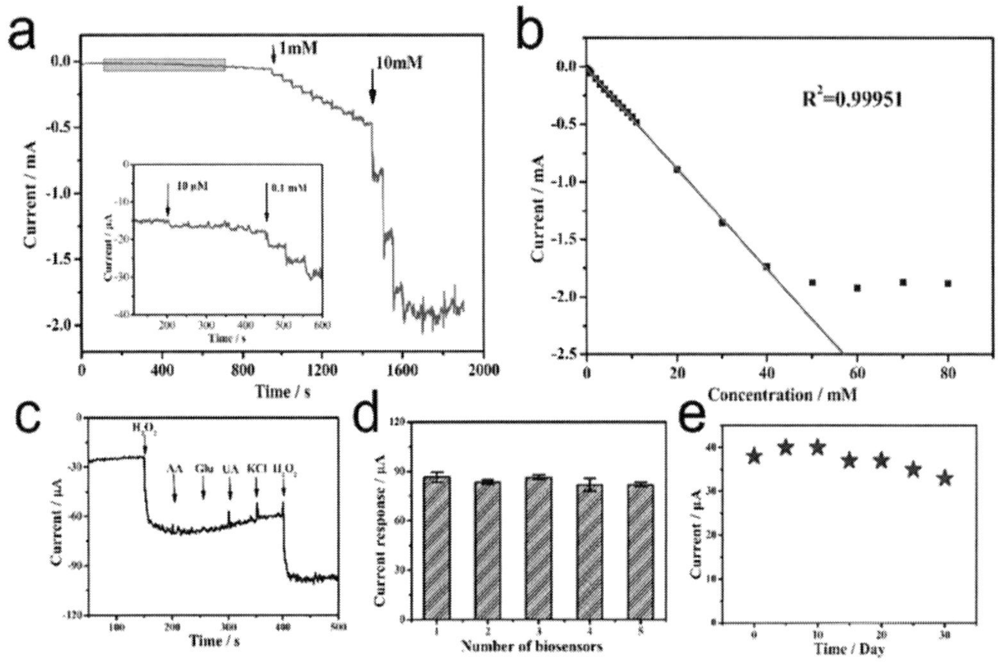

Figure 7. a I-t curve of PPy-Ag modified electrode with the subsequent addition of H_2O_2 into 0.1 M PBS (pH=6.5) at -0.3 V, and insert image is enlarged curve at low concentrations of H_2O_2. b Correlation between different H_2O_2 concentrations and current. c I-t curves of the PPy-Ag modified electrode by addition of 1 mM H_2O_2 and other interfering substances with the same concentration under stirring condition, applied potential at -0.3 V. d The current response of five different PPy-Ag modified electrodes with the addition of 2 mM H_2O_2, each test repeated five times. e The current response of sensors with 1 mM H_2O_2 every five days for one month.

Table 1. Comparison of typical H_2O_2 sensors.

Sensors	Sensitivity (μA mM^{-1} cm^{-2})	Linear range (mM)	LOD (μM)	References
Ag Nanowires	26.6	0.1~3.1	29.2	[43]
Ag-SBA-16/CPE	816.6	0.02~20	2.95	[44]
Ag NPs-PPy-Fe_3O_4/GCE	18.71	0.005~11.5	1.7	[45]
Ag-FeOOH/GCE	8.07	0.03~15	22.8	[46]
Ag-Cu_2O-polyimide films	null	0.05~0.5	4.0	[47]
Nf-Pd Ag-rGO-NH_2/GCE	1307.46	0.002~19.5	0.7	[48]
PPy-Ag/Au	621.43	0.01~43.55	4.52	This work

Moreover, the selectivity, reproducibility and time stability of the sensor were also examined. Figure 7c shows the influence to current response of interferents including ascorbic acid (AA), glucose, uric acid (UA) and KCl with the same concentration. Markedly, the PPy-Ag modified electrode displayed the highest electrochemical response to H_2O_2. The electrode reproducibility was assessed by measuring the current values of five different electrodes, and the RSD is 4.46% (Figure 7d). The time stability of the PPy-Ag modified electrode was examined every five days at the same condition for one month. The response current of the modified electrode still retained 86.84% after one-month test, indicating the electrode present a considerable long-term stability (Figure 7e). When a real sample of H_2O_2 disinfectant (3 wt%) was detected with the electrode, good recoveries (98.44~121%) were shown among different concentrations, and the RSD of the five measurements was less than 7.24% (Table 2). The above results indicate an excellent performance of the PPy-Ag sensor and thus provide a feasibility for practical applications.

Table 2. Determination of H_2O_2 in real sample.

Sample No.	Added (mM)	Measured by sensor (mM)	Recovery (%)	RSD (%)
1	0.1	0.121	121	1.18
2	1	1.026	102.6	4.51
3	5	4.972	98.44	7.24

4. Conclusions

This work presented an effective strategy to prepare PPy-metal (or metal oxide) nanocomposites on the surface of V_2O_5 coated cathode through one-step process. Essentially, the polymerization of pyrrole oxidized by V_2O_5 was synchronously proceed with electrodeposition of metal (or metal oxide) on the same cathode. As a result, PPy-Sn, PPy-Ag, PPy-ZnO, and PPy-NiO/Ni$(OH)_2$ nanocomposites were successfully prepared and proved the feasibility and versatility of this method. Moreover, the application of H_2O_2 non-enzyme detection was investigated on PPy-Ag modified electrode. The results exhibited a comparative or superior performance of H_2O_2-sensing to other parallel sensors. Predictably, such a fast, simple and efficient way for the preparation of PPy-based nanocomposites is expected to develop more competitive and promising nanocomposites in the future.

Acknowledgments

This research was finally supported by grand from the Program for New Century Excellent Talents in Fujian Province University, the National Nature Science Foundation of China (81271689, 31271009), and the Fundamental Research Funds for the Central Universities (20720150087). It also wishes to thank Prof. Yuan Jiang for his kind help in V_2O_5 preparing.

References

[1] Diaz AF, Castillo JI, Logan JA and Lee WY 1981 Electrochemistry of conducting polypyrrole films *J Electroanal Chem* vol 129 pp 115-132
[2] Gangopadhyay R and De A 2000 Conducting polymer nanocomposites: A brief overview *Chem Mater* vol 12 p 2064
[3] Tadesse Y, Brennan J, Smith C, Long TE and Priya S 2010 Synthesis and characterization of polypyrrole composite actuator for jellyfish unmanned undersea vehicle *Proc Spie* p 7642
[4] Rehman AU, Ikram M, Kan K, Zhao YM, Zhang WJ, Zhang JW, Liu Y, et al 2018 3D interlayer nanohybrids composed of reduced graphenescheme oxide/SnO_2/PPy grown from expanded graphite for the detection of ultra-trace Cd^{2+}, Cu^{2+}, Hg^{2+} and Pb^{2+} ions *Sensor Actuat B-Chem* vol 274 pp 285-295
[5] Fard LA, Ojani R, Raoof JB, Zare EN and Lakouraj MM 2017 Poly (pyrrole-co-aniline) hollow nanosphere supported Pd nanoflowers as high-performance catalyst for methanol electrooxidation in alkaline media *Energy* vol 127 pp 419-427
[6] Wang C, Liu H, Jiang M, Wang YD, Liu RN, Luo ZP, Liu XQ, et al 2017 Ammonium vanadate polypyrrole manganese dioxide nanowire arrays with enhanced reversible lithium storage *Appl Surf Sci* vol 416 pp 402-410
[7] Aravindan N and Sangaranarayanan MV 2016 Influence of solvent composition on the anti-corrosion performance of copper-polypyrrole (Cu-PPy) coated 304 stainless steel *Prog Org Coat* vol 95 pp 38-45
[8] Wu MX, Yan HJ, Gao J, Cheng Y, Yang J, Wu JR, Gong BJ, et al 2018 Multifunctional

supramolecular materials constructed from polypyrrole UiO-66 nanohybrids and pillararene nanovalves for targeted chemophotothermal therapy *Acs Appl Mater Inter* vol 10 pp 34655-34663

[9] Xu YD, Zhang X, LiY, Zhang YJ, He LJ and Zhao JW 2018 Effect of electrolyte cations on the capacitive performance of the hexagonal-MoO_3/polypyrrole hybrid *Mater Lett* vol 210 pp 314-316

[10] Pruna A, Shao Q, Kamruzzaman M, Li YY, Zapien JA, Pullini D, Mataix DB, et al 2017 Effect of ZnO core electrodeposition conditions on electrochemical and photocatalytic properties of polypyrrole-graphene oxide shelled nanoarrays *Appl Surf Sci* vol 392 pp 801-809

[11] Li YF and He GF 1998 Effect of preparation conditions on the two doping structures of polypyrrole *Synth Met* vol 94 pp 127-129

[12] Zhang J, Liu XH, Zhang LX, Cao BQ and Wu SH 2013 Reactive template synthesis of polypyrrole nanotubes for fabricating metal/conducting polymer nanocomposites *Macromol Rapid Comm* vol 34 pp 528-532

[13] Meenach SA, Burdick J, Kunwar A and Wang J 2007 Metal/conducting-polymer composite nanowires *Small* vol 3 pp 239-243

[14] Sonavane AC, Inamdar AI, Dalavi DS, Deshmukh HP and Patil PS 2010 Simple and rapid synthesis of NiO/PPy thin films with improved electrochromic performance *Electrochim Acta* vol 55 pp 2344-2351

[15] Vorotyntsev MA, Skompska M, Rajchowska A, Borysiuk J and Donten M 2011 A new strategy towards electroactive polymer-inorganic nanostructure composites: Silver nanoparticles inside polypyrrole matrix with pendant titanocene dichloride complexes *J Electroanal Chem* vol 662 pp 105-115

[16] Ali Y, Kumar V, Sonkawade RG, Shirsat MD and Dhaliwal AS 2013 Two-step electrochemical synthesis of Au nanoparticles decorated polyaniline nanofiber *Vacuum* vol 93 pp 79-83

[17] Kishore PS, Viswanathan B and Varadarajan TK 2008 Synthesis and characterization of metal nanoparticle embedded conducting polymer-polyoxometalate composites *Nanoscale Res Lett* vol 3 pp 14-20

[18] Jung YJ, Singh N and Choi KS 2009 Cathodic deposition of polypyrrole enabling the one-step assembly of metal-polymer hybrid electrodes *Angew Chem-Int Edit* vol 48 pp 8331-8334

[19] Hnida KE, Socha RP and Sulka GD 2013 Polypyrrole–silver composite nanowire arrays by cathodic co-deposition and their electrochemical properties *J Phys Chem C* vol 117 pp 19382–19392

[20] Cui X, Huang X, He Y, Dai L, Wang S, Sun Y, Shi W, et al 2014 Cathodic protection: A new strategy to enable the formation of nanostructured polypyrrole on magnesium *Synth Met* vol 195 pp 97-101

[21] Xiong CR, Aliev AE, Gnade B and Balkus KJ 2008 Fabrication of silver vanadium oxide and V_2O_5 nanowires for electrochromics *Acs Nano* vol 2 pp 293-301

[22] Wang JG, Liu HY, Liu HZ, Hua W and Shao MH 2018 Interfacial constructing flexible V_2O_5 polypyrrole core-shell nanowire membrane with superior supercapacitive performance *Acs Appl Mater Inter* vol 10 pp 18816-18823

[23] Bai MH, Bian LJ, Song Y and Liu XX 2014 Electrochemical codeposition of vanadium oxide and polypyrrole for high-performance supercapacitor with high working voltage *Acs Appl Mater Inter* vol 6 12656-12664

[24] Arora K, Chaubey A, Singhal R, Singh RP, Pandey M, Samanta S, Malhotra B, et al 2006 Application of electrochemically prepared polypyrrole–polyvinyl sulphonate films to DNA biosensor *Biosens Bioelectron* vol 21 pp 1777-1783

[25] Huguenin F, Girotto EM, Torresi RM and Buttry DA 2002 Transport properties of V_2O_5/polypyrrole nanocomposite prepared by a sol-gel alkoxide route *J Electroanal Chem* 536: 37-45

[26] Yang J, Lan TB, Liu JD, Song YF and Wei MD 2013 Supercapacitor electrode of hollow spherical V_2O_5 with a high pseudocapacitance in aqueous solution *Electrochim Acta* vol 105 pp 489-495

[27] Ardhi REA, Liu G, Tran MX, Hudaya C, Kim JY, Yu H and Lee JK 2018 Self-relaxant superelastic matrix derived from C-60 incorporated Sn nanoparticles for ultra-high-performance

Li-ion batteries *Acs Nano* vol 12 pp 5588-5604
[28] Shimizu M, Shimizu S, Katada A, Uejima M and Arai S 2018 Electrochemical preparation of free-standing carbon-nanotube/Sn composite paper *Mater Lett* vol 220 pp 182-185
[29] Nalage SR, Navale ST and Patil VB 2013 Polypyrrole-NiO hybrid nanocomposite: Structural, morphological, optical and electrical transport studies *Measurement* vol 46 pp 3268-3275
[30] Roy S, Mishra S, Yogi P, Saxena SK, Mishra V, Sagdeo PR and Kumar R 2018 Polypyrrole-vanadium oxide nanocomposite: polymer dominates crystallanity and oxide dominates conductivity, *Appl Phys a-Mater* vol 124 p 53
[31] Cho FH, Kuo SC and Lai YH 2017 Surface-plasmon-induced azo coupling reaction between nitro compounds on dendritic silver monitored by surface-enhanced Raman spectroscopy *Rsc Adv* vol 7 pp 10259-10265
[32] Qin X, Wang HC and Shan RF 2018 Morphology-controlled synthesis of Ag nanoparticle decorated glassy carbon electrode and its electrochemical performance *Ionics* vol 24 pp 1765-1772
[33] Wei GD, Nan CW, Deng Y and Lin YH 2003 Self-organized synthesis of silver chainlike and dendritic nanostructures via a solvothermal method *Chem Mater* vol 15 pp 4436-4441
[34] Hao H, Sheng QL and Zheng JB 2017 Ag C core-shell structure composites-decorated Ag nanoparticles: zero current potentiometry for detection of hydrogen peroxide *Chem Pap* vol 71 pp 535-542
[35] Qin X, Lu W, Luo Y, Chang G and Sun X 2011 Preparation of Ag nanoparticle-decorated polypyrrole colloids and their application for H_2O_2 detection *Electrochem Commun* vol 13 pp 785-787
[36] Wu CL, Qiao XL, Chen JG and Wang HS 2007 Controllable ZnO morphology via simple template-free solution route *Mater Chem Phys* vol 102 pp 7-12
[37] Xie J, Li P, Li YT, Wang Y J and Wei Y 2009 Morphology control of ZnO particles via aqueous solution route at low temperature *Mater Chem Phys* vol 114 pp 943-947
[38] Jia D, Ren Q, Sheng L, Li F, Xie G and Miao Y 2011 Preparation and characterization of multifunctional polypyrrole–Au coated NiO nanocomposites and study of their electrocatalysis toward several important bio-thiols *Sensor Actuat B-Chem* vol 160 pp 168-173
[39] Jiang L, Syed JA, Gao YZ, Lu HB and Meng XK 2018 Electrodeposition of $Ni(OH)_2$ reinforced polyaniline coating for corrosion protection of 304 stainless steel *Appl Surf Sci* vol 440 pp 1011-1021
[40] Ahn KS, Nah YC and Sung YE 2002 Surface morphological, microstructural, and electrochromic properties of short-range ordered and crystalline nickel oxide thin films *Appl Surf Sci* vol 199 pp 259-269
[41] Zhou C, Lin LN, Ma YZ, Huang LY, Li J, Liu ZP and Dong ZX 2017 Fabrication of amorphous mesoporous $Ni(OH)_2$ hollow spheres with waxberry-like morphology for supercapacitor electrodes *Chemelectrochem* vol 4 pp 2314-2320
[42] Asen P, Shahrokhian S and Zad AI 2017 One step electrodeposition of V_2O_5/polypyrrole/graphene oxide ternary nanocomposite for preparation of a high performance supercapacitor *Int J Hydrogen Energ* vol 42 pp 21073-21085
[43] Kurowska E, Brzozka A, Jarosz M, Sulka GD and Jaskula M 2013 Silver nanowire array sensor for sensitive and rapid detection of H_2O_2 *Electrochim Acta* vol 104 pp 439-447
[44] Azizi SN, Ghasemi S, Samadi-Maybodi A and Ranjbar-Azad M 2015 A new modified electrode based on Ag-doped mesoporous SBA-16 nanoparticles as non-enzymatic sensor for hydrogen peroxide *Sensor Actuat B-Chem* vol 216 pp 271-278
[45] Qi C and Zheng J 2015 Novel nonenzymatic hydrogen peroxide sensor based on Fe_3O_4/PPy/Ag nanocomposites *J Electroanal Chem.* vol 747 pp 53-58
[46] Zhang J and Zheng JB 2015 An enzyme-free hydrogen peroxide sensor based on Ag/FeOOH nanocomposites *Anal Methods-Uk* vol 7 pp 1788-1793
[47] Antink W H, Choi Y, Seong KD and Piao Y 2018 Simple synthesis of CuO/Ag nanocomposite electrode using precursor ink for non-enzymatic electrochemical hydrogen peroxide sensing *Sensor Actuat B-Chem* vol 255 pp 1995-2001

IOP Conf. Series: Materials Science and Engineering **611** (2019) 012037 doi:10.1088/1757-899X/611/1/012037

[48] Guler M, Turkoglu V, Bulut A and Zahmakiran M 2018 Electrochemical sensing of hydrogen peroxide using Pd Ag bimetallic nanoparticles decorated functionalized reduced graphene oxide *Electrochim Acta* vol 263 pp 118-126

IOP Conf. Series: Materials Science and Engineering **611** (2019) 012038 doi:10.1088/1757-899X/611/1/012038

Experimental and Numerical Analyses of the In-plane Permeability of 2.5D-woven Carbon Fabric Preforms with Compressive Deformation

Haoxin Liu, Pengfei He, Wenxiao Li*[*] and Shiqi Jin

School of Aerospace Engineering and Applied Mechanics, Tongji University, Shanghai, 200092, China

*Corresponding author

Abstract. The radial flow method is applied to obtain the in-plane permeability of a kind of 2.5D-woven carbon fabrics with various compression ratios. Geometrical parameters of weft, filling warp and binder warp yarns are measured by using computed tomography (CT) technology while the cross-sectional shape of fiber bundles and the orientation of yarns under certain compact effects are fully considered. Regarding the fiber bundle region as porous media, an unit cell model of the 2.5D-woven carbon fabrics is set up, and resin flow process is simulated by Computational Fluid Dynamics（CFD）method. The simulation results of the in-plane permeability are proved to be well consistent with the experimental data. Finally, using this method, the in-plane permeability of 2.5D-woven carbon fabrics with different fiber volume fractions (V_f) are predicted, and the changing tendency under different compressive deformation are analysed.

1. Introduction

Resin transfer molding (RTM) is an important processing technology of composite materials that is capable of fabricating extremely complex parts and three-dimensional (3D) reinforced composite structures. In RTM process, a dry preform is placed in a closed mold, impregnated with injected liquid resin and cured in the mold. To avoid the insufficient mold filling defects, finite element method (FEM) analysis tools is conventionally used to simulate the mold filling process and assist the mould design. The permeability of woven fabric preform is the key parameter for the processing simulation [1-3]. There are three main methods for obtaining permeability: experimental method, theoretical analysis method, and numerical simulation method. The permeability of fiber preform is generally obtained by experimental method and numerical simulation method [4]. The radial flow with flow visualization technique has become the main experimental method [5].

There are three basic types of numerical simulation method: Homogenization method, Lattice-Boltzmann method and Cell method [6]. More and more scholars obtain the permeability by using cell method.

In early researches, most scholars used the unit cell model with simplified cross-section of yarns. The influences of real texture of fabrics was neglected. To improve the accuracy of the simulation results, getting the actual feature size of the minimum periodic unit by image processing way is the most advanced method. Recently, by using the electron microscope, Endruweit [7] established the unit cell model which is similar to the actual fabric microstructure. Zeng [8] used the CT scan image to create a geometric unit cell model of the three-dimensional fabric and got a good result of the yarn cross-section. Isart [9] characterized the geometry of the fabric structure by using optical micrographs.

As for the numerical prediction of permeability, the Stokes Equation and the Brinkman Equation were used to describe the resin flow of unit cell respectively by Wu [10]. They used the finite element method to solve the permeability of the fabric. Verleye [11] used a meso-scale simulation tools to compute the permeability on the unit cell of textiles. Both He [12] and Wang [13] used the coupled flow model of resin flow with Darcy's law to establish a unit cell of plain fabric and predict the permeability. Based on Darcy Law, Dong [14] finished simulation of resin flow in the unit cell which was based on a two-scale porous medium. Zeng [8] used Darcy law and Navier-Stocks Equation to describe the resin flow of inter-yarn and intra-yarn respectively. The unit cell was considered as porous media to simulate the permeability. Kang [15] verified the reliability of porous medium and used the method of porous media replacing yarn bundle regions to calculate the permeability of unit cell.
It's found that during modeling, the unit cell model can't characterize accurate section structure and orientation of yarns so that the influence which was caused by these variations on the permeability was neglected. Besides, the two-scale flow in inter-yarn and intra-yarn has a great influence on prediction results. So, considering the accurate model of yarns and resin flow in intra-yarn is absolutely necessary. This paper is divided into three parts: (1) Radial flow experiments and CT tests were conducted. (2) Based on the CT image, the fabric's minimum periodic feature was obtained. Then, a nearly realistic unit cell model was established and Computational Fluid Dynamics was used for simulation of resin flow with the method of porous media model replacing fabric bundle regions model. (3) The actual cross-section feature size and orientation of yarns with different compression ratios were got and deformation of yarns was analyzed. The permeability of 2.5D fabrics under compression was obtained by dealing with experimental data, and the effects of compression degrees on the permeability were investigated. The simulation results were compared with experimental data and the feasibility of the modeling and simulation methods was verified. At last, the permeabilities of preforms with four kinds of fiber volume fractions were predicted by using this method.

2. Experiment

In order to get the geometrical parameters and permeability of 2.5D-woven fabric preforms with compressive deformation, the CT and radial flow experiment were conducted respectively.

2.1. CT Test

The interior geometry of the fabric was characterized at different compaction levels by Computed Tomography (CT) analysis. The fabric was compressed by customized acrylic fixture. The CT instrument is Precision II CT scanner of YXLON.
The ideal 2.5D-woven fabric preform consists of filling warp, weft and binder warp yarns which are shuttled between two adjacent weft yarns. The centerline of the filling warp and weft yarns is straight, without bending deformation, as shown in Figure 1. The x-direction is the direction of the centerline of weft yarn, and the y-direction is the direction of the centerline of the filling warp yarn. Then by using the software of Digimier to process CT images, structural parameters of yarns can be obtained.

2.2. Radial Flow Experiment

The experimental equipment with the specifications listed in Table 1 is characterized in this work. The method of radial flow is applied to obtain the in-plane permeability of 2.5D-woven fabric preforms. The vacuum pump provides a stable negative pressure so that the resin flows from the middle inlet and gradually wets the preform. At the same time, the flow visualization technology is used to monitor the resin flow front. The resin flow in the thickness direction will have a negative influence on the in-plane permeability measurement. In order to avoid this problem, the hollow drill is used to open a straight hole in the middle of the preform to ensure that the resin flow front in the thickness direction is uniform during the wetting process.

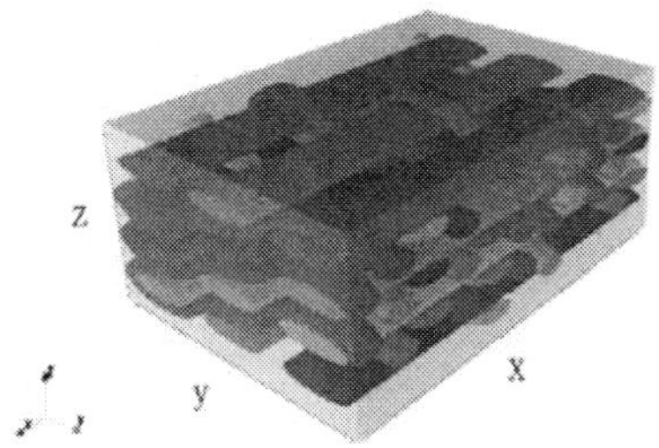

Figure 1. The ideal model of 2.5D carbon fabric.

Table 1. Specifications of experimental equipment.

Equipment	Specification
Vacuum Pump	Shanghai Vacuum Pump:2XZ
2.5D-woven preform	1/3 twill weave, T800 carbon fiber
Rotating viscometer	Bangxi Instrument: NDJ-8S
Flow monitoring	Apple Mngx2ch/a
Resin	Customization
Testing mold	Customization

Combined with the injection pressure and resin viscosity, the in-plane permeability and the deflection angle (δ) that between the X-direction of the principal permeability and the warp direction were calculated by using the two-dimensional in-plane permeability formula. The directions of principal permeability are defined as X-direction and Y-direction. The resin flow process is shown in Figure 2. The experimental parameters are shown in Table 2.

Table 2. Experimental parameters of the compressed preform.

Serial number	Original thickness	Thickness after compression	V_f (%)	Viscosity ($mP_a \cdot s$)
C0	H	H	48.4	50
C1	H	0.86H	54	50
C2	H	0.76H	61	50

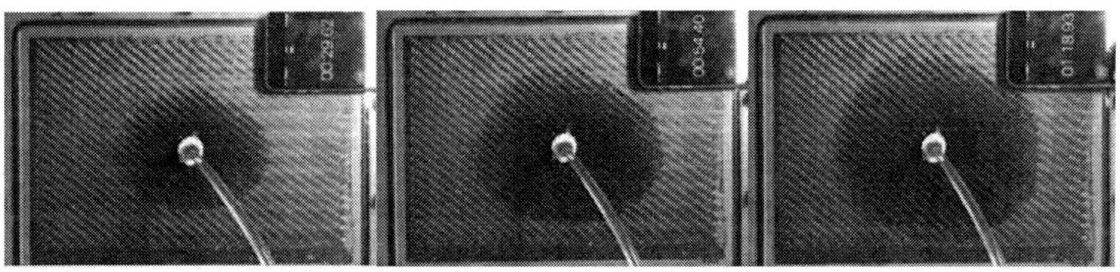

Figure 2. Typical resin flow process of the preform.

The time and distance by processing the resin flow image are obtained. According to the formula of two-dimensional in-plane permeability (1), (2), the in-plane permeability can be gained [16].

$$K_x = \frac{\mu\varepsilon}{4t\Delta P} x_f^2 \left[2\ln\left(\frac{x_f}{R_0}\right) - 1\right] \tag{1}$$

$$K_y = \frac{\mu\varepsilon}{4t\Delta P} y_f^2 \left[2\ln\left(\frac{y_f}{R_0}\right) - 1\right] \tag{2}$$

Here, K_x and K_y are the X-direction and Y-direction of principal permeability respectively. x_f and y_f are the positions of flow front in two principal directions. R_0 denotes the radius of the injection port. μ is the viscosity of the resin, ε is the porosity.

3. Numerical Analyses

3.1. Geometrical Modelling for 2.5D-woven Fabric Preforms

Structural parameters of yarns can be obtained by CT images. This paper uses the following assumptions when establishing the structure model of yarns:

(1) The cross-section of the warp yarn is rectangular;

(2) The weft yarns are divided into two types. Type I: The cross-section consists of two arcs which are at the top and bottom ends and the two arcs are connected by two straight lines respectively. Type II: The cross-section is a parallelogram. The cross-section area of these two kinds weft is identical. The transition area between the two yarn cross-sections is ignored;

(3) The centerline of the filling warp yarn is straight;

(4) The cross-section of all yarns is consistent along the centerline;

The simplified dimension and structure of preform are shown in Figure 3. The central part of the illustrated structure is weft yarn.

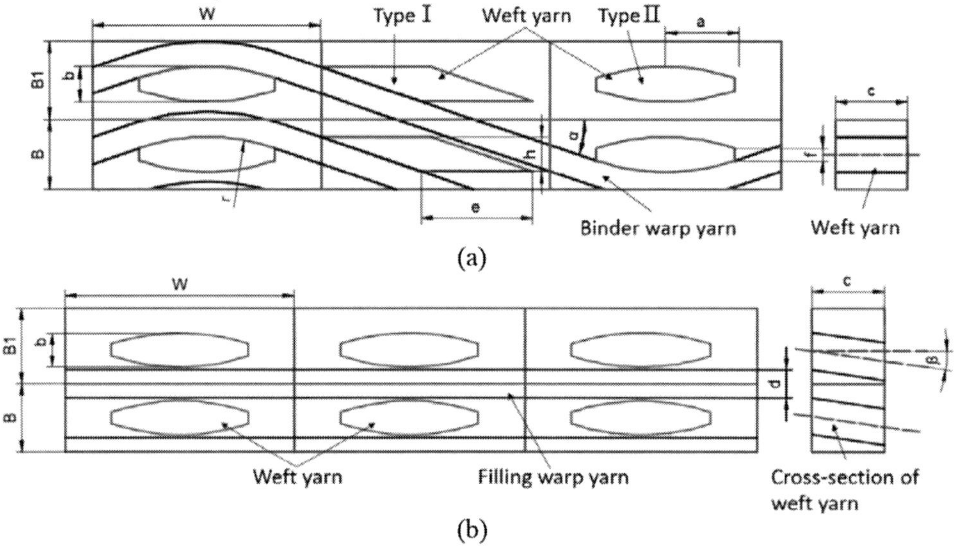

Figure 3. Structure parameters of 2.5D-woven fabric preforms. ((a) Cell structure in the position of binder warp yarn. (b) Cell structure in the position of filling warp yarn.)

Here, W is the cell length. c is the cell width, and it indicates the length of the warp cross-section. B is the height of the interior cell, B1 is the height of surface cell, a is the long semi-axle of Type I weft, and b is the height of the weft cross-section. f is the length of the straight line of the Type I weft. e and h are the length and height of the Type II bottom side respectively, and the cross-section areas of Type I and Type II are equal. d is the height of warp yarn, α is the orientation angle of binder warp yarn which characterizes its variation trend. β is the orientation angle of weft yarn, formed by the intersection of the centerline of weft yarn and the positive direction of horizontal line.

3.2. Unit Cell

According to the above structural parameters, the ideal model of 2.5D-woven fabric preforms (Figure 1) is transformed into the unit cell model which is shown in Figure 4(a). The minimum periodic unit cell is divided into eight parts which are shown in Figure 4(b).

Surface cells include SC1, SC2, SC3, and SC4. Interior cells include IC1, IC2, IC3 and IC4.According to the bending deformation, the weft yarns are divided into two types. One has a straight centerline, including SC1, SC2, SC3, IC1, IC2, and IC3, and the other has the orientation angle (β), including SC4 and IC4.

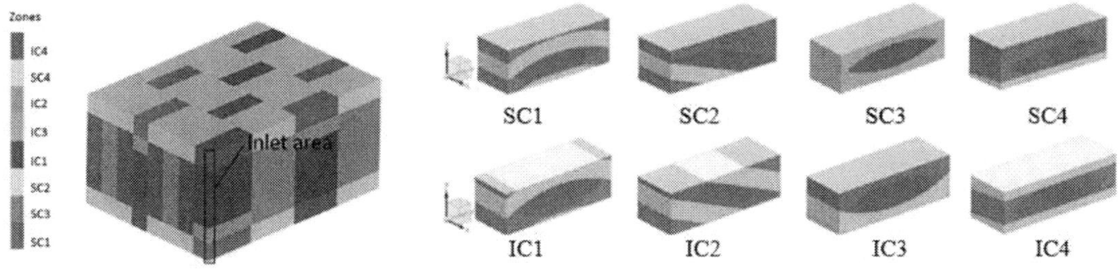

(a) (b)

Figure 4. Structure model of carbon fabrics. ((a)Structure of periodic unit cell in carbon fabrics,(b)Eight cell models, green: weft yarn, yellow: warp yarn, gray: the gap region of bundles.)

3.3. Computation of Numerical Permeability

Considering the two-scale structure in inter-yarn and intra-yarn, a porous medium model is adopted. The parallel and vertical porous media regions are used to replace the parallel and vertical fiber yarns respectively. The computational fluid dynamics (CFD) software is used to simulate the resin flow in eight unit cells to obtain the data of mass flow. Then, according to the Darcy formula [17], the permeability values of the eight cells are obtained.

The boundary nodes in the z-direction are set as the inlet, which is shown in Figure 4(a). The x-direction and y-direction permeability values of each cell are brought into the model. The resin flow is simulated by the software of PAM-RTM. According to the resin flow situation, the δ, flow velocity and filling time can be obtained. Then, the in-plane permeability of preform would be obtained by using the Darcy formula.

4. Results and Discussion

4.1. Deformation of Fabric Preforms with Compressive Deformation

The actual cross-section feature size and orientation of yarns with different compression ratios are got by CT. The yarn orientation and cross-section shape are shown in Figure 5.

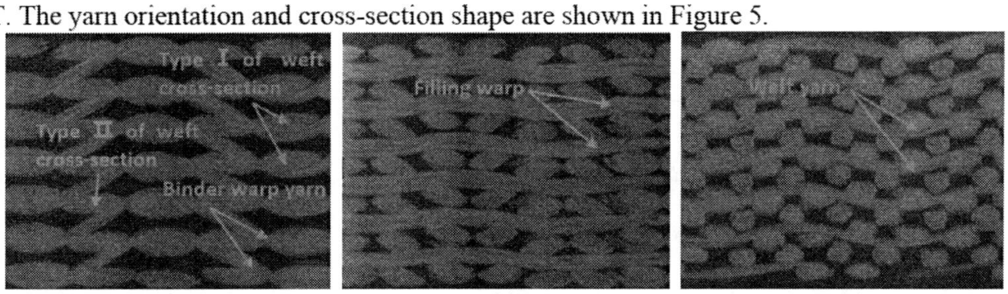

(a) (b) (c)

Figure 5. Fiber tow cross sections ((a) View from weft direction of binder warp zone, (b) View from weft direction of filling warp zone, (c) View from warp direction)

Compared with the ideal preform structure in Figure 1, it is found that the binder warp yarns between the upper and lower adjacent weft yarns, the filling warp yarn is not bent substantially and its centerline is straight. These two types of yarns are arranged at intervals. The weft direction and cross-section shape have changed, and micro-bending occurred at the same time. The cross-section shape and area of the weft and warp remain unchanged. The specific cell parameter information is shown in Table 3.

In these experiments, as the degree of compression increases, α, β, b, and d decreased. This is because after weaving, the weft yarn is compressed by warp yarn, resulting in slight deformation. And the heights of weft and warp are both reduced with the decrease of fabric thickness. The yarns of the preform are gradually compacted, and the remaining parameters of yarns remain unchanged.

Table 3. Main parameters of unit cell under different compression.

Serial number	Surface cell size(mm)			Interior cell size(mm)			α(°)	β(°)	b(mm)	d(mm)
	c/H	W/H	B1/H	c/H	W/H	B/H				
C0	0.119	0.355	0.147	0.119	0.355	0.141	26.9	11.0	0.74	0.67
C1	0.119	0.355	0.130	0.119	0.355	0.120	25.1	9.8	0.68	0.59
C2	0.119	0.324	0.118	0.119	0.324	0.105	23.2	8.5	0.61	0.54

4.2. In-plane Permeability

According to the formula of two-dimensional in-plane permeability, the in-plane permeability values can be gained. The calculation shows the time variation of in-plane permeability and deflection angle (δ) (Figure 6-7).K_a and K_b are reference standard value. T_0 is the total filling time, which represents the time from the beginning of the resin injection to the end. T is the time from resin injection to calculate permeability.

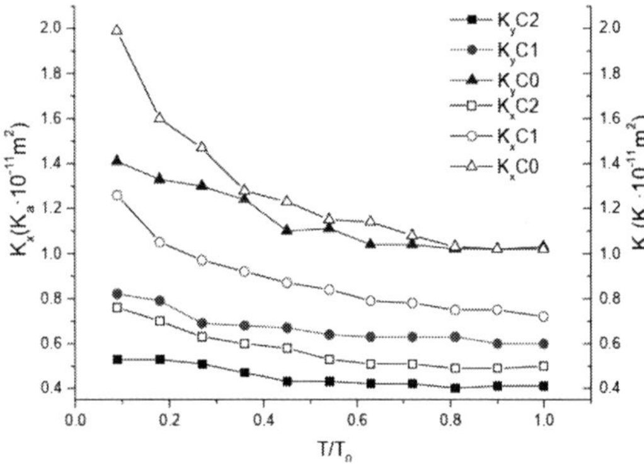

Figure 6. Principal permeability values of X and Y direction in three compression ratios.

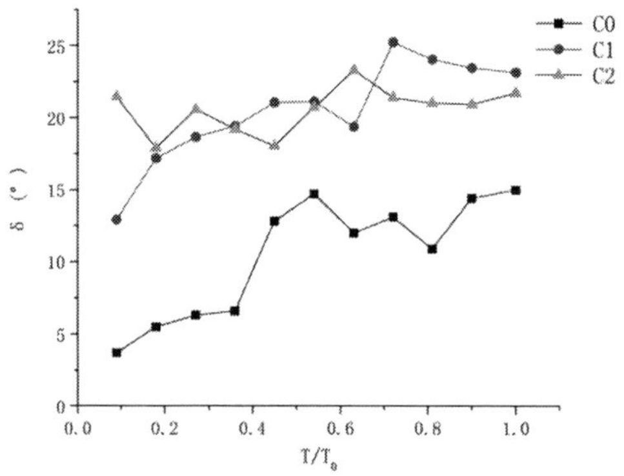

Figure 7. Deflection angle (δ) values in three compression ratios.

In the initial stage, due to the high flow rate and the big area ratio of the unsaturated region to the saturated region, the measured permeability values are large. As the resin flows continuously, the flow rate decreases and tends to be a constant value, the variation of measured permeability values has the similar tendency. The unsaturated flow in the 2.5D woven texture leads to the gradually expending rough boundary of the saturated region, combined with the deviation in setting the resin boundary by using the software of Digimier, the measured δ values have larger fluctuation. The time range in which the permeability values and δ are relatively stable is selected to calculate the value of principal permeability, as shown in Table 4.

Table 4. The permeability of 2.5D-woven fabric preforms with different compression ratios.

Serial number	V_f(%)	$K_x(K_a \times 10^{-11}\text{m}^2)$	K_y $(K_b \times 10^{-11}\text{m}^2)$	δ(°)
C0	48.4	1.07	1.12	13.6
C1	54	0.79	0.73	22.4
C2	61	0.51	0.42	21.2

It is found that when the V_f is 48%-61%, the principal permeability decreases with the increase of the compression degree. And the overall variation of permeability value is a linear tendency, while the value of δ is basically stable at 22 °after preform is compressed. As the degree of compression increases, the deformation of the preform mainly occurs in the z-direction. The yarns are gradually compacted, both of the cross-section height of the warp and weft yarns and the gap of inter-yarn and intra-yarn are reduced. These deformations are confirmed by CT, which gets the cross-section features of the preform. In summary, the compression leads to a reduction of the gap of inter-yarn and intra-yarn, which impeding the resin flow, resulting in the descent of the permeability.

4.3. Numerical Simulation

The in-plane permeability of preform are obtained by the Darcy formula, which is shown in Table 5.

Table 5. In-plane permeability of preform with different compression ratios.

Serial number	α (°)	K_x ($K_a \times 10^{-11}$m^2)		K_y ($K_b \times 10^{-11}$m^2)		δ (°)	
		Experiment	Simulation	Experiment	Simulation	Experiment	Simulation
C0	26.5	1.07	2.09	1.12	1.79	13.6	6.8
C1	25	0.79	1.50	0.73	1.10	22.4	9.4
C2	23	0.51	0.96	0.42	0.76	21.2	9.7

Comparing the numerical simulation results with experimental measurement results, it can be concluded that the simulated values of principal permeability are larger than the experimental values, but both of them are at the same order of magnitude. The simulated value of the δ is smaller than the experimental measurement, but the simulated value follows the same trend as the measured data. Besides the certain deviation of the experimental permeability to the actual permeability, the simulate setting of boundary conditions, in which the interaction between the two adjacent periodic unit cells is weaken, may mainly cause this problem. In the follow-up research, this aspect is meaningfully improved.

This numerical simulation method is used to predict the in-plane permeability of 2.5D-woven fabric preforms. The V_f of these four compressed fabrics are 50%, 52%, 56%, and 58%, respectively. Finally, the curve fitting results are shown in Figure 8 and Figure 9. It can be found that after the 2.5D-woven fabric preforms is compressed, the principal permeability decreases with the increase of the V_f, and the δ increases first and then gradually stabilizes, and its variation range is 6.8°-9.7°.

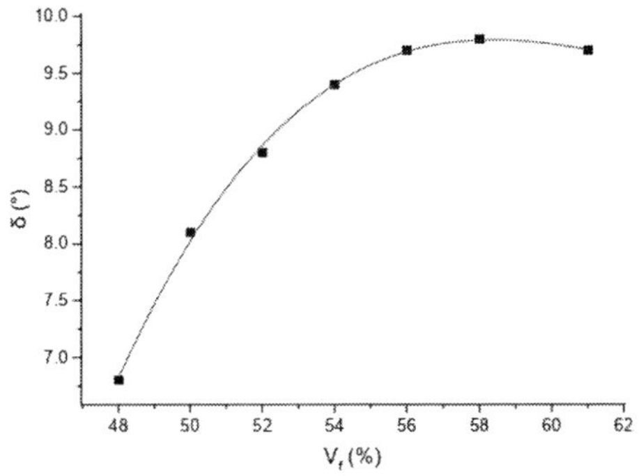

Figure 8. Prediction curve of deflection angle (δ) with fiber volume fraction.

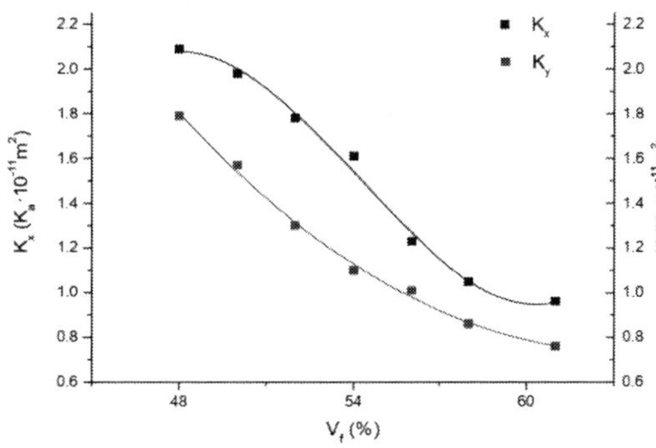

Figure 9. Prediction curves of X-direction and X-direction permeability in different V_f.

The major difficulty is that different regions of the carbon fiber fabric are subjected to different compressions during RTM molding, which makes it impossible to complete the experiment test and obtain the in-plane permeability by one test. However, the permeability values and δ can be obtained by the prediction curve, which can offer a reference for subsequent RTM molding simulation, optimize the design of molds, avoid manufacturing defects and reduce costs.

5. Conclusions

Compared with previous permeability researches, accurate section structure and orientation of yarns are taken into consideration in this work of 1/3 twill 2.5D-woven fabric preforms. Besides, priority consideration is given to the two-scale flow in inter-yarn and intra-yarn in numerical simulation. And good results are obtained. The following conclusions were drawn:

(1) The in-plane permeability coefficients of 2.5D-woven fabric preforms with various compression ratios are obtained by experiments. When the V_f is 48%-61%, the in-plane permeability decreases linearly with the increase of V_f, but the δ is stable at 22°.

(2) The accurate geometric modeling and numerical prediction method for 2.5D-woven fabric preforms are validated. The permeability of compressed 1/3 twill 2.5D-woven fabric preforms whose V_f is 48%-61% can be predicted and the deviation is within an acceptable range.

Acknowledgement
The research is supported by the Fundamental Research Funds for the Central Universities (Project No.22120180133).

References

[1] Song C 2014 Research on Design and Manufacture for Composites Blade in Aircraft Turbin. Thesis, Nanjing University of Aeronautics and Astronautics, China.
[2] Hurtado R 2008 Composites in Commercial Aircraft Engines. International Journal of Solids and Structures 9: 202-206.
[3] Luycker E, Moresin F, Boisse P, et al. 2009 Simulation of 3D interlock composite preforming. Composite Structures 88(4): 615-623.
[4] Kruckenberg T, Paton R, et al. 2009 Resin Transfer Moulding for Aerospace Structures. Aviation industry press.
[5] Chan A W and Hwang S 1991 Anisotropic in-plane permeability of fabric media. Polymer Engineering & Science 31(16): 1233-1239.
[6] Li J, Wu X, Feng C, et al. 2006 Research progress on the permeability prediction of fiber in RTM. Acta Materiae Compositae Sinica, 23(6):1-8.
[7] Endruweit A and Long A C 2010 Analysis of Compressibility and Permeability of Selected 3D Woven Reinforcements. Journal of Composite Materials 44(24):2833-2862.
[8] Zeng X, Brown L P, Endruweit A, et al. 2014 Geometrical modelling of 3D woven reinforcements for polymer composites: Prediction of fabric permeability and composite mechanical properties. Composites Part A 56(1):150-160.
[9] Isart N, Mayugo J A, Blanco N, et al. 2015 Geometric model for 3D through-thickness orthogonal interlock composites. Composite Structures 119:787-798.
[10] Wu Y, Yan S, Tan H, et al. 2008 Prediction of Permeability of Fiber Fabrics with two-scale void. Chinese Journal of Solid Mechanics, s1: 80-84.
[11] Verleye B, Lomov S V, Long A, et al. 2010 Permeability prediction for the meso-macro coupling in the simulation of the impregnation stage of Resin Transfer Moulding. Composites Part A 41(1):29-35.
[12] He H 2011 Structural Correlation Modeling of Permeability of Fiber Composite Preforms. Thesis, Shandong University, China.
[13] Wang W and Wu X 2014 Prediction Permeability of Glass Fiber Plain Fabric by Unit Cell Method. Fiber Composites 3:14-16.
[14] Dong S 2014 Permeability Prediction of Fiber Preform and Infiltration Process Simulation by Finite Element Method. PhD Thesis, Shandong University, China.
[15] Kang N and Tang Z 2016 Numerical Prediction on the Permeability of Non-crimp Fabrics. China Plastic Industry 44(1):44-47.
[16] Jin T, Wei Y, Yang B, et al. 2015 Permeability prediction of textile preform and influence of compression deformation. Acta Materiae Compositae Sinica 32(3):840-847.
[17] Ngo N D and Tamma K K 2001 Microscale permeability predictions of porous fibrous media. International Journal of Heat and Mass Transfer 44(16):3135-3145.

Preparation of Coral-like SnO_2 Hierarchical Nanostructures and Its Application in Ethanol Gas-sensing Performance

Shaohui Yan*, Jie Zhang, Yibo Huang and Yan Lu

College of Physics and Electronic Engineering, Key Laboratory of Atomic and Molecular Physics & Functional Materials of Gansu Province, Northwest Normal University, Lanzhou, Gansu 730070, China

*Corresponding author

Abstract. Coral-like SnO_2 hierarchical nanostructures were successfully fabricated by a simple hydrothermal route. By means of a comprehensive analysis of microstructures and morphologies of the hierarchical SnO_2, we found that the coral-like nanostructures were comprised of irregular nanorods and nanosheets. Considering the essential advantages of hierarchical architectures, its gas sensitivity and the formation mechanism have been further explored. The sensor response to 100 ppm ethanol is about 25 at the optimum working temperature of 240 °C, and the response and recovery time are about 17 s and 7 s, respectively. Based on the experimental results, a possible morphology-dependent sensing mechanism are speculated.
Keywords: coral-like SnO_2; hydrothermal; gas sensing; ethanol.

1. Introduction

Nanostructured semiconductor metal oxides have attracted considerable attention in many fields due to their large surface area, especially as gas sensors [1-3]. Over the past decades, quite a few efforts have been made to prepare many different nanostructures, such as nanoparticles, nanowires, nanorods, nanobelts and nanotubes, etc [4-9]. It is well known that the improvement of gas sensitivity mainly depends on crystalline structures, morphologies and dimensions. Nevertheless, not all nanostructures can effectively promot the gas sensitivity in the sensing films. The nanostructures are usually clustered together due to the natural reduction of the total surface energy. Undoubtedly such densely clustered structures are detrimental for the diffusion of the target gases into the surface of internal nanostructures, weakening their inherent sensing properties [10]. In particular, the oxides with three-dimension (3D) hierarchical architectures, which are assembled by the 1D or 2D nanoscale building blocks [11], often display novel characteristic [12]. For example, Xie et al. have synthesized porous SnO_2 based nanofibers by carbonization. The carbonized nanofibers had a hollow structure, which exhibited higher response and shorter response time toward toluene [13]. The three-dimensional SnO_2 nanostructures assembled with ultrathin shuttle-shaped nanosheets which showed excellent gas sensing performance to acetone at 500 °C [14]. The SnO_2 hierarchical microspheres consisted of two-dmensional nanosheets exhibiting high response and good selectivity to ethanol at 240 °C [15]. In this work, we report a simple hydrothermal method with the carbonization process for preparation of coral-like SnO_2 novel nanosructures. The SnO_2 sensor shows high sensitivity to ethanol at lower operating temperature than other studies. The possibility of gas sensing mechanism formation are also further discussed.

IOP Conf. Series: Materials Science and Engineering **611** (2019) 012039 doi:10.1088/1757-899X/611/1/012039

2. Experimental

2.1. Preparation of SnO_2 Hierarchical Nanostructures

All chemical reagents of experiments were analytical grade and used without further purification. The SnO_2 hierarchical nanostructures were fabricated via a simple hydrothermal process. The synthesis process was as follows: 0.460 g $SnCl_2$•$2H_2O$ powers were dissolved into distilled water (40 mL) with magnetic stirring for about 30 min. Meanwhile, 0.480 g NaOH was dissolved in 40 mL distilled water. Subsequently, both of them were mixed together by stirring at 35 °C for 2 h and then added into 0.740 g CTAB. Following, the resulting mixtures were transferred into 100 mL Teflon-lined stainless autoclave, sealed and maintained at 130 °C for 24 h. Finally, the weak yellow products of SnO_2 were obtained by centrifugation, followed by rinsing with distilled water and ethanol several times, and then dried at 80 °C for 12 h. The final SnO_2 nanostructures were obtained by calcining in a furnace at 600 °C for 3 h.

2.2. Materials Characterization

The microstructure and purity of the material was analyzed by X-ray diffraction (XRD, D/Max-2400) using Cu Kα radiation with λ = 0.15406 nm. The surface morphology was investigated by scanning electron microscope (SEM, S-4800). Transmission electron microscopic (TEM), high-resolution transmission electron microscopic (HRTEM), and selected area electron diffraction (SAED) image, as well as the energy dispersive X-ray (EDX) analysis results were obtained by USA FEI G2 TF20 transmission electron microscope.

2.3. Fabrication and Sensing Test of Gas Sensor

Gas sensor was fabricated as follows. The prepared SnO_2 power was mixed with ethanol to form a paste and coated onto an alumina ceramic tube with a pair of Au electrodes and four Pt wires. After sintering at 300 °C for 3 h in air to enhance the stability of material, a heating wire was inserting into the alumina tube as illustrated in figure 1(a). The sensor was then aged at 300 °C for one week in air to enhance its repeatability and stability.

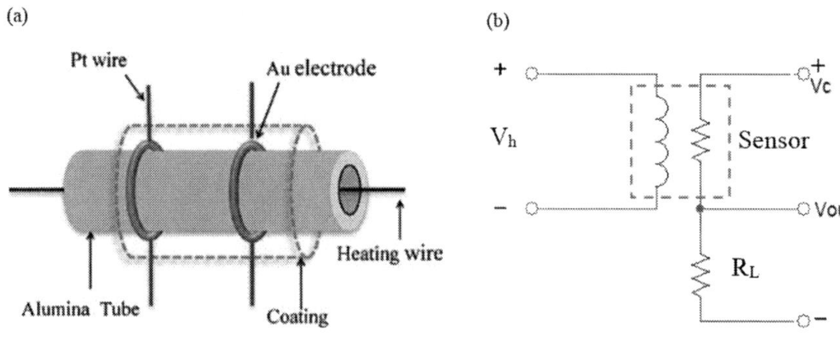

Figure 1. Schematic diagram of (a) the sensor device and (b) the gas sensing measurement circuit.

Gas-sensing properties were measured by WS-60A gas sensing measurement system (Wei Sheng Electronics Science and Technology Co., Ltd., China). The sensor response was defined as $S=R_a/R_g$, where R_a was the resistance of the sensor in air and R_g was the resistance of the sensor in the presence of the detected gas. The response time was defined as the time required reaching 90% of the steady response value and the recovery time was defined as the time attaining within 10% of the initial response value [16]. The schematic diagram of the gas sensing circuit is shown in figure 1(b). The input voltage was set tp 5 V during the gas sensing test. The test gas was formed by evaporating the liquid that injected on the bottom surface of gas chamber (the bottom contains a heating plate) by a microsyringe. The output voltage was measured by R_L, which varies with resistance of the sensor. The sensor resistance relied on test gas and its concentration.

IOP Conf. Series: Materials Science and Engineering **611** (2019) 012039 doi:10.1088/1757-899X/611/1/012039

3. Results and Discussion

3.1. Structure and Morphology of SnO_2 Hierarchical Nanostructures

XRD pattern in figure 2(a) shows that all diffraction peaks appearing from the as-prepared sample is completely indexed to a tetragonal rutile structure of SnO_2 (JCPDS No. 41-1445), and no other diffraction peak are observed, indicating that the prepared material has high purity. The corresponding EDX patterns of SnO_2 sample (figure 2(b)) further confirm that it is composed of Sn, and O (the presence of Cu signal is from the copper grid used in the TEM measurement).

It can be seen from a low-magnification SEM image shows in figure 3(a) that SnO_2 sample has coral-like nanostructures with a diameter of about 20-30 nm. High-resolution SEM image shown in figure 3(b) reveals that they consist of many nanorods and nanosheets. As can be seen from the typical TEM image in figure 3(c), the same results are also well supported. Obviously, the achieved hierarchical structures are all nanorods and nanosheets, randomly assembled together to form a coral-like morphology. The HRTEM image (figure 3(d)) exhibits well-defined lattice fringes, the interplanar distances of 0.33 and 0.26 nm are indexed to the (1 1 0) and (1 0 1) planes of the phase of tetragonal rutile SnO_2, respectively. This result also matches well with the previous XRD analysis. The SAED pattern (figure 3(e)) confirms the high-crystalline nature of the prepared SnO_2 material. The BET surface area of the SnO_2 hierarchical structures is calculated to be 36.2 m^2/g. Bigger surface area will provide more chance to adsorb more oxygen on the surface of SnO_2 nanostructures. It will improve the gas sensitivity of the sensor.

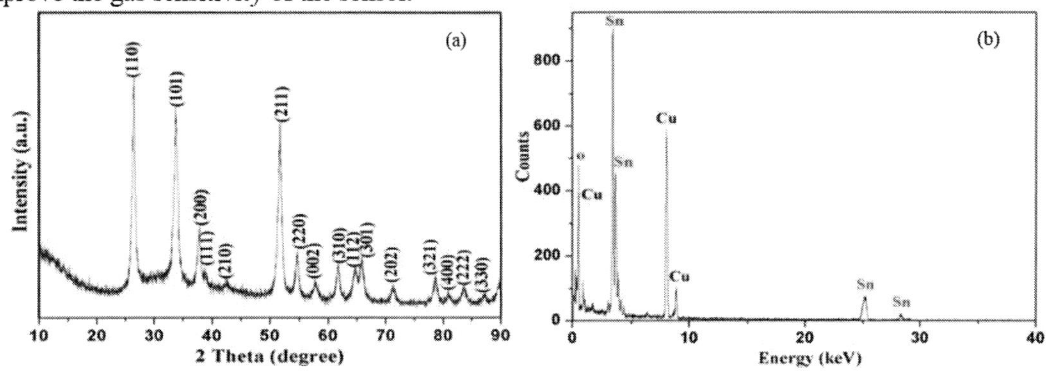

Figure 2. (a) XRD patterns and (b) EDX patterns of SnO_2 sample.

Figure 3. (a, b) Low-magnification and high-resolution SEM images of SnO_2 hierarchical nanostructures, (c) TEM images of SnO_2 hierarchical nanostructures, (d) HRTEM micrograph showing enlarged lattice images of SnO_2 hierarchical nanostructures, (e) SAED pattern of SnO_2 hierarchical nanostructures.

3.2. The Gas Sensing Characteristics

The operating temperature plays a crucial role in the adsorption/desorption of gas and the chemical reaction of surface adsorbed oxygen with ethanol molecules [17]. The response value of the sensor to 100 ppm ethanol at different operating temperatures were measured and shown in figure 4(a). The gas response initially increases with the increasing of operating temperature and shows a maximum response of 25 at 240 °C, and then decrease with further increase of the operating temperature. Therefore, we choose 240 °C as the optimal working temperature for subsequent gas-sensing measurements. Normally, the gas adsorption processes occurs at lower temperature, but it is weak for chemical activation. When the sensor work at higher temperature, some adsorbed gas molecules may escape before their reaction due to their enhanced activation [18].

The curve of response versus ethanol concentration is shown figure 4(b). The response of the sensor increases rapidly with increasing concentration of ethanol, and increases slowly above 2000 ppm, which indicates that the sensor becomes more or less saturated. The response shows a near-linear dependence with gas concentration in the range of 5-200 ppm (inset in figure 4(b)). The sensing response toward 5 ppm ethanol reaches 1.2, indicating that SnO_2 can be potentially used as sensing materials for detecting ethanol at low concentration.

Fast response and recovery times are essential for gas sensors. Figure 4(c) shows a single cycle response-recovery characteristics of the SnO_2 sensor operating at 240 °C to 100 ppm ethanol. The response and recovery time are about 17 s and 7 s, respectively. As shown in Fig. 2, the coral-like SnO_2 hierarchical nanostructures allows the gas molecules diffuse rapidly throughout the sensing film, resulting in its faster gas response speed. The short recovery time may be attributed to the hierarchical nanostructures which speeds up the desorption rate of the ethanol.

Figure 4(d) shows the gas sensing response of SnO_2 to different gases with concentration of 100 ppm at 240 °C. Obviously, the response toward ethanol is much higher than that toward methanol, acetone, DMF, ammonia, and acetic acid, which indicates that the SnO_2 nanostructures based sensor exhibits

IOP Conf. Series: Materials Science and Engineering **611** (2019) 012039 doi:10.1088/1757-899X/611/1/012039

prominent and high selectivity to ethanol. This result may be attributed to the fact that ethanol molecules are more susceptible to adsorb and react on the surface of SnO_2 hierarchical structures with defects and mismatches in contrast to others investigated volatile organic compounds (VOCs), leading to the large release of electrons to the conduction band of hierarchical nanostructured SnO_2. A comparison between this work and previous literature reports is summarized in Table 1. Therefore, coral-like SnO_2 hierarchical nanostructures could be used as a candidate material for effectively detecting ethanol due to the superior gas sensing properties.

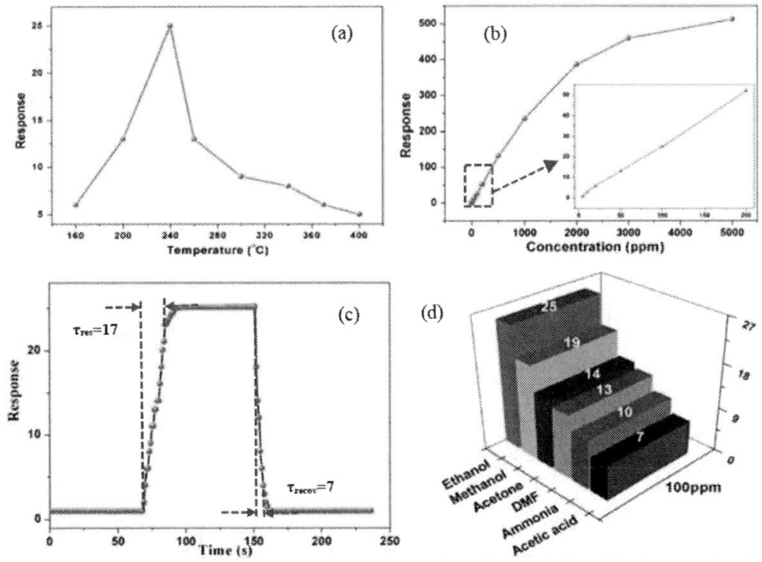

Figure 4. (a) The response of SnO_2 sensor to 100 ppm ethanol at different operating temperatures (160-400 °C), (b) The response of SnO_2 sensor to ethanol concentrations in the range from 5 to 5000 ppm at 240 °C (the inset shows the relationship in the range of 5-200 ppm), (c) The response-recovery curves of SnO_2 sensor to 100 ppm ethanol at 240 °C, (d) The response of SnO_2 sensor to 100 ppm different gases at 240 °C.

Table 1. Comparison of gas-sensing characteristics of other similar sensing materials based ethanol gas sensors.

Analyte gas	Gas-sensing materials	Synthesis route	Temperature (°C)	Response	Response/Revoery Time (s)	Reference
Ethanol (100 ppm)	Porous SnO_2 nanowires	Two-step method	380	17	22/18	[19]
Ethanol (100 ppm)	CuO nanosheets	Hydrothermal method	370	8.9	15/11	[20]
Ethanol (100 ppm)	α-Fe_2O_3 nanospheres	Laser irradiation	250	19.5	12/28.5	[21]
Ethanol (400 ppm)	ZnO nanoflowers	Hydrothermal method	350	30.4	10/4	[22]
Ethanol (100 ppm)	Horseshoe-shaped SnO_2	Self-assembly	225	17.3	8/780	[23]
Ethanol (100 ppm)	SnO_2 hierarchical nanostructures	Hydrothermal method	240	25	17/7	This work

3.3. The Gas Sensing Mechanism

It is well known that the gas sensing mechanism of metal oxide semiconductor is an adsorption-oxidation-desorption process that results in changes in conductance of sensor[24-27]. When the sensor is exposed to air ambient, the oxygen molecules are absorbed onto the surface of the SnO_2 nanostructures and ionized to O_2^- (ads) or O^- (ads) by capturing free electrons from the conduction band of SnO_2, which was highly dependent on the working temperatures [28,29]. As is shown in eqs 1 and 2

$$O_2\ (gas) + e^- \rightarrow O_2^-\ (ads) \quad (1)$$

$$O_2\ (gas) + e^- \rightarrow 2O^-\ (ads) \quad (2)$$

The adsorption of O_2^-(ads) and O^-(ads) ions on the surface of SnO_2 nanostructures improve the receptor function of the sensor. When the sensor is exposed to a reducing gas such as ethanol, these gas molecules could react with adsorbed O_2^- (ads) and O^-(ads) and release the trapped electron to the conduction band, resulting in the reduction of the resistance [30].

The enhanced ethanol gas sensing properties the present coral-like SnO_2 hierarchical nanostructures can be attribute to the following aspects. Firstly, the SnO_2 nanosructures are constructed by numerous nanorods and nanosheets, which possess a capacious space between adjacent neighbor nanorods and nanosheets. The nanorods and nanosheets array could increase the number of the gas channels leading to more effective surface areas. Secondly, hierarchical nanosructures are favorable for the sufficient diffusion of the gases detected among the sensing film. More importantly, the coral-like hierarchical nanostructures have the larger specific surface area will provide more contact chance between the SnO_2 nanostructures and target gas. Thirdly, in the as-prepared coral-like SnO_2 hierarchical nanostructures there are many nanojunctions between adjacent nanorods and nanosheets. These nanojuctions are similar to the grain boundaries of thin film gas sensor and might be considered as the active sites that can promote the response of gas sensor [31]. Thus, the enhanced sensor response can be attributed to the unique construction of SnO_2 hierarchical nanostructures.

4. Conclusions

In summary, we developed a simple hydrothermal growth route to prepare coral-like SnO_2 hierarchical nanostructures in absence of substrate. The sensor based on SnO_2 nanostructures displayed high response, good selectivity and quick response/recovery rate toward ethanol. The sensing response to 100 ppm of ethanol at 240 °C was up to 25. The improved ethanol gas sensing performance of the SnO_2 nanostructures can own to surface defects, unique structure and nanojunctions between contiguious nanorods and nanosheets.

Acknowledgements

This work was supported by China Postdoctoral Science Foundation (Grant NO. 2018M642426), The Scientific Research Project of Gansu Province (Grant NO. 17JR5RA072) and Natural Science Foundations of Gansu Province (Grant NO.1606RJYA263).

References

[1] Y.Q. Zhang, D. Li, L.G. Qin, P.L. Zhao, F.M. Liu, et al., Sens. Actuators B: Chem. 255 (2018) 2944-2951.
[2] X.L. Yang, S.F. Zhang, Q. Yu, L.P. Zhao, P. Sun, et al., Sens. Actuators B: Chem. 281 (2019) 415-423.
[3] G. Atanasova, A.O. Dikovska, T. Dilova, B. Georgieva, G.V. Avdeev, et al., Appli. Sur. Sci. 470 (2019) 861-869.
[4] N.B. Tanvir, O. Yurchenko, C. Wilbertz, G. Urban, J. Mater. Chem. A 4 (2016) 5294-5302.
[5] H.D. Chen, K.L. Jin, P.F. Wang, J.C. Xu, Y.B. Han, et al., J. Phys. Chem. Soli. 120 (2018) 271-278.
[6] T.V. Dang, N.D. Hoa, N.V. Duy, N.V. Hieu, ACS Appl. Mater. Interface 8 (2016) 4828-4837.
[7] I. Kortidis, H. C. Swart, S.S. Ray, D.E. Motaung, Sens. Actuators B: Chem. 285 (2019) 92-107.
[8] S. Yang, Y.L. Liu, T. Chen, W. Jin, T.Q. Yang, et al., Appl. Surface Science 393 (2017) 377-384.
[9] G.P. Evans, M.J. Powell, I. D. Johnson, D.P. Howard, D. Bauer, et al., Sens. Actuators B: Chem. 255 (2018) 1119-1129.
[10] Y.X. Li, Z. Guo, Y. Su, X.B. Jin, X.H. Tang, et al., ACS Sens. 2 (2017) 102-110.
[11] A. Dey, Mater. Sci. Eng. B 229 (2018) 206-217.
[12] Z.J. Wei, Q. Zhou, Z.R. Lu, L.N. Xu, Y.G. Gui, et al., Phys. E: Low-dimensional Sys. Nano. 109 (2019) 253-260.
[13] N. Xie, L.L. Guo, F. Chen, X.Y. Kou, C. Wang et al., Sens. Actuators B: Chem. 271 (2018) 44-53.
[14] Y.-X Li, Z. Guo, Y. Su, X.-B Jin, X.-H Tang, et al., ACS Sens. 2 (2017) 102-110.
[15] P. Sun, X.D. Mei, Y.X. Cai, J. Ma, Y.F. Sun, et al., Sens. Actuators B: Chem. 187 (2013) 301-307.
[16] S.H. Yan, X.Y. Liang, H.S. Song, S.Y. Ma, Y. Lu, Ceramics International. 44 (2018) 358-363.
[17] Y.V. Kaneti, Z. Zhang, J. Yue, Q.M.D. Zakaria, C. Chen, et al., Phys. Chem. Chem. Phys. 16 (2014) 11471-11480.
[18] Y.Q. Zhang, D. Li, L.G. Qin, P.L. Zhao, F.M. Liu, et al., Sens. Actuators B: Chem. 255 (2018) 2944-2951.
[19] R. Li, S. Chen, Z. Lou, L. Li, T.T. Huang, et al., Sens. Actuators B: Chem. 227 (2017) 79-85.
[20] A. Umar, A.A. Alshahrani, H. Algarni, R. Kumar, Sens. Actuators B: Chem. 250 (2017) 24-31.
[21] E.M. Dai, P.P. Wang, Y.X. Ye, Y.Y. Cai, J. Liu, et al., Mater Lett. 211 (2018) 239-242.
[22] L. Zhu, Y.Q. Li, W. Zeng, Appl. Surface Sci. 427 (2018) 281-287.
[23] Y.L. Wang, C. Liu, L. Wang, J. Liu, B. Zhang, et al., Sens. Actuators B: Chem. 240 (2017) 1321-1329.
[24] J. Luo, S.Y. Ma, A.M. Sun, L. Cheng, G.J. Yang, et al., Mater. Lett. 137 (2014) 17-20.
[25] F. Fan, Y. Feng, P. Tang, A. Chen, R. Luo, et al., Ind. Eng. Chem. Res. 53 (2014) 12737-12743.
[26] W. Guo, J. Mater. Sci.: Mater. Electron. 27 (2016) 7302-7310.
[27] J. Xu, Z. Xue, N. Qin, Z. Cheng, Q. Xiang, Sens. Actuators B: Chem. 242 (2017) 148-157.

[28] N.G. Shimpi, S. Jain, N. Karmakar, A. Shah, D.C. Kothari, et al., Appl. Surf. Sci. 390 (2016) 17-24.
[29] D.M. Han, Y. Li, F.B. Gu, Z.H. Wang, J. Coll. Inter. Sci. 531 (2018) 320-330.
[30] B. W. Zhang, W.Y. Fu, X.W. Meng, R. A., P.Y. Su, et al., Sens. Actuators B: Chem. 261 (2018) 505-514.
[31] C. Peng, J.J. Guo, W.K. Yang, C.K. Shi, M.R. Liu, et al., J. Alloys Compd. 654 (2016) 371-378.

Material Properties Analysis of Acid-alkali Resistance Geosynthetics for Highway

Ying Liu, Yongyan Yu and Lei Chen

Research Institute of Highway Ministry Oftransport, Beijing 100088, China

Abstract. In this paper, the engineering application characteristics for polypropylene filament geotextiles have been explored and analyzed. Besides, test about physical, chemical and mechanical properties has been conducted and conclusions have been drawn as below: engineering application characteristics for polypropylene geotextiles are good and corresponding physical parameters are excellent, especially good tensile property. The ageing resistance is outstanding, while breaking strength and elongation at break after aging test remain and show no obvious change. In addition, due to excellent acid and alkali resistant properties, the material can be applied to bituminous pavement and cement concrete pavement.
Keywords: geosynthetics; acid-alkali resistance property; engineering application; experimental analysis.

1. Introduction

The combined action of physical, chemical and biological factors results in damage of overall structure for pavement. The raw materials for overall structure of pavement usually include bituminous mixture and concrete which will result in acid or alkali pavement structure under various environmental factors. When geotextiles are used to prevent and control pavement crack, the engineering service life (i.e. engineering conformance) shall be considered. Chen Weidong[1] concluded good physical and chemical properties during disease control for pavement with geotextiles. Xie Jingkun[2] et al applied different testing factors and tested geotextiles with different specifications and then concluded good acid and alkali resistant properties of polypropylene fiber geotextiles under strong acid and alkali environment. Zheng Zhineng[3] explored acid-base properties of polypropylene reinforced belt and concluded its strong acid and alkali resistant properties. Chen Jie[4] studied chemical resistance of geosynthetics and indicated small mass change and high reduction in strength and high influences of temperature on strength of geotextile in alkaline solution. Shen Lei[5] implemented different mechanical property tests for different geosynthetics, applied saline solution to conduct indoor accelerated corrosion test for geosynthetics, conducted different tests for disposed geosynthetics, compared and analyzed data before and after test, and concluded the property of optimal engineering application. Zhang Baosen[6] et al tested and studied engineering application properties of geosynthetics of same and different varieties and concluded large differences in quality property of geosynthetics and that different geosynthetics should be selected upon engineering practice. Yang Guangqing[7] concluded production, actual application and property of geotextile and proposed that high-strength and environmental resistance type polypropylene filament needle punched geotextiles would be footstone for geotextiles in the future.

2. Measurement for Physical and Mechanics Parameters

2.1. Test Surface Parameter Specification
For functions of polypropylene filament geotextiles (hereinafter referred to as "geotextiles"), physical and mechanical parameters are measured to get 600g/m^2 in this test. According to related standards, physical and mechanical parameters such as area, thickness and breaking strength of monomer are measured for standard sample.

Table 1. 600g geotextile parameter table

Mass per unit area for monomer g/m^2		600
thickness mm		3.5
Breaking strength kN/m	Lateral	45
	Longitudinal	
Elongation at break %	Lateral	40-110
	Longitudinal	
Tearing strength kN	Lateral	1.3
	Longitudinal	

2.1.1. Test Deviation of Mass per Unit Area for Monomer of Geotextiles. The typical samples shall be selected when monomer area of geotextiles is measured. For engineering materials such as geotextile with spare property, ensuring mechanical parameters under maximum mass per unit area for monomer will reduce production cost and also cost for engineering materials. When mass deviation for geotextiles is measured, the following formula shall be applied to calculate and test:

$$\rho_A = \frac{m \times 10000}{A} \tag{1}$$

Where, ρ_A refers to mass per unit area (g/m^2); m refers to mass of sample (g); A refers to area of sample (cm^2).
According to sampled data, the average value and deviation of mass per unit area can be calculated.

2.1.2. Measure Thickness of Geotextile. When geotextiles are used as reinforced cloth and laid between pavement surface and basement layer to prevent and control reflection crack, thickness of geotextiles has relatively large influences on anti-cracking effect. If thickness of geotextiles is large, effect of combination between surface layer and base layer will be influenced, and then geotextiles will act as isolated layer to influence cementation condition. The proper thickness gauge shall be selected to measure thickness. At least 10 samples shall be taken and designed with proper size based on related regulations of thickness gauge. The diameter of sample shall be 1.75 times of diameter of pressure foot for thickness gauge. The average value shall be calculated based on measured results.

2.1.3. Test Related Parameters of Breaking Strength for Geotextiles. The strength value of geotextiles depends on roughness of monofilament and raw material and is closely related to structural form of geotextile knitmesh and combined technology. Therefore, the property of strength can be reflected when geotextiles resist changes of external force. At least five pieces of lateral and longitudinal geotextiles shall be selected to measure parameters. When samples are held, test length and tensile direction shall be in parallel in lateral and longitudinal test. The breaking strength shall be calculated upon formula as below:

$$\alpha_f = \frac{F_f}{B} \tag{2}$$

Where, α_f refers to breaking strength (kN/m); F_f refers to maximum load (kN); B refers to width of sample(m).
The average value can be calculated based on measured lateral and longitudinal breaking strength.

2.1.4. Test Tearing Strength of Geotextiles. When geotextiles are used on roads as anti-cracking cloth to prevent and control extension of reflection crack, the anti-tearing property has played an important role. The standard sample shall be selected and then a isosceles trapezoid shall be drawn in cloth with trapezoid sample plate to test anti-tearing property of geotextiles. Besides, the sample shall be held along both unparalleled sides of isosceles trapezoid.

Table 2. Geotextiles physical parameter checklist

Mass deviation per unit area for monomer, %		3
Thickness, mm		3.83
Breaking strength, kN/m	Lateral	51.6
	Longitudinal	54.2
Elongation at break, %	Lateral	84
	Longitudinal	68
Tearing strength, kN	Lateral	1.65
	Longitudinal	1.28

2.2. Aging Test for Xenon Lamp

In consideration of durability of geotextiles, xenon lamp is applied to conduct aging test and simulate and accelerate influences of light and heat in nature on geotextiles. Actually, climate and environmental factors have influences on aging of geotextiles during application. At present, it has been known that spectral energy distribution of xenon arc lamp among artificial light sources is most similar to ultraviolet and visible light. Therefore, we have selected xenon lamp to conduct ageing test for geotextiles. When ageing test is conducted for xenon lamp, control for various factor variables shall be paid attention to and then breaking strength and retention rate of elongation at break for 500 hours shall be calculated. The breaking retention rate has been calculated as below:

$$\mathrm{R}_F = \frac{F_e}{F_c} \tag{3}$$

Where, R_F refers to retention rate of strength for sample (%); F_e refers to breaking strength of sample (N); F_c refers to breaking strength of reference sample (N).

$$\mathrm{R}_\varepsilon = \frac{\varepsilon_e}{\varepsilon_c} \tag{4}$$

Where, R_ε refers to retention rate of elongation (%); ε_e refers to breaking elongation of sample (mm); ε_c refers to breaking elongation of reference sample (mm).

Table 3. 500 hours xenon lamp aging data

Retention rate for breaking strength, %	Lateral	96
	Longitudinal	94
Retention rate of elongation at break, %	Lateral	80
	Longitudinal	82

3. Measurement for Chemical Properties

From the view of current wide application of geotextiles, geotextiles are usually applied to strengthen roadbed, protect slope, coast and river embankment, control water and soil loss, and prevent leakage of landfill, hazardous wastes treatment plant and mineral tailing pond. The properties of solid and liquid wastes in different landfill are different and properties of mineral waste residue from various mines are different. Therefore, liquid waste from landfill and tailing pond has different acid and alkaline properties. For example, due to organic degradation, sanitary landfill usually generates a mass of acidic materials, resulting in acidic liquid waste. Due to application of strong alkaline sodium hydroxide as dissolution agent during smelting process for aluminium oxide for red mud tailing pond,

IOP Conf. Series: Materials Science and Engineering **611** (2019) 012040 doi:10.1088/1757-899X/611/1/012040

red mud and liquid waste show strong alkaline property. During road engineering application, due to mixture of various engineering application materials, different materials have various acid and alkali properties. Besides, pH is 10-12 for gold tailings and 1.5-4.5 for ardealite. During engineering application, different acid and alkali environment usually results in decline of physical and chemical properties of geotextiles to some extent. From the view of protection, strengthening, isolation, filtration and drainage for geotextiles, geotextiles shall directly contact with acid and alkali in soil or DO water solution. The geotextiles can resist erosion of such chemical substances, on the one hand, depending on composition, production technology, formed structure and current damage of raw materials, on the other hand, depending on composition of liquid and environmental conditions.

3.1. Acid and Alkali Resistant Property of Geotextiles
The acid-base property test shall be conducted for standard geotextile samples in accordance with related standards. The sulfuric acid with concentration of 10% and sodium hydroxide solution shall be prepared to soak for 168 hours. Then retention rate of breaking strength and elongation at break can be measured.

Table 4. Data sheet for acid and alkali resistance of Geotextiles

Acid resistant property (10%, H_2SO_4, 168 hours)		
Breaking strength, kN/m	Lateral	49.0
	Longitudinal	50.4
Retention rate for breaking strength, %	Lateral	95
	Longitudinal	93
Elongation at break, %	Lateral	79
	Longitudinal	65
Retention rate of elongation at break, %	Lateral	94
	Longitudinal	95
Alkali resistant property (10%, NaOH, 168 hours)		
Breaking strength, kN/m	Lateral	47.0
	Longitudinal	49.9
Retention rate for breaking strength, %	Lateral	91
	Longitudinal	92
Elongation at break, %	Lateral	77
	Longitudinal	62
Retention rate of elongation at break, %	Lateral	92
	Longitudinal	91

3.2. Influences of Monofilament Geotextile on Acid-base Properties
Due to weak polarity, low surface energy and bad hydrophilic performance of polypropylene, formed polypropylene monofilament shows excellent properties and acid and alkali resistant property can be understood through electron microscopy observation.

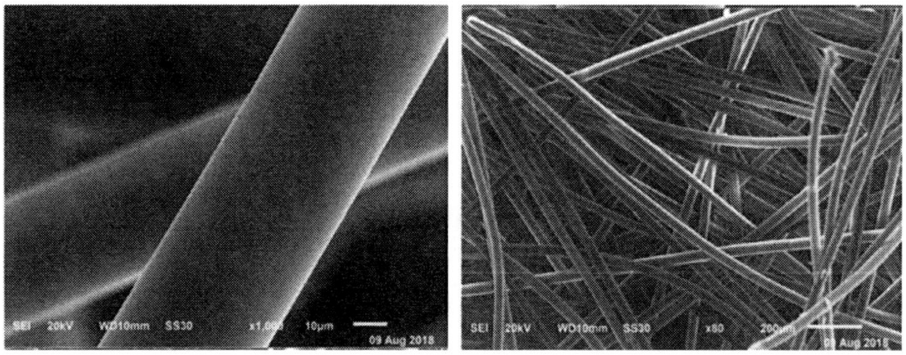

Figure 1. Surface structure display of geotextiles under electron microscope

According to electron microscopy observation for polypropylene filament geotextiles, polypropylene monofilament has smooth surface and strong hydrophobic property and no obvious roughness, which will be beneficial to flow of waste acid and alkaline liquid. Therefore, there is no obvious waste acid and alkaline liquid under acid and alkaline liquid environment, and acid and alkali resistant property for polypropylene geotextiles is greater than polyester geotextiles based on comparison with polyester filament geotextiles.

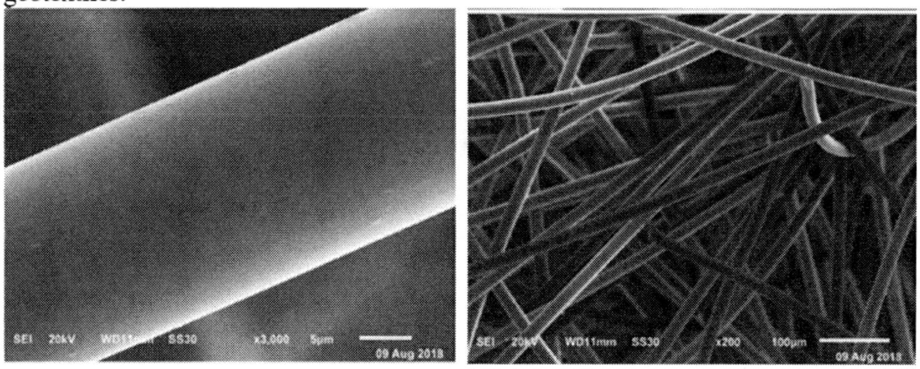

Figure 2. Electron microscope photographs of polyester filament Geotextiles

According to surface observation for monofilament of polyester filament geotextiles, surface distribution is small and raised, which is not beneficial to flow of waste acid and alkali liquid. Therefore, liquid locking capacity of polyester filament geotextile is stronger than polypropylene filament geotextiles, which is also one of acid and alkali resistant properties for polypropylene filament geotextiles.

3.3. Micro Exploration for Alkali Resistance of Polypropylene and Polyester

Due to national administration for strong corrosive solution, the polypropylene and polyester filament geotextiles are soaked in 100% strong alkaline solution for 180 days at normal temperature under current conditions for electron microscopy observation.

After 180 days for polypropylene geotextiles soaked in strong alkaline solution, according to surface corrosion of polypropylene monofilament through electron microscopy observation, strong alkaline solution has influences on polypropylene monofilament, but surface of monofilament remains to be complete and has no obvious corrosive characteristics.

After 180 days for polyester geotextiles soaked in strong alkaline solution, according to surface corrosion of polypropylene monofilament through electron microscopy observation, strong alkaline solution has influences on polyester monofilament, but surface of monofilament has been damaged and has obvious pit slot and corrosive characteristics.

4. Conclusions

Through exploration for engineering properties of polypropylene filament geotextiles, conclusions can be drawn as below:

Strong engineering application property and good tensile property; Excellent anti-ageing performance, over 90% retention rate for breaking strength and over 80% retention rate for elongation at break; Strong acid and alkali resistant property, excellent acid resistant property, over 90% retention rate for breaking strength of acid resistant property and over 60% retention rate for elongation at break; over 60% retention rate of elongation at break; over 90% retention rate of breaking strength and over 60% retention rate of elongation at break for alkali resistant property; According to electron microscope observation for polypropylene and polyester filament geotextiles, alkali resistant property for polypropylene geotextiles is greater than polyester geotextiles.

References

[1] CHEN Weidong. Exploration and Analysis on Application of Reflection Crack Technology for Bituminous Pavement with Geotextile Prevention and Control [J]. Development Guide To Building Materials. 2011, (11).

[2] XIE Jingkun, LI Jianxin, CHU Zhaojing. Exploration on Acid-base Resistance Property for Non-woven Textiles [J]. China Mine Engineering, 2017, 46(01): 61-65.

[3] ZHENG Zhineng. Research on Durability Test for Geosynthetics [D]. Chongqing Jiaotong University, 2003.

[4] CHEN Jie. Research on Durability Test for Railway Geosynthetics [D]. Shijiazhuang Raiway University, 2017.

[5] SHEN Lei. Research on Composite Geotextile for Sacked Cement Cast-in-place Concrete Pile in Saline Soil Region [D]. Chang'an University, 2013.

[6] ZHANG Baosen, ZHANG Xiquan, ZHANG Junxia, LI Li. Research on Engineering Characteristics Test for Geosynthetics [J]. Journal of Gansu University of Technology, 2002(02): 103-107.

[7] YANG Guangqing. Outlook on Current Development Situation and Trends for Geotextile [J]. China Textile Leader, 2017(05): 20+22+24-25.

[8] Prambauer M, Wendeler C, Weitzenböck J, et al. Biodegradable geotextiles - An overview of existing and potential materials[J]. Geotextiles and Geomembranes, 2019, 47: 48-59

[9] GB/T 17632-1998, Geotextiles and geotextile-related products - Test method for determining the resistance to liquids of acids and bases [S].

[10] Karademir T, Frost J D. Micro-scale tensile properties of single geotextile polypropylene filaments at elevated temperatures [J]. Geotextiles and Geomembranes, 2014, 42(3):201-213.

[11] José Ricardo Carneiro, Morais M, Lopes M D L. Degradation of polypropylene geotextiles with different chemical stabilisations in marine environments [J]. Construction & Building Materials, 2018, 165:877–886.

Optimization of Reaction Conditions for Preparing Carboxymethyl Cellulose from Bamboo Scraps by Response Surface Methodology

Weiqing Chen* **and Ying Wen**

College of Biological and Environmental Engineering, Zhejiang Shuren University, Hangzhou, Zhejiang, 310015, China

*Corresponding author

Abstract. Sodium carboxymethyl cellulose (CMC) is mainly prepared from cotton linter and wood pulp. The bamboo scraps is a kind of natural lignocellulosic material with rich cellulose content. In this paper, the manufacture process of CMC from pre-treated bamboo scraps was reported. Box-Behnken design and response surface methodology were implemented for optimizing the reaction conditions in the light of single factor experiments. The results proved that the optimal factors of carboxymethyl reaction were: pre-treated bamboo powder (contained 91.7% α-cellulose) was added into aqueous ethanol with sodium hydroxide solution, the alkalization kept at 33.6℃ for 120min; subsequently, etherifying agent chloroacetic aicd was put into the system and etherification carried out at 54.6℃ for 138min. After this preparation process, the product's viscosity of its 1% water solution was 233 mPa·s.

1. Introduction

Considered as one sort of crucial cellulose derivatives, sodium carboxymethylcellulose (CMC), due to its thickening, viscosity and some other properties, is widely applied in textiles, daily chemical industry, medicine, food, building materials and other fields[1, 2]. Currently, most of the raw materials for CMC preparation is cotton linter and wood pulp. The waste biomass, such as bamboo scraps (bamboo powder and shaving), bagasse and corn-cob, is a kind of natural lignocellulosic materials with rich cellulose content. The manufacture method of CMC made of varieties of cellulose sources could be detail explored. In recently years, it became an interesting topic and many achievements in related field were reported [3, 4, 5].

The aim of this research was to get optimal reaction factors of carboxymethylation for CMC manufacture from pretreated bamboo scraps, then obtain the increased viscosity value of CMC. During the research, response surface methodology (RSM), which known as a valid statistical method with more precise and economical than the other ways, was applied to optimize the synthesis process. RSM could survey assorted factors simultaneously and obtain data about the interaction between variates, so as to get rational optimal results [6, 7].

2. Materials and Methods

2.1. Reagents

All of the chemical reagents and enzyme applied in this study, including ethanol, sodium hydroxide, monochloro acetic acid, acetic acid, cellulose, were purchased by Zeheng Biotechnological Co. Ltd.. All of these reagents were applied without further purification.

2.2. Bamboo Scraps and Its Pretreatment Method

Bamboo scraps was gathered from a company of bamboo products manufacturing located in Anji, Zhejiang Province. Bamboo scraps should be baked at 70℃ by a vacuum oven after washing, subsequently powdered and sieved by 0.5 mm mesh sieve. Afterwards, bamboo powder was soaked into 6.1% sodium hydroxide liquor with the ratio of 1:10, strring, maintained 90min at 50℃, then disposed the sample with cellulase solution (200U/mL) for 30 min at 49.0℃. After these procedures, the α-cellulose content of pretreated bamboo powder enhanced from 45.2% to 91.7%.

2.3. CMC Preparation and Optimizing Study by RSM

CMC synthesis by refined bamboo powder (contained 91.7% α-cellulose) was implemented by alkalization and etherification in the system successively. Based on the single-factor test what was done before, we got the ranges of each factors. The powder was added into aqueous ethanol with sodium hydroxide solution to start the reaction. Then Plackett-Burman design (PBD) was employed to screen the remarkable effect factors. Table 1 demonstrated the design of steepest ascent path and corresponding conclusions. RSM was applied to optimise the most remarkable variable factors and analyze the relationship between the factors and the product's viscosity value. Box-Behnken design (BBD) was employed for further studied on these independent factors, and table 2 showed the design and results. The model's analysis was assessed by analysis of variance (ANOVA) with Design Expert version 7.0.0. The model's quality was evaluated by the determination coefficient, R^2. The significance of the polynomial equation for the response variables was estimated by p value. In this experiment, the selected response value was CMC's viscosity of its 1% aqueous solution.

3. Results and Discussion

Based on PBD experiment, it showed alkalization temperature (X_1), etherification temperature (X_2) and etherification time (X_3) were significant variables ($p<0.05$), accompanied with an extensive variation of viscosity. Therefore, X_1, X_2 and X_3 were chosen for further study to obtain the optimal response.

Table 1 showed the path of steepest ascent experiments and the consequences, it demonstrated Run 2 would be close to the scope of the maximum response, which could be selected for the next step optimal test.

Table 1. Steepest ascent design and the consequences

Run	alkalization temperature (℃)	etherification temperature (℃)	etherification time (min)	Viscosity (mPa·s)
1	37	115	60	90.9
2	35.5	120	58	107
3	34	125	56	148
4	32.5	130	54	186
5	31	135	52	117

BBD for RSM was adopted to analyze the optimized values of the studied variables which influenced samples' viscosity value remarkably. Table 2 illustrated the Box-Behnken design and the consequences. The latter was confirmed to a second-order polynomial equation:

$$Y=+194.70+23.15*X1+14.95*X2+9.04*X3+1.75*X1*X2+1.00*X1*X3+5.25*X2*X3-28.28*X12-31.47*X22-6.94*X32 \quad (1)$$

Table 2. Box-Behnken design and consequences of CMC synthesis conditions

Run	X_1 alkalization temperature (℃)	X_2 Etherification temperature(℃)	X_3 Etherification time (min)	Viscosity (mPa·s)
1	32.5	54.0	120.0	167
2	35.0	52.0	140.0	144
3	35.0	52.0	140.0	193
4	32.5	54.0	130.0	192
5	32.5	54.0	130.0	183
6	28.3	54.0	130.0	89.0
7	32.5	54.0	130.0	206
8	36.7	54.0	130.0	139
9	30.0	52.0	140.0	89.0
10	32.5	54.0	130.0	201
11	32.5	54.0	130.0	187
12	32.5	50.6	130.0	77.0
13	30.0	56.0	120.0	101
14	30.0	56.0	140.0	122
15	35.0	56.0	140.0	187
16	30.0	52.0	120.0	86.0
17	32.5	54.0	130.0	207
18	35.0	52.0	120.0	140
19	35.0	56.0	120.0	159
20	32.5	57.4	130.0	133

Table 3 showed the analysis of variance (ANOVA) for this model. R-Squared, the determination coefficient value was 0.9626. It meant 96.26% of the total variance could be illuminated according to this model. Generally, a regression model of R^2 exceeds 0.9 clarifying a high correction. The adjusted determination coefficient (Adj R^2=0.9289) was high as well, it was enough to demonstrate the significance of the model. Simultaneously, in this model, p-value was less than 0.0001, showed the highly significant fitness. Additionally, the lack of fit score was only 0.2807, which meant non-significant, displayed it was suited to forecast the response and explicate the influence of variables on this response.

The adopted three-dimensional response surface diagrams (Figure 1, Figure 2 and Figure 3) confirmed the interaction among the variable factors. Every 3D diagram displayed associations between two experimental variables, while the other one was at the central level. An oval-shaped contour or saddle nature of the contour plots implied there was a statistically significant interaction between two variables. Based on the quadratic model, the maximum viscosity value was predicted as 234·4mPa.s, while the alkalinity temperature was 33.6 ℃, and etherification temperature and etherification time were 54.6℃ and 138min, respectively.

Table 3. The analysis of variance of quadratic model for CMC synthesis

Source	Sum of Squares	df	Mean Square	F Value	p-value Prob > F
Model	35995.28	9	3999.48	28.60	< 0.0001 significant
X_1	7315.94	1	7315.94	52.31	< 0.0001
X_2	3052.65	1	3052.65	21.83	0.0009
X_3	870.55	1	870.55	6.22	0.0317
$X_1 X_2$	24.50	1	24.50	0.18	0.6844
$X_1 X_3$	8.00	1	8.00	0.057	0.8158
$X_2 X_3$	220.50	1	220.50	1.58	0.2378
X_1^2	11625.30	1	11625.30	83.13	< 0.0001
X_2^2	14388.21	1	14388.21	102.88	< 0.0001
X_3^2	436.51	1	436.51	3.12	0.1077
Residual	1398.52	10	139.85		
Lack of Fit	886.52	5	177.30	1.73	0.2807 not significant
Pure Error	512.00	5	102.40		
Cor Total	37393.80	19			
R-Squared	0.9626				
C.V. %	7.88				
Adj R^2	0.9289				

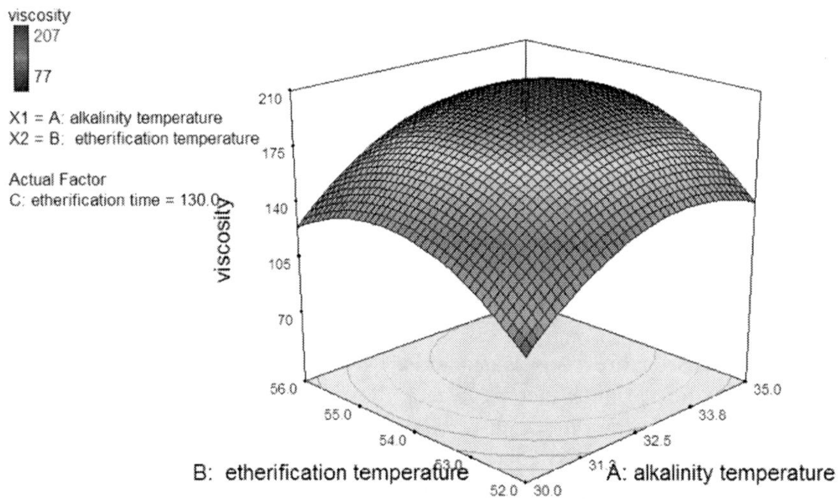

Figure 1. Three-dimensional response surface diagram: the interaction of alkalization temperature and etherification temperature

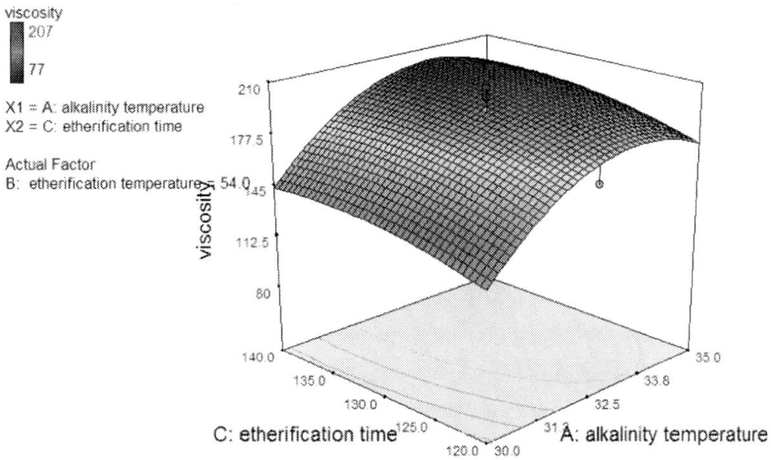

Figure 2. Three-dimensional response surface diagram: the interaction of alkalinity temperature and etherification time

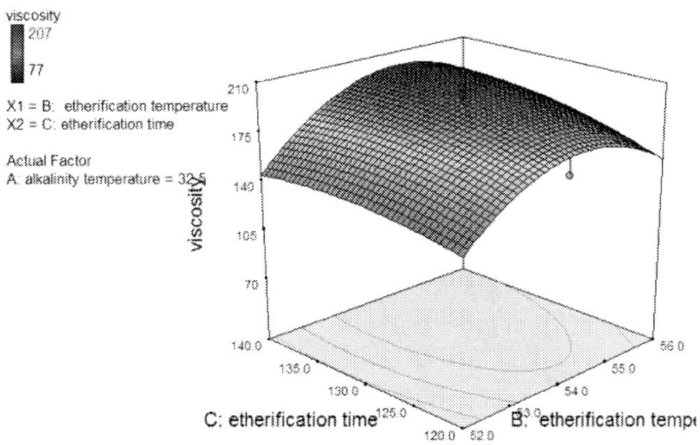

Figure 3. Three-dimensional response surface diagram: the interaction of etherification temperature and etherification time

For purpose of verifying the model fitness for predicting the maximum response, a probative test was carried out for five times. With the predicted optimal conditions, the average experimental value was 233 mPa·s, which was conform to the model predicted. The viscosity of CMC was improved by 194% compared with that of original sample.

4. Conclusions

Importantly, our results provided evidence of synthesis conditions for CMC made from refined bamboo scraps. The effective mathematical model illuminated RSM could be applied to obtain the optimal factors about carboxymethyl reaction. The viscosity of CMC product increased significantly. This research could contribute to waste management and circular using of resource.

References

[1] Dolz M, Jiménez J, Hernández M J, *et al.* 2007 J petrol Sci Eng. 57(3-4), pp 294-302.
[2] Pushpamalar V, Langford S J, Ahmad M, *et al.* 2006 *Carbohyd Polym.* 264(2), pp 312-318.
[3] Singh R K, Singh A K. 2013 *Waste Biomass Valori.* 4, pp 129-137.
[4] Ibrahim A A, Adel A M, El–Wahab Z H A, *et al.* 2011 *Carbohyd Polym.* 83(1), pp 94-115.
[5] Agil R, Oomah D B, Mazza G, *et al.* 2012 *Food Bioprocess Tech.* 5, pp 2655-2664.
[6] Sun Y G, Ma Y L, Wang L Q, *et al.* 2015 *Carbohyd Polym* 117, pp 486- 493.
[7] Vaheed H, Shojaosadati S A, Galip H. 2011 *J Ind Microbiol Biot.* 38, pp 101-111.

Applying Fractal Dimension to the Effects of Root System on Slope Soil Displacement

Xiaolei Ji*

College of Construction Engineering, Jiangsu Open University, Nanjing, Jiangsu, China

*Corresponding author

Abstract. Based on the box-counting dimension method of fractal theory, a set of computing image fractal dimension software Fractal 1.0 was developed which can be used to determine the root fractal dimension of plants. By combining the theory of fractal with the finite element theory, the effects of root system fractal dimension on the amount of displacement taking place on the surface of slope were studied. By analyzing the relationship between them from different aspects, the results indicate that fractal dimension can be regarded as the parameter of root system, the effects of different root systems on the stability of soil in slope surface were quantitatively characterized through integrating fractal dimension with finite element method.
Keywords: fractal dimension; root form; slope stability

1. Introduction

Fractals are used to describe objects that are so irregular that they are not considered as classical geometry, through chaotic phenomena and irregular configuration, it tries to reveal the intrinsic relation and motion law between the local and the whole which is hidden behind the phenomenon, fractals can be measured with fractal dimension (Falconer 2007; Li and Huang 2001; Li and Ruan 2004). Mandelbert proposed the concept of fractal dimension firstly, and established the fractal geometry (Liao and Yu 2001). By using fractal dimension, the geometry complexity and capacity of fulling space can be quantified, as well as, the complexity of geometry which is characterized by the irregularity and self-similarity can be compared quantitatively (Xia et al. 2011).

Root system is a complex and irregular geometry structure, it is often considered to be a disordered meridional structure in euclidean geometry (Fattet et al. 2011; Hu et al. 2008; Hu et al. 2016). As a time-consuming job, it is difficult to be measured. Since now, an appropriate scheme of quantitatively describing the morphological features of the root system has not been propose (Bast et al. 2015). Aiming at the growth law and branch feature of plant root system, a set of studies is conducted, It is found that the root morphology has obvious statistical self-similarity on a certain scale, so the fractal geometry theory can be applied to explore the root morphology, then the parameter of root morphology will be measured and analyzed rapidly and accurately so that the reliability of parameter of root morphology will be promoted effectively (Eab et al. 2014; Gary and Sotir 1996). In this paper, the computer aided image analysis system and fractal theory are combined to be applied to the root morphology analysis. Basing on the Basing on the box dimension method of fractal theory, a software package (Fractal1.0) for calculating the fractal dimension of plant roots is developed, this software can be used to determine the dimension of plant roots while the effects of different root systems on the stability of soil in slope surface were quantitatively characterized through integrating fractal dimension with finite element method. The displacement of slope soil is closely related to the stability of slope. The slope soil produces displacement with rainfall, and when the displacement increases to the limit value of soil, the slope slide occurs. Root-planted slope can effectively reduce the soil displacement and therefore improve the stability of the slope.

2. Development of Fractal Software

2.1. Development Thoughts

Fractal dimension computing software (Fractal1.0) is developed based on the method of box-counting dimension. The definition of box-counting dimension is, setting assembly A∈F(x), F(x) represents a set of all compact support sets on the metric space x, ρ is a kind of measure, (x, ρ)is a metric space. As for every $\delta > 0$, the $N_\delta(A)$ represent the minimum amount of closed ball whose radius is $\delta > 0$. If the $\lim_{\delta>0} \frac{\log N_\delta(A)}{-\log\delta}$ is available, then this limit is box counting dimension od assembly A, signed as $\dim_B^A$[1-3].

2.2. Software Algorithm Implementation

The original grayscale image of root system $f(x,y)$was obtained by digital image acquisition device, the image is converted into a binary image $f_2(x,y)$ by using a gray threshold valueς so that every pixel point on the image is divided into black and white, in the process of binarization, try to keep the details which is relevant to the application in the image. There are many ways to choose the threshold value of binarization, the average gray value method which sets the average of all gray value as the threshold value is used in this software, then the binary image is represented by matrix of m by n ($n \geq m$),the matrix *F* is showed as follows:

$$F = \begin{bmatrix} F_{11} & F_{12} & \cdots & F_{1n} \\ F_{21} & F_{22} & \cdots & F_{2n} \\ \cdots & \cdots & \cdots & \cdots \\ F_{m1} & F_{m2} & \cdots & F_{mn} \end{bmatrix}$$

Where $F_{ij} \in \{0,1\}$, when $F_{ij} = 1$,it is the pixel of root system; when $F_{ij} = 0$,it is background pixel. As for the matrix of *F*, it is covered by a square with length of L_{max} to assure all pixels are located in the square.By dividing the length L_{max} into equal proportions at the ratio of *r*, multiple squares whose length od side is $L = rL_{max}$of the grid.It is assumed that the number of square which includes at least one prexel point in it is *N*, the fractal dimension of root system is *D*, then

$$N = r^{-D} \tag{1}$$

So that the total area of all squares which includes at least one pixel point in it is:

$$S = N * L^2 \tag{2}$$

By applying the equation $L = rL_{max}$ into equation (2) yields:

$$S = N * L^2 = \mathrm{r}^{-D} * r^2 L_{max}^2 \tag{3}$$

Take the logarithm of both ends of equation (3) yields:

$$\log S = (2 - D)\log r + 2\log L_{max} \tag{4}$$

Then:

$$D = 2 + \frac{2\log L_{max} - \log s}{\log r} \tag{5}$$

According to the equation (4), there is a linear relationship between $\log S$ and$\log r$, so the least square method is employed to estimate fractal dimension in practice, not use the equation (50) directly. By fitting multiple data points$\{\log S, \log r\}$, the estimate of fractal dimension is obtained by the slope of the fitting line (Ghestem et al. 2011; Sonnenberg et al. 2010; Abe and Ziemer 2011).

2.3. Data Analysis and Experiment Result

The soil used in research was silty clay collected from field. The indexes are as followed:

Table 1. Physical and mechanical properties of testing soil

Index	Value
Density (g/cm3)	1.80
dry density (g/cm3)	1.41
specific gravity of solids	2.72
water content (%)	27.7
liquid limit (%)	38.2
plastic limit (%)	17.5
Best water content (%)	16.9
maximum dry density (g/cm3)	1.72
void ratio (%)	0.93
Saturation (%)	81

The depth of root system influencing the surface soil of slope is related to the length of root system. The bermuda root influenced soil at the depth of 200mm, 300mm and 400mm respectively, and the oleander root at the depth of 600mm.

The root samples were treated by using fractal calculation software, the root of bermuda grass and the root of oleander are showed in Figure.1 and Figure.2. Taking the root distribution in consideration, the test categorized the bermuda grass root system as 200mm, 300mm and 400mm according to the fibrous root amount and length, as showed in Figure.3. The root system of oleander was categorized as 30°, 45°, 60°, 90° according to the angle between main root and lateral root, as showed in Figure.4. Then fractal dimension of root system is calculated and the collected data is counted. By choosing the average of fractal dimension of root system, the fractal dimension of bermuda grass and oleander are 1.56753,1.76572 respectively.

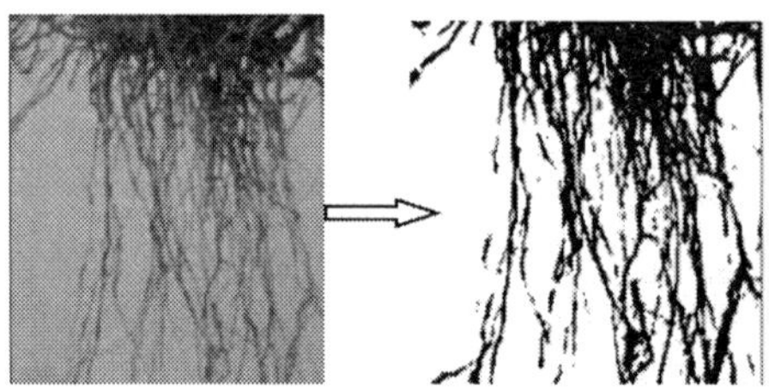

Figure 1. Binary processing diagram of bermuda grass root system

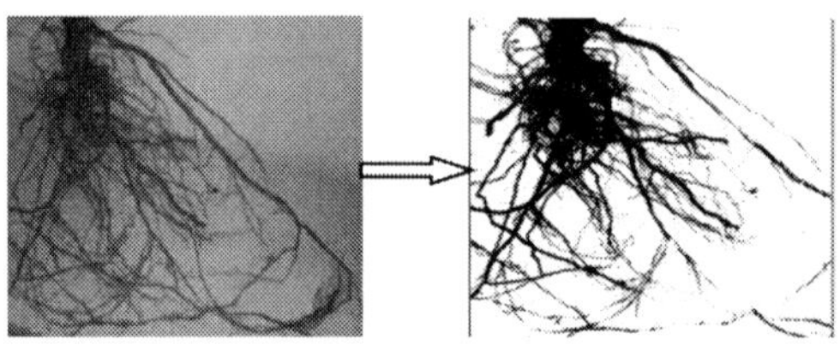

Figure 2. Binary processing diagram of oleander root system

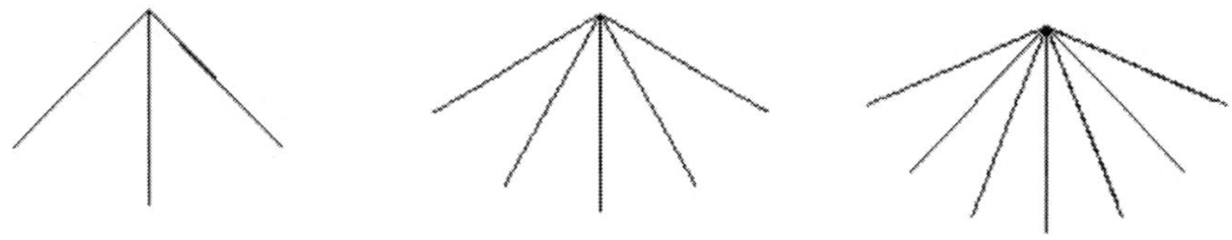

Figure 3. Bermuda grass root system distribution

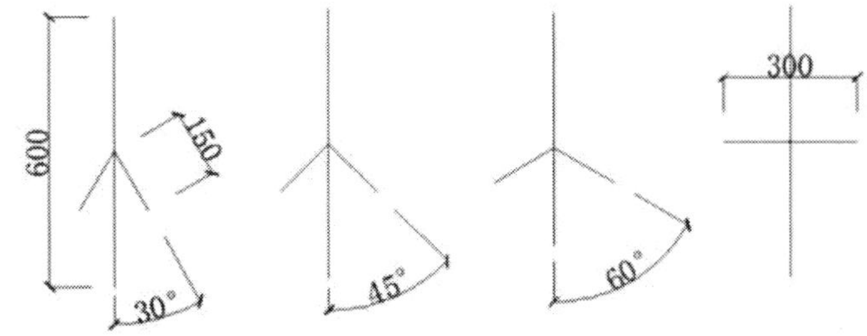

Figure 4. Simplified oleander root forms with different main-lateral root angles

3. Application of Fractal Software

3.1. The Relationship between the Fractal Dimension of Root Model and Soil Displacement

Fractal dimension can show root system morphology and reflect the growth of root system. The root system with large fractal dimension is widely distributed and has strong ability of dividing. The larger the fractal dimension, the stronger the root growth; the stronger the entanglement of roots and soil, the better the solid soil effect (Ghestem et al. 2013; Ha et al. 2014). Bermuda grass and oleander, two widely used slope-protecting plants were focused as the example, and the paper explored how the root models related to slope soil displacement by means of numerical simulation software Abaqus (Hardin and Drnevich 1972), the quantitative calculation theory of the effect of root fractal dimension on Slope solid soil performance, this has significant meaning for the choice of plant in the process of ecological slope protection.

In numerical calculation, the root system was simplified in the process of model establishing. The actual morphology is much more complicated than modelling, but as there exists the intrinsic geometric regularity in root form which accord with the proportional self-similarity of fractal theory, fractal dimension can be used to measure and reflect the actual form of root system. The influence of plant roots on slope stability mainly depends on the rooting ability of plants. The greater root abundance, the more branches of plant roots, the greater contribution to soil stability. Based on the numerical calculation of the influence of different root systems on the surface displacement, the relationship curves between the fractal dimension and displacement field are established. The effect of root model fractal dimension on soil displacement is showed in Table.2 and Table.3.

Table.2 Fractal dimension of numerical model of bermuda grass root system

Fibrous root amount	3			5			7		
Length of root (mm)	200	300	400	200	300	400	200	300	400
Fractal dimension	1.003	1.015	1.032	1.109	1.117	1.032	1.153	1.155	1.171
Horizontal displacement(mm)	1.252	1.246	1.237	1.241	1.230	1.219	1.232	1.218	1.202
Vertical displacement(mm)	5.276	5.264	5.254	5.270	5.256	5.253	5.267	5.255	5.245
Total displacement(mm)	5.374	5.361	5.355	5.368	5.351	5.347	5.353	5.349	5.337

Table.3 Fractal dimension of numerical model of oleander root system

Main lateral root angle	Lateral root amount	Fractal dimension	Horizontal displacement (mm)	Vertical displacement (mm)	Total displacement (mm)
30°	2	0.986	1.867	6.639	6.808
	4	1.064	1.860	6.632	6.798
	6	1.088	1.852	6.626	6.788
45°	2	0.97	1.859	6.633	6.803
	4	1.027	1.843	6.621	6.787
	6	1.704	1.826	6.609	6.772
60°	2	0.976	1.859	6.635	6.804
	4	1.054	1.841	6.624	6.790
	6	1.097	1.822	6.613	6.776
90°	2	1.005	1.874	6.645	6.807
	4	1.050	1.869	6.644	6.815
	6	1.112	1.865	6.643	6.805

In order to show the relation between the fractal dimension of root system and the displacement of surface soil mass, according to the data in Table.1 and Table.2, curves of differences of horizontal slope soil displacement, vertical displacement and total displacement vary with fractal dimension were plotted out. The curves of effect of bermuda grass root system on soil displacement are showed in Figure.5, Figure.6, Figure.7, The curves of oleander root effects on soil displacement are demonstrated in Figure.8, Figure.9, Figure.10.

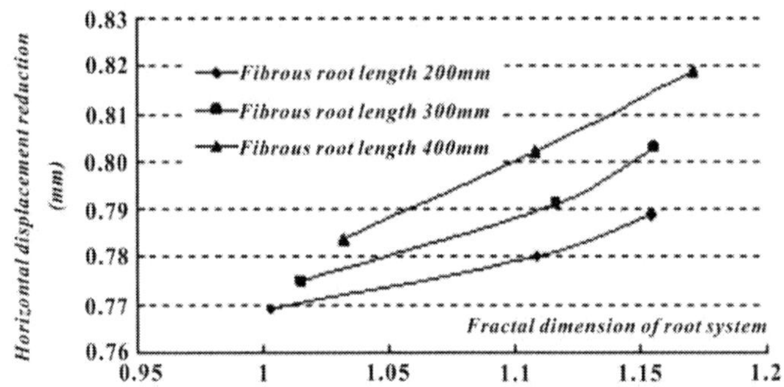

Figure 5. The effect of fractal dimension of bermuda root on horizontal displacement of soil

By data fitting and analyzation, it was of more accuracy to regress the relationship between the bermuda root fractal dimension and slope horizontal displacement through polynomial fitting method. The following lists the fitting formula:

Fibrous root length 200mm: $y = 0.6339x^2 - 1.2339x + 1.3689 \quad R^2 = 1$ (6)

Fibrous root length 300mm: $y = 1.1008x^2 + 2.1888x + 1.8625 \quad R^2 = 1$ (7)

Fibrous root length 400mm: $y = 0.256x^2 - 0.3111x + 0.8324 \quad R^2 = 1$ (8)

Where: x means fractal dimension of root system, y means horizontal displacement reduction (mm)

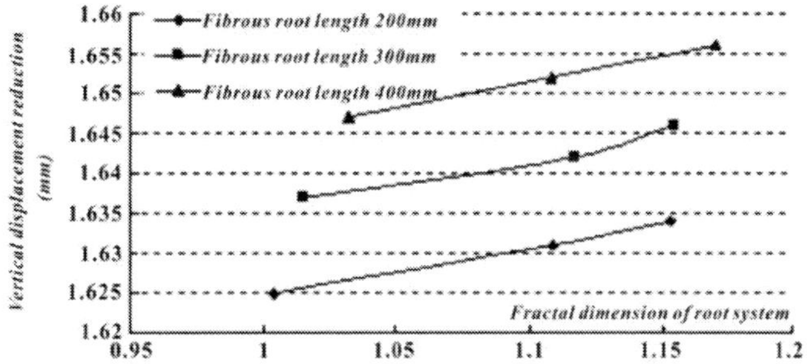

Figure 6. The effect of fractal dimension of bermuda root on vertical displacement of soil

By data fitting and analyzation, it was of more accuracy to regress the relationship between the bermuda root fractal dimension and slope horizontal displacement through polynomial fitting method. The following lists the fitting formula:

Fibrous root length 200mm: $y = 0.0629x^2 - 0.0756x + 1.6375$ $R^2 = 1$ (9)

Fibrous root length 300mm: $y = 0.3905x^2 - 0.7831x + 2.0295$ $R^2 = 1$ (10)

Fibrous root length 400mm: $y = -0.0124 + 0.0923x + 1.565$ $R^2 = 1$ (11)

Where: x means fractal dimension of root system, y means the vertical displacement reduction (mm)

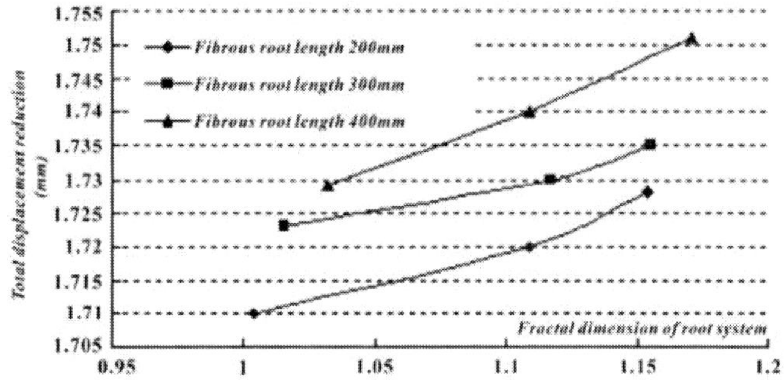

Figure 7. The effect of fractal dimension of bermuda root on total displacement of soil

By data fitting and analyzation, it was of more accuracy to regress the relationship between the bermuda root fractal dimension and slope horizontal displacement through polynomial fitting method. The following lists the fitting formula:

Fibrous root length 200mm: $y = 1.3964x^2 - 2.8922x + 3.2062$ $R^2 = 1$ (12)

Fibrous root length 300mm: $y = 0.4351x^2 - 0.8585x + 2.1461$ $R^2 = 1$ (13)

Fibrous root length 400mm: $y = -0.2354x^2 + 0.6464x + 1.3105$ $R^2 = 1$ (14)

Where: x means fractal dimension of root system, y means the total displacement reduction (mm)

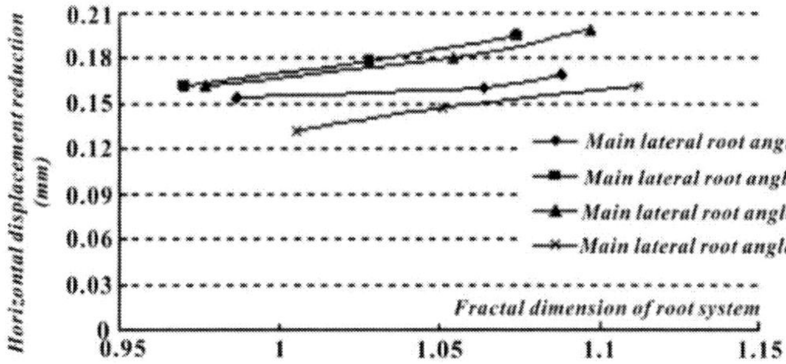

Figure 8. The effect of fractal dimension of oleander root on horizontal displacement of soil

By data fitting and analyzation, it was of more accuracy to regress the relationship between the bermuda root fractal dimension and slope horizontal displacement through polynomial fitting method. The following lists the fitting formula:

Main lateral root angle 30: $y = 0.1331x + 0.0221$ $R^2 = 0.9844$ (15)

Main lateral root angle 45: $y = 0.3161x - 0.1454$ $R^2 = 0.9936$ (16)

Main lateral root angle 60: $y = 0.2975x - 0.1298$ $R^2 = 0.9885$ (17)

Main lateral root angle 90: $y = 0.2765x - 0.145$ $R^2 = 0.9933$ (18)

Where: x means fractal dimension of root system, y means the horizontal displacement reduction (mm)

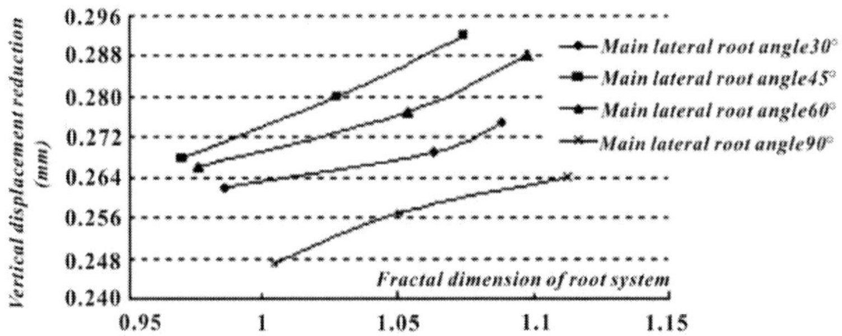

Figure 9. The effect of fractal dimension of oleander root on vertical displacement of soil

By data fitting and analyzation, it was of more accuracy to regress the relationship between the bermuda root fractal dimension and slope horizontal displacement through polynomial fitting method. The following lists the fitting formula:

Main lateral root angle 30: $y = 1.5243x^2 - 3.0346x + 1.7722$ $R^2 = 1$ (19)

Main lateral root angle 45: $y = 0.4906x^2 - 0.772x + 0.5552$ $R^2 = 1$ (20)

Main lateral root angle 60: $y = 0.933x^2 - 1.7527x + 1.0879$ $R^2 = 1$ (21)

Main lateral root angle 90: $y = -0.9527x^2 + 2.1747x - 0.9763$ $R^2 = 1$ (22)

Where: x means fractal dimension of root system, y means vertical displacement reduction (mm)

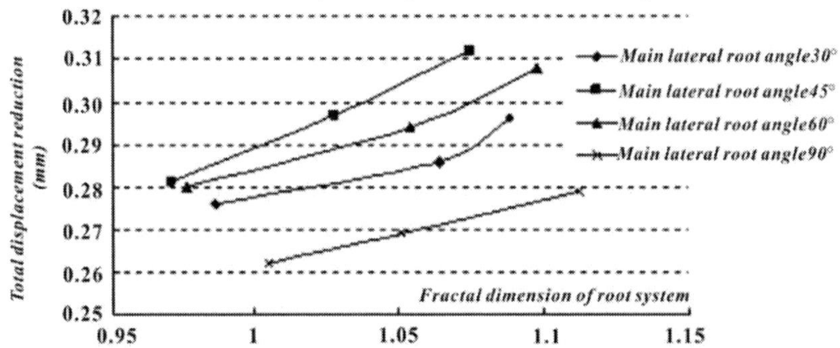

Figure 10. The effect of fractal dimension of oleander root on total displacement of soil

By data fitting and analyzation, it was of more accuracy to regress the relationship between the bermuda root fractal dimension and slope horizontal displacement through polynomial fitting method. The following lists the fitting formula:

Main lateral root angle 30: $y = 2.7507x^2 - 5.51x + 3.0346$ $R^2 = 1$ (23)

Main lateral root angle 45: $y = 0.4461x^2 - 0.6137x + 0.4564$ $R^2 = 1$ (24)

Main lateral root angle 60: $y = 1.1875x^2 - 2.2308x + 1.326$ $R^2 = 1$ (25)

Main lateral root angle 90: $y = 0.1017x^2 - 0.05577x + 0.2172$ $R^2 = 1$ (26)

Where:x means fractal dimension of root system, y means total displacement reduction (mm).

Based on the curves of the root fractal dimension and the slope soil displacement, both kinds of root systems showed the consistent relationships of root fractal dimension and displacement reduction, appearing polynomial fitting curve. The fractal dimension of bermuda grass root system increases with the increase of root length, for the plant root system with long root, the variation of slope displacement reduction with fractal dimension of root system is more obvious. The horizontal, vertical and total displacement of surface soil decrease as the fractal dimension of root system increases. The variation of fractal dimension impact the horizontal displacement less than vertical displacement and total displacement, this suggests the vertical soil displacement can be limited effectively with increase of fractal dimension of root system.

According to the variation curves of the root fractal dimension and the slope surface soil displacement, for different main lateral root angle of plant root system, the soil-reinforcement of root system is different. When the main lateral angle of oleander root system is 45 degrees and 60 degrees, the performance of reducing slope surface displacement is decent with the increase of fractal dimension of root system, and this change feature is corresponding to the result of digital simulation. The form of oleander root system with lateral root angle of 90 degree is single, the arrangement is regular, it is not conducive to forming a winding network structure in soil and the performance of reducing surface displacement is not so obvious (Danjon et al. 2008; Kaestner et al. 2006; Tasser and Tappeiner 2005). For oleander root system with lateral root angle of 30 degree, because of the angle is small, the distribution in the soil is small, the contact with soil is not sufficient, so the performance of soil-reinforcement is bad (Liang 2015). According to the curve of relationship between soil-reinforcement performance and fractal dimension, the relation between fractal dimension of root system and stability of slope surface can be identified, it also can be assured that the performance of reducing slope surface displacement become better with the increase of fractal dimension of root system (Pierret et al. 1999). It is further verified that fractal dimension can be regarded as a parameter to scale root morphology through the relationship between slope surface displacement and fractal dimension of root morphology.

3.2. The Relationship between the Fractal Dimension of Natural State and Soil Displacement

The root system at natural state is more complex than that of numerical simulation, the fractal dimension is bigger, so the performance of reducing slope soil displacement and stabilizing soil is more obvious. According to the feature of bermuda grass root morphology and relationship between the fractal dimension of root digital model and slope soil displacement, the fitting formula of variation of displacement difference and fractal dimension of bermuda grass root model with root length of 300mm. was chosen to calculate the effect of fractal dimension of natural root system on the slope surface displacement. The equation of horizontal displacement with fractal dimension is: $y = 1.1008x^2 + 2.1888x + 1.8625$, The equation of vertical displacement with fractal dimension is: $y = 0.3905x^2 - 0.7831x + 2.0295$, The equation of total displacement with fractal dimension is: $y = 0.4351x^2 - 0.8585x + 2.1461$. The fractal dimension of bermuda grass root was calculated by using Fractal1.0 software, the value was 1.56753.After the root fractal dimension of the natural state is substituted into the fitting formula, it was observed that the root of bermuda grass can reduce the horizontal displacement of the surface soil by 0.975mm and vertical displacement by1.41mm and total displacement 1.48mm ,as well as, by implanting a naturally growing bermuda grass root, the horizontal displacement of slop surface can be reduced by 41.2%,vertical displacement can be reduced by 20.4%.total displacement can be reduced by 20.9%.

The fitting formula of displacement variation and fractal dimension of oleander root model with main lateral root angle of 45 degree was chosen as formula that is used to calculate impact of fractal dimension of natural root system on slope surface displacement. The equation of horizontal displacement with fractal dimension is: $y = 0.3161x - 0.1454$, The equation of vertical displacement with fractal dimension is: $y = 0.4906x^2 - 0.772x + 0.5552$, The equation of total displacement with fractal dimension is: $y = 0.4461x^2 - 0.6137x + 0.4564$. he fractal dimension of oleander root system was calculated by using Fractal1.0 software,the value was 1.76572.After the root fractal dimension of the natural state is substituted into the fitting formula, it was observed that the root of oleander can reduce the horizontal displacement of the surface soil by 0.42mm and vertical displacement by0.726mm and total displacement 0.767mm ,as well as, by implanting a naturally growing bermuda grass root, the horizontal displacement of slop surface can be reduced by 20.7%,vertical displacement can be reduced by 10.5%.total displacement can be reduced by 10.8%.

4. Conclusion

In this paper, a software of root fractal dimension calculation Fractal1.0 was developed basing on morphological characteristics of plant root system. Theory of fractal with the theory finite element were combined together innovatively, he growth morphology of root system was characterized by fractal dimension. The relationship between fractal dimension of root system and displacement of surface pf slopes was studied, by analyzing this relationship, the impact of morphological characteristics of plant root system to displacement of slope surface was reflected truly (Reubens et al. 2007).Research result indicates fractal dimension is capable to be regarded as parameter to describe growth morphology of root system, It is feasible to combine the fractal dimension of root system with the finite element method, but it need to be further investigated to be applied in practice. The influence factors of plant root to solid slope are multifaceted (Lin et al. 2010), but the root morphology of plants was simplified in the study (Stokes et al. 2009) , it is a new attempt to apply the method in quantitative analysis of plant slope, further research is needed on the theory and practice of soil reinforced by root.

Acknowledgments

This research was supported by Science and Technology Project Plan of Ministry of Housing and Urban-rural Development of the PRC [2016-K4-020]; "Qing Lan Project" of universities in Jiangsu province; Jiangsu province "Six Talents Peak" project; Natural Science Foundation of the Jiangsu Higher Education Institutions [18KJB220001]; Jiangsu Open University "The 13th Five-year Plan" Project [17SSW-Z-Y-024].

References

[1] Abe, K., Ziemer, R. R. 1991. Effect of tree roots on a shear zone: modeling reinforced shear stress. Can. J. For. Res. 21, 1012–1019. Alpandino, 2011. Alpandino by Institute of Botany, University of Basel, Switzerland.

[2] Bast, A., Wilcke, W., Graf, F., Lüscher, P. Gärtner, H. 2015. A simplified and rapid technique to determine an aggregate stability coefficient in coarse-grained soils. Catena 127, 170–176.

[3] Danjon, F., Barker, D. H., Drexhage, M. & Stokes, A. 2008. Using three-dimensional plant root architecture in models of shallowslope stability. Ann. Botany 101, No. 8, 1281–1293.

[4] Eab, K., Takahashi, A. , Likitlersuang, S. 2014. Centrifuge modelling of root-reinforced soil slope subjected to rainfall infiltration. Géotechnique Lett. 4, No. 3, 211–216.

[5] Falconer K. Translated by Zeng Wenqu. 2007. Fractal geometry:mathematical foundations and applications (2nd ed). Beijing,China: People's Posts and Telecom Press.

[6] Fattet, M., Fu, Y., Ghestem M., et al. 2011. Effects of vegetation type on soil resistance to erosion: Relationship between aggregate stability and shear strength. Catena, 87(1), 60-69.

[7] Gary, D. H. , Sotir, R. B. 1996. Biotechnical and soil bioengineering slope stabilization: a practical guide for erosion control. New York, NY, USA: John Wiley & Sons.

[8] Ghestem, M., Sidle, R.C., Stokes, A. 2011. The influence of plant root system on subsurface flow: implications for slope stability. Bioscience 61, 869–879.

[9] Ghestem, M., Veylon, G., Bernard, A., Vanel, Q. & Stokes, A. 2013. Influence of plant root system morphology andarchitectural traits on soil shear resistance. Plant and Soil 377, No. 1–2, 43–61.

[11] Ha, J. G., Lee, S. H., Kim, D. S. & Choo, Y. W. 2014. Simulation of soil–foundation–structure interaction of Hualien large-scale seismic test using dynamic centrifuge test. Soil Dynamics Earthquake Engng 61–62, 176–187.

[12] Hardin, B. O. & Drnevich, V. P. 1972. Shear modulus and damping in soils: design equations and curves. J. Soil Mech. Found. Div. 98, No. 7, 667–692.

[13] Hu, P., Song, X.G., Wu, D.G. 2008. Experimental research on reinforcement mechanism of

expressway slope protection with greensward. Rock and Soil Mechanics, 29(2), 442-444.

[14] Hu, X., Li, Z.C., Li, X.Y., Liu, L.Y. 2016. Quantification of soil macropores under alpine vegetation using computed tomography in the Qinghai Lake Watershed, NE Qinghai-Tibet Plateau. Geoderma 264, 244–251.

[15] Kaestner, A., Schneebeli, M., Graf, F. 2006. Visualizing three-dimensional root networks using computed tomography. Geoderma 136, 459–469.

[16] Li, H.G., Huang, M.R. 2001. Fractal theory and its application to plants researches. Chinese Bulletin of Botany, 18(6), 684-690.

[17] Li, H.G., Ruan, X.G., Wang, Y.Q., et al. Development and application of software for fractal dimension calculation (FDC1.0) of complex traits in forest trees. 2004. Journal of Nanjing Forestry University (Natural Sciences Edition), 28(3), 5-8.

[18] Liang, T. 2015. Seismic performance of vegetated slopes. PhD thesis, University of Dundee, Dundee, UK.

[19] Liao, C.Z., Yu, X.H. 2001. Application of fractal theory on studies of the root structure of plant. Journal of Jiangxi Agriculture University, 23(2), 192-196.

[20] Lin, D. G., Huang, B. S. & Lin, S. H. 2010. 3-D numerical investigations into the shear strength of the soil–root system of Makino bamboo and its effect on slope stability. Ecol. Engng 36, No. 8, 992–1006.

[21] Pierret, J.S., Prasher, S.O., Kantzas, A., Langford, C. 1999. Three-dimensional quantification of macropore networks in undisturbed soil cores. Soil Sci. Soc. Am. J. 63, 1530–1543.

[22] Reubens, B., Poesen, J., Danjon, F., Geudens, G., Muys, B, 2007. The role of fine and coarse roots in shallow slope stability and soil erosion control with a focus on root system architecture: a review. Trees 21, 385–402.

[23] Sonnenberg, R., Bransby, M. F., Hallett, P. D., Bengough, A. G., Mickovski, S. B. & Davies, M. C. R. 2010. Centrifuge modelling of soil slopes reinforced with vegetation. Can. Geotech. J. 47, No. 12, 1415–1430.

[24] Stokes, A., Atger, C., Bengough, A.G., Fourcaud, T., Sidle, R.C. 2009. Desirable plant root traits for protecting natural and engineered slopes against landslides. Plant Soil 324, 1–30.

[25] Tasser, E., Tappeiner, U. 2005. New model to predict rooting in diverse plant community compositions. Ecol. Modell. 185, 195–211.

[26] Xia, Z.Y., Zhou, Z.J., Huang, X.L, et al. 2011. Preliminary study of relationship between shallow soil reinforcement and fractal characteristics of vegetation roots in biotechnical slope protection. Chinese Journal of Rock Mechanics and Engineering, 30(Supp.2), 3641-3647.

IOP Conf. Series: Materials Science and Engineering **611** (2019) 012043 doi:10.1088/1757-899X/611/1/012043

Fractal Analysis on Damage of the Aggregate Gradation Crumb Rubber Concrete Beam with MATLAB

Hongquan Sun*, **Jing Wang and Yunxue Yang**

Silicon Lake College of Vocational Technology, Suzhou, Jiangsu, China

*Corresponding author

Abstract. The fractal dimensions of the surface crack of the three concrete beams with different gradations coarse aggregate crumb rubber in damaging state are calculated. The research indicates that the damage evolution process of the test beams is a process of the fractal dimension increasing. The research result reveals the relationship between damage variables and fractal dimension of the test beams. The small particle size of coarse aggregate gradation has the best mechanical characteristics with the increasing load. It provides a new method for studying the mechanical properties of crumb rubber concrete beams.

1. Introduction

Elastic rubber concrete is made by the concrete cement mortar mixing rubber powder ground from waste tires [1]. By studying the material mechanical properties of waste tire rubber particle, it is found that the addition of rubber powder can reduce the occurrence of cracks in the mortar's plastic shrinkage, and the cracks also decrease in length and width. Studies have shown that the mechanical properties of rubber concrete are greatly affected by the particle size, shape and surface roughness of rubber powder. Some papers present the fractal analysis of effect of coarse aggregate gradation on damage of rubber concrete [2].

The damage and fracture of materials have a large number of micro-damage nucleation, expansion and connection, which eventually lead to the fracture and failure of materials. In this paper, MATLAB is used to calculate the fractal dimension of the box image. A program is written to calculate the fractal dimension of damage and fracture by the pixel point covering method [3]. We use square boxes with different side lengths δ to cover the graphics, and count the number N_δ of boxes that cover the graphics' pixels. The fractal dimension of surface crack development CRC test beam is calculated, and the quantitative relationship between CRC damage and fractal dimension is obtained.

2. The Definition of Box Dimension

In practical engineering, box dimension is one of the most widely used dimensions [4], which is relatively easy to calculate and estimate empirically. Let $F \subset R^n$ be any non-empty bounded subset, and let $N_\delta(F)$ be the number of grids intersecting the set F with the coordinate network, the definitions of the lower box dimension $\underline{dim}_B F$ and the upper box dimension $\overline{dim}_B F$ of the set F are respectively:

$$\underline{dim}_B F = \underline{lim}_{\delta\to 0} \frac{\log N_\delta(F)}{\log 1/\delta} \tag{2.1}$$

$$\overline{dim}_B F = \overline{lim}_{\delta\to 0} \frac{\log N_\delta(F)}{\log 1/\delta} \tag{2.2}$$

When these two values are equal, the common value F is called the Box Dimension, denoted as $dim_B F$, where

$$dim_B F = \lim_{\delta \to 0} \frac{\log N_\delta(F)}{\log 1/\delta} \qquad (2.3)$$

In order to calculate the box dimension of the plane set, we can first draw the coordinate square grid on the plane and calculate the number $N_\delta(F)$ of intersections of F with the grid square [5] (the square can be regarded as a box, hence the box dimension is named).

3. Test Beam Design

In this test, three rubber concrete beams with different coarse aggregate gradations were made [6], all of which were made of fine aggregate with substitution ratio 7.5% of rubber powder. The design strength grade of the crumb rubber concrete (CRC) is C30. The specimen numbers are CRC-small, CRC-mixed and CRC-large representing the rubber concrete with coarse aggregate gradations of 5 mm -16mm, 5 mm -31.5mm and 16 mm -31.5mm respectively. In the case of the same substitution rate, by changing the coarse aggregate gradation and using the fractal method, the damage effect of the rubber concrete beams with different gradations was studied. In order to form a 600mm pure bending section in the middle span of the test beam, the three-point method was adopted for test loading. The dimension and reinforcement of the test beam are shown in figure 1.

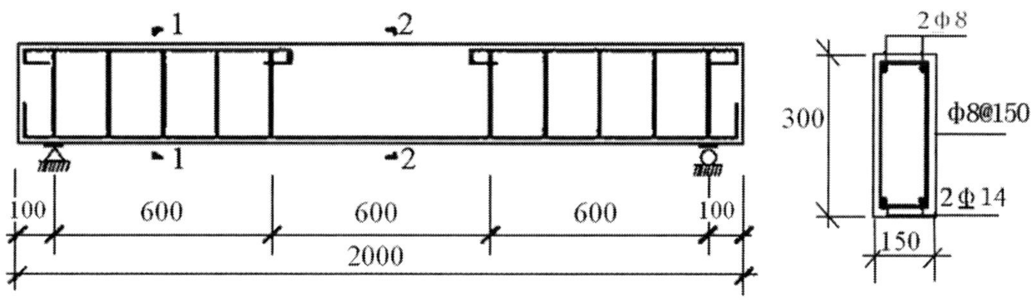

Figure 1. Reinforcement dimension diagram of test beam (unit: mm)

4. Programming with MATLAB

Read the pictures taken in the experiment into the computer and program with MATLAB (MATLAB 7 R14) [7]. The program with MATLAB is following.

```
D = imread; % read the image in the Matlab working folder;
A = rgb2gray (D); % convert the original image to a grayscale image;
G = im2bw (A); % Gray image to binary image;
P = size (G); % to see the size of the binary image;
G = imresize; % Resize the matrix;
Function t = js(n,G); % writing statistical functions
X = [ ]; Y = [ ]; % sets two empty matrices to hold statistical data;
For m = 1: n; %traverse the binary image matrix
    B = blockproc (G, (2^m 2^m], @ (bs) any (bs) data (:) = = 0)); % counts the number of sub
        blocks that contain 0,
    X = [x 2^m]; % traverse the size of the box and put it into x
    Y = [y sum (B (:))]; % count the number of boxes and put it into y
The end
X = x/(2 ^ n); % unitize the original binary image matrix, and the box size is adjusted accordingly
X = - log (x); % take the logarithm of the inverse of the box size
Y = log (y); % take the logarithm of the number of boxes
T = (x, y), % outputs the statistical and processed value
Jsl = js (n, G); % the box width and number of boxes per count.
```

IOP Conf. Series: Materials Science and Engineering **611** (2019) 012043 doi:10.1088/1757-899X/611/1/012043

5. Fractal Analysis of Damage of Rubber Concrete Beams

The fractal dimension of cracks of each test beam from loading to failure is calculated by using the fractal dimension calculation method provided in this paper, and the stiffness change of the test beams under load is listed in the following table 1.

Table 1. Relationship Between Fracture Fractal Dimension of Test Beam under Different Loads and Damage Variables (*D-S*)

Specimen Numbers	Loads (KN)	Fractal Dimensions (D)	Initial Stiffness	Post Damage Stiffness	Damage Variable (S)
CRC-small	35	1.3750	40000.0	31372.6	0.216
	53	1.4445		25906.7	0.352
	73	1.4571		23411.4	0.415
	103	1.4932		18975.3	0.526
CRC-mixed	35	1.4185	41666.7	26890.8	0.355
	53	1.4441		23041.5	0.447
	73	1.4501		21084.3	0.494
	118	1.4602		14180	0.660
CRC-large	35	1.2908	30769.2	27826.1	0.096
	53	1.3830		23041.5	0.251
	73	1.3851		20958.1	0.319
	108	1.4134		12152.8	0.605

From the table 1, we know that in the process of test of rubber concrete beam loading, the fractal dimension of the surface crack and the damage variable are increased with the increasing of the load. And the damage variable is proportional to the fractal dimension. Therefore, the fractal dimension value can be used to characterize the damage degree of rubber concrete beam.

The fractal dimension values of three test beams with different grades of coarse aggregate rubber-concrete can be calculated by using the fractal dimension calculation method provided in this paper. In the failure state, the fractal dimension value of the crack in the test beam is shown in table 2. Table 2 shows that the fitting degree of the fractal dimension of rubber concrete beams with different grades of coarse aggregate is more than 99%, which further verifies that the development of the surface crack of the test beam has good fractal characteristics.

Table 2. Fracture fractal dimension of test beam in failure state

Specimen Numbers	Fractal Dimensions (D)	Maximum Crack Width (mm)	Correlation Coefficient (R^2)
CRC-small	1.4932	0.4	0.9972
CRC-mixed	1.4602	1.2	0.9976
CRC-large	1.4134	0.6	0.9948

From the table 2, It can also be seen that, in the failure state, the fractal dimension value of CRC-small is the largest and the maximum crack width of CRC-small is the smallest. So CRC-small is the best in the three test beams. It can be explained as follows: as the rubber powder is mixed into the ordinary concrete, the aggregate gap is effectively filled and the mechanical properties of the ordinary concrete are changed. The grading of CRC-small is the smallest, so that the development and distribution of surface cracks is the most uniform. From loading to failure, CRC-small has excellent crack resistance, the maximum damage variable, and sufficient energy consumption.

6. Conclusions

1) The damage variables and fractal dimension values of the test beam show an increasing trend as the load increases, and the qualitative relationship between them can be fitted by MATLAB.
2) The damage evolution process of the rubber concrete beam is actually a process of increasing surface fractal dimension. The specimens at different damage stages have the corresponding fractal dimension value. Fractal dimension value can be used to characterize the damage degree of rubber concrete beam.
3) Due to the addition of rubber powder, the mechanical properties of ordinary concrete are improved. Among the three rubber concrete test beams with different coarse aggregate gradations, the CRC-small has the largest fractal dimension, the most uniform crack development and the least damage variable, so it has high crack resistance and sufficient energy consumption. Among the three test beams, CRC-small is the best one.

References

[1] Hong-quan SUN 2016 Computer Simulation on Surface Crack Growth of Recycled Concrete Beams with Fractals, 2016 6th International Conference on Mechanics, Biotechnology and Environment (ICMMBE 2016), pp 630-635.
[2] Hong-quan SUN, Wen-hui ZHOU and Chao-yi WEI 2014 Experimental study on the mechanical properties of rubber concrete. 2014 2nd International Forum on Mechanical and Material Engineering, IFMME 2014. Zhuhai, China. Advanced Materials Research. Volume: 915-916. pp 685-689.
[3] Hong-quan SUN 2011 Fractal Geometry and Fractal Interpolation [M], science press.
[4] Genady P Cherepanov 1995 Alexander S Balankin, Vera S Iavnova. Fractal Fracture Mechanics-A review [J]. Engng. Fract. Mech, (51) pp 997-1033.
[5] Hua ZHU and Cui-cui JI 2011 Fractal Theory and Its Application [M], science press.
[6] Hong-quan SUN, Jun DING, Jian GUO and Dong-liang FU 2011 Fractal Research on Cracks of Reinforced Concrete Beams with Different Aggregates Sizes [J].Advanced Building Materials (Advanced Materials Research), pp 1818-1822.
[7] Jun DING and Hong-quan SUN 2011 Optimization of Fractal Dimension Calculation Method [J], Sichuan Architectural Science Research. 37 (5), pp 54-56.

Effect of Stratum on the Interfacial Stress and Structural Integrity of Cement Sheath

Jingfu Zhang, Jiancheng Wang, Shuting Wang, Shuai Shao, Wenhui Xiao and Ziyang Lin

Enhanced Oil and Gas Recovery Key Laboratory, Ministry of Education, Northeast Petroleum University, Daqing 163318, Heilongjiang, China

Abstract. In order to determine the method of protecting the cement sheath structural integrity in oil and gas wells, the finite element mechanics models of the Casing-Cement sheath-Stratum Combination Body (CCSCB) were established based on different stratum action mechanism. The effect of stratum type, well depth, crustal stress and stratum elastic modulus on the interface stress and structural integrity of cement sheath were researched. The results show that, in the creep stratum, the interface contact pressure of cement sheath increase and the circumferential stresses reduce with the increase of crustal stress. There is no effect of the crustal stress on the interface stress in the elastic and rigid stratum. In elastic stratum and low crustal stress creep stratum, circumferential tensile stress would be generated at cement sheath when loading. The cement sheath is easy to crack by this tensile stress and the structural integrity is destroyed when the load was higher. The effect of stratum elastic modulus on the interface stress is related to the stratum and stress type. There is no influence of the elastic modulus on the interface stresses in the rigid stratum. The influence of the elastic modulus on the interface contact pressure is obvious in the elastic stratum, and the effect on the circumferential stress is obvious in the creep stratum.
Keywords: creep stratum; elastic stratum; rigid stratum; cement sheath; structural integrity.

1. Introduction

In recent years, some progress has been made in the mechanics analysis and research on the deformation and bearing capacity of cement sheath under the action of construction load, focusing on the problem of annular seal leakage in oil and gas wells or gas storage wells [1-15]. The mechanic model of the Casing-Cement sheath-Stratum Combination Body (CCSCB) were established. The influence on the crustal stress and the mechanic parameters of the cement sheath on the stress of the CCSCB were analyzed. The adaptability relationship between the cement sheath performance and its load were studied. The optimization method of mechanic parameters of cement sheath was put forward. The above research results have laid a foundation for the further study of the mechanic mechanism and safeguard of the structural integrity of cement sheath.

In the CCSCB, the mechanic response of the cement sheath under the load is closely related to the loading mode, load size, borehole geometry, casing properties, stratum properties, crustal stress and the mechanic properties of the cement sheath itself etc. Therefore, the key to the accurate analysis of the stress and deformation of the cement sheath is to establish the mechanic model of the CCSCB which is suitable for the mechanic characteristics of the stratum, based on the correct analysis of the interaction between the stratum and the cement sheath.

However, in the mechanic analysis and model building, most of the related achievements are set the stratum as creep stratum. It is considered that stress in the far field can always be loaded into the cement sheath. The mechanic properties that well rock can bear part or all crustal stress under steady state of the wall of oil and gas wells are not taken into account. Therefore, based on the mechanics analysis of literature [3] and [4], three kinds of stratum with different material properties such as creep, elasticity and rigidity are selected to establish the mechanic model of the CCSCB. The influence of stratum environmental condition such as strata type, well depth, crustal stress, stratum elastic modulus on the interface stress and structural integrity of the cement sheath was analyzed, which can provide the basis for the reasonable design of methods and approaches for ensuring the reliability of the annular seal of cement sheath according to the stratum conditions or different well depths.

2. Mechanic Action between Stratum and Cement Sheath

Because oil and gas wells have different strata environments in project, the mechanic properties of strata of borehole wall will be different. Even strata at different depths in the same well could have different mechanic properties because of their different structures, compositions and microstructures. As a result, it shows different deformation characteristics and bearing capacity under the load, and then affects the mechanic interaction between the stratum and the cement sheath. The types of strata can be classified into creep stratum, elastic stratum and rigid stratum by the mechanics of materials and the theory of elastic-plastic mechanics.

2.1. Mechanic Action of Crustal Stress on Cement Sheath

Creep stratum. Creep rock is mainly composed of strata of rock salt, mudstone and shale. Under the action of far-field stress, creep deformation can occur. In cementing wells, the creep stress under long-term steady state can continue to act on the cement sheath and casing. That is, the stress in the far field can always be loaded on the cement sheath.

Elastic stratum. Under the action of external load, the rock stratum has the characteristics of elastic-plastic deformation. However, when the load size is within a certain range, the rock strata only produce elastic deformation or small amount of plastic deformation can be ignored and the elastic deformation is dominant. And its deformation is stable under long-term load. After unloading, it can completely (or basically) restore its original shape and size. At this time, the mechanic properties of strata can be idealized as elastic stratum. Under normal circumstances, these kinds of stratum such as tight sandstone and limestone can be roughly determined into the elastic stratum.

During well construction, the borehole wall formed by elastic stratum reaches the mechanic equilibrium under the effect of liquid column pressure and crustal stress to keep borehole wall stable. The stratum will not actively load crustal stress onto cement sheath when no load is applied in the casing after cementing. When construction load is applied in the casing, the stress and strain transfer of the casing and cement sheath, the surrounding rock of the well bore will have corresponding stress and strain changes. When the load in the casing is removed, the elastic deformation generated by the stratum will be restored to its original state. The stress between the borehole wall and the cement sheath occurs only at the initial stage of unloading. But there is no interaction between stratum and cement sheath when the borehole wall is restored to its original position.

Rigid stratum. Rigid stratum refers to the rock which only stress response is produced without any deformation under load. The mechanic properties of hard rock formation in deep depth can be generally idealized as a rigid stratum. In view of the mechanic characteristics of rigid strata, the crustal stress is borne by the borehole wall stratum. When the casing is added load or unloaded, the borehole wall stratum always generates no strain and does not exert any stress on the cement sheath.

2.2. The Finite Element Mechanic Model of Combination Body

According to the above analysis, the CCSCB are selected as figure 1 and the PLANE183 plane strain element is used to discrete the elements of the CCSCB of figure 1. The 1/2 model (Figure 2) is selected to analyze the stress and strain mechanics problems of the CCSCB under load such as loading and unloading in casing. The finite element mechanic model of the CCSCB is established for different kind of stratum. In the figure 1, the casing and borehole annulus are eccentric annulus. p_H, p_h are respectively the original maximum, minimum horizontal crustal stress in the far-field.

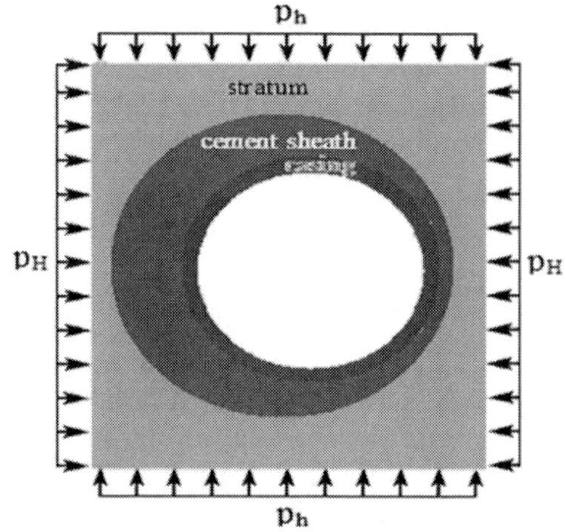

Figure 1. Casing-Cement sheath- Stratum combination body

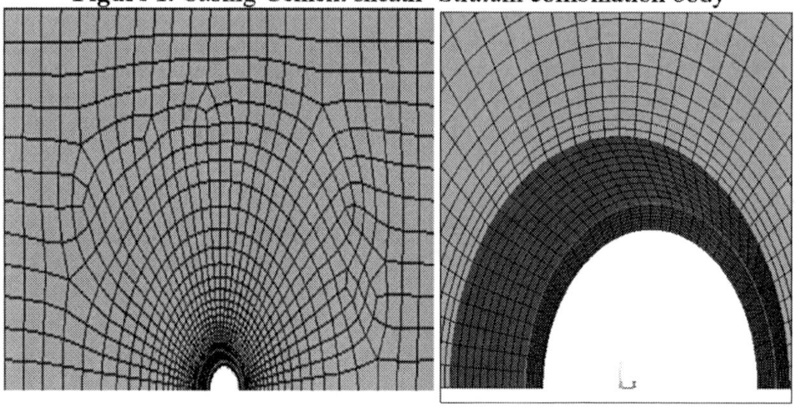

Figure 2. Finite element model of the combination body

The followings should be considered when the CCSCB was analyzed by mechanic analysis:(1)The borehole wall strata are selected as creep, elastic and rigid stratum respectively; (2)Non-uniform ground stress;(3)eccentric casing;(4)The two load action form for the load and unload during casing pressure test construction are considered at the same time;(5)The bearing capacity of the cementation interface which formed by cement sheath and casing, cement sheath and stratum and the state of contact or dispersion of CCSCB are considered.

3. Results and Discussion

3.1. Effect of Crustal Stress on the Interfacial Stress of Cement Sheath in Different Stratum

Above analysis show that different stratum has different mechanic properties and bearing capacity, which affects the mechanic effect of crustal stress on the cement sheath. Figure 3 and figure 4 show the simulation results of the interface contact pressure and circumferential stress of cement sheath at narrow gap in eccentric annulus under the different crustal stress in different stratum.

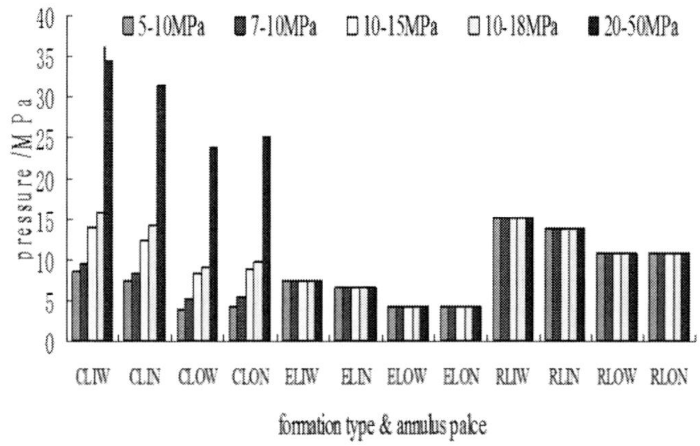

Figure 3. Effect of formation type and crustal stress on interface contact pressure

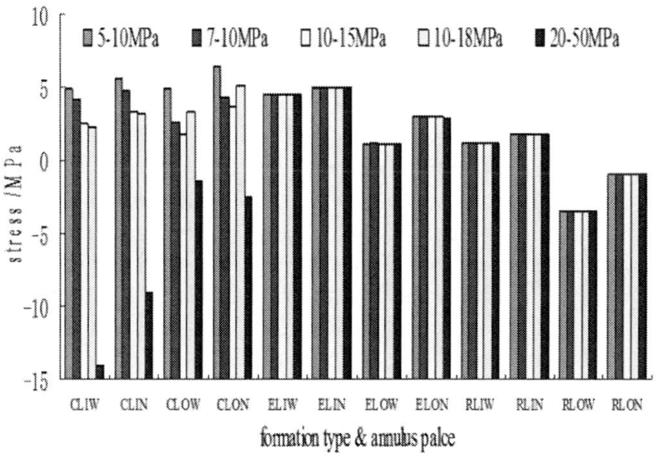

Figure 4. Effect of formation type and crustal stress on interface circumferential stress

In the figure 3 and figure 4, the stratum type, the cement sheath inner or outer interface, loading or unloading conditions are indicated by symbolic combination respectively. Where C, E and R respectively represent the creep, elastic and rigid stratum. L or U represent the loading or unloading state. I or O respectively represent the inner or outer surface of the cement sheath (the same below). In the simulation calculation, the selected basic condition parameters are as follows: Borehole diameter is 241.3 mm, casing diameter is 177.8 mm, casing thickness is 12.65 mm and casing eccentricity is 0.2. The fluid density in the borehole is 1200kg/cm^3. The depth of calculate point is 850m. For casing the elasticity modulus is 210 GPa and the Poisson's ratio is 0.3. The surrounding rock stratum radius takes 1m, and the stratum elastic modulus is 15GPa, the Poisson's ratio is 0.2. For cement sheath, the elastic modulus is 7.86GPa, the Poisson's ratio is 0.178, the yield strength is 20.3MPa. The inner interface cementation strength is 0.6MPa, and the outer interface cements strength is 0.18MPa. The load added at well head is 30MPa. During the process of calculating, the condations of five combination of the min and max crustal stress are selected as: 5,10MPa, 7,10MPa, 10,15MPa, 10,18MPa, 20,50MPa.

The calculation results in figure 3 and figure 4 show that: Under the conditions of non-uniform crustal stress and eccentric annulus, the interface contact pressure (positive value is compressive stress, negative value is tensile stress) at the wide gap of the cement sheath is higher than that at the narrow gap during the loading process in various kinds of strata. However, the circumferential stress (positive is tensile stress and negative is compressive stress) of the inner interface at the narrow gap of the cement sheath is always higher than that at the wide gap, which is the weak point of brittle tensile failure of the cement sheath.

The effect of the crustal stress on the cement sheath is only limited to the creep stratum, and there is no influence of crustal stress in the elastic and rigid stratum. For creep stratum, the overall influence trend of crustal stress on the interface stress of cement sheath is as follows: the crustal stress increases (minimum or maximum crustal stress increases), the contact pressure on the inner and outer surfaces of cement sheath increases, and the circumferential stress decreases.

High circumferential tensile stress appears in cement sheath under the conditions of partial strata and crustal stress according figure 3. When the tensile stress exceeds the tensile strength of the cement sheath, local or integral tensile cracks will appear in the cement sheath and the structural integrity is damaged. This situation mainly occurs in elastic stratum and low-stress creep stratum. For example, when the crustal stress in elastic stratum are 5,10MPa and the cement sheath tensile strength is 4.0MPa, the whole body of cement sheath produces tensile cracks at the wide and narrow gaps. Therefore, full attention should be paid to the high circumferential tensile stress of the cement sheath at the narrow gap in the elastic stratum under the conditions of eccentric annulus when evaluating the structural integrity of the cement sheath and designing the mechanic parameters of the cement sheath.

3.2. Effect of Stratum Types on the Interfacial Stress of Cement Sheath

Figure 5 and figure 6 show the curves of the interface contact pressure and circumferential stress of the cement sheath during the loading and unloading under different stratum and well depth. In order to compare and analyze the influence of stratum type on the interface stress of cement sheath, the borehole wall stratum is assumed to be creep, elastic and rigid stratum respectively, and the variation of stratum type with depth is not considered

The basic condition parameters selected in the calculation are as follows: casing eccentricity is 0. Elastic modulus is 15GPa and Poisson's ratio is 0.25 of the well surrounding rock. For cement sheath, elastic modulus is 4.75GPa, yield strength is 14.8MPa, Poisson's ratio is 0.178 and tensile strength is 3.1MPa. The remaining conditions are consistent with the corresponding parameters of the figure 3.

According to the engineering practice, the variation of hydrostatic pressure of working fluid and crustal stress with the depth of well is considered. Figure 5a shows that when there is a fluid with a specific density in the casing, the interface contact pressure values of the cement sheath are the same for the creep and elastic stratum at the surface wellhead, which both are lower than that of the rigid stratum. With the increase of well depth, the casing load and crustal stress increases, and the ninner interface contact pressure of cement sheath increase under different geological conditions. But the increase degree is different, the stress increased obviously in the creep stratum. This is mainly due to that the cement sheath interface contact pressure is only depended on the casing load in the elastic and rigid stratum, but the contact pressure is generated both by the casing load and crustal stress CCSCB in the creep stratum.

At the beginning of well depth increase, the interface stress of cement sheath of creep stratum is lower than that of rigid stratum. When the depth of a well reaches a given deep, with the depth increase continually, the effect of crustal stress increase, and the interface stress of cement sheath of creep stratum is higher than that of rigid stratum. When the well depth reaches 3000m, the inner interface of the cement sheath in the creep and rigid stratum are all yield. Within the well depth given in this paper, the interface contact pressure of the cement sheath in the elastic stratum is the lowest.

When load at the wellhead is withdrawn after operation and the hydrostatic pressure and cruatsl stress are still maintained in well, interface contact pressure of cement sheath with elastic deformation in the loading process is restored to the original loading state in the well. During the process of loading, the cement sheath occurs yield local or overall. Cement sheath deformation cannot fully recover after unload. The interface contact pressure will reduce. At this point, if the crustal stress is low (or no pressure) and in the situation of no crustal stress action in elastic strata, rigid stratum, the interface contact stress of cement sheath will turn to be tensile stress from compressive stress. There is a potential risk of tearing apart the bond interface of the cement sheath.

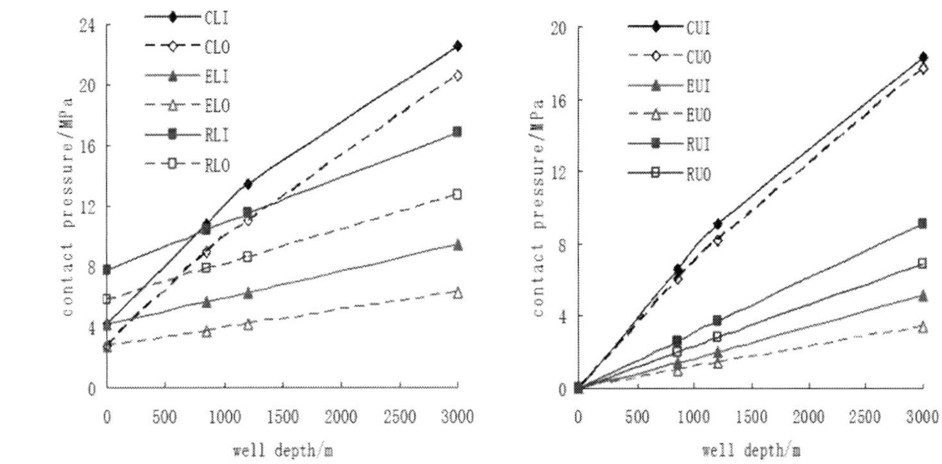

(a)loading (b)unloading

Figure 5. Change of cement sheath interface contact pressure

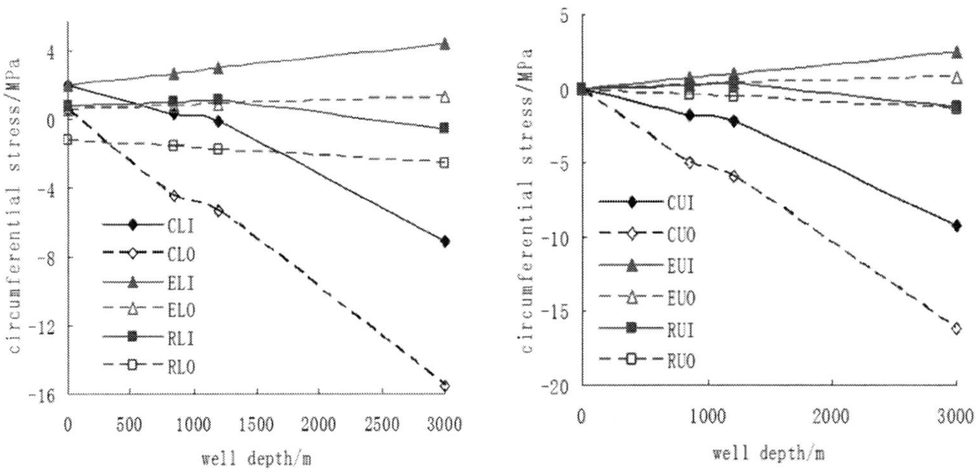

(a)loading (b)unloading

Figure 6. Change of cement sheath interface circumferential stress

The circumferential interface stress of the cement sheath varies with the well depth and stratum types during loading and unloading (Figure 6a). In case of loading, circumferential tensile stress is generated in both the inner and outer surfaces of the cement sheath under the condition of creep formation and elastic formation at the wellhead, and the values are the same. While in the case of rigid stratum, the inner interface circumferential stress is tensile stress, and the outer is compressive stress. Under the confining pressure and crustal stress in the creep formation, the inner and outer interface circumferential stress of the cement sheath decreases with the increase of well depth and the stress state gradually changes from tensile to compression deformation. The crustal stress cannot be acted on cement sheath in the elastic stratum due to the borehole wall can bear the crustal stress. Under the action of internal pressure in casing, the inner and outer intrface circumferential stresses of cement sheath are tensile stress. The deeper the well depth, the higher the tensile stress is. For a certain well depth, the tensile crack could be made at the interface or in the internal part of cement sheath when the inner interface tensile stress is higher than the tensile strength, so the structure integrity of cement sheath is destroyed. Under the condition of rigid stratum, the interface circumferential stress of cement sheath increases first and then decreases with the increase of well depth, while the outer circumferential stress of cement sheath always shows a downward trend.

Figure 5b and figure 6b are obtained with hydrostatic pressure or crustal stress remaining in the well after the wellhead load is removed. When only elastic deformation occurs to the cement sheath in loading process, the interface circumferential stress of the cement sheath reverts to the loading state of hydrostatic-liquid column pressure or crustal stress in the well after unload in various kinds of stratum. The above analysis results show that, during the process of loading or unloading in the oil and gas well construction, the well depth will have an impact on the interface stress and structural integrity of the cement sheath, and the degree and law of the impact is closely related to the stratum type. In fact, with the change of well depth, the composition and mechanic properties of surrounding rocks in wells will change, and even the material mechanic constitutive relationship will also change, which made the stratum to belong to creep, elasticity, rigidity respectively. Therefore, the effect of well depth on stratum type and rock properties must be considered in the mechanic analysis and structural integrity evaluation of cement sheath in the wells.

3.3. Effect of the Stratum Elastic Modulus on the Interfacial Stress of the Cement Sheath

Using the calculation situation parameters of figure 3 and figure 4, make minimum and maximum horizontal crustal stress in inhomogeneous stress stratum for 10MPa and 15MPa, figure 7 and figure 8 show the influence of elastic modulus on the interface contact pressure and circumferential stress at narrow gap of cement sheath in different stratum after load and unload, which the elastic modulus are selected as 5,7,10 and 20GPa respectively.

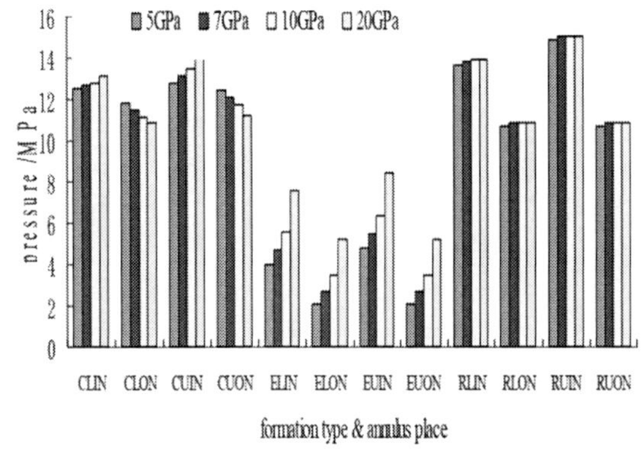

Figure 7. Effect of formation type and elasticity modulus on interface contact pressure

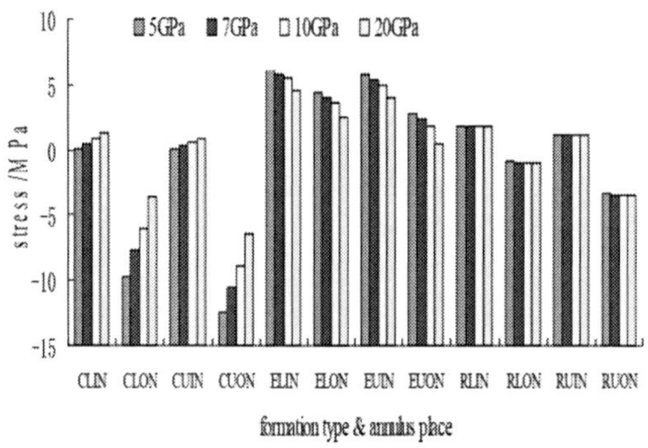

Figure 8. Effect of formation type and elasticity modulus on interface contact pressure

The results show that, there is no influence of elastic modulus on the stress of cement sheath in the rigid stratum. The elastic modulus of the elastic stratum has an obvious influence on the interface contact pressure of the cement sheath while the elastic modulus of the creep stratum has a more obvious influence on the interface circumferential stress. For creep stratum, with the increase of elastic modulus, the inner interface contact pressure of cement sheath increase and the outer's decrease, the interface circumferential stress are both increase. For elastic stratum, the interface contact pressure increase and the circumferential stress decrease with the increase of elastic modulus.

The key basic condition is that the stratum type determines the interaction mode between crustal stress and cement sheath, and restricts the effect of parameters such as the stratum elastic modulus on the interface stress and structural integrity of the cement sheath. Therefore, when carrying out mechanic analysis of cement sheath, structural integrity evaluation and mechanic parameter design of cement sheath, it is necessary to accurately evaluate and determine the stratum type of specific depth stratum based on accurately determining the crustal stress and stratum rock mechanic properties.

4. Conclusion

- In creep stratum, with increase of crustal stress, the contact pressure of cement sheath increase, but circumferential stress decrease; In elastic stratum and rigid stratum, there is no influence of crustal stress on interfacial stress of cement sheath.
- In elastic stratum and creep stratum with low crustal stress, when the load or cement sheath elastic modulus is high, the cement sheath is cracked easily due to high circumferential tensile stress so that broke the integrity of cement sheath structure. Under the condition of inhomogeneous crustal stress and eccentric annulus, annular narrow gap is the weakest part of the cement sheath.
- With the increase of well depth, the inner and outer interface contact stress of the cement sheath increase obviously in the creep stratum, and the inner and outer interface circumferential tensile stress of the cement sheath increase in elastic stratum. There will be a potential risk that the inner interface of cement sheath is cracked and structure integrity is broken in elastic stratum.
- There is no influence of elastic modulus on the stress of cement sheath in the rigid stratum. The elastic modulus of the elastic stratum has an obvious influence on the interface contact pressure of the cement sheath while the elastic modulus of the creeping stratum has a more obvious influence on the interface circumferential stress.

Acknowledgment
The authors wish to thank National Natural Science Foundation of China and National Natural Science Foundation of Heilongjiang Province for supporting the research. The supporting research projects were respectively: Mechanic analysis and research on annular sealing reliability of cement sheath (51474074); Analysis on structure integrity of casing-cement sheath-stratum cementing combination body (E201334).

References

[1] Rodriguez W J , Fleckenstein W W, Eustes A W. Simulation of collapse loads on cemented casing using finite element analysis .SPE 84566, 2003 :1-9.
[2] Li Zifeng, Zhang Yonggui. Yang xinjun. Mechanics model for interaction between creep stratum and oil well casing. Acta Petrolei Sinica, 2009, 30(1): 129-131.
[3] Li Guoqing. Analysis of boundary condition of stress calculation on casing/cement-sheath.Petroluem Drilling Techniques, 2012,40(2):20-24.
[4] Zhang Jingfu, Zhang Debing, Zhang Qiang, et al. Impact of elastic parameters of cement ring on structural integrity of casing-cement ring-strata cementing combination. Oil Drilling & Production Technology, 2013, 35(5):43-46.
[5] Zhang Jingfu, Lv Yingbo, Liu Shuoqiong, et al. Adaptability among loads and mechanical parameters of cement sheath. Oil Drilling & Production Technology, 2016,38(5):594-600.
[6] Li Jun, Chen Mian. Liu Gonghui, et al. Elastic-plastic analysis of casing-concrete sheath-rock combination. Acta Petrolei Sinica, 2005,26(6):99-103.
[7] Li Jun, Chen Mian, Zhang Hui, et al. Effect of cement sheath elastic modulus on casing external collapse load. Journal of the University of Petroleum, China: Edition of Natural Science, 2005, 29(6): 41-44.
[8] Yin Youquan, Chen Chaowei. Li Enping, et al. Theoretical solutions of stress distribution in casing-cement and stratum system. Chinese Journal of Theoretical and Applied Mechanics, 2006,38 (6): 835-842.
[9] Yin Youquan, Li Zhiming. Zhang Guangqing, et al. Study on Casing Loads in Creep Stratums. Chinese Journal of Rock Mechanics and Engineering, 2004,23 (14): 2381-2384.
[10] Wang Yaofeng, Li Junqiang, Yang Xiaohui, et al. Research on maching cement's elastic modulus and poisson's ratio. Petroleum Drilling Techniques, 2008, 36 (1):25-29.
[11] Zeng Dezhi, Lin Yuanhua, Li Shuanggui, et al. Study of stress distribution of cement sheath cross section non-uniform stratum stress. Petroleum Drilling Techniques, 2007, 35(1):32-34.

[12] Fang Jun, Yue Boqian, Zhao Huaiwen, et al. Analysis of surface loading on casing and cement sheath under nonuniform geologic stress. Journal of the University of Petroleum, China: Edition of Natural Science, 2005, 29(6): 46-48.
[13] G. Alfano, M. A. Crisfield. Finite element interface models for the delamination analysis of laminated composites: mechanical and computational issues. Int. J. Numer. Meth. Engng, 2001,50:1701-1736.
[14] Liu Yang, Yan Haibing, Yu Xin, et al. Negative impacts of borehole pressure change on cement sheath sealing integrity and countermeasures. Natural Gas Industry, 2014, 34(4):95-98.
[15] X.P. Xu, A. Needleman. Numerical simulations of fast crack growth in brittle solids. J. Mech. Phys. Solids, 1994, 42(9): 1397-1434.

A Study on Analysis of Long Span Continuous Rigid Frame Bridge

Guihong Xu[1,*], Bin Yang[2], Chuang Chen[2], Shirong Wei[2] and Zhenxue Li[3]

[1]Guizhou institute of technology, GuiZhou, GuiYang, 550001, China
[2]China Communications Technology Co., Ltd. Beijing, 100121, China
[3]Xingtai Zhengyang building and installation Co., Ltd., Xingtai, Hebei, 054500, China

*Corresponding author

Abstract. A case study of a large span pre-stressed concrete continuous rigid frame bridge (Ba-Mao-Chong bridge) of GUI DU high speed railway. Finite element program Midas/Civil is used to simulate the construction process and operation phase, the influence of concrete overweight, pre-stressing loss, concrete shrinkage and creep, stiffness loss and other factors that related the long-term deflection of bridge are analyzed. The calculation results show that: (1) The overweight concrete and Bridge deck pavement construction error could cause the bridge long-term deflection increases. (2) The creep coefficient, environment humidity, and other factors on the bridge have significantly influence on long-term mid-span deflection, (3) The beam stiffness reduction could lead to the bridge long-term mid-span deflection increase.
Keywords: Continuous Rigid Frame Bridge.

1. Introduction

The GUI Du highway is a fast way between Guiyang and Du-yun city, it is an common segment channel of *National Highway Planning Network* of Xia Men-Cheng Du, Lan Zhou - Hai Kou,Shang Hai to Kun Ming, in the middle part of the Gui-Zhou province. The highway starting point in the Huo-Shi-Po of Du-yun City, connected with Xia-Rong expressway, through the Xiao-Wei-zhai, Da-Tie-Xiang of Gui-ding County, Chang Ming Town, Long-li county and other places, The end is at Lan-Ni-Gou of Gui-yang Expressway, Cross connecting with the Guiyang city highway and the south exit road of Guiyang, the route is 80.969 kilometers.

Ba-Mao-Chong bridge is located in Gui-du expressway line, the left line of bridge beginning and end were ZK218+768 and K219+764. The span arrangement is 4×40+5×40+5×40+(80+150+80)+3×40m, bridge length is 996m; the right line of bridge beginning and end were YK218+747.792 and K219+764. The span arrangement is 5×40+5×40+5×40+(80+150+80)+3×40m, the bridge length is 1032.92m. The main span is 80+150+80m, with variable cross-section of pre-stressed concrete continuous rigid frame structure, Pre-stressed concrete T beam are used on both sides.

2. Layout of Height Measurement Point of Bridge Deck

Optical fiber sensors, demodulator, precision leveling instrument, total station instrument, dynamic load data acquisition instrument are used to monitor the bridge of Ba-Mao-Chong. The instrument and equipment are shown in table 1.

In order to ensure the continuity and effectiveness of the monitoring data of the bridge, a permanent control network for the Ba-Mao-Chongbridge is established. Point arrangement, as shown in Figure1.

Table 1. Main instrument and equipment.

Number	Name	Model	Quantity
1	Precision electronic level	Leica DNA03	1
2	total station	Leica TPS1201+	1
3	Grating surface strain gauge	FBGTECH FSS0221DS	54
4	Grating thermometer	FBGTECH FST2111DS	16
6	Bridge modal testing system	DH5907	9

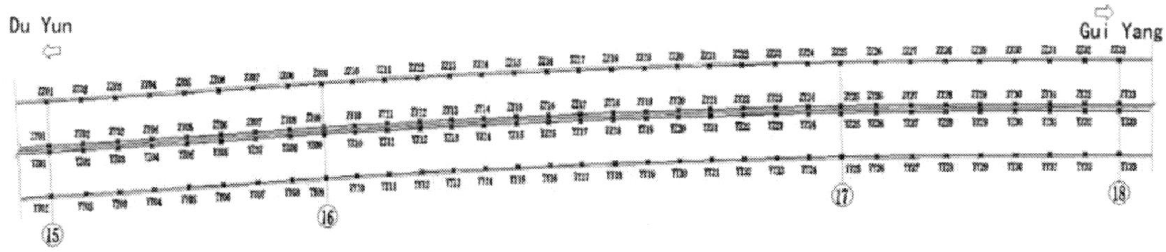

Figure 1. Layout of bridge elevation survey points.

Through the past 4 years (2012.10--2016.10) elevation monitoring of the Ba-Mao-Chong bridge, There is the cumulative variation curve of the bridge elevation in different observation periods was shown in Fig.2 to Fig.5.

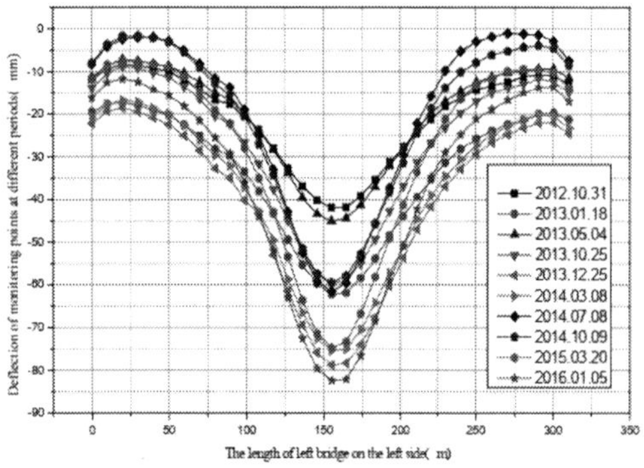

Figure 2. Monitoring results of the left side of the left bridge.

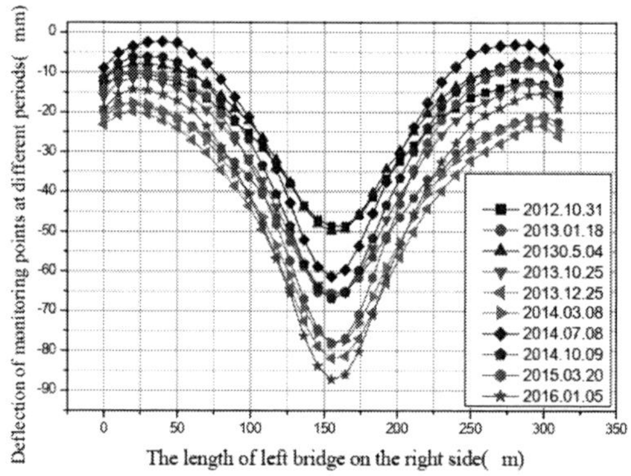

Figure 3. Monitoring results of the right side of the left bridge.

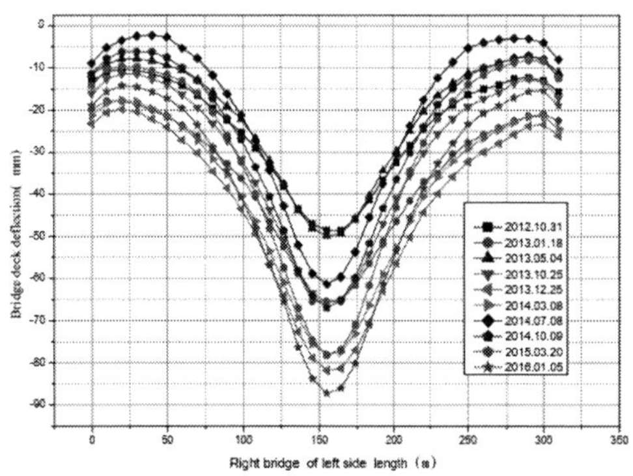

Figure 4. Monitoring results of the left side of the right bridge.

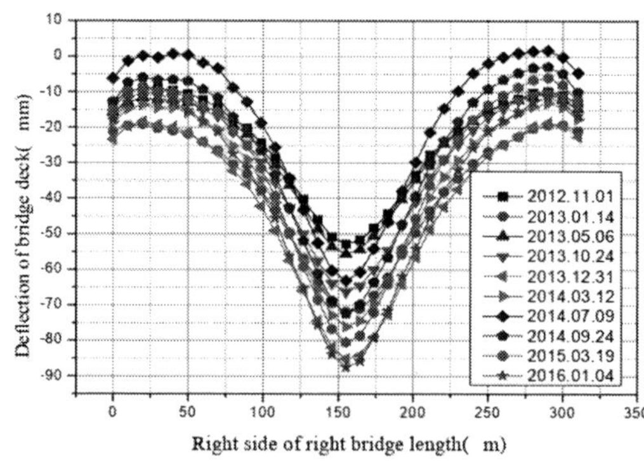

Figure 5. Monitoring results of the right side of the right bridge.

It can be seen from the monitoring results that the deflection value of the Z17 point at the mid-span monitoring point of the Ba-Mao-Chong bridge is the largest. As the time increases, the cumulative value of the deflection increases gradually at the Z17 point.

The results of the Fourth measurement is different from the third for the left side of the left bridge, is 14.75mm. The result of the ninth measurement is quite different from the result of the eighth measurement, which is 15.22mm. The results of the first and third monitoring were similar. And the results of tenth, ninth, fifth and sixth were similar. The difference between the third and the fourth measurements on the right side of the left side is very different, and the ninth and the eighth monitoring results is large difference.The maximum deflection value of the cumulative of Ba-Mao-Chong bridge is 87.26mm, The results of the third and fourth monitoring were difference , The difference between the two is 7.5mm. The difference between the eighth measurements and the ninth measurements is 6.06mm, and the first times and second times are similar, second times, eighth times and sixth times are similar, fifth times and tenth times are similar.

The maximum ZY17 value of the right side bridge deflection is 87mm, the difference between the third and the fourth monitoring results is larger, 15.83mm, and the deflection and time of the other points increase linearly.

3. Numerical Simulation

3.1. Calculation Model

The finite element program Midas/CiviL is used to establish the finite element model, as shown in Fig.6. The three main girders of the "T structure"consist of 28 cantilevers. The whole bridge is divided into 139 units and 144 nodes. The main girder is divided into 123 units, the main span piers are double thin-walled piers, the 1 pier is divided into 8 units, and 2 pier is divided into 8 units.

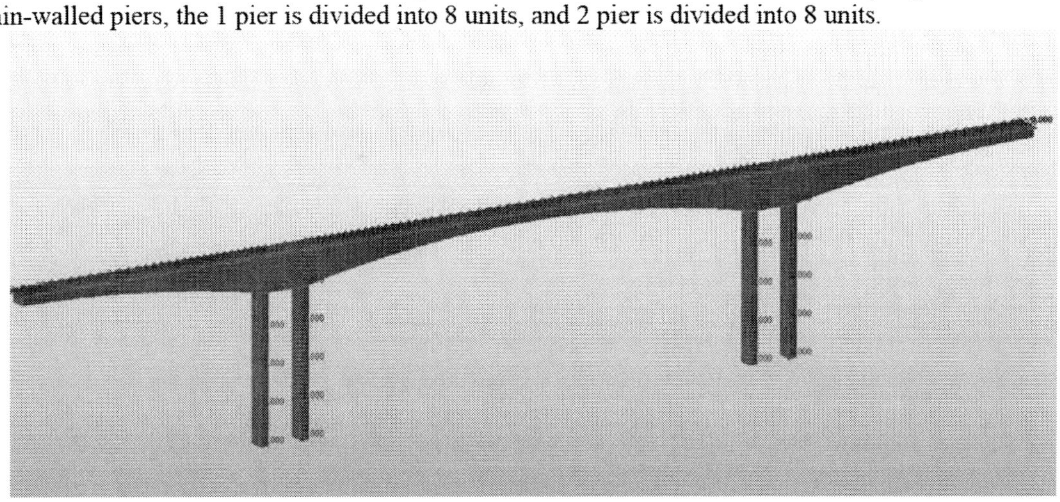

Figure 6. Calculation model of Ba Mao River Bridge.

The constraints of the calculation model of the pendulum bridge are shown in Table 2.

Table 2. Computing model constraints.

Constraint node number	Constraint position	Dx	Dy	Dz	Rx	Ry	Rz
1	Support of side span	×	√	√	×	×	×
124	Support of side span	×	√	√	×	×	×
141	Bottom of piers of 1#	√	√	√	√	√	√
142	Bottom of piers of 1#	√	√	√	√	√	√
143	Bottom of piers of 2#	√	√	√	√	√	√
144	Bottom of piers of 2#	√	√	√	√	√	√

In Table 1, Dx, Dy, and Dz are translational degrees of freedom in the direction of the X, Y, and Z axis, Rx, Ry, Rz are rotational degrees of freedom in the direction of the X axis of the whole coordinate system, respectively.

3.2. Comparison between Numerical Simulation Results and Measurement Results

In order to analyze the change characteristics of the monitoring points along the time, ZZ17 (middle span) ,ZZ29 (third span) and ZZ33 (endpoint) are selected for analysis. Fig 7 is the deflection characteristic diagram of the left Ba-Mao-Chong bridge of the left side.

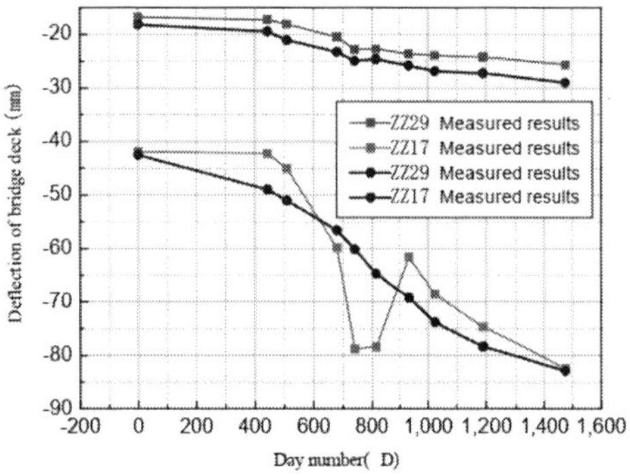

Figure 7. The relationship of Vertical deflection and time for left side of left bridge.

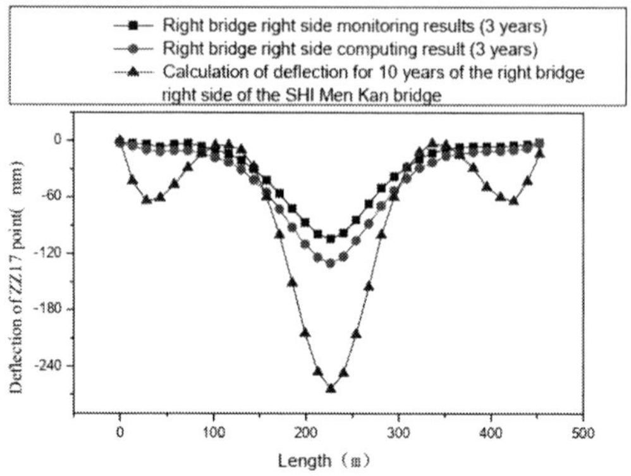

Figure 8. The deflection of Measured results and calculation results of Ba-Mao-Chong bridge.

It can be seen from fig.8, that the vertical displacement increases with time, and generally shows a downward trend. The monitoring point ZZ17 vertical displacement increases with the time, in general, the decline trend. The vertical displacement of the second cross span is the largest. The measured deflection is 82.42mm. and The calculated results are in good agreement with the measured results. The comparison of ZZ29 points shows that the calculated results are slightly larger than the measured results, but the deformation trend is consistent. It can be seen from 9 that the calculated results are larger than the monitoring results by 25mm in the middle span of Shi Men Kan bridge. Up to the time of measurement, the maximum deflection is 151mm of the left bridge in the middle span ,and the maximum deflection is 148mm in the middle span of right. Both of the two are less than the theoretical deflection value 264mm.

The comparison between Fig. 7and Fig.8 shows that the calculated results are slightly larger than the measured value. Therefore, the latter analysis is more secure.

3.3. The Influence of Self Weight

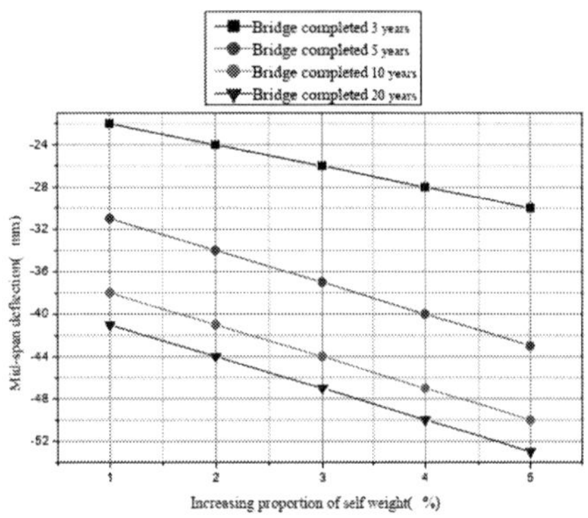

Figure 9. Influence of bridge weight increase on long term deflection of bridge.

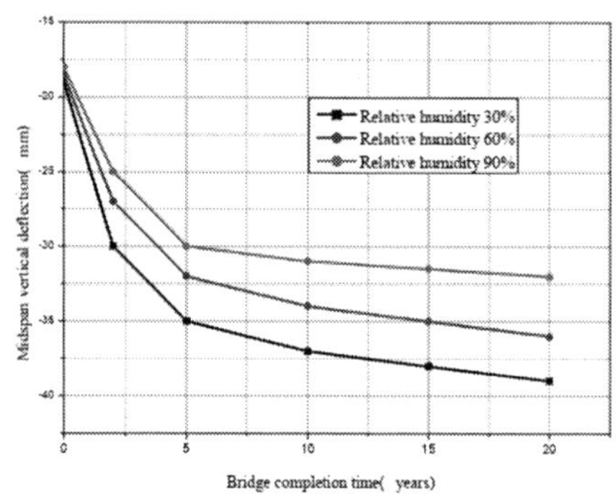

Figure 10. Effect of humidity on deflection of bridge in different stages.

Based on the calculation model of the Ba-Mao-Chong Long Bridge, assuming that the weight of the main girder is 1%, 2%, 3% and 4% respectively, the deflection under different overweight ratios is calculated as shown in Fig.9.

As shown in Figure 9, the Maximum deflection caused by deadweight of main girder In the early stages of bridge construction With the increase of the time, the deflection of bridge caused by deadweight is reduce. 20 years after the bridge was built, When the weight of the main beam increases by 4%, the increase of the span deflection of the bridge is 51mm.

3.4. Humidity Influence

After the concrete is poured, it has been exposed to the nature for a long time, and humidity has a significant influence on the performance of concrete. According to the climatic conditions in Gui-zhou area, the relative humidity of 30% , 60% and 90% are calculated in this paper. The effect of humidity on the mid-and long-term deflection of Ba-Mao-Chong Bridge at different times was studied.

From figure 10, it is known that the greater the relative humidity, the smaller the long span deflection of the bridge. When the bridge is 20 years old, and the humidity is 30% and 60%, the deflection is bigger 17.94% and 11. 1 % respectively than the relative humidity is 90%. Under the same conditions, with the passage of bridge completion time, the larger the humidity, the smaller the ratio of deflection increase.

3.5. Impact of Pre-stressed Loss

In order to investigate the effect of pre-stress loss on bridge span deflection, in this paper the pre-stressed loss is simulated by the pre-stress reduction in the software. Assuming that the longitudinal pre-stressed of the bridge is reduced by 5%, 10%, 15%, 20% respectively, the maximum deflection of bridge in different service time is calculated. The calculation results are shown in Figure 11.

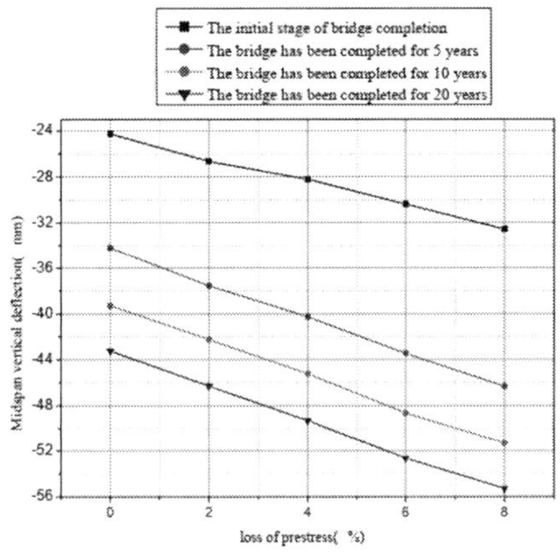

Figure 11. Relationship between pre-stress loss and mid-span deflection.

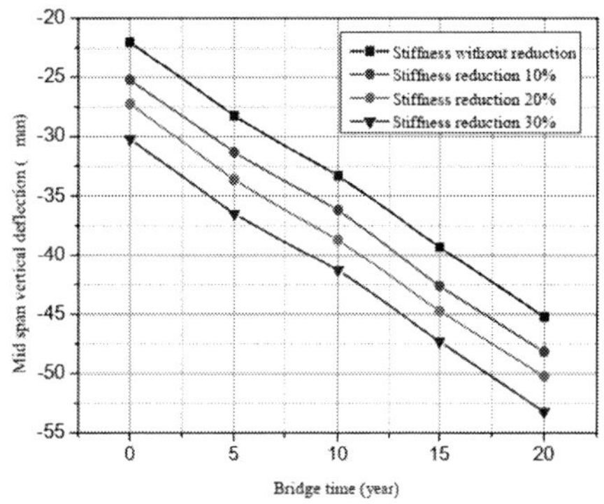

Figure 12. Effect of stiffness reduction on mid-span deflection of bridge.

The completion of the bridge from 5 years to 20 years, when the pre-stress decreases by 2%, the mid span deflection increases 3.3mm of the Ba-Mao-Chong bridge. After 20 years of the bridge foundation, the pre-stress loss is 2%, the maximum deflection of the middle span of the bridge increasing is 4~5mm.

3.6. Effect of Beam Cracking on Long Term Deflection of Bridge

When the pre-stressed can't counteract the tensile stress produced by external loads, the beam will experience cracking, and the bending moment of inertia will be reduced after the cracking of the beam, so the stiffness of the beam will reduce. The stiffness is proportional to the strength of the concrete, so the effect of the cracking of the beam on the long-term deflection of the bridge is calculation through the reduction of the compressive strength. The stiffness reduction of the main beam is 10%, 20%, 30%, and the deflection of the bridge in different periods, as shown in Figure 12.

From Figure 12, it can be seen that with the increase of main girder cracks, the overall stiffness gradually decreases, and the maximum deflection of the bridge increases gradually. The increase of deflection is basically linear relationship with the time and stiffness.

4. Conclusion

In order to analyze the growth of the deflection of the Ba-Mao-Chong bridge, The finite element program Midas/CiviL is used to simulate the construction process and the operation stage. The influence of overcapacity, pre-stressed loss, shrinkage, creep and stiffness loss on deflection of long-span concrete continuous rigid frame bridge is analyzed. The following conclusions are obtained:(1)In the process of bridge construction, the increase of self weight caused by the overweight of concrete and the error of bridge deck pavement construction will cause the long-term deflection to increase. Therefore, it is necessary to avoid the construction irregularity in bridge construction, resulting in the excessive deflection of the middle part of the girder and the excessive deflection.(2)The influence of ambient humidity on the Long Span Rigid Frame Bridge is remarkable. Pre-stress loss has significant influence on long term deflection of long-span continuous rigid frame bridge. From 5 to 20 years, the mid-span deflection of Bam Chong Bridge increases about 3.3 mm when the pre-stress decreases by 2 mm.(3)The girder beam stiffness reduction has great influence on the long-term mid-span deflection of the bridge, from the Bridge construction completion to the bridge 20 years of service, When the stiffness of the main beam is reduced by 10%, the mid span deflection is increased by about 5mm.

Acknowledgements

This work was performed under grants from the Natural Science Foundation of China (grants no.51508121 and 51868009) and Natural Science Foundation of Gui-zhou (grants no. J [2015]2063, and No. LH [2014] 7374). These foundations are greatly appreciated.

References

[1] Ma Cheng, Chen Wei zhen. PC bridge pre-stressed and creep coupled time-varying effect numerical simulation [J], bridge construction, 2013.43 (3): 77-82.
[2] Liu Wen hua, Yan Fanghua. Seismic response analysis of long-span continuous rigid frame bridge. [J], Highway No. 23 / 2: 107-110
[3] Xie WZ, Li ZT, Xu YZ. Effect of concrete creep on long-term deflection of continuous rigid frame bridge. [J] .Journal of Henan City Construction University,2012, 21(1): 4-7.
[4] Xiong Ying. Analysis of the factors affecting the middle and lower deflection of continuous rigid frame bridge and its prevention and control measures [J]. highway traffic technology, 2011 (8): 262-264.
[5] Zheng Yuanxun, Yang Peibing, hanging basket design and preloading test of rigid frame bridge cantilever construction [J]. Journal of Zhengzhou University (SCIENCE EDITION), 2015.47 (4): 108-112.
[6] Meng Xin Qi, Wei Lunhua, Zhang Jinchen and so on. Study on the problem of mid and lower deflection of Large Span Rigid Frame Bridges [J], world bridge, 2013.41 (2): 76-80
[7] Zhou Yongjun, Zhao Yu, he Quanhai. Rigid frame continuous composite bridge impact coefficient [J] multi factor sensitivity analysis, vibration and shock, 2012.31 (3): 97-82
[8] Ma Ming. (108+2*185+115) experimental study on shrinkage and creep of M continuous rigid frame bridge [J], Journal of railway engineering, 2014 (10): 67-72
[9] Li Yong, Yan Weiming, Chen Yanjiang. Vibration and impact study on real time monitoring of [J]. bridge inclination deformation test based on 2013.32 (5): 65-70.
[10] Lining, study on the creep effect of prestressed concrete bridge based on finite element method [J]. highway engineering. 2017.42 (3):58-63
[11] Ma Zhendong, Liu Anshuang, Technical measures for controlling excessive deflection of Long-Span continuous rigid frame Bridge. [J]. Bridge construction. 2015.45(2):71-78

Experimental Research on Seismic Behaviour of Railway Bridge Pier

Guocheng Fu[1], Yuntian Hu[2], Lei Liu[2], Shanzhong Liu[1] and Xiaoxue Wang[2,*]

[1]Chenglan railway co. LTD, Chengdu, Sichuan Province, China
[2]School of civil engineering, Beijing Jiaotong University, Beijing, China

*Corresponding author

Abstract. The shear failure of railway bridge piers has become the main feature of modern bridges. In order to study the shear resistance of railway bridge piers, three models were designed in this paper, including two piers model with a ratio of 1:5 based on Chenglan Railway bridge prototype piers and a test research model. Through the pseudo-static test study on the shear resistance of piers, the failure process of the specimen was given. It was indicated that that A1 and A2 specimen were typical bending failure and possessed enough ductility. And A3 specimen was typical flexure-shear failure. Before loading to yield displacement, stirrup provided little shear strength. After loading to yield displacement, the shear strength provided by stirrup increased gradually with the development of cracks. When the stirrup yielded, its restraint to the core concrete was weaken rapidly, which resulted in the rapid destruction of the bridge pier.

1. Introduction

The earthquake damage of the bridge structure mainly includes the crash of the superstructure, the damage of the supporting connector, the destroyed of the pier and the abutment, the failure of the foundation etc[1]. The pier is the most vulnerable member during the earthquake. There are three kinds of typical failure modes of piers: flexure failure, shear failure and flexure-shear failure[2]. The flexure failure belongs to the ductile failure, which is manifested by the crushing and spalling of concrete, bare and buckling of the steel bar, and even breaking. There is a large plastic deformation when it is broken. It usually occurs in piers with large shear span ratio or sufficient restriction hoops. Shear failure belongs to the brittle failure, which is mainly manifested as a small amount of spalling of concrete, hoops are broken, and forming obvious shearing inclined cracks. There is small plastic deformation and a sharp drop in strength and stiffness when damage occurs. The ductility and energy dissipation capacity are very poor. It often occurs in piers with small shear span ratio or insufficient restriction hoops. Flexure-shear failure is a failure form between flexure and shear failure. Its ductility and energy dissipation capacity are poor, which has certain brittle characteristics. For pier with shear failure and flexure-shear failure, its deformability and energy dissipation capacity are poor, and it is difficult to repair after damage.

With the development of the research on the ground motion and the structural dynamic characteristics, the seismic design method has changed from the strength theory to the ductility seismic theory. It was found that using the plastic deformation of the structure under seismic action to consume seismic energy can better against the seismic action, and it is easy to repair after the event. However, in the ductile seismic design of piers, sufficient shear strength must be obtained in the ductile stage to ensure that brittle shear failure will not occur under strong earthquakes[3]. Therefore, two actual piers of

Chenglan Railway Bridge with a ratio of 1:5 were selected as prototype piers. A test research model was designed for studying the shear performance of pier[5] and there are three models in total. And these models were used to conduct pseudo-static test.

2. Design Parameters of Trial Piers

Respectively, A1, A2 and A3 are the test models of the three piers. A1 and A2 are the prototype test models of Chenglan Railway Bridge piers, and A3 is the test model to study the shear performance. During the test, on the one hand, the rationality of the design of the prototype bridge pier is evaluated, on the other hand, the shear failure mechanism of the plastic hinge area is studied, and the influence of design parameters on the shear capability of the pier is analyzed. The specific model parameters of the three piers are shown in **Table 1**.

Table 1. Model parameter.

Model number	Section shape	Section size(mm)	Effective height (mm)	Longitudinal reinforcement ratio	Stirrup ratio	Axial pressure (kN)	Axial load ratio	Shear span ratio
A1	Rectangle	710×600	2200	0.47%	0.16%	120	0.70%	3.7
A2	Rectangle	846×446	1200	0.21%	0.13%	240	1.59%	2.7
A3	Rectangle	720×420	630	0.93%	0.12%	120	0.99%	1.5

3. Test Scheme

3.1. Layout of Strain Gauge

The length of plastic hinge area at the bottom of pier is calculated and predicted at first. The strain gauge with longitudinal reinforcement tries to cover the height of possible plastic hinge region along the height direction of the specimen, and 1~2 times the height range of the section is taken. The strain gauge with longitudinal reinforcement is arranged symmetrically on both sides of main reinforcements and perpendicular to the loading direction, five main bars are arranged on each side. The strain gauge with hoops is arranged on both sides and parallel to the loading direction and in a continuous manner along the direction of height. The range of layout is consistent with the strain gauge of the main reinforcement.

For catching the strain variation regularity of concrete before cracking, three strain gauges are arranged on both sides of the pier along the plastic hinge area, and the direction perpendicular to the loading direction. Three groups of strain rosette are arranged on one side of the specimen which parallel to the loading direction. The static strain testing system was used to collect the strain values of steel bars and concrete at each peak of load.

3.2. Sensor Placement

The displacement of pier crown under horizontal load mainly includes bending deformation, shear deformation and slip deformation of longitudinal reinforcement[8]. In order to investigate the deformation components of the bridge pier in the loading process, displacement sensors were placed in some positions to record the deformation of the pier at different positions in the loading process.

In this test, 1) MTS actuator[9] sensor was used to record the lateral force of pier crown. The lateral displacement of pier crown is modified by the measured value of displacement sensor at effective height as the actual side displacement of pier crown, while the recorded value of displacement sensor on the base is used for aided-accuracy. 2) In the plastic hinge region on both sides of the pier, the recorded value of the linear differential displacement sensor along the height of the pier is used to calculate the slippage component of the longitudinal reinforcement and curvature characteristics of the section. 3) Three groups of linear differential displacement sensors arranged on the side that parallel to the loading direction are used to calculate the shear deformation component of the pier. The actuator and sensor layout diagram are shown in **Figure 1**.

3.3. Loading System

Variable displacement loading is adopted in this test. A small displacement grade is used at the preliminary stage of loading, and the step loading is carried out with an increment of 1mm until the pier yields (In this test, the longitudinal strain strain value at 5 cm from the bottom of the pier is 2000 ε for the first time.) In order to observe the development and damage changes of the pier before yielding crack. After the pier yields, the yield displacement is the increment of each stage, and the cycle of each stage is three times. According to the code for seismic design of railway engineering[6], when the loading displacement of the bridge pier at a certain level is first circulated, the lateral force drops below 80% of the maximum resistance on the same side, which is defined as the failure of the bridge pier and the termination of the test. The test load waveform[7] is as shown in **Figure 2**.

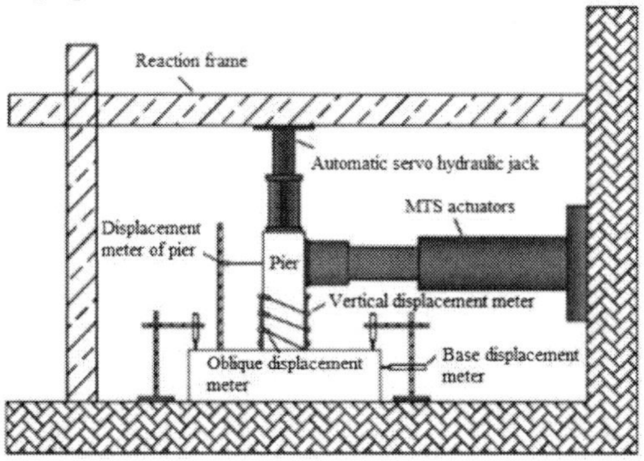

Figure 1. Test setup and instrumentation of pier specimens.

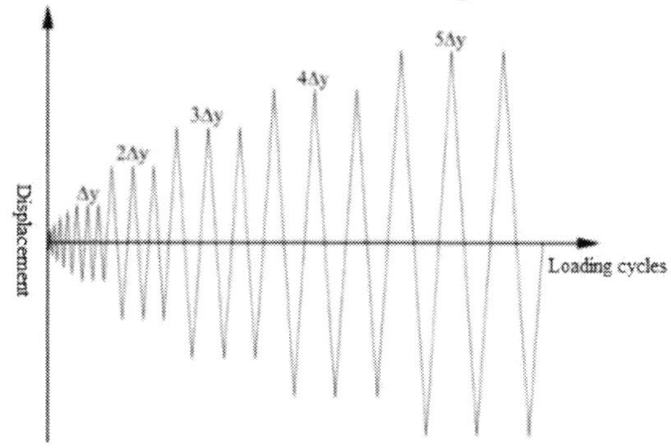

Figure 2. Load pattern.

4. The Experiment Results and Analysis

4.1. Summary of Test Results

According to the measured data in the test, the parameters of specimen seismic performance are summarized as shown in **Table 2**.

4.2. Description of Test Phenomenon

According to **Table 2**, the failure types of the test pier are mainly divided into flexure failure and flexure-shear failure. A1 and A2 piers are flexure failure, and A3 piers are flexure-shear failure. When the actuator is pushed forward, the direction is positive, and when pulled back, it is negative.

Table 2. Summary for the parameters of specimen seismic performance.

Model number	Damage direction	Equivalent yield displacement	Ultimate displacement	Ultimate shear strength (kN)	Displacement ductility ratio	Failure mode
A1	negative	7.21	70.18	139.01	9.74	Flexure
A2	positive	3.04	35.75	147.58	11.76	Flexure
A3	negative	5.46	24.60	382.33	4.51	Flexure-shear

4.2.1. Flexure Failure. At the initial stage of loading, transverse cracks of the pier with flexure failure appear at the bottom of the pier front (plane A and C) and extend to the two sides. With the continuous loading, the cracks at the bottom of the pier are constantly widened, other pre-existing cracks develop slightly, and fewer new cracks are generated. In the process of further loading, transverse cracks at the bottom of the pier continue to widen, and other cracks widen slightly. The concrete at the bottom of the pier begins to spall gradually, and part of the reinforcement is bare. In the final stage of loading, the cracks at the bottom of the pier are rapidly developed, the steel bars are cracked, the bearing capacity is obviously decreased, and the specimens are destroyed. The destruction features of this kind of specimen are: there are few cracks in the pier body, and the width is small. The crack at the bottom of the pier develops rapidly and the crack width is large, and the final failure is the longitudinal reinforcement fracture. The prototype pier is the failure mode. **Figure 3** and **Figure 5** show the ultimate failure cracks of bridge piers A1 and A2.

IOP Conf. Series: Materials Science and Engineering **611** (2019) 012046 doi:10.1088/1757-899X/611/1/012046

The hysteretic curves of flexure failure types are plump and show good ductility. The hysteretic curves and skeleton curves of A1 and A2 piers are shown in **Figure 4** and **Figure 6** respectively.

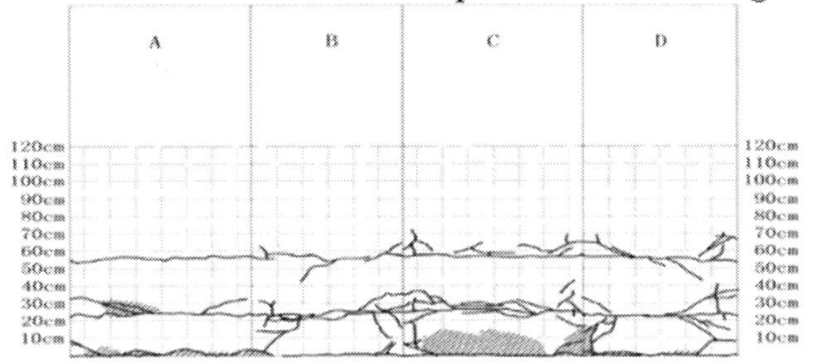

Figure 3. Cracks for A1.

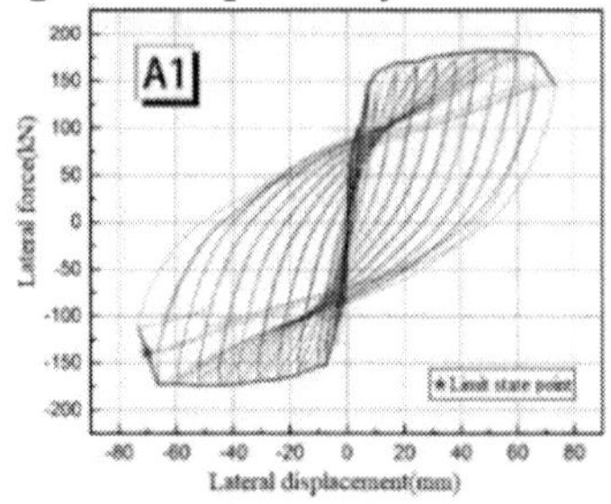

Figure 4. Skeleton curves for A1.

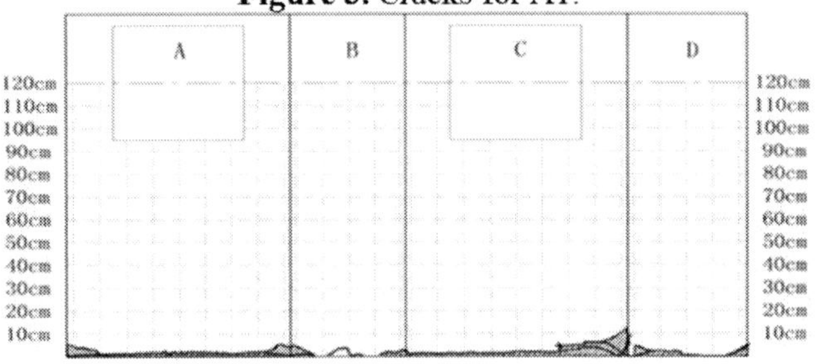

Figure 5. Cracks for A2.

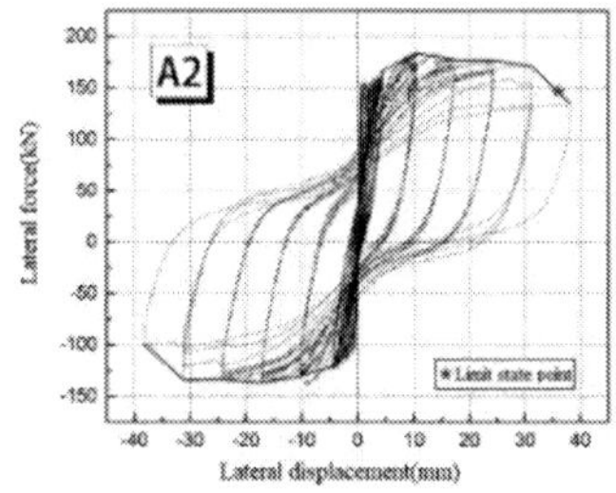

Figure 6. Skeleton curves for A2.

4.2.2. Flexure-shear Failure. At the initial loading stage of the flexure-shear failure type model, transverse cracks were generated in the front of the pier (plane A and C), then the cracks continued to develop and extend to the side (plane B and D) of the pier. With the continued loading, there are only a few penetrating transverse cracks on the front of the pier, and the inclined cracks on the side of the pier gradually increase and widen. In the later stage of loading, the number of transverse cracks in the front of the pier body tends to be stable, but the crack width continues to increase and the concrete gradually bulges. The side of the pier body forms-shaped main oblique crack with an obvious "X", and the concrete has an outer drum, peeling off, the main steel bar buckling, the hoop outer drum, the bearing capacity of the pier is greatly reduced, and the specimen is destroyed. The destruction features of this kind of model are as follows: a large number of oblique cracks appear on the side when the transverse crack appears on the front of the pier body, and, the concrete in the plastic hinge area is crushed and peeled with hoops are bulging outside when the failure occurs. **Figure 7** shows the ultimate failure cracks of bridge piers A3. The hysteretic curve of the flexure-shear failure type is obviously pinched, with poor ductility and a fusiform shape. The hysteretic curves and skeleton curves of A3 pier are shown in **Figure 8**.

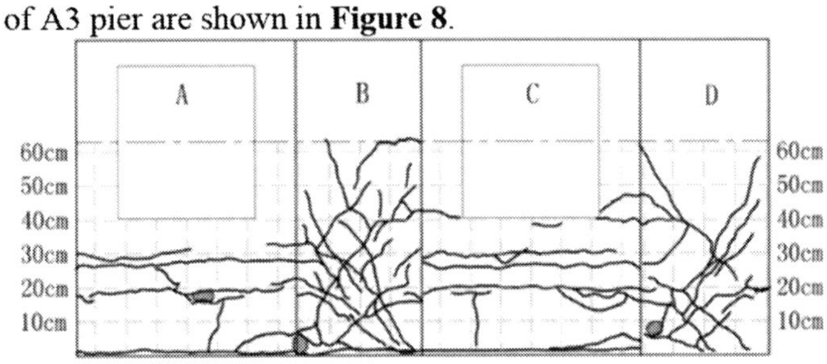

Figure 7. Cracks for A3.

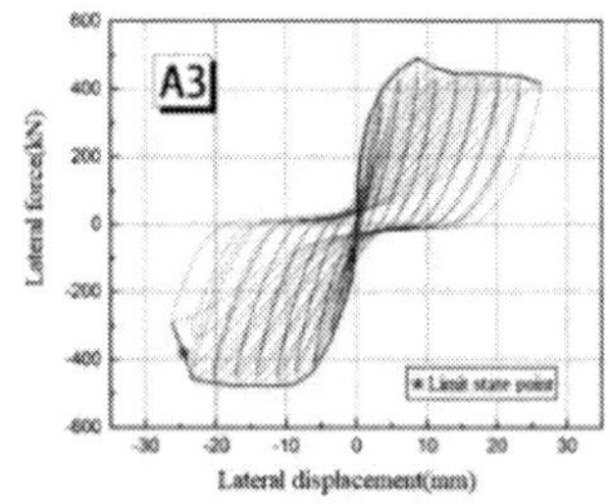

Figure 8. Skeleton curves for A3.

4.3. Analysis of Shear Failure Mechanism

In this paper, the shear failure mechanism is obtained by analysing the shear behaviour of hoop and concrete. In the process of cyclic loading, there are two main roles of stirrup. One is to bear the shear force and improve the shear behaviour of the structure. The other is formed reinforcement cage with the longitudinal reinforcement. It can not only effectively prevent the buckling of longitudinal reinforcement, but also effectively restrain the core concrete, so as to improve the mechanical behaviour of concrete.

Figure 9 and **Figure 10** show the change law of stirrup strain with loading grade for B and D of specimen A3 (typical shear failure) under forward and reverse loading respectively. According to the data in the figure and the failure phenomenon analysis of shear failure, the strain value of stirrup is relatively small or even negative before the longitudinal reinforcement yield. In other words, stirrups provide less shear capacity, or even coordinate with concrete compression. Now the shear capacity of piers is mainly provided by concrete. When the longitudinal reinforcement yields, with the displacement increasing continuously, the oblique crack develops rapidly, and the strain value of the stirrup which through the oblique crack increases rapidly and reaches the yield. While the restraint effect of core concrete decreases rapidly due to the sudden increase of stirrup strain, which leads to brittle shear failure of the specimen.

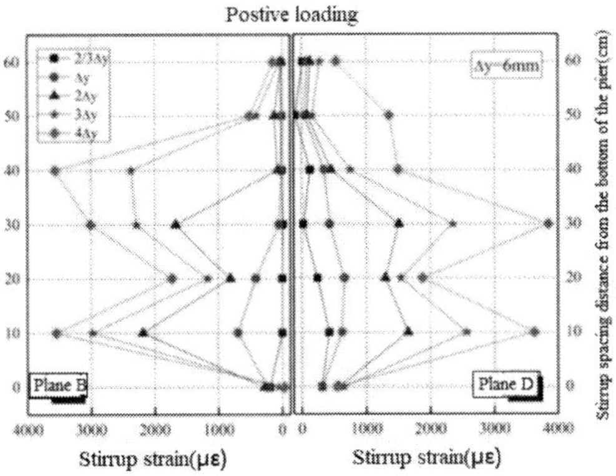

Figure 9. Stirrup steel tension strains for A3 (positive loading).

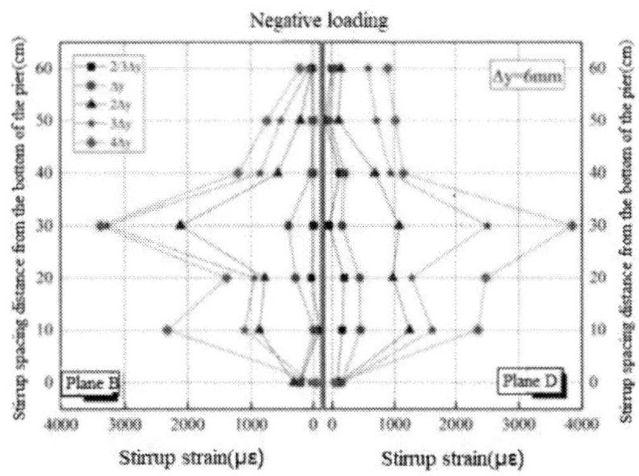

Figure 10. Stirrup steel tension strains for A3 (negative loading).

In the test, the total shear force on the pier is divided into two parts. One part is the shear bearing capacity provided by stirrup, which can be calculated from the data of strain gauge. The other part is the shear capacity of concrete, including the shear capacity provided by concrete, axial pressure and longitudinal reinforcement. The total shear force can be readout by the horizontal actuator, so the shear capacity of concrete can also be calculated. **Figure 11** shows the total shear of A3 and the change law of shear bearing capacity with displacement ductility ratio of each part. The same rule can be obtained for other specimens.

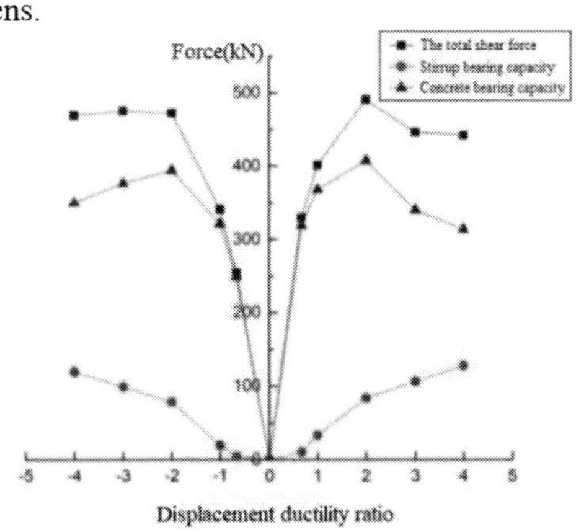

Figure 11. The variation of stirrups and concrete shear contribution for A3.

The epigraph vividly illustrates the change mechanism of shear capacity of pier specimens under repeated loading. Before the longitudinal bars yield, the inclined cracks are less developed, the stirrups hardly bear the shear force, or even coordinate with concrete compression, and the concrete bears almost the shear force. When the longitudinal bars yield, the cracks develop rapidly. When the inclined crack which intersecting with stirrup is formed, the shear contribution provided by stirrup increases gradually, while the area of compression zone of core concrete decreases gradually due to the generation and development of cracks. Repeated shear slip makes the aggregate occlusion and frictional resistance at the crack reduce continuously, and with the gradual crush of concrete, the bolt removal effect of longitudinal reinforcement also decreases. All these make the shear contribution of concrete decrease gradually. Once the stirrup yields, the restraint on the core concrete is greatly weakened, and the shear contribution of concrete begins to decrease obviously. Meanwhile, the stirrup has reached the shear bearing capacity, which can no longer make up for the lack of shear bearing capacity of concrete. As a result, the bearing capacity of specimens decreases rapidly, and the failure of specimens presents brittle feature.

5. Conclusion

Based on the research background of "The seismic comprehensive test of Chenglan railway pier", this paper described the pseudo-static test scheme of reinforced concrete pier in detail, and then the results of the experiment were analyzed and summarized.

(1) According to the analysis of the pseudo-static test results of the Chenglan railway prototype piers (A1 and A2), it was indicated that both piers were typical bending failure, with sufficient ductility reserve. When the pier body cracked, the crack at the bottom of the pier developed rapidly and the crack width was large, and the final failure was the longitudinal reinforcement fracture.

(2) According to the analysis of A3 specimen, it was indicated that pier was typical flexure-shear failure. A large number of oblique cracks appeared on the side when the transverse crack appeared on the front of the pier body, and the concrete in the plastic hinge area produced crush and peel as well as bulging outside of hoops when the failure occurred.

(3) The mechanism of shear failure under cycling loading is as follows: Before loading to yield displacement, shear strength was provided by stirrup little, which was mainly provided by concrete. After loading to yield displacement, the shear strength provided by stirrup increased gradually with the development of cracks. When the stirrup yields, its restraint to the core concrete was weaken rapidly, which resulted in the rapid destruction of the bridge pier.

References

[1] Fan, L. C. and W. D. Zhuo. *Bridge ductility seismic design*. Beijing: China Communications Press 2001.

[2] Ye, A. J. and Z. G. Guan. *Seismic design of bridges*. Beijing: China Communications Press 2011.

[3] TANABE, Tadaaki. "Comparative performances of seismic design codes for concrete structures." *Publication of Japanese Society of Civil Engineers* 1 (1999).

[4] Tang, G. W., J. Z. Li, and X. X. Tao. "Guidelines for seismic design of highway bridges." *Ministry of Transport of the People's Republic of China, Beijing* (2008).

[5] WANG, Dong-sheng, et al. "Experiment on shear strength of reinforced concrete bridge column in plastic hinge zone under seismic effect." *China Journal of Highway and Transport* 24.2 (2011): 34-41.

[6] Xu, Z. Z., S. Q. Liao, and Y. P. Ni. "Code for seismic design of railway engineering." *Ministry of railways of the People's Republic of China, Beijing* (2009).

[7] Li, G. Q. "Experimental study and numerical analysis on seismic performance of reinforced concrete bridge columns." *Chongqing Jiaotong University, Chongqing* (2010).

[8] Zheng, G. and G. Q. Li. "The effective rigidity of reinforced concrete bridge piers." *China Civil Engineering Journal* 46 6 (2013): 44-52.

[9] Pan, P. and Y. Y. Song. "Multi-dimensional pseudo-static loading experimental method and control." *China Civil Engineering Journal* 34 2 (2001): 26-32.

Analysis of Factors Affecting Performance of Cement-Bentonite Mud Impervious Wall

Jihui Ding[1], Yu Gan[1,*], Tianyu Zhang[1,2], Meng Yan[1] and Qi Zhao[2]

[1]College of Civil Engineering Hebei University, Baoding, Hebei, China
[2]China Academy of Building Research, Beijing, China

*Corresponding author

Abstract. The triaxial unconsolidated undrained test, unconfined compressive strength test and variable head penetration test were used to study the water-cement ratio (R_{wc} between 3.85 and 5.88) and water-bentonite ratio (R_{wb} between 3.85 and 7.14) in different indoor mix ratios. Effect on mechanical properties and permeability of cement-bentonite mud impervious wall. The test results showed that:(1)The unconfined compressive strength of cement-bentonite anti-seepage wall has a double exponential function relationship with water-cement ratio and water-bentonite ratio.(2)When the water-bentonite ratio is constant, the deformation modulus increases with the decrease of water-cement ratio. When the water-cement ratio is constant, the influence of water-bentonite ratio on deformation modulus is not obvious, but too much bentonite will reduce deformation modulus.(3)As the strength of the cut-off wall increases, the modulus of deformation also increases, and there is a good linear relationship between the two.(4)The permeability coefficient of cement-bentonite anti-seepage wall has a double exponential function relationship with water-cement ratio and water-bentonite ratio.

1. Introduction

Cement-bentonite mud is a kind of self-hardening mixed material synthesized from cement, bentonite, water and external admixture , with low compressive strength (not more than 5MPa) and low elastic modulus (not more than 1000MPa)[1]. It has the characteristics of large ultimate strain and high impermeability, and is often used in environmental protection projects. Considering that the cement has a stronger cementation effect through the hydration reaction, the formed solid body has higher strength and can achieve better deformation control. Therefore, the cement-bentonite mud anti-seepage wall has higher practicability. At present, the development of cement-bentonite slurry cutoff wall material in China is slow, and the stress state and anti-seepage effect of vertical cutoff wall used in environmental protection engineering are different from other walls. Therefore, studying the mechanical properties and anti-seepage performance of cement-bentonite mud anti-seepage wall is of great significance in the rational use and promotion.

Evans et al. found that the cement has a stronger cementation effect through the hydration reaction, and the formed solid body has higher strength and can achieve better deformation control [2]. Royal found that the cement-bentonite material did not stabilize its physical parameters after 28 days, but it took 60 days or more [3]. Grant found that the solidified body obtained by the agitated mud on site was low in strength and high in permeability [4]. Jiao et al. studied the influence of the ratio of bentonite to cement on the strength characteristics of plastic concrete under low confining pressure[5] . Deschenes conducted a mechanical test on cement-bentonite mud for 3 to 103 days and found that the strength of the mud increased significantly in the early stage of curing [6]. Huang et al found that the ultimate strain of cement bentonite materials cured for 28d decreased with the increase of cement content by testing the influence

of cement content on ultimate strain [7]. Dai et al passed the test to show that the cement-bentonite under certain ratio can meet the requirements of the impervious property of the anti-seepage slurry[8]. Jefferis obtained from the cement-bentonite consolidation test, which can significantly improve the impermeability of the cut-off wall by replacing the cement with a small amount of slag[9]. According to Khera, considered that the optimum content of bentonite should be controlled to about 18%[10]. Zhang measured the permeability of the soil-bentonite anti-seepage wall through the flexible wall penetration test, and found that the consolidation pressure increased from 30 kPa to 100 kPa, and the permeability coefficient would be 5.21×10^{-8} cm/s to 3.78×10^{-8} cm/s. The permeability coefficient of the rigid wall penetration test was measured to be 7.14×10^{-8} cm/s under the 10 kPa consolidation pressure [11].
The triaxial unconsolidated undrained test, the unconfined compressive strength test and the variable head penetration test are used to study the mechanical properties and permeability factors of the cement-bentonite mud cutoff wall under different ratios.

2. Test Materials and Methods

2.1. Test Materials
Based on the anti-seepage construction of a sanitary landfill in hebei, the permeability coefficient is required to be less than 1×10^{-7}cm/s. The cement used in this test is ordinary Portland cement, with a density of 3.08g/cm^3, a final setting time of 259min, a flexural strength of 8.9mpa and a compressive strength of 59MPa. Bentonite is an artificial sodium bentonite with a colloidal value of 200-500ml/15g and an expansion ratio of 6-8 times.
In order to study the influence of a small amount of soil mixed into the process of mud mixing and transportation, the soil in this area was selected as silty clay with a natural moisture content of 16.7%, a density of 1.87g/cm^3, a liquid limit of 22.2% and a plastic limit of 14.2%, clay particles of <0.005mm accounting for 16%, powder particles of 0.005-0.075mm accounting for 69.4%, and sand particles of >0.075mm accounting for 14.6%.

2.2. Test Methods
The mass ratio of water to cement is defined as $R_{wc} = m_w / m_c$; the mass ratio of water to expansive soil is defined as $R_{wb} = m_w / m_b$; the mass ratio of water to soil is defined as $R_{ws} = m_w / m_s$ .Where m_w is the mass of water, g; m_c is the mass of cement, g; m_b is the mass of bentonite, g; m_s is the mass of the soil, g.
The experiment was designed with 1000g water, bentonite: 140g-260g, cement: 170g-260g, and dispersant Na_2Co_3:2g. 16 mixture schemes were used. Two groups of three test blocks were made for each mixture scheme.R_{wc}=3.85~5.88, R_{wb}=3.85~7.14.

3. Test Result Analysis
Taking into account the strength of the cement-bentonite mud after the solidification of the sample is less than 1MPa, and the shear rate and other factors, the unconfined compressive strength test and the triaxial unconsolidated undrained test are all based on the TSZ automatic triaxial instrument. The head change test was carried out using a TST-55 permeameter.

3.1. Influence of R_{wc} and R_{wb} on Unconfined Compressive Strength of Samples
The relationship between unconfined compressive strength and R_{wc} and R_{wb} is fitted in a double exponential relationship:

$$f(q_u)=C_1\exp[(-R_{wb}/t_1)+(-R_{wc}/t_2)]+C_2\exp(-R_{wc}/t_2)+C_3\exp(-R_{wb}/t_1)+C_4 \text{ (kPa)} \tag{1}$$

The values of the fitting parameters are shown in Table 1. The square of the correlation coefficient is greater than 0.9, and the curve has a high degree of fit. The fitted cloud image is shown in Figure 1.

Table 1. Fitting parameter values

age	R-Square	C1	C2	C3	C4	t1	t2
28d	0.96998	15917.3	-714.35	-2706.74	177.505	4.49762	2.00746
60d	0.97745	-215.452	4.44844	6524.814	-53.3155	-2.47948	1.64702

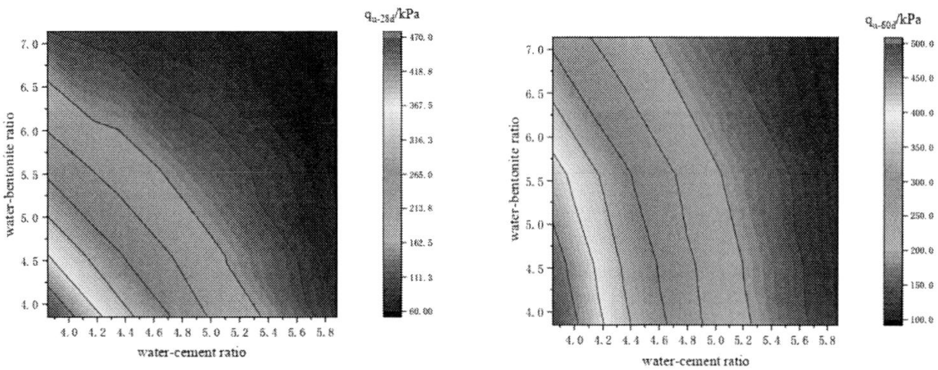

Figure 1. Fitting the relationship between q_u and R_{wc} and R_{wb}

Figure 1. is a cloud plot of unconfined compressive strength and R_{wc} and R_{wb} at different ages. As shown in Figure 1., the unconfined compressive strength decreases with increasing R_{wc} and R_{wb}, but with a different degree of reduction. The increase in R_{wc} has a significant effect on the reduction of strength and has a strong regularity.The increase of R_{wb} will reduce the strength, but the regularity is not strong. At 28 days, R_{wb} increases from 4.7 to 5.5, and the strength decreases from 400 kPa to 200 kPa. The regularity is strong.

Figures 2. and 3. are the fitted curves of R_{wc} and R_{wb} and unconfined compressive strength at different ages. It can be seen from Fig.2. that when the proportion of other raw materials is constant, the intensity values of all the curves are increased by the decrease of the R_{wc}. After fitting, it is found that when R_{wb} is constant, the R_{wc} is 3.85 to 7.14, and its corresponding unconfined compressive strength is a single exponential function, and the correlation coefficient is above 0.97.

As shown in Figure 3., the R_{wc} is 5.88, and the R_{wb} is reduced from 7.14 to 5.56. The unconfined compressive strength of the consolidated bodies 28d and 60d is increasing. The R_{wb} is 5.56. After that, the strength decreased significantly; when the R_{wc} was equal to 5.00, 4.35 and 3.85, the unconfined compressive strength of the R_{wb} decreased from 7.14 to 3.85, and it was found to be a single exponential function after fitting. And the correlation coefficient is greater than 0.93.

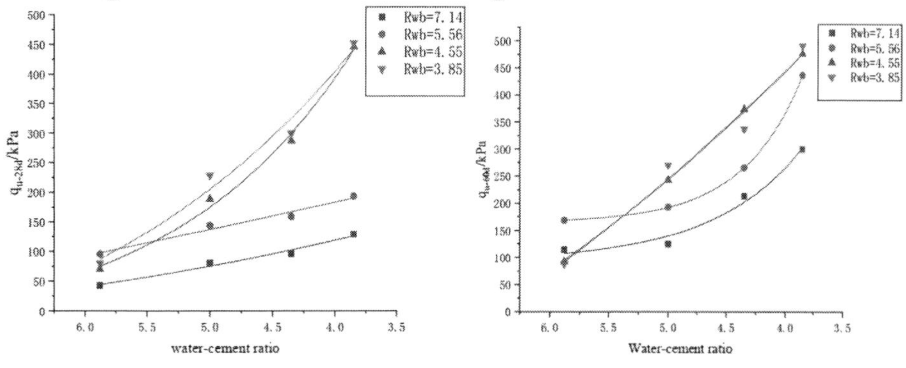

Figure 2. Fitting the relationship between q_u and R_{wc}

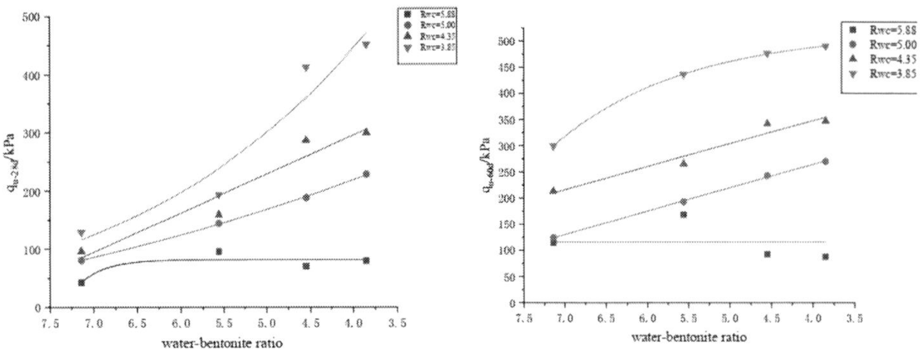

Figure 3. Fitting the relationship between q_u and R_{wb}

3.2. Influence of R_{wc} and R_{wb} on Deformation Characteristics of Specimens

Figure 4. shows the relationship between the deformation modulus and R_{wc} and R_{wb}. It can be seen from Figure 4. that as the R_{wc} decreases, the deformation modulus gradually increases. It can be seen that the cement plays a major role in the strength of the consolidated body. With the increase of the cement content, the plasticity of the solid body is unrecoverable, and the deformation modulus is continuously increased.When the age is 28 days and the R_{wb} is 7.14 and 5.56, the total increments of deformation modulus are 10.31 MPa and 2.94 MPa, respectively, which are relatively small. When the age is 60 days and the R_{wb} is 4.55 and 3.85, the increments of deformation modulus are relatively large, reaching 19.96 MPa and 23.62 MPa.When the amount of cement, dispersant and water is constant, the effect of R_{wb} on the modulus of deformation is considered. The deformation modulus of the two ages is not obvious with the R_{wb}.

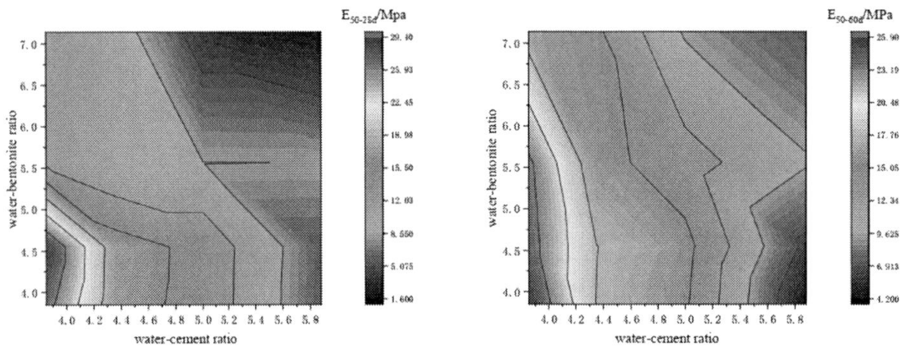

Figure 4. Relationship between deformation modulus and R_{wc} and R_{wb}

3.3. Relationship between Strength and Deformation Characteristics

As the strength of the cut-off wall increases, the modulus of deformation increases, and there is a good linear relationship between the two. The 28d and 60d ages are shown in the figure below. The correlation coefficient of 28d is equal to 0.9406, and the fitting straight line equation of $E_{50\text{-}28d}$ and $q_{u\text{-}28d}$ is as follows:

$$E_{50\text{-}28d} = (59.75 \pm 1.60) q_{u\text{-}28d}\ (\text{kPa}) \tag{2}$$

The correlation coefficient of 60d is equal to 0.9464, and the fitting linear equation of $E_{50\text{-}60d}$ and $q_{u\text{-}60d}$ is as follows:

$$E_{50\text{-}60d} = (47.55 \pm 2.92) q_{u\text{-}60d}\ (\text{kPa}) \tag{3}$$

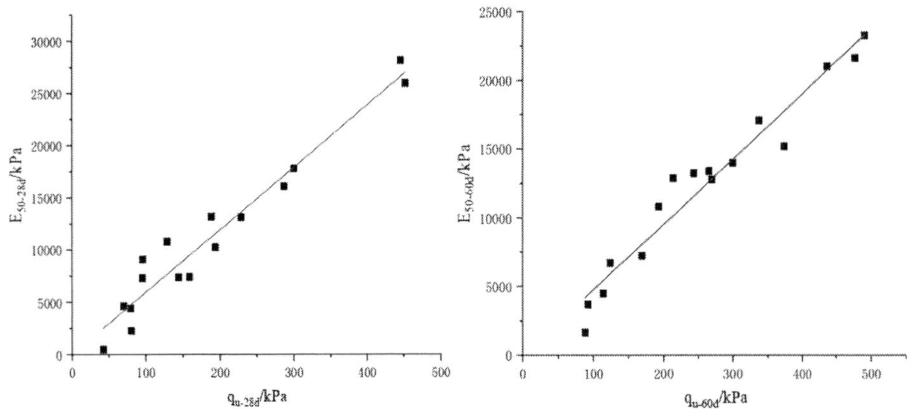

Figure 5. Curve of unconfined compressive strength and deformation modulus

3.4. Effect of R_{wc} and R_{wb} on the Permeability Coefficient of the Sample

The permeability coefficient of the cut-off wall is fitted with R_{wc} and R_{wb} according to a double exponent:

$$f(k)=C_1\exp[(-R_{wb}/t_1)+(-R_{wc}/t_2)]+C_2\exp(-R_{wc}/t_2)+C_3\exp(-R_{wb}/t_1)+C_4 \quad (\times 10^{-7}\text{cm/s}) \qquad (4)$$

The values of the fitting parameters are shown in Table 2. The correlation coefficients are all greater than 0.96, and the fitting degree of the curve is higher. Figure 6. shows the fitting relationship cloud image.

Table 2. Fitting parameter values

Test	R-Square	C1	C2	C3	C4	t1	t2
28d variable head test	0.96161	0.00321	0.87752	-0.00365	-0.96422	-3.34841	-0.94918
60d variable head test	0.96728	0.00646	0.16779	-0.00036	0.04159	-2.44943	-1.25803

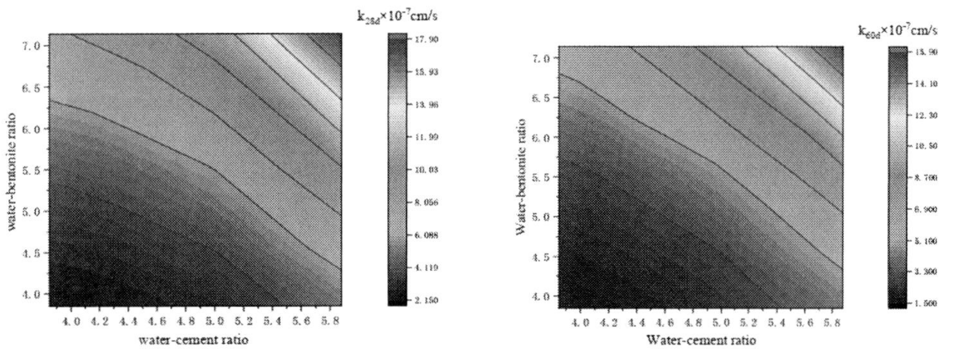

Figure 6. Relationship between the fitted permeability coefficient and R_{wc} and R_{wb}

It can be seen from Fig.6. that when the R_{wb} is between 3.85 and 7.14, the permeability coefficient increases gradually with the increase of the R_{wc}, but the increase range is different. When the cement ratio is between 3.85 and 5, the permeability coefficient increases as the R_{wb} increases. When the R_{wc} is between 5 and 5.88, the permeability coefficient decreases first and then increases with the increase of R_{wb}. When the R_{wc} is equal to 5.88 and the R_{wb} is equal to 7.14, the seepage coefficient of the cut-off wall is the largest, and the permeability coefficients of 28d and 60d are 1.82×10-6cm/s and 1.62×10-6cm/s, respectively.

4. Conclusion

(1) The unconfined compressive strength of cement-bentonite anti-seepage wall has a double

exponential function relationship with R_{wc} and R_{wb}. The R_{wb} is between 3.88 and 7.14, and the unconfined compressive strength of the cut-off wall is a single exponential function relationship with the R_{wc}. When cement ratio equals 5.0, 4.35 and 3.85, the relationship between the unconfined compressive strength of cut-off wall and R_{wb} is a single exponential function.
(2) When the ratio of water to bentonite is constant, the modulus of deformation increases with the decrease of R_{wc}. When the R_{wc} is constant, the effect of water-bentonite on the deformation modulus is not obvious. When the peaks of deformation modulus were reached in each group, the deformation modulus mostly decreased with the decrease of R_{wb}.
(3) The deformation modulus has a good linear relationship with the unconfined compressive strength, and the ratio between them is 45 to 60.
(4) The permeability coefficient of cement-bentonite anti-seepage wall has a double exponential function relationship with R_{wc} and R_{wb}. The permeability coefficient of the mud-solidified body increases as the R_{wc} increases. When the R_{wc} is relatively high, the permeability coefficient first decreases with the increase of the R_{wb}, and then the increase is larger. When the R_{wc} is relatively low, the permeability coefficient increases with the increase of the R_{wb}.

5. Acknowledgements
This work was supported by Post-graduate's Innovation Fund Project of Hebei University. The item number is hbu2019ss016.

References
[1] Evans J C, Dawson A R. 2014 Slurry walls for control of contaminant migration a comparison of UK and US practices: Geo-Engineering for Underground Facilities, [C]. 105-120.
[2] Evans J C, Dawson A R. 2014Slurry walls for control of contaminant migration a comparison of UK and US practices: Geo-Engineering for Underground Facilities, [C]. 105-120.
[3] Royal A C D,Makhover Y,Moshirian S,et al.2013Investigation of Cement-Bentonite slurry Samples Containing PFA in the UCS and Triaxial Apparatus[J].Geotechnical & Geological Engineering ,31(2):767-781.
[4] Grant W. Hal Jr, Rutiedge John M, Gardner Craig A. 1990 Quality of bentonite and its effect on cement-slurry performance. SPE Production Engineering,, 5(4): 411-414.
[5] Jiao Kai, Dang Faning, Xie Kaijun. 2016Effect of Bentonite and Cement Mixing Ratio on Strength Properties of Plastic Concrete[J]. Journal of Hydroelectric Engineering,, 35(3): 11-18.
[6] D Jean-Hugues, M Michel, Tournier, et al. 1995Testing of a cement-bentonite mix for a low-permeability plastic barrier. ASTM Special Technical Publication,, 1293: 252-270.
[7] Huang Liang, Xu Chao, Wu Fang. 2009Laboratory Experimental Study on Mechanical Properties of Cement-Bentonite Mud Consolidation[J]. 2018Investigation Science and Technology. (6): 3-8.
[8] Dai Guozhong, Sheng Yanmin, Li Shujin, Zhang Yaxing. Experimental Study on the Ploughability of PBFC Anti-seepage Slurry in Landfill Site[J]. Journal of Silicate Bulletin. (02)
[9] Jefferis S A. 1981 Bentonite-cement slurries for hydraulic cut-offs. Proc., 10th Int. Conf., International Society for Soil Mechanics and Foundation Engineering, 1: 425-440.
[10] Khera R P. 2014Calcium Bentonite, Cement, Slag, and Fly Ash as Slurry Wall Materials; Geoenvironment 2000@sCharacterization, Containment, Remediation, and Performance in Environmental Geotechnics, [C]. 1237-1249.
[11] Zhang Wenjie, Gu Chen. 2017Test of permeability and diffusion coefficient of soil-bentonite anti-seepage wall under low consolidation pressure[J]. Chinese Journal of Geotechnical Engineering, 10: 1915-1921.

Simulation Analysis on Pile-driving Process of Anti-flood Bionic Screw Pile

Fanfei Jia, Yanqin Tang, Kun Hu and Yuxing Wang*

School of Engineering, South China Agricultural University, Guangzhou, China

*Corresponding author

Abstract. In order to reduce the dynamic energy consumption of pile-driving of the anti-flood screw piles, the unsmoothed surface is applied to the screw pile tip to form an anti-flood bionic screw pile. In ANSYS, the FEM-SPH coupling algorithm is used to model and simulate the pile-driving process of bionic screw piles, and the simulation results of soil stress distribution, radial displacement of soil particles and pile-driving torque are obtained. The results of the pile-driving test show that the test torque is less than the simulated torque, and the soil pressure is consistent with the simulation results. The correctness of the simulation model is verified, which provides a theoretical basis for further improving the bionic screw pile and the pile-driving equipment.
Keywords: bionic; screw pile; pile-driving; simulation.

1. Introduction

The screw pile pile-driving is a new type of anti-flood pile-driving. Compared with traditional impact pile-driving, screw pile pile-driving has the advantages of the faster pile-driving and pile-up speed, low vibration and noise, and no secondary disaster etc.

The unsmoothed surface of the body surface of soil animals has the function resistance reduction [1], which has already formed a consensus in the academic circle. In order to reduce the pile-driving torque, the unsmoothed surface is applied to the tip of screw pile to form the anti-flood bionic screw pile. When the bionic screw pile is driving, it is very difficult to theoretically analyze the pile-soil effect. It is necessary to combine the numerical simulation with the field test to carry out the study of the pile-driving process.

Poulos studied the bearing capacity characteristics of pile foundation by finite element method and theoretical analysis [2]. Chen Jing used ABAQUS to construct pile-soil friction piles and analyzed the bearing capacity transmission of screw piles [3]. Zhang Mingyi proposed the displacement penetration method, and used ANSYS to simulate the pile-driving process of static pressure piles [4]. Wei Xieen optimized the pitch and width of the helical blade by ANSYS finite element software and designed the geometric parameters of the screw pile [5]. Pang Hongchen took measures such as reducing the diameter of the pile body and the length of the helical blade. The results show that the pile-driving resistance can be reduced [6], but at the cost of reducing the firmness of the pile.

The predecessors are all researches on the pile-driving process of ordinary screw piles, but there is no progress in the research on the pile-driving process and the pile-driving test of the bionic screw piles. This paper is to study the pile-driving process of the anti-flood bionic screw piles.

IOP Conf. Series: Materials Science and Engineering **611** (2019) 012048 doi:10.1088/1757-899X/611/1/012048

2. Numerical Simulation

2.1. Basic Assumption

The pile-soil model is a point-to-face contact method in which the pile is used as the main contact surface and the soil is used as the contact surface. During the pile-driving process of the bionic screw pile, the friction coefficient between pile-soil is constant [7].

The bionic screw pile is a rigid body without any strain and stress, and adopts the tetrahedral solid element. The soil is an isotropic and continuous homogeneous elastoplastic material, adopting the Mohr-Coulomb model [7].

2.2. Parameters of Bionic Screw pile and Soil

Bionic Screw Pile Parameters. The screw pile consists of a tapered pile tip, cylinders of different diameters, a trapezoidal convex, a round shaft and helical blades. According to the unsmoothed surface study, a certain size of the bionic surface has a function resistance reduction superior to the smooth surface [8]. The biggest difference between a bionic screw pile and a common screw pile is that the pile tip is uniformly bionic convexes. The bionic screw pile includes 1 meter pile and 1.5 meter pile. This paper refers to the 1 meter pile.

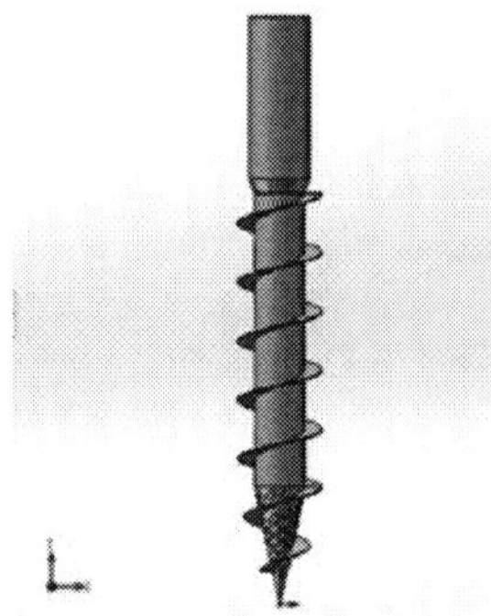

Figure 1. 3D model of bionic screw pile

Soil Parameters. The Mohr-Coulomb model is suitable for yielding under maximum shear stress, but the maximum shear stress depends only on the maximum principal stress and the minimum principal stress, while the second principal stress does not affect the yield; it is more suitable for loose or cemented granular materials [9] (Soil, rock and concrete). Therefore, the soil selects the Mohr-Coulomb model and sets the material parameters. The specific values are shown in Table 1 [10].

Table 1. Soil parameters

Parameter name	Selection
Density(g•cm^{-1})	1.85
Cohesion(KPa)	100
Poisson's ratio	0.3
Elastic modulus(MPa)	20
Unconfined compressive strength(KPa)	400
Shear modulus(MPa)	7.7
Bulk modulus(MPa)	16.7

2.3. Simulation Model Establishment

2.3.1. Finite Element Model of Bionic Screw Pile. The tetrahedral mesh is used to divide the bionic screw pile to improve the problems of mesh fracture and distortion, and reduce the possibility of penetration. The model has 6494 nodes and 20881 units. In order to solve the rust problem of common screw pile, the bionic screw pile is made of 304 stainless steel, and the mechanical parameters are shown in Table 2.

Table 2. Mechanical parameters

Parameter name	Material	Density	Elastic modulus	Poisson's ratio
Selection	304 stainless steel	7930Kg/m^3	193000MPa	0.31

2.3.2. Soil Finite Element Model. The sinking pile of the bionic screw pile belongs to the process of large deformation of the soil, and the SPH meshless algorithm is suitable for the deformation of the continuous structure such as rupture, dissolution and brittle fracture. In this paper, the SPH soil model is established, and particles are arranged uniformly and regularly. Each particle is isotropic, and the total number of particles is 83,300.

2.3.3. Pile-Soil Simulation Model Constraint Setting. In contact, the contact of the bionic screw pile with the soil is of the point-to-face type. Set the contact penetration factor of 0.1. For better contact astringency, set the time step factor for the simulation calculation to 0.31. Apply a 400N vertical uniform load to the pile end. Ignore the gravitational acceleration of the bionic screw pile and set the gravitational acceleration of the soil model to 9.8m/s2.
For the sport, the X and Z axes of the bionic screw pile are fully constrained. Constrains the degree of freedom of the peripheral particles of the SPH soil model. In order to shorten the simulation calculation time, the speed of the bionic screw pile is increased from the actual 2 rad/s to 8 rad/s, and the whole simulation time is 4 s.

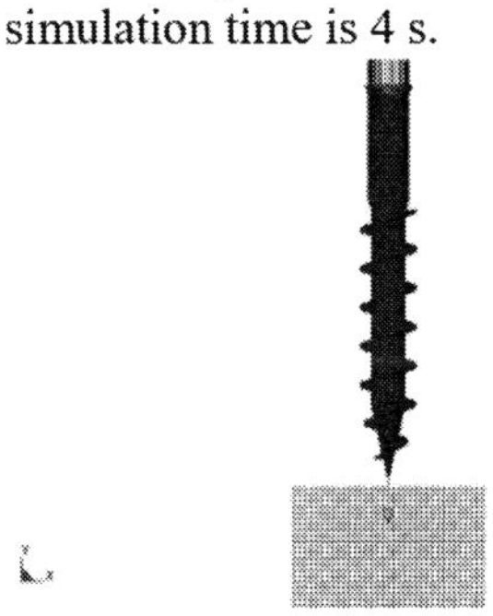

Figure 2. Pile-soil simulation model constraints

Figure 3. Partial magnification of the pile tip

3. Simulation Results

3.1. Soil Stress Distribution
Comparing the simulation results of the pile-driving processes of common screw piles and bionic screw piles, the soil stresses distribution in the two models are analyzed, as shown in Figure. 4.

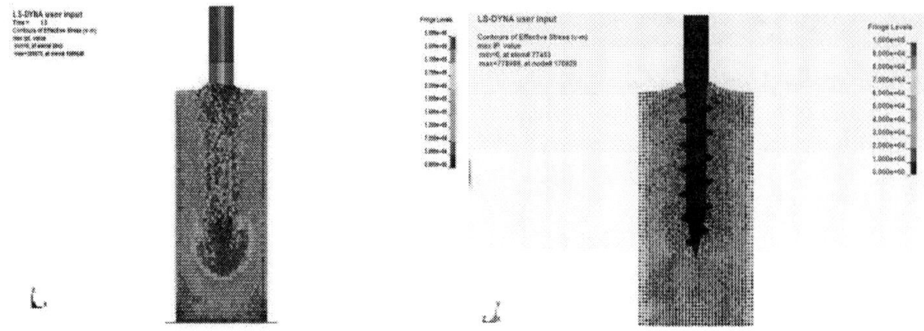

(a) Common screw pile (b) Bionic screw pile

Figure 4. Soil stress distribution

The surface soil stress of common screw piles is very large, and there is almost no change in soil stress from top to bottom. The surface soil stress value of the bionic screw pile is normal, and the soil stress increases gradually from the top to the bottom about the pile. The soil stress at the pile tip reaches the maximum and is hemispherical. The soil stress at the bionic pile tip is 100KPa, which is smaller than the soil stress at the common pile tip 400KPa, and the pile-driving resistance of the bionic screw pile is small. Considering the difference of structure and parameters between the two simulation models, the soil stress difference is large, which has certain guiding significance for the comparison of pile-driving resistance.

3.2. Radial Displacement of Soil Particles

Four groups of soil particles with a radial distance of 150 mm and a depth of 300 mm were selected, and the angle between adjacent particles was 90°. Figure 5 shows the radial displacement of soil particles.

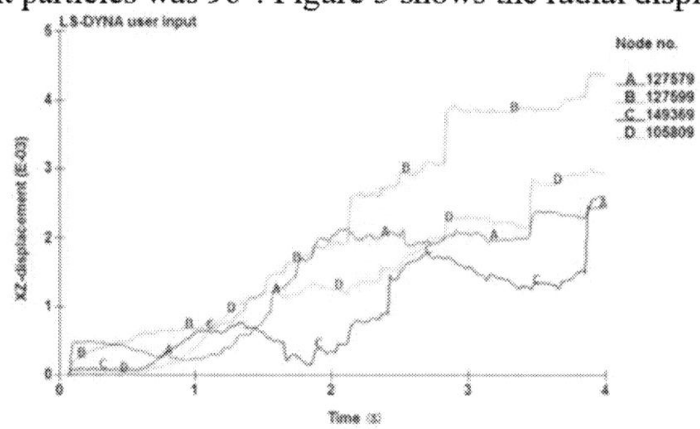

Figure 5. Radial displacement of soil particles

The soil particles are all squeezed during the pile-driving process of the bionic screw pile and forced to radially displace. The displacement change of A particle is large, but the total displacement is not large, indicating that it is subject to large disturbance. The displacement of B particle increases linearly, and the total displacement is relatively large, indicating that it is strongly squeezed. For C and D particles, the effect of extrusion is small. In summary, the radial displacement of B particles is the largest, and it is easier to detect the stress change of soil, which provides a theoretical basis for the pile-driving test of bionic screw piles.

4. Pile-driving Test

The bionic screw pile is made of 304 stainless steel, as shown in Figure 6. The test soil is a naturally deposited laterite. Select the four corners and the center with a side length of 1.7m square as the pile-driving point. The upper end of the pile is equipped with strain gauges for measuring the strain of the bionic screw pile, and measuring with the soil pressure sensor at the position of the B particles, and conducting the pile-driving test, as shown in Figure7.

Figure 6. Bionic screw pile **Figure 7.** Pile-driving test

The test data was taken into the equations of the torque calibration and the soil pressure to obtain the average value. The pile-driving torque curve of the bionic screw pile was obtained, as shown in Figure8; the pile-driving torque curve of the simulation model is shown in Figure9.

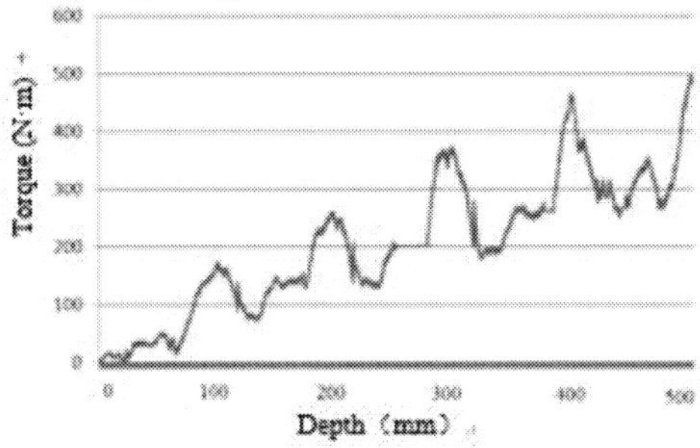

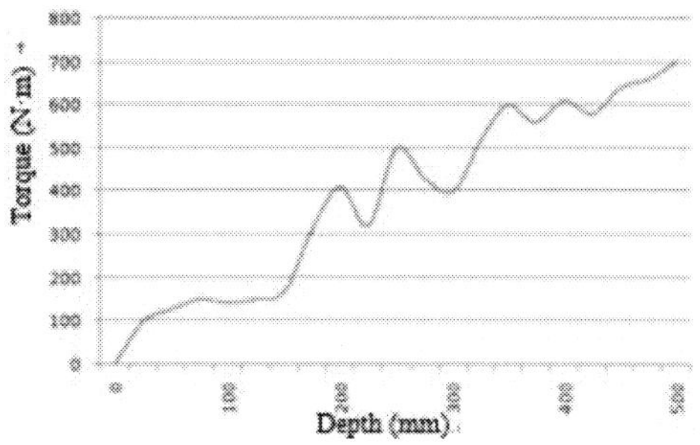

Figure 8. Test torque **Figure 9.** Simulation torque

During the pile-driving process, the overall test torque increases, accompanied by the large fluctuation of the torque. It indicates that the bionic pile tip pierces the soil, and the helical blade shears the soil, then the soil has large deformation and weakens the effect of adhering the pile, and the pile-driving torque reduces. At the same time, the helical blade sinks and has a pressing effect on the soil. The soil is gathered and compacted, and has a reaction force on the helical blade. It and the bionic screw pile generate the large friction, then the pile-driving torque increases.

Both the test torque and the simulation torque are increased, and the overall variation is not much different, and the test maximum torque is 500 N•M less than the simulation maximum torque of 700 N•M. The pile-driving test cannot be consistent with the simulation, resulting in errors, and the correctness of the simulation model is verified to some extent.

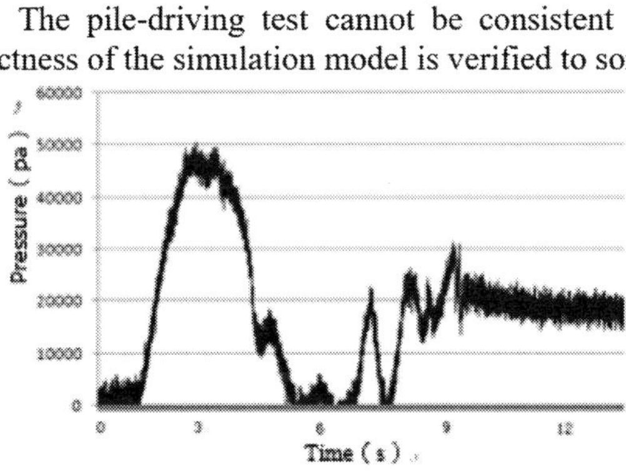

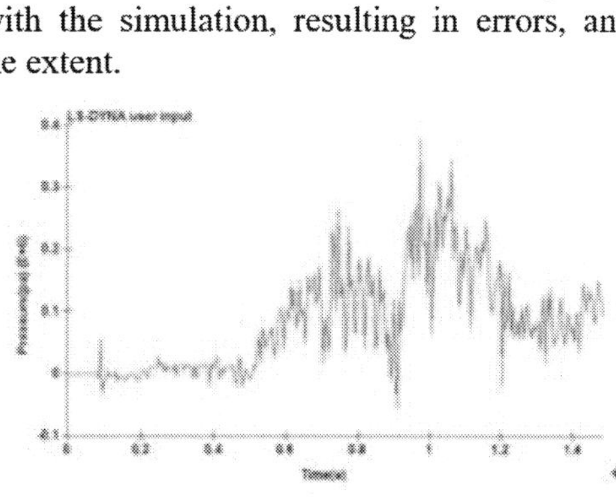

Figure 10. Soil pressure test curve **Figure 11.** B particle simulation soil pressure curve

From the soil pressure test, the maximum value of 50KPa is smaller than the soil stress of the simulation model. The embedding of the soil pressure sensor causes extrusion hardening of the soil, and the sensitivity of the sensor also has the error. It is impossible to analyze the soil pressure through accurate values.

Due to the embedding of the sensor, the initial pressure of the soil is so small. Between 1s and 9s, the pressure of the soil fluctuates sharply. When the bionic pile tip sinks and squeezes the soil, the soil pressure increases linearly, then the pressure drops and fluctuates slightly, showing the state of tension timely. When the bionic pile tip passes here, the soil pressure is basically stable. The test soil pressure and the B particle simulation soil pressure have similar changes, which realizing the same process of bionic screw pile-driving, and also verifies the correctness of the simulation model to some extent.

5. Conclusion

(1) SPH-FEM coupling simulation can better simulate the deformation of soil due to pile extrusion and helical blade shearing.

(2) During the pile-driving process, the bionic pile tip penetrates and destroys the soil, and the large deformation of soil weakens contaction with the pile tip, then the pile-driving resistance reduces. The helical blade sinks to gather and compact the soil, resulting in huge friction, and pile-driving resistance increases. The bionic pile tip crushes and destroys the soil, and the soil pressure fluctuates sharply, showing the state in which the soil is pressed and pulled. After the bionic pile tip passes through the soil, the soil pressure drops but remains at a certain value.

(3) Through the pile-driving test, the change of torque and soil pressure can show the same trend with the simulation results, and the correctness of the simulation model is verified.

Acknowledgments

Subject was supported by Guangdong water conservancy science and technology innovation project (NO.2017-31).

References

[1] Ren Luquan, Cong Wei, Chen Bingcong, et al. Study on the anti-stick properties of geometric non-smooth typical organisms [J]. Transactions of the Chinese Society of Agricultural Machinery, 1992, (02): 29-35.

[2] POULOS H G. Analysis of the settlement of pile groups [J]. Geotechnique, 1968, 18 (4) :449-471.

[3] Chen Jing. Numerical simulation of pile-soil interaction based on ABAQUS [J]. Henan Science, 2009, 27(8): 974-976.

[4] Zhang Mingyi, Deng Anfu, Gan Tengjun. Displacement penetration method for numerical simulation of static piles [J]. Rock and Soil Mechanics, 2003, 24 (1): 113-117.

[5] Wei Xieen, Wang Yuxing, et al. Simulation of anti-flood spiral pile driving process based on ANSYS/LS-DYNA [C]. 2010 World Automation Congress, 2010:239~243.

[6] Pang Hongchen, Wang Yuxing, Tang Yanqin, et al. Comparative analysis of SPH-FEM coupling simulation of anti-mite and rescue screw piles [J]. Chinese Agricultural Mechanization, 2015, 36(04): 271-274.

[7] Luo Zhanyou, Gong Xiaonan, Wang Jianliang, et al. Numerical simulation of squeezing effect of static pressure piles and analysis of influencing factors [J]. Journal of Zhejiang University (Engineering Science), 2005, 39(7): 92-96.

[8] Dian Wu Zhang,Wang Yu Xing,Tang Yan Qin, et al.Experimental Study on Bionic Surface Based on the Compound Eye of Ant[C]. Applied Mechanics and Materials.2010, 26-28:579-584.

[9] Qiao Hongjun, Peng Liyun, Yang Chengcai, Wang Jinyong. Static Load Test and Numerical Simulation of Blade Screw Pile [J]. Journal of Beijing University of Civil Engineering and Architecture, 2017, 33(04): 12-17.

[10] Cui Kerui. Practical Manual of Geotechnical Engineering, Volume III [M]. Beijing: Chemical Industry Press, 2007.

Structural Optimization of Flood Control Screw Piles: A Review

Kun Hu[1], Yanqin Tang[2], Fanfei Jia[3] and Yuxing Wang[4,*]

[1,2,3,4]School of Engineering, South China Agricultural University, Guangzhou, China

*Corresponding author

Abstract. This paper discusses the importance of research on flood control piling equipment, and lists the existing domestic flood control piling equipment. The research status of screw piles at home and abroad in recent ten years is summarized from three aspects: bearing capacity, installation torque, stress and strain of soil. This paper introduces some optimization methods closely related to screw piles, such as evolutionary algorithm, neural network, numerical simulation optimization method, and makes a comparative analysis of the optimization methods that can be adopted for the screw pile.at last Several directions which need to be further studied in the future are put forward.
Keywords: screw piles; optimization algorithm; numerical simulation; pile driving; flood control.

1. Introduction

There are many rivers and lakes in China, and the total amount of water conservancy projects built on that is huge. Some shortcomings in design, construction, maintenance of water conservancy projects will lead to various kinds of dangers in flood season. Common dangers include overflow, wave washout, bank crack and collapse. [1] piles and its supporting pile drivers are essential for flood control emergency, which can be mainly used for embankment reinforcement, curtain hanging to plug loophole, barrier setting in fissure, ship and cable hanging operation and so on.

At present, there are mainly the following methods for flood control piles: manpower hammer piles, mechanical hammer piles and mechanized sink screw metal piles. However, the manpower hammering piles need high labor intensity, but pile sinking speed still slow and waste serious of human resources; mechanical hammering piles are difficult to display in flood disaster areas, and it is easy to cause secondary disasters such as cracks and collapses on the dam. [2]

The research results of the existing flood control pile machines mainly include: ZJ100A type flood control pile driver [3] adopts the principle of impact pile sinking, which has the advantages of flexible maneuverability and convenient operation, but the hitting noise is very large. DZF-120 portable flood control pile driver [4]sink piles by periodic impact due to vibration. The main engine and power device are separated from each other, and are connected by a transmission flexible shaft, which makes the operation of pile driver more complicated. The anti-mite and rescue screw pile [5] has the advantages of high efficiency, low vibration and noise, no secondary hazard and corrosion resistance, and is a new direction worth investing energy and time.

Based on the structural optimization of flood control screw piles, this paper summarizes the research status related to screw piles, and combines various optimization design methods to explore the feasible design scheme of screw pile structure, so that the screw piles can better meet the actual working conditions of flood control and rescue.

2. Research on Screw Piles at Home and Abroad

The researches on flood control piles and their associated machines at home and abroad are not too much, but domestic and foreign scholars have already had a lot of understanding of the characteristics of screw piles in pile foundation engineering, including the installation torque, bearing capacity and group pile effect of screw piles.[6] Therefore, we can refer the study of screw piles used in pile foundation engineering to solve the problem of structural optimization of flood control screw piles. For the study of screw piles, different scholars tend to apply different research methods, which can be roughly divided into theoretical analysis, numerical simulation, indoor model test and field test research. In the following, the research of screw piles by previous scholars will be summarized from three aspects: bearing capacity, installation torque and stress and strain on soil.

2.1. Research on Bearing Capacity of Screw Piles

The role of the screw pile is generally subjected to compression or tension. Therefore, the bearing capacity must be considered in the design process of the screw pile. A large number of experts and scholars have analyzed the screw pile from the perspective of bearing capacity.

The systematic research of screw steel piles abroad began in 1950s, and the research of bearing capacity of screw steel piles have made great progress in the past ten years. M. Hesham El Naggar et al.[7-8] began to study the monotonic and cyclic characteristics of screw pile foundation system in 2007, developed a new type of screw pile foundation system suitable for seismic transformation of existing foundation and new structure, and conducted some static and dynamic load tests about it. Mohammed Sakr et al [9-10] carried out full-scale pile load tests using single pile and double screw piles, and studied the performance of single pile and double screw pile under axial compression, tension and lateral loads. Amy B. Cerato et al. [11] compared the uplift prediction methods used in screw pile design as well as discussed the effects of long-term dynamic loads and water level fluctuations on the performance of screw piles used as the foundation of small wind tower stay cables. Hamed Niroumand et al. [12] carried out laboratory tests of small-scale models with sand paving technique and measured uplift load and uplift displacement. Yong-Seone KIM et al. [13] compared the laboratory model tests of pre-drilled screw piles and smooth surface piles with different diameters, pitches and screws, and then combined with the finite element method to find appropriate parameters of screw pile mainly based on ultimate bearing capacity. Adnan Anwar Malik et al. [14] carried out laboratory model tests on pile loads and straight piles, and tested the displacement of piles and the stress changes of soils. The results show that screw piles in similar pile shaft diameter showed higher end bearing capacity than straight piles. Lisa n. Wheeler et al. [15] proposed the use of screw pile to reduce the displacement of railway track and play a role in strengthening the foundation of railway track, at the end conducted experiments for nearly a year, showing that the effect of screw pile on foundation reinforcement is obvious.

Although the research on domestic screw piles started later, due to the large number of applications of pile foundation engineering in China, domestic scholars have already accumulated a lot of research on the bearing capacity of screw piles. In recent ten years, the research direction of domestic scholars tends to be more and more diversified, and there are many researches on specific engineering practical problems. Dong Tianwen et al. [16] systematically conducted theoretical and experimental research on screw steel pile. Through the theoretical analysis, the failure mode of the screw pile foundation and the calculation method of the bearing capacity of the screw steel pile are proposed, and the influence of several geometric parameters of the pile on the bearing capacity of the pile is analyzed by model test and field test. Wang Zhao et al. [17] summarized the application of screw steel pile manufacture of screw pile, conducted structural design, and carried out anchoring and field drawing tests on anchor rods with different soil penetration depth, overlying soil thickness, standing time and whether grouting. Liu Jiankun et al. [18] investigated the photovoltaic support screw pile, through the tension and compression test, the pile type parameters (blade spacing, number of blade paths, blade diameter, pile length, etc.) of the screw steel pile and its ultimate pullout resistance The relationship between the compressive bearing capacity is studied, and the recommended values of the pile parameters are given combined with test results. Zhou Yang [19] compared the bearing characteristics of variable-section threaded piles with ordinary threaded straight piles and ordinary cylindrical smooth piles by laboratory model test and finite element method, and analyzed the influence of pitch, varied-section and thickness of screw blade on the bearing capacity of threaded pile.

2.2. Research on Screw Piles about Installation Torque
The flood control screw pile requires that the pile body should reduce the torque as much as possible on the basis of meeting the bearing capacity requirements, so that the supporting pile-sinking equipment can be more simple and efficient, and easy to carry. Since the installation and settlement of the screw pile in the pile foundation engineering can be equipped with sufficient power, there is generally no need to worry that the pile cannot provide sufficient torque and pressure, so there is relatively little research on the pure installation torque of the screw pile. However, many scholars have found that there is a certain relationship between the installation torque and the bearing capacity after the pile is installed, so that a lot of researches have been carried out on that.
El Naggar et al. [20-21] tested the installation torque and bearing capacity of screw piles, compared and obtained the torque bearing relationship applicable to large-diameter screw piles, and conducted finite element analysis to study the ultimate load criterion and load transfer mechanism of screw piles. Nelson Aoki [22] theoretically analyzed the relationship between the installation torque and uplift capacity of deep screw pile in sand layer, verified the correctness of this expression through centrifugal and direct shear interface tests, and finally compared the torque correlation factor calculated in different literatures. Mohammed Sakr [23] also presented a theoretical model developed to estimate the torsional resistance of cohesionless soils, and verified the theoretical torque model using some installation records. In order to study axial capacity and end installation torque of piles in cohesive soils, Zhengyang Guo et al. [24] conducted field tests on full-size screw piles and established a theoretical model of end installation torque. Umair Khan and Sumi Siddiqua [25] continue to study the relationship between the installation torque and bearing capacity of screw piles, focusing their research on the settlement of ultimate load and laying a foundation for the wide application of screw piles.

2.3. Analysis of Soil Stress and Strain by Screw Piles and Pile Group
The The dams need flood control and rescue operations are generally fragile, and it is very likely to cause secondary damages to the dam during pile installation. Therefore, it is necessary to consider the influence of the soil structure during the process of sinking piles into the soil. Many studies have been done on soil stress and strain analysis at home and abroad, but most of the tests are complicated and the equipment is expensive. CC HIRD et al. [26] carried out a model test of screw piles in beds of transparent synthetic soil simulating clay. Reflected particles were mixed into the clay, and the soil displacement could be recorded under laser irradiation, at the end vertical axis of the screw axis was measured by particle image velocimetry. This method can reveal the real mechanism of soil deformation around the screw pile by comparing and analyzing the deformation field under the influence of different pullout forces. Hamed Niroumand et al. [12] analyzed the measurement of drawing load and vertical displacement, and studied the failure mechanism of soil.
In the rescue process, the flood-prevention screw pile generally needs a lot of piles to sink into a certain area at the same time, so group pile analysis is also very necessary. There are many researches on pile group theory in pile foundation engineering: Dong Tianwen and Zheng Yingren [28] Based on a lot of theories established a bearing capacity function of pile group foundation, and the concept of ultimate load criterion and safety reserve factor of group pile foundation is proposed. C.C. Mendoza et al. [29] extended the knowledge of shear mechanism, stress generated, displacement field and the contribution of supporting raft to the overall system performance by combining finite element analysis with field test. Bushra s. Albusoda et al. [30] studied 36 kinds of screw pile group models embedded in expansive soil, which was formed by four piles arranged in a square way.

2.4. Research on Soil Replacement Screw Piles
Although the soli displacement screw piles are only used to squeeze soil and drilling, ultimately what created bearing force is the concrete which have been poured in the hole after drilled, the interaction between soil and drill bit during drill the hole has a high reference significance for the pile installation modeling of flood control screw piles.

T. Pucker [27] conducted numerical simulation of screw pile installation process based on Coupled Eulerian–Lagrangian approach, analyzed the influence of different ratios between the rotational and the penetration velocity , and verified the numerical model with experimental data of others. A. Krasin[31] conducted experiments and numerical simulation on the interaction between screw pile and non-cohesive foundation in the process of compression load transfer. Deng Yibing et al. [32-34] conducted discrete element analysis based on PFC software on the pile installation process of soil displacement screw piles, and compared it with indoor model test and field test to verify the effectiveness of numerical simulation. Different parameters of screw piles are taken for finite element analysis and indoor model test analysis, which provided a basis for optimal design of screw pile's drill bit. The influence of rotation speed ratio on the pore formation characteristics of loose and dense soil was analyzed by means of macroscopic model test and microscopic digital image analysis.

3. Research on Optimization Method of Screw Pile Structure

The structure of screw pile is mainly composed of central cylindrical pile and screw blades, so the structural optimization of screw pile is based on the interaction between cylindrical pile and soil and the parameter optimization of screw blade. The optimization methods commonly used in mechanical design include tabu-search algorithm, simulated annealing algorithm, genetic algorithm, ant colony optimization algorithm, artificial neural network, Lagrange relaxation algorithm and so on. In the published literature, four methods are mainly used for pile optimization and screw optimization: fmincon function optimization method, evolutionary algorithm, neural network, numerical simulation combined with experimental.

3.1. Fmincon Function Optimization Method

Fmincon function is widely used in various fields because it can directly call the command of the mathematical calculation software MATLAB after building the model, which is convenient and simple to use.

Ma Aili [35] and Xue Zixuan [36] respectively constructed the mathematical model of screw cutter optimal design for vertical rotary-cultivated ditcher. Zhang Chengxin et al. [37] established the mass objective function of the helix and calculated the optimal solution of the mathematical model by using the fmincon function. Wu Yinliang et al. [38] used fmincon function to optimize the plane layout, spacing, anchoring depth and section shape of the pile, which has many references for the layout of the flood control screw piles.

3.2. Evolutionary Algorithm

Evolutionary algorithms cover a wide range, including genetic algorithm, ant algorithm [39], particle swarm optimization, harmony search algorithm [40] and so on. At home and abroad, there are many examples of applying evolutionary algorithms to the optimization of screws.

Cristina Teixeira [41] adopted multi-objective ant colony optimization algorithm (MOACO) to deal with the twin-screw configuration problem, and compared the performance with the existing multi-objective evolutionary algorithm and two-stage local search algorithm. Ching-chi Hsu et al. [42] adopted genetic algorithm for multi-objective optimization in the investigation of bone screws to study the optimal design of bending strength and bone holding power. Chian-her Lee et al. [43] established a three-dimensional numerical model of locking compression plate femoral fracture in order to find the optimal number and position of locking compression plate screws, and used particle swarm optimization algorithm based on simulation to determine the position and number of screws. Kyunguk Na et al. [44] on the basis of predicting the bearing capacity of screw piles, optimized the components of screw piles and adopted harmony search algorithm to minimize the material cost of screw piles. Li Xiaohuo and other scholars used evolutionary algorithms to optimize the structural parameters of the shearer auger drill and drum, the optimization variables were the drill pipe inner diameter, blade screw angle, blade height, blade thickness, blade pitch and motor speed: hybrid algorithm based on genetic algorithm and ant algorithm (GAAA) on the coal productivity and conveying efficiency optimization [45];Evolutionary algorithm based on strength pareto evolutionary algorithm (SPEA) optimizes transportation productivity, machine quality and energy consumption [46].The improved genetic algorithm was used to carry out multi-objective optimization design for the conveying power, secondary crushing rate, coal conveying quantity and drill pipe quality of the screw coal drilling and mining machine [47]. Qin Dagong [48] used particle swarm optimization algorithm and annealing algorithm to optimize the cutting performance of shearer drum.

With the increasing maturity of the evolutionary algorithm to solve multi-objective optimization problems [49], the evolutionary algorithm plays an increasingly important role in the screw optimization problem. Therefore, the optimization problem of flood control screw piles needs further study of evolutionary algorithm in the future. The optimization objectives of flood control and emergency spiral pile include the quality of spiral pile, driving torque and bearing capacity. For the problem of four optimization targets, there are many classic second-generation multi-objective optimization algorithms, such as SPEA-II, PAES-II,NSGA-II, which have good optimization effect. These algorithms have many successful cases in engineering optimization, among which NSGA-II is the most widely used.

3.3. Artificial Neural Network

The neural network is mainly used to predict the bearing capacity of pile, and can also be used for inverse prediction of soil parameters [50], multi-objective optimization [51] and other problems, so it has a wide range of applications. Neural networks are suitable for nonlinear systems and can be used for systems that are difficult to be described by mathematical expressions, but the neural network needs a large amount of data to perform the operation.

Irshad Ahmad et al. [52] established six artificial neural network models for predicting various response parameters of soil pile interaction. F. PooyaNejad [53] established a neural network model to predict pile settlement according to standard penetration test (SPT) data. Mansour Mosallanezhad [54] used competitive algorithm (ICA) to train, test and optimize the artificial neural network, and the ICA-ANN model established has a good prediction effect on the ultimate bearing capacity of screw piles. Hossein Moayedi [55] used classical support vector machine and its two variants, namely, normalized generalized proximal support vector machine and double support vector machine, adaptive neural fuzzy inference system and genetic programming, to model and predict the friction resistance of clay foundation piles. H. I. PARK et al. [56] introduced the application of ANN in dynamic load test of pile foundation, predicted pile tip resistance and pile body resistance, and measured these values accordingly. FatehniaMilad [57] established a model to predict the bearing capacity by using the data of the last meter hammer strike times of 100 driven piles, the basic properties of surrounding soil, pile geometry and friction Angle of pile and soil, etc., combined with the artificial neural network. Then, using genetic algorithm program and linear regression method, the influence equations of plate number, soil parameters and pile geometry on pile bearing capacity are established. Xu Lina [50] used the combination of finite element and experiment to analyze the bearing characteristics of large-diameter pile foundation, and applied the BP neural network to back-analyze soil parameters.

3.4. Numerical Simulation Combined with Experimental

When using numerical simulation to optimize parameters, relevant verification tests are usually designed to verify the accuracy of numerical simulation optimization solutions.

Ma Aili [35] and Xue Zixuan [36] respectively carried out numerical simulation of soil cutting with screw cutting ditcher based on ANSYS/ ls-dyna, which was used to calculate the maximum power of soil cutting. Wang Shaowei et al. [58] established a simulation model for optimization of structural parameters of tooth cutter by using response surface analysis method and EDEM software, and obtained the structural parameters of ditcher through optimization. Gao Wenying et al. [59] used orthogonal experiments to optimize the deep burial rate of corn straw, the power consumption of furrow cutting and the maximum deviation of unit linear driving. Sebastian lobo-guerrero et al. [60] used discrete element method to analyze and visualize the influence of pile type on the permeability resistance of driving pile to crushing particles. Hong-bin Xiao et al. [61] analyzed the interaction between pile and soil in expansive land by numerical simulation. NuoDuan et al. [62] used discrete element method to study the effect of foundation pile installation.

3.5. Other Optimization Methods

Michal Lisicki et al. [63] applied bayesian optimization method to optimize the design and operation of Archimedes screw turbine (AST), so as to maximize the total return rate of specific devices. KyeongEop Kang [64] used response surface method to optimize the pretreatment parameters when designing the double-screw extruder, and evaluated the pretreatment effect through relevant tests.

4. Conclusions and Prospects

This paper discusses the importance of flood control and emergency rescue, enumerates some existing pile driving equipment for flood control and emergency rescue, summarizes and analyzes the research status of spiral pile used in pile foundation engineering, and explores the optimization methods related to spiral pile. Through the summary of the problems related to the optimization of screw pile structure parameters, it is found that the optimization of screw pile in the future exploration should pay attention to the following aspects:

- At present, there is little research on the theory of spiral pile sinking, so it is necessary to fully learn from the mature pile-soil interaction mechanism, combine the pile-soil friction theory, comprehensively consider the parameters of spiral pile and soil, and make the theoretical model of spiral pile sinking into soil more conform to the actual situation. When the theoretical model and bearing capacity model of screw pile installation are established, each component of the theoretical model is verified by combining the test and finite element method. For example, the pile head is isolated to test its installation torque, and the pressure required for pile installation is tested after removing screw blades of the screw pile, so as to test and modify the theoretical model.
- When using finite element model simulation, especially when using existing commercial software, such as ANSYS, to do screw pile installation and load analysis, there is some gap between the pile-soil interaction model used and the actual pile installation process, so the simulation results may not very consistent with the actual degree. Therefore, it is necessary to combine the finite element method with others, such as: study the dynamic contact of piles and soil by combining the finite element method and boundary element method, and investigate the coupling of finite element method and smooth particle method. In this way, part of operational errors which are difficult to overcome by the finite element method can be corrected.
- In the optimization design of flood control and emergency helical pile, the optimization objective of screw pile can be extended from the single objective optimization of installation torque of screw pile to multiple objectives optimization: mass of screw pile, installation torque, bearing capacity, etc. The second-generation multi-objective optimization algorithm such as the widely used NSGA-II, can be used to calculate the theoretical model of screw pile installation and bearing capacity. Under certain soil parameters, an optimal set of structural parameters of screw pile is obtained.

Acknowledgments

Subject was supported by Guangdong water conservancy science and technology innovation project (NO.2017-31).

References

[1] Liu Zheng. Common dangerous situations and emergency methods of embankment project [J]. Science and technology information,2011(23):44.

[2] Pang Hongchen. Simulation and structural improvement of screw pile installation process [D]. Guagnzhou: South China agricultural university, 2014.

[3] Wang Yuxing, Luo Xiwen, Zhou Guangde, et al. Development of gasoline internal combustion pile driver for flood control and emergency rescue [J]. Transactions of the Chinese Society for Agricultural Machinery, 2000, 31(5):123-125.

[4] Zhou Linsen, Tang zhenke, Han Linshan, et al. DZF-120 portable flood prevention and emergency pile driver[J]. Engineering Machinery, 2002(04):12-13.

[5] Lǘ Youjie, Wang Yuxing, Tang Yanqin, et al. Application of bionic surface on blades of screw pile[J]. Transactions of the CSAE, 2007,23(4): 134-137.
[6] Cai Jiazhe. Study on field test and numerical simulation of vertical bearing capacity of spliced screw steel piles[D]. Southwest Jiaotong University,2018.
[7] Elkasabgy M, El Naggar M H. Dynamic response of vertically loaded helical and driven steel piles[J]. Canadian Geotechnical Journal, 2013, 50(5):521-535.
[8] Zeyad H. Elsherbiny, M. Hesham El Naggar. Axial compressive capacity of helical piles from field tests and numerical study[J].Can. Geotech. J,2013,50(9):1191-1203.
[9] Mohammed Sak. Performance of helical piles in oil sand[J]. Canadian Geotechnical Journal,2009,46(8):1046-1061.
[10] Sakr M. Comparison between high strain dynamic and static load tests of helical piles in cohesive soils[J]. Soil Dynamics & Earthquake Engineering, 2013, 54(11):20-30.
[11] Amy B. Cerato, P.E., A.M.ASCE. Effects of Long Term Dynamic Loading and Fluctuating Water Table on Helical Anchor Performance for Small Wind Tower Foundations[J]. Journal of Performance of Constructed Facilities, 2009, 23(4): 251-261
[12] Nazir R, Chuan H S, Niroumand H, et al. Performance of single vertical helical anchor embedded in dry sand[J]. Measurement, 2014, 49:42-51.
[13] Lee C W, Kim Y S, Park S Y. Development of Prebored Screw Pile Method and Evaluation of Its Bearing Characteristics[J]. Marine Georesources& Geotechnology, 2016, 34(1):42-56.
[14] Adnan Anwar Malik, Jiro Kuwano, Shinya Tachibana, Tadashi Maejima. End bearing capacity comparison of screw pile with straight pipe pile under similar ground conditions[J].Acta Geotechnical, 2017, 12(4):415-412.
[15] Hendry, Take, Hoult. Field performance of a peat railway subgrade reinforced with helical screw piles[J]. Canadian Geotechnical Journal, 2018.
[16] Dong Tianwen, Liang Li, Wang Wei, Wang Mingshu. Experimental analysis on the lamina-soils interaction in pullout screw pile foundation[J] Engineering mechanics, 2008(08):150-155+163.
[17] Wang Qian, Zhou Hongan, Li LIhua, Chu Kaiming, Cui Bojun. Design and trial production of fiber reinforced plastic screw anchor[J]. Rock and Soil Mechanics, 2007(11):2325-2328+2332.
[18] Wang Tengfei, Liu Jiankun, Tai Bowen, Lǘ Peng. Model tests on frost jacking behaviors of helical steel piles[J]. Chinese Journal of Geotechnical Engineering, 2018,40(06):1084-1092.
[19] Zhou Yang, Xiao Shiguo, Xu Jun, Hu Yayun. Model test on vertical bearing capacity of variable cross-section thread piles[J]. Rock and Soil Mechanics, 2017,38(03):747-754+783.
[20] Harnish J, El Naggar M H. Large Diameter Helical Pile Capacity - Torque Correlations[J]. Canadian Geotechnical Journal, 2017, 54(7): 968-986.
[21] Harnish J L. Helical Pile Installation Torque and Capacity Correlations[J]. Journal of Cutaneous Pathology, 2015, 24(2):457-60.
[22] Tsuha C D H C, Aoki N. Relationship between installation torque and uplift capacity of deep helical piles in sand[J]. Canadian Geotechnical Journal, 2010, 47(6):635-647.
[23] SakrM. Retraction: Relationship between installation torque and axial capacities of helical piles in cohesionless soils[J]. Canadian Geotechnical Journal, 2017, 54(1):44-58.
[24] Zhengyang (Moira) Guo, Lijun Deng. Field behaviour of screw micropiles subject to axial loading in cohesive soils[J].Canadian Geotechnical Journal, 2019, 55(1): 34-44.
[25] Khan U, Siddiqua S. Study of compressive loading capacities of helical piles using torque method and induced settlements[J]. Environmental Earth Sciences, 2018, 77(1):22.
[26] C. C. HIRD, Q. NI, I. GUYMER. Physical modelling of deformations around piling augers in clay[J].Geotechnique,2011, 61(11),:993–999.
[27] Pucker T, Grabe J. Numerical simulation of the installation process of full displacement piles[J]. Computers and Geotechnics, 2012, 45(2):93-106.
[28] Dong T W, Zheng Y R. Limit analysis of vertical anti-pulling screw pile group under inclined loading on 3D elastic-plastic finite element strength reduction method[J]. Journal of Central South University, 2014, 21(3):1165-1175.

[29] Mendoza C C, Cunha R, Lizcano A. Mechanical and numerical behavior of groups of screw (type) piles founded in a tropical soil of the Midwestern Brazil[J]. Computers and Geotechnics, 2015, 67:187-203.
[30] Albusoda B S, Abbase H O. Performance assessment of single and group of helical piles embedded in expansive soil[J]. International Journal of Geo-Engineering, 2017, 8(1):25.
[31] Krasiński, A. Numerical simulation of screw displacement pile interaction with non-cohesive soil[J]. Archives of Civil and Mechanical Engineering, 2014, 14(1):122-133.
[32] Deng Yibing, Zhou Jian, Liu Wenbai, Liu Zhong, Li Zhiyi. PFC numerical simulation of augered piling of soil displacement screw piles[j]. Chinese Journal of Geotechnical Engineering, 2014, 14(1):122-133.
[33] Li Zhiyi, Liu Zhong, Zhao Yanfei, Zhang Huidong. Optimization Analysis on Drill Auger of Soil Displacement Screw Pile[J]. Journal of Tongji University(Natural Science), 2011,39(08):1145-1149.
[34] Shi danda, Yang Yancheng, Deng Yibing, Liu Wenbai. Experimental study of the effect of drilling velocity ratio on the behavior of auger piling in sand[J]. Rock and Soil Mechanics, 2018,39(06):1981-1990+1998.
[35] Ma Aili, Liao Qingxi, Tian Boping et al. Numerical Simulation of Soil Cutting by Spiral Cutter on the Basis of ANSYS/LS DYNA[J].Journal of Huazhong Agricultural University, 2009,28(02):248-252.
[36] Xue Zixuan, Lv Xinmin, Tang Weiwei. The simulation analysis of soil cutting with screw cutter based on ANSYS/LS-DYNA971[J]. Research on Agricultural Mechanization, 2011,33(04):13-16.
[37] Zhang Chengxin, Cui Zongke, Wang Song, Leng Luyou. Optimal design of Screw conveyor Based on MATLAB[J]. Engineering Machinery, 2017,48(06):34-36+6-7.
[38] Wu Yinliang, Chen Lin, Guo Libo. Optimal design of anti-slide pile based on nonlinear Fmincon method[J]. Railway Engineering,2011(05):81-84.
[39] Merkle D, Middendorf M. An Ant Algorithm with a New Pheromone Evaluation Rule for Total Tardiness Problems[J]. Proceedings of Evo Workshops , 2000:287--296.
[40] Geem Z W. Music-Inspired Harmony Search Algorithm[J]. Studies in Computational Intelligence, 2009, 191:163-172.
[41] Teixeira C, Covas J A, Stützle, Thomas, et al. Multi-objective ant colony optimization for the twin-screw configuration problem[J]. Engineering Optimization, 2012, 44(3):351-371.
[42] Hsu C C, Chao C K , Wang J L , et al. Multiobjective optimization of tibial locking screw design using a genetic algorithm: Evaluation of mechanical performance[J]. Journal of Orthopaedic Research, 2006, 24(5):908-916.
[43] Lee C H , Shih K S , Hsu C C, et al. Simulation-based particle swarm optimization and mechanical validation of screw position and number for the fixation stability of a femoral locking compression plate[J]. Medical Engineering & Physics, 2014, 36(1):57-64.
[44] Na K., Lee D., Lee H., Jung K., Choi H. Optimum Configuration of Helical Piles with Material Cost Minimized by Harmony Search Algorithm. In: Kim J., Geem Z. (eds) Harmony Search Algorithm. Advances in Intelligent Systems and Computing, vol 382. Springer, Berlin, Heidelberg,2016.
[45] Li Xiaohuo, Lin Qiyue, He Yang. Parameter optimization of spiral conveyor for coal auger based on GAAA algorithm[J]. Journal of China Coal Society, 2010,35(03):498-502.
[46] Zhang Qiang, Fu Yunfei, Han Su, Yuan Zhi. Multi-objective Fuzzy Reliability Optimization for Auger Coal Miner Based on SPEA Algorithm[J].Machine Design and Research, 2012,28(05):117-119.
[47] Song Yuning, Song Yejiao, Xu Xiaochen. Application of Coal Auger Enchantment and Optimum Design Under Improved Genetic Algorithm[J]. Coal Mining Technology, 2018,23(05):29-33.

[48] Qin Datong, Wang Zhen, Hu Minghui, et al. Dynamic matching of optimal drum movement parameters of shearer based on multi-objective optimization [J]. Journal of China Coal Society, 2015, 40(S2) : 532 - 538.
[49] Gong MG, Jiao LC, Yang DD, Ma WP. Research on evolutionary multi-objective optimization algorithms. Journal of software, 2009,20(02):271-289.
[50] Xu Lina. Research on numerical simulation of the bearing performance of large-diameter bored piles[D]. Jilin University,2014.
[51] Konak A, Coit D W, Smith A E. Multi-objective optimization using genetic algorithms: A tutorial[J]. Reliability Engineering & System Safety, 2006, 91(9):992-1007.
[52] Ahmad I, Naggar M H E, Khan A N. Artificial neural network application to estimate kinematic soil pile interaction response parameters[J]. Soil Dynamics and Earthquake Engineering, 2007, 27(9):892-905.
[53] Nejad F P ,Jaksa M B , Kakhi M , et al. Prediction of pile settlement using artificial neural networks based on standard penetration test data[J]. Computers and Geotechnics, 2009, 36(7):1125-1133.
[54] MosallanezhadM, Moayedi H. Developing hybrid artificial neural network model for predicting uplift resistance of screw piles[J]. Arabian Journal of Geosciences, 2017, 10(22):479.
[55] Hossein M, Sajad H. Artificial intelligence design charts for predicting friction capacity of driven pile in clay[J]. Neural Computing and Applications, 2018.
[56] Park H I, Cho C W. Neural Network Model for Predicting the Resistance of Driven Piles[J]. Marine Geotechnology, 2010, 28(4):21.
[57] Milad F, Kamal T, Nader H, et al. New method for predicting the ultimate bearing capacity of driven piles by using Flap number[J]. KSCE Journal of Civil Engineering, 2015, 19(3):611-620.
[58] Wang Shaowei, Li Shanjun, Zhang Yanlin, Zhang Chi, Chen Hong, Meng Liang. Design and optimization of inclined helical ditching component for mountain orchard ditcher[J].Transactions of the Chinese Society of Agricultural Engineering, 2018,34(23): 11-22.
[59] Gao Wenying, Lin Jing, Li Baofa, Ma Tie. Parameter Optimization and Experiment for Spiral Type Opener Device of Maize Straw Deep Bury and Returning Machine[J]. Transactions of the Chinese Society for Agricultural Machinery, 2018,49(09):45-54.
[60] Lobo-Guerrero S, Vallejo L E. Influence of pile shape and pile interaction on the crushable behavior of granular materials around driven piles: DEM analyses[J]. Granular Matter, 2007, 9(3):241-250.
[61] Xiao H B , Zhang C S , Wang Y H , et al. Pile-Soil Interaction in Expansive Soil Foundation: Analytical Solution and Numerical Simulation[J]. International Journal of Geomechanics, 2011, 11(3):159-166.
[62] NuoD, Pik C Y, Wei L J. DEM analysis of pile installation effect: comparing a bored and a driven pile[J]. Granular Matter, 2018, 20(3):36-.
[63] Lisicki M, Lubitz W, Taylor G W. Optimal design and operation of Archimedes screw turbines using Bayesian optimization[J]. Applied Energy, 2016, 183:1404-1417.
[64] Kang K E , Han M , Moon S K , et al. Optimization of alkali-extrusion pretreatment with twin-screw for bioethanol production from Miscanthus[J]. Fuel, 2013, 109:520-526.

Research on Shear Lag Effect of Three-span Continuous Curved Steel Box Girder Bridge

Kongliang Chen[1,*], Liankun Wang[1] and Wenzhi Zhang[1]

[1]School of Civil and Architectural Engineering, Wuyi University, Jiangmen, Guangdong, 529000

*Corresponding author

Abstract. The normal stress of box-section beam under symmetrical load is the non-uniform characteristic called shear lag effect. Based on actual engineering project, the plate-beam finite element method is used to study the shear lag effect of the three-span continuous-curve steel box-girder bridge under the variety conditions of the curve radius, and the width span ratio, and the distribution of the shear lag effect in the longitudinal and horizontal direction under the symmetrical load are summarized in this paper, providing references for design of steel box-girder bridge. The results show that the shear lags of steel box girders are all positive shear lags under symmetrical loads. Under the conditions of three parameters, the curve radius and the width-span ratio have a great influence on the shear lag effect of steel box girders, and the high-span ratio change has no obvious effect on the shear lag effect, and the shear lag effect of bottom plate is far less severe than the top plate.
Keywords: steel box girder; shear lag effect; curvature radius.

1. Introduction

Box section is a commonly used section form in bridge design. Under the action of symmetrical bending load, considering the shear deformation of box girder, the distribution of the normal stress shows unevenness along the lateral direction, which is called shear lag effect [1-6]. In the structural design, the neglect of the shear lag effect will underestimate the deflection and stress at the joint between the web and the wing of the box girder, resulting in unsafe structural design. Several foreign countries have not considered the "shear lag effect" and caused the instability of the structure or a partial damage accident. At present, as the rapid development of steel structure technology, the use of steel box girder bridges is more and more extensive. Due to its advantages of light weight, high bending stiffness and torsional stiffness, good integrity and quick installation, steel box girder bridges are widely used in overpass bridges and long-span bridges. Some workers have been working on steel box girder bridges. The research work on the shear lag is carried out[7]. The shear lag effect of box girder is analyzed by constructing the warping displacement mode of cross section, and the energy variation method is based on the principle of minimum potential energy in paper[8]. Based on the deformation coordination and equilibrium conditions of the composite wing plate micro-element, the normal stress differential equation of the cross-section wing is established, and the shear lag coefficient of the composite box girder wing is derived in paper[9]. Through the calculation and analysis of the shear lag effect of curved box girder bridges with different parameters, the shear lag effect variation law of curved box girder continuous bridge with various influencing factors is summarized in paper[10], but the research results of the shear lag effect of the curved continuous steel box girder bridge still need further research.

2. The Shear Lag Effect of Curved Steel Box Girder Bridge

The curved continuous beam bridge generates torque under load. The "bending and torsion coupling" phenomenon is very important in the curved beam bridge, which will cause the structural stress of the curved bridge to be inconsistent with the force of the linear bridge. Therefore, the inner and outer forces of the curved bridge are inconsistent with the deformation and deformation under the action of the symmetrical load, leading to the difference of the shear lag between the inner and outer sides of the curved beam bridge.

$$\Pi=\int (M(x)+M'(x))(\frac{d^2w}{dx^2})^2 dx+\frac{1}{2}\int EI_w(\frac{d^2w}{dx^2})dx+\frac{1}{2}\iint t_u(E\varepsilon_{xu}^2+G\gamma_u^2)dxdy$$
$$+\frac{1}{2}\iint t_b(E\varepsilon_{xb}^2+G\gamma_b^2)dxdy+\frac{1}{2}\int (GK_TK_\phi^2+EI_wK_\phi^2)dz$$
$$=\int (M(x)+M'(x))(\frac{d^2w}{dx^2})^2 dx+\frac{1}{2}\int EI_w(\frac{d^2w}{dx^2})dx+\frac{1}{2}I_s\int E\left\{\left[(w'')^2+\frac{3}{2}w''u'+\frac{9}{14}(u')^2\right]+\frac{9Gu^2}{5b^2}\right\}dx \quad (1)$$
$$+\frac{1}{2}\int (GK_TK_\phi^2+EI_wK_\phi^2)dz$$

$$\varepsilon_{xu}=-h_b\left[w''+(1-\frac{y^3}{b^3})u'(x)\right] \qquad \varepsilon_{xb}=h_b\left[w''+(1-\frac{y^3}{b^3})u'(x)\right] \quad (2)$$

Where: $I_{su}=2t_ubh_u^2+2abt_uh_u^2 \quad I_{sb}=2t_bbh_b^2 \quad I_s=I_{su}+I_{sb}$

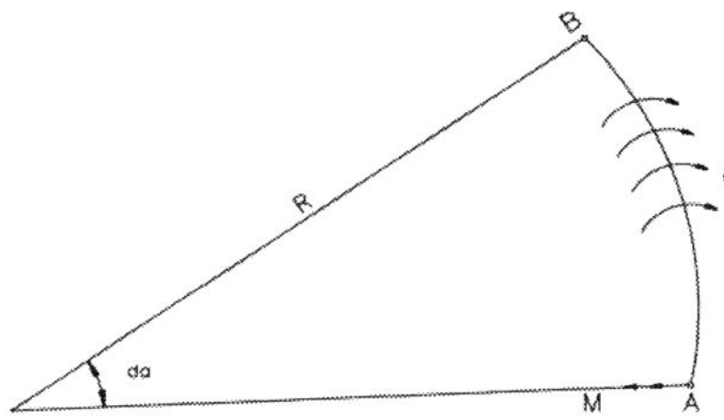

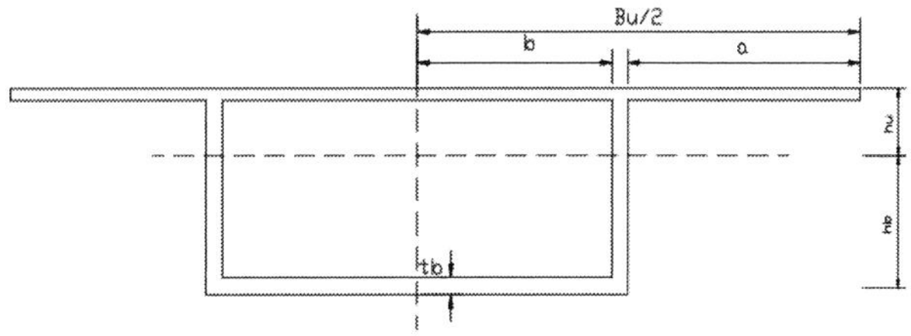

Figure 1. Shear stagnation point arrangement **Figure 2.** Loading diagram

From the above curve steel box beam elastic control differential equation, it can be seen that the influencing factors of the curved steel box girder have an important relationship between the radius curvature, the high span ratio and the width span ratio, which will be discussed for steel in this paper.

3. Project Background

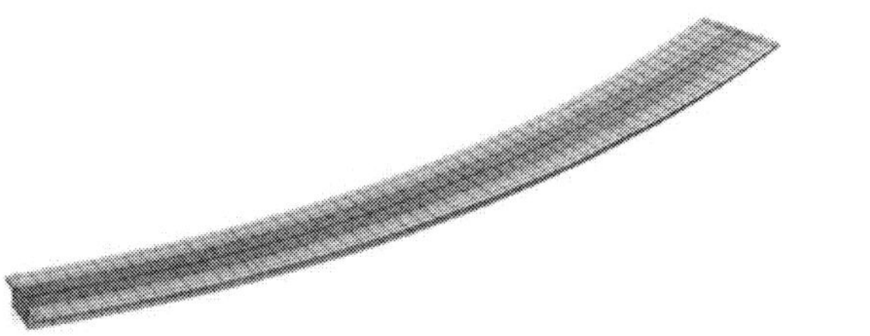

Figure 3. A steel box girder bridge of some Expressway **Figure 4.** R=160m steel box girder model

A continuous curved steel box girder of Guang-Zhong-Jiang Expressway is taken as the research object. The bridge span is 28m+52m+26m, and the upper structure is a single-chamber steel box girder with curvature radius R=160m, and the width of bridge deck is10.5m, and the beam height is 2.5m, and there is a horizontal slope on the inner side of the main beam, shown in Figure 3. In addition to the two-way 3m spacing stiffeners in the span, and the rest are strongly and weak stiffened every 2m along the longitudinal direction. In this paper, the Midas finite element program is used to establish the spatial plate beam combination model. The top plate, bottom plate, wing plate, web and diaphragm of the box girder are simulated by the plate element, and the longitudinal stiffeners are simulated by the beam element. Figure 4 shows the radius of curvature R = 160m calculation model.

4. Analysis of Shear Lag Effect

Figure 5 is the main section of the main beam, and the points 1 and 9 of the top plate are located on the inner side and the outer side of the cross section respectively, and the points 1 and 4 on the bottom plate are located on the inner side and the outer side of the cross section respectively. The type of load is the full-span symmetric uniform load q=10kN/m is mainly considered in the calculation.

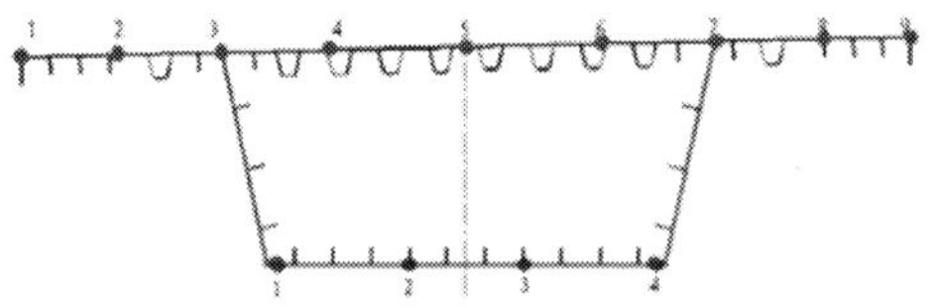

Figure 5. Shear stagnation point arrangement

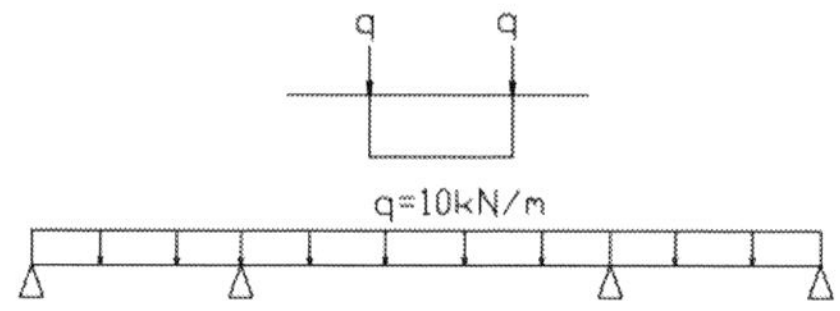

Figure 6. Loading diagram

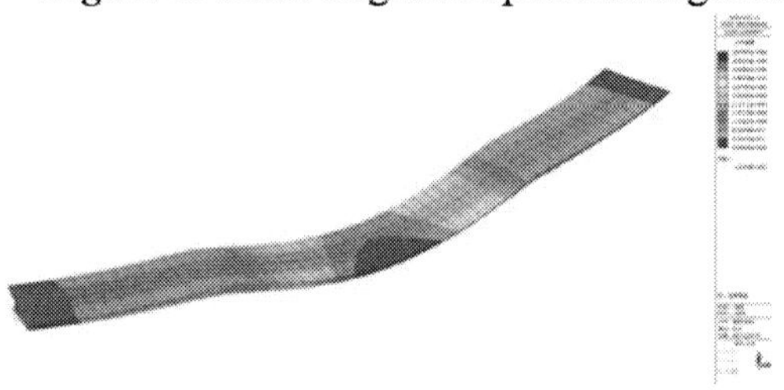

Figure 7. The structure displacement of symmetric load

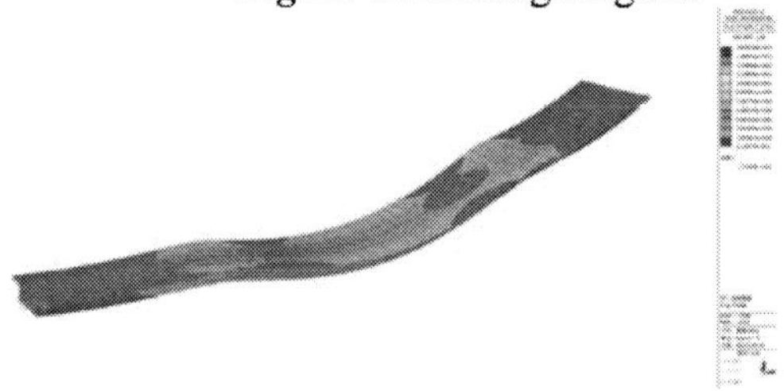

Figure 8. The plate stress of symmetric load

4.1. Analysis of Shear Lag Effect of Steel Box Girder

The shear lag results of the R=160m model under symmetric load are shown in Figures 9-11. Figure 9 is the shear lag effect of the roof with uniform load, and the Figure 10 is the shear of the bottom plate, and Figure 11 is the longitudinal shear lag effect, from which can lead to the following conclusions:

1) Under the action of symmetric load, the shear lag curve of the roof is similar, and both are positive shear lag effect, and the shear lag coefficient of the inner and outer sides of the roof is different, and the shear lag effect of the shear lag is inconsistent.

2) There is also shear lag in the steel box girder bottom plate, but the shear lag coefficient is smaller than the top plate. The bending and torsion effect of the curved bridge leads to inconsistent shear lag in the inner and outer sides of the bridge.

3) It can be known from the longitudinal shear lag of the graph that as the radius of curvature becomes smaller, the longitudinal shear force hysteresis effect of the curved steel box girder increases, but the increase is about 20%; The load is still concentrated, and there is no negative shear lag in any section along the longitudinal section of the bridge span. This may be due to the strong stiffening and weak stiffening in the longitudinal direction; the shear lag effect near the support is better than the rest.

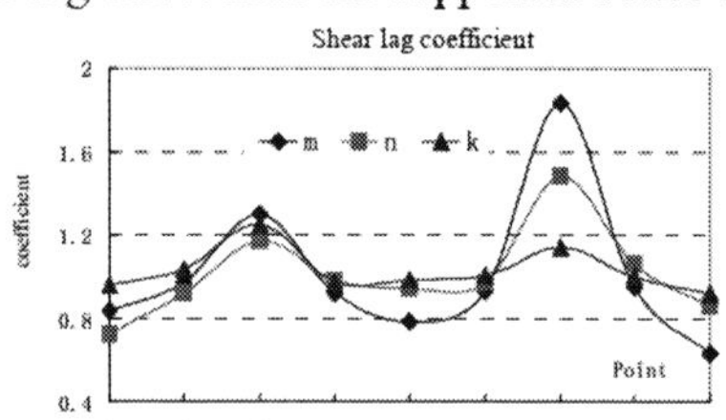

Figure 9. The shear lag of the Roof

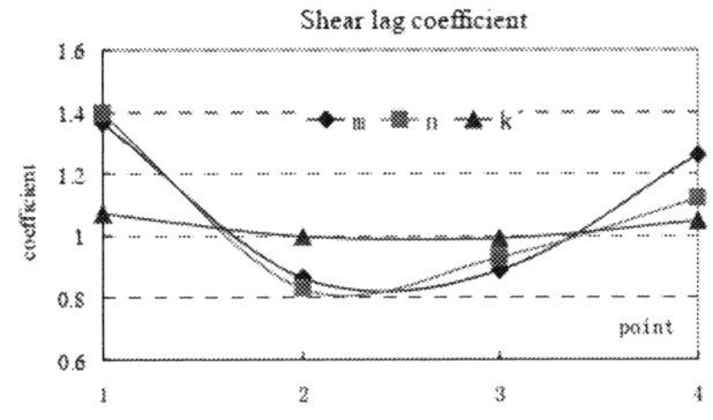

Figure 10. The shear lag of Bottom plate

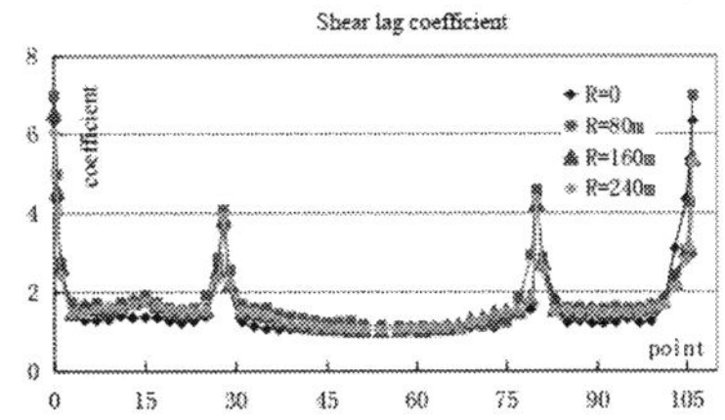

Figure 11. The longitudinal shear lag effect

4.2. The Parameter Variation of The Shear Lag

For the curvature radius R=80, 160m, 240m, R=0 four different curvature radius of bridge, and The width span ratio is defined as the ratio of bridge deck width to the length of the beam axis (2b/L). Keep the width of the section unchanged, while the span L changes, the ratio of 2b/L is 0.1, 0.2, and 0.3 respectively. The variation law of shear lag is summarized.

1) As can be seen from figures 12-13, the peak value of the shear lag coefficient of the outer web across the mid-span and the fulcrum section is smaller than the peak value of the shear lag coefficient of the inner web. This is due to the inconsistent shear lag of the inner and outer sides caused by the combination of bending and torsion of the steel box girder. The peak value of shear lag coefficient increases with the decrease of curvature radius. When the curvature radius is 160m<R<240m, the shear lag curve is closer; when the curvature radius is much larger than 240m, the curved steel box beam can be The shear lag coefficient is calculated approximately as a straight line beam without a large error. Under the action of symmetric load, the shear lag coefficient of the steel box girder bottom plate is a downward concave curve. Under the conditions of curvature radius, the shear lag curve of the bottom plate is not obvious, which also indicates that the shear lag effect of the bottom plate is far from the top plate.

2) It can be seen from the data in Table 1 and Figure 14 that the law of shear lag is a quadratic parabola with increasing width-to-span ratio. The width-to-span ratio has a greater influence on the shear lag coefficient. As the width-to-span ratio increases, the shear lag effect increases, but the degree of aggravation gradually decreases.

3) It can be seen from the Figures 15-17 that the shear lag curve is basically bonded when the high-span ratio changes, indicating that the high-span ratio change has no obvious effect on the shear lag coefficient of the top and bottom plates.

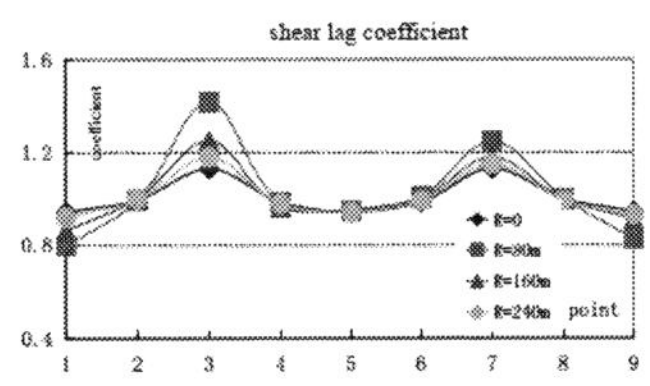

Figure 12. The shear lag of the roof with curvature radius

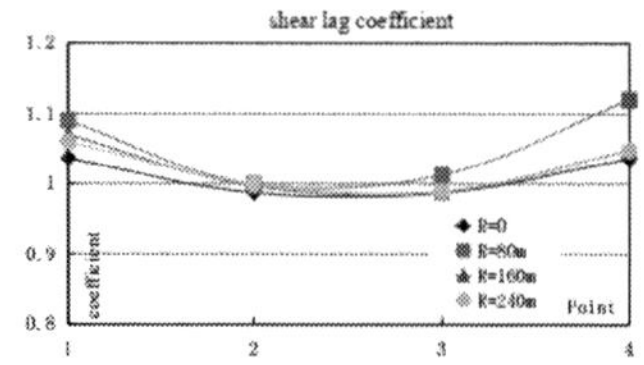

Figure 13. The shear lag of bottom plate with curvature radius

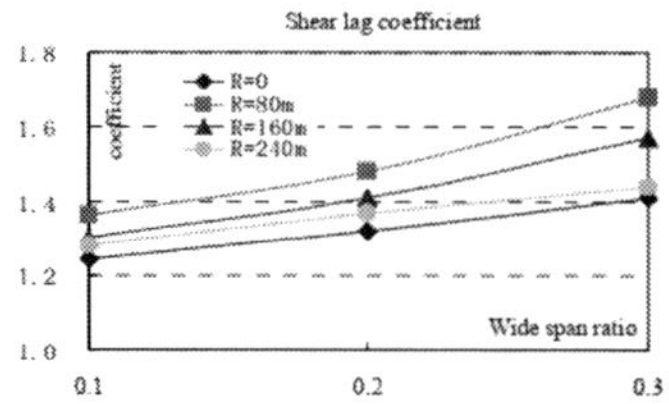

Figure 14. The shear lag effect with width span ratio

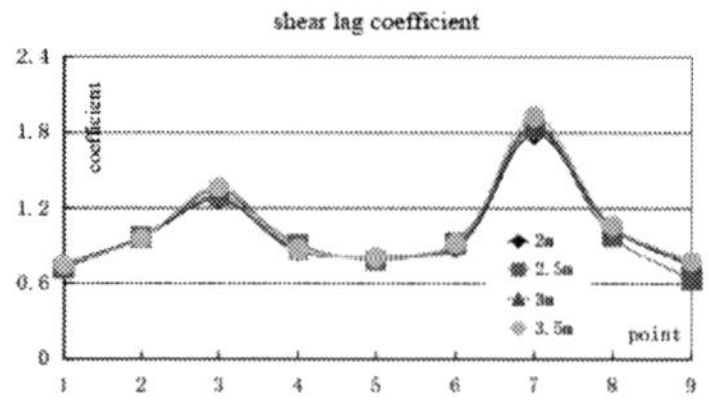

Figure 15. The shear lag of the roof with high span ratio (1)

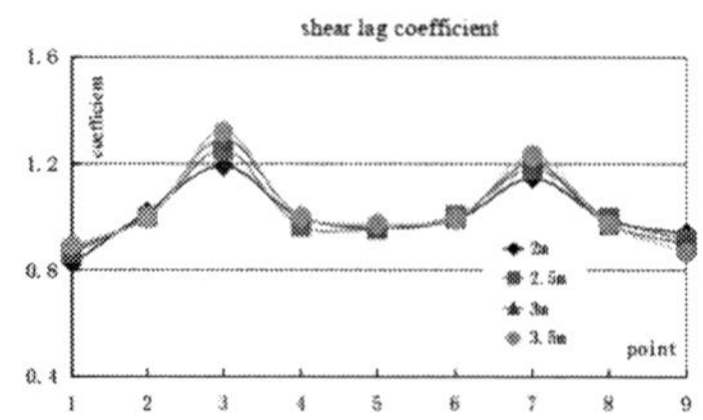

Figure 16. The shear lag of the roof with high span ratio (2)

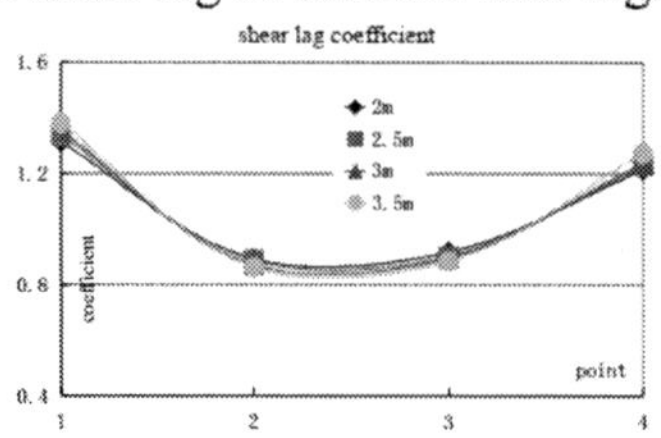

Figure 17. The shear lag of bottom plate with high span ratio

Table 1. Shear lag coefficient of No. 3 point on the top plate

section	radius	width span ratio			Fitting formula
		0.1	0.2	0.3	
Side-span	160m	1.3	1.41	1.57	$y = 2.5223x^2 + 0.3388x + 1.2413$
Near the pivot	160m	1.17	1.26	1.39	$y = 1.9245x^2 + 0.3378x + 1.1155$
Mid-span	160m	1.25	1.26	1.39	$y = 6.0834x^2 - 1.7417x + 1.365$

5. Conclusion
Through the above calculation and analysis, the variation law of shear lag effect of continuous curved steel box girder bridge under symmetric load is summarized:
1) Under the action of symmetrical load, the shear lag of steel box girder is positive shear lag, and the curve radius and width span ratio have relatively large influence on the shear lag effect of steel box girder, the variation law of shear lag curve increases with the width span ratio and is similar to the quadratic parabola with irregularity, and the high-span ratio change has no obvious effect on the shear lag effect.
2) Under the action of symmetric load, the shear lag coefficient at the boundary between the inner and outer webs of the steel box girder and the roof is the peak, and the variation of the shear lag coefficient of the bottom plate is the downward concave curve, and the shear lag effect is far less severe than the shear lag of the top plate.
3) The curvature radius is an important parameter for the analysis of shear lag effect of continuous curved steel box girder bridge. Due to the different inner and outer radii of curvature, the shear lag coefficient of the inner and outer sides of the steel box girder is inconsistent.
4) Under the action of symmetrical load, there is no negative shear lag in the longitudinal direction of the bridge. This may be due to the strong stiffening and weak stiffening in the longitudinal direction.

Acknowledgements
The paper supported by the Special Funds of the National Natural Science Foundation of China (11502172) and Guangdong Science and Technology Department Project (2016A040403125).

References
[1] Zhang Yuan-hai, Su Yan-dong, Lin Li-xia.2010.Finite Beam Element Analysis on Shear Lag Effect of Skewly Supported Continuous Box Girder. Journal of the China Railway Society.pp 44-50.
[2] Li Yun-sheng, Zhang yan-ling, Fan Jian-sheng. 2010. Shear lag effect study of cracked steel-concrete continuous beams.Journal of Building Srtuctures. pp 390-397.
[3] Li Xia-yuan, Wan Shui, Chen Jian-bing, Mo Yi-lung. 2018. Analysis on shear lag effect in thin-walled box girders based on modified warping displacement function.Journal of Southeast University.pp 851-856.
[4] Stuart S Chen, Amjad J. Aref, Methee Chiewanichakorn, Il-Sang Ahn. 2007. Proposed Effective Width Criteria for Composite Bridge Girders. Journal of Bridge Engineering. pp 325-338.
[5] Aref Amjad J, Methee Chiewanichakorn, Chen Stuart S. 2007.Ahn Il-Sang. Effective slab width definition for negativemoment regions of composite bridges. 2003. Journal of Bridge Engineering.pp425-426.
[6] Li Li-feng, Shao Xu-dong, Yi Wei-jian, Zhang Xin.2007. Model Test on local stability of flat steel box girder. China Journal of Highway and Transport. pp 60-65.
[7] Zhang Yuan-hai, Li Lli-xia. 2013. Initial parameter method for analyzing shear lag effect of thin-walled box girders. Engineering Mechanics. pp 205-211.
[8] Luo Qi-zhi, Wu You-ming, Liu Guang-dong. Shear lag of the thin-wall box girder with varying depths[J]. Journal of the China Railway Society.pp 81−87.
[9] Luo Qi-zhi. 1998. Analysis of the shear lag effect on continuous box girder bridges with variable depth[J]. China Journal of Highway and Transport, pp 63−70.
[10] Chen Yu-ji, Luo Qi-zhi. 2007. The analysis of large deflection of curved box girders in considering three shear lag functions. China Railway Science, pp13−18.

IOP Conf. Series: Materials Science and Engineering **611** (2019) 012051 doi:10.1088/1757-899X/611/1/012051

A Laboratory Analysis of Chloride Ions Penetration in Recycled Aggregates Concrete Admixed with Ground Granulated Blast Furnace Slag

Weichao Hu[1,2], Shengxin Li[1], Chunmei Song[1], Zheng Chen[1,*], Liyuan Chen[1], Yang Yang[1] and Weilong Luo[1]

[1]Key Laboratory of Disaster Prevention and Structural Safety of China Ministry of Education, School of Civil Engineering and Architecture, Guangxi University, Nanning 530004, China
[2]Hunan Electric Power Design Institute, Changsha 410007, China

*Corresponding author

Abstract. The corrosion of reinforcements in recycled aggregates (RA) structures could be triggered by chloride attack. Ground granulated blast furnace slag (GGBS) has been recognized a significant cement replacement to improve chloride resistant of concrete. However, the study aimed at time-dependent chloride diffusion of recycled aggregates concrete (RAC) with GGBS was still scarce. Thus, with the data generated at varying immersion durations from long-term bulk immersion tests, the effects of the content of ground granulated blast furnace slag (GGBS) and immersion time on time-dependent chloride diffusion were studied. The results indicated that the apparent diffusion coefficient (D_{app}) of recycled aggregates concrete decreases with immersion time. Moreover, the total chloride concentration decrease with the increase of GGBS replacement. When increasing the content of GGBS from 0% to 30%, the value of D_{app} decreases from 28.9% to 52.2%, while 67.2%~40.6% of reduction was observed on D_{app} with the addition of 60% GGBS at different immersion time.

1. Introduction

In recent years, construction waste has increasingly become a potential threat to environmental pollution, and treatment of it seems to be a worldwide problem. Currently, construction waste accounts for about 40 percent of solid waste discharged by urban residents, making it one of the biggest wastes in Chinese rapid urbanization [1]. More and more attention has been paid to it, resulting from that the recycling of construction waste can reduce the discharge of solid waste and the consumption of natural resources. However there are wide variations caused by the differences between natural aggregate and recycled coarse aggregate (RCA) in particle shape, surface structure, water absorption and many other aspects, which restricts the application of RAC in marine environments [2].

Nowadays, much attention has been devoted to studying the durability of RAC in the marine environments. Hu [3] investigated the effects of recycled aggregate, sand ratio, and water-cement ratio on chloride transport in RAC from an unsteady rapid chloride migration test (RCM), which the decrease in the water cement ratio lead to the development in the chloride penetration resistance of RAC. Zhang et al. [4] found that the chloride diffusivity of RAC with high quality recycled fine aggregates seems to be better than that of natural aggregates concrete (NAC), while the impermeability of the RAC with high quality course aggregates is close to that of NAC by using RCM. The existence of recycled aggregate may give rise to concrete having inherent pores and worsen durability actually [5].

Nonetheless, GGBS is capable of improving it by reducing micro pores and cracks, i.e. pozzolanic effect[6]. Du et al. [7] had made some efforts to explore the chloride permeability of RAC containing different pozzolans, such as fly ash and slag through ASTM C1202 test method. It showed that the mineral admixture can improve the chloride diffusivity of RAC effectively, and its chloride resistance gradually increases with the increase of curing age. However, in engineering practice, the time-dependent characteristics of erosion environment and material properties are usually found in NAC, which reflected in the chloride diffusion coefficient, surface chloride ion concentration and other parameters of concrete changing with time [8]. The correlation between chloride diffusion coefficient and time was first discovered by Takewake [9]. Wang et al. [10] analyzed a series of test data of NAC from different long-term fields and laboratory experiments exposed to chloride environment. The results showed that the apparent diffusion coefficient presents an exponential decay with time and tend to be stable at 30 years exposure time. Nevertheless, previous studies focused on the time-dependency of NAC and its influencing factors. Due to the existence of the double interface transition zone, the chloride diffusion of RAC is quite different from that of NAC. Therefore, the time-dependency of chloride ions in RAC needs further study.

In this paper, the effects of the content of GGBS and immersion duration on time-dependent chloride diffusion was studied through the long-term bulk immersion tests. Finally, a theoretical gist for the research on the chloride permeability of RAC Structure was provided.

2. Modeling of Chloride Transport Based on Fick's 2nd Law

The transport of chloride in interior concrete is dominated by diffusion and the concentration of chloride ions decreases with the increase of depth, while there exist the complex chloride transport methods including diffusion, capillary adsorption and convection in the surface of concrete [11]. By assuming the concrete as a semi-infinite homogeneous medium, the transmission of chloride can be expressed by Fick's 2nd law of diffusion [12]:

$$\frac{\partial C}{\partial t} = D \cdot \frac{\partial^2 C}{\partial x^2} \tag{1}$$

Where C represents the chloride concentration; t represents the exposure time; D represents the chloride diffusion coefficient in concrete; and x represents the distance from concrete surface.

By solving Eq. (1) in a simple mathematical way, the error function solution based on Fick second law can be obtained [13]:

$$C(x,t) = C_0\left[1 - erf\left(\frac{x}{2\sqrt{Dt}}\right)\right] \tag{2}$$

Where $C(x, t)$ represents the concentration of chloride ion at exposure time t and depth x; C_0 represents the equilibrium chloride concentration on concrete surface; D represents the chloride ion diffusion coefficient of concrete; and $erf(x)$ represents the error function.

3. Experiment

3.1. Test Materials and Mix Proportion Design

In this study, P•II 42.5 ordinary portland cement supplied by Guangxi Huarun Cement Corporation of China was used, and its apparent density was 3020 kg/m^3. A GGBS was obtained from Guangxi Yuansheng Corporation of china and it has a density of 2910 kg/m^3. Crushed concrete (C30) was used as the coarse aggregate and the composition of different gradation in RAC was shown in Table 1. Fine aggregate, having a density of 2613.65 kg/m^3 and a fineness modulus of 2.725, came from Wuning, Guangxi, China.

Three sets of RAC samples were produced with different content of GGBS (0%, 30% and 60%), as shown in Table 2. The replacement rate of recycle aggregate of all RAC specimens were 100%.

IOP Conf. Series: Materials Science and Engineering **611** (2019) 012051 doi:10.1088/1757-899X/611/1/012051

Table 1. The composition of different gradation in recycled aggregate.

Aggregate type	Gradation(mm)	Volumetric proportions
	5-10	0.25
RA	10-15	0.5
	15-20	0.25

Table 2. Mix proportion design of recycled aggregates concrete.

Group	Sample number	Water (kg/m^3)	Cement (kg/m^3)	Slag (kg/m^3)	Sand (kg/m^3)	Recycle aggregate (kg/m^3)
1	RA-0%	203.7	452.67	0	701.86	931
2	RA-30%	201.77	313.86	134.51	701.78	931
3	RA-60%	199.84	177.63	266.45	701.85	931

3.2. Experimental Method

The RAC mixtures were prepared in the laboratory using a $0.2m^3$ pan mixer. Cubes of 100×100×100mm in dimension were cast in mold and vibrated until suitably consolidated. Then, they were demoulded after 24h and further cured at 20±3℃ and 95% relative humidity until testing. Two faces of cubes exposed to test solution was remained, and other faces were coated with epoxy resin after 28d. Then, samples were immersed in the 165g/L NaCl solution according to NT Build 443[14]. For obtaining the chloride concentration profiles, related powder was collected from cubes specimens at different depths after the immersion period (30d, 60d and 180d). Layers of thickness would be adjusted from 2~3mm to 4~5mm with the increasing of immersion time. Acid-soluble chlorides were taken out from obtained powder by using Improved Volhard Method [15], which giving an indication of the total chloride concentration.

4. Results and Discussion

4.1. Chloride Penetration

The evolution of total chloride concentration for different GGBS contents at immersion ages of 30, 60 and 180 days is represented in Fig.1, and it is found that the total chloride concentration decreases as the content of GGBS increases at different immersion ages. This is primarily attributed to hydration products of C_3A and C-S-H gels which enhancing the physical-chemistry adhesion of concrete to chloride ions, thus decreasing the diffusion coefficient significantly. Also it can be found from Fig.1 that the effect of GGBS on chloride resistance of RAC decreases with time. This may be due to the pozzolanic reaction of GGBS starts at very early age, and GGBS significantly decreases the content of $Ca(OH)_2$ crystals in the aggregate–mortar ITZ [16], which in turn leads to the insufficient of $Ca(OH)_2$ for pozzolanic reaction with the increase of immersion time [17].

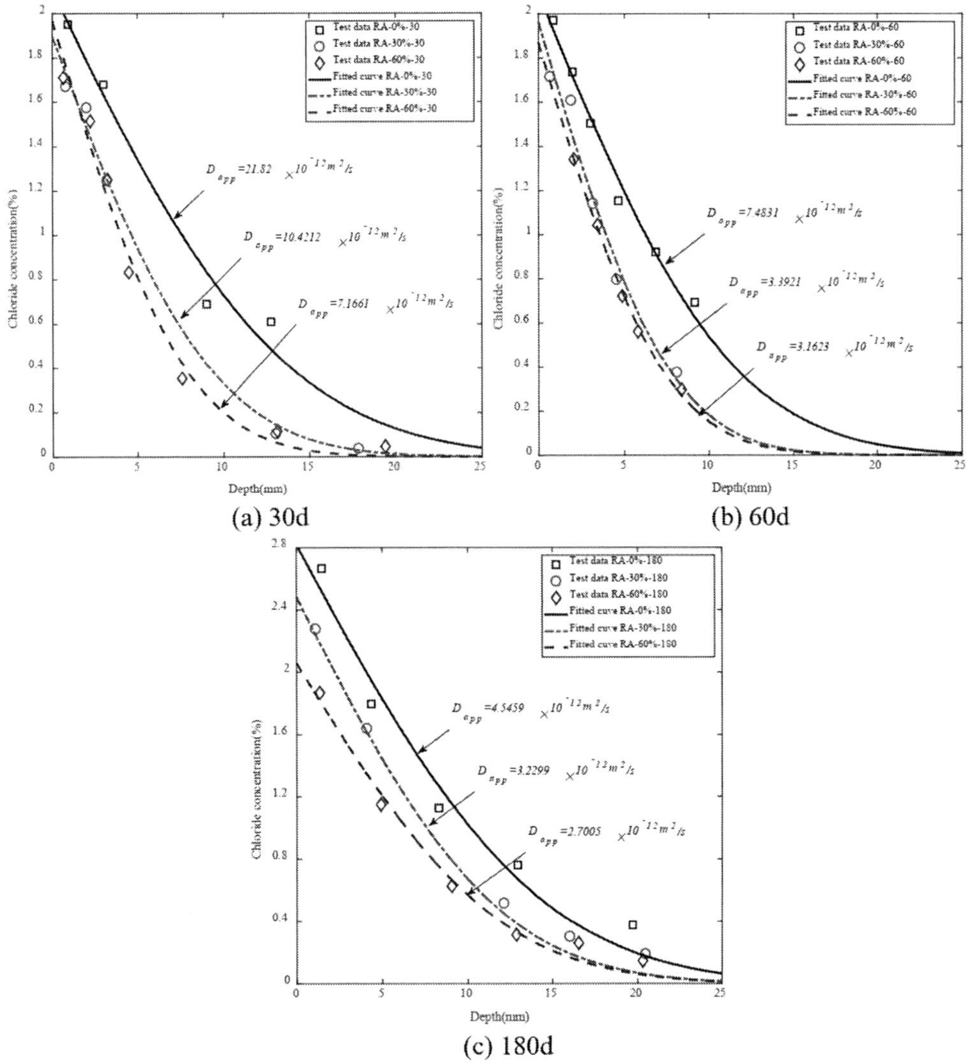

Figure 1. Distribution of total chloride ion concentration of recycled aggregates concrete at different immersion time.

4.2. Apparent Chloride Diffusion Coefficient

Based on the experimentally obtained chloride concentrations profile, a non-linear regression analysis by means of the least squares fit method (by using Eq. (2)) is performed so as to determine the apparent diffusion coefficient D_{app}, as shown in Fig. 2. We can see that D_{app} is a time-dependent parameter, and decreases gradually with the increase of immersion time. Besides, significantly decrease at the early immersion stage can be observed. It can be observed that the increase of immersion time from 30d to 60d produces a decrease of 65.71%, 67.45% and 55.87% on D_{app} for mixes with the content of GGBS equal to 0%, 30% and 60%, respectively, while the change on the immersion time from 60d to 180d results in a reduction of 39.25%, 4.78% and 14.6% on D_{app} with the various GGBS content. The main reason is that RAC mixtures have a high a capillary porosity and larger pore size at early age, while the pores are interconnected. Then, as the development of hydration degree, the pore size of RAC will be decreased with the capillary pores being blocked with hydration products [18]. In addition, it can be noticed from Fig.2 that the D_{app} of RAC incorporation 30% GGBS decreases 28.9%-52.2% compared to the control RAC without GGBS at different immersion time, while decreasing 67.2%-40.6% with the addition of 60% GGBS in RAC mixtures.

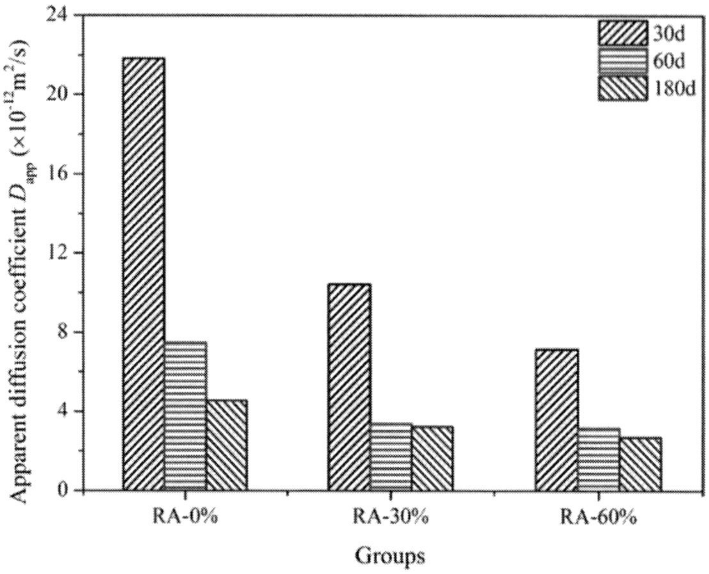

Figure 2. D_{app} of RAC with different immersion time.

5. Conclusion

Long-term bulk immersion test is operated to research the time-dependent chloride diffusion coefficient of RAC with different GGBS replacement. The relationship between immersion time and chloride coefficient is investigated. The effect of GGBS replacement on chloride coefficient is derived. The conclusion obtained from the experimental findings are followed:

● Replacing partially the cement with GGBS can reduce the chloride concentration in RAC effectively. However, the improvement of GGBS on impermeability of RAC decreases with time.

● The apparent diffusion coefficient of RAC of long-term bulk immersion tests decreases with immersion time. When the immersion time increases from 30d to 60d, the values of D_{app} decreases by 65.71%-55.87% with the GGBS content, while an increase of immersion time from 60d to 180d resulted in a reduction of 39.25%-4.78%.

● The reduction of 28.9%-52.2% in D_{app} can be observed with the increase of GGBS content from 0% to 30% at different immersion time. However, the addition of 60% GGBS resulted in a reduction of 67.2%-40.6% in apparent diffusion coefficient compared to the RAC without GGBS.

Acknowledgement

This research was funded by National Natural Science Foundation of People's Republic of China (NSFC 51468004, 51762004) and the Guangxi Special Project for Innovation-driven Development (GKAA18242007, GKAA18118029). The authors are thankful for these supports.

References

[1] Xiao J, Li W, Fan Y, et al. An overview of study on recycled aggregate concrete in China (1996–2011) [J]. *Construction & Building Materials*, 2012, 31(6):364-383.

[2] Sun Y, Xiao J, Aggregate of recycled concrete [J]. *Concrete*, 2004(6):33-36. (Chinese)

[3] Hu B, Liu B and Zhang L, Chloride ion permeability test and analysis for recycled concrete [J]. *Journal of Hefei university of technology (natural science)*, 2009, 32(8):1240-1243. (Chinese)

[4] Zhang J, Li Q, Du J, et al, Experimental study on mineral admixture and recycled aggregates affecting the rapid chloride permeability of high-performance recycled concrete [J]. *Concrete*, 2009(2):94-97. (Chinese)

[5] Thomas C, Setién J, Polanco J A, et al. Durability of recycled aggregate concrete[J]. *Construction and Building Materials*, 2013, 40:1054-1065.

[6] Wang H, Sun X, Wang J, et al. Permeability of concrete with recycled concrete aggregate and

pozzolanic materials under stress[J]. *Materials*, 2016, 9(4):252.

[7] Du T, Li H, Guo T, et al, Test Study on the Resistance of Chloride ion Penetration of Recycled Aggregate Concrete [J]. *Journal of Wuhan University of Technology*, 2006, 28(5):33-36. (Chinese)

[8] Yang L, Hu C, Chen Z and Hong B, Stochastic and Time-Dependent Diffusion of Chloride Ion in Concrete and Its Concentration Distribution [J]. *Journal of Building Materials*, 2013, 16(2):210-216. (Chinese)

[9] Takewaka K, Mastumoto S, Quality and cover thickness of concrete based on the estimation of chloride penetration in marine environments, in: V.M. Malhotra (Ed.), Proc. 2nd Int. *Conf. Concr. Marine Envir*, ACI SP-109, 1988, pp. 381–400.

[10] Wang S, Su Q, Z. Fan, et al, Durability design principle and method for concrete structures in Hong Kong-Zhuai-Macau sea link project [J]. *China Civil Engineering Journal*, 2014(6):1-8. (Chinese)

[11] Jin W, Xue W and Chen J, Effecting coefficients for concrete structure durability design index [J]. *Journal of Building Structures*, 2011, 32(12):86-97. (Chinese)

[12] Mangat P S, Molloy B T, Prediction of long term chloride concentration in concrete, *Mater. Struct.* 27 (1994) 338–346.

[13] Jin Z, Sun W, Zhang Y, et al. Study on damage of HPC under the corrosion of chloride and sulfate [J]. *Industrial Construction*, 2005, 35(1):5-7. (Chinese)

[14] NORDTEST, NT BUILD 443:1995. hardened concrete: accelerated chloride penetration, *Espoo*; 1995.

[15] Yang L F, Chen Z, Feng Q G, et al., “A method for determining chloride content in high performance concrete”, *CN101614718*, 2009. (Chinese)

[16] Gao J M , Qian C X , Liu H F , et al. ITZ microstructure of concrete containing GGBS[J]. *Cement and Concrete Research*, 2005, 35(7):1299-1304.

[17] Villagrán-Zaccardi Y A, Zega C J, Di Maio Á A. Chloride penetration and binding in recycled concrete[J]. *Journal of materials in civil engineering*, 2008, 20(6): 449-455.

[18] Kuma mehta, Paul J M montero. Concrete: microstructural properties and materials [M]. *China Electric Power Press*, 2008.

IOP Conf. Series: Materials Science and Engineering **611** (2019) 012052 doi:10.1088/1757-899X/611/1/012052

Effects of Superfine Grinding on the Structure, Bioactive Components and Antioxidant Activity of Pineapple Leaf Fiber

Lihong Cui[1,2,*], Jihua Li[1], Ze Wang[3], Xiaoyi Wei[1], Wen Xia[1] and Wei Zhou[1]

[1]Agriculture Products Processing Research Institute, Chinese Academy of Tropical Agricultural Sciences, Zhanjiang, China
[2]Chinese Society for Tropical Crops, Haikou, China
[3]College of Food Science and Technology of Huazhong Agricultural University, Wuhan, China

*Corresponding author

Abstract. To investigate the influence of particle size of pineapple leaf fiber (PALF) powders on their structure, bioactive components and antioxidant activities, different sized PALF powders were prepared using superfine grinding. Particle sizes and surface morphology of PALF powders were examined with laser diffraction instrument and scanning electron microscopy (SEM) respectively. Meanwhile, the contents of polyphenols and flavonoids in PALF powders were determined by UV method, and the antioxidant capacity was investigated. During the superfine grinding process, as the time prolonged, the particle size of PALF powders sharply decreased and the surface area increased. When the grinding time reached to 60 min, PALF superfine powder having a median diameter of 18.209 μm was obtained. Scanning electron micrograph displayed that the shape and particle size of superfine powders were uniform. After superfine pulverization, the bioactive components of PALF can be well preserved. The total polyphenols content of PALF superfine powder was 2.347 mg GAE/g DW and the total flavonoids content was 0.227 mg RE/g DW. Therefore, the antioxidant capacity of PALF superfine powder remains high level, including DPPH radical scavenging rate (0.167 mg GAE/g DW), ABTS radical scavenging rate (0.105 mg GAE/g DW) and total antioxidant activity (67.379 unit/g DW). The results demonstrated that PALF superfine powder has good potential to be used in functional food, cosmetics and biomedical materials.

1. Introduction

Pineapple is one of the most popular and delicious perennial fruit. With the rapid development of pineapple cultivation, a large number of pineapple leaves are disposed as a waste after fruit harvesting annually. Recently, due to the concerns of sustainable development, more and more pineapple leaves as an alternative fiber resource has become increasingly attractive[1]. Many attemps has been made to develop the potential applications of this agricultural residues [2].

Pineapple leaf fiber (PALF) is a natural and environmentally friendly fiber extracted from the pineapple leaves, which presents superior mechanical properties[3]. It also possesses some better characteristics such as vascular bundle system and hygroscopicity compared to other natural fibers [4]. In recent years some researchers have isolated nanofibers from PALF, expanding their applications in the field of textile, food, drug delivery, medicinal implants and green packaging materials [5].

Superfine grinding technology is a physical processing approaches to make superfine powder with superior properties [6-7]. It has been adopted in food material industries and biotechnology companies nowadays, but far from prevalent [7]. Many scholars have reported the effects of this technology on the physicochemical properties of fruit, vegetable and grain powders, such as pear pomace powders [6], sugar beet pulp powders [7], red grape pomace powders [8], ginger powders [9] and red rice powders [10]. However, to our knowledge there is no report of PALF superfine powders. This study aims to evaluate the influences of superfine grinding on the surface structure, polyphenol content and antioxidant activities of PALF. We hope this research could provide a useful basis for green utilization of PALF.

2. Materials and Methods

2.1. Materials and Regents

Pineapple (*Ananas comosus* cv. Cayenne) leaf fibers were supplied by Agricultural Machinery Institute, Chinese Academy of Tropical Agricultural Sciences. Gallic acid (≥98%), rutin (≥98%), 1,1-diphenyl-2-picryl-hydrazyl (DPPH), Folin-Ciocalteu reagent and 2,2'-azinobis (3-ethylbenzothiazoline-6-sulfonic acid) (ABTS) were bought from Merck Co., Ltd. Total antioxidant capacity (T-AOC) assay kit were bought from Institute of Jiancheng Biology Engineering, Nanjing.

2.2. Preparation of PALF Powders

Dried PALF was treated with a two-roll mill, and then screened through 40-mesh sieve to obtain coarse powder. The superfine powders were prepared by a vibrating micronizer (WZJ-12B1, China) for 15, 30 and 60 min, respectively. Finally, four sizes of PALF powder samples were obtained, numbered 0, 1, 2 and 3.

2.3. Particle Size Determination

The particle size determination of PALF powders were carried out on Mastersizer 2000 laser particle analyzer manufactured by Malvern, UK. All samples were measured three times.

2.4. Surface Morphology Analysis

The surface morphology of PALF particles was analyzed by using a scanning electron microscope (Hitachi S-4800, Japan) with 0.5-30.0 kV accelerated voltage. All samples were coated with gold of 20 nm before analysis.

2.5. Polyphenols and Flavonoids Content Analysis

The free phenolic, bound phenolic and flavonoid fractions in PALF powder were extracted and determined by our previously reported methods [11]. All samples were extracted three times.

The free and bound phenolic content of PALF powder extracts were determined at 760 nm using Folin-Ciocalteu reagent in a UV–visable spectrophotometer (Shimadzu UV-1780, Japan). Total flavonoid content in PALF powder extracts were measured at 510 nm.

2.6. Antioxidant Capacity Analysis

Extracts of PALF powders were prepared as described in section 2.5 and stored at 4 °C for further use. All samles were measured for three times.

The activities of PALF powder extracts to scavenge DPPH and ABTS free radicals were measured by the method of Wang *et al* [11].

T-AOCs of different sized PALF powder extracts were investigated using a commercial assay kit. Measurements were performed according to the procedure of the assay kit.

3. Results and Discussion

3.1. Particle size of PALF Powders

Table 1 shows the influence of superfine grinding on PALF particle size. The $D_{0.5}$ value represents the median diameter of the particles and can be used to compare the size of different PALF powder samples. Comparing to the coarse powder (0), the diameter of the PALF powder treated by superfine grinding for 15 min (1) was significantly reduced. Then the particle size decreased gradually with longer treating times. The $D_{0.5}$ of the sample obtained after 60 min of superfine grinding was only 18.209 μm.

According to table 1, as the paticle size decreased, the specific surface area increased from 0.173 to 0.474 m^2/g. It was clear that micronization treatment effectively increased the specific surface area of PALF powder by breaking down it into smaller particles. The superfine powders would successfully achieved homogeneity when they are mixed with other powder additives [12].

Table 1. Micromeritic parameters of different size PALF particles[a]

Samples	Particle size $D_{0.5}$ (μm)	Surface-weighted mean diameter (μm)	Volume-weighted mean diameter (μm)	Specific surface area (m^2/g)
0	116.385±0.172	34.699±0.006	168.898±0.092	0.173±0.0001
1	41.510±0.081	18.883±0.002	56.963±0.036	0.318±0.0004
2	29.630±0.066	17.409±0.004	38.329±0.042	0.345±0.0004
3	18.209±0.003	12.647±0.004	22.435±0.022	0.474±0.0005

[a] Each data in the table is expressed as the mean ± standard deviation of three replicates.

3.2. Surface Morphology of PALF Powders

The surface structural changes of PALF powders after superfine grinding treatment were shown in figure 1. The PALF coarse powder retained with original fiber structure and irregular shape. The size of fine powder (treated for 15, 30 min) obviously reduced, but not uniform, still existing partial long fibers. And the size of superfine powder (treated for 60 min) was small and uniform, almost not existing long fiber, which has better structural property than other superfine grinding samples.

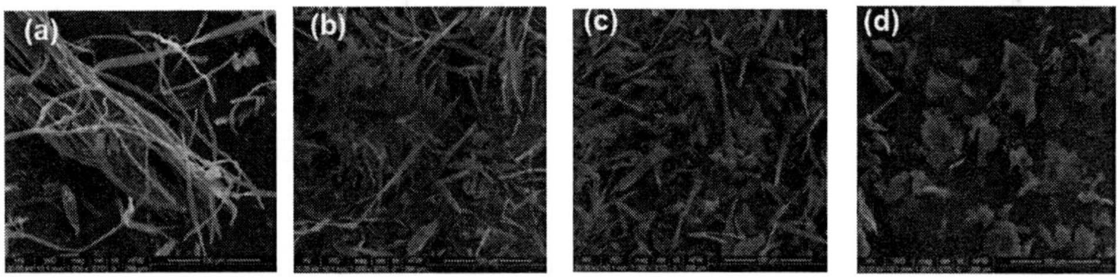

Figure 1. Morphology of different sizes of PALF powders: Coarse powder (a), superfine grinding for 15 (b), 30 (c) and 60 min (d).

3.3. Polyphenols and Flavonoids Content in PALF Powders

It has been reported that the grinding-fractionation process has a great effect on chemical composition separation of different particle size fractions [13]. In our previous studies [11], PALF was found to contain polyphenols and flavonoids. Therefore, the influence of size variation of PALF powder on these bioactive components was investigated. As shown in figure 2, the free phenolic content of PALF powder decreased from 1.752 to 1.344 mg GAE/g DW after 15 min of superfine grinding, and then gradually increased to 1.615 mg GAE/g DW as the grinding time increased from 15 to 60 min. While in the same process, the bound phenolic content of PALF powder increased slightly as the size reduction. For all superfine grinding treated PALF powders, the free phenolic content increased with decreasing particle size. This might be due to the finer the powders and the larger the specific surface area, making it easier to release the souble free phenolic component [14]. The variation trend of total phenolic content was consistent with that of free phenolic content. Total phenolic content of PALF coarse powder and superfine grinding 60 min powder could reach 2.369 and 2.347 mg GAE/g DW, respectively, the difference being not significant. It implied that the bioactive polyphenols of PALF could be preserved well after treated by superfine grinding.

In addition, the total flavonoids content decreased with the size reduction of PALF powders, but increased slightly when the superfine grinding time reached to 60 min. Interestingly, it was not consistent with the changing trend of total polyphenols. The results indicated that the total flavonoids content in PALF powder was more sensitive to temperature and O_2 than total polyphenols. During the process of superfine grinding, the oxidation of total flavonoids might be accelerated by the increasing temperature and the contact area with O_2 along with the treating time.

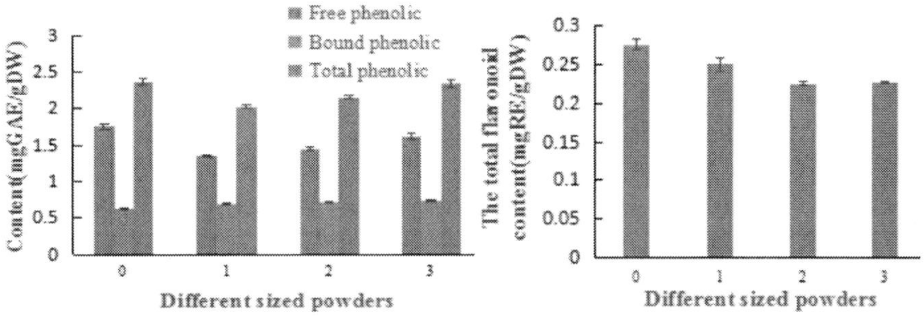

Figure 2. The free phenolic, bound phenolic and total flavonoids content of different particle-sized PALF powders.

3.4. Antioxidant Activity of PALF Powders

The DPPH and ABTS radical models were the basis of antioxidant assay. According to figure 3, the polyphenols in PALF coarse powder have higher values both of DPPH and ABTS radical scavenging capacity than superfine grinding treated powder, reaching to 0.169 and 0.129 mg GAE/g DW, respectively. For all samples treated by superfine grinding, the DPPH and ABTS scavenging activity increased with the size reduction of PALF powder, and these capacities of ultrafine powder obtained at 60 min of superfine grinding were 0.167 and 0.105 mg GAE/g DW, respectively. The changing trend of these two free radical scavenging capacities were nearly identical, and was in agreement with the total polyphenols content.

Furthermore, the T-AOCs assay of different sized PALF powders were carried out, and the change of T-AOC was also in accord with the total polyphenols content. It might be suggested that the antioxidant activity of PALF powder extracts has a great correlation with the total polyphenols content.

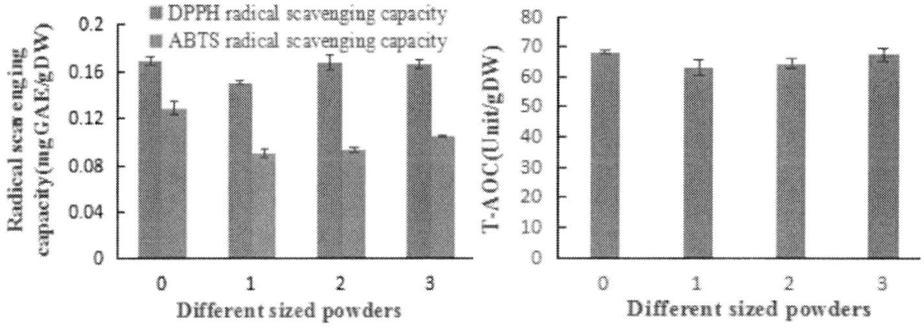

Figure 3. Antioxidant activities of different sized PALF powders extracts

4. Conclusion

Superfine grinding was employed in this work to prepare finer PALF powders. After 60 min of treatment, the particle size of PALF powders drastically decreased from 116.385 to 18.209 μm. According to our results, the content of free and total phenolic increased along with the particle size reduction in the superfine grinding process, as well as DPPH free radical scavenging activity and T-AOC. Compared with the coarse powder, the content of total flavonoids in treated PALF powder decreased slightly, but the superfine PALF powder also had some characteristics keeping high level, including free and total phenolic content, DPPH free radical scavenging activity and T-AOC. In view of these structural advantages and bioactive components, superfine PALF powder can be used in functional food, cosmetics, medicines and biomaterials.

Acknowledgements
We are grateful to the financial support from Zhanjiang Science and Technology Guidance Special Project (No. 2017A03016).

References

[1] Tanpichai S and Witayakran S. 2018. *Polymer Composites*, vol 39, pp 895-903.
[2] Tanpichai S, Witayakran S and Boonmahitthisud A. 2019. *Journal of Environmental Chemical Engineering*, vol 7, p 102836.
[3] Todkar S S and Patil S A. 2019. *Composites Part B*, vol 174, p 106927.
[4] Shamsulalam M, Arifuzzamankhan G M and Abdurrazzaque S M. 2009. *Journal of Natural Fibers*, vol 6, pp 138-150.
[5] Ravindran L, Sreekala M S and Thomas S. 2019. *International Journal of Biological Macromolecules*, vol 131, pp 858-870.
[6] Yan L, Li T, Liu C and Zheng L. 2019. *LWT - Food Science and Technology*, vol 107, pp 171-177.
[7] Huang X, Dou J and Li D. 2018. *LWT - Food Science and Technology*, vol 87, pp 203-209.
[8] Zhao X, Zhu H, Zhang G and Tang W. 2015. *Powder Technology*, vol 286, pp 838-844.
[9] Zhao X, Yang Z, Gai G and Yang Y. 2009. *Journal of Food Engineering*, vol 91, pp 217-222.
[10] Chen Q M, Fu M R, Yue F L and Cheng Y Y. 2015. *Food and Nutrition Sciences*, vol 6, pp 1277-1284.
[11] Wang Z, Cui L, Tang B, Wang F and Li J. 2016. *Advances in Engineering Research*, vol 63, pp 267-272.
[12] Riley C K, Adebayo S A, Wheatley A and Asemota H N. 2008. *Powder Technology*, vol 185, pp 280-285.
[13] Maaroufi C, Melcion J P, Monredon F, Giboulot B, Guibert D and Le Guen M P. 2000. *Animal Feed Science & Technology*, vol 85, pp 61-78.
[14] Zhong C, Zu Y, Zhao X, Li Y, Ge Y, Wu W, Zhang Y, Li Y and Guo D. 2016. *International Journal of Food Science and Technology*, vol 51, pp 212-221.

IOP Conf. Series: Materials Science and Engineering **611** (2019) 012053 doi:10.1088/1757-899X/611/1/012053

Adsorption and Co-precipitation of Metoprolol with Struvite Recovered from Synthetic Source Separated Black Wastewater

Jiangjiang Wang [1], **Qixing Zhou** [2,*] **and Yang Liu** [3]

[1]College of Environmental Science and Engineering, Nankai University, Tianjin 300350, China

[2]College of Environmental Science and Engineering, Nankai University, Tianjin 300350, China

[3]Department of Civil and Environmental Engineering, 7-263 Donadeo Innovation Centre for Engineering, University of Alberta, Edmonton, Alberta, Canada T6G 1H9

*Corresponding author

Abstract. Due to the residues of metoprolol (MET) in black wastewater, recovering phosphate as struvite ($MgNH_4PO_4 \cdot 6H_2O$) from black wastewater may pose threats to the utilization of struvite in the agricultural planting and human health. In this study, the adsorption and co-precipitation experiments of MET in the formation of struvite from synthetic black wastewater will be performed and some factors like initial MET concentration, pH values, temperatures were investigated. Results revealed that the adsorption process can be divided into rapid adsorption, fluctuation stage and steady stage, and the maximum MET adsorption quantity was 0.05029mg/g under 80ppb, 20℃. Chemical adsorption and pseudo second-order kinetics were evident at the low concentration level. Neither Langmuir nor Freundlich isotherms fitted for the adsorption. The presence of $Mg[H_2O]_6^{2+}$ (Hexahydrate and magnesium ions) interfered with MET adsorption through recombining with MET, which was also determined by pH variation.
The existence of MET in black wastewater would not obstruct the formation of struvite in alkaline condition generally and serious terrible influences would not be generated when using struvite as fertilizer. A vast majority of MET will stay in the solution after 3 hours' reaction.

1. Introduction

A large amount of pharmaceuticals have been produced and applied in the treatment of human diseases. In the treatment of high blood pressure (hypertension), angina, arrhythmias and heart rhythm disturbances [1-3], β-blockers have been widely used for many years as a series of effective compounds. According to Ramil et al. [4], metoprolol was the most frequently used β-blocker, and can be detected in the environmental matrix and domestic black water frequently [2].

Like other pharmaceutically active compounds, β-blocker will be excreted through urine and faces as a mixture of metabolites, unchanged substances or conjugated with inactivating substances [5]. No matter in surface water, sewage treatment plants or blackwater, β-blockers and their metabolites can be detected in the concentration range from μg/L to mg/L [1,6]. Although no comprehensive study about the acute aquatic toxicity of β-blockers has been conducted, a significant cumulative effect on the metabolism of non-target organisms and the ecosystem can be achieved by them [7].

Nutrients like ammonia and phosphorus are still exist in sewage waste and black water. Recovering ammonia and phosphorus from wastewater in the form of struvite ($MgNH_4PO_4 \cdot 6H_2O$) was an ideal

approach which had been applied in Japan, Netherlands, Italy and so on [8-11]. Struvite was an excellent slow release fertilizer in agriculture industry because of its composition and physicochemical characteristics. Few researchers have studied the issues of micro-pollutants adsorption or coprecipitation in the process of struvite crystallization in source separated black water. Inestimable effects will be achieved on the soil and even the ecosystem when such struvite fertilizer used in agriculture.
Therefore, the objectives of our work are: firstly, stimulating the main components of black wastewater to recover struvite and discussing the adsorption effects. Secondly, considering the coprecipitation mechanisms of metoprolol (MET) with struvite formation.

2. Materials and Methods

2.1. Materials

Inorganic salts in this experiment including $MgCl_2 \cdot 6H_2O$, $Na_2HPO_4 \cdot 12H_2O$, NH_4Cl, $CaCl_2$, KCl, KNO_3, Na_2SO_4, $NaHCO_3$, and ammonia hydroxide etc. All of them are guaranteed reagent and purchased from Sinopharm Chemical Reagent Co., Ltd. L-Ascorbic Acid (≥99.99%), KH_2PO_4 (≥99%) are bought from Acros Organic (USA). Potassium antimonyl tartrate sesquihydrate and ammonium molybdate are guaranteed reagent and bought from Energy Chemical (Shanghai, China). (±) metoprolol (+) tartrate salt (MET, CAS NO.56392-17-7) is bought from Sigma-Aldrich Inc. (USA) and dissolved in ultrapure water. Methanol and formic acid are UPLC/LC-MS and UPLC grades, both of them are bought from ANPEL Laboratory Technologies Inc. (Shanghai, China). The pH of metoprolol (MET) solution are adjusted to the required values with diluted 0.5mol/L NaOH or HCl.

2.2. Metoprolol Dsorption Experiment

2.2.1. Struvite Recovery from Synthetic Blackwater. Inorganic salts are added into distilled water to simulate black wastewater matrix after treated by wastewater treatment plants. The initial phosphate concentration is set at c_0=204mg/L as the baseline of other ions. Stimulation of ionic concentrations in black water after treated by wastewater treatment plants are presented in Table.1. Magnesium ion concentration is a limiting factor, but excessive ammonium is highly beneficial to the precipitation of struvite [12]. In order to get the maximum degree of phosphate recovery, the ratios of $C_{Mg^{2+}}$:$C_{NH_4^+}$:$C_{PO_4^{3-}}$ are adjusted as 0.8:1:1 to 2.0:1:1. The pH value of struvite precipitation is maintained in 8.0 according to previous literature [13]. Meanwhile, struvite formation process will pass through two stages, including nucleation and crystal growth. Nearly 3 hours will be taken to complete struvite precipitation [12,14]. Therefore the mixture will be shaken for 180 min on a rotary shaker at 120rpm under room temperature (298K). The concentrations of phosphate in different samples (c_e) are measured according to ascorbic acid colorimetric method [15]. And the removal (%) is calculated by

$$\frac{c_0 - c_e}{c_0} \times 100\% \tag{1}$$

After that, the precipitates will be filtered with 0.45μm cellulose acetate filter, washed with deionized water, and dried under 40℃ in oven for 48 hours.

Table 1. Stimulation of ionic concentrations in black wastewater after treated by wastewater treatment plants.

Ionic species	HCO_3^{2-}	NH_4^+	PO_4^{3-}	Ca^{2+}	K^+	Na^+	Cl^-	SO_4^{2-}	NO_3^-
Concentration (mg/L)	1040.4	1414.0	204.0	39.0	424.0	604.0	695.0	157.0	26.0

After finding the optimum ratio of $C_{Mg^{2+}}$:$C_{NH_4^+}$:$C_{PO_4^{3-}}$, a large amount of struvite will be produced and stored in a sealed container for the subsequent experiments.

2.2.2. Struvite Characterization. Surface morphologies and microstructure of struvite samples were examined by environmental scanning electron microscope with FEG (ESEM, Quanta FEG 250, FEI,

USA) and transmission electron microscopy (TEM, JEM-2800, 200kV, Japan). X-ray powder diffraction (XRD) patters were recorded in the 2θ range of 10-80° at a scan speed of 10°/min by using a diffractometer (Ulitma IV, Rigaku, Japan) with Cu Kα radiation (40mA,40kV). Nitrogen-adsorption-desorption isotherms were measured at the liquid nitrogen temperature 77K using ASAP 2460 (Micromeritics Inc., USA). The struvite samples were degassed at 50℃ for 5h under vacuum prior to analysis. The specific surface area (S_{BET}) was determined from the linear part of the BET plot ($P/P_0 = 0.01 - 0.15$). The pore size (d) was calculated from the adsorption branch of isotherm by using Barrett-Joyner-Hallenda (BJH). The total pore volume (V_{total}) was determined from the adsorbed nitrogen amount at P/P_0 =0.99.

2.2.3. Adsorption Processes. A series of batch experiments were performed to investigate the metoprolol adsorption performances of struvite adsorbents. Metoprolol solution with different concentrations were prepared by dissolving (±) metoprolol (+) tartrate salt (MET) in ultrapure water.

Adsorption kinetic experiment was performed with metoprolol initial concentrations in 35, to 300 ppb. pH values of solution were adjusted to 8.0. The process was performed on a thermostat shaker under 120rpm, 298K. Samples were withdrawn from the flask at pre-determined time intervals from 15 to 180 min and filtered by 0.22μm syringe filters immediately before further analysis. The MET concentration was detected by ultra-performance liquid chromatography (UPLC, Waters, USA.)/quadrupole-time-of-flight tandem mass spectrometry (MS/MS, Xevo TQ-S, USA.) with a 100mm (1.7μm particle size) UPLC BEH C18 column. The UPLC-MS/MS was operated in ESI positive mode with methanol (solvent A, 65%) and 0.2% formic acid (solvent B, 35%) at a flow rate of 0.3mL/min and an injection volume of 10μL. The adsorption capacities (q_t) were calculated by the following equation:

$$q_t = \frac{(C_0 - C_t) \cdot V}{m} \tag{2}$$

where C_0 and C_t were the MET concentrations in solution at initial and different time periods (μg MET /L), respectively; V was the volume of solution (L) and m was the mass of struvite adsorbent (g).

In order to analyse the kinetic mechanism of adsorption process, the experimental data were fitted in the pseudo first-order, pseudo second-order models which were calculated as (3) [16,17], (4) [16,17].

Pseudo first-order equation:

$$e^{-k_1 t} = \ln(1 - \frac{q_t}{q_e}) \tag{3}$$

Pseudo second-order equation:

$$\frac{t}{q_t} = \frac{1}{k_2 q_e^2} + \frac{t}{q_e} \tag{4}$$

where q_t and q_e were the amount of MET adsorbed over a given period of time t (mg MET/g) and at equilibrium (mg MET/g), respectively; t was the adsorption time (min); k_1 (1/min), k_2 (g/mg/min) were the adsorption rate constants of the pseudo first-order adsorption, pseudo second-order adsorption, respectively.

Equilibrium experiment was conducted with different metoprolol concentrations and temperature respectively. The pH value of solution was adjusted to 8.0. Solutions were filtered through 0.22μm syringe filters after 1.5h immediately prior to further analysis. The MET concentration was detected by UPLC-MS/MS (Waters UPLC Xevo TQ-S, USA.).

The adsorption isotherms were fitted by Langmuir and Freundlich models which were described as (5) [17,18] and (6) [19].

$$\text{Langmuir model: } q_e = \frac{q_m k_l c_e}{1 + k_l c_e} \tag{5}$$

$$\text{Freundlich model: } q_e = k_F c_e^{\frac{1}{n_F}} \tag{6}$$

where C_e (μg MET/L) was the concentration of MET solution at equilibrium; q_e (mg MET/g) was the corresponding adsorption capacity; q_m (mg MET/g) and k_F (mg/g) were the constants in the

IOP Conf. Series: Materials Science and Engineering **611** (2019) 012053 doi:10.1088/1757-899X/611/1/012053

Langmuir and Freundlich isotherm models. k_L (L/mg) was the constant in Langmuir isotherm model which related to the energy of adsorption. n_F was the constant in Freundlich isotherm model which indicated the adsorption intensity.

2.2.4. Effects of Solution pH on Adsorption. Considering struvite was very sensitive to pH changes [20] and mainly existed in alkaline condition [13,14]. The interaction between struvite and metoprolol was only possible at pH levels between 8.0 and 10.5.
This part experiment was performed with pH values from 8.0 to 10.5. After 1.5 hours, filtering the solution through 0.22 μm syringe filters immediately and analysed by UPLC-MS/MS.

2.3. Struvite Precipitation from Metoprolol Contaminated Blackwater

2.3.1. Effects of Solution pH on Co-precipitation. Adding 60ppb MET solution into black wastewater matrix based on Table.1. And pH values are from 8.0 to 11.0. After 3 hours, separating solution and solid compounds by filtering through 0.22μm syringe filters immediately. Testing the concentration of phosphate and MET. Meanwhile, the solid compounds were tested by XRD spectrogram.

2.3.2. Effects of MET Concentration and Contact Time on Co-precipitation. Adding 60ppb, 80ppb, and 100ppb MET solution into blackwater matrix under 298K, 120rpm. Adjusting pH value was 8.0. Sampling the solution in 20min intervals until 3 hours and testing MET concentration by UPLC/MS-MS after filtering through 0.22μm cellulose acetate membrane. The sediments were separated and tested by XRD spectrogram, which was compared with standard struvite.

3. Results and Discussion

3.1. Phosphate Removal and Struvite Characterization
Phosphate removal (%) from synthetic blackwater under different $C_{Mg^{2+}}:C_{NH_4^+}:C_{PO_4^{3-}}$ ratio were presented in Figure 1. When $C_{Mg^{2+}}:C_{NH_4^+}:C_{PO_4^{3-}}$ were 1.6:1:1, 1.8:1:1, and 2.0:1:1, phosphate removal (%) would reach to more than 95.5%, which was higher than other ratios. Meanwhile, no significant differences among these three results.

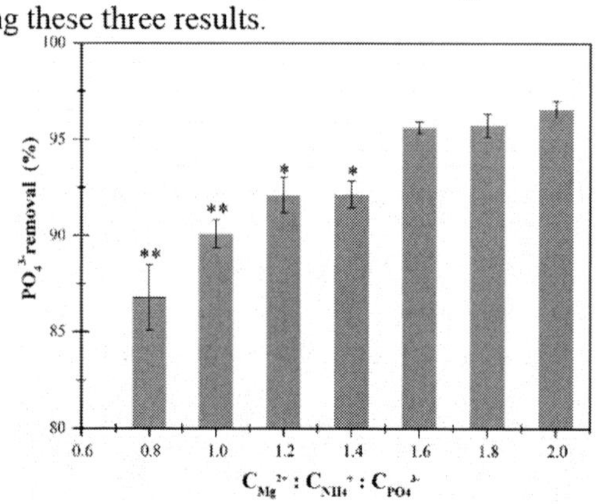

Figure 1. Phosphate removal (%) under different $C_{Mg^{2+}}:C_{NH_4^+}:C_{PO_4^{3-}}$ ratio.
"**"indicates $P < 0.01$ compared to 1.6, "*"indicates $0.01 < P < 0.05$ compared to 1.6.

XRD and SEM images were also used to analyse the synthetic struvite samples. Figure 2 was the comparison of synthetic struvite samples with standard substance by XRD spectrogram. There were no significant differences among synthetic and standard struvite. The SEM images of synthetic struvite samples were presented in Figure 3. In Figure3 (a) and (b), a large amount of transparent rod-like crystals were observed which can be considered as struvite [13]. But in Figure3 (c), circular coacervates impurities were appeared.

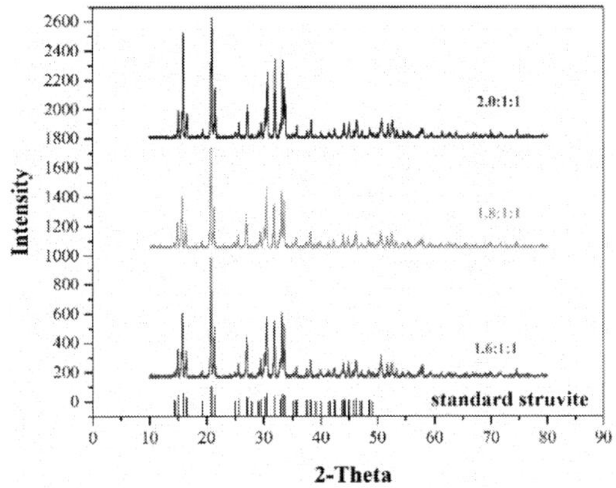

Figure 2. XRD graph of synthetic and standard struvite.

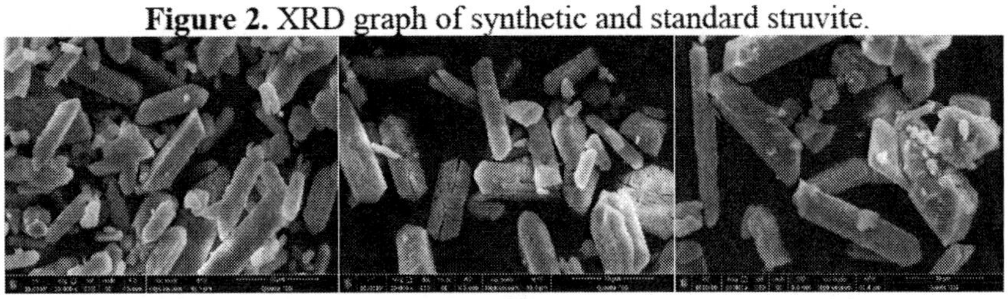

(a) (b) (c)

Figure 3. SEM images under different $C_{Mg^{2+}}:C_{NH_4^+}:C_{PO_4^{3-}}$ conditions
(a) $C_{Mg^{2+}}:C_{NH_4^+}:C_{PO_4^{3-}}$ =1.6:1:1, (b) $C_{Mg^{2+}}:C_{NH_4^+}:C_{PO_4^{3-}}$=1.8:1:1, (c) $C_{Mg^{2+}}:C_{NH_4^+}:C_{PO_4^{3-}}$ =2.0:1:1.

In spite of phosphate removal (%), the specific surface area of generated struvite samples would be considered at the same time. Table.2 was the BET surface area of different struvite samples.

Table 2. BET surface area of struvite produced from different $C_{Mg^{2+}}:C_{NH_4^+}:C_{PO_4^{3-}}$ ratio.

$C_{Mg^{2+}}:C_{NH_4^+}:C_{PO_4^{3-}}$	1.6:1:1	1.8:1:1	2.0:1:1
BET surface area (m2/g)	97.4460	83.3276	70.5147

The surface area of struvite sample produced under the $C_{Mg^{2+}}:C_{NH_4^+}:C_{PO_4^{3-}}$ ratio was 1.6:1:1 can reach to 97.4460m^2/g, which was much larger than any other ratios. N_2 adsorption-desorption curve and pore size distribution for struvite powder were shown in Figure 4.

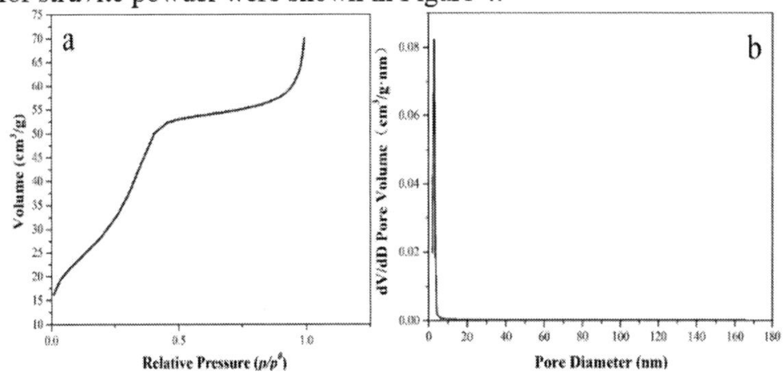

Figure 4. Nitrogen adsorption-desorption curves (a) and pore size distribution (b) for struvite powder.

Therefore, 1.6:1:1 of $C_{Mg^{2+}}:C_{NH_4^+}:C_{PO_4^{3-}}$ ratio was chosen as the best one to generate struvite, which used in the next adsorption experiments. Its TEM images was shown in Figure5.

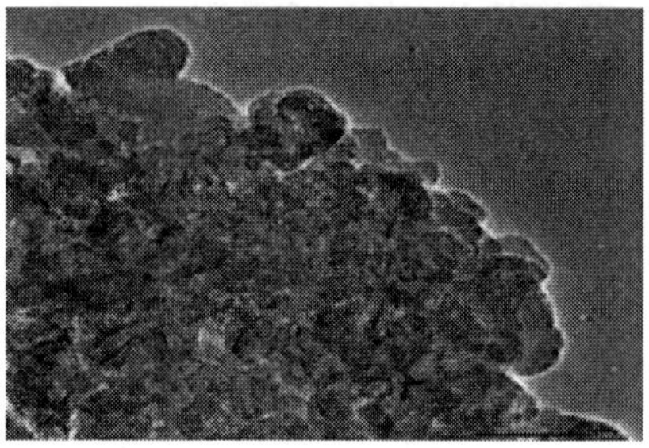

Figure 5. TEM image of struvite sample under $C_{Mg^{2+}}:C_{NH_4^+}:C_{PO_4^{3-}}$ =1.6:1:1.

3.2. Metoprolol Adsorption Experiment

3.2.1. Metoprolol Adsorption Kinetics. The adsorption quantity by adding struvite in the solution with different MET concentrations, e.g. 35ppb, 60ppb, 80ppb, 100ppb, 200ppb, and 300ppb respectively, was shown in Figure6 (a). The adsorption kinetics graphs were shown in Figure 6 (b) and Figure 6 (c), and the correlation coefficients and parameters of adsorption kinetics were presented in Table. 3. As seen from Figure6 (a), when the MET concentrations maintained in relative low levels, e.g. 35ppb, 60ppb, and 80ppb, and the solution pH was 8.0, chemisorption was predomination. Meanwhile, the adsorption efficiency improved with the MET concentration increased. In these concentrations, the whole adsorption processes can be divided into three stages, rapid adsorption stage (0-15min), fluctuation stage (15-120min), and steady stage (120-180min). The maximum adsorption quantity was 0.05029mg/g in 80ppb MET concentration.

While, physical adsorption, e.g. Van der Waals' force, preponderated when the concentrations of MET reached to relative high levels, like 100ppb, 200ppb, and 300ppb under pH 8.0. It was highly notable that MET concentrations can affect Van der Waals' force. The adsorption efficiency maintained in low level when the concentration of MET was in high level. Overall, the physical adsorption efficiency was decreased when the MET concentration increased. High concentrations of MET will resign adsorption between MET and struvite.

Under practical conditions, the average concentration of MET in source separated black wastewater was 45ppb, and the maximum was 68ppb [6]. Thus, MET adsorption in struvite will not affect its quality theoretically. The environmental risks generated from MET adsorbed in struvite can be ignored.

From Table.3, the correlation coefficients of pseudo first-order were not good enough. But the pseudo second-order correlation coefficients were better when MET concentration was lower than 100ppb. These suggested that there were chemical forces by sharing or exchanging electrons at the solid/solution interface [16,21].

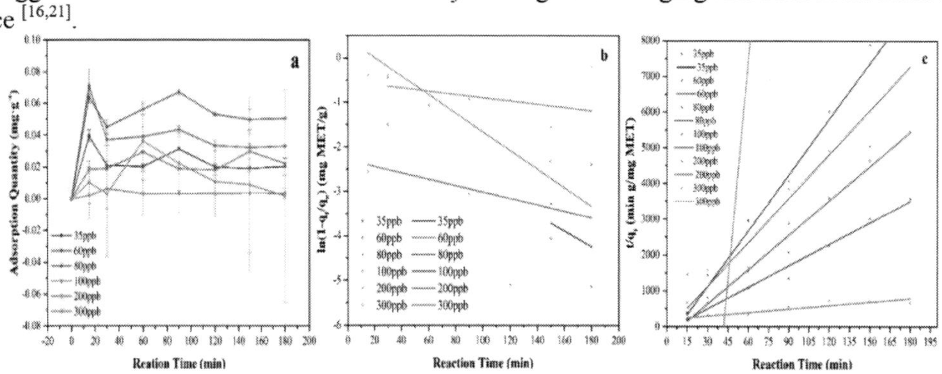

Figure 6. MET adsorption quantity graph (a), pseudo first-order model (b), and pseudo second-order model (c).

Table 3. Pseudo first-order model and pseudo second-order model parameters under different

Pseudo first-order model	MET concentration	35ppb	60ppb	80ppb	100ppb	200ppb	300ppb
	R^2	-	-	-	0.1312	0.60711	0.06188
Pseudo second-order model	MET concentration	35ppb	60ppb	80ppb	100ppb	200ppb	300ppb
	R^2	0.96334	0.98726	0.98102	0.93722	0.82181	0.50744
	q_e (mgMET/g)	0.0192	0.031213	0.049843	0.024402	0.305119	0.002422
	k_2 (g/mg/min)	-5.95843	-3.0675	-3.64009	-14.919	0.056011	-9.85424

3.2.2. Metoprolol Adsorption Isotherms. MET removal rate graphs under 20℃, 35℃, 40℃, and 60℃ were shown in Figure7 (a), Langmuir and Freundlich isotherms were shown in Figure7 (b) and (c), respectively.

From Figure7 (a), under 20℃ condition, MET removal rate was the highest among the other temperatures no matter in any concentration level. The highest removal rate reached to over 70% when MET concentration was 50ppb, and in other temperatures, the removal rate of MET were lower than 35%. A conclusion can be drawn from Figure 7(b), (c) and Table.4, neither Langmuir isotherm nor Freundlich isotherm fitted for MET adsorption. So the adsorption between MET and struvite was not monolayer, homogenous or multilayer, heterogeneous adsorption [2,17,22,23].

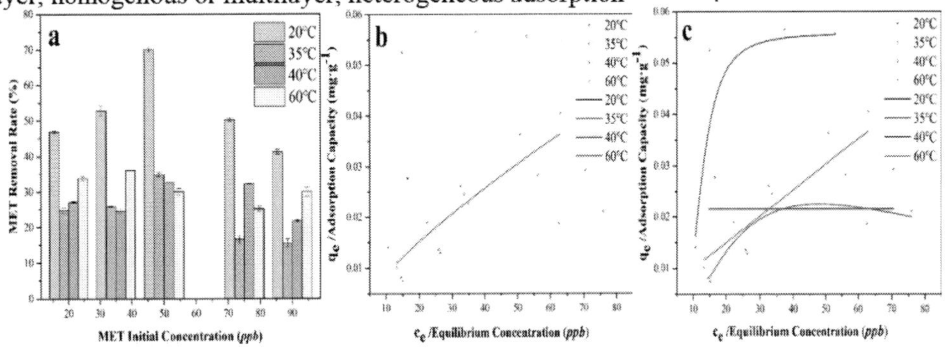

Figure 7. MET removal rate under 20℃, 35℃, 40℃, and 60℃ (a), Langmuir isotherm (b), and Freundlich isotherm (c).

Table.4 Correlation coefficients of Langmuir and Freundlich isotherms under 20℃, 35℃, 40℃, and 60℃.

	Temperature (℃)	Langmuir isotherm	Freundlich isotherm
R^2	20	-	0.7083
	35	-	0.71651
	40	-	-
	60	0.90292	0.90057

3.3. The Effect of pH

Solution pH was another important factor which may affect the adsorption between struvite and MET. Both MET residual rate and removal rate under pH 8.0 to 10.5 were shown in Figure8. In the pH ranges from 8.0 to 9.5, the removal rate of MET was decreasing gradually. While the removal rate was increasing sharply after pH 9.5.

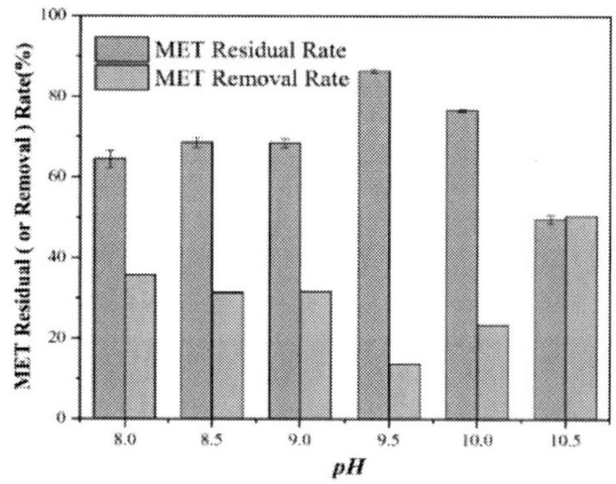

Figure 8. MET residual and removal rate under pH 8.0-10.5.

It has been known that struvite crystal consists of PO_4^{3-} and NH_4^+ tetrahedral and $Mg[H_2O]_6^{2+}$ octahedral, and the rectangular ac or bc crystallite facets (a, b, c, are the three-dimension coordinate axes) are rich in $Mg[H_2O]_6^{2+}$ or positive charge as compared to ab crystal facets [18,24]. Considering that pk_a of MET was 9.68 [25], it would protonate when the pH range was 8.0 to 9.5 and deprotonate at pH 10.0 and 10.5. Therefore struvite crystallite facets in $Mg[H_2O]_6^{2+}$ were prone to reject MET at pH 8.0, 8.5, and 9.0, and the adsorption of MET would occur when pH values were 10.0 and 10.5.

3.4. Effects of pH on Co-precipitation

Under different pH values, phosphate removal rate and MET residual rate in solution were shown in Figure 9(a), and the XRD spectrograms under different pH conditions after 3 hours were compared with standard struvite samples presented in Figure 9(b).

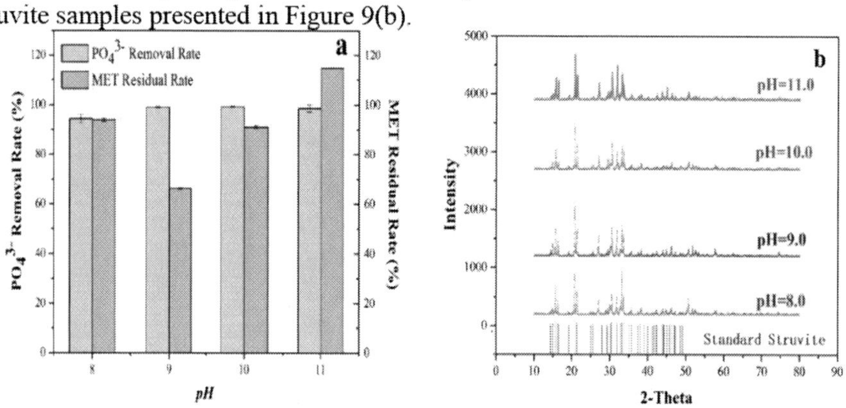

Figure 9. under pH value was 8.0, 9.0, 10.0, and 11.0, PO_4^{3-} removal rate and MET residual rate (a), and XRD graphs of struvite samples compared with standard struvite (b).

In Figure 9(a), phosphate removal rate was in the range of 95% to 100%. Comparing the sediments with standard struvite, there were no differences between their XRD graphs (Figure 9(b)). We can conclude from Figure 9(a) and (b) that the existence of MET will not obstruct the formation of struvite. On the other hand, adding MET into the black wastewater matrix at the pH value was 8.0, 9.0, and 10.0 will increase the residual of MET in solution significantly and this conclusion can be demonstrated by comparing Figure 9(a) with Figure 8. In addition, the residual rate of MET under different pH value was in the range of more than 65% to nearly 100%, most of MET will be maintained in the solution rather than getting into the crystal of struvite. Although when pH was 9.0, residual rate of MET was only 66%, the phosphate removal rate was close to 100% and XRD graph was very similar to standard sruvite. A conclusion still can be drawn that phosphate recovery from blackwater will not be affected by the existence of MET obviously. This conclusion was similar to the research taken by Abel-Denee, Abbott and Eskicioglu [26] who studied the disturbance of struvite formation when triclosan and triclocarban existed.

3.5. Effects of MET Concentration and Contact Time on Co-precipitation

MET residual rate under the concentrations of MET were 60ppb, 80ppb, and 100ppb from 20min to 180min at 8.0 pH value was shown in Figure 10. XRD spectrograms of sediments after 3 hours' reaction and compared with standard struvite was shown in Figure 11.

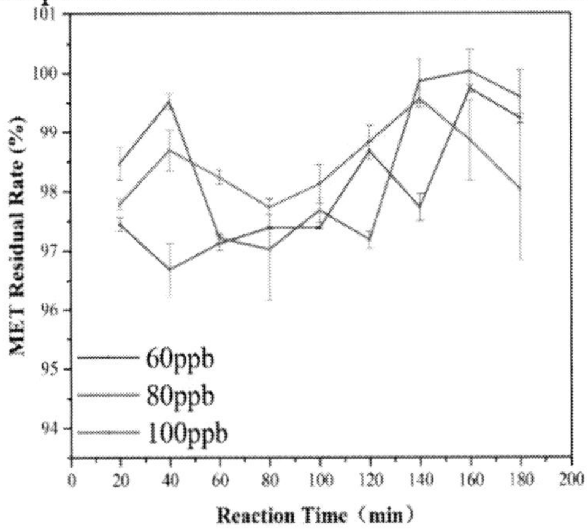

Figure 10. MET residual rate under time intervals

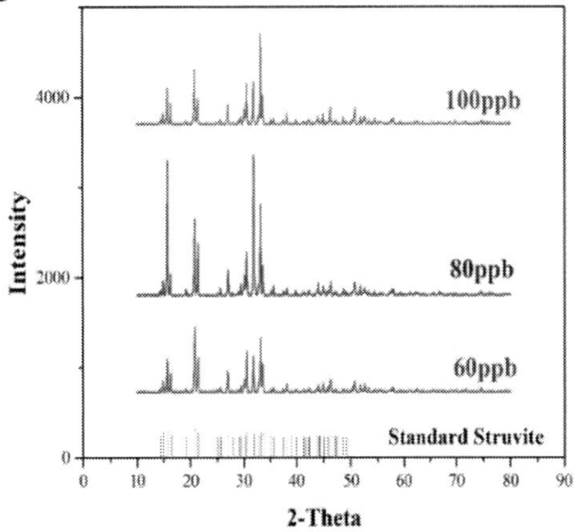

Figure 11. XRD graphs of sediments.

From Figure 10, there was no significant increase or decrease trend of MET residual rate in solution when MET added into black wastewater matrix. But a conclusion can be drawn that no matter in low or high MET concentration conditions, the final residual rate was higher than the initial state. The residual rate of MET in any time intervals were higher than 96.5%. Figure 11 indicated that the sediments of reaction were struvite.

Thus, we can conclude that, MET molecule shuttled back and forth in the interface of solid-liquid during the process of struvite crystal, and the vast majority of MET will stay in the solution in the end. This phenomenon may be explained by the two steps of struvite formation, which called nucleation and crystal growth [12,14], both of them would complete no more than 45min [18]. So after 60min, MET molecule will get in touch with struvite in the interface of solid-liquid and be rejected by struvite, which was consistent with 3.3.

4. Conclusion

This study gave information about the adsorption and co-precipitation of MET in the formation of struvite from synthetic black wastewater. In adsorption experiment, the whole process can be divided into three stages: rapid adsorption, fluctuation stage, and steady stage. The maximum adsorption quantity was 0.05029mg/g (20 ℃, pH was 8.0). Chemical adsorption predominated in the low concentration condition, but physical adsorption preponderated in high concentration of MET and decreased with MET concentration increased. Pseudo second-order kinetics fitted for MET adsorption when the concentration was lower than 100ppb. Neither Langmuir nor Freundlich adsorption isotherms fitted for the adsorption experiments.

The pH value was another significant factor which will affect the adsorption efficiency. $Mg[H_2O]_6^{2+}$ in the surface of struvite crystal will adsorb deprotonated MET molecule when pH > pk_a, and reject protonated MET molecule when pH < pk_a.

The existence of MET in blackwater matrix would not obstruct the formation of struvite in alkaline codition generally. But MET molecule will shuttle back and forth in the interface of solid-liquid during the formation of struvite and the vast majority of MET will stay in the solution finally.

No matter in the adsorption or co-precipitation experiments, the existence of MET would not affect the struvite recovery from black wastewater. A very majority of MET will be maintained in the solution. The adsorption quantity of MET in the surface of struvite was very few which will not generate serious influences when using struvite as the fertilizer in the agriculture area.

Acknowledgements
This work is financially supported by the Ministry of Education, People's Republic of China, as special funds for basic scientific research services of central colleges and universities, a 111 program (grant No. T2017002) and an innovative team rolling project (grant No. IRT_17R58).

References
[1] Maurer M, Escher B I, Richle P, et al. Elimination of beta-blockers in sewage treatment plants [J]. Water Research, 2007, 41(7): 1614-1622.
[2] Maszkowska J, Stolte S, Kumirska J, et al. Beta-blockers in the environment: part I. Mobility and hydrolysis study [J]. Maszkowska J , Stolte S , Kumirska J , et al. Beta-blockers in the environment: Part I. Mobility and hydrolysis study[J]. Science of The Total Environment, 2014, 493:1112-1121.
[3] Maszkowska J, Stolte S, Kumirska J, et al. Beta-blockers in the environment: part II. Ecotoxicity study [J]. Science of the Total Environment, 2014, 493: 1122-1126.
[4] Ramil M, El A T, Fink G, et al. Fate of beta blockers in aquatic-sediment systems: sorption and biotransformation [J]. Environmental Science & Technology, 2010, 44(3): 962-970.
[5] Hallingsørensen B, Nors N S, Lanzky P F, et al. Occurrence, fate and effects of pharmaceutical substances in the environment--a review [J]. Chemosphere, 1998, 36(2): 357-393.
[6] de Graaff M S, Vieno N M, Kujawa-Roeleveld K, et al. Fate of hormones and pharmaceuticals during combined anaerobic treatment and nitrogen removal by partial nitritation-anammox in vacuum collected black water [J]. Water Research, 2011, 45(1): 375-383.
[7] Bendz D, Paxeus N A, Ginn T R, et al. Occurrence and fate of pharmaceutically active compounds in the environment, a case study: Hoje River in Sweden [J]. Journal of Hazardous Materials, 2005, 122(3): 195-204.
[8] Giesen A. Crystallisation Process Enables Environmental Friendly Phosphate Removal at Low Costs [J]. Environmental Technology Letters, 1999, 20(7): 769-775.
[9] Ueno Y, Fujii M. Three years experience of operating and selling recovered struvite from full-scale plant [J]. Environmental Technology, 2001, 22(11): 1373-1381.
[10] Battistoni P, Boccadoro R. Struvite crystallisation in sludge dewatering supernatant using air stripping: the new full-scale plant at Treviso (Italy) sewage works [J]. 2001.
[11] Battistoni P, Boccadoro R, Fatone F, et al. Auto-Nucleation and Crystal Growth of Struvite in a Demonstrative Fluidized Bed Reactor (FBR) [J]. Environmental Technology, 2005, 26(9): 975-982.
[12] Stratful I, Scrimshaw M D, Lester J N. Conditions influencing the precipitation of magnesium ammonium phosphate [J]. Water Research, 2001, 35(17): 4191-4199.
[13] Hao X D, Wang C C, Lan L, et al. Struvite formation, analytical methods and effects of pH and Ca^{2+} [J]. Water Science & Technology, 2008, 58(8): 1687-1692.
[14] Corre K S L, Valsamijones E, Hobbs P, et al. Phosphorus recovery from wastewater by struvite crystallization: a review [J]. Critical Reviews in Environmental Science & Technology, 2009, 39(6): 433-477.
[15] Federation W E, Association A P H. Standard methods for the examination of water and wastewater [J]. American Public Health Association (APHA): Washington, DC, USA, 2005.
[16] Wang H, Wang X, Ma J, et al. Removal of cadmium (II) from aqueous solution: A comparative study of raw attapulgite clay and a reusable waste-struvite/attapulgite obtained from nutrient-rich wastewater [J]. Journal of Hazardous Materials, 2017, 329: 66-76.
[17] Huang W-Y, Li D, Liu Z-Q, et al. Kinetics, isotherm, thermodynamic, and adsorption mechanism studies of La(OH) 3 -modified exfoliated vermiculites as highly efficient phosphate adsorbents [J]. Chemical Engineering Journal, 2014, 236: 191-201.
[18] Ye Z-L, Deng Y, Lou Y, et al. Adsorption behavior of tetracyclines by struvite particles in the process of phosphorus recovery from synthetic swine wastewater [J]. Chemical Engineering Journal, 2017, 313: 1633-1638.

[19] Ivanets A I, Kitikova N V, Shashkova I L, et al. Removal of Zn2+, Fe2+, Cu2+, Pb2+, Cd2+, Ni2+ and Co2+ ions from aqueous solutions using modified phosphate dolomite [J]. Journal of Environmental Chemical Engineering, 2014, 2(2): 981-987.
[20] Başakçılardan-Kabakcı S, Thompson A, Cartmell E, et al. Adsorption and Precipitation of Tetracycline with Struvite [J]. Water Environment Research, 2007, 79(13): 2551-2556.
[21] Lalhmunsiama, Tiwari D, Lee S-M. Surface-functionalized activated sericite for the simultaneous removal of cadmium and phenol from aqueous solutions: Mechanistic insights [J]. Chemical Engineering Journal, 2016, 283: 1414-1423.
[22] Xia P, Wang X, Wang X, et al. Synthesis and Characterization of MgO Modified Diatomite for Phosphorus Recovery in Eutrophic Water [J]. Journal of Chemical & Engineering Data, 2016, 62(1): 226-235.
[23] Chao Y, Zhu W, Wu X, et al. Application of graphene-like layered molybdenum disulfide and its excellent adsorption behavior for doxycycline antibiotic [J]. Chemical Engineering Journal, 2014, 243: 60-67.
[24] Ye Z, Shen Y, Ye X, et al. Phosphorus recovery from wastewater by struvite crystallization: Property of aggregates [J]. Journal of Environmental Sciences, 2014, 26(5): 991-1000.
[25] Butkovskyi A, Hernandez Leal L, Rijnaarts H H, et al. Fate of pharmaceuticals in full-scale source separated sanitation system [J]. Water Research, 2015, 85: 384-392.
[26] Abel-Denee M, Abbott T, Eskicioglu C. Using mass struvite precipitation to remove recalcitrant nutrients and micropollutants from anaerobic digestion dewatering centrate [J]. Water Research, 2018, 132: 292-300.

IOP Conf. Series: Materials Science and Engineering **611** (2019) 012054 doi:10.1088/1757-899X/611/1/012054

Study on Catalyst Performance of Compact Methane Reformer with Nickel Loading

Wenrui Teng, Zhiyu Yao and Guogang Yang*

Henan Vocational College of Applied Technology, Zhengzhou 450042, Henan, P.R. China

*Corresponding author

Abstract. The compact methane reformer has the advantages of small size and high efficiency, making it possible to produce hydrogen on the fuel cell operating site. The development of a reforming catalyst with good selectivity and high stability is of great significance for the industrialization of the reformer. Among them, the nickel content is an important factor affecting the performance and cost of the catalyst. A compact methane reformer experimental platform was designed and built, and Ni/CeO_2 catalyst was prepared by equal volume impregnation method. The optimum loading of Ni was determined by experiments to be 10%. The mechanism of the effect of Ni loading on the catalytic performance was studied by XRD and SEM crystal phase analysis.

1. Foreword

In recent years, with the increasing attention paid to energy conservation and environmental protection, designing and developing new energy sources has become an important research project in the domestic and international scientific and technological circles. Among many options, fuel cells have broad application prospects in the transportation industry and distributed energy because of its advantages of high efficiency and cleanliness[1,2]. Hydrogen is the main raw material of the fuel cell, which has low density and high risk and is not conducive to transportation and storage. Therefore, using a wide range of sources and low cost of raw materials to produce hydrogen on the fuel cell work site has become an urgent problem to be solved[3,4].

Natural gas (methane generally accounts for more than 98%) plays an important role in China's energy structure, which is low in price and easy to obtain. Using methane to produce hydrogen in the industry has a long history, and the research on production processes and catalysts has been relatively mature. However, the oversized equipment of industrial product is the biggest obstacle of hydrogen production on the fuel cell work site[5]. In recent years, the emerging compact methane reformer has the advantages of small size and high efficiency, which makes using methane to produce hydrogen on-site come ture[6].

2. Experimental System and Experimental Scheme

2.1. Compact Methane Reformer Structure

Figure 1 is a schematic diagram of a compact methane reformer. The reformer is connected in parallel by a plurality of units, each unit consisting of a combustion duct and a reforming duct, the two ducts being separated by a thin metal plate. In the combustion pipeline, when methane gas and air flow in the main flow direction, a catalytic combustion reaction occurs in the catalytic layer sprayed on the surface of the metal plate, and the heat of reaction is transmitted to the catalytic layer of the reforming pipe through the metal plate to supply methane and water vapor. A reforming reaction takes place. The thickness of the metal sheet reaches the order of mm, which effectively increases the heat transfer efficiency and also greatly reduces the size of the reformer.

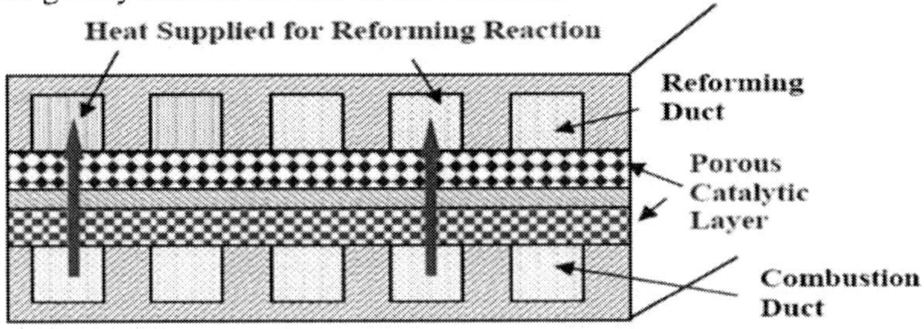

Figure 1. Schematic diagram of Marine compact methane reformer.

2.2. Experimental System Schematic

The design diagram of the experimental bench is shown in Figure 2. The device is divided into three parts: a feeding device, a methane steam reforming reactor and an analysis device.

Methane, N_2 (protective gas) and H_2 are controlled by the mass flow controller and read by the mass flow meter. The water vapor is generated by the water pump and the steam generator. The two raw materials are passed through the three-way valve and all flow into the same pipe. Mix evenly in the subsequent lines. The heat of the methane steam reforming reactor is supplied by a tube furnace which is placed in a constant temperature zone of the tube furnace to maintain a constant reaction temperature. The raw gas flows into the reactor through the preheating line to a temperature close to the reaction to carry out a reforming reaction. The effluent mixed gas is first condensed and dedusted by a pretreatment device, and the remaining reaction products are passed into an online infrared gas component and a calorimeter to analyze the volume fraction of each component, and the results are recorded every second, and the analyzer is connected to a computer for data analysis.

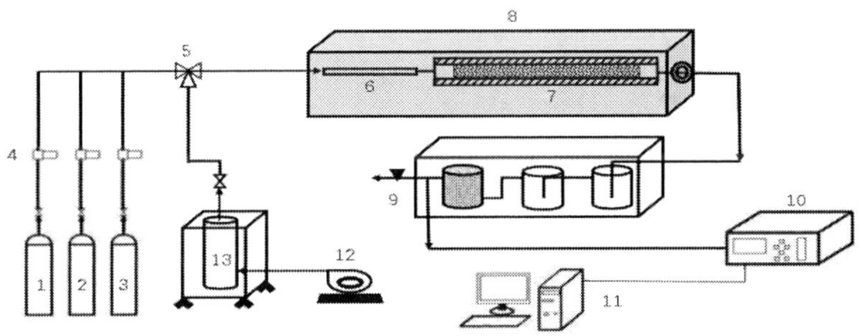

(1. Methane 2. Nitrogen 3. Hydrogen 4. Mass flow meter 5. Three-way valve 6. Preheating pipe 7. Reactor 8. Tube furnace 9. Exhaust gas treatment device 10. Online infrared gas composition and calorimeter 11 .Computer Control 12.Water Pump 13. Steam Generator)

Figure 2. Experimental flow chart.

2.3. Preparation of Catalysts with Different Ni Loadings Using CeO2 as Carrier by Equal Volume Impregnation

According to the Ni loading amount of 6.0wt%~12.0wt% (mass fraction, the same below), the desired $Ni(NO_3)_2$ was respectively weighed. H_2O was dissolved in n (ml) deionized water. After being sufficiently dissolved, the $Ni(NO_3)_2$ solution was immersed on the CeO_2 support and aged at room temperature for 12 hours to produce more uniform crystals. Then, the oil was evaporated in an 80°C oil bath in an evaporator, and the evaporated mixture was dried in a blast drying oven at 120°C for 4 hours. After drying, the sample was ground into a powder and placed in a tube furnace and calcined for 4 hours with the heating rate that was 2°C / min (the same as the calcination rate) was from room temperature to 500°C. Then a Ni/CeO_2 catalyst sample (hereinafter abbreviated as Ni/CeO_2) was prepared.

2.4. Catalyst Characterization Instrument

2.4.1. Scanning Electron Microscope (SEM). The morphology of the catalyst was examined using a scanning electron microscope of the model of SUPRA 55 SAPPHIRE from Carl Zeiss, Germany, with a scan voltage of 3-20 kV.

2.4.2. X-ray Diffraction (XRD) Analysis. The catalyst crystal phase was characterized using an X-ray diffractometer model of D/MAX-Ultima+ by Nippon Ricoh Co., Ltd. with a scanning step size of 0.02° and a scanning range of 10°-90°.

2.5. Experimental Result Evaluation and Analysis Method

The experiment investigated the catalytic performance and working conditions of the catalyst by investigating the conversion rate of CH_4, the amount of H_2 produced, and the ratio of H_2 to CO in the synthesis gas. The calculation formula is as follows:
Methane conversion rate expression:

$$CH_4(\%) = \frac{CH_4(\text{inlet}) - CH_4(\text{outlet})}{CH_4(\text{inlet})} * 100\%$$

Hydrogen generation rate expression:

$$H_2(\%) = \frac{H_2(\text{outlet})}{CH_4(\text{inlet}) - CH_4(\text{outlet})} * 100\%$$

The hydrogen production rate is the hydrogen production compared to the methane consumption, which means the hydrogen production corresponding to the methane per unit molar flow. Therefore, when the conversion rate of methane is lowered, the hydrogen production rate remains unchanged, but the actual hydrogen production decreases, mainly due to the decrease in the activity of the catalyst (carbon deposition, poisoning, etc.).
Carbon monoxide selective expression:

$$CO_{\text{selectivity}}(\%) = \frac{CO(\text{outlet})}{CO(\text{inlet}) + CO_2(\text{outlet})} * 100\%$$

3. Experimental Results and Analysis

3.1. Experimental Results of Catalytic Activity of Ni/CeO_2 under Different Ni Loadings

Under the conditions of a reaction temperature of 700°C, a pressure of 1 standard atmosphere, a water-carbon ratio of n (H_2O) / n (CH_4) of 3, and a GHSV of 1000 h-1 , the Ni loading is 6% to 12%. Ni/CeO_2 was tested and its catalytic performance is shown in Figure 3. It can be concluded from Figure 3 that when the Ni loading is 6%~10%, the conversion rate of methane increases with the increase of Ni loading, and the hydrogen production rate also increases greatly. When the Ni loading reaches 9%, it is produced. The hydrogen rate reached the highest value of 182.1%, and the methane conversion rate reached 73.6%. When the Ni loading continued to increase, the change of methane conversion rate was no longer obvious, and it decreased at 12%, and the hydrogen production rate also. From the CO selectivity analysis, it can be seen that as the Ni loading increases, the CO selectivity decreases first and then increases, and the minimum value is 22.97% at 9% (23.1% at 10%). At this time, the rate of water vapor reaction is relatively small.
The reasons for affecting Figure 3 are analyzed, mainly Ni content and dispersion. XRD analysis showed that the loading of Ni affected the dispersion of NiO on Ni/CeO_2. At the 6% Ni content, the increase in the loading amount causes the NiO grains to become larger, the active component and the active center to increase, and the methane conversion rate and the hydrogen production rate are greatly increased. However, when the Ni loading reaches 10%, since Ni itself has the property of easy aggregation and easy sintering, the NiO which continues to increase is sintered, and a large amount of

active components cannot uniformly cover the surface of the carrier, but gather in one place. The catalytic activity is decreased. At the same time, it is suspected that an excessive Ni loading will increase the carbon deposition reaction, thereby increasing the selectivity of the reaction to CO, which may affect the activity of the catalyst. Studies have shown that there is an optimum value for the Ni content in the nickel-based catalyst, which allows all active components to be uniformly dispersed at the active sites on the surface of the support, and neither the aggregation phenomenon nor the active center is obtained as much as possible.

Combined with experimental data analysis, comprehensive consideration can be concluded that the most reasonable Ni loading in Ni/CeO_2 is 10% under suitable conditions.

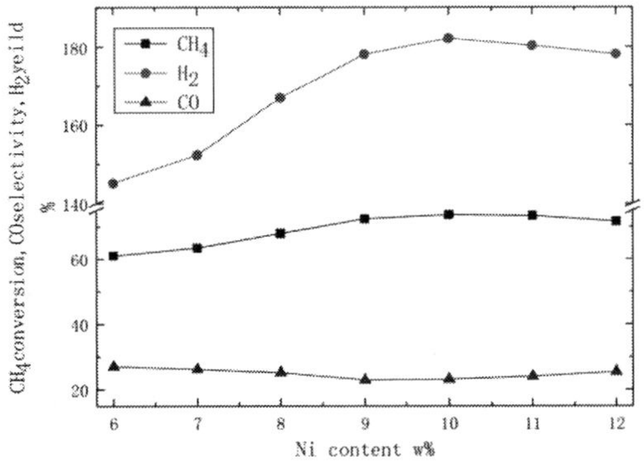

Figure 3. Catalytic properties of catalysts with different Ni loads for methane steam reforming.

3.2. Crystal Phase and Structure of Ni/CeO2 under Different Ni Loadings

3.2.1. XRD Analysis. The results of XRD analysis of each Ni loading amount of Ni/CeO_2 can be compared by Figure 4. It can be seen that there are characteristic peaks of NiO at 2, 43.5°, 50.8°, and 75°, and characteristic peaks of CeO_2 exist at 2, 3, 3, and 38.5°, and no characteristic peak of Ni is found. It is indicated that the active material Ni is all present in the form of NiO in the Ni/CeO_2 catalyst. Longitudinal contrast, as the Ni loading increases from 6% to 10%, it can be seen that the characteristic peak of NiO is gradually becoming stronger, reaching a relative maximum value of 10%. At this time, the characteristic peak is high and sharp, and then the characteristic peak The change in intensity is no longer noticeable even at 43.5°. This is because at the beginning of Ni/CeO_2, as the Ni content increases, the grain size of NiO gradually increases, so that the characteristic peak is remarkably enhanced, and the NiO crystal plane of the surface is the most. However, when the Ni loading reaches 10%, the NiO characteristic peak no longer becomes strong and the shape is no longer sharp, and the peak width begins to increase. According to the Scherrer formula: the degree of broadening of the X-ray diffraction band and the grain size of the crystal In inverse proportion, it can be inferred that the continued increase in Ni content may not have a positive effect on the catalytic activity of Ni/CeO_2. At the same time, an excessive Ni content tends to cause aggregation of NiO particles, which will seriously affect the catalytic activity of Ni/CeO_2.

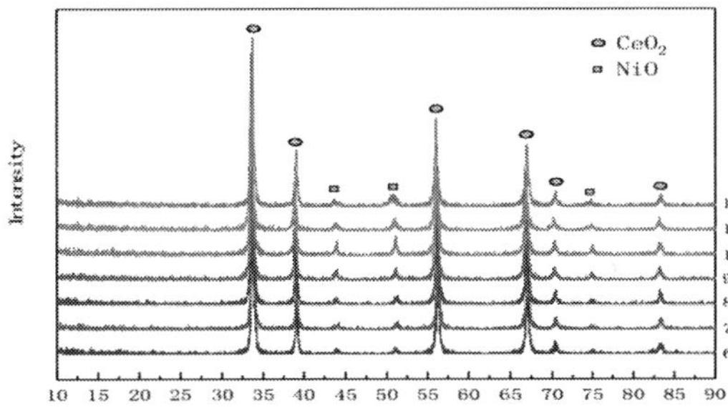

Figure 4. XRD patterns of catalysts with different Ni loads.

3.2.2. SEM Analysis. From the results of EDS spectrum in Figure 5, it can be roughly seen that the NiO crystal is uniformly distributed on the surface of the carrier CeO_2, and there is an active site on the surface of CeO_2, and the content of NiO around it is relatively high around this active site.

Figure 6 and Figure 7 show the morphology of the 10% Ni/CeO_2 catalyst particles. It can be seen that the surface morphology of the carrier CeO_2 is porous, and NiO is a small particle with a size of about 50 nm embedded in the surface of CeO_2. It is relatively high, and there is no phenomenon of agglomeration and sintering.

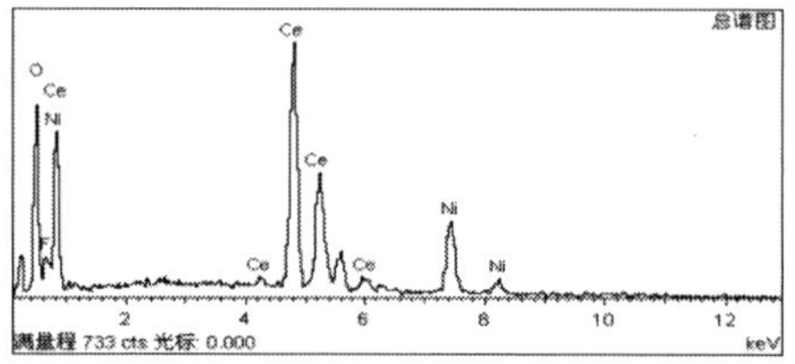

Figure 5. EDS spectrum of 10%Ni/CeO_2.

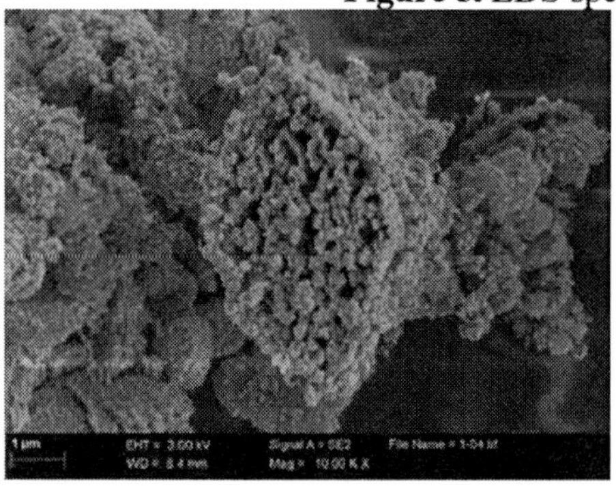

Figure 6. Results of Ni/CeO_2 scanning electron microscopy with 10%Ni load.

Figure 7. Results of Ni/CeO_2 scanning electron microscopy with 10%Ni load.

4. Conclusion

CeO_2 was used as the carrier in the nickel-based catalyst under the condition of compact methane reformer. The optimum loading of Ni was 10%, the conversion of methane was 73.6%, the hydrogen production rate was 182.1%, and the carbon monoxide selectivity was 23.1%. The catalyst was characterized by XRD and SEM, and the main components, NiO dispersion and surface morphology of the prepared catalyst were obtained.

Acknowledgments

This work was financially supported by NSFC fund(51779025).

References

[1] Yousri M. A.Welaya, M. Mosleh, Nader R. Ammar. J. *Energy Analysis of a Combined Solid Oxide Fuel Cell with a Steam Turbine Power Plant for Marine Applications.* J. Marine Sci. Appl. 23(2013)12: 473-483.

[2] H. Patel, T. Woudstra, P. Aravind, *Thermodynamic analysis of solid oxide fuelcell gas turbine systems operating with various biofuels.* J. Fuel cells 12 (6)(2012) 1115-1128.

[3] ZhangLiang, ZhangWeijing. *The Key Technical Issues of SOFC as a Power Source for Ship Application.* J. Chinese Journal of Ship Research, 6(2011), pp. 109-113.

[4] PanYongqi, ZhuZiwen, ZhengQingrong, JiHouzhi. *Research of Typical Marine Propulsion System Powered by PEMFC.* J. Ship Engineering, 38(2016), pp. 36-38.

[5] Bhattacharya A. K., Breach J. A., Chand S., etc. *Selective oxidation of methane to carbon monoxide on supported palladium catalyst.* J. Applied CatalysisA: General, 1992, 80(1): 1-5.

[6] Feng Wanga, Bo Qi, Guoqiang Wang, Longjian Li (2013). *Methane steam reforming: Kinetics and modeling over coating catalyst in micro-channel reactor*. J. International Journal of Hydrogen Energy, 38, pp. 5693-5704.

IOP Conf. Series: Materials Science and Engineering **611** (2019) 012055 doi:10.1088/1757-899X/611/1/012055

Four Amino Acid Residues in Acyltransferase Affect the Production of Epothilone A to Mixed Epothilones A and B

Xin Sun[1], Pengfei Li[2], Yuezhong Li[2] and Lin Zhao[1,*]

[1]State Key Laboratory of Biobased Material and Green Papermaking, school of bioengineering, Qilu University of Technology, Shandong Academy of Sciences, Jinan, 250353, China

[2]State Key Laboratory of Microbial Technology, College of Life Science, Shandong University, Jinan 250100, P. R. China

*Corresponding author

Abstract. Surveying more than 100 native epothilone producing strains were isolated from different locales revealed that the production ratio of epothilones A and B was sorted into two groups, almost single epothilone A or approximate 2:1 ratio of A and B. Sequencing the second acyltransferase in the *epoC* module (AT_{ModC2}) of different *Sorangium* producers indicated that four amino acid residues at sites 90, 91, 95 and 196 in the enzyme active centre altered in parallel to the production ratio shift. In those producers with single epothilone A, AT_{ModC2} had almost the same sequence and computational structure as AT_{ModC1}, which selectively accepts malonyl-CoA as the extender units.
Keywords: epothilone; acyltransferase; aminoacid.

1. Introduction

Epothilones are taking many attentions because of their Taxol-like microtubule stabilization activity and potential large-scale fermentation production [1]. At present, there are at least six epothilones used in anti-cancer trials [2], and ixabepilonehas been approved by the USFDA. The reported three strains, i.e. So ce90, SMP44, and So0157-2 [3, 4, 5] produce a mixture of epothilones A and B (approximately 2:1 in ratio) as the major product. There are differences of between epothilones A and B, and people studies to increase B ratio by either optimizing fermentation procedures, or engineering the producing strains and the biosynthetic genes[1].

Production of epothilones A and B is decided by the second acyltransferase domain of module C (AT_{ModC2}), via promiscuously accepting malonyl-CoA or methylmalonyl-CoA as the extender units [6, 7]. Although it is practical in other systems to swap AT domains to alter the substrate selection of acyltransferases, selective producing single epothilone B has never been realized in *Sorangium* strains [8, 9, 10].

In our screening work, we obtained many epothilone producing *Sorangium* strains from different locales. Although the yields of epothilones varied in different *Sorangium* producers, the ratios of epothilones A and B were sorted into two groups, either almost single epothilone A or approximate 2:1 ratio of A and B. Interestingly, the sequence and computational structure of the active centre of AT_{ModC2} was significantly changed in parallel to the ratio shift. We amplified the AT_{ModC2} of the strain So0157-2 which can produce both epothilone A and B. Then a four points mutation was introduced into this enzyme which made it transform into the AT_{ModC2} of the strains with single epothilone A just like So0007-12.

IOP Conf. Series: Materials Science and Engineering **611** (2019) 012055 doi:10.1088/1757-899X/611/1/012055

2. Experimental Procedure

2.1. Strain and Condition

The *Sorangium* strains were in Table 1. *Sorangium* strains were cultured on CNST agar as previously described [11]. For DNA extraction, the *Sorangium* cells from 4-day CNST cultures were transferred into M26 liquid medium and incubated for 7 days [12]. For the production of epothilones, strains were inoculated on M26 for 4 or 5 days. Then XAD-16 resin were added, and incubated for additional 9 to 10 days as previously described [5]. *E. coli* cells were cultured in LB with Ampicillin. The *Sorangium* strains were incubated at 30°C, while *E. coli* was at 37°C.

Table 1. Production of epothilones in different Sorangium cellulosum strains.

Isolate	Accession number of AT_{ModC2}	Epothilone A (mg/L)	Epothilone B (mg/L)	B/A
So007-12-2A	EU554397	1.52	0	–
So007-12-2B	EU554398	1.51	0	–
So0087-2-1	EU864255	0.47	0	–
So0087-2-2	EU554399	0.19	0	–
So0087-2-3	EU554400	0.54	0	–
So0087-7-1	EU554401	0.94	±	–
So0087-14	EU554402	1.71	0	–
So0157-2	EU554406	1.7	0.8	0.47
So ce90[a]		22	11	0.5
SMP44[b]		12.3	6.7	0.5

a. cited from the reference of Gerth et al. 1996.
b. cited from the reference of Bollag et al. 1995.

2.2. Detection and Identification of Epothilones

Resins were extracted with methanol. After dried at 40°C, the extracts were stored at -20°C. For a typical HPLC-MS analysis, samples from 50 ml cultures were redissolved in 500 μl methanol. HPLC interfaced with a Finnigan MSQ classic quadruples mass spectrometer and MRC-ODS columnwith the mobile phase of 65% methanol and 35% buffer. Epothilone A and B were eluted at 12.5 and 14.2min. To quantify the production titers, 10 μL aliquots of the samples were analyzed on Surveyor HPLC under the same chromatography condition, detected at 249 nm. They were quantified based on purified epothilone A and B.

2.3. Nucleic Acid Extraction and PCR Amplification

The genomic materials of *Sorangium* cells were extracte. The AT_{ModC2} domain sequences in different *Sorangium* strains were amplified using the nest PCR protocol to avoid unspecific amplifications. The first route of PCR amplification was performed with the primer pair of 5'-AGTGGACGCATGACGCTGAC-3' and 5'-ACGTCGATTTCGTGGAATGC-3', targeting the KS and KR domains. The product is 2.7 kb, containing the AT_{ModC2} domain. After electrophoresis, they were incubated for 15 min, rinsed with Milli-Q water for 10 min, and photographed using a BioRad GelDoc station (BioRad, California, USA). The target bands were excised and extracted using a Gel extract kit (Bioflux), and dissolved in 30 μL Milli-Q water. The second PCR route used a primer pair of 5'-CTG CGC GAG CAC CTG GAC ATG C-3' and 5'-GCT GCC GCT GCC ACG GAT AGG T-3'. The PCR fragments were extracted and ligated into T-vecto rand transformed into DH-5α.

2.4. Nucleic Bioinformatics Analysis

The epothilone biosynthetic gene sequences of So ce90 (AF210843) and SMP44 (AF217189) were obtained from the GenBank databases. The sequences were completely aligned using the ClastalW programs. Phylogenetic reconstruction was carried out in MEGA 3D structure of AT_{ModC2} domain sequences was on-line predicted using swissmodel (http://swissmodel.expasy.org/), based on the acyltransferase of the DEBS biosynthetic gene cluster for erythromycin in the SWISS PDB databases

[13]. The similarity of the sequences of the DEBS acyltransferase and AT_{ModC2} is about 44%. The logP value and MR value were calculated by ChemDraw Pro6.0 (http://www.cambridgesoft.com/software/ChemDraw/).

3. Results

3.1. Epothilones A/B Ratios in Sorangium Producers

There have been reported three epothilone producers, So ce90, SMP44 and So0157-2 [3, 4, 5]. These strains all had an approximately 2:1 ratio of mixed epo A and B. To obtain mutants that were able to selectively produce epothilone B, So0157-2 strain was mutated [5]. Although mutants with increased proportion of epothilone B (but in low productivity of epothilones) were occasionally obtained, no strains were able to selectively produce epothilone B only, or had normal productivity with higher proportion of epothilone B than A. Our screening projects obtained more than 100 epothilone producing Sorangium strains. These native producers showed great differences in epothilone production, but the ratios of epothilones A and B in these strains were only sorted in two groups. Some strains produced a mixture of epothilones A and B with about 2:1 ratio (named $epoB^+$ in this paper), while others had extremely low or even undetectable epothilone B (named $epoB^-$, which means they lost the ability of epothilone B production). However, none of these Sorangium strains were able to selectively yield epothilone B without A, or even higher proportion of B than A. In this study, seven natural $epoB^-$ strains, together with three $epoB^+$ strains were analyzed (Table 1 showed the productions of epothilones A and B in these strains).

3.2. AT_{ModC2} Domain Sequences Analysis

Epothilones are biosynthesized in *Sorangium* cells by seven multifunctional proteins[14, 15]. The acyltransferases (ATs) are responsible for the choice of extender units. Based on the incorporation of radioactively labeled precursors and the composition of biosynthetic enzymes [6, 7], the production of epothilone A or B is suggested to be attributed to the promiscuous substrate selection with malonyl-CoA or methylmalonyl-CoA. Besides AT_{ModC2}, the other eight acyltransferases had a strict selection of malonyl-CoA or methylmalonyl-CoA. Fig. 1 exhibited the model of epothilones A and B biosynthesis. Because epothilone producing *Sorangium* strains are divided into $epoB^+$ and $epoB^-$ groups, it is then interesting to see changes of the AT_{ModC2} gene in those $epoB^-$ strains.

We amplified the AT_{ModC2} of all strains. All of the ATs in nine modules of the three reported strains were phylogenetic analyzed and the results indicated that all sequences are clearly sorted into two groups: the malonyl-CoA accepting AT (MCAT) and the methylmalonyl-CoA accepting AT (MMCAT) (Fig. 2A). The AT_{ModC2} domain belongs to the malonyl-CoA type, showing high identity to the first acyltransferase domain in epoC module (AT_{ModC1}), which is only able to accept malonyl-CoA,[6, 7]. The AT_{ModC2} sequences from different *Sorangium* strains were highly conserved with more than 98.6% similarity in nucleotide acid sequences and 97.7% similarity in amino acid sequences. AT_{ModC2} were all located in the AT_{ModC1}/AT_{ModC2} subgroup (Fig.2A).

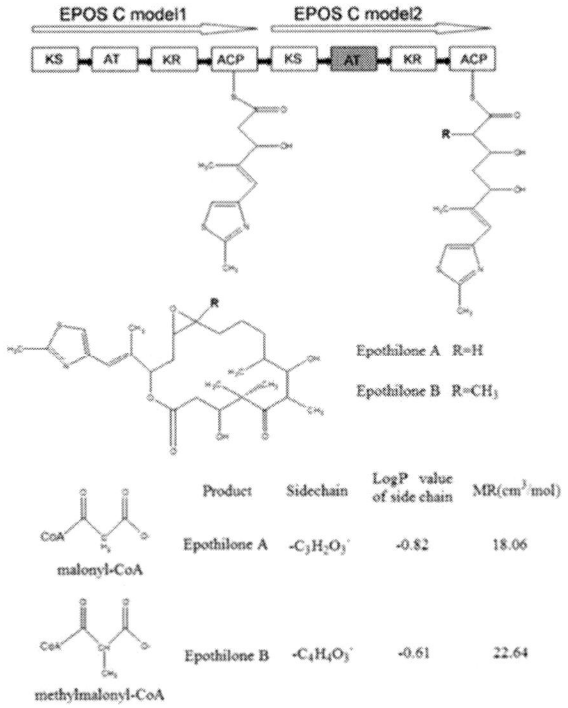

Figure 1. Substrate selection of AT_{ModC2} domain (gray) in the biosynthesis of epothilones in *Sorangium* strains.

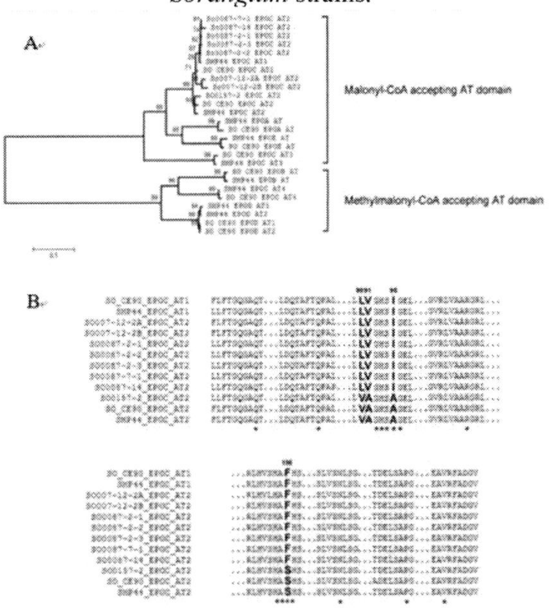

Figure 2. (A) Phylogenetic analysis of the AT_{ModC2} domains from different *Sorangium* strains, with all of the acyltransferases in the biosynthetic gene clusters from So ce90 and SMP44 strains. (B) Alignment of AT_{ModC2} sequences, with AT_{ModC1} sequences from So ce90 and SMP44 strains. The active sites are marked with '*'; four sites that we analyzed are marked bold black).

Some particular amino acids (or motifs) of ATs are responsible for the substrate selection. We compared the 13 amino acids might be active sites of all the AT sequences in PKSs. (highlighted in Fig.2B) They were essential for the substrate selectivity and reaction[11]. However, 11 of the 13 sites were the same in both epoB$^+$ and epoB$^-$ *Sorangium* strains, except the 95th and 196th residues. In epothilone native producers, this putative 4-aa decision motif in the epoB$^+$ strains is HASH, but HAFH in all epoB$^-$ strains (Fig. 2B). The results suggested that, besides the HASH/HAFH/YASH motif, there were some other potential residues cooperating with them for the substrate selection.

3.3. 3D Model of AT_{ModC2} Domain and Substrate Selection Bioinformatics' Analysis

In acyltransferases, the sites of Ser_{94} and Arg_{119} are responsible for the catalysis (the process was shown in Fig.3A). The computational 3D structure model of AT_{ModC2} was constructed with the acyltransferase of DEBS in *Streptomyces erythreus* as template (pdb code: 2hg4a), which was highly consistent with the previous report of malonyl-CoA type acyltransferase [16]. Comparing the 3D structures of AT_{ModC2} domains from different *Sorangium* producers revealed that four amino acid residues in the active centre (marked in the sequence alignment in Fig. 2B) were altered significantly. Two (residues 95 and 196) are among the 13 previously putative active sites in PKS acyltransferases [11], while the other two (residues 90 and 91) have not been mentioned. These four amino acids were close to the two catalyzing sites (Ser_{94} and Arg_{119}) (Fig. 3B), suggesting their potential roles in the substrate selection. The four amino acids in epoB$^-$ (*i.e.* Leu_{90}, Val_{91}, Ile_{95} and Phe_{196}) were much bigger than those in epoB$^+$ (Val_{90}, Ala_{91}, Ala_{95}, and Ser_{196}), increasing 29% (90th site), 146% (91th site), 218% (95th site), and 276% (196th site) in molecular volume. In addition, these four amino acids also showed more hydrophobic in epoB$^-$ strains than in epoB$^+$ strains. The substitution of the amino acid residues at these four sites formed a complicated stereo-hindrance for the entrance of methylmalonyl-CoA into the catalyzing centre of epoB$^-$ AT_{ModC2}. When we constructed the 3D-structure with acyltransferase of FADs in *Saccharopolyspora erythraea* as template (pdb code: 2qo3a), it showed that the substitution of 90, 91 site changed the position of neighboring Ser_{94}, which made the catalyzing sites of Ser_{94} and Arg_{119} to be closer in epoB$^-$ AT_{ModC2} than in epoB$^+$ AT_{ModC2} (Fig. 3B). The stereo-hindrance in the active centre of the epoB$^-$ AT_{ModC2} had less effect for accepting Malonyl-CoA, which is smaller and less hydrophobic than methylmalonyl-CoA. Interestingly, the four amino acids of AT_{ModC2} in those epoB$^-$ strains were identical to those of AT_{ModC1} (Fig. 2B). Thus, it is suggested that these four amino acids are crucial and cooperative in the substrate selection of AT_{ModC2} domain.

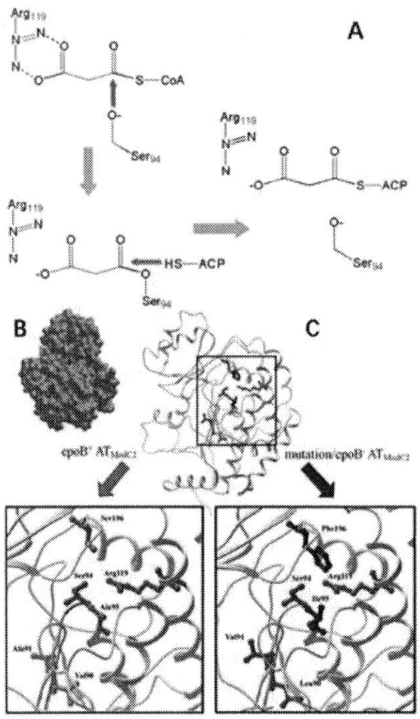

Figure 3. (A) Pathway of the transacylation. (B) The skin representation of epoB$^+$ AT_{ModC2} and the catalyzing region (red is Ser_{94}; purple is Arg_{119}). (C) 3D structure of epoB$^+$ and epoB$^-$ constructed with template Pdb code: 2qo3a.

4. Conclusion

Myxobacteria are famous not only for their complicated multi-cellular social lifestyle [17], and their diverse secondary metabolites [18, 19]. The ratio of epo A and B in native *Sorangium* strains suggested that the biosynthetic genes are highly conserved. Acyltransferase is responsible for the choice of extender units for the PKS. In the enzymatic complex, there are nine acyltransferases either selectively accepting malonyl-CoA or methylmalonyl-CoA, except AT_{ModC2}. However, AT_{ModC2} had about 98% sequence identity to the malonyl-CoA type acyltransferase AT_{ModC1}. Paralleled to the production ratio shift, the amino acid residues at the sites 90, 91, 95 and 196 in the active region differed between $epoB^-$ and $epoB^+$ strains. The changes led to a high stereo-hindrance of AT_{ModC2} for the acceptance of methylmalonyl-CoA in $epoB^-$. In cells, malonyl-CoA is more easily accessible than methylmalonyl-CoA, because of their importance for cellular basic metabolisms. It is a disappointing suggestion that natural *Sorangium* producers probably cannot produce epothilone B without A. However, we assumed that it may be practical to swap the complete AT_{ModC2} domain with some other tolerable methylmalonyl-CoA type acyltransferases for the selective epothilone B without A. It will significantly improve the value of the strain for epothilone B.

Acknowledgments

The work was financially supported by Shandong Provincial Natural Science Foundation (NO. ZR2017BH021).

References

[1] Rowinsky E K, and Calvo E. Novel agents that target tubulin and related elements. *Semin Oncol.* 2006, 33: p421-435.
[2] Larkin J M, and Kaye S .B Epothilones in the treatment of cancer. *Expert. Opin Investig Drugs.* 2006, 15: p691–702.
[3] Gerth K, Bedorf N, Höfle G, Irschik H, and Reichenbach H. Epothilones A and B: Antifungal and cytotoxic compounds from *Sorangium cellulosum* (Myxobacteria) -production, physico-chemical and biological properties. *J Antibiot.* 1996, 49: p560-564.
[4] Bollag D M, McQueney P A, Zhu J, Hensens O, Koupal L, Liesch J, Goetz M, Lazarides E, and Woods C M. Epothilones, a new class of microtubule-stabilizing agents with a taxol-like mechanism of action. *Cancer Res.* 1995, 55: p2325-2333.
[5] Gong G L, Sun X, Liu X L, Hu W, Cao W R, Liu H. Mutation and high-throughput screening method for improving the production of Epothilones of *Sorangium. J Ind Microbiol Biotechnol. 2007.* 34: p615-623.
[6] Gerth K, Steinmetz H, Höfle G, and Reichenbach H. Studies on the biosynthesis of epothilones: the biosynthetic origin of the carbon skeleton. *J Antibiot.* 2000, 53: p1373-1377.
[7] Molnár I, and Schupp T, Zirkle R, Milnamow M, Nowak-Thompson B, Engel N. The biosynthetic gene cluster for the microtubule-stabilizing agents epothilones A and B from *Sorangium cellulosum* So ce90. *Chem Biol. 2000,* 7: p97-109.
[8] Petković H, Lill R E, Sheridan R M, Wilkinson B, McCormick E L, McArthur H A. A novel erythromycin, 6-desmethyl erythromycin D, made by substituting an acyltransferase domain of the erythromycin polyketide synthase. *J Antibiot. 2003,* 56: p543-551.
[9] Regentin R, Cadapan L, Ou S, Zavala S, and Licari P. Production of a novel FK520 analog in Streptomyces hygroscopicus: improving titer while minimizing impurities. *J Ind Microbiol Biotechnol.2002,* 28: p12-16.
[10] Patel K, Piagentini M, Rascher A, Tian Z Q, Buchanan G O, Regentin R. Engineered biosynthesis of geldanamycin analogs for Hsp90 inhibition. *Chem Biol. 2004,* 11: p1625-1633.
[11] Yadav G, Gokhale R S, and Mohanty D. Computational approach for prediction of domain organization and substrate specificity of modular polyketide synthases. *J Mol Biol.* 2003, 8: p335-363.
[12] Nguimbi E, Li Y Z, Gao B L, Li Z F, Wang B, Wu Z H. 16S-23S ribosomal DNA intergenic spacer regions in cellulolytic myxobacteria and differentiation of closely related strains. *Syst Appl Microbiol. 2003,* 26: p262-268.

[13] Tang Y, Kim C Y, Mathews I I, Cane D E and Khosla C. The 2.7-Angstrom crystal structure of a 194-kDa homodimeric fragment of the 6-deoxyerythronolide B synthase. *Proc Natl Acad Sci.* 2006, 03: p11124-11129.
[14] Julien B, Shah S, Ziermann R, Goldman R, Katz L, and Khosla C. Isolation and characterization of the epothilone biosynthetic gene cluster from *Sorangium cellulosum. Gene. 2000,* 249: p153-160.
[15] Molnár I, and Schupp T, Zirkle R, Milnamow M, Nowak-Thompson B, Engel N. The biosynthetic gene cluster for the microtubule-stabilizing agents epothilones A and B from *Sorangium cellulosum* So ce90. *Chem Biol. 2000,* 7: p97-109.
[16] Oefner C, Schulz H, D'Arcy A, and Dale G E. Mapping the active site of Escherichia coli malonyl-CoA-acyl carrier protein transacylase (FabD) by protein crystallography. *Acta Cryst.2000,* 62: p613-618.
[17] Whitworth D E. Myxobacteria: multicellularity and differentiation. ASM Press, 2007.
[18] Reichenbach H. Myxobacteria, producers of novel bioactive substances. *J Ind Microbiol Biotechnol.* 2001, 27: p149-156.
[19] Gerth K, Pradella S, Perlova O, Beyer S, and Müller R. Myxobacteria: proficient producers of novel natural products with various biological activities--past and future biotechnological aspects with the focus on the genus *Sorangium. J Biotechnol.* 2003, 106: p233-253.

Sub-chronic Exposure to Tris(1,3-dichioro-propyl) Phosphate Induces Hepatotoxicity in Rats

Simin Wang and Qixing Zhou*

Ministry of Education Key Laboratory of Pollution Processes and Environmental Criteria / Tianjin Key Laboratory of Environmental Remediation and Pollution Control, College of Environmental Science and Engineering, Nankai University, Tianjin 300350, China

*Corresponding author

Abstract. Tris(1,3-dichloro-2-propyl) phosphate (TDCIPP), one of the most widely used organophosphorus flame retardants (OPFRs) is frequently detected in various environment and biota. Thus, its potential risk on wildlife and human health should be concerned. It has been implicated to induce hepatotoxicity, but the toxicological mechanism still remains unclear. Thus, Sprague Dawley (SD) male rats were administrated TDCIPP at 125, 250, or 500 mg/kg/d for 8 weeks in the present study. And hepatic histopathology and enzyme activities in serum and livers were analyzed to determine the molecular mechanisms of hepatotoxicity induced by TDCIPP. The histopathology showed the hepatocytes of rats exposed to TDCIPP were damaged, and the necrosis was more severe in 500 mg/kg/d TDCIPP-exposed group. And the decrease of transaminases in serum and livers were observed after TDCIPP exposure for 8 weeks, which indicated that severe hepatotoxicity was induced by TDCIPP. The abnormalities of oxidative stress makers and inflammation factors also indicated that TDCIPP caused dysfunction of oxidative system and severe inflammation response in livers of rats. Collectively, the results in our study demonstrated that TDCIPP induced hepatic oxidative stress and inflammatory response, and caused hepatotoxicity in rats.
Keywords: tris(1,3-dichloro-2-propyl) phosphate; hepatotoxicity; autophagy; oxidative stress; inflammation.

1. Introduction

As the phasing out of polybrominated diphenyl ethers (PBDEs), the production and application of OPFRs have thus been increasing in recent years.[1] OPFRs are not chemically bonded to products and can leach into various environments easily.[2] TDCIPP has been used as the main replacement of PBDEs and constantly detected in indoor air,[3] water,[4] dust[5] and biota.[6] Furthermore, TDCIPP and its metabolite have been detected in human breast milk, urine, and seminal fluid.[1] Although the usage of TDCIPP is increasing rapidly, there is limited toxicological information available for TDCIPP. Recent studies showed that TDCIPP could adversely affect the nervous system in zebrafish and induce neurotoxicity in cultured cells.[7-8] Based on several animal studies, TDCIPP is regarded as a probable human carcinogen.[9] It has also been reported that TDCIPP could pervert endocrine system.[10] Some studies indicated that exposure to TDCIPP could alter the genes involved in growth and the thyroid hormone (TH) pathway in avian hepatocytes.[11] It has been suggested that TDCIPP exposure could also disturb the expression of hepatic genes involved in inflammation significantly in zebrafish.[12] In chicken embryos, studies showed that hepatic genes containing immune response and xenobiotic may be changed after TDCIPP exposure.[13] The lack of reports about hepatotoxicity in mammals is surprising since the studies above demonstrated that liver is a critical target for TDCIPP toxicity. Consequently, in the present study, the hepatic histopathology and enzyme activities in serum and livers were analyzed in SD rats to obtain reliable risk assessment and explore underlying molecular mechanisms of TDCIPP-induced hepatoxicity.

2. Materials and Methods

2.1. TDCIPP Treatment of Rats

TDCIPP (CAS#13674-87-8; >95% purity) was purchased from Tokyo Chemical Industry (Tokyo, Japan) and primarily dissolved in extra virgin olive oil (Abril, Spain). SD rats were from Beijing Vital River Laboratory Animal Technology and raised under standard animal housing condition (12-hr light/dark cycle, 20 ± 2 °C, and 50 ± 5% humidity). After acclimation for 2 weeks, 48 male rats were then divided into 4 groups. The rats in TDCIPP-exposed groups were administered TDCIPP in olive oil at dose of 125, 250, and 500 mg/kg/d for 8 weeks. The control group was exposed to the same volume olive oil per day. Plasma sample from each group were collected by retro-orbital bleeding to detect hepatic function related biochemical indicators at the end of 2nd, 4th, 6th, and 8th week. After TDCIPP exposure for 8 weeks, 3 rats from each group were sacrificed. The liver tissues were excised for histopathology and TEM analysis. All experimental procedures were conducted in accordance with Chinese legislation and approved by Animal Ethical committee of Nankai University.

2.2. Histopathological and TEM Analysis

The liver tissues were fixed in 10% buffered formalin for no less than 24 h. Fixed samples were embedded in paraffin wax, sectioned to 5 μm and then stained with hematoxylin and eosin staining (H&E staining). The sections were examined by brightfield microscope (IX71, Olympus, Japan). For TEM, liver tissues were quickly excised and then fixed in 2.5% glutaraldehyde for 48 h, followed by osmium tetroxide at 4 °C. Samples were then sectioned ultrathin which were analyzed using TEM (H-7650, Hitachi, Japan).

2.3. Biochemical Analysis

The levels of ALT, AST, GSH, SOD, IL-1β and TNF-α in hepatic homogenate were detected using commercial kits according to the manufacturers' instructions (Nanjing Jiancheng Bioengineering Institute, Nanjing, China). And the activities of ALT and AST in serum were analyzed with an automatic biochemical analyzer (MD-2880, U.S.A).

2.4. Statistical Analysis

Data were analyzed using one-way ANOVA with Tukey's test. Differences were accepted as significant when P-value < 0.05 compared with control. SPSS 21.0 was used for statistical analyses. All experiments were repeated with no less than 3 biological replicates.

3. Results and Discussion

3.1. Validation of Hepatotoxicity by Histological and TEM Examination

To explore detailed hepatic damages, focal inflammation and hydropic degeneration were examined by examining 30 images from four hepatic sections for three rats in each group. H&E staining of liver slices revealed no evident abnormal pathological in the livers of control group (Fig.1a). Focal inflammation were both detected in the livers of all TDCIPP-exposed rats, with mild to moderate hydropic degeneration observed in 250 and 500 mg/kg/d TDCIPP-exposed rats (Fig.1b, 1c and 1d). In addition, the hepatic ultrastructural analysis was evaluated using TEM. No abnormality in hepatocellular organelle morphology was detected in the livers of control group (Fig.1e). Nuclear condensation was observed in the livers of 500 mg/kg/d TDCIPP-exposed group compared with control, which is a sign of apoptosis. And the results also showed increasing autophagy formation in the livers of 500 mg/kg/d TDCIPP-exposed rats (Fig.1f, 1g and 1h). Taken together, histopathological examination and TEM analysis confirmed that TDCIPP exposure can cause pathology damages and hepatocellular injuries of rats.

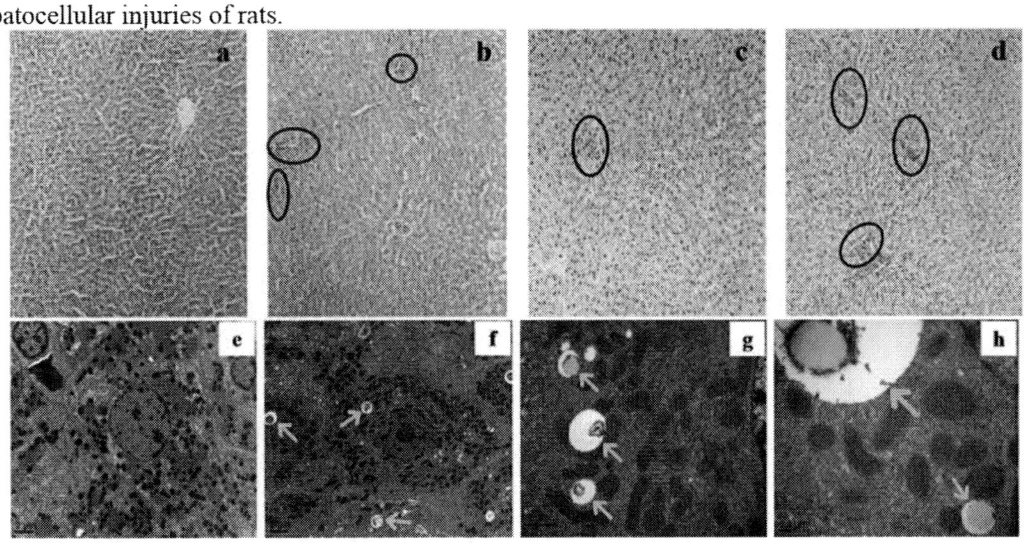

Figure 1. a–d, H&E stained hepatocytes. (a) control group; (b) 125 mg/kg/d TDCIPP-exposed group; (c) 250 mg/kg/d TDCIPP-exposed group; (d) 500 mg/kg/d TDCIPP-exposed group. e–h, TEM ultrastructure of rats hepatocytes. (e) control group; (f-h) 500 mg/kg/d TDCIPP-exposed group.

3.2. Enzymatic Activity Analysis

In serum, the activities of ALT and AST were frequently considered to be the clinical and laboratory biomarker of hepatotoxicity. The activities of ALT (Fig.2a) and AST (Fig.2b) in serum were measured every two week during TDCIPP administration for 8 weeks. The results showed that no statistically differences were observed in ALT and AST activities between TDCIPP-exposed groups and control group from 2nd week to 4th week. At the end of 6th week, the levels of ALT and AST in 500 mg/kg/d TDCIPP-exposed rats were significantly decreased compared with control. And after TDCIPP exposure for 8 weeks, the activities of ALT and AST in serum of TDCIPP-exposed groups were found to be significantly lower than control group. We further determined the levels of ALT (Fig.2c) and AST (Fig.2d) in livers of rats after TDCIPP exposure for 8 weeks. The levels of ALT in livers of TDCIPP-exposed groups were statistically decreased compared with control. And the activities of AST in livers of TDCIPP-exposed groups were lower than that of control and the data showed a statistically decreased in 250 and 500 mg/kg/d group. Elevated levels of ALT and AST in serum and livers are correlated with hepatotoxicity, while reductions of ALT and AST have been demonstrated in some hepatic disease.[14] Consequently, the data above indicated that a potential liver injury was induced by TDCIPP.

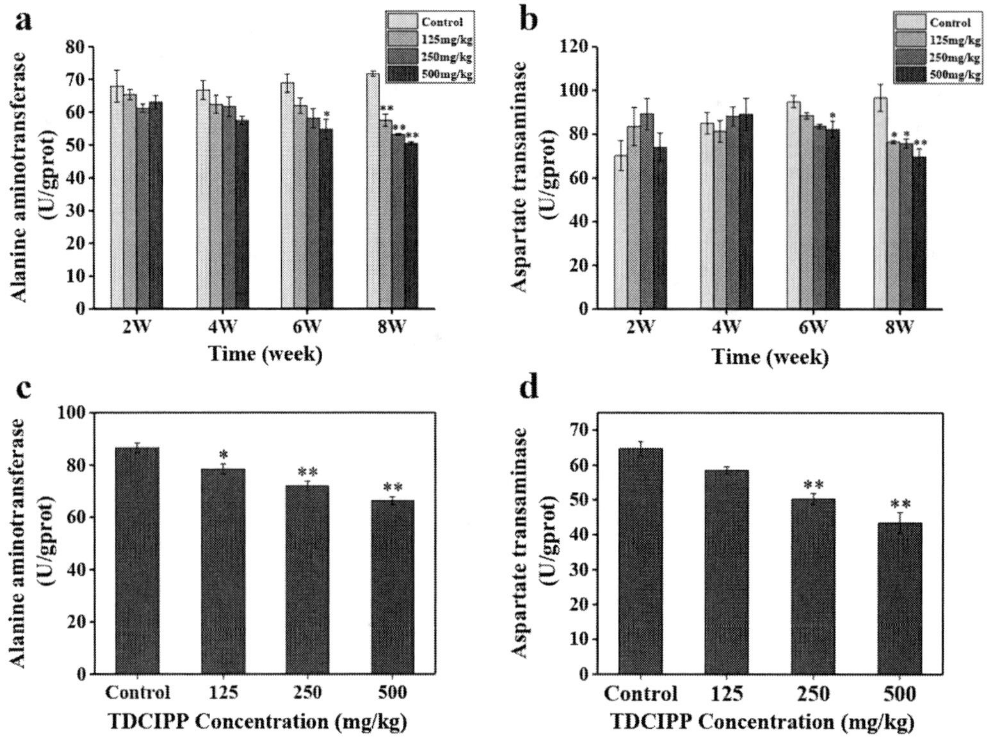

Figure 2. The levels of ALT (a) and AST (b) in serum of rats at the end of 2nd, 4th, 6th and 8th week. The activities of ALT (c) and AST (d) in the livers of rats. **p<0.01, *p<0.05

3.3. Biochemical Indicators Involved in Oxidative Stress and Inflammation

To further confirm the TDCIPP-induced hepatotoxicity in rats at molecular level, the activities of GSH, SOD, IL-1β and TNF-α in livers of rats were determined. In TDCIPP-exposed groups, the levels of GSH (Fig.3a) were significantly decreased dose-dependently compared with control. The levels of SOD (Fig.3b) were statistically decreased in the livers of 500 mg/kg/d TDCIPP-exposed rats while there was no significantly changes in any other TDCIPP-exposed groups. The results revealed that oxidative damages in the livers of rats can be induced by TDCIPP. The data indicated that the level of IL-1β (Fig.3c) can be increased after TDCIPP exposure and it was significantly increased in 500 mg/kg/d TDCIPP-exposed group. The levels of TNF-α (Fig.3d) were increased in the livers of TDCIPP-exposed groups dose-dependently. And the activities of TNF-α were statistically increased in 250 and 500 mg/kg/d TDCIPP-exposed groups. The results above verified that TDCIPP induced inflammation response in the livers of rats. Recent studies also indicated that TDCIPP exposure can cause inflammation response and change inflammatory genes in zebrafish and chicken embryos.[12-13]

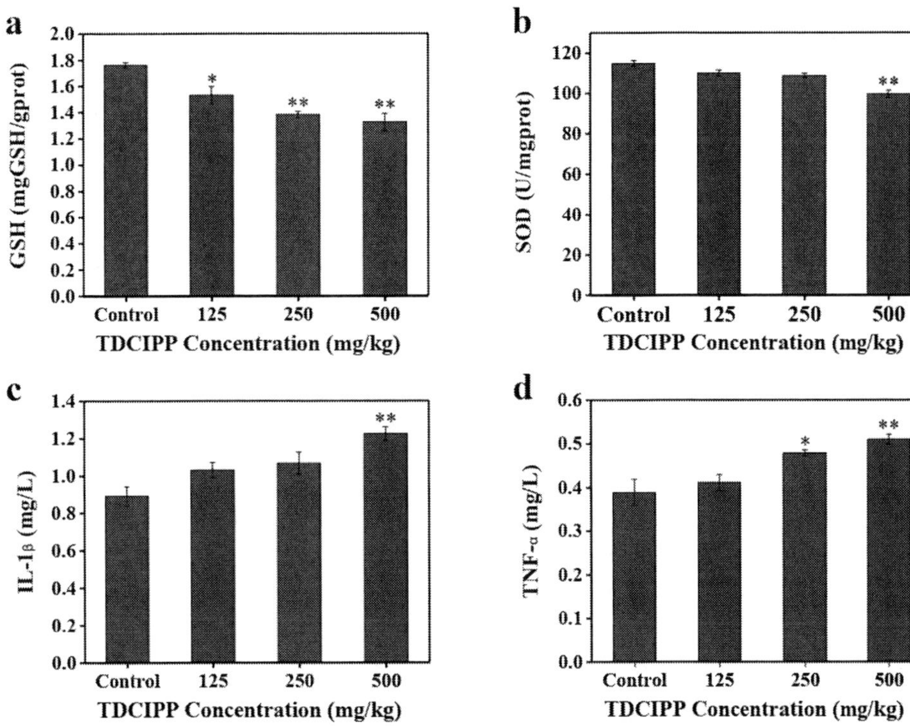

Figure 3. (a-b) Biochemical indicators of oxidative stress: (a) GSH and (b) SOD; (c-d) Biochemical indicators of inflammation: (c) IL-1β and (d) TNF-α in the livers of rats. $^{**}p<0.01$, $^{*}p<0.05$

4. Conclusions

In the present study, hepatotoxicity induced by TDCIPP was evaluated due to its increasing detection in environment and potential risk on human health. The observation of hepatic pathology damages, hepatocellular injury, abnormality of transaminase, oxidative stress and inflammation indicated the hepatotoxicity in rats can be induced by TDCIPP. And our study further added the toxic data of TDCIPP in mammals.

Acknowledgments

This work was supported by the National Natural Science Foundation of China (Grant Nos. 31670508) and the 111 program (T2017002), special funds for basic scientific research services of central colleges and universities.

References

[1] van der Veen, I., and de Boer, J. Phosphorus flame retardants: Properties, production, environmental occurrence, toxicity, and analysis. Chemosphere, 2012, 88 (10), 1119−1153.

[2] Betts, K. S. New Details on Organophosphate Flame Retardants Exposure in Men Appears Stable over Time. Environmental Health Perspectives, 2013, 121(5), A168-A168.

[3] Stapleton, H. M., Klosterhaus, S., Eagle, S and et al. Detection of Organophosphate Flame Retardants in Furniture Foam and US House Dust. Environmental Science & Technology, 2009, 43 (19), 7490-7495.

[4] Martinez-Carballo, E., Gonzalez-Barreiro, C., Sitka, A. and et al. Determination of selected organophosphate esters in the aquatic environment of Austria. Science of the Total Environment, 2007, 388 (1-3), 290–299.

[5] van den Eede, N., Dirtu, A.C., Neels, H. and et al. Analytical developments and preliminary assessment of human exposure to organophosphate flame retardants from indoor dust. Environment International, 2011, 37 (2), 454–461.

[6] Leonards, P., Steindal, E.H., van der Veen, I. and et al. 2011, Screening of Organophosphor Flame Retardants 2010. SPFO-Report 1091/2011. TA-2786/2011.

[7] Jarema, K.A., Hunter, D.L., Shaffer, R.M. and et al. Acute and developmental behavioral effects of flame retardants and related chemicals in zebrafish. Neurotoxicology and Teratology, 2015, 52, 194–20.

[8] Li, R.W., Zhou, P.J., Guo, Y.Y. and et al. Tris (1,3-dichloro-2-propyl) phosphate-induced apoptotic signaling pathways in SH-SY5Y neuroblastoma cells. Neurotoxicology, 2017, 58, 1–10.

[9] OEHHA. Evidence on the Carcinogenicity of Tris(1,3-Dichloro-2-Propyl) Phosphate. Reproductive and Cancer Hazard Assessment Branch, Office of Environmental Health Hazard Assessment, California Environmental Protection Agency: Sacramento, CA, 2011.

[10] Farhat, A., Buick, J. K., Williams, A. and et al. Tris(1,3-dichloro-2-propyl) phosphate perturbs the expression of genes involved in immune response and lipid and steroid metabolism in chicken embryos. Toxicology and Applied Pharmacology, 2014, 275 (2), 104–112.

[11] Crump, D., Chiu, S. and Kennedy, S. W. Effects of tris(1,3-dichloro-2-propyl) phosphate and tris(1-chloropropyl) phosphate on cytotoxicity and mRNA expression in primary cultures of avian hepatocytes and neuronal cells. Toxicological Sciences, 2012, 126 (1), 140–148.

[12] Chunsheng Liu, Guanyong Su, John P. Giesy and et al. Acute Exposure to Tris(1,3-dichloro-2-propyl) Phosphate (TDCIPP) Causes Hepatic Inflammation and Leads to Hepatotoxicity in Zebrafish. Scientific reports, 2016, 6, 11.

[13] Farhat, A. et al. In ovo effects of two organophosphate flame retardants-TCPP and TDCIPP-on pipping success, development, mRNA expression, and thyroid hormone levels in chicken embryos. Toxicological Sciences, 2013, 134 (1), 92–104.

[14] Eric J. Ditzel, Thu Nguyen, Patricia Parker and et al. Effects of Arsenite Exposure during Fetal Development on Energy Metabolism and Susceptibility to Diet-Induced Fatty Liver Disease in Male Mice. Environmental Health Perspectives. 2016, 124(2), 201-209.

Analysis of Immunological Activity of Recombinant Outer Membrane Protein of *Fusobacterium Necrophorum*

Hang Jia[1], Shengxin Wang[1], Xiaoli Zhang[1], Yang Guo[1] and Jing Xu[2,*]

[1]Medical Laboratory Technology, College of Medical Technology, Qiqihar Medical University, Qiqihar 161006, China

[2]Department of Biochemistry and Molecular Biology, Qiqihar Medical University, Qiqihar 161006, China

*Corresponding author

Abstract. To investigate the prokaryotic expression of 93 ku outer membrane protein (OMP) obtained from *Fusobacterium necrophorum* and the response to immunization and cytotoxicity in mice. The recombinant plasmid pET-28a-OMP was transformed into *E. coli* BL21 (DE3) competent form, and a recombinant outer membrane protein containing 2 His tags (label size 3.8 ku) was expressed. The mice were immunized as immunization antigens by Freund's incomplete adjuvant. SDS-PAGE and Western blot analysis showed that the recombinant protein had a molecular weight of about 93 ku. After the immunization, the serum antibody titer of the experimental group was higher, at the same time, in the challenge experiment, the liver has certain toxicity, and the protection rate is 83.3%. The recombinant 93 ku outer membrane protein of *Fusobacterium necrophorum* has a good immune response level and provides basic data for the study of the subunit vaccine of *Fusobacterium necrophorum*.
Keywords: *Fusobacterium necrophorum*; Outer membrane proteins; Immunoprotective.

1. Introduction

Fusobacterium necrophorum (FN) is a zoonotic pathogen, which is generally chronic. Infected animals often experience weight loss, miscarriage and death. The bacteria are also main pathogens of inflammation liver hepatic abscess, rot disease, and human acute and chronic throat etc [1-4]. FN is a Gram-negative anaerobic bacterium, no spores, no capsules, no flagella, often found in the intestines, mouth of animals and humans. Usually FN is divided into 2 sub-species, namely Fnn and Fnf. Fnn mainly infects livestock and has low virulence to humans, while Fnf mainly infects humans, including A, B and AB genotypes. The bacterium is the main pathogen of ruminants, such as cattle, sheep, deer, etc. Infected animals can cause cross-contagion if not treated in time, seriously affecting the development of animal husbandry, and even cherish the breeding of animals, etc. If the infection continues to be frequent, it will cause great economic losses. This infection of zoonotic bacteria is also worthy of the attention of many researchers. At present, vaccination is the most direct and effective method to prevent FN infection. Outer membrane proteins (OMP) are unique structures in the cell wall of Gram-negative bacteria, which maintain the structure of bacteria, transport of substances and transport agents and antibacterial agents. Regulation and other plays a vital role, and it is also one of the main virulence factors of bacteria, and has good immunological activity. It is an important potential target for the development of vaccines [5-9].

Based on the prokaryotic expression of *Fusobacterium necrophorum* and the primary extraction of OMP, this study determined that the recombinant protein 93ku OMP was expressed in 44.5 ku OMP

[10-11]. The immunogenicity study of recombinant 93 ku OMP provides basic data for the screening of new FN subunit vaccines.

2. Materials and Methods

2.1. Primary Reagent
Restriction enzymes Xho I, EcoR I (BioLabs); ProteinMolecular Weight Standard-high (TaKaRa); Coomassie Brilliant Blue R-250 (Sigma); Protein Pre-stained Marker(Fermentas); HEPEs(Beijing Huamei Bioengineering Co., Ltd); HRP-labeled goat anti-mouse IgG, DAB chromogenic kit (Beijing Zhongshan Jinqiao Biotechnology Company)

2.2. Strains and Vectors
Fusobacterium necrophorum (FN) AB strain, a virulence strain isolated, identified and preserved by the Institute of Special Products of the Chinese Academy of Agricultural Sciences; *E. coli* TOP10 competent strain, deposited by the Institute of Special Products of the Chinese Academy of Agricultural Sciences; *E.coli* BL21 (DE3), pET-28a was purchased from Novagen; the recombinant plasmid pET28a-OMP was constructed by our laboratory.

2.3. Experimental Animal
Kunming BALB/c mice, clean grade 8 weeks mice (25g) were purchased from the Experimental Animal Center of Jilin University and passed the test of Jilin Provincial Laboratory Animal Quality Testing Center.

2.4. Recombinant Outer Membrane Protein Expression
Construction of recombinant plasmid pET28a-OMP, expression of the outer membrane protein of *Fusobacterium necrophorum* 93ku[11].

2.5. MTS Test
The mouse macrophage (AnA-1) was adjusted to a cell concentration of 10^4-10^5 cells/mL with a culture medium, and inoculated into a 96-well cell culture plate, 100 μL of the above-mentioned concentration cells were added to each well, and a control empty (medium, MTS, PBS, cells), zero-well (medium, MTS), about 20 h, discard the culture medium in the well, and wash the surface of the cells once with PBS. Adjust the expressed OMP concentration to 100 μg/mL, filter with 0.45 μm, preset 6 gradients (2-fold gradient dilution), 3 replicate wells per gradient, and add it to the well-packed 96-well plate .After 16 h of incubation, 120 μL of MTS dilution (100 μL + 20 μL MTS) per well was added to a 96-well plate in the dark, and cultured in a cell culture incubator for about 3.5 h. The light absorption value of each well was measured by a microplate reader under OD_{490} absorption light.

2.6. Preparation of Immune Serum
Sixty Kunming mice were randomly divided into 4 groups, 15 in each group. Group 1 blank control, multiple HEPEs (20 mM), 0.2 mL/n in the back and abdominal cavity; 2nd group negative control, 0.2 mL/F incomplete Freund's adjuvant (IFA); Group 3 experiment Group, back and abdominal multi-point injection of recombinant OMP 100 μg/protein (protein antigen 0.1 mL, IFA 0.1 mL) emulsified with Freund's complete adjuvant, group 4 positive control, multi-point inoculation of super-breaking bacteria in the back and abdominal cavity Body pellet (20 mL culture, 2 mL 20 mM HEPEs resuspended), 0.2 mL/only. Each group was immunized once every other week for a total of three immunizations. In the case of booster immunization, the antigen was emulsified with Freund's incomplete adjuvant. After 10 days of the last immunization, fasting for 10 hours or so, blood collection of 3 /group, 4,000 r/min centrifugation. Serum was prepared at 10 min and stored at -80 °C until use.

IOP Conf. Series: Materials Science and Engineering **611** (2019) 012057 doi:10.1088/1757-899X/611/1/012057

2.7. Western-blotting Test
SDS-PAGE electrophoresis, antigen is recombinant OMP, primary antibody is mouse anti-OMP (FN strain) serum (1:200), secondary antibody is HRP-labeled goat anti-mouse IgG (1:3000), DAB chromogenic kit color.

2.8. Protection Rate Detection
The remaining mice after the above three exemptions were tested for protection rate, and the back and abdominal cavity were inoculated with the super-degraded Bacillus nebula bacterial cell pellet (100 mL culture solution, 10 mL 20 mM HEPEs resuspended), 0.2 mL/only, note Observe the mice's eating, metabolism, coat color, limbs, weight, activity, death and so on.

3. Results and Analysis

3.1. SDS-PAGE Verification
SDS-PAGE identified that the recombinant OMP of Fusarium oxysporum had a molecular weight of 93 ku and was stained with Coomassie Brilliant Blue R-250. A clearly visible band appeared at approximately 93 ku, as shown in Figure 1.

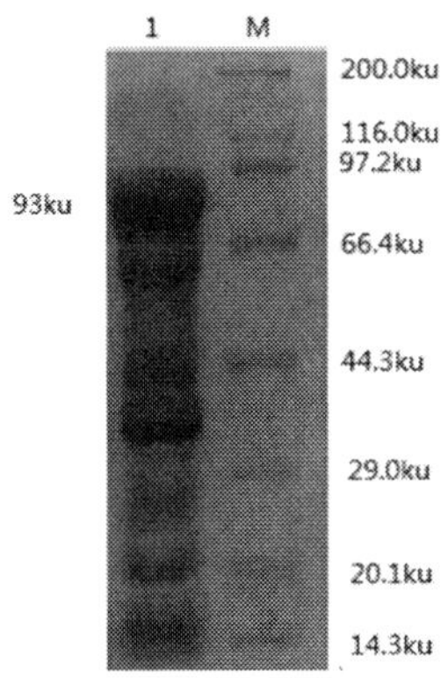

M, Protein Molecular Weight Marker; 1, Recombinant OMP fragmentation
Figure 1. Identification of OMP by SDS-PAGE

3.2. Western-blotting Identification
Western blotting showed that a band appeared at 93 ku, indicating that the expressed protein has good specificity, as shown in Figure 2.

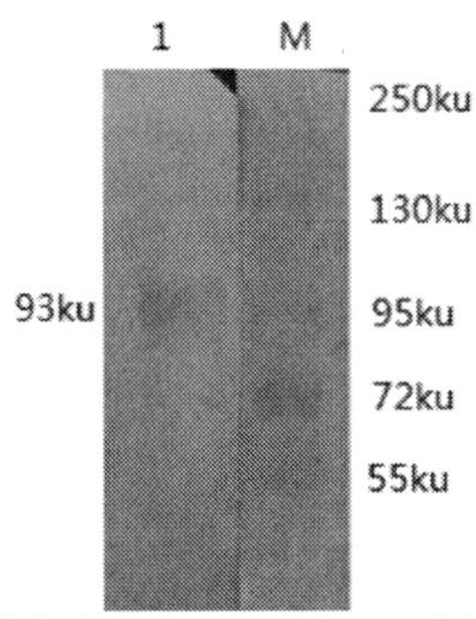

M, Protein Molecular Weight Marker; 1, Recombinant OMP fragmentation
Figure 2. Identification of OMP by Western-blotting

3.3. MTS Test
Recombinant OMP 100 μg/mL, preset 6 gradients (2-fold gradient dilution: 1/2, 1/4, 1/8, 1/16, 1/32, 1/64) to the small paved MTS dilutions were added to the mouse macrophage (AnA-1) plate in the

dark. The OD490 light absorption value of each well was measured by a microplate reader, and the results showed that the absorbance values were lower at 1/2 and 1/4 of the dilution factor, as shown in Figure 3.

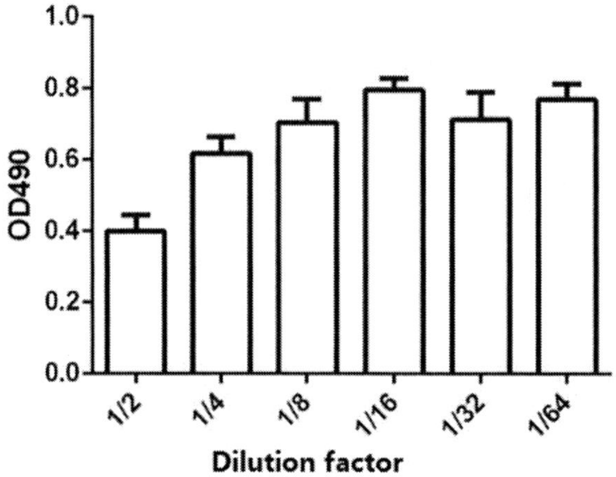

Figure 3. MTS reduction test of the AnA-1 cell

3.4. *Immune Serum Antibody Titer Test*

The serum antibody titer was detected. The serum of the mice immunized with recombinant OMP (3 in each group) was determined by ELISA. The results showed that the serum titer of the experimental group was 1:320000. The control group and the blank group were not detected. Antibody production indicates that the method yields high serum.

3.5. *Mouse Protection Rate Test*

After the mice were exempted, the remaining 12 groups in the first group of blank group, 3 had blood collection test; the remaining 2 groups in the second group of negative control group, 3 had blood test; the remaining 10 groups in the third group, 3 had blood test 2 had died; the remaining 4 in the 4th positive control group, 3 had blood test and 5 had died. The super-degraded Bacillus necrotic bacteria (100 mL culture medium, 10 mL 20 mM HEPEs were resuspended), 0.2 mL/only, after multiple inoculation in the back and abdominal cavity, after two weeks, the results showed the first and second. All the mice in the group died, and the remaining mice in the third and fourth groups were 7 (3 deaths) and 5 (2 deaths), and the protection rates were 70% and 71.4%, respectively (protection rate = number of surviving mice / The total number of test mice × 100%), as shown in Table 1.

Table 1. Rate test of mouse protection

Group	Group 1 blank group	Group 2 negative control	Group 3 experimental group	Group 4 positive control
Total	15	15	15	15
Remaining	12	12	10	7
Protection	12	12	3	2
Remaining	0	0	7	5
Protection rate	0%	0%	70%	71.4%

4. Discussion

In recent years, necrobacillus is a zoonotic disease that is widely concerned by scholars at home and abroad. It is widely distributed, causing many types of diseases, the treatment effect is not obvious, and it is easy to relapse. Therefore, prevention and control are very difficult [13-15]. The immunological activity and immunoprotection of outer membrane proteins have been studied in many bacteria and been the target of subunit vaccine research. The outer membrane protein is the main structure of the bacterial cell wall, which directly contacts the external environment and plays an important role in the

invasion, adhesion, and prevention of host serum defense and killing [16-17]. The study found that many human diseases are related to the outer membrane proteins of bacteria, so the outer membrane protein is concerned by many scholars, such as Oprl (a major OMP of Pseudomonas aeruginosa), which can induce small mice after immunization with recombinant protein against *Pseudomonas aeruginosa* [18].The pathogenic role of pathogenic bacteria is achieved by immune evasion to adapt to the host's defense mechanism. When the pathogen is adsorbed to the susceptible part of the organism, it is easy to cause disease; the pathogen that adsorbs to the epithelial cells of the respiratory tract plays a key role. , avoiding the mechanical removal of the host, thus achieving the first step of the invasion [19-22].
In this study, the virulence factor outer membrane protein of *Fusobacterium necrophorum* was selected as the research object. The prokaryotic expression of *Fusobacterium necrophorum* recombinant 93 ku OMP was emulsified by incomplete Freund's adjuvant, and then it was used as an antigen to immunize Kunming mice to detect serum antibody effect. The price is higher, and it has better protection in the experiment. The Western blotting shows that it has good reaction specificity, and it is the same protein as the crude protein of 44.5 ku. OMP lays the foundation for the development of FN subunit vaccine.

Acknowledgments
This study was supported by Heilongjiang Provincial Department of Education Project; National Key Project of Heilongjiang University Students Innovation and Entrepreneurship Training Program (201811230011).

References
[1] Chen L Z. Study on isolation, identification and its leukotoxin immunological characteristics of bovine origin pathogenic *fusobacterium necrophorum* [D]. Changchun: Jilin University, 2008.
[2] Miao L G. The study on identification of protective antigen in fusobacterium necrophorum and gene selection [D]. Beijing: Graduate School of Chinese Academy of Agricultural Sciences, 2006.
[3] Panchavati P K , Kar B , Hassoun A ,et al. Fusobacterium necrophorum tonsillitis with mild case of Lemierre's syndrome [J]. Anaerobe, 2017, 43:102-104.
[4] Farooq S, Wani S A, Hassan M N, et al. The detection and prevalence of leukotoxin gene variant strains of Fusobacterium necrophorum in footrot lesions of sheep in Kashmir, India [J], Anaerobe, 2018, 51
[5] Ann M G K, Marianne K T, Therese O, et al. Clinical and biochemical characteristics of patients with Fusobacterium necrophorum -positive acute tonsillitis [J].European Archives of Oto-Rhino-Laryngology, 2015, 272(6):1-7.
[6] Clifton R, Greeen L E, Purdy K J. Development and validation of a multiple locus variable number tandem repeat analysis (MLVA) scheme for Fusobacterium necrophorum [J], Veterinary Microbiology, 2018, 213:108-113.
[7] Klug T E, Rusan M, Fuursted K. A systematic review of Fusobacterium necrophorum -positive acute tonsillitis: prevalence, methods of detection, patient characteristics, and the usefulness of the Centor score [J]. European Journal of Clinical Microbiology, 2016, 35(12):1903-1912.
[8] He M Y. Application of genetic engineering vaccine in animal disease prevention and treatment [J]. China Animal Husbandry and Veterinary Medicine, 2015, 31(11):222+175.
[9] Xiao X X,Liu J S, Yin X P. Prokaryotic expression and purification of flagellins from salmonella trphimurium [J].China Animal Husbandry and Veterinary Medicine,2015,42(4):871-876.
[10] Meng F, Li S C, Yang W N, et al. Research progress on biofilms in pseudomonas aeruginosa [J]. China Animal Husbandry and Veterinary Medicine, 2016,43(11):2976-2981.
[11] Xu J, Chen L Z, Liu X Y, et al. Preliminary Extraction and Identification of the 44.5 kDa Outer Membrane Proteins Isolated from Bovine *Fusobacterium necrophorum* (AB) [J]. Indian J Microbiol (Oct–Dec 2013) 53(4):395–399.
[12] Xu J, Chen L Z, Liu X Y, et al. Preliminary Extraction and Identification of the 44.5 kDa Outer Membrane Proteins Isolated from Bovine *Fusobacterium necrophorum* (AB) [J]. Indian J

Microbiol (Oct–Dec 2013) 53(4):395–399

[13] Kulkarni H M, Nagaraj R M V. Protective role of E.coli outer membrane vesicles against antibiotics [J]. Microbiological Research, 2015, 181:1-7.

[14] Breda L C, Hsieh C L, Valencia M M, et al. Correction: Fine Mapping of the Interaction between C4b-Binding Protein and Outer Membrane Proteins LigA and LigB of Pathogenic Leptospira interrogans [J]. Plos Neglected Tropical Diseases, 2015, 9(12):e0004286.

[15] Kleanthous C, Rassam P, Baumann C G. Protein-protein interactions and the spatiotemporal dynamics of bacterial outer membrane proteins [J]. Current Opinion in Structural Biology, 2015, 35:109-115.

[16] Jyoti S S and James E Peacock. Fusobacterium necrophorum endocarditis case report and review of the literature [J]. Diagnostic Microbiology & Infectious Disease, 2011, 69(2):192-195.

[17] Cao Y, Johnson H M, Bazemore-Walker C. R. Improved enrichment and proteomic identification of outer membrane proteins from a Gram-negative bacterium: focus on caulobacter crescentus [J]. Proteomics, 2012, 12(2): 251–262.

[18] Hua C Z, Hu W L, Shang S Q, et al. Serum concentrations of antibodies against outer membrane protein P6, Protein D and their T- and B-cell combined antigenic epitopes of nontypeable Haemophilus influenzae in children and adults of different ages [J]. Clinical and Vaccine Immunology, 2015, pii: CVI.00506-15.

[19] Cattelan N, Villalba M I, Parisi G, et al. The outer membrane protein OmpQ of Bordetella bronchiseptica is required for mature biofilm formation [J]. Microbiology, 2015, doi: 10.1099/mic.

[20] Urfer M, Bogdanovic J, Lo Monte F, et al. A Peptidomimetic Antibiotic Targets Outer Membrane Proteins and Disrupts Selectively the Outer Membranein Escherichia coli [J]. Journal of Biological Chemistry, 2015, pii: jbc.M115.691725.

[21] Li S G,Zhang N, Chen L K, et al. Construction and immunogenicity analysis of streptococcus suisrGDH prokaryotic expression vector [J].China Animal Husbandry and Veterinary Medicine, 2015,42(10):2594-2599.

[22] Xu J, Chen L Z, Liu X Y, et al. Cloning and expressin of out membrane protein (OMP) of *Fusobacterium necrophorum* (AB) [J].China Animal Husbandry and Veterinary Medicine, 2013, 40(6): 57-59.

Application of Aeration Spray Technology in China and Its Reduction Effects in VOCs Control

Zhida He[1], Maoshen Ye[2,*], Qingwei Ni[2] and Junqi Ding[2]

[1]Tongzhou Municipal Commission of Development and Reform, Beijing, China
[2]China Eastern Route Corporation of South-to-North Water Diversion, Beijing, China

*Corresponding author

Abstract. The impression of paint spraying operation is that the paint mist is flying and it can pollute the environment and endanger the human health. The Airmix® spray technology invented by Chemlin completely changed the working conditions of the painting operation. This painting technology is called "mixed air spraying". Although the technology is still little known in China, aeration spray has become a newcomer in the field of spray technology. It not only has the advantages of good atomization effect of traditional air spraying, high surface quality of paint film, convenient use, etc., but it also has many excellent process characteristics: it can spray paint with high viscosity, high spraying efficiency, save paint and reduce cost, and get a thicker film. In particular, "mixed air spraying" inhibits the scattering of paint mist; it not only improves the spraying environment, occupational health and safety production guarantee, but also effectively reduces the pollution of VOCs and greatly promotes the development of productivity. The China's government should promote the application of this new technology and gradually phase out some backward technology equipment.

1. Introduction

The earliest painting process was to apply a brush to paint or dip the work piece directly into a paint-filled container. In order to improve the coating quality and work efficiency, various coating processes have emerged such as air spraying, HVLP (low pressure and large flow) spraying, electrostatic spraying, electrostatic powder coating, electrophoretic coating, high pressure airless spraying, aeration spraying, and electrostatic mixing spraying. The research objective is comparing aeration spray technology to the normally used air spray methods and point out its superiorities.

Table 1. Process parameters and transmission efficiency of different paint spray guns

Painting type	Gas pressure (Bar)	Paint pressure (Bar)	Paint transmission efficiency[1]
Air spray gun	2.5～6.0	0.5～3.5	30～45%
Environmentally friendly air spray gun	0.5～1.5	0.1～2.0	45～75%
Airless spray gun	0	100～500	40～60%
Aeration spray gun	0.5～1.5	60～120	50～85%
Electrostatic spray gun	2.0～4.0	0.5～3.5	70～90%
Aerated electrostatic spray gun	0.5～1.2	60～120	75～95%

Table 2. Parameters and efficiency of different spray atomization technology equipment

Atomizing spray gun	Maximum sprayable consistency (Ford Cup 4)	Paint thickness	Painting efficiency[2]
Environmentally friendly air spray gun	15 sec	Very thin	Very slow
Air spray gun	30 sec	Thin	Slow
Aeration spray gun	90 sec	Thick	Fast
Airless spray gun	> 90 sec	Very thick	Very fast

Table 1 compares the process parameters and transmission efficiency of different paint spray guns, and Table 2 lists the parameters and efficiency of different spraying atomization technology equipment. The main advantages of the air-mixing spray gun are as follows.

(1) Controllable spray effect makes the paint quality higher;

(2) Higher paint efficiency and improved production efficiency;

(3) Reduced overspray and save up to 40% of paint;

(4) It can use high viscosity paint to save 80% of thinner material;

(5) To control the scattering of paint mist, and greatly improve the working environment and reduce pollution.

If the rules and new products or technologies work together, it will have a very large impact on the industry; if there is sufficient market capacity (to support his R&D investment), all companies will follow the rules. When the rules are strictly enforced, only those companies with long-term goals will get the most profit and get a lasting development.

2. Theory and Methodology

As early as 1975, the German company Kremlin created the Airmix ® aeration spray technology, which is still considered to be one of the most outstanding inventions in the field of spray application technology. Aerated spray began to become a new type of spray technology recognized in the industry. The inspiration for creating a gas-blown spray technology comes from combining the advantages of traditional air spray and airless spray, complementing the blank areas that are applied to the paint output.

The aeration spray technology combines the technical advantages of traditional air spray and airless spray. First, at a low paint pressure of 30 to 100 bar, the paint is subjected to preliminary atomization through a nozzle similar to airless spray. Second, through the positioning of the precise air outlet hole on the air cap, the pressure of the atomizing air is very low, usually 0.5 to 1.2 bar; this process makes the fan shape uniform and the paint particles become fine. Third, because the speed of the coating particles is very low at only 0.7 m/s, which is less than one tenth of the speed of the conventional spraying particles, the scattering and rebound of the coating are effectively suppressed, the pollution hazard of VOCs is greatly reduced, and the production environment is improved. Reduce the investment and management costs of the company for occupational health, safety and environmental protection.

3. Experiments and Results

3.1. Experimental Design of Aeration Spray Technology

The test used a full set of imported air-mixing equipment: two Xcite spray guns from Kremlin and two paint supply devices from Wagner, Germany. In the engine-painting booth, the air-mixing spray engine test was carried out. The Beijing Municipal Institute of Labor Protection and the German

Wagner Corp. conducted on-site inspection and training. Weichai Power Corp. provided more than 150 engines and related technicians and operators.

3.2. Test Results

3.2.1. Spray quality. After repeated trials, the viscosity of the paint was set at 37 to 43 seconds, which was higher than the paint viscosity of 22 ± 5 seconds required by the air spray process, which created favorable conditions for improving the thickness of the single spray paint film.

A. Film thickness

Table 3. Comparison of the average thickness of air spray and air spray paint film

Part type	Paint film thickness process requirements	Average thickness of air spray paint film	Average value of mixed spray paint film thickness	Difference	Thickness increase[3]
Primer casting parts	≥90μm	97.8	147.5	49.7	50.8%
Primer ribbed parts	≥60μm	80.4	117.5	37.1	46.1%
Primerless parts	≥55μm	73.75	89.75	16	21.7%
		Average percentage of thickness			39.5%

It can be seen from Table 3 that the paint film thickness of the two sprayed diesel engines meet the process requirements because the operation mode of the new spray gun is different from that of the air spray gun. The operation habit of the paint sprayer is difficult to change at one time, and the paint film thickness is mixed. The average value is thicker than the average thickness of the air spray paint film of 39.5%. That is, the performance of the aeration spray method is better than the air spray method. Moreover, no flow lacquering occurs at this thickness, mainly because the painting equipment can accommodate higher paint viscosities.

B. Appearance quality of paint

Due to the small particle size and less diffusion of the mixed spray paint, the appearance of the sprayed surface in the diesel engines was more delicate than the air sprayed ones, and the gloss is better. In particular, peripheral components such as a cylinder head cover and a supercharger are more obvious.

Figure 1. Comparison of the appearance of air-sprayed(left) and aerated-sprayed(right) diesel engine cylinder head covers

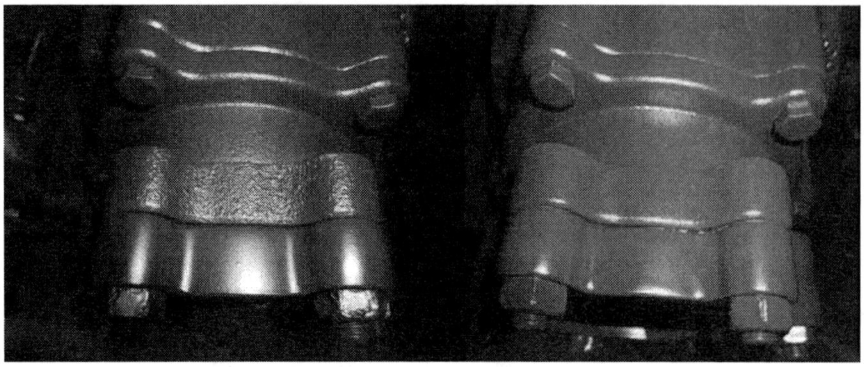

Figure 2. Comparison of appearance of air spray(left) and aerated spray(right) diesel exhaust pipe

3.2.2. Analysis of spray material usage and production cost

Table 4. Paint, curing agent, thinner consumption

	Air spray gun	Aeration spray gun	Difference	Price per kilogram(¥)	Saving price per unit[4]
Thinner consumption of single diesel engine (kg)	0.95	0.47	0.48	13	6.24
Single diesel engine paint consumption (kg)	1.37	1.31	0.06	29.5	1.77
Single diesel engine curing agent consumption (kg)	0.1377	0.1311	0.0066	63	0.4158
Total cost per unit					8.4258

It can be seen from Table 4. If compared with air spraying method, the air-mixing spray method can save 8.4258 yuan per unit. Considering an annual output of 500,000 diesel engines and the film thickness factor, using the current paint film thickness requirements, the total annual savings is about 8.4258*500000*(1+39.5%)=5876995.5 yuan, or 5.877 million yuan.

3.2.3. Improvement of air quality in the workplace. During the test, the BMILP was commissioned to test the ambient air quality of the two painting equipment. In the manual painting room, paint mist concentration sample in the air spray site and the sample in aeration spray site were collected. The concentrations of benzene, toluene, xylene, butyl acetate, and non-methane total hydrocarbons (VOCs) in the air were tested. The test results are shown in Table 5.

From the above comparison, the aeration spray gun reduces the pollutant emissions of the air spray gun by more than 50%; the air quality of the workshop is greatly improved. In addition, the paint mist image generated from two different spraying methods on the spot can also visually compare the difference in paint mist concentration [6-7].

On-site robots and artificial air spray guns are used to spray paint. The paint mist is flying, and the pollution is serious. The difficulty of harmful substances is harmful to the health of the operators on the site, and it incurs the fire safety risks of the workshop[8-9]. See Fig. 3, 4.

Table 5. Comparison of workshop air quality test

Serial number	Test items	Concentration value(mg/m^3)		Comparative Results[5]		Remarks
		Air spray gun	Aeration spray gun	Decrease the number	Reduction rate	
1	Non-methane total Hydrocarbon	175.3	-	-	-	Robot
2	Non-methane total Hydrocarbon	79.4	9.7	69.7	87.78%	Artificial
3	Butyl acetate	7.58	1.92	5.66	74.67%	Artificial
4	Xylene	1.60	0.30	1.30	81.25%	Artificial
5	Toluene	<0.03	<0.01	0.02	66.67%	Artificial
6	benzene	<0.02	<0.01	0.01	50.00%	Artificial

Figure 3. Robot spraying on-site paint fog condition (air spray)

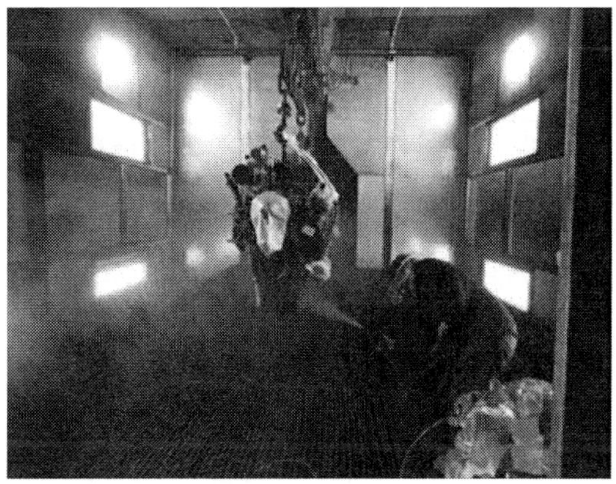

Figure 4. Artificial air spray on-site paint fog condition

Spraying with a mixed air spray gun, the paint rebounds little, the paint mist has no obvious traces, and it is beneficial for on-site equipment maintenance and workers health. As shown in Fig. 5.

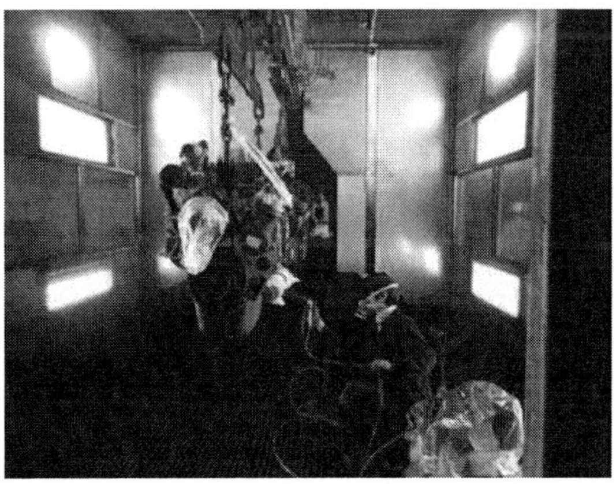

Figure 5. Artificial aeration spray painting scene paint fog condition

It can be seen from the above figure. Due to the technical characteristics of the air-mixing spray gun, there is a paint mist restraint zone formed by the air-assisted pressure. So, the paint spray concentration on the air-mixed spray site is weakened much compared to the air spray method, and reduced the damage to the field operator. There is a sharp difference or promotion between air-mixing spray method and the air spray gun in the environmental pollution and the paint waste. The operator's experiences of the operation process were collected on site. The operator was very satisfied with the improvement of the working environment brought about by the reduction of the paint mist. The reduction in the concentration of flammable substances in the air also reduces the risk of safety accidents such as fires [10-12].

3.2.4. Comparison of spraying operation speed. The average time taken by a skilled painter to use a gas-blown sprayer and the average time spray a diesel engine using a conventional air spray gun are shown in Table 6 [13-15].

Table 6. Average time of air spray gun and air spray gun painting

Gun type	Aeration spray gun	Air spray gun
Spraying time	4′47″	5′46″

The fastest and slowest spraying times of the two painters are shown in Table 7:

Table 7 Comparison of Spraying Time of Mixing Air Guns of Different Operators

Painter	Fastest spraying time	Slowest spraying time
Liu guoqiang	3′20″	5′06″
Ma Ruiwei	4′20″	5′50″

In addition, in order to compare the paint spraying time efficiency of different painters, we selected two spray painters Liu and Ma on May 16,2018 to compare the spray time of 10 diesel engines sprayed with a mixed air spray gun. The contrast curve is shown in Figure 6[16-17].

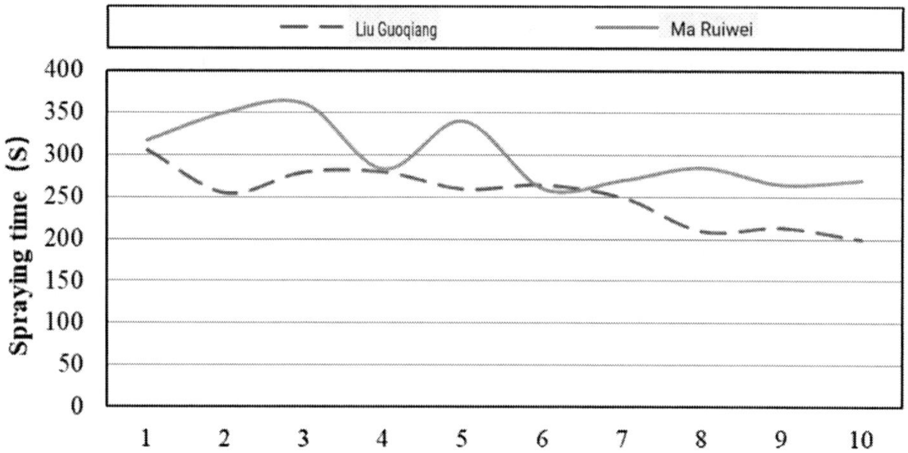

Figure 6. Spraying time curve of different sprayers

In table 6, 7 and figure 6, from the data obtained from 40 diesel engines, regardless of the skill difference of the painter, the spray rate of the aeration spray gun is higher than that of the air spray gun. The spraying rate of different painters is different. In this test, under the premise that the thickness meets the requirements of the process quality, the painter Liu Guoqiang sprays a diesel engine for a maximum of 3 minutes and 20 seconds, which is 1 minute faster than the last time for Ma Ruiwei to spray one. The painting speed is faster. Significantly higher than air spray. It can be seen from the spray time curves of the two painters in Figure 6. As the painter gradually becomes familiar with the operation of the spray gun, the spraying time tends to be shortened.

Through the case study of the paint production line of the engine of Weichai Power Co., Ltd., the aeration spray technology has the following obvious technical advantages.

First, the increase of the viscosity of the paint reduces the use of organic solvents and reduces the pollution of harmful substances; the measured use of thinners is reduced by 50%, and the total amount of VOCs emissions is reduced by 75%;
Second, it reduced the rebound of paint fog, improved the utilization of paint, saved paint (paint, thinner, etc.) 20%; according to the annual output of 50 units, it can save 5.777 million yuan;
Third, it significantly reduced the scattering of paint mist and reduced noise; effectively reduce the health hazards of operators. The VOCs content of the spray booth is more than 100 mg/m^3 when air sprayed and less than 10 mg/m^3 when air spray is applied;
Fourth, the dust and aerosol in the discharge gas of the spray booth are greatly reduced, the use time of the filter material is prolonged, and the total amount of organic gas emissions (VOCs) is reduced several times;
Fifth, the surface quality of the engine spray paint is smooth and smooth, and the thickness of the paint film is generally improved, which increases the market competitiveness of the engine products.

4. Conclusion

Recently, the Ministry of Environmental Protection and the State Administration of Work Safety of China have attached great importance to the occupational health hazards and safety hazards of organic gas (VOCs) emissions and organic waste gases. Cities such as Beijing and Shanghai have developed more stringent VOCs emission standards. In the spring of 2014, the surrounding eastern part of Beijing caused persistent severe smog weather, forcing the Beijing Municipal Government to propose the corresponding measures for VOCs reduction. In the future, it is necessary to limit the backward spraying process of the paint spraying industry, and eliminate the backward spraying process and device. The technical advantage of the aeration spray technology is that it can suppress the occurrence of paint mist, improve the use efficiency of paints (paints, thinners, additives), reduce the consumption of compressed air, improve the surface quality of spray products, lift production efficiency, depress emissions of VOCs, and cut down the production cost and environmental protection cost of an enterprise. It is an effective technical means and worth promoting.

References

[1] Buzcu B, Fraser M P. Source Identification and Apportionment of Volatile Organic Compounds in Houston, TX. Atmospheric Environment, 2006, 40(13): 2385-2400.
[2] CHEN Hui, LIU Xianjun. Development and Application of Arc Coating Technology in China, Chongqing Car and Motorcycle, Welding, Painting Academic Symposium Proceedings, 2000.
[3] DU Yanling. The Coating Interface Structure under the Action of Medium Temperature, South China University, 2005.
[4] Guenther A, Geron C, Pierce T, et al. Natural Emissions of Non-methane Volatile Organic Compounds, Carbon monoxide, and Oxides of Nitrogen from North America. Atmospheric Environment, 2000, 34(12-14): 2205-2230.
[5] EPA. EPA 8260C-2006 Volatile Organic Compounds by gas chromatography/mass spectrometry, 2006.
[6] European commission joint research centre. Total Volatile Organic Compounds (TVOC) in Indoor Air Quality Investigations. 1997.
[7] HAO Jianjun. Research on Sandblasting and Arc Coating Technology for Iron Parts Repair, Heibei Agriculture University, 2001.
[8] Jamrog Robert. Automotive Water-borne Coating: Clean Air Legislation is Pushing Automakers Toward Water-Borne basecoat, Products Finishing, October, 1993.
[9] Kraul, Chris. A Warmer Climate for Furniture Makers, The Los Angeles Times, May 14, 1990.
[10] LI Heqi, LIU Jia, CHEN Aijun. Development on the Equipment of Arc Coating with Computer Controlling, The Proceedings of The 9th National Conference on welding(Book 1),1999.
[11] LI Hongsheng, WANG Wenli. The Equipment Application of Airless Technology, China Construction News, 2002.
[12] Randall Paul. Pollution Prevention Methods in the Surface Coating Industry, Journal of Hazardous Materials, 29(1992): 275-295.

[13] Robinson A L, Donahue N M, Shrivastava M K, et al. Rethinking Organic Aerosols: Semi Volatile Emissions and Photochemical Aging. Science, 2007, 315(5816):1259-1262.
[14] Schulenberg Fred. America's Great Coating Centres, American Paint and Coatings Journal, January 5th, 1987.
[15] Sillman S. The Relation Between Ozone, NOx, and Hydrocarbons in Urban and Polluted Rural Environments// Developments in Environmental Science. Environmental Science & Technology, 2002:1821-1845.
[16] Tan SMeiller JMahaffey K. Critical Reviews in Toxicology. Informa Healthcare, (2009), 228-269.
[17] YU Meijie. Research on High-temperature Abrasion Resistant Coatings of Circulating Fluidized Bed Boiler. Shangdong Technology University, 2004.

IOP Conf. Series: Materials Science and Engineering **611** (2019) 012059 doi:10.1088/1757-899X/611/1/012059

Biomimetic Synthesis of Calcium Carbonate with the Assistance of Amino Acids

Yanru Liu[1], Lei Yang[1,*], Zhiyuan Ou-Yang[1] and Jin Chen[2]

[1]Faculty of Chemistry and Environmental Science, Guangdong Ocean University, Zhanjiang 524088, China

[2]Zhanjiang Jiahui Pearl Ltd, Zhanjiang 524088, China.

*Corresponding author

Abstract. Two acidic amino acids (aspartic acid, glutamic acid) and four neutral amino acids (cysteine, serine, tyrosine, glycine) were used as inducers for calcium carbonate biomimetic synthesis. The results showed that aspartic acid introduced is more effective for the growth of aragonite, while glutamic acid used is more effective for the growth of vaterite. Compared to those acidic amino acids, neutral amino acids used lead to less aragonite or vaterite produced, and tyrosine and serine show stronger effects on the growth of aragonite and vaterite than cysteine and glycine. On the other hands, it was found that the increased concentration of amino acid shows positive effects on the growth of aragonite and vaterite. In addition, the enhanced concentration of calcium chloride results in the increased crystalline volume of calcite significantly, as well as the more orderly structure as shown in scanning electron microscopy investigation. Meanwhile, the high calcium chloride concentration indicates weak inhibitory effect on aragonite growth but no significant influence on the growth of vaterite.

1. Introduction

In the process of pearl mineralization, organic matter is one of the important factors that regulate the growth and nucleation of inorganic minerals[1]. Studies have shown that the organic matters secreted by cells include chitin proteins and some biomacromolecules[2], which have the functions of regulating the directional nucleation growth and provide a template for their synthesis to form nano-structures of special assembly methods[3]. Feng's Research Group extracted the water-soluble matrix (WSM), acid-soluble matrix (ASM) and acid-insoluble matrix (AIM)[4] from the pearl to carry out in vitro mineralization simulation experiment. WSM has an effect on crystal morphology and size[5]. WSM can induce aragonite in solution with different pH values, and the pH value has a significant effect on the morphology of calcite. WSM at different solution temperature has a small effect on the growth of calcium carbonate crystals, while has a significant effect on the morphology of calcium carbonate[6]. It was also found that WSM could control grains which smaller than ASM, and some organic substrates could induce the production of certain crystals[7]. Picker[8] found that specific amino acids have multiple effects on the early stage of $CaCO_3$ crystallization. Chuajiw[9] showed that the interaction between COO^- groups and Ca^{2+} cations in $CaCO_3$/amino acid precipitation samples is polarity or hydrophilic interaction.

Qiao[10,11] proved that acid-soluble organic matters could induce the morphology and crystal type of calcium carbonate mineralized crystals. Yang[12-15] found that aspartic and glutamic could induce the formation of calcium carbonate into vaterite crystals from studying the influence of acid amino acids on the crystal structure and morphology of calcium carbonate. Chen[16] found that the negative charge of amino acids after pH adjustment could attract Ca^{2+} ions, and the change of amino acid concentration would affect the number and size of crystals. Chen[17] found that the products induced by L-phenylalanine were totally calcite, while L-phenylalanine inhibited the growth of calcite. Guo[18] studied the regulation of L-valine, L-arginine and L-serine on $CaCO_3$ morphology. Wang[19] found that the concentration and temperature of L-arginine both affect the growth of calcium carbonate crystals. Xiao[20] synthesized acicular nanometer calcium carbonate crystals using L-histidine as an organic template. Wu[21], Ren[22], and Liu[23] studied the synthesis of calcium carbonate induced by L-lysine.

In our research group, we already have relevant research foundation. Yang[24], Wu[25, 26] discovered the effect of organic matter on pearl growth, and studied the induction effect of amino acids on calcium carbonate at different deposition times. Tang[27] studied the induction of calcium carbonate crystal morphology by different amino acid systems: various amino acids can induce the formation of calcite, but alanine, leucine, tyrosine and glutamic acid cannot induce the formation of aragonite. Under the conditions of histidine, isoleucine, valine, methionine and proline, more aragonite was found, and the interaction structure of aragonite layer and calcite was obtained under the action of mixed amino acids. Ou-yang[28] found that different reaction temperatures and amino acid contents can regulate the morphology and composition of calcium carbonate based on the research of Tang[27]. It is easier to prepare the nacre layer when the same volume of amino acids is mixed and the temperature is higher. This study is based on the changes in calcium chloride and amino acid concentrations that were not involved in previous studies, to study the induction effect of acidic amino acid and neutral amino acid on calcium carbonate and the influence of different calcium chloride and amino acid concentrations on the mineralization of calcium carbonate.

2. Exprerimental Section

2.1. Amino Acid Selection

In this experiment, two acidic amino acids and four neutral amino acids were selected, among which the acidic amino acids were aspartic and glutamic, and the neutral amino acids were glycine, tyrosine, serine and cysteine.

2.2. Experimental Reagents and Instruments

Anhydrous calcium chloride (AR), ammonia carbonate (AR) L-aspartic, L-glutamate, L-glycine, L-cysteine, L-serine, L-tyrosine (BR)

Fourier-transform infrared spectroscopy (perkin-elmer spectrum, USA, 100 resolution, better than 0.2 cm-1, wave number range 4000 ~ 400cm^{-1}), X-ray diffraction (Empyrean, copper target K ray, voltage 40KV, current 40 mA, emission slit 1/2°, anti-scattering slit 1°, anti-scattering slit 8.0 mm, 2θ :20°~ 60°, step length 0.02°, residence time 80s, scanning electron microscope (HITACHI S-4800, Japan).

2.3. Experimental Procedure

$CaCl_2$ solution and amino acid solution at the concentration of 10 mmol/L and 20 mmol/L were respectively prepared. Group A was formed with the amino acid solution (10 mmol/L) and $CaCl_2$ solution (20 mmol/L). Group B was formed with amino acid solution(10 mmol/L) and $CaCl_2$ solution (10 mmol/L). Group C was formed with amino acid solution (20 mmol/L) and $CaCl_2$ solution (10 mmol/L). Calcium carbonate mineralization was carried out under the closed condition for 30 days and the $(NH_4)_2CO_3$ was changed every 10 days during mineralization to ensure that the experiment was carried out under CO_2 saturation.

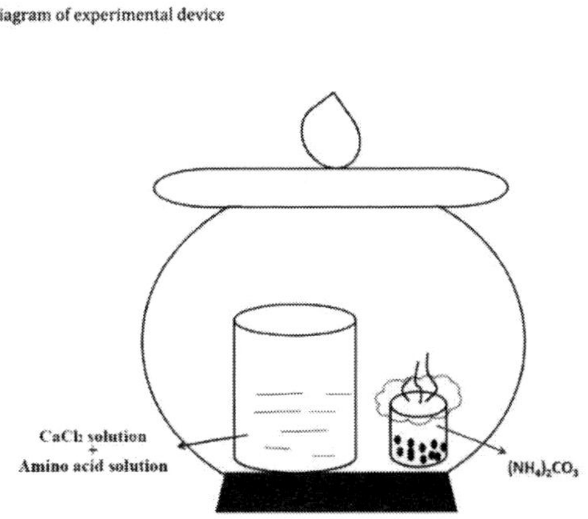

Figure 1. Diagram of experimental device.

3. Results and Discussion

3.1. FT-IR

In the infrared spectrogram (Figure 2), the calcium carbonate crystal of the blank group in group B was significantly absorbed at 1475 cm^{-1},856 cm^{-1}, and had a weak peak at 1083 cm^{-1}, 710 cm^{-1}. It was speculated that there were both aragonite and vaterite in the blank group. The absorption peak of aragonite induced by aspartic was obvious, but that of vaterite was weakened. The absorption peak of glutamate globular aragonite was obvious and that of calcite disappeared. The crystal characteristic peak of glycine and cysteine didn't change obviously. The red shift of serine, tyrosine and aragonite absorption peak indicates that these two amino acids have certain induction effect, and the CO_3^{2-} generated is unstable.

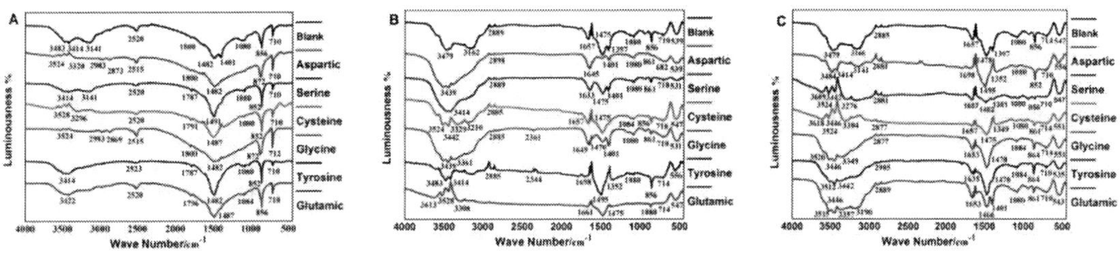

A:10 mmol/L of amino acid solution and 20 mmol/L of $CaCl_2$ solution;

B:10 mmol/L of amino acid solution and 10 mmol/L of $CaCl_2$ solution;

C:20 mmol/L of amino acid solution and 10 mmol/L of $CaCl_2$ solution;

Figure 2. Infrared spectrum corresponding to group A, B and C.

1: Blank; 2: L-aspartic; 3: L-glutamic; 4: L-glycine; 5: L-cysteine; 6: L-serine; 7: L-tyrosine

The absorption peak of cysteine in the blank group disappeared and the characteristic peak of calcite appeared by comparing group A with group B. The absorption peak of globular aragonite in aspartic, glutamate, glycine, tyrosine and serine weakened and even disappeared, while the existing aragonite absorption peak appeared blue shift, indicating that the stability of aragonite was enhanced. The absorption peak corresponding to other functional groups has a certain movement, and it is presumed that the concentration of Ca^{2+} is increased, and the intermolecular forces such as hydrogen bonds are enhanced, so that the increase of the half-peak width causes the peak shift. At the same time, the absorption peak of group A in 3200~3500 cm^{-1} showed obvious weakening compared with group B. It may be that the increase of calcium chloride concentration made the mineralization of calcium carbonate more tight, and the space of -OH stretching and bending vibration are weakened.

The comparison between group C and group B showed that the intensity of characteristic peaks of each group increased after increasing the concentration of amino acids, indicating that the dipole moment in the crystal of group C changed greatly, that is, the crystal instability generated increased. The aspartic acid and glutamic acid in group C showed obvious absorption peaks of aragonite and calcite, the absorption peaks of glycine aragonite and vaterite increased, the absorption peak of serine and cysteine aragonite increased, while tyrosine was not obvious. It can be speculated that the increasing concentration of amino acids is conducive to the formation of aragonite and calcite. At the same time, the absorption peaks of other functional groups all show small fluctuations, which may be due to the amino acid groups, not involved in the reaction, affect the fluctuation of the absorption peak with the increasing concentration of amino acid.

3.2. XRD

In the XRD pattern (Figure 3), the blank group of group B has a strong peak of calcite at 29.8°. Calcite was induced by aspartic, serine and tyrosine. The crystals induced by cysteine and glutamate had strong peaks near 26.5° and weak peaks between 40°~50°, indicating the presence of a large amount of calcite and a small amount of aragonite mixed with vaterite. There was such a weak peak of glycine at 26.5°, and it is speculated that a small amount of aragonite exists.

The comparison between group A and group B showed that aragonite and vaterite absorption peak, induced by six kinds of amino acid, weakened and even disappeared while calcite absorption peak become obviously with the increasing concentration of calcium chloride. It can be speculated that increasing the concentration of calcium chloride is conducive to the formation of calcite, but reduces the induction effect of amino acids on aragonite and vaterite. The possible reason for this phenomenon is that there is a saturation ratio between calcium chloride and amino acid. When calcium chloride is supersaturated, some Ca^{2+} occupying the amino acid template are removed, and Ca^{2+} are still present. These Ca^{2+} are free from the constraints of the template and form a crystal of calcite freely and rapidly with carbonate. The crystal has a spatial conformation with a coordination number of six, the lowest energy and stable structure.

After increasing the concentration of amino acid in group C, the XRD peaks were significantly changed, aspartic showed the absorption peak of aragonite, the absorption peak of glutamic was weakened, and the absorption peak of aragonite was enhanced. The absorption peak of aragonite appeared in serine, the absorption peak of vaterite appeared in tyrosine, and the three absorption peaks of cysteine were enhanced. The comparison of group B and group C can be found that glutamate has strong ability to induce aragonite, aspartic acid has strong ability to induce aragonite, glycine and cysteine have weak induction ability, tyrosine and serine have the weakest induction ability. After the concentration of amino acids increased, the ability of six amino acids to induce aragonite and vaterite was significantly enhanced.

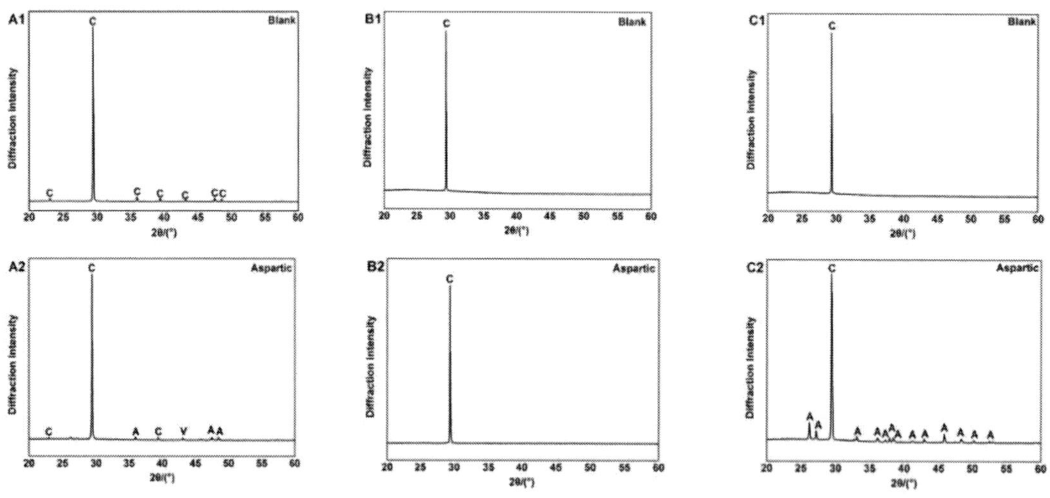

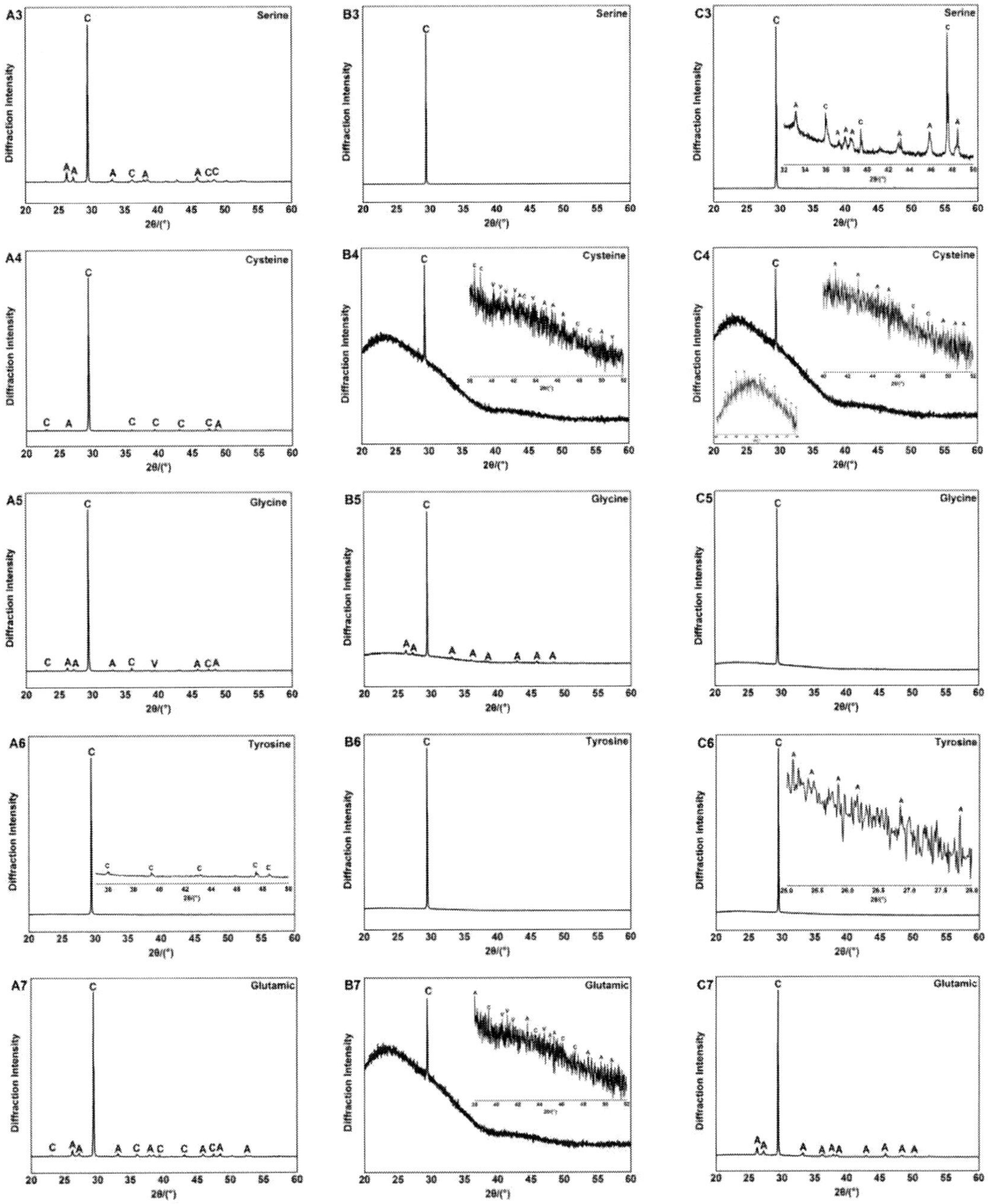

A: 10 mmol/L of amino acid solution and 20 mmol/L of $CaCl_2$ solution;
B: 10 mmol/L of amino acid solution and 10 mmol/L of $CaCl_2$ solution;
C: 20 mmol/L of amino acid solution and 10 mmol/L of $CaCl_2$ solution;

Figure 3. XRD pattern corresponding to group A, B and C

1: Blank; 2: L-aspartic; 3: L-glutamic; 4: L-glycine; 5: L-cysteine; 6: L-serine; 7: L-tyrosine

3.3. SEM

In group B, a large number of regularly structural hexagonal and quadrilateral calcites (2~3μm) were observed in the blank group, but the arrangement was disordered. After the addition of amino acids, the crystals induced by glutamate and aspartic acid were observed to be scale-like, closely arranged and in the same direction, mostly pentagonal and quadrilateral aragonite (1~2 μm), and there was vaterite in glutamate. The crystals induced by serine became smaller and disordered, and there were vaterite particles (100~200 nm) and quadrilateral aragonite (1 μm). Cysteine induces the synthesis of aligned aragonite (100 nm) and a small amount of calcite (2 μm). Glycine induced flaky calcite (2~3 μm) with large size and different shapes. There was a small amount of columnar aragonite in the calcite gap, and the overall arrangement of the calcite was orderly compared with the blank group. Tyrosine crystal induced by separate, mineralization in the same direction, there is irregular calcite needle with small particles aragonite and small vaterite.. By comparing the six amino acids with the blank group, it can be found that glutamate crystasl induced by aspartic are ordered and regular in shape, and the aspartic acid induces the best effect of aragonite, and the glutamic acid induces the the best effect of vaterite, followed by tyrosine and cysteine, and serine and glycine are the weakest.

With the increasing of calcium chloride concentration in group A, it was observed that the amount of calcite in blank group A1 increased, the crystals were more regular, and the arrangement was tight. The number of A7 calcite quantity increased, but there were still obvious spherolite and a small amount of aragonite. A3 calcite has irregular shape, increased size and more regular in arrangement than B3. A4 has a large amount of calcite arranged in a scale-like order and a small amount of vaterite. The calcite of A5 was obviously increased by twice as much as that of B5, and aragonite was found in disordered arrangement. A6 calcite is highly ordered in arrangement and its size is twice as large as that of B6. The presence of acicular aragonite indicates that the increasing concentration of calcium chloride leads to the increase in size, quantity and order of calcite, which is conducive to the mineralization and crystallization of calcite, but has little effect on aragonite and vaterite.

After increasing the concentration of amino acid in group C, the comparison with group B showed that the size of group C was significantly reduced and the structure was more compact and neat. In group C2, there were a large number of acicular clusters and particle aggregation of nano-aragonite, and the number of C7 vaterite was significantly increased. C3 crystals become smaller and arranged in order. C4 calcite becomes larger, and spherolite occurs at the same time. Vertical aragonite appeared in C5 and C6, with a small amount of spherolite. Therefore, it can be concluded that the size of the crystals induced by the increase of the concentration of amino acids is smaller and more orderly, which is conducive to the production of aragonite and vaterite.

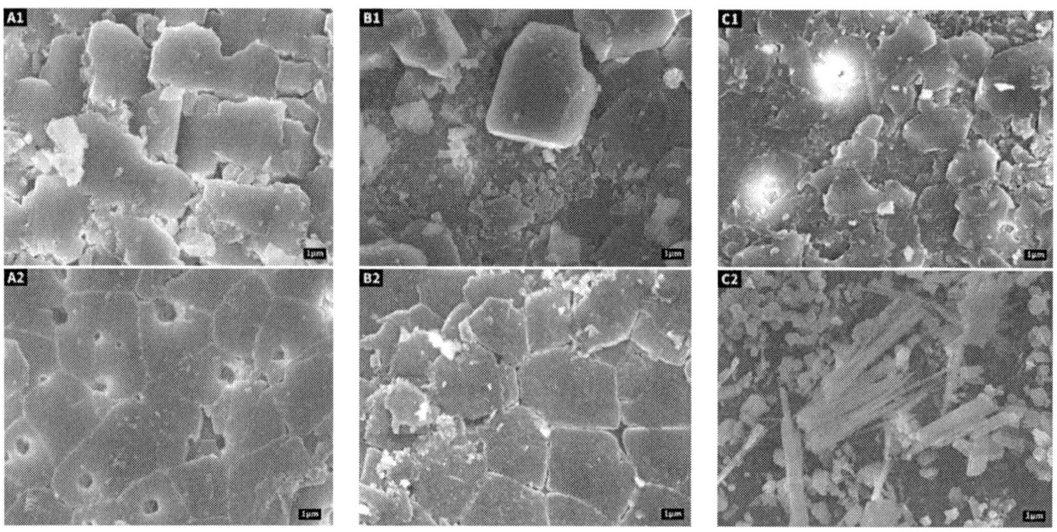

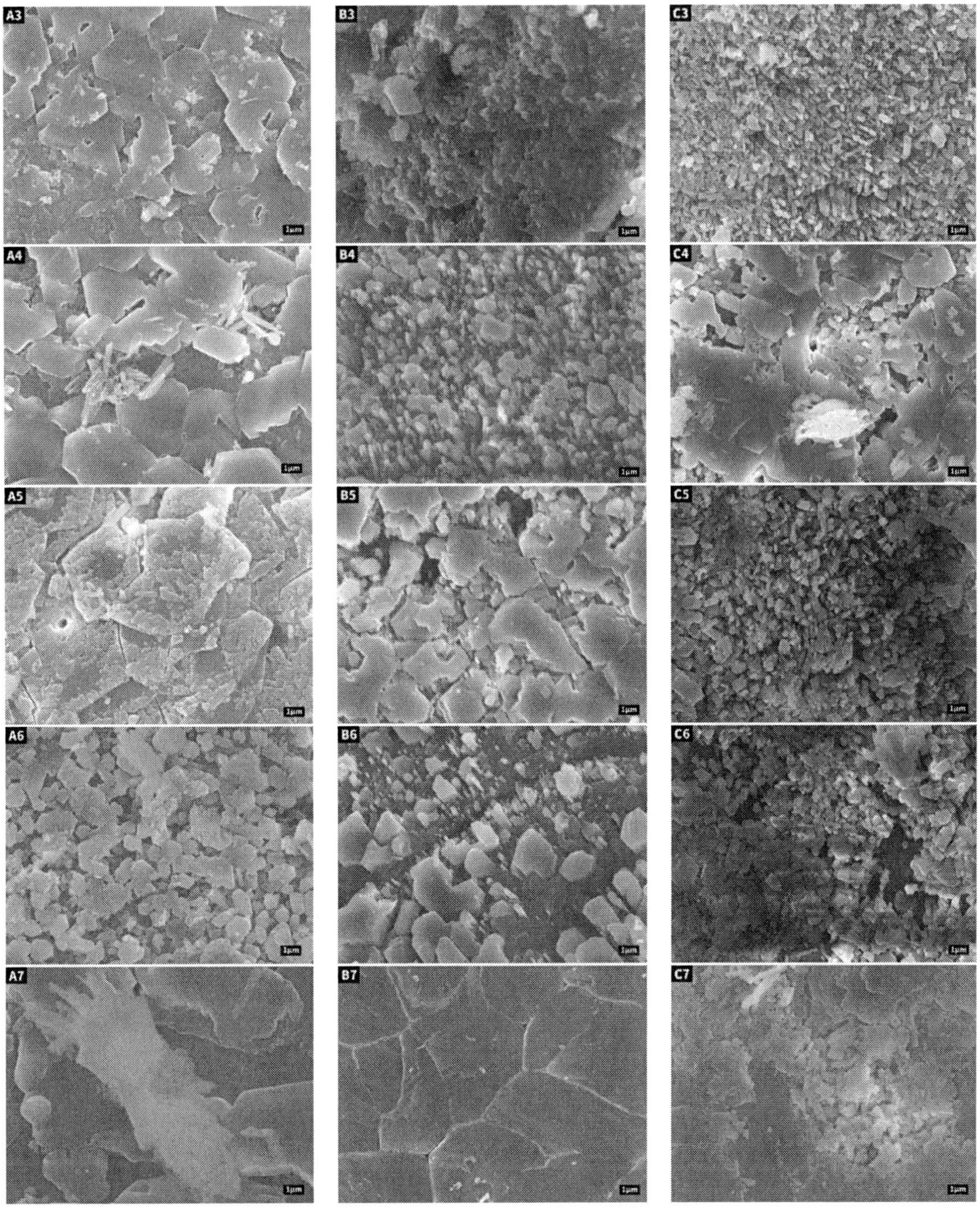

A: 10 mmol/L of amino acid solution and 20 mmol/L of $CaCl_2$ solution;
B: 10 mmol/L of amino acid solution and 10 mmol/L of $CaCl_2$ solution;
C: 20 mmol/L of amino acid solution and 10 mmol/L of $CaCl_2$ solution;

Figure 4. SEM images corresponding to group A, B and C

1: Blank; 2: L-aspartic; 3: L-glutamic; 4: L-glycine; 5: L-cysteine; 6: L-serine; 7: L-tyrosine

4. Summary

Based on the analysis of FT-IR, XRD and SEM, there are three main conclusions. Firstly, aragonite and vaterite can be induced by adding acid amino acid during the mineralization of calcium carbonate. And aspartic acid introduced is more effective for the growth of aragonite, while glutamic acid used is more effective for the growth of vaterite. Compared to those acidic amino acids, the neutral amino acids used lead to less aragonite or vaterite produced. The induced effects of four neutral amino acids from strong to weak were tyrosine, serine, cysteine and glycine. Secondly, the increased amino acid concentration leads to more aragonite and vaterite produced under the experimental conditions. Last but not least, it is found that the calcium chloride concentration investigated is highly positively correlated with calcite growth, but negatively correlated with aragonite growth. As for vaterite growth, there is no significant influence when calcium chloride concentration was changed under the experimental investigation.

Acknowledgement
This research was financially supported by the Zhanjiang Ocean and Fisheries Bureau Project (zj [2018] 008) and Guangdong Provincial Department of Science and Technology Project (No.2016A020222013).

References
[1] M. Chuanbin, L. Hengde, C. Fuzhai, Bionic synthesis of inorganic materials, Progress in Chemistry. 10(1998)246-254.
[2] Z. Jianmin, M. Kangsen, Z. Wenbing, Shell and pearl layer and its bionic application, High Technology Communication. 11(2003)94-98.
[3] Z. Xuewu, W. Jianfang, W. Wenjian, Biomineralization of shell and pearl layers and its implications for biomimetic materials, Journal of Inorganic Materials. 21(2006)257-266.
[4] M. Yufei, G. Yonghua, F. Qingling, Characterization of organic matrix extracted from fresh water pearls, Materials Science and Engineering C. 31(2011)1338-1342.
[5] M. Yufei, G. Yonghua, R. Dongni, Effect of soluble organic matter on CaCO3 crystallization in freshwater aragonite pearls, Journal of Rock Mineralogy. 28(2009) 605-610.
[6] M. Yufei, G. Yonghua, F. Qingling, Effects of pH and temperature on CaCO3 crystallization in aqueous solution with water soluble matrix of pearls, Journal of Crystal Growth. 312(2010)3165-3170.
[7] M. Yufei, Q. Li, F. Qingling, In-vitro study on calcium carbonate crystal growth mediated by organic matrix extracted from fresh water pearls, Materials Science and Engineering C. 32(2012) 1963-1970.
[8] A. Picker, M. Kellermeier, J. Seto, The multiple effects of amino acids on the early stages of calcium carbonate crystallization, Zeitschrift für Kristallographie - Crystalline Materials.227(2012) 744-757.
[9] W. Chuajiw, K. Takatori, A. Koiwaib, Interaction between Aliphatic Organic Molecules (Di-amine, Mono-amine and Amino Acid) with Calcium Carbonate, The Clay Science Society of Japan., A10: 52-53.
[10] Q. Li, F. Qingling, Alternate Deposition of Oriented Calcite and Amino Acid Layer On Calcite Substrates, The Journal of Physicial Chemistry. 112(2008)13635-13640.
[11] Q. Li, F. Qingling, L. Shanshan, In Vitro Growth of Nacre-like Tablet: Forming From Amorphous Calcium Carbonate, Nanostacks to Hexagonal Tablets,Crystal Growth and Design. 8(2007)1509-1514.
[12] Y. Shilin,G. Qian,W. Haitang, Study on the biomimetic mineralization of calcium carbonate regulated by acidic amino acids, Journal of natural science, heilongjiang university. 4(2010)495-499.
[13] S. Hui, Study on the bionic synthesis of calcium carbonate by polyaspartic acid, Heilongjiang University. 2013.
[14] H. Yaodong, Study on the preparation of calcium carbonate by polyaspartic acid preassembly biomimetic mineralization, Heilongjiang University. 2014.
[15] L. Yue, Study on the regulation of calcium carbonate precipitation by polyaspartic acid and magnesium ion, Heilongjiang University. 2016.
[16] C. Xin, R. Dongni, F. Qingling, Effects of several amino acids on the mineralization of calcium carbonate, Acta Mineralogica Sinica. 31(2011)692-697.
[17] C. Chao, L. Qiongfang, Z. Qingming, Effects of two amino acids on the mineralization of

calcium carbonate at low temperature, Geological Journal of China Universities. 23(2017)606-614
[18] G. Yuming, Wang, Feifei, Zhang, Jie, Biomimetic synthesis of calcium carbonate with different morphologies under the direction of different amino acids, Research on Chemical Intermediates. 39(2013)2407-2415.
[19] W. Xiaofeng, C. Changyun, W. Gang, Preparation of calcium carbonate crystal in amino acid mixed solvent system, Journal of Nanjing normal university (natural science edition). 33(2010)54-57.
[20] X. Yupeng, Z. Xu, Z. Lina, Acicular calcium carbonate was synthesized by bionics with L-histidine as template, Journal of chemistry of colleges and universities. 30(2009)530-533.
[21] W. Gang, Z. Shouquan, L. Jin, Study of L-lysine on crystal growth of calcium carbonate, Journal of Nanjing Xiaozhuang University. 6(2009)33-35.
[22] R. Liying, Z. Qun, Z. Wanhua, Bionic synthesis of calcium carbonate microrings, Journal of Synthetic Crystals. 44(2015)250-255.
[23] L. Xiaohong, P. Yifan, M. Yanhong, Synthesis of l-lysine grafted nanometer calcium carbonate, China Powder Science and Technology. 22(2016)86-95.
[24] X. Jing, D. Yali, B. Huili, Study on calcium carbonate deposition induced by urea-lysis microorganism, Function Materials. 4(2016)4001-4005.
[25] W. Han, Y. Lei, O. Qianqian, Study on microstructure and formation mechanism of cultured pearls in liusha bay, Journal of Chinese Electron Microscopy Society. 36(2017)153-160.
[26] W. Han, Composition and structure of mariculture pearls and in vitro mineralization based on nacre, Guangdong Ocean University. 2017.
[27] T. Hena, Y. Lei, O. Zhiyuan, A preliminary study on the mineralization mechanism of seawater pearls based on in vitro bionics, Journal of Chinese Electron Microscopy Society. (2017)589-597.
[28] O. Zhiyuan, Y. Lei, T. Hena, Preliminary study on the synthesis of nacre-like substances induced by amino acids, Journal of Chinese Electron Microscopy Society. 38(2019)34-42

Properties and Extraction of Crude Pectin Esterase from Potato (*Solanum Tuberosum*)

Jing Li, Xiaowan Li, Shanshan Li, Guanghui Shen, Qingying Luo, Hejun Wu, Anjun Chen and Zhiqing Zhang*

College of Food Science, Sichuan Agricultural University, Ya'an China

*Corresponding author

Abstract. In order to study the enzymatic properties of potato esterase, the crude enzyme of potato esterase was extracted with phosphate buffer solution, and the extraction conditions were optimized. The optimal condition is: the optimal condition for phosphate buffer solution pH 7.0, water bath time for 30 min, extraction temperature 35℃, and solid-liquid ratio 1:4 (g/mL). The property analysis showed that the potato esterase extracted from phosphate buffer solution was not a single esterase, but a kind of esterase isozyme. At 273nm and 325nm, the potato esterase crude enzyme had two characteristics uv absorption peak, its isoelectric point was near pI 4.4, and its molecular weight was roughly distributed at 16.4, 20.866, 22.73 and 40.286 KDa. In 40℃ ~ 45℃, under the condition of pH 7.0, better able to maintain its activity, and seven kinds of metal ions are inhibitory effect to it. This study provides a theoretical basis for further understanding the enzymatic characteristics of potato esterase and its application in pesticide residue detection.
Keywords: *solanum tuberosum*; esterase; extraction conditions; properties.

1. Introduction

Esterase is a general term for catalyzing the hydrolysis of ester bonds and synthetases, which is widely found in animals, plants and microorganisms. Most esterases are encoded by multiple gene families, and there are differences in esterase isoenzymes between species and subspecies, so they are important target enzymes for evolutionary relationship analysis and cultivar identification [1].

Organophosphorus insecticides can form covalent bonds with serine residues at the active site of human acetylcholinesterase (AChE)[2], inactivating acetylcholinesterase, and inhibiting normal nerve conduction. Thus causes the entire physiological and biochemical process maladjustment, produces the toxicity to the organism[3]. Based on this toxicity mechanism, a method for detecting organophosphorus pesticide residues by acetylcholinesterase inhibition method has emerged, such as test paper method[4], electrochemical method[5] and so on. This method is different from chromatography and mass spectrometry relying on large analytical instruments[6]. Using the strong affinity of pesticides and acetylcholinesterase, it can quickly and sensitively detect pesticide residues.

The commonly used AChE is extracted from animal and insect tissues or blood. The extraction process requires complex purification techniques and low temperature treatment[7], which means that the widespread use of this enzyme in rapid detection is limited. Relatively speaking, plant esterase is economically easy to obtain, extract and preserve, which has a broad application prospect. The extracted esterase activity is different, and the affinity for pesticides will be different. Therefore, finding a suitable plant esterase source is important for detecting pesticide residues by enzyme inhibition.

We tried to extract plant esterase from fresh potato tubers, and explored the relevant properties of the crude enzyme before purifying the enzyme. This research would provide a reference for further comparative analysis of enzymatic properties of purified esterases and further search for the general rule of potato esterase activity and its foundation in the application of pesticide residue detection.

2. Materials and Methods

2.1. Materials and Instruments

Potatoes were acquired from Dingxi city (Gansu, China), and potato cultivar is Qingshu9. α-naphthyl acetate, α-naphthol and Fast blue B salt were purchased from Shanghai Yuanye Biotechnology Co., Ltd. (Shanghai, China). Disodium hydrogen phosphate, sodium dihydrogen phosphate, $CaCl_2$, $BaCl_2$, $MgCl_2$, $AlCl_3$, $CuSO_4$, $FeSO_4$, $MnSO_4$ and citric acid were purchased from Chengdu Kelong Chemical Reagent Factory (Chengdu, China). All of these reagents were analytical grade and used as received.
UV-vis spectra were recorded using a UV-3100PC type spectrophotometer, which purchased from Shanghai Mepuda Instrument Co., Ltd. (Shanghai, China). The gel imaging using a Gel Doc2000 gel imaging system (Bio-Rad, USA).

2.2. Analytical Methods

2.2.1. Method for Extracting Potato Esterase Crude Enzyme. Wash the potatoes and peel them then broken with a broken machine. Weigh 5g and add 25 mL of phosphate buffer solution (PBS, 0.2mol / L, pH 6.5), and let it stand for 30 minutes in a water bath at 30℃, then filtered with gauze. The filtrate was centrifuged at 4000 r/min and 4℃ for 15 min. The supernatant was taken as a crude enzyme solution and stored at 4 °C for use.

2.2.2. Determination of Potato Esterase Activity. Refer to the method of Ye et al. [8] with slightly modify. Prepare 0.4, 0.5, 0.6, 0.7, 0.8, 0.9, 1 mmol/L α-naphthol standard solution to obtain a linear regression equation between α-naphthol concentration (Y) and absorbance (X) as Y=2.1144X - 0.0388 (R2 = 0.999). The linear relationship of the regression equation is good which can be used to test the determination of esterase activity. The unit enzyme activity is calculated by the follow formula:

$$EAs=\frac{k\times OD+b}{5\times V}\times N$$

Where *EAs*--unit enzyme activity of the enzyme solution, U/mL (1 U is defined as the amount of enzyme required to catalyze 1 μmol of α-naphthol per minute under certain conditions); *N*--dilution of the enzyme solution; *k*--the slope of the alpha-naphthol standard curve; *b*--the intercept of the alpha-naphthol standard curve; *OD*--absorbance of the reaction solution; *V*--the volume of the reaction enzyme solution, mL; 5--the reaction time was determined to be 5 min.

2.2.3. Optimization of Extraction Conditions for Potato Esterase. (1) Single factor experimental design.
When the pH of the PBS is 6.5, the temperature is 30℃, and the time is 20 min, the ratio of material to liquid (g/mL) is set to 1:3, 1:4, 1:5, 1: 6,1:7. When the pH is 6.5, the ratio of material to liquid is 1:5, and the temperature is 30 °C, the time is set at 10, 20, 30, 40, 50 min. When the temperature is 30 °C, the time is 40 min, and the ratio of material to liquid is 1:5, the pH is set to 5.5, 6.0, 6.5, 7.0, 7.5. When the pH is 6.5, the time is 20 min, and the ratio of material to liquid is 1:5, the temperature is set at 25, 30, 35, 40, 45 °C. Taking the unit enzyme activity as the evaluation index of the extraction effect. Analyzing the effect of the various factors on the extraction of potato esterase.
(2) Orthogonal design
According to the results of single factor test, the pH (A), extraction time (B), extraction temperature (C) and ratio of material to liquid (D) were selected as the investigation factors, and the unit enzyme activity of potato esterase was taken as the investigation index. L_9 (3^4) orthogonal experiments were performed, each factor was paralleled 3 times, and the results were averaged.

2.2.4. Study on Properties of Potato Esterase. (1) Determination of ultraviolet absorption spectrum
Analyzed the peak shape and peak position of the ultraviolet absorption curve at 200~800 nm.
(2) Determination of isoelectric point of potato esterase
Adjust the crude enzyme solution pH to 2.0, 3.0, 4.0, 5.0, 6.0, 7.0, 8.0, 9.0, 10.0, 11.0, and let it stand at 4℃ for 30 min. Then centrifuge at 4000 r/min for 15 min at 4℃. Take the supernatant, determine the protein content[9]. Then select 11 points (take a point every 0.4) near the lowest point of the protein content, repeat the above steps to determine the isoelectric point of potato esterase.
(3) SDS-PAGE gel electrophoresis analysis
The method of SDS-PAGE gel electrophoresis were reference to Laemmli et al. [10]. The electrophoresis conditions were: the concentration of upper gel is 4%, voltage of 90 V, 30 min; separation gel concentration of 12%, voltage of 120 V, 120 min.
(4) Determination of optimum temperature of potato esterase
Take a certain amount of potato esterase crude enzyme solution in a 25, 30, 35, 40, 45, 50, 60 °C water bath, respectively, after 15 min incubation, determine the unit enzyme activity of the crude enzyme solution. The enzyme activity of the unheated crude enzyme solution is 100%. Calculate the relative enzyme activity, and the optimum temperature of potato esterase was investigated.
(5) Determination of optimum pH of potato esterase
Take a certain amount of potato esterase crude enzyme solution, add 0.2 mol/L buffer solution with pH values of 3.0, 4.0, 5.0, 6.0, 7.0, 8.0, 9.0 respectively, and incubate in a 30°C water bath for 15 min. Then measure the crude enzyme solution. The enzyme activity per unit enzyme was 100% of the enzyme activity of the unadjusted pH crude enzyme solution, and the relative enzyme activity was calculated to determine the optimum pH of the potato esterase.
(6) Effect of metal ions on potato esterase
Take appropriate amount of potato esterase crude enzyme solution, add 1 mL of 2, 6, 10 mmol/L Ca^{2+}, Ba^{2+}, Mg^{2+}, Al^{3+}, Cu^{2+}, Fe^{2+}, Mn^{2+} solution, respectively, and measure the unit enzyme activity. The enzyme activity of the crude enzyme solution without adding metal ion solution was 100%, and the relative enzyme activity was calculated.

2.3. Statistical Analysis
The experimental data was analyzed by IBM SPSS Statistics data editor and plotted with Origin Pro 9.0 software. Electropherogram acquisition and molecular weight calculation were performed using Quantity one software.

3. Results and Analysis

3.1. Effects of Extraction Conditions on Crude Enzyme Activity of Potato Esterase
In Fig. 1a, the activity per unit of enzyme is continuously increased, reaching a maximum at 1:5 (g/mL). And then gradually reduced. When the optimum ratio of material to liquid is reached, the esterase dissolution in the solution tends to be balanced, and too much solvent has a dilution effect on the system, resulting in a decrease in unit enzyme activity. In Fig. 1b, the enzyme activity reached the maximum at 20min, and the unit enzyme activity decreased gradually after 20min. When the extraction time is short, the esterase is not easy to separate from other components [11], and the enzyme activity is low. As the extraction time is extended, the esterase is fully dissolved and reaches equilibrium at 20 min. Due to the conformational instability of the esterase in the solution state [12], the extraction time is too long and the enzyme activity is likely to decrease. In Fig. 1c, the optimum pH of the phosphate buffer solution is 6.5. At the same time, as shown in Fig. 1c, the effect of pH on the enzyme is significant, and the unsuitable pH has a great influence on its stability. In Fig.1d, the enzyme activity reached the maximum at 40 °C. The higher of temperature, the higher activity of enzyme, which may be caused by the high temperature to increase the solubility of the enzyme protein, but too high temperature will cause the hydrophobic group to be exposed, and the solubility will decrease [13], resulting in lower esterase activity. At lower temperature, the enzyme is difficult to release completely, resulting in lower enzyme activity.

IOP Conf. Series: Materials Science and Engineering **611** (2019) 012060 doi:10.1088/1757-899X/611/1/012060

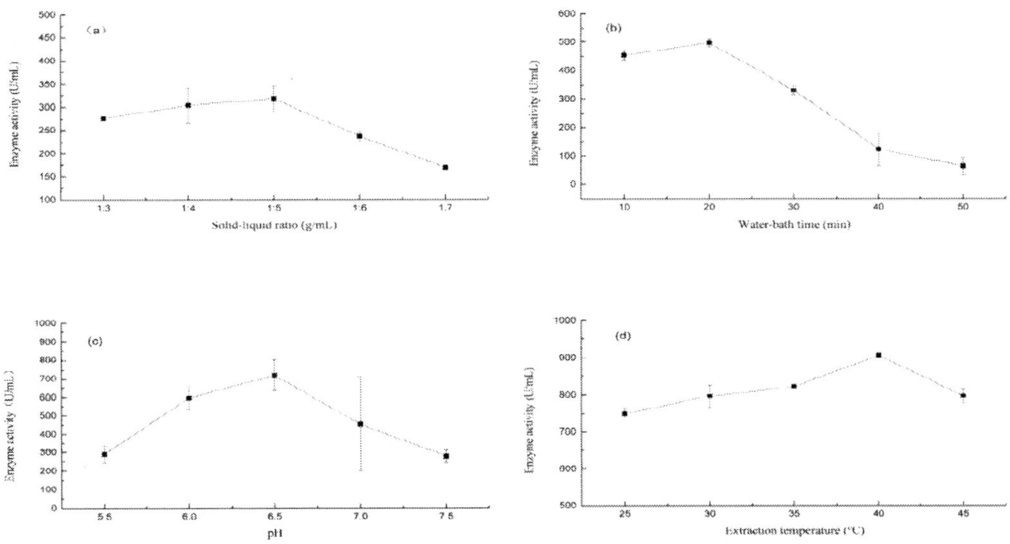

Figure 1. Effect of extract conditions on enzyme activity: (a) material - liquid ratio, (b) water-bath time, (c) pH, (d) extraction temperature

3.2. Orthogonal Optimization Experiment of Crude Enzyme Extraction Conditions of Potato Esterase

The results of orthogonal enzyme experiment to optimize the extraction of potato esterase crude enzyme are shown in Table 1.

Table 1. Results of orthogonal experiment

No.	Factors				Enzyme activity(U/mL)
	pH A	Time B(min)	Temperature C(°C)	Solid-liquid D	
1	6.0	10	35	4	709.846
2	6.0	20	40	5	646.696
3	6.0	30	45	6	533.928
4	6.5	10	40	6	741.985
5	6.5	20	45	4	806.262
6	6.5	30	35	5	860.391
7	7.0	10	45	5	881.817
8	7.0	20	35	6	860.391
9	7.0	30	40	4	973.723
K1	1890.469	2333.647	2430.628	2489.831	
K2	2408.638	2313.349	2362.403	2388.904	7015.038
K3	2715.931	2368.042	2222.007	2136.303	
k1	630.156	777.882	810.209	829.944	
k2	802.879	771.116	787.468	796.301	
k3	905.310	789.347	740.669	712.101	
Range	275.154	18.231	69.540	117.843	
Priority			A>D>C>B		
The optimal level	A_3	B_3	C_1	D_1	
Optimal combination			$A_3B_3C_1D_1$		

It can be seen from Table 1. that the range of pH is the largest, and the second is the ratio of material to liquid, temperature and water bath time. The optimal combination of extraction conditions is: the pH was 7.0, the time was 30 min, the temperature was 35 °C, and the ratio of material to liquid was 1:4. The verification experiments were carried out according to the optimal extraction conditions ($A_3B_3C_1D_1$). The unit enzyme activities of the measured potato esterase were 1053.224 U/mL, 961.882 U/mL, and 958.442 U/mL, respectively. The results showed that the optimized extraction conditions were stable and could be used as the best extraction conditions for extracting potato esterase.

3.3. Properties of Potato Esterase

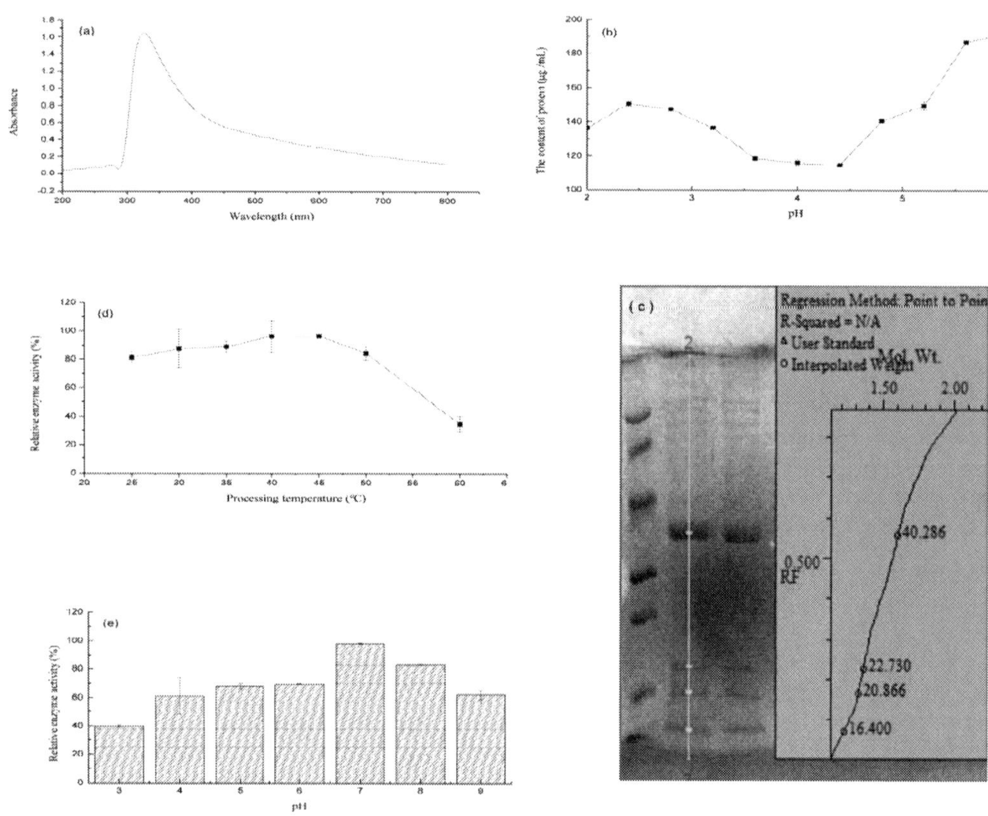

Figure 2. Properties of crude pectin esterase from *Solanum tuberosum* (a UV- spectrometry, b pI analysis value, c SDS-PAGE, d temperature, e pH)

As shown in Fig. 2a, the crude enzyme solution of potato esterase has a major UV characteristic absorption peak at 325 nm and an insignificant absorption peak at 273 nm. In the near ultraviolet region (200~400 nm), aromatic amino acids have the ability to absorb light, and the maximum absorption peaks of tyrosine, phenylalanine and tryptophan are between 260 and 280 nm [14]. It is indicated that potato esterase is within the characteristic range of protein ultraviolet absorption spectrum.

When the isoelectric point of the protein is the same as the pH of the solution, the positive and negative charges generated by the dissociation of the protein molecules are equal, and at this time, the protein molecules will agglomerate and precipitate. Therefore, the protein has the least solubility at the isoelectric point and is most likely to form precipitates [15]. Fig. 2b shows that the protein content in the supernatant is the lowest when the pH is 4.4 (115.119 g/mL). It is concluded that the pI of potato esterase is about 4.4. At pH 2, probably due to the peracid environment leading to denaturation of the enzyme protein, thereby changing the conformation, affecting and resulting in lower content of

measured protein [16].
SDS-PAGE analysis showed (Fig. 2c) that there were 4 clear bands in the enzyme solution, indicating that the potato esterase extracted by phosphate buffer solution is not an enzyme, and its active component is esterase isoenzyme. Calculated the molecular weight distribution of potato esterase isoenzyme, and the results indicated that the molecular weight of potato esterase isoenzyme were 16.4, 20.866, 22.73, and 40.286 KDa, which is consistent with the previous report of most plant esterase molecular weights from 20kDa to 60kDa [17~18].
The relative enzyme activity of potato esterase tends to be stable when kept at 25~50 °C. The relative enzyme activities at 40°C and 45°C were 96.13% and 96.24%, respectively. It's indicating the optimal temperature of potato esterase between 40~45 ℃. When the temperature exceeds 50 ℃, the enzyme activity begins to decrease sharply. It's indicating that the excessive temperature has a destructive effect on the specific structure of the enzyme active center, which denatures the enzyme protein [19] and the activity is reduced. The relative enzyme activity of potato esterase was stable between pH 4~6. The relative enzyme activity at pH 7 was 98.11%, which was higher than other groups. The active center of the esterase is located in a canyon area, and the bottleneck of the canyon area is a constricted area. The environment of strong acid and alkali may cause the amino acid conformation of this area to change, thereby inhibiting the activity of the enzyme [20].
Some metal ions can act as activators or inhibitors of the enzyme. By changing the configuration of the enzyme protein, the active site of the enzyme is more suitable for binding to the substrate [21]. In Table 2, Ca^{2+} and Ba^{2+} have inhibitory effects on potato esterase activity. Low concentration (2 mmol/L) of Mg^{2+} , Al^{3+} and Fe^{2+} on potato esterase activity was greater than that of the three metal ions (6 mmol/L, 10 mmol/L); while Cu^{2+} and Mn^{2+} inhibited potato esterase with increasing concentration.

Table 2. Relative enzyme activity of potato esterase with different metal ions

Metal ions	**Metal ion concentration (mmol/L)**			
	0	**2**	**6**	**10**
Ca^{2+}	100±5.66a	64.47±2.83b	66.85±0.18b	71.73±1.42b
Ba^{2+}	100±5.66a	52.96±8.85c	66.35±1.59bc	69.35±0.53b
Mg^{2+}	100±5.66a	31.19±5.31c	64.97±1.06b	73.35±1.24b
Al^{3+}	100±5.66a	47.58±0.18c	77.86±1.95b	78.98±0.35b
Cu^{2+}	100±5.66a	44.20±2.12d	58.71±2.83c	73.60±1.59b
Fe^{2+}	100±5.66a	39.20±4.60c	59.96±7.08b	86.49±2.48a
Mn^{2+}	100±5.66a	47.45±0.71c	53.71±1.42c	89.24±4.60b

Control: Relative activity of untreated potato esterase. Different letters following date at same line mean significant difference among treatments (P<0.05).

4. Conclusion

In this study, the optimal extraction conditions of potato esterase were obtained, and the properties of crude enzymes of fresh potato tuber esterase were preliminarily determined. It was found that its active component consisted of potato esterase isoenzyme, rather than a single esterase. The optimal temperature of potato esterase was in the range of 40~45 ℃. The optimum pH is 7, and several metal ions selected in the experiment have inhibitory effects on potato esterase. The enzyme solution used in this study is a crude enzyme that has not been purified, and its related properties will be different in the enzymatic properties after purification. The results of this study will guide the targeted purification of the ester. The enzyme provides more reference, which can provide a theoretical basis for further in-depth research and application.

References

[1] Kim M S , Seo J S, Ahn S J , et al. Duplication of phospholipase C-δ gene family in fish genomes[J]. Genomics, 2008, 92(5):0-371.
[2] Aurilia V, Parracino A, D Auria S. Microbial carbohydrate esterases in cold adapted

environments[J]. Gene, 2008, 410(2):0-240.
[3] Yuan-ping Pang. Species marker for developing novel and safe pesticides[J]. Bioorganic & Medicinal Chemistry Letters, 2007, 17(1):197-199.
[4] Yan Wu, Yong-fang Sun, Fu-bing Xiao, et al. Sensitive inkjet printing paper-based colormetric strips for acetylcholinesterase inhibitors with indoxyl acetate substrate[J]. Talanta, 2017, 162(Complete):174-179.
[5] Hui-ying Wan, Mingyue Liu, Xinxi Hu, et al. Electrochemical Determination of Glycoalkaloids Using a Carbon Nanotubes-Phenylboronic Acid Modified Glassy Carbon Electrode[J]. Sensors, 2013, 13(12):16234-16244.
[6] Liang Dong, Yan Ren, Jun-jie Li, et al. Detection of Carbendazim Residues in Aqueous Samples by Fluorescent Quenching of Plant Esterase[J]. Journal of Applied Spectroscopy, 2018, 85(3):535-542.
[7] Dyk J S V, Pletschke B. Review on the use of enzymes for the detection of organochlorine, organophosphate and carbamate pesticides in the environment[J]. Chemosphere, 2011, 82(3):291-307.
[8] Lin Ye, Xue Yang, Zhi-qing Zhang, et al. Optimization of Extraction Process for Wheat Bran Esterase by Response Surface Methodology[J]. Journal of Triticeae Crops, 2017, 37(3):420-427.
[9] Lin Gou, Zhi Shan. Biochemical experiment[M]. Chengdu: Southwest Jiaotong University Press, 2009: 72-73.
[10] U. K. Laemmli. Cleavage of structural proteins during the assembly of head of bacteriophage T4. Nature 227: 680–685(15 August 1970).
[11] Jun-Peng K , Chang-Qing R , Li-Dong W , et al. Optimization of Extraction Conditions for Protein from Barley Malt Roots[J]. Food Science, 2012, 33(22):95-98.
[12] Rahman M A, Culsum U, Kumar A, et al. Immobilization of a novel cold active esterase onto Fe_3O_4~cellulose nano-composite enhances catalytic properties[J]. International Journal of Biological Macromolecules, 2016, 87:488-497.
[13] Paraman I, Hettiarachchy N S, Schaefer C, et al. Hydrophobicity, Solubility, and Emulsifying Properties of Enzyme-Modified Rice Endosperm Protein[J]. Cereal Chemistry, 2007, 84(4):343-349.
[14] Ji Zheng, Junhui Chen. General Biochemistry (Fourth Edition) [M]. Beijing: Higher education press, 2007: 76.
[15] Qing-song Wang, An-hong Xiao, Li-juan Han, et al. Extraction and properties of protein from radish seed meal[J]. China Oils and Fats, 2018, 43(03): 26-29+34.
[16] JIANG Bin, FENG Zhi-biao, LIU Chun-hong, et al. Extraction and purification of wheat-esterase using aqueous two-phase systems of ionic liquid and salt[J]. Journal of Food Science and Technology, 2014, 52(5), 2878-2885.
[17] Veerabhadrappa P S, Montgomery M W. Polyacrylamide-gel electrophoresis of pea and green bean carboxylesterases[J]. Phytochemistry, 1971, 10(6):1171-1174.
2.2.1. [18] Bartley I M, Stevens W H. Carboxylic Ester Hydrolases of Apple[J]. Journal of Experimental Botany, 1981, 32(129):741-751.
[19] Jia-jia Zhang. Study on the extraction of Acetylcholinesterase from chicken brain organ and its characteristics[D]. Guangzhou: South China University of Technology, 2016.
[20] Koellner G, Kryger G, Millard C B, et al. Active-site gorge and buried water molecules in crystal structures of acetylcholinesterase from Torpedo californica[J]. Journal of Molecular Biology, 2000, 296(2):0-735.
[21] Kai-Ping K, Sheng-Gui Z, Hog L I, et al. Detection of lipase activity in malt and its enzymatic properties research[J]. Science & Technology of Food Industry, 2013, 34(6):190-193.

IOP Conf. Series: Materials Science and Engineering **611** (2019) 012061 doi:10.1088/1757-899X/611/1/012061

Browning and Flavor Changing of Cloudy Apple Juice during Accelerated Storage

Danshi Zhu, Yusi Shen, Lingxia Xu, Xuehui Cao, Changxin Lv and Jianrong Li*

College of Food Science and Technology, Bohai University; Food Safety Key Lab of Liaoning Province; National & Local Joint Engineering Research Center of Storage, Processing and Safety Control Technology for Fresh Agricultural and Aquatic Products; Jinzhou, Liaoning, 121013, China

*Corresponding author

Abstract. In this paper, the browning and flavor changing of cloudy apple juice during accelerated storage were investigated with storage temperature at 37 °C. These indicators, such as, soluble solids content, titratable acid content, Vc, color difference, and turbidity of cloudy apple juice were measured every 1 or 2 hours during storage. Meanwhile, the volatile flavor of juice were also analyzed using electronic nose. The results showed that L^* and ΔE^* values decreased during storage, while a^* and b^* values increased. The electronic nose could distinguish the apple juice stored for 0 hour from other storage time, and the sensor with the highest contribution rate was W1S, followed by W1C and W2S, which were sensitive to hydrocarbons, aromatics, and alcohols, respectively. The values of Vc content, sugar-acid ratio, and turbidity fluctuated slightly during storage. In general, the content of Vc decreased significantly, and the ratio of sugar to acid also decreased, while, turbidity increased during storage.

1. Introduction

Apple is one of the four most widely grown fruits and one of the most commonly consumed fruits [1]. Apple juice is the second most consumed juice among U.S. consumers due to its valuable nutrional properties and highly appreciated flavor[2]. However, long-term storage of apple juice at room temperature conditions lead to the accelerated development of brown color and off-flavors, which results in decreases of the commercial value of products[3]. Browning can be divided into enzymatic browning and non-enzymatic browning. The enzymatic browning reaction in fruits are mainly catalyzed by polyphenol oxidase (PPO) in the presence of oxygen and reducing substances[4]. However, the browning of juice after pasteurization is mainly due to non-enzymatic browning, which is the result of the Maillard reaction, including condensation between reducing sugars and amino acids, caramelization, ascorbic acid degradation and pigment damage[5].

This work was to explore browning and flavor changes during accelerated storage of cloudy apple juice. This could provide some theoretical references for understanding the relationship between browning and flavor changing in cloudy apple juice, and then better for preservation of the flavor components in the future.

2. Methods

2.1. Sample Preparation

Rall's ripe apple fruit (*Malus pumila*) were purchased from the fruit market in Jinzhou, Liaoning

Province, China. All apples had reached commercial maturity. Cored apples were pulped using DS-1 tissue blender (Wanhua experimental instrument factory, Jintan, China), with ascorbic acid (1g/kg) was added as antioxidant. Then, juice was filtered through a 300-mesh filter cloth to remove impurities. After that, the cloudy apple juice was placed in an open glass bottle and stored in a constant temperature incubator at 37 °C for 9 hours, and every 1 or 2 hours for sampling analysis.

2.2. Color Measurement

The color of cloudy apple juice was measured using a CR-400 chroma meter (Konica Minolta Sensing, Ink., Tokyo, Japan). The investigated parameters were as follows[6]: L^* (brightness), a^* (red-green component), b^* (yellow-blue component), ΔE^* (the total color difference) (Eq. (1)), H^0(hue angle) (Eq. (2)), and C^* (chroma, quantitative attribute for colorfulness) (Eq. (3)). The values were measured by 3 repetitions.

$$\Delta E^* = \sqrt{(\Delta L^*)^2 + (\Delta a^*)^2 + (\Delta b^*)^2} \tag{1}$$

$$H^0 = \arctan(b^* / a^*). \tag{2}$$

$$C^* = \left[(a^*)^2 + (b^*)^2\right]^{\frac{1}{2}}. \tag{3}$$

2.3. Electronic Nose

The smell characteristics of apple juice were analyzed using a portable PEN3 E-nose (Insent Company, Japan). The equipment of E-nose mainly consists of a pattern-recognition software, a sampling apparatus and a sensor array unit. The sensor array were composed often metal oxide semiconductor (MOS) sensors[7].

2.4. Sugars and Acidity

The soluble solid content (SSC) of apple juice was determined by PAL-3 hand-held sugar meter (ATAGO, Japan). The content of titratable acid was titrated with NaOH, and phenolphthalein solution was used as indicator[8].

2.5. Turbidity and Vitamin C

Turbidity was directly measured in a transparent glass tube of 30 mL using the WGZ-2000 nephelometer (INESA physico Optiacal Instrument Co, Ltd, Shanghai, China) at room temperature. Measurements were repeated for 3 times, and the results were expressed as nephelometric turbidity units (NTU). Vitamin C content was measured by the method of 2, 6-dichlorophenol indophenol titration.

2.6. Statistical Analysis

The values were expressed as mean ± standard deviation. Data were analyzed by one-way analysis of variance test using SPSS Statistics 20.0 followed by drawing with Origin 9.0. Signifcance was determined at 95% confdence level.

3. Results

3.1. Color Difference

Table 1 showed the results of the change in color difference of cloudy apple juice. L^* and ΔE^* values decreased while a^* and b^* values increased during storage. This indicated the loss of the typical apple juice's color (yellowness) and formation of browning (enhanced dark and redness) during storage at 37 °C.

Table 1. Changes in color difference of apple juice during storage at 37 °C.

Time/h	L^*	a^*	b^*	ΔE^*
0	38.69 ± 0.08^a	-7.49 ± 0.04^f	14.25 ± 0.13^b	36.21 ± 0.10^a
1	38.33 ± 0.07^a	-7.10 ± 0.06^e	14.87 ± 0.40^b	35.98 ± 0.15^a
2	38.36 ± 0.14^a	-6.93 ± 0.04^e	15.23 ± 0.45^a	36.09 ± 0.20^a
3	38.28 ± 0.13^a	-6.52 ± 0.04^d	15.49 ± 0.16^a	36.03 ± 0.17^a
4	38.26 ± 0.39^a	-6.36 ± 0.05^d	15.42 ± 0.68^a	35.98 ± 0.18^a
5	38.59 ± 0.13^a	-6.07 ± 0.09^c	15.21 ± 0.24^a	36.19 ± 0.11^a
7	36.67 ± 0.41^b	-4.38 ± 0.18^b	15.75 ± 0.68^a	34.37 ± 0.23^b
9	34.87 ± 0.17^c	-2.73 ± 0.37^a	15.28 ± 0.59^a	32.42 ± 0.33^c

Values in the same column with different lower-case letters (a- f) are significantly different at $p < 0.05$.

The results of hue angle (H^o) and chroma (C^*) of cloudy apple juice were shown in Figure 1. H^o and C^* are vital parameters for appraisal of fruit juice color. From Figure 1 (A), the value of hue angle dropped significantly ($p < 0.05$). It showed that cloudy apple juice continuous browning visually during 9 hours accelerated storage. This indicated that, hue angle could be used to describe the changes of fruit juice browning. However, the chroma increased slowly and then decreased (Figure 1 (B)) during storage, and this might not be a good indicator for browning.

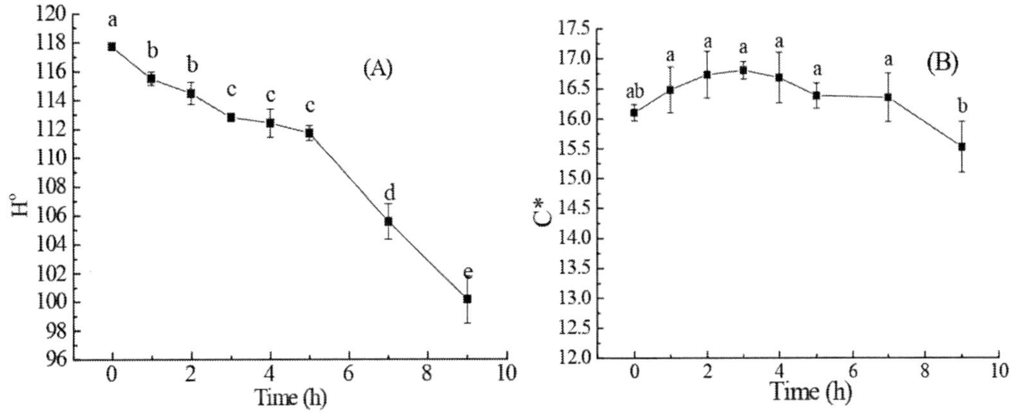

Figure 1. Changes of hue angle (A) and chroma (B) of cloudy apple juice during accelerated storage. Data were the means of three replicates ± SD. Different letters (a-e) indicate significant differences ($p < 0.05$).

3.2. Volatile Flavor Change

Electronic nose is a common method for detecting juice flavor. The detection limit of the electronic nose can reach 10^{-9}. It has strong sensitivity to odor, and has the advantages of short time and fast detection speed[9]. Figure 2 showed the linear discriminant analysis (LDA) and sensor load analysis (LA) plots of cloudy apple juice during storage. From Figure 2(A), the first linear discriminant (LD1) and the second linear discriminant (LD2) explained 90.79% of the total variance (LD1 for 57% and LD2 for 33.79%). From the LD1, samples that stored 0 hour and other storage times could be well distinguished. And from the LD2, the flavor substances at different storage time were well differentiated except 5h and 7h. From Figure 2(B), the first sensor load accounted for 98.52%, and the second sensor load accounts for 0.72%. From the first main axis, the highest contribution sensor was W1S, followed by W1C and W2S. The sensors W1S, W1C, and W2S were sensitive to hydrocarbons, aromatics, and alcohols, respectively.

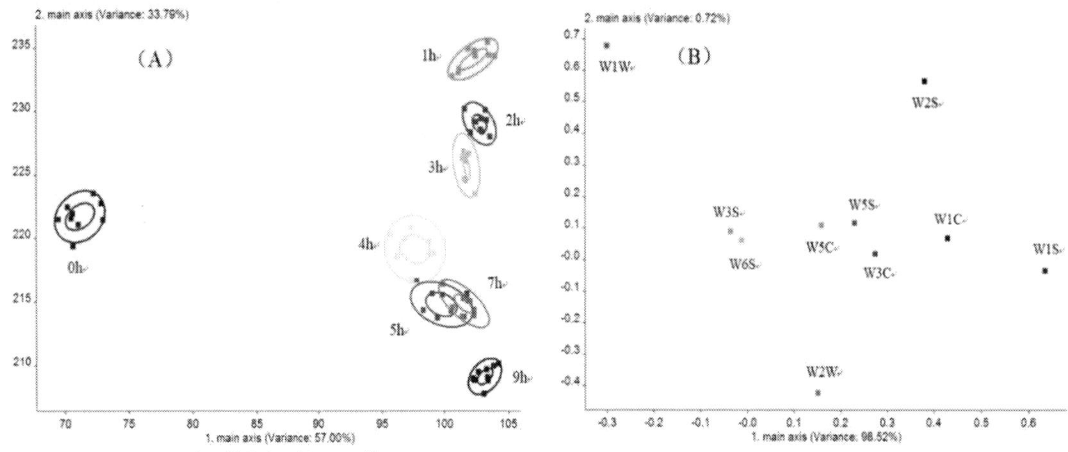

Figure 2. Linear discriminant analysis, LDA (A) and sensor load analysis, LA (B) of cloudy apple juice analyzed using electronic nose.

3.3. Taste Change

The changes of soluble solids content, titratable acid content, and sugar to acid ratio in cloudy apple juice during accelerated storage were shown in Figure 3. From 0 to 9 hours of storage time, the SSC scarcely changed except for a slight decrease in storage for 4 hours. In addition, the titratable acid appeared rising trend, and the ratio of sugar to acid showed a downward trend. Usually, the taste dropped during the browning process of juice, and generally this taste change will not be favored by consumers [2,10].

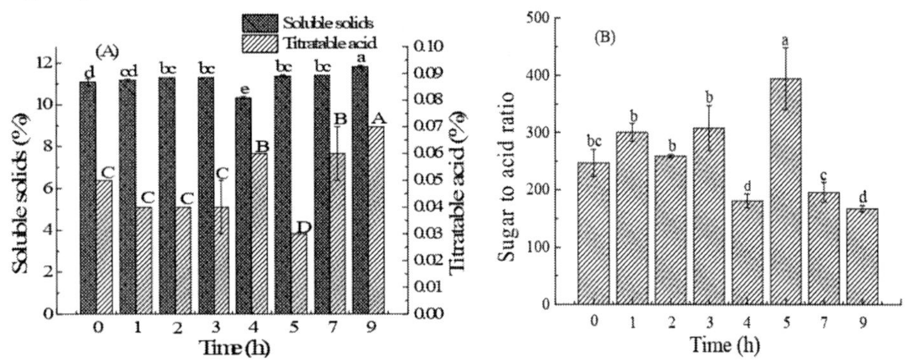

Figure 3. Changes of soluble solids content, titratable acid content (A) and sugar-acid ratio (B) in apple juice during storage. Different letters (A-D) and (a-d) meant significant difference ($p < 0.05$).

3.4. The Change of Vc Content and Turbidity

Turbidity is an index to evaluate the degree of light scattered by particles (suspended solids) in water or other liquids. To some extent, it can indicate the degree of browning under certain circumstances. The changes of Vc content and turbidity in cloudy apple juice during storage were shown in Figure 4. The Vc content decreased firstly, and then increased gradually after storage for 2 hours, and dropped again after storage for 4 hours. The change of turbidity was exactly contrary to the change of Vc. There are many ascorbic acid and dehydroascorbate reductase in apple fruit. The partially dehydroascorbic acid could be reduced to ascorbic acid, and it led to the transitory increase of Vc content. However, the oxidation of Vc was the main reaction during juice storage, and the total content of Vc decreased finally. For turbidity, particles in cloudy juice aggregation and degradation caused by various oxidation-reduction reaction, which resulted in the decline and increase of turbidity[11].

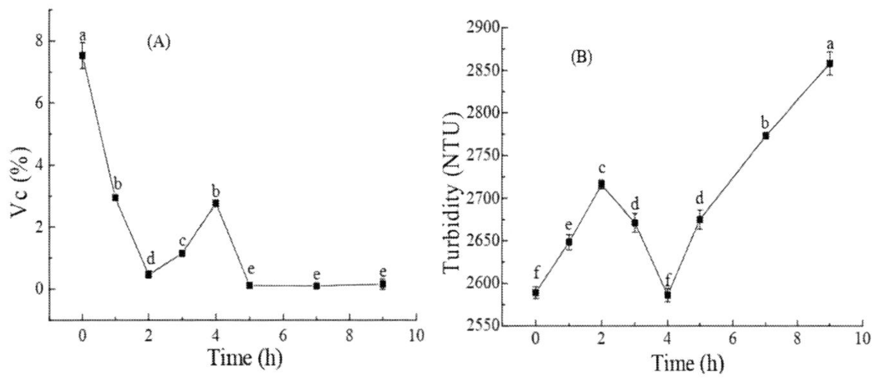

Figure 4. Changes of Vc (A) and turbidity (B) content in cloudy apple juice during accelerated storage. Values in the pictures with different lower-case letters (a- f) are significantly different at $p < 0.05$.

4. Conclusions

Cloudy apple juice stored at accelerated conditions resulted in the color changes from the yellowness to red-brown, which apparently browning. The electronic nose could distinguish the apple juice stored at 0 hour from other storage time. And the flavor substances at different storage time were well differentiated except 5h and 7h. The highest contribution sensor was W1S, which was sensitive to hydrocarbons. During the browning process, the taste of apple juice became sour and the turbidity was increased, and also, the Vc loss was severe. The quality of cloudy apple juice deteriorated significantly during accelerated storage. The browning lead to losses of desirable flavour and nutritional value of juice.

Acknowledgements

This work was financed by National Key Research and Development Program (2017YFD0400704) and National Natural Science Foundation of China (31701618) and Natural Science Foundation of Liaoning province, china (20180551103). We thank Dr. Li Jianrong (Bohai University, CN) for his critical review of the manuscript.

References

[1] Huang Z, Hu H, Shen F, et al 2018 Relatively high acidity is an important breeding objective for fresh juice-specific apple cultivars[J] *Scientia Horticulturae* vol 233, pp 29-37
[2] Paravisini L, Peterson DG 2018 Role of Reactive Carbonyl Species in non-enzymatic browning of apple juice during storage[J] *Food Chemistry* vol 245, pp 1010-7
[3] Lu S, Luo Y, Turner E, Feng H 2007 Efficacy of sodium chlorite as an inhibitor of enzymatic browning in apple slices[J] *Food Chemistry* vol 104, pp 824-9
[4] López-Nicolas JM, Pérez-López AJ, Carbonell-Barrachina A, García-Carmona F 2007 Use of natural and modified cyclodextrins as inhibiting agents of peach juice enzymatic browning[J] *Journal of Agricultural & Food Chemistry* vol 55, pp 5312
[5] Burdurlu HS, Karadeniz F 2003 Effect of storage on nonenzymatic browning of apple juice concentrates[J] *Food Chemistry* vol 80, pp 91-7
[6] Yi J, Kebede BT, Grauwet T, et al 2016 A multivariate approach into physicochemical, biochemical and aromatic quality changes of purée based on Hayward kiwifruit during the final phase of ripening[J] *Postharvest Biology and Technology* vol 117, pp 206-16
[7] Ortega JB, Garcia JG, Ortega JG, editors 2015 Precision of volatile compound analysis in extra virgin olive oil: The influence of MOS electronic nose acquisition factors. *IEEE International Conference on Industrial Technology* vol 36, pp 72-6
[8] Cao J, Jiang W, Zhao Y 2007 *Experiment Guidance of Postharvest Physiology and Biochemistry of Fruits and Vegetables* (Beijing, China Light Industry Press) vol 2, pp 28-30
[9] Zhou Z, Zhu Y, Cheng C, Chen G, Qiong MA, Pan S 2014 Analysis and Evaluation on Overall

Flavor of Rosa roxburghii Juice by Electronic Nose[J] *Journal of Hubei University for Nationalities* vol 87, pp 33-9

[10] Klimczak I, Gliszczynska-Swiglo A 2017 Green tea extract as an anti-browning agent for cloudy apple juice[J] *Journal of the science of food and agriculture* vol 97, pp 1420-6

[11] Ma, YuHua, Ma, FengWang, Wang, YongHong, et al 2011 The responses of the enzymes related with ascorbate–glutathione cycle during drought stress in apple leaves[J] *Acta Physiologiae Plantarum* vol 33, pp 173-80

IOP Conf. Series: Materials Science and Engineering **611** (2019) 012062 doi:10.1088/1757-899X/611/1/012062

Effects of Application of BB Fertilizer on the Yield, Quality and Economic Benefits of Citrus

Dongfeng Huang[1,2]*, Zhucai Pan[3] and Xiaoxuan Qiu[1,2]

[1]Institute of Soil and Fertilizer, Fujian Academy of Agricultural Sciences
[2]Fujian Key Laboratory of Agro-products Qualitiy & Safety
[3]Nan'an Institute of Agricultural Sciences, Fujian Province

*Corresponding author

Abstract. The effects of four different fertilization treatments [i.e., Conventional fertilization, 45% imported compound fertilizer, Sacofu compound fertilizer (N-P_2O_5-K_2O=20-10-10), BB fertilizer for citrus (N-P_2O_5-K_2O=19-8-13)]on yield, quality and economic benefit of Newhall navel orange were studied through three-year field plot experiment. The results showed that compared with conventional fertilization treatment, the application of Citrus BB fertilizer could not only increase the yield of Newhall navel orange by 13.65%, increase the soluble solids (TSS) content by 51.50%, soluble sugar (WSS) content by 56.64%, vitamin C (Vc) content by 8.87%, reducing the titratable acid (TA) content by 22.83%, increase the solid-acid ratio of navel orange fruit by 96.31%, sugar-acid ratio by 102.97%, but also enhance the economic benefit by 15496 yuan/hm^2. Therefore, the special BB fertilizer for citrus is suitable for further popularization and application in citrus orchards on hilly red soil hills in Fujian Province and even in southern China, where the climate and soil conditions are similar to those in Fujian Province.

1. Introduction

China is one of the major citrus producing areas in the world. According to the data of *China Statistical Yearbook 2018*, in 2017, the citrus planting area of China was 2.587 million hm^2, the total yield reached 38.168 million tons, which was 1.03% higher than that of 2016, and the yield increased by 6.27%. In 2017, the citrus planting area of Fujian Province was 18.57 million hm^2, and the output was 3.15 million tons, which was one of the nine major citrus producing areas in China [1].

Fertilization is one of the most important agronomic measures for citrus trees to obtain high yield and quality. At present, there are many studies on the effects of nitrogen, phosphorus and potassium fertilizers on the yield and quality of citrus fruits (such as soluble solids, titratable acids, vitamin C, etc.) at home and abroad, while there are few studies on the effects of nitrogen, phosphorus and potassium fertilizers on the yield and quality of Newhall navel oranges (such as soluble sugar and main organic acids)[2]. Moreover, due to the obvious differences in climate, soil and management strategies in different regions, we can not copy the rigid [3].

BB Fertilizer, namely Bulk Blending Fertilizer, is several single granular Fertilizer and compound Fertilizer in according to certain proportion mixing and become a kind of compound Fertilizer material[4]. The main advantages of BB fertilizer are: comprehensive nutrients, high concentration, significant yield increase effect; Simple processing, low production cost, no pollution; The formula is flexible and can realize small batch production as low as several hundred kilograms[4]. At present, the production of BB fertilizer in the world accounts for about 40% ~ 50% of total fertilizer production, while the production of BB fertilizer in developed countries such as the United States, Japan and

Canada accounts for more than 70% of total fertilizer consumption[5]. However, the development of BB fertilizer in China is still in its infancy, with a very low market share[6].

BB fertilizer for citrus (N-P_2O_5-K_2O=19-8-13) was developed by Institute of Soil and Fertilizer, Fujian Academy of Agricultural Sciences, based on the nutrient characteristics of red soil orchard soil in southern China, and combined with the demand rule of Citrus for nitrogen, phosphorus and potassium. After many years of experiments, the suitable nutrient formula of nitrogen, phosphorus and potassium was put forward and entrusted with Fujian Longdebao Geodetic Formula Technology Development Co., Ltd.

In order to verify the field application effect of BB fertilizer for citrus, a field trial was conducted with four treatments, such as Habits fertilization, 45% imported compound fertilizer, SaKeFu compound fertilizer (N-P_2O_5-K_2O = 20-10-10), and Citrus BB fertilizer (N-P_2O_5-K_2O= 19-8-13), to test the effect on yield, quality and economic efficiency for Newhall navel orange. And the results would provide scientific basis for the promotion and application of special BB fertilizer products in hilly citrus gardens in southern China's red soil and similar climate and soil conditions.

2. Materials and Methods

2.1. Basic Conditions for Trial Field

The test site is located at the umbilical orange planting base of Meixian Town, Youxi County, Fujian Province. It is located in the northeast of Youxi County, Sanming City, Fujian Province. The terrain is high in the West and low in the East. The mountains in the territory are undulating, and the small valleys and small basins fall between them. Between 90 and 760 m. It is a subtropical maritime monsoon climate with an average annual temperature of 19.6 °C, a frost-free period of 300 d, an average annual rainfall of 1600 to 1800 mm, and a sunshine of 1864.6 h. The basic properties of the tested soil: pH 4.81, organic matter 23.7 g/kg, total nitrogen 1.34 g/kg, alkalysis 68.5 mg/kg, effective phosphorus 177.89 mg/kg, available potassium 268.82 mg/kg, exchange magnesium 0.191 cmol/kg, effective sulfur 34.61 mg/kg.

2.2. Trial Design

The test consists of 4 treatments, namely: treatment 1, Customary fertilization(urea 1 kg/plant, potassium chloride 0.18 kg/plant, 45 % imported compound fertilizer 0.45 kg/plant); Treatment 2, 45% imported compound fertilizer(3.53 kg/plant); Treatment 3, Common compound fertilizer(2.65 kg/strain); Treatment 4, Special BB fertilizer for citrus(2.79 kg/plant). Repeat 3 times per treatment.

The citrus varieties tested were Newhall umbilical orange with a tree age of 9a and a planting density of 675 plants per hm^2. Four umbilical orange trees with relatively uniform growth were selected for each test cell, and random group arrangement was adopted for each test cell. The types of fertilizers tested were as follows: dedicated BB fertilizer for citrus(N-P_2O_5-K_2O= 19-8-13), produced by Fujian Longdebao Soil Survey Formula Technology Development Co., Ltd.; Ordinary compound fertilizer, using "SaKeFu mixed fertilizer(N-P_2O_5-K_2O=20-10-10)", produced by Pengwei Agricultural Resources Co., Ltd. of Longkou City; Urea(containing N 46%), potassium chloride(60% containing K_2O), and 45% imported compound fertilizer(N-P_2O_5-K_2O= 15-15-15).

The fertilization method is: fertilization 4 times a year. Among them, winter fertilizer(January) was applied uniformly to fermented chicken dung 5kg/plant, spring fertilizer(April) accounted for 40 % of the total fertilizer application, young fruit fertilizer(May) accounted for 30% of the total fertilizer application, and strong fruit fertilizer(July) accounted for 30% of the total fertilizer application. Before fertilization, the topsoil of the drip water line along the edge of the crown is about 20 cm wide and deep, and then the fertilizer is evenly spread in the ditch, mixed with the soil, and finally covered with soil. Harvest and measure production in early November of each year. The experiment began in January 2015 and ended in December, 2017. In addition to different treatments, other field cultivation management measures are completely consistent.

2.3. Methods of Test and Analysis

Every year before the mature umbilical orange is harvested in early November, there are 3 fruits with the same random picking orientation, size, maturity and other indicators from the middle and upper parts of each plant for the test tree crown. Each treatment takes 12 fruits for fruit quality measurement. Moreover, Soluble Solids (TSS) content was determined by hand-held sugar meters, Titrable Acid(TA) was determined by NaOH neutralization titration, Vitamin C(Vc) content was determined by 2,6-indiphol sodium, Soluble Sugar(WSS) was determined by ferrocyanic oxidation[6,7] .And data processing uses Excel 2003 software, variance analysis uses SPSS19 .0 software, and LSD multiple comparison method.

3. Results and Analysis

3.1. Effects of Different Fertilization on Fruit Yield of Newhall Cord Orange

The experimental results(Figure 1.) showed that: compared with the customary fertilization treatment(treatment 1), the application of citrus special BB fertilizer(treatment 4) and the application of 45 % imported compound fertilizer(treatment 2) could be significantly ($P<0.05$) to increase the production level of Newhall navel oranges, the average increase rate for the three years was 13.65 % and 11.09 %, respectively, equivalent to 4208 kg/hm^2 and 3420 kg/hm^2. However, there was no significant difference between the yield of umbilical orange and that of conventional fertilizer treatment(treatment 1) with common compound fertilizer treatment(treatment 3)($P>0.05$).

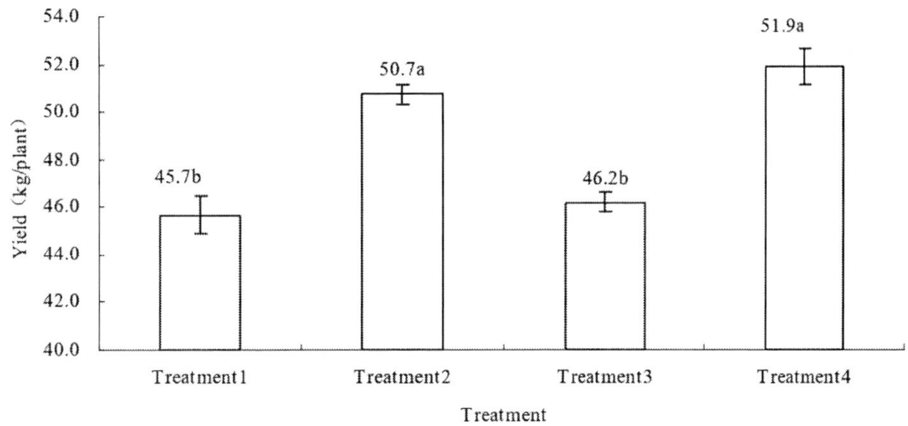

Figure 1. Effects of different fertilization treatments on fruit yield of Newhall navel orange

3.2. Effects of Different Fertilization Treatments on Fruit Quality of Newhall Navel Orange

The results of the experiment (Table 1) showed that compared with the conventional fertilization treatment (Treatment 1), the fruit quality of Newhall navel orange could be improved to some extent by applying several different compound (mixed) fertilizers. Among them, the content of soluble solids (TSS), soluble sugar (WSS) and vitamin C (Vc) increased by 20.53%~51.50%, 31.06%~56.64% and 1.50%~8.87% respectively, and the content of titratable acid (TA) decreased by 7.61%~22.83%, and increased the solid-acid ratio of navel orange fruit 30.46%~96.31%, sugar-acid ratio 41.85%~102.97%. Among them, the application of BB fertilizer for citrus (treatment 4) had the best effect on improving the quality of navel orange fruit. The content of soluble solids (TSS) was significantly higher ($P < 0.01$) than that of conventional fertilization (treatment 1) and common compound fertilizer (treatment 3), and significantly ($P < 0.05$) than that of 45% imported compound fertilizer (treatment 2); the content of soluble sugar (WSS) was significantly higher ($P < 0.01$). The content of vitamin C (Vc) and titratable acid (TA) in conventional fertilization treatment (treatment 1) was significantly higher than that in conventional fertilization treatment (treatment 3), but the difference was not significant ($P > 0.05$) with 45% imported compound fertilizer (treatment 2). The content of vitamin C (Vc) and titratable acid (TA) in conventional fertilization treatment (treatment 1) was significantly higher than that in conventional fertilization treatment (treatment 1). The ratio of solid to acid and sugar to acid was extremely significant or significantly better than that in conventional fertilization treatment 1. So, in those designed treatments, the quality of fruit of treatment 4 is the best.

Table 1. Effects of different fertilization treatments on fruit quality of Newhall navel orange

Treatment number	Treatment contents	TSS (%)	TA (%)	WSS (%)	Vc (mg/100ml)	solid-acid ratio	sugar-acid ratio
Treatment 1	conventional fertilization	9.01Bc	0.92a	8.21 Bc	63.36b	9.79Dd	8.92Cd
Treatment 2	45%imported compound fertilizer	12.63ABb	0.78b	12.28 Aa	68.65a	16.19Bb	15.74Ab
Treatment 3	conventional fertilization	10.86Bc	0.85ab	10.76 ABb	64.31b	12.78Cc	12.66Bc
Treatment 4	Citrus BB Fertilizer	13.65Aa	0.71b	12.86 Aa	68.98a	19.23Aa	18.11Aa

Note 1: Each data in the table is the average of three repetitions, the same below.
Note 2: For the different letters after the same column values in the table, the difference is very significant in capital letters ($P < 0.01$), and significant in lowercase letters ($P < 0.05$), the same below.

3.3. Effect of Different Fertilization Treatment on Economic Benefit of Newhall Navel Orange

From the analysis results of economic benefits of Newhall navel orange under different fertilization treatments (Table 2), it can be seen that the order of economic benefits of different fertilization treatments is: BB fertilizer for citrus (Treatment 4) > 45% imported compound fertilizer (Treatment 2) > conventional fertilization (Treatment 1) > common compound fertilizer (Treatment 3). Among them, compared with the conventional fertilization (treatment 1), the application of BB fertilizer for citrus (treatment 4) could increase the economic benefit by 15496 yuan/hm^2 with an increase rate of 13.09%, and the application of 45% imported compound fertilizer (treatment 2) could increase the economic benefit by 9146 yuan/hm^2 with an increase rate of 7.72%. However, the economic benefit of applying common compound fertilizer (treatment 3) was slightly lower than that of conventional fertilization (treatment 1).

Table 2. Effects of Different Fertilization Treatments on Economic Benefits of Newhall Navel Orange

Treatment number	Yield (kg/hm^2)	output value (Yuan/hm^2)	Fertilizer cost (Yuan/hm^2)	Economic Benefits (Yuan/hm^2)	Ratio increase (%)
Treatment1	30825	123300	4882	118418	\
Treatment2	34245	136980	9416	127564	7.72
Treatment3	31185	124740	7806	116934	-1.25
Treatment4	35033	140130	6216	133914	13.09

Note: fermented chicken manure 670 yuan / t; urea 2100 yuan / t; 45% imported compound fertilizer 3000 yuan / t; 45% imported compound fertilizer 2600 yuan / t; citrus BB fertilizer 2100 yuan / t; Sacofu compound fertilizer (20-10-10) 3100 yuan / t; potassium chloride 2400 yuan / t; navel orange 4 yuan / kg.

4. Discussion and Conclusion

Crop specialized fertilizer product is the embodiment of comprehensive balanced fertilization technology, and an important link to realize the one-stop agrochemical science and technology service of "determining soil nutrients, determining fertilizer formula, producing fertilizer products and guiding scientific fertilizer use". Shu Changcheng[8] studied the fertilizer effects of different formulations of special citrus fertilizer in Qujiang District of Zhejiang Province. The results showed that formula 1 (N-P_2O_5-K_2O=10-4-6, Organic matter content 25%) and formula 2 (N-P_2O_5-K_2O=8-3-4, Organic matter content 20%) of special citrus fertilizer could be obtained by applying 3000 kg/hm^2 and 3975 kg/hm^2

respectively in the whole year. The fruit quality is better with higher yield. Liang Suiquan *et al.*[9] and Hu Yun *et al.*[10] studied the application of Xingnong brand special citrus fertilizer ($N-P_2O_5-K_2O$=11-7-7) mixed equal amount cake fertilizer as summer citrus fertilizer in citrus orchard of Yixing County, Hubei Province, and achieved better yield increase and better fruit quality. Therefore, there are great differences in climate, soil, management strategy and fruit varieties in different regions. There are also great differences in the distribution of nitrogen, phosphorus and potassium nutrients recommended for citrus special fertilizer products.

This experiment was carried out on hilly red soil orchards in Fujian Province to test the application effect of BB fertilizer for citrus ($N-P_2O_5-K_2O$=19-8-13).The results showed that compared with the conventional fertilization treatment (treatment 1), the application of BB fertilizer for citrus (treatment 4) not only increased the yield of Newhall navel orange by 13.65%, increased the content of soluble solids (TSS) by 51.50%, soluble sugar (WSS) by 56.64%, vitamin C (Vc) by 8.87%, decreased the content of titratable acid (TA) by 22.83%, and increased the ratio of solid to acid and sugar to acid by 96.31% and 102.97%, respectively. And the economic benefit was increased by 15496 yuan/hm^2. Therefore, the special BB fertilizer for citrus is suitable for further popularization and application in citrus orchards on hilly red soil hills in Fujian Province and even in southern China, where the climate and soil conditions are similar to those in Fujian Province.

Acknowledgments

This work was supported by several projects, which included Fujian Provincial Public Welfare Research Institute Basic Research Project(2018R1022-4), Fuzhou Science and Technology Project(2018-G-65), and Fujian Provincial Academy of Agricultural Sciences Science and Technology Innovation Team Project(STIT2017-2-10).

References

[1] Shen Z M. Current Situation and Future Development Forecast of Citrus Production in China. *Friends of Fruit Growers*,2019(2):1~4.
[2] Zhu Z Y, Li M, Zhang C M, *et al.* Optimum P and K ratio for high yield and good quality of 'Newhall'navel orange. *Journal of Plant Nutrition and Fertilizers*, 2018, 24(4): 1105~1112.
[3] Fan W G, Yang S A, Liu G Q, *et al.* Efects of fertilizer application level on the yield, quality and economic return of navel orange.Journal of Mountain Agriculture and Biology, 2006, 25(3): 194~196.
[4] Yao B Q, Li Y. Production Technology and New Product Development of Blended Fertilizer (BB Fertilizer).*Journal of Min Xi Vocational College*, 2003(1):74~76.
[5] Xu P Z. Present Situation of Blended Fertilizer (BB Fertilizer) in China. *Tropical and Subtropical Soil Science*, 1997,6(1): 67~69.
[6] Ru J W, Chen F, Wan Y F, *et al.* Effect of Potassium Fertilizer Application Rate on Yield and Quality of Navel Orange. *Journal of Fruit Tree*, 2001,18(6): 272~275.
[7] Zhou J F, Yang Y T, Li X F, *et al.* On Influence of Different Percentage of N, P, K and Ca On Navel Orange Yield and Fruit Quality. *Journal of Southwest China Normal University (NaturalScienceEdition)*, 2015,40(2):54~58.
[8] Shu C C. Research on the effect of different formulas of orange plants specia-purposel fertilizer. *Soil and Fertilizer of China*, 2006 (3): 35, 36, 43
[9] Liang S Q, Hu Y, Fu G Z, *et al.* The experiment of Yichang Xingnong special citrus fertilizer mixed with equal amount of cake fertilizer for citrus summer fertilizer. *Guangxi Horticulture*,2000 (2): 5-6
[10] Hu Y, Fu G Z, Liang S Q, *et al.* Experiments on the application of special citrus fertilizer mixed with equivalent cake fertilizer to citrus summer fertilizer. *Hubei Agricultural Science*, 2000 (5): 38-39

Taste Correction and Safety Evaluation of Rhodiola Extraction

Shouqing Zhao[1], Shanshan Wang[2], Changsong Xue[3] and Guangqing Xia[4,*]

[1]Shengli Hospital Jilin Province
[2]Cancer institute of Ailuokang
[3]School of pharmaceutical science and medicine, Tonghua Normal University
[4]School of life science, Tonghua Normal University

*Corresponding author

Abstract. Objective To explore tannins removal of *Rhodiola* extraction and evaluate its safety. **Method:** Content of Tannins was treated with cheese, tannins removal was used by chitosan embedding, safety evaluation was tested by acute toxicity. The pharmacodynamics were detected by in vivo and in vitro antioxidant methods. **Results:** 20g/L *Rhodiola* extraction added 1g chitosan under 50°C embedding 2h, tannis was removed effectively and taste improved greatly. The results of acute toxicity tests in mice showed that the daily dose of 29.73 g was safe on the basis of 70kg adult weight. Antioxidant activity in vitro was evaluated by DPPH, super-oxide anion and hydroxyl radical scavenging assays showed at concentration of 0.1% indicated better results, while results in vivo showed at concentration of 1.2% decreased p53, p21 gene expression. **Conclusion**: A reasonable process for removing tannins from were determined *Rhodiola* extraction and taste of *Rhodiola* extraction is safe under the removal extraction technique. Results of antioxidant proved that *Rhodiola* extraction is effective.

1. Introduction

Rhodiola is a plant of *Rhodiola* family, the current Pharmacopoeia variety is Rhodiola crenulata. It is widely believed that it has the effects of anti - ischemia, anti - hypoxia, anti - fatigue and improving athletic ability[1]. *Rhodiola* mainly contains rhodiola glycosides, tannins and polysaccharides[2-3]. The tannins have seriously affected the clarity and taste of *Rhodiola* rosea extract preparations, especially its strong bitter taste, which has affected the development and utilization of *Rhodiola rosea* in the health products market. In order to correct its bitter taste, it is necessary to remove the enamel components as much as possible. At present, the methods for removing tannin in China are generally classified into alkaline alcohol precipitation method, dialysis method, gelatin precipitation method and chitosan precipitation method[4-6]. This study investigated the effects of different chitosan dosage, agitation temperature and embedding time on the taste of rhodiola rosea. On this basis, the safety and pharmacodynamics of the modified rhodiola rosea extract were evaluated to provide theoretical basis for the further development and application of the health food with *Rhodiola* rosea as the main component.

2. Materials and Methods

2.1 Materials

Rhodiola (Chinese medicinal materials wholesale market purchased), approved by Professor Qin

Jiamei from the School of Life Sciences of Tonghua Normal University as *Rhodiola crenulata*; mice(SPF kunming mice, purchased from changchun gaoxin animal center, male and female, healthy and age-appropriate); zebrafish is provided by Metabolism and Aging Laboratory of Tonghua Normal University. All the drugs used in the experiment were analytically pure.

2.2 Methods

2.2.1 Determination of tannin content. Operate in accordance with Appendix XB of the 2015 Edition (Part 1) of the Pharmacopoeia of the People's Republic of China.

2.2.2 Method for the removal of Rhodiola tannin. The dried *Rhodiola* was cut into pieces and crushed through a 30-mesh sieve. 20g rhodiola powder was added to 400ml distilled water and soaked for 24h. The mixture was heated and extracted in a water bath at 85°C for 1.5 h, suction filtered, and the filtrate was removed, the remaining part was adjusted to a volume of 1000 ml to prepare *Rhodiola rosea* extract. Divided the *Rhodiola* extract into 4 groups, investigate different doses of chitosan (0.5g, 1.0g, 1.5g, 2.0g), different mixing times (0.5h, 1.0h, 1.5h, 2.0h) and stirring temperature (30°C, 40°C, 50°C, 60°C) effect on its tannin content. Each group of experiments was repeated 3 times to determine the average.

2.2.3 Modified Rhodiola Extract Safety Test. Maximum toxicity test for acute toxicity test: 20 mice, 10 in each group, were randomly divided into the drug-administered group and the control group. Administration group: given the maximum volume (0.4 mL/kg) available for this product(the average weight of the mice was 20 g, and the maximum dosage was about 9.8g/kg). One night before the administration, the mice were fasted for 12 hours, and the next day, the administration was started. The administration interval was 3 hours. During the interval, the mice were fasted but allowed to drink, and the drug was administered continuously for 7 times. Control group: The same amount of distilled water was administered, and the time, duration, recovery period, number of deaths, etc. of the mice were recorded at 12 h after administration[8].

2.2.4 Study on pharmacodynamics of Rhodiola extract. Experimental study on antioxidant activity of *Rhodiola* extract
The 0.005%-0.1% *Rhodiola* solution was prepared separately. The superoxide anion radical scavenging rate was determined by pyrogallol auto-oxidation method, and the hydroxyl radical scavenging rate was determined by FeSO4 salicylic acid method.
Experimental study on the antioxidant activity of *Rhodiola* extract
The control group (CK), low concentration group (0.6%), medium concentration group (1.2%), and high concentration group (1.8%) were prepared respectively to treat zebrafish embryonic cells. The culture medium was changed once every day, and the cells were continuously treated for 3 days. Samples were collected for gene expression analysis[10].

2.2.5 Statistical analysis. Data of experimental results were expressed as $\bar{x} \pm s$, and SPSS16.0 software was used for one-way analysis of variance, and $p<0.05$ was considered to be statistically significant.

3. Results and Analysis

3.1. Effect of Different Chitosan Dosage on Tannin Removal
Tannins are widely found in the plant kingdom, and more than 70% of Chinese herbal plants contain tannins. In the extraction process of Chinese herbal medicines, especially in the extraction process of the solution, the tannin is often removed as an impurity. Under the experimental conditions, the effects of different chitosan on tannin removal were shown in figure 1. As can be seen from figure 1, with the increase of chitosan dosage, the change trend of tannin content first decreases, then increases and then

decreases. When chitosan content reached 1.0g, the tannin content was the lowest, and then increased with the increase of chitosan content.

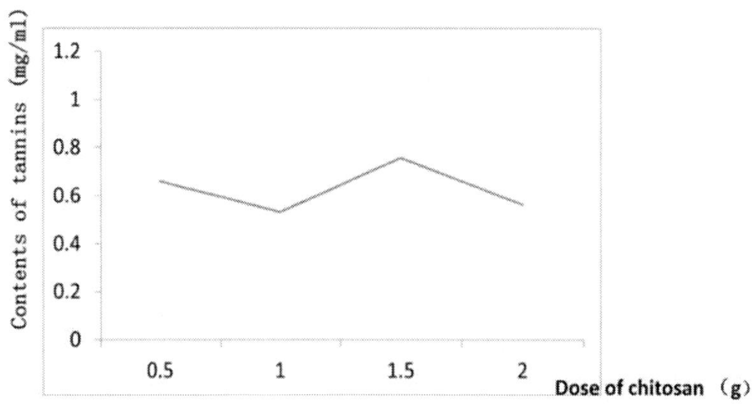

Figure 1. Effect of different chitosan dosage on tannin removal

3.2. Effects of Different Agitation Time on Tannin Removal

Different agitation times have a greater impact on the embedding of tannins, as shown in figure 2. It can be seen from the figure that the maximum tannin content was 0.693mg/ml when the agitation time was 1.5h. When the embedding time reached 2h, the tannin content reached a low value of 0.425 mg/ml. Subsequent experiments showed that the content of tannin remained at about 0.425 mg/ml as the agitation time increased.

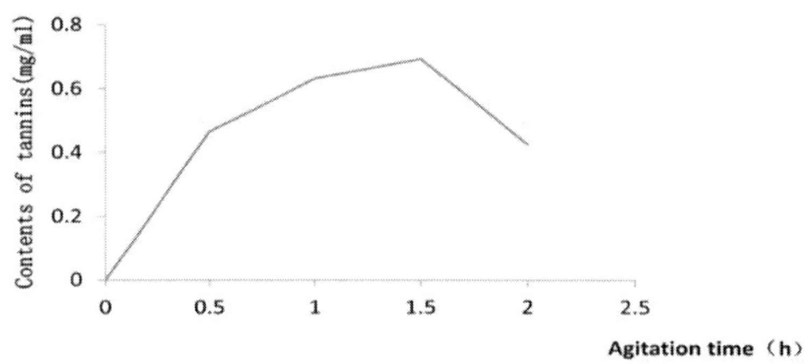

Figure 2. Effects of different agitation time on tannin removal

3.3. Effects of Different Agitation Temperatures on Tannin Removal

After adding the same chitosan, the influence of different stirring temperatures on tannin content was shown in figure 3. It can be seen from the figure that in the temperature range of 30°C to 60°C, the content of tannin increases with the increase of temperature. When the temperature reached 50°C, the content of tannin was the lowest, and then the content of tannin showed an increasing trend with the increase of temperature.

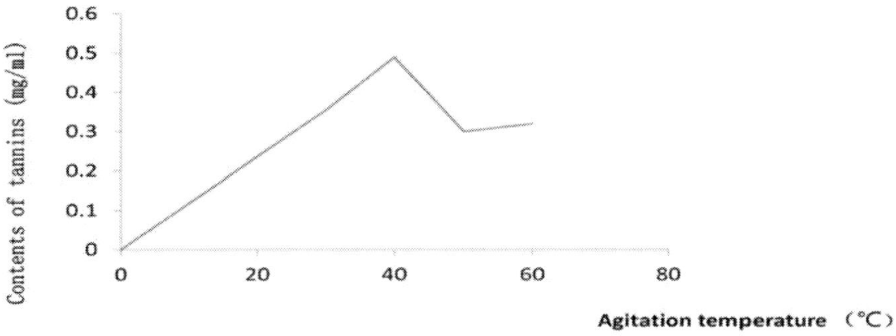

Figure 3. Effects of different agitation temperatures on tannin removal

3.4. Effect of Modified Rhodiola Extract on Toxicity and Hypoxia Tolerance in Mice

The product shall be given in the maximum available volume (0.4ml/kg)(The average weight of the mice was 20g, and the maximum dosage was about 9.8g/kg). After continuous administration for 7 days (Table 1), no abnormality or death was found in mice after intragastric administration, indicating that the modified *Rhodiola* extract used in this study was actually non-toxic. Subsequently, the effects of different concentrations of *Rhodiola* extract on survival time of mice under normal pressure were investigated (Table 2). The results showed that the survival time of mice could be prolonged by different concentrations of *Rhodiola* extract. Compared with the control group, the survival time of mice in the high-dose group reached a significant level ($p<0.05$), indicating that *Rhodiola* extract could improve the anti-hypoxia ability of mice.

Table 1. Effect of Rhodiola extract on toxicity in mice.

Times of administration	The first administration	The second administration	The third administration	The fourth administration	The fifth administration	The sixth administration	The seventh administration
Dosage (mL)	0.4	0.4	0.4	0.4	0.4	0.4	0.4
Deaths	None	None	None	None	None	None	None

Table 2. Effects of different concentrations of *Rhodiola* extract on hypoxia survival time of mice under normal pressure($\bar{x}$ ±s).

Groups	Dosage (g/kg)	Survival time (min)
Control group		28.42 ±6.25
Rhodiola extract group	Low dose group	26.76±8.35
	Medium dose group	27.08±6.56
	High dose group	30.63±5.87*

*.The significance level was 0.05

3.5. Experimental Study on Antioxidant Activity in Vitro of the Modified Rhodiola Oral Solution

Different concentrations of modified *Rhodiola* have certain scavenging effects on DPPH, superoxide anion radicals and hydroxyl radicals, as shown in Table 3. It can be seen from Table3 that the removal rate of superoxide anion reached 96.33% when the concentration of modified *Rhodiola* oral solution was 0.1%. At the concentration of 0.05%, the DPPH clearance rate was the highest at 89.01%. Although this experiment has removed the tannin of *Rhodiola*, it still contained a certain amount of tannin, so the scavenging rate of hydroxyl radical was relatively low.

Table 3. Effects of *Rhodiola* Oral Liquid on Hydroxyl Radical, Superoxide Anion Radical and DPPH Clearance.

Rhodiola oral liquid concentration	Hydroxyl radical (%)	Superoxide anion free radical(%)	DPPH(%)
0.1%	10.06	96.33	83.67
0.05%	9.52	91.33	89.01
0.01%	4.79	91.0	77.67
0.005%	1.06	87.65	51.33

3.6. *Experimental study on antioxidation of modified Rhodiola oral liquid in vivo*

In combination with the in vitro antioxidant experiment, the zebrafish embryo cells developed for 8h were treated by low-concentration group, medium-concentration group and high-concentration group of *Rhodiola*. And after 24 hours of drug treatment, the effects of each treatment on the survival rate and deformity rate of zebrafish were valued(Figure 4). After 3 days of treatment, the effects of aging-related gene expression were examined(Table 4). The results showed that the treatment of medium concentration drug had little effect on the development of zebrafish, but the treatment of high concentration group affected the survival and development of zebrafish, and the mortality and deformity rate were higher. In addition, low concentration treatment can significantly reduce the expression of p53 and p21 genes. Compared with the control group, there was no significant difference in the expression levels of p53 and mdm2. High concentration group treatment increased the expression of p53 and p21 genes.

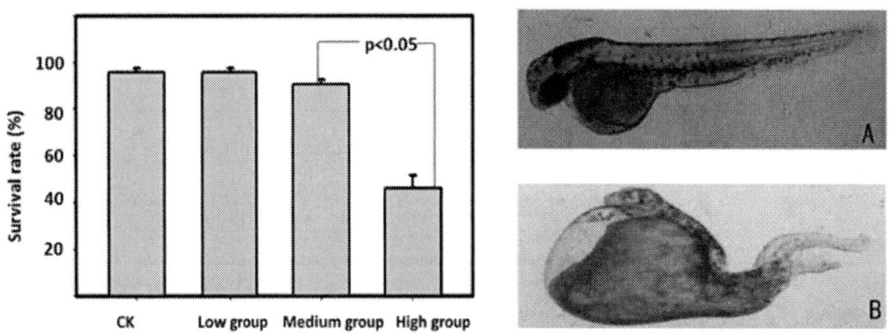

Figure 4. Effect of modified Rhodiola Oral Liquid on Survival Rate and Deformity Rate of Zebrafish Embryos(Figure A normal developing zebrafish, B deformed zebrafish)(The data in the figure is the standard deviation of 30 zebrafish embryos).

Table 4. Effect of the treatment of modified *Rhodiola* Oral Liquid on the expression of aging-related genes in zebrafish (The data in the table is the standard deviation of 30 zebrafish embryos).

gene	control group	low concentration	medium concentration	high concentration
mdm2	1.0012±0.0599	1.0000±0.0080	1.0915±0.1706	1.0057±0.0061
p53	1.0008±0.0492	0.9377±0.0110	1.0118±0.0736	1.7252±0.0206
p21	1.5301±0.0370	1.4242±0.0262	1.4579±0.0273	1.8095±0.0473

4. Discussion

In this paper, the single factor experiment method was used to investigate the addition amount, agitation time and agitation temperature of chitosan. The final determined rationality of *Rhodiola* extract taste correction is that the dried *Rhodiola* was cut into pieces and crushed through a 30-mesh sieve. 20g *Rhodiola* powder was added to 400ml distilled water and soaked for 24h. The mixture was heated and extracted in a water bath at 85℃ for 1.5 h, suction filtered, and the filtrate was removed, the remaining part was adjusted to a volume of 1000 ml to prepare *Rhodiola rosea* extract. Take 100 ml of *Rhodiola* extract, add 1.0 g chitosan and stir at 500/min in an electromagnetic heating stirrer, heat at 50℃ for 3 h, filter and dilute to 100 ml. This study laid a foundation for the orthogonal test in the future to optimize the taste correction process of *Rhodiola* extract.

At the same time, according to the standard requirements of the Food Safety Toxicology Evaluation Procedure, the safety test and hypoxic tolerance experiment were conducted on the modified *Rhodiola* extract. Under the experimental conditions, with the increase of the concentration of *Rhodiola rosea* extract, the hypoxia tolerance of the experimental mice increased accordingly. According to the adult body weight of 70 kg, the daily dose was 29.73g, with no adverse effects. Therefore, the modified *Rhodiola* extract used in this study was actually non-toxic.

The main reason for human aging is that with the increase of age, the ability to scavenge free radicals in the body decreases, and free radicals have become the root cause of 85% of the human body's diseases[10]. Looking for low-cost and suitable antioxidants for the human body has been a hot topic of scientific research. Compared with synthetic antioxidants, natural antioxidants have more extensive development and utilization prospects[11].

Higher organisms are basically developed from single cell fertilized eggs. Many scholars believe that life is most sensitive to exogenous chemicals in the embryonic period, so it is of great significance to monitor them. The zebrafish can be observed directly under microscope because of its in vitro fertilization, in vitro development and transparent embryo body. In addition to being a tumor suppressor gene, p53 also has many important functions, including cell apoptosis and cell senescence. When subjected to stress signals such as DNA damage, hypoxia and radiation, p53 exerts its transcriptional activation function, regulates the transcriptional expression of a series of target genes, and then causes stress reactions such as DNA damage repair, cell cycle arrest and apoptosis[12]. P21 is activated by p53 at the transcriptional level and mainly mediates aging caused by telomere dependence and various emergency conditions such as DNA damage[13]. Mdm2 is currently the most important intracellular negative regulator of p53. Mdm2 contains a p53 gene binding site that binds to p53 to form a complex that inhibits the transcriptional activity of p53. Overexpression of mdm2 can block the trans-activation mediated by p53, resulting in loss of p53 function, gene instability and cell proliferation. DNA damage leads to mdm2 inactivation and elevated p53 levels[14].

Based on the above analysis, it can be seen that the modified *Rhodiola* Oral Liquid has certain antioxidant activity of scavenging free radicals in vitro. And this effect may be mediated by the p53 signal transduction pathway, which provides experimental basis for further research and development of *Rhodiola* natural antioxidant drugs.

Acknowledgement

This work was supported by Jilin Province Pharmaceutical Industry Development Guidance Fund Project (20150311061YY).

References

[1] Zhang Linan, Li Yiliu, et al. Progress on modern pharmacology of Rhodiola[J].Clinical Focus, 2011, 26(21):31-33

[2] Li Jun, Chen Zhi, Li Jianmin, et al. The Review of Rhodiola[J].Journal of Agricultural University, 2007, 22(1):61-64

[3] Ji Yubin, Geng Xin, Ji Chenfeng. Progress on Rhodiola[J].Tianjin Journal of Traditional Chinese Medicine, 2007, 24(1):81-85

[4] Li Wei, Huang Qinni. Advances in Studies andApplications on Rhodiola Rosea L[J].Journal of Capital Normal University, 2003, 24(1):56-59

[5] Jia Guofu. GENERAL STUDY AND ITS PROSPECT ON RHODIOLA ROOT [J].Si Chuan Prarire, 1997, 3:38-40
[6] Lin Junzhi. Evaluation on Pre-Preparation of Rhodiola extract [D].Chengdu: Chengdu University of Chinese Medicine, 2013
[7] National Pharmacopoeia Committee. Pharmacopoeia of Republic of China [S].Beijing: Chemical Industry Press, 2015
[8] Wang Qingli, Peng Jian. Reflections on Acute Toxicity Tests of Chemical Drugs [J].Chinese Journal of New Drugs, 2004,13(11):961-964
[9] Xia Guangqing, Yao Huimin, Dong Lihong, Liu Wei. Effect of ganoderma polysaccharide on survival, development and senescence of zebrafish [J]. Chin J Pharmacol Toxicol, 2014(4):491-496
[10] Halliwell B, Gutteridge J M C. Free radicals in biology and medicine[J]. NY: Oxford University Press. 2011
[11] Peng X, Xiong Y L, Kong B. Antioxidant activity of peptide fractions from whey protein hydrolysates as measured by electron spin resonance[J]. Food Chemistry, 2009,113(1):196-201.
[12] Lahav G, Rosenfeld N, Sigal A. Dynamics of the p53-mdm2 feedback loop inindividual cells[J]. Nat Genet, 2004, 36(2):147-50
[13] Jaskelioff, M. Muller, F. L. Paik, E. Depinho, R. A. Telomerase reactivation reverses tissue degeneration in aged telomerase-deficient mice[J]. Nature. 2011, 69:102-106
[14] Lessel, D., Wu, D., Trujillo, C., , Goebel, I., et al. Dysfunction of the MDM2/p53 axis is linked to premature aging[J]. J Clin Invest, 2017, 127, 3598-3608.

Edible Quality and Safety Evaluation of *Zanthoxylum* from Western China

Jingxuan Ke[1], Yuan Qu[1], Guanghui Shen[1], Shanshan Li[1], Biao Pu[1], Meng Ye[2] and Zhiqing Zhang[1,*]

[1]College of Food Science, Sichuan Agricultural University, Ya'an, China
[2]Collge of Forestry, Sichuan Agricultural University, Chengdu, China

*Corresponding author

Abstract. In order to investigate the edible safety of Chinese prickly ash (*Zanthoxylum*), the contents of four inorganic contaminants (lead, cadmium, mercury, and arsenic) and two pesticide residues (permethrin and cypermethrin) in seventeen prickly ash samples were detected, which gathered from the main plant region of Western China. In addition, potential risks due to the contaminants which exposure to it had been evaluated. Eestimated daily intakes (EDIs) and hazard quotient (HQ) of the contaminants were analyzed too. The data showed that EDIs of inorganic contaminants were lower than the acceptable daily intake (ADI), and their HQ values, ranging from 2.65% (As) to 19.05% (Pb), were considerably <1. In addition, HQ values for permethrin and cypermethrin were determined to be 1.83% and 2.09%, respectively. These findings demonstrated a high level of edible quality and safety of *Zanthoxylum* from western China.
Keywords: *Zanthoxylum*; inorganic contaminants; pyrethroids; safety evaluation.

1. Introduction

Zanthoxylum (Prickly ash) is an important industrial woody plant of rutaceae. It includes more than 200 kinds of prickly ash distributed around the world[1]. Its fruit is a famous spicery in Chinese cooking culture, offering a unique numb taste especially in Sichuan cuisine. In the past decade, the planting area of Chinese prickly ash has exceeded 500,000 ha and the output has exceeded 100,000 tons, with increased agricultural inputs such as fertilizers and pesticides. *Zanthoxylum* is rich in nutrients, such as volatile oil[2], numb-taste composition[3], phenols[4], etc. Many researches have studied *Zanthoxylum* in terms of its organic compositions, such as volatile oil[5], numb-taste composition[3] and so on. Currently, most processed products of *Zanthoxylum* are food products. In addition to condiment oil, prickly ash powder, essential oil and other traditional food ingredients in Sichuan hot pot, highly processed products such as prickly ash sauce, have gradually entered the market. Furthermore, other highly processed products in medicine and commodity of *Zanthoxylum* become more abundant, with one example being prickly ash essential oil soap[6]. Nevertheless, few researches studied its edible quality and safety, especially focus on potential contaminations and this quality thus merits further research.

Inorganic contaminants, e.g., Pb, Cd, Hg, As, etc., in food can arise through many ways, such as contaminated soil, irrigation water, industrial emissions, and harvesting process[7]. In agriculture, farmers use pesticides (e.g., pyrethroid, organophosphorous, etc.) to obtain high yield and quality of agricultural products[8]. Insecticides, and herbicides are widely used especially in developing countries, including China[9]. This can lead to contamination of pesticides in food. Consumption of pesticide-contaminated food can cause great harm to human health[10], leading to various symptoms, such as, headache, nausea, declined reproductive function, endocrine system disorders, etc. [11]. Therefore, whether or not the amount of contaminants in food exceed the maximum allowable concentrations set by the relevant institutions has raised public and scientific concerns [12]. A few studies have reported the detection of pesticide residues in other condiments, such as white and black pepper [13]. Furthermore, there is no definitive standard of maximum allowable concentrations of pesticide residue in *Zanthoxylum* domestically or worldwide.

The aims of this work are to determinate the presence and content of inorganic contaminants and pesticide residues in two varieties *Zanthoxylum bungeanum* Maxim. and *Zanthoxylum armatum* DC.of *Zanthoxylum* with 17 acessions gathered from main planting distrcts in China, such as Jinyang, Hanyuan in Sichuan province and Hancheng, Fengxian in Shanxi province; and to comprehensively evaluate their edible safety. The results will provide new information regarding the edible quality of *Zanthoxylum*.

2. Materials and Methods

2.1. Sample Collection and Preparation

Seventeen samples of *Zanthoxylum* were collected from seventeen districts in the region of Western China (Table 1). The samples were crushed and homogenized before passing through a 40-mesh screen. The resultant powder samples were dried at 60 °C for 48 h for further analysis.

2.2. Standards and Reagents

Stock standard solutions of Pb, Cd, Hg, As (1000 μg/mL each), and pyrethroid pesticides (permethrin and cypermethrin in n-hexane, 100 μg/mL each) were purchased from Bejing Tanmo science and technology of quality control Co., Ltd. (Bejing, China). These solutions were of a purity range of 99.71% to 99.99%, and were stored in a freezer at -18 °C prior to use.

The reagents were purchased from Chengdu Jere Technology Co., Ltd. (Chengdu, China). Deionized water was purified by a Milli-Q water purification system (Millipore, Milli-Q, USA).

2.3. Analytical Methods

2.3.1. Determination of Contaminations. The determination of Pb and Cd was carried out according to the National Food Safety Standard of China: determination of lead (cadmium) in foods [14, 15]. The quantification of Pb and Cd was conducted using a flame atomic absorption spectrometer (FAAS) equipped with an air-C_2H_2 flame and a polarizer for Zeeman background correction (Z-2000, Japan). FAAS was operated with the instrument recommended procedure.

The digestion and determination of Hg and As were conducted according to the National Food Safety Standard of China: Determination of mercury (arsenic) in foods [16, 17]. The quantification of Hg and As, was carried out using an AFS-3000 (Beijing Haiguang Instrument Co., Ltd., Bejing, China) and operated with the instrument recommended procedure.

The Determination of pyrethroids in *Zanthoxylum* were carried out according to the Agricultural Standards of China [18].

2.3.2. Comparion content of Contaminants with MRLs. Maximum Residue Limits (MRLs) are not the toxicological limits ensuring that food is safe for consumption. The obtained data were compared with the Chinese established MRLs, according to the Safety Standard of *Zanthoxylum* concentration limits for Pb, Cd, Hg, and As [19]. There are no standard MRLs for permethrin and cypermethrin in *Zanthoxylum*. Therefore, MRLs for permethrin and cypermethrin were compared to MRLs of malathion in *Zanthoxylum* [19].

2.3.3. Risk Evaluation of Intake Contaminants. MRLs represent the highest concentration of contaminants allowed in food, they however lacks food consumption information, which is an essential data for food safe assessment. Eestimated daily intakes (EDI) is used to accurately identify whether or not the food is sufficiently safe for consumption [20]. In addition, it is possible to compare the exposure levels to the chronic ADI levels [21]. The dietary exposure was analyzed based on an intergration of the analytical data concerning the contaminants of interest (obtained from this study) and the food consumption data [22]. The computing method of EDIs, hazard quotient (HQ) and hazard index (HI) were referred to [9]. The commodity can be deemed safe when the value of HI is lower than 1. Conversely, food may contain high health hazard or consumer is not be adequately protected when the value of HI exceed 1 [23].

3. Results and Discussion

3.1. Quality Assurance of Contaminants Quantitative Determination
DL (defensive lineman) of Pb, Cd, Hg, and As were 0.0180, 0.0021, 0.000375, and 0.000167 mg·kg^{-1}, respectively. These values were obtained with correlation coefficients, $R^2 \geq 0.999$. In addition, LODs (Limit of detection) of permethrin and cypermethrin were 0.016 and 0.011 mg·kg^{-1}, respectively. The precision of results were described in terms of relative standard deviation (RSD) [24]. RSDs were between 0.3% and 5.5% for the quantification of inorganic contaminants, between 2.12% and 6.0% for the quantification of pesticides.

3.2. The Content Level of Contaminants in Zanthoxylum
Concentrations of Pb and As in all analyzed samples were within the safty range (<MRLs), and all four inorganic contaminants were detected in almost all samples (Table 1). Concentrations of Cd and Hg were above MRLs in more than half of samples. As indicated by mean values, Pb was the highest, while Hg was the lowest, in consistent with the previously reported data [25]. Moreover, *Zanthoxylum* belongs to Rutaceae, is the perennial plants that are enriched with harmful elements. The sources of heavy metals, especially Cd and Hg should be focus and controlled, the measured content of which were above MRLs. Two types of pyrethroids found in analyzed samples exceeded MRLs; cypermethrin was the majority (found in 94.11% of samples). Half of samples (50%) contained permethrin of concentration lower than the LOD. Concentration range of the detected permethrin were from under LOD to 0.6126 mg·kg^{-1} (Table 1). Mean values, ranges, and total number of samples containing four inorganic contaminants and two pesticides above MRLs are presented in Table 2.

3.3. Evaluation of Risk Based on Amount of Intake Contaminants
Food and Agriculture Organization and World Health Organization (FAO/WHO) and Joint Expert Committee on Food Additive have recommended that the Provisional Tolerated Weekly Intake (PTWI) of Pb, Cd, Hg, and As are 0.025, 0.007, 0.005, and 0.015 mg·kg^{-1}bw, respectively[26]. In the present study, ADI of Pb, Cd, Hg and As were calculated for adults with a mean body weight of 60 kg. The ADI of pesticides are in reference to the standards established by Joint Meeting of Pesticide Residues (JMPR)[21]. A list of ADI values used for the assessment of chronic exposure as well as the contribution of inorganic contaminants and pyrethroids to the calculated total EDI and HI are tabulated in Table 3. For inorganic contaminants, HQ values ranged from 2.65% (for As) to 19.05% (for Pb). HQ values for permethrin and cypermethrin were 1.83% and 2.09%, respectively. HI values of lower than 1 also show that these inorganic contaminants as well as pyrethroids are at the safe levels.
Because *Zanthoxylum* can easily suffer from pest attack, farmers often spray pesticides on the plant to prevent such attack. Unfortunately, MRLs for permethrin and cypermethrin in prickly ash are currently not available. According to the data in this work, permethrin and cypermethrin were detected in most analyzed samples. It concluded that MRLs for permethrin and cypermethrin in *Zanthoxylum* should urgently be investigated and made available to ensure food safety of consumers.

Table 1. Mean content of the monitoring contaminants (Pb, Cd, Hg, As, permethrin and cypermethrin) in samples

No.	Samples	Plant district	Inorganic contaminants (mg·kg^{-1})				Pesticides (mg·kg^{-1})	
			Pb	Cd	Hg	As	Permethrin	Cypermethrin
1	Hancheng Dahongpao	Hancheng of Shanxi province	1.537±0.020	0.1188±0.0035	0.0448±0.0024	0.0752±0.0034	0	0.0123±0.0006
2	Fengxian Dahongpao	Fengxian of Shanxi province	0.504±0.019	0.1200±0.0035	0.0524±0.0007	0.0333±0.0006	0.0712±0.0030	0.6552±0.0336
3	Maoxian Dahongpao	Maoxian of Sichuan province	0	0.0125±0.0000	0.0388±0.0020	0.0567±0.0014	0.0391±0.0015	0.6234±0.0132
4	Zhaotong Qinghuajiao	Zhaotong of Yunnan province	0.860±0.014	0.0869±0.0027	0.0378±0.0008	0.0934±0.0013	0.0217±0.0103	0.6207±0.0199
5	Yuexi Gongjiao	Yuexi of Sichuan province	0.816±0.030	0.0496±0.0013	0.0192±0.0004	0.0695±0.0027	0.3065±0.0126	0.7723±0.0356
6	Henan Dahongpao	Jiyuan of Henan province	1.744±0.068	0.1355±0.0031	0.0299±0.0009	0.0627±0.0009	0	0.7823±0.0255
7	Jiulong Huajiao	Jiulong of Sichuan province	0.785±0.019	0.0288±0.0007	0.0200±0.0011	0.0076±0.0005	0.3062±0.0625	0.5952±0.0328
8	Wudu Dahongpao-1	Majie town ,Wudu of Gansu province	0.356±0.016	0.0813±0.0031	0.0304±0.0004	0.0577±0.0021	0.1601±0.0073	0.4639±0.0161
9	Wudu Dahongpao-2	Longfeng town, Wudu of Gansu province	1.231±0.062	0.0410±0.0020	0.0361±0.0012	0.0475±0.0026	0	0.3472±0.0080
10	Wudu Dahongpao-3	Wuyuhusng town, Wudu of Gansu province	0.439±0.020	0.0369±0.0018	0.0394±0.0012	0.0948±0.0010	0	0.4276±0.0172
11	Wudu Jiu yejiao	Majie Seedling Base of Gansu province	1.035±0.057	0.0315±0.0007	0.0500±0.0018	0.1619±0.0004	0.6126±0.016	0.7171±0.0172
12	Jinyang Qinghuajiao	Jinyang of Sichuan province	0.981±0.044	0.1031±0.0035	0.0359±0.0016	0.0852±0.0018	0	0.8354±0.0189
13	Hanyuan Qinghuajiao	Yidong town, Hanyuan of Sichuan province	1.264±0.038	0.1181±0.0009	0.0406±0.0002	0.0490±0.0015	0	0.2854±0.0564
14	Hanyuan Zhenglujiao	Sanjiao town, Hanyuan of Sichuan province	0.923±0.031	0.0750±0.0029	0.0409±0.0009	0.0693±0.0016	0	0
15	Hanyuan Dahongpao	Sanjiao town, Hanyuan of Sichuan province	0.559±0.023	0.0633±0.0029	0.0400±0.0001	0.0922±0.0006	0.0220±0.0011	0.3236±0.0178
16	Santai Qinghuajiao-1	Santai town, Mianyang of Sichuan province	0.846±0.031	0.1063±0.0043	0.0443±0.0013	0.0749±0.0014	0.4958±0.0268	0.8786±0.0485
17	Santai Qinghuajiao-2	Santai town, Mianyang of Sichuan province	0.576±0.023	0.0967±0.0014	0.0420±0.0018	0.0793±0.0015	0.1753±0.0071	0.3914±0.0097

Table 2. 4 inorganic contaminants and 2 pyrethroids detected in samples from Western China.

The monitoring contaminants	MRL (mg·kg^{-1})	Mean value (mg·kg^{-1})	Range (min-max)	No.of samples> MRL	No.of samples < DL
Inorganic contaminants					
Pb	1.86	0.8504	0 -1.7440	0	1(5.88%)
Cd	0.05	0.0768	0.0125-0.1355	11(64.71%)	0
Hg	0.03	0.0380	0.0192-0.0524	14(82.35%)	0
As	0.30	0.0712	0.0076-0.1619	0	0
Pyrethroids					
Permethrin	8.00	0.1147	0-0.6126	0	7(50%)
Cypermethrin	8.00	0.5136	0-0.8786	0	1(5.88%)

Table 3. EDI of the monitoring contaminants (Pb, Cd, Hg, As, permethrin and cypermethrin) in all samples and risk assessment.

The monitoring contaminants		ADI (μg·kg^{-1}bw day^{-1})	EDI (μg·kg^{-1}bw day^{-1})	HQ(%)
Inorganic contaminants				
Pb		3.571[a]	0.6803	19.05
Cd		1[a]	0.0614	6.14
Hg		0.7142[a]	0.0304	4.25
As		2.1428[a]	0.0569	2.65
Total EDI	0.8290μg·kg^{-1}bw day^{-1}			
Hazard index (HI)	32.09%			
Pyrethroids				
Permethrin		50	0.0917	1.83
Cypermethrin		20	0.4108	2.09
Total EDI	0.5025μg·kg^{-1}bw day^{-1}			
Hazard index (HI)	3.92%			

[a] PTWI/7

3.4 Estimation Intake and Risk of the Monitoring Contaminants Based on Samples

Contributions of *Zanthoxylum* samples to the total EDIs and HQs were calculated, and the data are shown in Table 4. HQ values for inorganic contaminants were from 7.46% for sample No. 3 (*Maoxian Dahongpao*) to 55.6% for sample No. 6 (*Henan Dahongpao*). For pyrethroids, pesticide residues were not detected in sample No. 14 (*Hanyuan Zhenglujiao*), and their HQ values were between 0 to 4.31%. In the present work, all samples had HQ of lower than 1, a safe level of *Zanthoxylum* that the presence of these inorganic contaminants and peticides is indicated.

It is worth noting that the dietary intakes of inorganic contaminants and pyrethroids estimated in the present study only showed possible exposures from *Zanthoxylum* but did not imply any other condiments, such as, black pepper, salt, amomum, and so on. In turn, they cannot be considered as 'total' dietary exposures for inorganic contaminants (pyrethroids) because fruits, vegetables, or grains were not considered. Moreover, other factors, such as the dietary or digestion process, were neglected. Chinese prickly ash seed is generally used as spice rather than direct consumption. Other uses of Chinese prickly ash can be in the form of powder or oil. Thus, HQ values presented in this study can be the overestimates of actual exposures to inorganic contaminants and pyrethroids.

The aim of this paper is to measure the content of four inorganic contaminants (including Pb, Cd, Hg, and As) and two pyrethroids (permethrin and cypermethrin) in Chinese prickly ash, as well as to assess the edible safety of Chinese prickly ash. The results show that nearly all samples contained inorganic contaminants. As indicated by mean values, Pb was the highest, while Hg was the lowest, in consistent with the previously reported data [25]. It is known that heavy metals have impacts on the human body once they enter the food chain [27]. Polluted water and soil produce a greater effect on the content of heavy metals in crops [28]. Moreover, Chinese prickly ash, family Rutaceae, is the perennial plants that are enriched with harmful elements. We should therefore control the sources of heavy metals, especially Cd and Hg, the measured content of which were above MRLs.

In addition to a few other research that have studied pesticide residuals in Chinese prickly ash, the measured contents of two pyrethroids presented in this work were lower than the reference MRLs. Additionally, more than half of samples were positive for such peticides. Although the HQ values of Chinese prickly ash were low, the residual pesticides in Chinese prickly ash should not be ignored. This study only present two common pyrethroid pesticides, the assessment of edible safe is therefore partial. Potential exposure to other hazards may also include organophosphorus and carbamate pesticide residuals, and so on. Nowadays, consumers advocating pollution-free food urge that use of pesticides should be reduced, while use of a more biological pest control should be implemented.

It is worth noting that the dietary intakes of inorganic contaminants and pyrethroids estimated in the present study only showed possible exposures from *Zanthoxylum* but did not imply any other condiments, such as, black pepper, salt, amomum, and so on. In turn, they cannot be considered as 'total' dietary exposures for inorganic contaminants (pyrethroids) because fruits, vegetables, or grains were not considered. Moreover, other factors, such as the dietary or digestion process, were neglected [29]. *Zanthoxylum* fruit is generally used as spice rather than direct consumption. Other products of *Zanthoxylum* can be in the form of powder or oil. Thus, HQ values presented in this study can be the overestimates of actual exposures to inorganic contaminants and pyrethroids.

Table 4. DI and risk assessment for each samples.

No.	**4 Inorganic contaminants**		**2 Pyrethroids**	
	EDI[a] (μg·kg^{-1}bw day^{-1})	HQ (%)	EDI[b] (μg·kg^{-1}bw day^{-1})	HQ (%)
1	1.4206	51.76	0.0098	0.49
2	0.5677	28.00	0.5811	2.73
3	0.0864	7.46	0.5300	2.56
4	0.8624	33.93	0.5139	2.52
5	0.7634	26.99	0.8630	3.58
6	1.5776	55.60	0.6258	3.13
7	0.6731	22.41	0.7211	2.87
8	1.0056	20.03	0.4992	2.11
9	1.0844	36.67	0.2777	1.39
10	0.4880	20.73	0.3420	1.71
11	0.9987	33.39	1.0637	3.85
12	0.9641	37.42	0.6683	3.34
13	1.1773	44.14	0.2283	1.14
14	0.8865	33.84	0	0
15	0.6036	25.50	0.2764	1.33
16	0.5872	35.21	1.0995	4.31
17	0.6352	28.30	0.4533	1.85

[a]EDI of each sample was calculated as the sum of the EDI for 4 inorganic contaminants in that sample.
[b]EDI of each sample was calculated as the sum of the EDI for 2 pyrethroids in that sample.

4. Conclusion

The work demonstrated that although Pb, Cd, Hg, and As were detected in almost all *Zanthoxylum* samples in Western China with high frequencies, their EDIs values were lower than ADIs, and HI values were well below 1. Such values are adequately safe for consumers, not at the hazard levels that can cause harm to health. Nonetheless, the risk due to contaminated food should be monitored, and the pollution of soils and irrigating water should be regulated so that inorganic contaminants or pesticide residues in food (including *Zanthoxylum*) remain at the safe levels. Based on the results for pesticides determination, the MRLs for permethrin and cypermethrin in *Zanthoxylum* should be urgently pay more attention in food security.

Acknowledgements

This work was financial supported by the Scientific Fund of the Application Fundamental Project [2016JY0118], the Key Research & Development Project [18ZDYF1175] of the Science & Technology Department of Sichuan Province, China.

References

[1] KUMAR A, NEGI N, HAIDER S Z, et al. Composition and Efficacy of *Zanthoxylum alatum* Essential Oils and Extracts Against Spodoptera litura [J]. Chemistry of Natural Compounds, 2014, 50(5): 920-923.
[2] LI K, ZHOU R, WANG JIA W, et al. *Zanthoxylum bungeanum* essential oil induces apoptosis of HaCaT human keratinocytes [J]. Journal of ethnopharmacology, 2016, 186: 351-361.
[3] KE J, QU Y, LI S, et al. Application of HPLC fingerprint based on acid amide components in Chinese prickly ash (*Zanthoxylum*) [J]. Industrial Crops and Products, 2018, 119: 267-276.
[4] KE J X, QU Y, LI S S, et al. Application of HPLC fingerprint based on acid amide components in

Chinese prickly ash (*Zanthoxylum*)[J]. Industrial Crops & Products, 2018, 119:267-276.

[5] KE J, QU Y, LI S, et al. Establishment of GC-MS Fingerprint Based on Essential Oil Components in *Zanthoxylum* and Application on Hanyaun *Zanthoxylum bungeanum* Maxim. [J]. Journal of the Chinese Cereals and Oils Association, 2018, 33(11): 116-126.

[6] YIN H, WANG X, DOU X, et al. Review and Out look of *Zanthoxylum bungeanum* Processing for Medicinal Use [J]. Shanxi Forest Science and Technology, 2018, 46(4): 102-105.

[7] FAN Y, MA H. Determination of 20 kinds of pesticide residues in honey by QuEChERS- gas chrom atography-tandem mass spectrometry[J]. Journal of Food Saftry and Quality, 2019, 10(10): 3106-3113.

[8] AN M,YE H, XIE Z,WEI J, ZHANG Q, HUANG J. Safety Assessment of Agricultural Residues and Heavy Metals in Wine-making Grain Bases[J]. Liquor-making technology, 2019, 2(26).

[9] FENG C Y, BAO J Q, FANG X M, et al. Investigation and analysis of pesticide residues for commercially available fruits and vegetables in Shanghai[J]. Acta Agriculturae Zhejiangensis, 2015, 27(3).

[10] JUAN-BORR S M, DOMENECH E, ESCRICHE I. Mixture-risk-assessment of pesticide residues in retail polyfloral honey [J]. Food Control, 2016, 67:127-134.

[11] SHAO J M, YANG P, FAN Z Z, FENG H K, CHEN B, YE J M. Fabrication and analytical application of centrifugal microfludic chip for rapid detection of pesticide residues[J]. Food science, 2017, 12: 229-304.

[12] AKBARI B, GHARANFOLI F, KHAYYAT M H, et al. Determination of heavy metals in different honey brands from Iranian markets [J]. Food additives & contaminants Part B, Surveillance, 2012, 5(2): 105-111.

[13] OKUDA T , KOSHI N , MATSUMURA A , et al. Validation Study on a Multi-Residue Method for Determination of Pesticide Residues in Agricultural Products by New Automatic Pretreatment Equipment (FASRAC) and GC-MS/MS[J]. Food Hygiene and Safety Science (Shokuhin Eiseigaku Zasshi), 2014, 55(5):216-229.

[14] CHINA. N H O. National food safety standard of China [M]. Determination of lead in foods. Beijing, China; National Health of China. 2010.

[15] CHINA N H O. National food safety standard of China [M]. Determination of cadmium in foods Beijing, China; National Health of China. 2014.

[16] NATIONAL HEALTH OF CHINA. National food safety standard of China [M]. Determination of arsenic in foods. Beijing, China; National Health of China. 2014.

[17] CHINA N H O. National food safety standard of China [M]. Determination of mercury in foods. Beijing, China; National Health of China. 2014.

[18] CHINA S A O. Determination of organochlorines and pyrethroid pesticide multiresidues in vegetable foods [M]. Beijing, China.; Standardization Administration of the People's Republic of China. 2008.

[19] CHINA S A O. Prickly ash [M]. Beijing, China.; Standardization Administration of China. 2013.

[20] YUN D , CHENG Q , JIAN L I, et al. Organophosphorus pesticide residues in Hainan bitter gourd,Chinese flowering cabbage and cowpea and dietary exposure risk assessment[J]. Journal of Southern Agriculture, 2014, 45(9): 1594-1598.

[21] JMPR. Inventory of evaluations performed by the Jion Meeting on Pesticide Residues(JMPR) [M]. Availableonline:<http://apps.who.int/pesticide-residues-jmpr-database> JMPR. 2012.

[22] JIN S. Survey report on the nutrition and health of China residents:Date set on the status of nutrition and health in 2002 [M]. People's Hygiene Press. Beijing,China; Nutrition and health data set. 2008.

[23] SALUNKHE V P, SAWANT I S, BANERJEE K, et al. Enhanced Dissipation of Triazole and Multiclass Pesticide Residues on Grapes after Foliar Application of Grapevine-Associated Bacillus Species [J]. J Agric Food Chem, 2015, 63(50): 10736-10746.

[24] ACAR O, TUN ELI A, T RKER A R. Comparison of Wet and Microwave Digestion Methods for the Determination of Copper, Iron and Zinc in Some Food Samples by FAAS [J]. Food Analytical Methods, 2016, 9(11): 3201-2308.

[25] ZHENG Y N, FAN M T, GUO S N, et al. Determination of Toxic and Essential Metal Element Contents in Chinese Prickly Ash of Shaanxi [J]. China Condiment, 2012,

[26] ORGANIZATION F A, ORGANIZATION W H. Summary of evaluations performed by the joint FAO/WHO Expert Committee on Food Additives (JECFA) [J]. Nutrition Reviews, 2010, 58(3): 90-90.
[27] HUANG Z, PAN X-D, WU P-G, et al. Heavy metals in vegetables and the health risk to population in Zhejiang, China [J]. Food Control, 2014, 36(1): 248-252.
[28] ZHOU Y, XUE L B, YI S M. Reusing of the proficiency testing results of organophosphorus pesticide residues in food [J]. Journal of Food Safety and Quality, 2019, 10(7): 2043-2046.
[29] TIAN J R, CHEN W, ZHANG C J, SUN X S, TAN T T, REN J. Effect of Different Pre-treatment Methods on Detection Results of Organophosphorus Pesticide Residue Vegetables [J]. Modern Food Science and Technology, 2013, 29(3): 664-667.

IOP Conf. Series: Materials Science and Engineering **611** (2019) 012065 doi:10.1088/1757-899X/611/1/012065

Research on the Integration Design of Double Well Head Drilling System for Semi-submersible Drilling Platform

Weixu Wang[1,2], Guangchan Ma[1], Xiuju Yang[1,2], Tianming Zhou[1,2], Peng Zhang[1,2], Yunpeng Li[1,2], Nan Hu[1,2] and Anyi Wang[1,2]

[1]Baoji Oilfield Machinery Co., LTD, Baoji, China
[2]National Oil&Gas Drilling Equipment Research Center, Baoji, China

Abstract. In order to upgrade the capability in offshore drilling and exploring, it is imperative to design and study deep water drilling system. Firstly, surveying and researching the present situation and development status of deep water drilling system. Secondly, analysing the characteristics and configuration of the deep water drilling rig used for semi-submersible drilling platform, also selecting key equipment. Finally, designing and coming up with a dual-frame derrick drilling system with 3660 meters water depth and 15250 meters drilling depth, compared with the trade technology, its advantage: achieving top drive core and realizing continuous drilling, allowing duplex main wellhead operation, the two wellhead could be working at the same time, and a well is being used to piping, and another well is being used for drilling. As a result of saving the non-drilling time and increasing the drilling efficiency. It is of great significance to the development of marine energy sources.
Keywords: double wellhead operation; ocean drilling system; double derrick drilling rig; drilling rig selection; semi-submersible drilling platform.

1. Introduction

With the rapid development of global economy, the demand for Marine oil and gas is increasing sharply, but nearly 20 years, with the exhaustion of the oil and gas resources of the land, the world's deepwater oil and gas field exploration and development results emerge in endlessly, of which more than 70% in the northern gulf, southeast of Brazil and west Africa three deep waters, the deepwater oil and gas recoverable resources in three major areas of global deepwater oil and gas is 40%~50% of the total recoverable resources[1].

As the development of deepwater oil and gas field size and water depth increasing, the deep ocean engineering equipment and technology rapid development, accelerating the development process of Marine resources. For a long period of time, Marine oil and gas resources development around the world has become the important development goals and it is effective powerful weapon and the important sign of Marine oil and gas resources development. And the shallow sea exploration and development means has become increasingly mature, the development of deepwater drilling equipment is imperative [2][3].

Deepwater drilling equipment can provide technical support for the design, construction and application of deepwater drilling platforms, and the main factor affecting the efficiency of deepwater drilling system is the form of derrick. Currently, most ultra-deepwater drilling platforms with operating water depth greater than 5000ft in the world adopt semi-derrick and dual-derrick rigs, while those with operating water depth less than 5000ft basically adopt single derrick. This drilling system with double derrick drilling rig can improve drilling efficiency [4].

IOP Conf. Series: Materials Science and Engineering **611** (2019) 012065 doi:10.1088/1757-899X/611/1/012065

2. Drilling System Design Process

Taking the needs of ultra-deep water drilling operations as the goal, the key parameters of drilling modules, circulation systems, solid control systems, etc. of the drilling system and the matching calculation of the system are carried out, then the configuration and selection of drilling systems are completed. Establish modular dynamic system, circulation system, solid control system and other equipment performance, structure parameter configuration database. Combined with the analysis of the mechanical behavior of the ultra-deep water riser under various working conditions, it provides theoretical basis for the type selection, design and operation of the drilling system equipment [5]~[7].
Calculating the equipment layout and completing the layout design of the drilling system. According to the optimization strategy of function zone module planning, operation cost planning and reducing the barycenter of the platform.
Carrying out analysis of the interface system between drilling systems and drilling platforms, and comprehensively analysing the man-machine interface, electro-mechanical interface, intelligent interface, and gas-liquid interface between drilling systems and drilling platforms or drilling vessels.
Achieving the matching analysis and control channel requirement design for power module and hydraulic module of drilling system. In order to determine the demand of power, hydraulic and control signals for the equipment matching under different operating conditions, analysing the correlation between the input interface of drilling system and the output interface of drilling platform, and studying the interconnection relationship.
Finishing the functional design of the drilling system input interface and the drilling platform output interface, including the performance design of the power interface, human-machine interface, and gas-liquid interface, and the study of the communication interface protocols and standards.
The overall design technical route of the drilling system is shown in Figure 1.

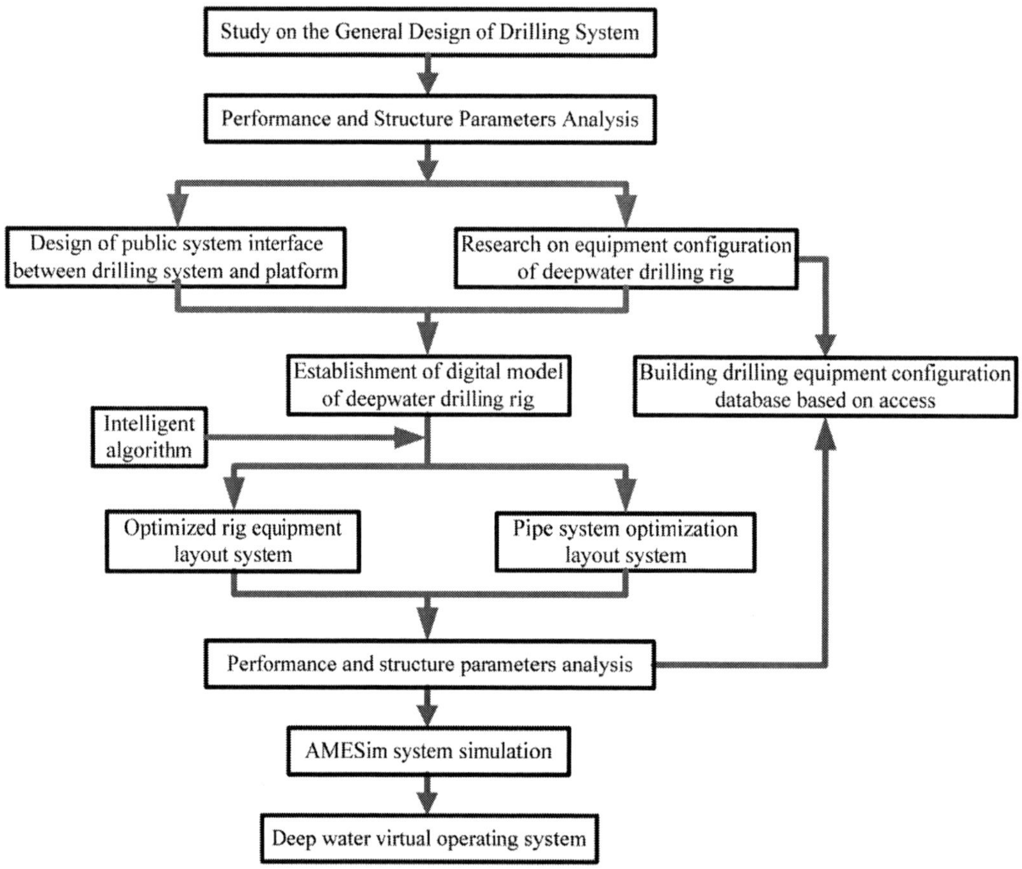

Figure 1. General design technical route of drilling system.

3. Drilling Rig System Configuration and Equipment Options

At present, the ocean deep water rig refers to a floating ocean platform rig that is used for operating water depth of more than 500 meters. At present, there are three types of deep water rig [7]: AC inverter rig, Ram rig (hydraulic rig) and DMPT rig (double column multifunctional tower rig). The DMPT rig is a new type of rig.

Due to the high power of deep water rig equipment and the limitation of deck load and deck space, the low volume and high power AC frequency converter equipment is used in deep water rig. For AC frequency conversion drilling rig, the compensation device (such as crown block, travelling block, drawworks, dead rope compensation device, etc.) must be used to compensate the sinking motion of the drill string. The AC frequency rigs are widely used [8] [9]. Most of the new ocean rigs use this type and have good experience in installation, commissioning, use and maintenance. The first deep-water semi-submersible drilling rig HYSY981 also adopts this type of rig to adapt to the working water depth of 3,660 meters. A semi-submersible drilling rig with a drilling depth of 15,250 metres is recommended for this type of rig.

The main systems of the rig include: lifting system, compensation system, circulatory system, rotating system, underwater system and solid control system. The selection of mast type and compensation device has a relatively large impact on the drilling system and is the main influence factor for the rig configuration selection, especially the mast type. It is one of the most important effecting factors on the cost, weight and working efficiency of deep water drilling rig.

3.1. Hoist System

Used for completing drilling, lifting, rising and dropping systems such as drilling, casing, tubing, riser, BOP(blowout preventer) or pumping rod string, including key equipment such as derrick, travelling system, and drawworks, etc.. The derrick used on the semi-submersible drilling platform is an ocean dynamic derrick. The derrick is use of placing crown block, travelling block, hook, and top drive, installing top drive guide rail and storing pipe string.

At present, there are three types of derrick, such as single derrick, double derrick and one half derrick, which are suitable for deep water semi-submersible drilling rig. Among them, the double derrick rig on the deep water semi-submersible drilling platform has two types of operating system: main auxiliary wellhead and double main wellhead. Deeper water drilling platforms should use double derrick deep water drilling rigs. For example, for deep water semi-submersible drilling platforms with a maximum operating depth of 3,660 meters, the time for landing riser and BOP are longer. In order to improve operational efficiency, double derrick should be used.

The vital equipment of the lifting system is the derrick and drawworks. The selection of the height of the derrick and the input power of the drawworks is the most critical choice index of the system.

3.2. Rotation System

Used for rotating the drilling tool in the well to continuously break the rock, including a rotary table and top drive device. A new generation of high torque AC frequency top drive is commonly used in deep water drilling rigs. The top drive lifting load is matched with the maximum hook load of the drill rig, and the scooter can have telescopic function. The diameter of the rotary table opening shall not be less than 1536.7 mm, so that when the pipe is decentralized, the unilateral gap between the rotary table opening and the outer wall of the pontoon shall not be less than 76.2 mm to 127 mm. For the seventh-generation deep-water semi-submersible drilling platform, due to its operating depth of 3,660 m, riser with a diameter greater than 1371.6 mm will be deployed, so an open diameter 1917.7 mm rotary table should be used.

3.3. Drilling String Compensation System

The semi-submersible drilling platform must be equipped with compensation system to eliminate the impact of the platform sinking movement on the drill string to ensure constant drilling pressure and ensure that the drill tool can continue to drill during the platform sinking movement and improve drilling efficiency. The compensation system is divided into passive compensation type and active compensation type. In addition, according to the different installation locations of the compensation mechanism, the drilling compensation is divided into crown block model, travel block model and hoisted model compensation. At present, there are two kinds of commonly used lifting compensation methods for deep-water drilling rigs: active lifting compensation for sky models, and compensation for hoisted models (active compensation drawworks). Both ultra-deep water semi-submersible drilling platforms can choose two compensation types of them. When using a double derrick, if you need to reduce the barycenter of the entire platform, you can use an active compensation drawworks to compensate for the trip.

3.4. Circulatory System
The circulatory system mainly includes high-pressure mud pumps and low-pressure centrifugal pumps. Deep water drilling systems are generally equipped with 3 to 4 high-pressure mud pumps with 2,200 hp, which are driven by AC frequency Motors. Deep water drilling systems use mud pumps with a power of 2,200 hp and a pump pressure of 7500 psi (mud pumps with gemstone machinery can be used to realize equipment localization). The deep water semi-submersible drilling platform has large operating depth and hook load, and is equipped with dual-frame derrick. Equipped 5 to 6 mud pumps. The semi-submersible drilling platform with working water depth of 3,660 meters and drilling depth of 15,250 meters is recommended to be equipped with 6 mud pumps. Four pumps used and two pumps reserved.

3.5. Pipe String Handling System
Composed of a number of mechanized, automated and intelligent equipment. It completely or partially replaces manual operations during the drilling string processing operation. It is used to complete all or part of the transportation, connection, placement, etc. of drilling pipe string such as drill pipe, drill collar, or casing, and is divided into pile field area equipment and drill platform area equipment. Pipe string yard equipment mainly includes pipe crane, power cat walk, and requires maximum operating area of the crane cover pipe pillar yard area. And its maximum processing capacity should be extended to be suit for riser. The equipment of the drilling area includes iron roughneck, racking pipe devices, pipe string lifting devices, support devices, and clamping devices.

3.6. BOP and X-mas Tree Handling System
At present, the diameter of the deep-water blowout preventer group (following abbreviation, BOP) is 476.25 mm, and the rated working pressure is (15000 to 20000) psi. Deep water semi-submersible drilling platforms can be equipped with universal BOP with 10000 psi pressure level or 7 to 8 gate blowouts with 20000 psi pressure. The BOP includes the LMRP (lower marine riser package), the middle connector, the lower blowout preventer group, and the blowout preventer combination frame.
According to the different spatial arrangements, BOP and the X-mas tree processing system can share the same processing system, or use two separate systems. When a common system takes into account, its maximum processing capacity is designed according to the weight of BOP. When two sets of independent systems are used, one set is used for processing BOP and another set is used for processing X-mas trees.

3.7. Solid Control
Deep water drilling rigs generally adopt solid control systems of level 5 or above: scrapers, shale shaker, degasser, mud cleaner desander (generally configured according to specific circumstances), desilter, centrifugal pumps, etc.. The parameters and configuration of the solid control system are mainly determined by the drilling rig cycle displacement (mud pump maximum displacement) and mud treatment process.

4. Designing a Deep Water Drilling System

4.1. Major Equipment Arrangement
The drilling system of the platform is centered on the rig module, and the entire rig module is arranged in the central position of the platform. For dual-frame derrick rig, the center position of the platform adopt the center position of the derrick or the center position of the main rotary table. The central part

of the platform is monthly pool, for the decentralization and recovery operation of underwater equipment. The mud pump, daily mud pool, pulp distribution equipment, bag store, BOP control system equipment, etc. are set in the special cabin of the hull on the platform.
Drilling equipment is mainly arranged in several areas [10] ~[13]: tube yard deck (upper deck), drilling platform area, moon pool area, lower deck and other areas.

4.1.1. Pipe Deck /Upper Deck Area
Upper deck area have a large number of drilling equipment [14], including riser crane, BOP crane, X-mas tree crane, solid control equipment rooms, power catwalk (should be part of the automated pipe handling system), folded crane, piping machine, and pipe yards, etc..

4.1.2. Drill Floor Area
The drilling floor area include drillers' cabin, local instrument rooms, derrick frame, lifting systems, rotary table, chock and kill manifold, standpipe manifold, and drill platform equipment such as iron roughneck, Iron derrick man, power Slip and mud stuffing box, etc.. The drawworks is also generally arranged in the drill floor area.

4.1.3. Moon Pool Area
Moon pool area placement of moon pool equipment: riser tensioner, high pressure air accumulator, accumulator, nitrogen tank, control skid, moon pool winch, blowout preventer and X-mas tree pulley, slide and lower blowout preventer guide device (mounted on the bottom of the moon pool).

4.1.4. Lower Deck
Lower deck set up high pressure mud pumping room, small pit, mud purification equipment, power system equipment are all arranged on the.

4.2. Drilling Pipe Pile Layout

4.2.1. Drilling Pipe and Casing
The horizontal pipes are laid on the upper deck, and the verticals are on the setback area. The pipe string pile is placed in the stern area of the main deck. Both the pipes are stacked on the pile in a horizontal storage manner. When used, the pipes can be transported, connected and placed into the setback area by using the horizontal and vertical string processing system equipment[10].

4.2.2. Riser
The marine riser has three arrangement ways for semi-submersible drilling platforms: horizontal separately, both of horizontal and vertical, vertical separately. Different placement methods affect the layout of the platform, the riser handling project and its flowing and operating efficiency.
Firstly, the riser is level. The horizontal riser need to be changed to the vertical state by the conveyor device before it will be sent to the rotary table wellhead. The horizontal placement riser is generally located on the main deck, and its area can be used in combination with the casing storage area. Secondly, some is horizontal and another is vertical. The horizontal risers are placed on the main deck pipe yard. When used, they need to be connected and transported to the wellhead through treatment equipment. The vertical risers can be placed in the vertical placement area, when used, they are directly transported to the wellhead. Finally, the riser is vertical, it can improve the efficiency of the riser landing and even the overall operation, so that the riser can be laid down the upper deck below, is placed on the lower deck, reducing the barycenter. Recommending to use the finally arrangement for the semi-submersible platform.

4.3. The Drilling System Configuration Designing
Based on the above, designing the configuration of the deep-water drilling system for ultra-deep water semi-submersible drilling platforms with a depth of 3,660 meters and a depth of 15,250 meters, Compared with trade technology as shown in table 1, layout scheme as shown in Figure 2.

Table 1. Comparison table with designing and trade technology.

Name	Parameters		Comparative result
	Designing schema	others	
Projected depth	3660 m	3660 m	equivalence
Depth-drilled	15250 m	15250 m	equivalence
Maximum hook load	1150 t	1134 t	equivalence
Table opening Dia	1917.7mm(75.5 in)	1917.7mm(75.5 in)	equivalence
Mud pump	quantity and power: 4+2(reserved), 2200 hp	quantity and power: 4+2(reserved), 2200 hp	equivalence
BOP/X-tree handling system	BOP crane: 600 t X-tree crane: 150 t	BOP crane: 550 t X-tree crane: 120 t	precedence
Drill string handling system	Drill string transport capacity, 2-7/8 in≤D≤36 in Piping capacity, 2-7/8 in≤D≤14 in	Drill string transport capacity, 2-7/8 in≤D≤36 in Piping capacity, 2-7/8 in≤D≤14 in	equivalence
Riser handling system	Verticality place, single height 130 ft	Verticality place, single height 110 ft	precedence
Subsea BOP stack	Rated working pressure: 20000 psi	Rated working pressure: 15000 psi	precedence

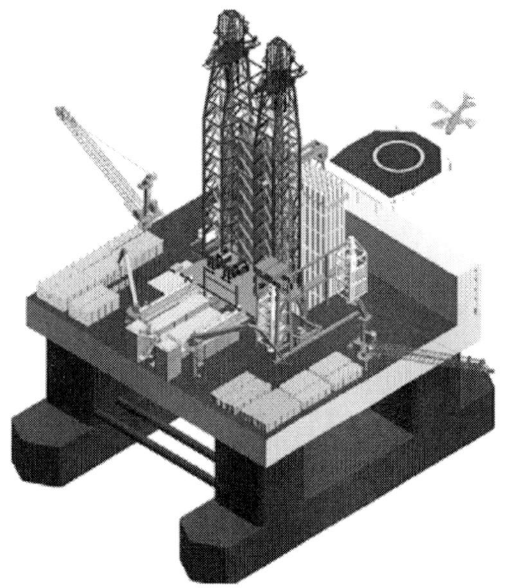

Figure 2. Distribution of drilling system.

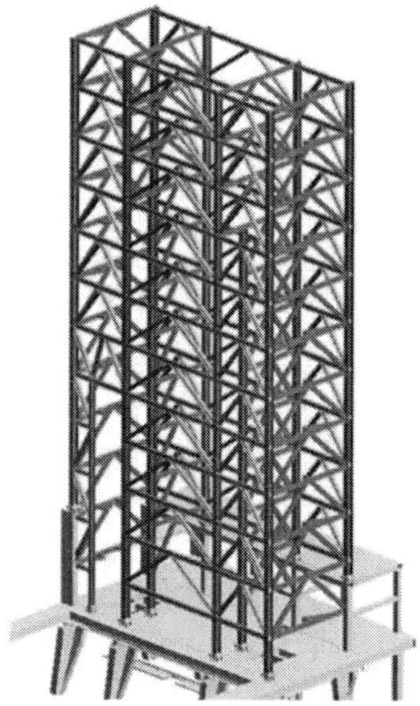

Figure 3. Dual-frame derrick.

4.3.1. Dual-frame Derrick

The derrick adopts a dual-frame, which adopts a large-span structure to separate the drilling operation area on both sides of the derrick from the maintenance of the equipment in the middle. This facilitates drilling operations and equipment maintenance. Its structural performance is approved, and the dual-frame derrick structure is as shown in Figure 3.

As equation (1), the effective height H_M of the derrick mainly depends on the drilling operation of the rig. When the travelling block is up to the highest position, it can still ensure sufficient safety height, which is closely related to the height of the standpipe, the length of the travelling block and the top drive, the compensation displacement, and the collision protection distance.

$$H_M = h_{TP} + l_{TL} + h_{CPD} + h_{CPH} \tag{1}$$

IOP Conf. Series: Materials Science and Engineering **611** (2019) 012065 doi:10.1088/1757-899X/611/1/012065

In the above formula:

h_{TP}, The effective height of the standpipe, take 41.15 metres.

h_{TL}, The actual height between travelling block and top drive, take 13 metres.

h_{CPD}, Top safety anti-collision space, take 4.5 metres (including 3.81 metres compensation trip).

h_{CPH}, The bottom safety anti-collision space, take 4.5 metres (including 3.81 metres compensation trip).

So, $H_M = 63.15\ m$, determine the effective height of the derrick 64 metres.

4.3.2. Substructure Height

The substructure height is one of the most important parameters for substructure design. The factors to be considered include: mud backflow pipe angle, riser tensioner swing angle, the BOP and X-max tree handling net space, etc.. At the same time, the substructure height has a greater influence on the barycenter for the drilling system. As equation (2), the substructure height H can be obtained.

$$H = h_1 + h_2 + h_3 \tag{2}$$

In the above formula:

h_1, The height of mud diverter inlet, $h_1 = 3\ m$ in figure 2,

h_2, Vertical height difference between the mud outlet and the mud diverter inlet. The length of the backflow pipe between the mud outlet and the mud diverter inlet is L, and the tilt angle φ is generally not less than 5°, so, $h_1 = L \times sin\varphi = 2.62\ m$.

h_3, The height difference from Mud outlet to the drill floor. Taking into account the actual dimensions of rotary table, rotary table beam and diverter, the height difference h_3 is not less than 3 m.

Therefore, the drill floor height $H = 3\ m + 3\ m + 3\ m = 8.62\ m$, so, it is reasonable to determine a 9 metres high drill floor height.

In addition, need to further verify when the riser tensioner is fully extended and the swing angle is 3°, guarantee to have sufficient safety space between the tensioner and the side wall of the moon pool.

4.3.3. Lifting System

The lifting system adopts variable frequency drawworks and travelling system with compensation, and the drawworks placed on the outside of the derrick, and its performance is reliable.

The main technical parameters of Semi-active crown block compensation device for drilling system include maximum static load and maximum compensation load. The maximum static load of the drilling system should not be less than the maximum hook load of the drilling system.

The maximum compensation load of the crown block compensation device during drilling as equation (3) below.

$$Q_{\max} = Q_{ST} + Q_{TR} - F_{FL} \tag{3}$$

In the above formula:

Q_{ST}, maximum string weight, 7250 kN

Q_{TR}, travelling system weight, 700 kN

F_{FL}, pipe string flotation, $F_{FL} = \rho g V = 1500 \times 9.8 \times 70.24 \approx 1030$ kN

So, $Q_{\max} = 7250 + 700 - 1030 = 6920 kN$

So, The maximum static load of the crown block compensation device is 11500 kN and the maximum compensation load is 6920 kN.

4.3.4. Rotation System

Mainly includes rotary table and top drive.

- Rotary table

The rotary table is the main load-bearing and rotating part of ocean rig equipment. Bearing the weight of all the drill pipe string when lifting the drill pipe string. Bearing the weight of BOP and underwater equipment When decentralizing BOP and underwater equipment. It can be used for the rotation of drilling tools when dealing with underground accidents in emergency situations. When choosing a rotary table, it mainly considers the opening diameter and the maximum static load of the rotary table.
The rotary table opening diameter must ensure the smoothly handling the riser. At present, the rotary table diameter of the deep water drilling platform is more than 60.5 in (1536.7 mm), with the operation depth increasing, the rotary table opening diameter will continue to increase. The outer diameter of the floating material of the riser supporting exceeds 60 in(1524 mm). The safe operating gap is usually 5"(127 mm) when decentralizing the riser. So the rotary table opening diameter is 75.5 in (1727.2 mm).
The maximum static load of the rotary table is the maximum axial load that the rotary table can withstand, and its value should not be less than the maximum hook load 11,500 kN.
The rotary table is driven directly by a hydraulic motor, which has the characteristics of integrating mechanical, electrical and hydraulic design, small size, light weight and strong function.

- Top drive

The cost of offshore drilling is high. In order to improve drilling efficiency, the top drive is used instead of the swivel and rotary table. And the main parameters of the top drive include continuous working torque and maximum hook load. Having telescopic capability.
Continuous operating torque of the top drive should be equation as (4).

$$M_{\max} \geq 10^{-2} \cdot L_{\max} - 5 \tag{4}$$

In the above formula, L max is the upper limit of the nominal well depth, which is 15250 metres. So, Continuous operating torque of the top drive should be equation as (5).

$$M_{\max} \geq 10^{-2} \times 15250 - 5 = 147.5 \ kN \cdot m \tag{5}$$

The top drive maximum static load is the maximum lifting capacity that the top drive can withstand. The maximum static load of the top drive is the same as the hook load, which is 11,500 kN.
In order to further improve the working efficiency of the rig travelling system, the pipe string automatic processing system can realize the continuous operation. During the top drive moving upwards from the drill floor, the top drive needs a retractable guide rail frame to effectively avoid the wellhead.

4.3.5. BOP and X-max Tree Handling Systems

The equipment of the BOP and X-max tree handling system is deployed on the upper deck.

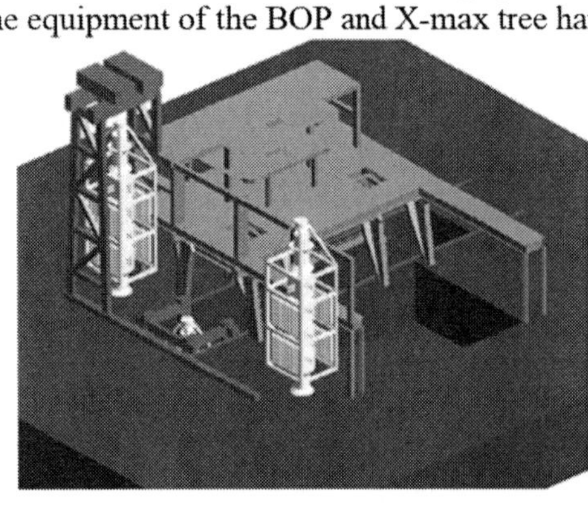

Figure 4. BOP handling system layout.

Figure 5. X-max tree handling system layout.

Equipped with two sets of BOP, sharing one LMRP, reserving with each other, with a maximum working pressure of 20000 psi, using the BOP platform design with self-guided function and hydraulic step slider, having the advantages of simple operation and high reliability. The layout is as shown in figure 4.
Equipped with 5-6 sets of X-max trees, X-Y two-way sliding and steering systems, hydraulic systems, electronic control systems, etc. to achieve a series of operations such as removing and transporting ultra-

large X-max trees from the storage area to the wellhead. The X-max tree treatment equipment can realize the X-Y two-way slip. The layout plan is as shown in figure 5.

4.3.6. Automatic Handling System

Specially equipped with fingerboard for storing riser, bridge crane and pallet for transporting riser. Two risers for one string, through the cabinet standing on the lower deck.

Drilling standpipe placed on the inside of the derrick, it can save space and reduce the piping requirements.

The drilling system is efficiently operated by the automatic treatment system and drill integrated control system, etc. Realizing automatic handling system.

5. Conclusion

Describing the general design technical route of the drilling system, analysing the deep water drilling rig, determining the type of deep water drilling rig, and expounding the characteristics of the key equipment, finally determining a drilling system design and layout scheme as shown in figure 2. To design and deploy the drilling system of the deep water rig in the platform precise shift and steering ability, riser column safety monitoring ability, automatic pipe processing and storage capacity, monthly pool opening and heavy equipment transportation capacity, mud pool storage capacity, etc.. The development of marine exploration is changing rapidly, and the localization of marine equipment highlights its importance. In the design of offshore platform drilling systems, the superiority of high-end localized equipment is highlighted. The innovation and upgrade of localized equipment will lead to a new technological breakthrough in marine oil exploration and development.

Acknowledgments

Fund projects: Ministry of Industry and Information Technolog " Seventh Generation Super Deep Water Drilling Platform(Ship) Innovation Special"-"Application Research of Drilling Package Integration and Some Key Equipment" Project(Ministry of Industry and Information Technology Joint installed[2016]24).

References

[1] Li Zhang. Research on Key Technologies and Support Technologies of Marine Engineering Equipment [J]. Chemical Industry Management, 2019 (17): 120-121.

[2] Yuqing Huang, Hai Lan,Yingyi Hong, Shuli Wen, He Yin. Optimal generation scheduling for a deep-water semi-submersible drilling platform with uncertain renewable power generation and loads[J]. Energy,2019,181.

[3] Lina Shao. Research on the Development Trend of Semi-submersible Platform [J]. Ship Standardization and Quality, 2018 (06): 42-47.

[4] Yvette. Bluewhale 1 Ultra-Deep Water Double-Tower Water Semi-submersible Drilling Platform [J]. Design, 2019,32(02): 32-35.

[5] Anyi Wang, Xiuju Yang, Chunming Fan, Peng Zhang, Zhitong Liu. Analysis of key operation technology of double derrick drilling rig for ultra-deep water semi-submersible drilling platform [J]. Ship, 2019,30 (01): 69-73.

[6] Nan Hu, Anyi Wang, Xiuju Yang, Zhitong Liu, Peng Zhang. Overall scheme study of drilling rig module for ultra-deep water semi-submersible drilling platform [J]. Ship, 2018, 29 (03): 73-79.

[7] Bin Xie, Yang Li, Wei Zhang, Wenhui Xie, Junrong Wang. Ultra-deep water semi-submersible drilling platform design technology innovation and application [J]. Marine Engineering Equipment and Technology 2015, 2 (6): 353-360.

[8] Yi Wang, Menglan Duan, Xiaonan Jiao. Current status and outlook of deep-water oil and gas development equipment development [J]. Petroleum Machinery, 2013, 41 (10): 51-55.

[9] Anyi Wang, Nan Hu, Xiuju Yang, Yuanbo Liu, Peng Zhao. Study on configuration and layout of pipe string processing system for ultra-deep-water marine drilling rig [J]. Ship, 2017, 28 (05): 87-91.

IOP Conf. Series: Materials Science and Engineering **611** (2019) 012065 doi:10.1088/1757-899X/611/1/012065

[10] Wensheng Xiao, Lei Wu, Xue Tian, Keke Wei. Study on the layout of drilling equipment on the upper deck of deep-water semi-submersible drilling platform [J]. Petroleum Machinery, 2014, 42 (5): 44-48.
[11] Yong Wang, Shulin Liu, Wei Du, etc.. Marine drilling platform tube transfer manipulator [J]. Machinery manufacturing, 2014 (01): 14 -17.
[12] Qiuhong Song, Haohui Wang, Shaorong Xu, Jinni Lu. The status quo and outlook of the design and development of the crane [J]. Machinery Manufacturing, 2015, 03:72 -76.
[13] Xuezheng Cui, Pingquan Liu, Wensheng Xiao, etc.. Marine drilling platform automated cat path design [J]. Oil Mine Machinery, 2011, 40 (2): 20-23.
[14] Xiuju Yang, Anyi Wang, Nan Yang, Nan Hu, Yunpeng Li. Study on Pipe Stack Layout Scheme of Deepwater Semi-submersible Drilling Platform [J]. Ship, 2018, 29 (05): 64-74.

Nonlinear Dynamics and Stability Analysis of Gas Turbine Rotor System Considering Eccentricity

Shiwei Wu, Rui Zhu, Guangchao Wang, Anlei Zhao and Xin Xia

Energy and Mechanical Engineering, Shanghai University of Electric Power, Shanghai, China

Abstract. According to the actual model of the gas turbine rotor system, a double-disc rotor-bearing system model is established. The appropriate parameters are selected, and the integral is solved by numerical integration method. The effects of rotational speed, unbalance and oil film thickness on the dynamics and stability of the rotor system are analyzed by combining illustration, axis trajectory and Poincaré. The results show that the rotor system experiences a single cycle, and the double cycle finally enters a chaotic state; with the increase of the unbalance, the oil film stiffness is improved, and the rotor system become much more stable; and the oil film thickness is too large to increase the oil film whirl and reduce the stability of the system. If the thickness is too small, whirl and oscillation will not occur. Finally, the experiment has been performed, and the experimental results well illustrate the influence of the increase of eccentricity on the stability.

1. Introduction

Rotor stability directly affects the safety of gas turbine rotor operation. As a fault of a rotor system, oil film instability is likely to cause rubbing and other regenerative faults. Therefore, it is of great significance to study the stability, improve the stability of the rotor system to ensure the normal and stable operation of the unit, and optimize the design system. Gas turbine rotor is a typical rotor-bearing system. The stability of sliding bearing-rotor system has been a hot topic in rotor dynamics research.

In the retrieved articles, the investigations of stability of the rotor-bearing system are mainly focused on the nonlinear oil-film model. Tiago[1] compared the numerical results for rotors based on nonlinear with linear. Sun[2] et al considered the temperature influence, while temperature made great impacts on the abrasion. Li[3] study a tilting pad bearing, and meanwhile, because of the particularity of turbo-expander, temperature was also considered. Aiming at different locations of unbalance and considering the nonlinear oil film behavior of FRBs, Wang[4] investigate how the unbalance locations affect the nonlinear response characteristics of turbocharger. Sghir[5] studied the nonlinear stability of a balanced flexible rotor considering the equilibrium point, and found that the progressive decrease of rotor stiffness reduced the stable speed range. Cao[6] et al analyzed the rub-impact and rotor focused on the effects of fractional order damping, while Okabe[7]analyzed the turbulence effect through a simulation of a Jeffcott/Laval rotor supported by tilting pad bearing. Tian[8]calculated the nonlinear bearing force by a new method, investigate the unbalance engine excitation effects.

In this paper, a nonlinear oil film force model of the rotor system has been considered under the action of gas turbine rotor. According to this model, it chooses appropriate unbalance, and oil film thickness as the main research object, by the Runge-Kutta method to solve the function. Combined with the bifurcation diagrams and other graphic, it conducts a qualitative analysis of the gas turbine rotor system stability under nonlinear oil film force. It provides the design of rotor bearing system in practice.

IOP Conf. Series: Materials Science and Engineering **611** (2019) 012066 doi:10.1088/1757-899X/611/1/012066

2. Double Disc Rotor-bearing Model

Figure1 is a schematic diagram of a rotor-bearing system of a gas turbine. The rotor is composed of a compressor, a turbine and sliding bearings at both ends.

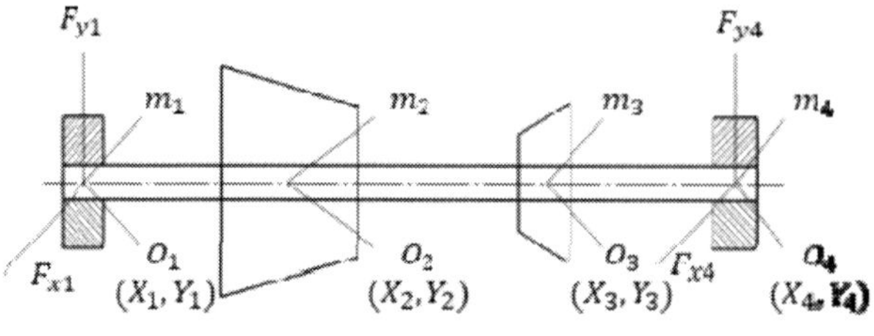

Figure 1. Double disk rotor-bearing system

2.1. Model of Oil-film Force

This paper studies the stability of gas turbine rotor under the oil film force. The dynamic pressure oil film model adopts Capone model based on the short bearing theory [9].The data expression of this model is as follows. Herein, μ is oil viscosity, L and R is the length and radius of bearing, while c is mounting clearance.

$$\begin{cases} F_X = \sigma f_x \\ F_Y = \sigma f_y \end{cases} \tag{1}$$

$$\sigma = \mu\omega RL\left(\frac{R}{c}\right)^2\left(\frac{L}{2R}\right)^2 \tag{2}$$

$$\frac{f_x}{f_y} = -\frac{\left[(x-2\dot{y})^2+(y+2\dot{x})^2\right]^{1/2}}{1-x^2-y^2}\times\begin{bmatrix}3xV(x,y,\alpha)-\sin\alpha G(x,y,\alpha)-2\cos\alpha S(x,y,\alpha)\\3xV(x,y,\alpha)+\cos\alpha G(x,y,\alpha)-2\sin\alpha S(x,y,\alpha)\end{bmatrix} \tag{3}$$

$$\begin{cases} V(x,y,\alpha)=\dfrac{2+(y\cos\alpha-x\sin\alpha)G(x,y,\alpha)}{1-x^2-y^2} \\ S(x,y,\alpha)=\dfrac{x\cos\alpha+y\sin\alpha}{1-(x\cos\alpha+y\sin\alpha)^2} \\ G(x,y,\alpha)=\dfrac{2}{(1-x^2-y^2)^{1/2}}\times\left[\dfrac{\pi}{2}+\arctan\dfrac{y\cos\alpha-x\sin\alpha}{(1-x^2-y^2)^{1/2}}\right] \\ \alpha=\arctan\dfrac{y+2\dot{x}}{x-2\dot{y}}-\dfrac{\pi}{2}sign\left[\dfrac{y+2\dot{x}}{x-2\dot{y}}\right]-\dfrac{\pi}{2}sign\left[y+2\dot{x}\right] \end{cases} \tag{4}$$

Where: f_x and f_y is the dimensionless nonlinear oil film force component in the x direction and y direction of the bearing, respectively, and σ is the number of sommerfeld.

2.2 Model of Rotor-Bearing System

The gas turbine rotor is a typical double-disk rotor system, where the concentrated masses of bearing are m_1 and m_4, and the concentrated masses of compressor and transparency are m_2 and m_3.The radial displacement of the bearing is x_1, y_1, x_4 and y_4, and the radial displacement of the compressor and turbine is x_3, y_3, x_4 and y_4.Ignoring the gyroscopic moment and torsional vibration. Equation(5) show the describes the nondimensional rotor dynamics equations. Herein, $\tau = \omega_r t$, $\overline{e} = e/c$, $x_i = X_i/c$, $y_i = Y_i/c$, $i = 1,\dots,4$, ζ stand for the damping ratio, and F_{bx1}, F_{by1}, F_{bx4} and F_{bx4} is the forces witch produced by the bearing.

$$\begin{cases} x_1'' + \dfrac{2\zeta}{\lambda} x_1' + \dfrac{k_{11}}{\omega_1^2 \lambda^2}(x_1 - x_2) = -\dfrac{F_{bx1}}{cm_1\omega_1^2\lambda^2} \\ x_2'' + \dfrac{2\zeta}{\lambda} x_2' + \dfrac{k_{21}}{\omega_1^2 \lambda^2}(-x_1 + x_2) + \dfrac{k_{22}}{\omega_1^2 \lambda^2}(4x_2 - 4x_3) = \dfrac{e_2}{c}\cos\tau \\ x_3'' + \dfrac{2\zeta}{\lambda} x_3' + \dfrac{k_{32}}{l_2^3\omega_1^2 \lambda^2}(-4x_2 + 4x_3) + \dfrac{k_{33}}{\omega_1^2 \lambda^2}(x_3 - x_4) = \dfrac{e_3}{c}\cos\tau \\ x_4'' + \dfrac{2\zeta}{\lambda} x_4' + \dfrac{k_{43}}{\omega_1^2 \lambda^2}(-x_3 + x_4) = -\dfrac{F_{bx4}}{cm_4\omega_1^2\lambda^2} \\ y_1'' + \dfrac{2\zeta}{\lambda} y_1' + \dfrac{k_{11}}{\omega_1^2 \lambda^2}(y_{11} - y_2) = -\dfrac{F_{by1}}{cm_1\omega_1^2\lambda^2} - \dfrac{g}{c\omega_1^2\lambda^2} \\ y_2'' + \dfrac{2\zeta}{\lambda} y_2' + \dfrac{k_{21}}{\omega_1^2 \lambda^2}(-y_1 + y_2) + \dfrac{k_{22}}{\omega_1^2 \lambda^2}(4y_2 - 4y_3) = \dfrac{e_2}{c}\sin\tau - \dfrac{g}{c\omega_1^2\lambda^2} \\ y_3'' + \dfrac{2\zeta}{\lambda} y_3' + \dfrac{k_{32}}{\omega_1^2 \lambda^2}(-4y_2 + 4y_3) + \dfrac{k_{33}}{\omega_1^2 \lambda^2}(y_3 - y_4) = \dfrac{e_3}{c}\sin\tau - \dfrac{g}{c\omega_1^2\lambda^2} \\ y_4'' + \dfrac{2\zeta}{\lambda} y_4' + \dfrac{k_{43}}{\omega_1^2 \lambda^2}(-y_3 + y_4) = -\dfrac{F_{by4}}{cm_4\omega_1^2\lambda^2} - \dfrac{g}{c\omega_1^2\lambda^2} \end{cases} \quad (5)$$

The basic parameters of a gas turbine rotor are as follows: the total mass of the rotor is about 16000kg, the mass of the compressor is 12000kg, the mass of the turbine is 4000kg, the working speed is 3000r/min, and the first-order critical speed is 1700r/min. Therefore, the parameters of the gas turbine system are selected, as shown in Table 1.

Table 1. Parameters of gas turbine system

Parameters	Numbers
Mass m_i(i=1,2,3,4)kg	500,12000,4000,460
Equivalent stiffness K_i(i=1,2,3)N/m	3.78e8,3.63e8,3.73e8
Radius of bearing R_i(i=1,2)m	0.5,0.45
Length of bearing L_i(i=1,2)m	0.7,0.63
Mounting clearance c/mm	1.75
Oil viscosity μ/(Pa·s)	0.0373
Damping ratio ζ	0.036

3. Nonlinear Analysis

In order to obtain the dynamic characteristics of the gas turbine rotor-bearing system, Runge-Kutta method should be used for numerical simulation to solve the ordinary differential equation (5), with the step size is 2π/100.

3.1. Influence of Rotate Speed

Figure.2 shows the bifurcation diagram of the system response with time when the rotor-bearing system e_2=0.23, e_3=0.25 and the oil film gap is 1.75mm. As can be seen from the figure, the system mainly went through three stages as the rotation speed increased. When the system speed ω_r<1700r/min, the system maintains a stable single-period motion, which is shown as an isolated point in the Poincare cross section, and the system is in a stable state. With the increase of rotation speed, when the rotation speed is 1700r/min< ω_r<2550r/min, the system moves from a single period to a double period, and the bifurcation occurs. At this point, the Poincare cross section is shown as two isolated points. There are vortex frequency, fundamental frequency and multiple frequency in the spectrum diagram, and the vortex frequency is about 0.5 times of the fundamental frequency, and the half-speed oil film vortex appears in the system. When the rotation speed ω_r =2550r/min, oil film vorticity develops into oil film oscillation through accumulation, and Hopf bifurcation occurs in the system. Poincare diagram shows an approximate closed circle, showing the quasi-periodic motion state.

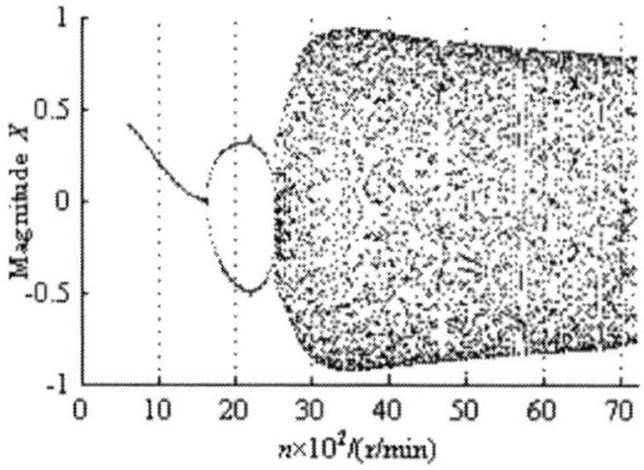

Figure 2. Bifurcation diagram at e_2=0.23 and e_3=0.25

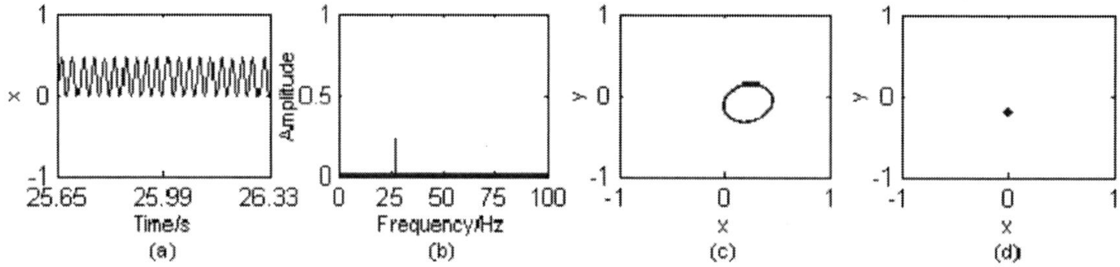

Figure 3. System response at 1620r/min(a) signal of time,(b) frequency,(c) trajectory of rotor axis ,and (d) Poincaré diagram

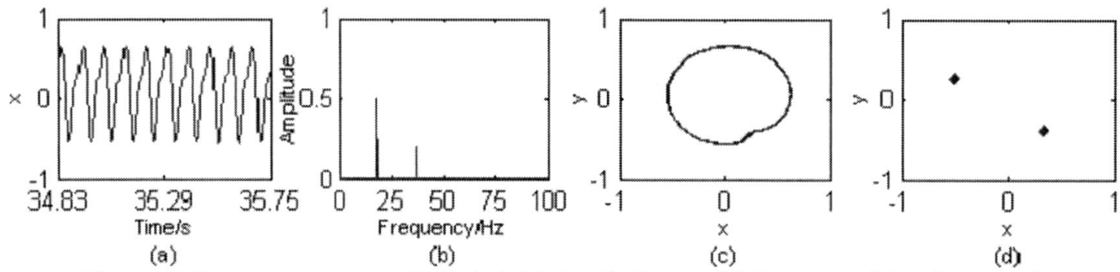

Figure 4. System response at 2200r/min(a)signal of time, (b) frequency,(c) trajectory of rotor axis ,and (d) Poincaré diagram

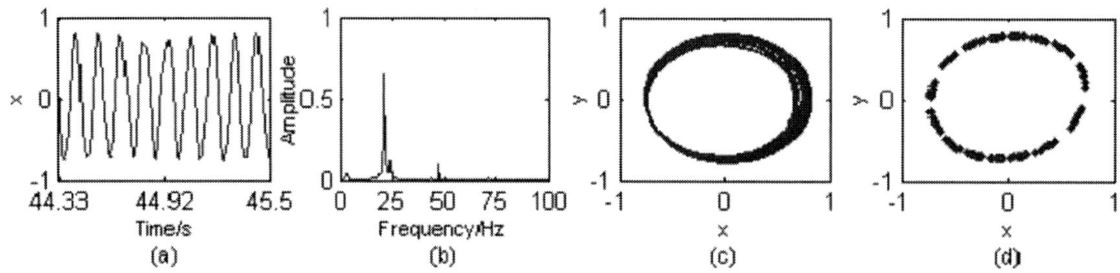

Figure 5. System response at 2800r/min(a)signal of time, (b) frequency,(c) trajectory of rotor axis ,and (d) Poincaré diagram

3.2. Influence of Eccentricities

In practice, due to the rotor machining error and installation eccentricity, the rotor is inevitably unbalanced. The results show that the magnitude of unbalance has a certain influence on the stability of rotor. Figure 6 is the bifurcation diagram when e_2=0.46 and e_3=0.48.

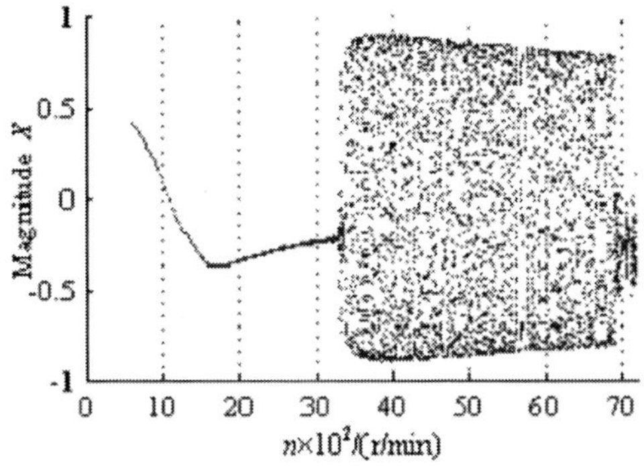

Figure 6. Bifurcation diagram at e_2=0.46 and e_3=0.48

As can be seen from Figure 2 and Figure 6 of the bifurcation diagram, with the increase of eccentricity, the unbalanced force becomes larger. The larger unbalanced force simplifies the motion form of the system, and the bifurcation state of the system becomes simpler. The system is in a stable state under the second critical speed. In addition, it can be seen that after the increase of unbalance, the critical rotational speed of oil film vortexes increases, and the increase of unbalance makes the system instability lag, which indicates that the increase of unbalance improves the stability of the system. This is because when the unbalance increases, the clearance of the bearing oil film decreases, the stiffness of the oil film increases, and the bearing capacity of the bearing increases. This result corresponds to Capone short bearing oil film force model.

3.3 Influence of Clearance

The stability of sliding bearing system mainly depends on the stability of oil film, and the thickness of oil film is an essential factor to determine the stability of oil film. Figure 7 shows the bifurcation diagram of the rotor system with corresponding oil film thickness. When the oil film thickness is 1.25mm, the rotational speed of the generated oil film vortex increases and is not obvious, and the oil film vortex rapidly turns to the oil film oscillation. This can be explained that when the oil film thickness is small, the oil film thickness is insufficient, thus affecting the formation of oil film vortexes. However, when the oil film thickness is 2.75mm, the dynamic characteristics of the rotor become complex, and both oil film vortexes and oil film oscillations occur in advance. And a similar phenomenon occurs when thethickness is 4.25mm. This phenomenon indicates that the larger oil film thickness will reduce the stability of the oil film, so that the oil film vortex intensifies, leading to the reduction of system stability.

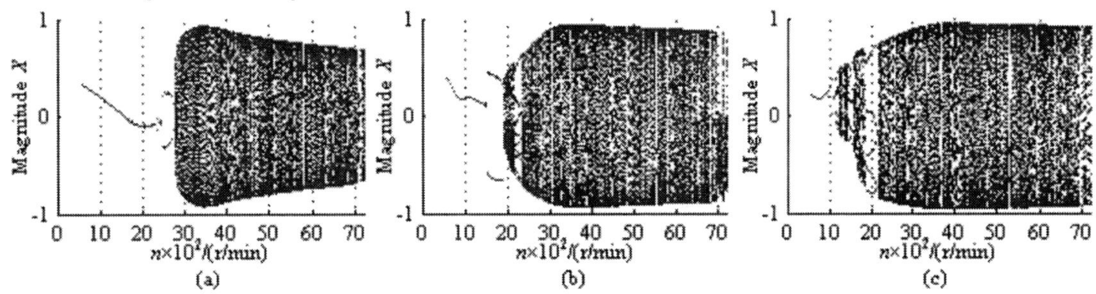

Figure 7. Bifurcation diagram with different clearances:
(a)c=1.25mm(b)c=2.75mm(c)c=4.25mm

4. Experimental Analysis

It is essential to do the experimental analysis. It is effective to correct the deficiency of the theoretical model through experimental study.Figures8 to 10 show the whole experimental system. The system consists of a rotor, motor control and the signal collecting.

During the experimental research, the response of the system under different eccentricity and rotor speed at ω_r =2000r/min, ω_r =4000r/min andω_r =6000r/min is tested respectively. Eccentricity is realized by the eccentricity mass installed on the disk.

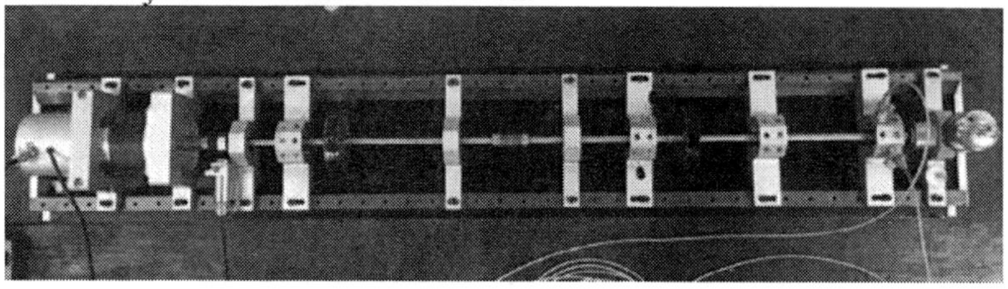

Figure 8. Motor and rotor system

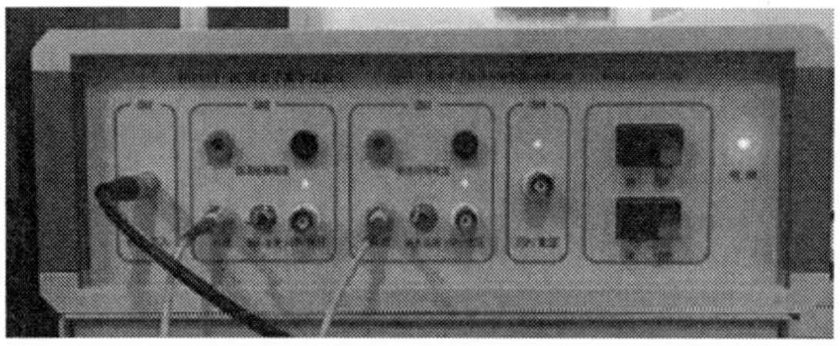

Figure 9. Signal collecter

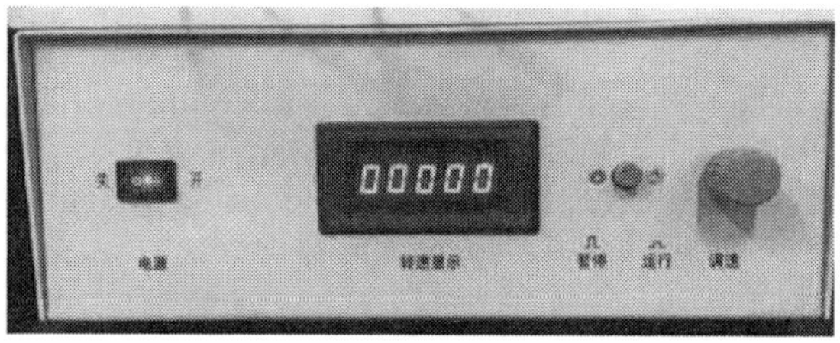

Figure 10. Motor controller

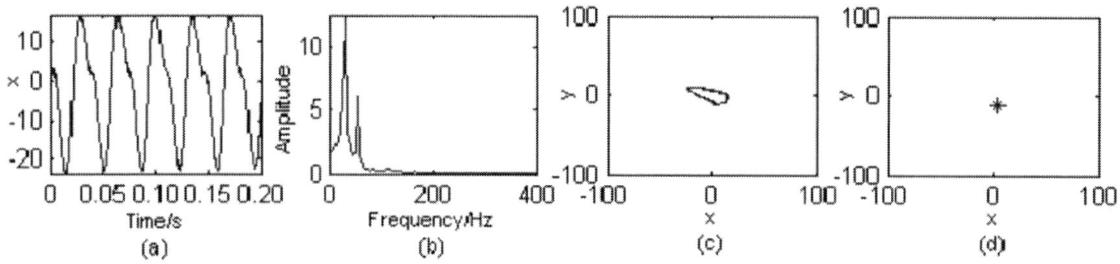

Figure 11. Response of expander Impeller at 2000r/min(a)signal of time, (b)frequency(c)trajectory of rotor axis ,and (d)Poincaré diagram

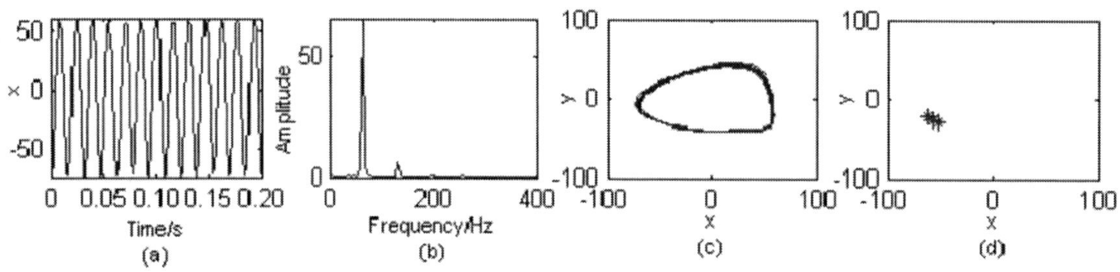

Figure 12. Response of expander Impeller at 4000r/min(a)signal of time,(b)frequency(c)trajectory of rotor axis ,and (d)Poincaré diagram

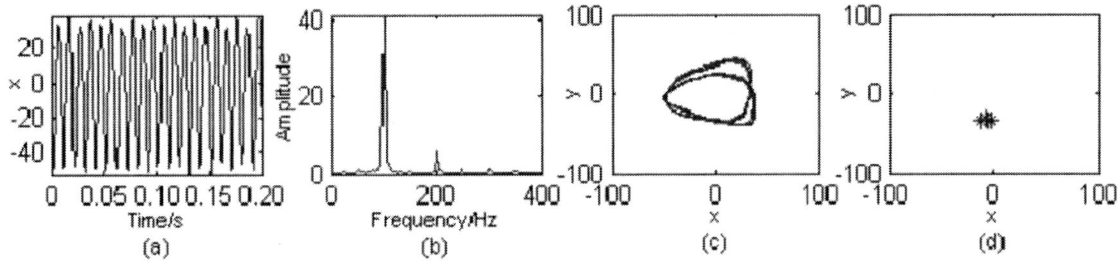

Figure 13. Response of expander Impeller at 6000r/min(a)signal of time,(b)frequency(c)trajectory of rotor axis ,and (d)Poincaré diagram

Figures11 to 13 show the system responses at different speeds in the process of speed rising of the non-eccentric rotor. When ω_r=2000r/min, due to the low rotation speed,stable oil film force has not yet been formed. The axis track diagram presents as a rough oval. The system is dominated by the fundamental frequency, with a small amount of double frequency.When ω_r=4000r/min, it can be found from the axis track that: after passing the first critical speed, the rotor gradually transits from non-stationary to stationary state, and the axis track presents as a more rounded oval. Poincare chart shows an isolated point. The frequency is based on the fundamental frequency, with a small amount of double frequency.When ω_r=6000r/min, P2 motion of the rotor can be seen from the axis track diagram. Poincare diagram shows two fuzzy points. In the spectrum diagram, in addition to the fundamental frequency, there are half-frequency and high-frequency multiples, indicating that the system has oil film vorticity at this time.

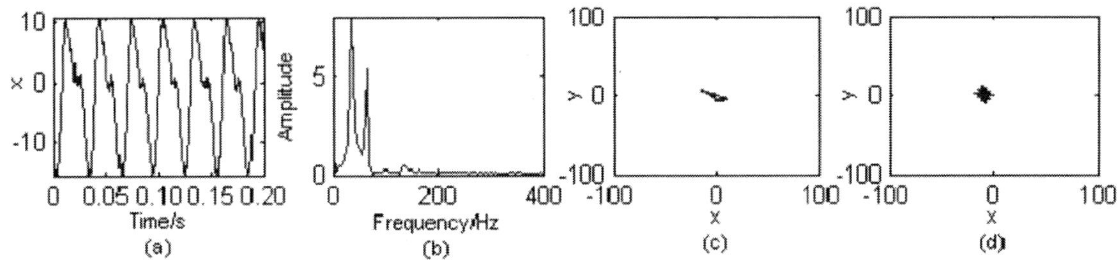

Figure 14. Response of expander Impeller at 2000r/min(a)signal of time,(b)frequency(c)trajectory of rotor axis ,and (d)Poincaré diagram

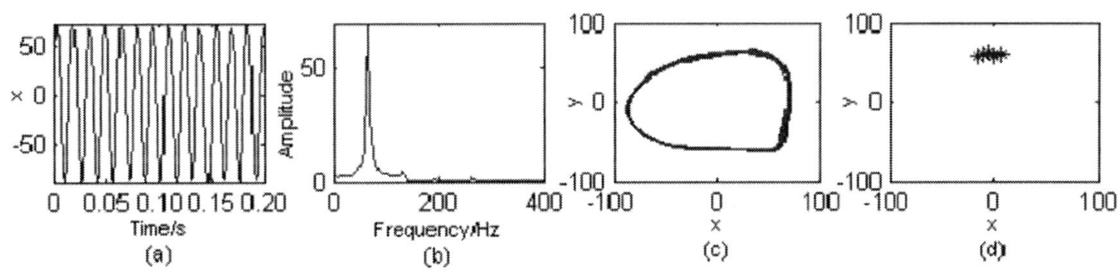

Figure 15. Response of expander Impeller at 4000r/min(a)signal of time,(b)frequency(c)trajectory of rotor axis ,and (d)Poincaré diagram

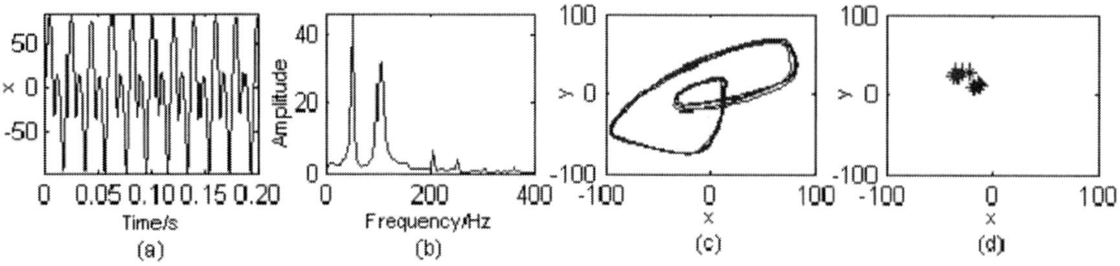

Figure 16. Response of expander Impeller at 6000r/min(a)signal of time,(b)frequency(c)trajectory of rotor axis ,and (d)Poincaré diagram

The eccentric mass block with a mass of 0.6351g was mounted on the disc for experiment.Figures14 to 16 show the response of a rotor system with an eccentric mass of 0.6351g in the acoustic velocity process. Compared with Figure 11-13, it can be seen that with the increase of eccentricity, the dynamic characteristics of the rotor system change greatly. When ω_r=2000r/min, the axis trajectory is rougher, and the half-frequency amplitude in the frequency is significantly increased, indicating that the excessive eccentricity has inhibited the formation of the dynamic pressure oil film of the sliding bearing. When the speed rises to 4000r/min, the system runs stably, but the motion amplitude increases significantly. When the rotation speed rises to 6000r/min, an obvious beat vibration phenomenon appears in the time-domain waveform. Poincare chart now shows two isolated points. In the spectrum diagram, except the base frequency, there are half frequency and double frequency of higher amplitude.This result corresponds to the fact that the increase of unbalance in the simulation results will improve the stability of the rotor system and affect the formation of oil film vortexes. The results show that too much unbalance will seriously affect the formation of bearing dynamic pressure oil film, making the rotor movement more complex. Blind increase of unbalance will not improve the stability of the rotor, but affect the formation of bearing dynamic pressure oil film, and even cause damage to the rotor.

5. Conclusion

In this paper, adual-disc rotor-sliding bearing model is adopted to analyze the influence of rotor unbalance and oil film thickness on the stability of the rotor subsystem in the process of rotor speed rise, and the following conclusions are drawn:

(1) In the process of rising speed, the gas turbine rotor system goes through a single period, and the period doubling finally enters a chaotic state.After the system passes the first critical speed, the oil film vortexes accumulate continuously until the oil film oscillates.

(2) The increase of unbalance will affect the stability of the system, and simplify its dynamic behavior, The appropriate increase of unbalance can improve the oil film stiffness of the bearing to enhance the stability of the system.

(3) Too much oil film thickness will reduce the stability of the system.Excessive oil film thickness will increase the oil film vorticity and reduce the stability of the system.

(4) The experimental results verify that the increase of eccentricity will improve the stability of the rotor system.

References

[1] Machado T H Alves,D. S.and Cavalca, K. L 2018 *Discussion about nonlinear boundaries for hydrodynamic forces in journal bearing.* (Nonlinear Dynamics)

[2] Sun, W, Yan, Z, Tan and Luo X. 2018 *Nonlinear characterization of the rotor-bearing system with the oil-film and unbalance forces considering the effect of the oil-film temperature.* (Nonlinear Dynamics)

[3] Li H. Li, M. Zhu R and Bai H.2016 *Rotor dynamics behavior of turbo-expander involving droplet impact.* (Mathematical and Computer Modelling of Dynamical Systems) pp 1-19.

[4] Wang, L. Bin, G. Li, X. and Liu, D. 2016. *Effects of unbalance location on dynamic*

characteristics of high-speed gasoline engine turbocharger with floating ring bearings. (Chinese Journal of Mechanical Engineering) 29 (2) pp 271-280.

[5] Chouchane M and Sghir R 2012 *Stability and bifurcation analysis of a flexible rotor-bearing system by numerical continuation.*

[6] Cao J, Ma C,Jiang Z and Liu S. 2011. *Nonlinear dynamic analysis of fractional order rub-impact rotor system.* (Communications in Nonlinear Science and Numerical Simulation) 16 (3) pp 1443-1463.

[7] Okabe, E. P, & Cavalca, K. L. 2009. *Rotordynamic analysis of systems with a non-linear model of tilting pad bearings including turbulence effects.* Nonlinear Dynamics, 57(4), 481-495.

[8] Tian, L.Wang, W. J. Peng, Z. J. 2011. *Dynamic behaviours of a full floating ring bearing supported turbocharger rotor with engine excitation.* (Journal of Sound & Vibration) pp 4851-4874.

[9] Adiletta G., Guido A. R.and Rossi C 1996 *Chaotic motions of a rigid rotor in short journal bearings.* 10 (3) pp 251-269.

Material Analysis and Design of New Household Humidifier

Zhiwei Zhu

Hangzhou Normal University Qianjiang College, Hangzhou, Zhejiang, China

Abstract. With the continuous improvement of people's living standards, more and more people have higher requirements for health and quality of life. Humidifier as an indispensable small household appliances product in dry areas, it has become part of people's life gradually. This paper investigates the market of humidifiers, and material analysis of different humidifier products, and design and launch a new type of household humidifier. The positioning of this design is to meet the user's purchasing needs for modelling, Requirements for modelling not only highlight the simplicity and atmosphere of the product, but also meet the needs of consumers. While improving the quality of life and family environment, it becomes part of the happy family memory.

1. Introduction

Humidifier can be divided into ultrasonic humidifier, net humidifier, heating humidifier immersion electrode humidifier and cold fog humidifier in principle. As a new process product, humidifier brings a comfortable, enjoyable and healthy environment to people in humidification, bacteriostasis, cosmetic function and so on. However, the development history of humidifier is relatively short, Humidifier has many shortcomings in shape, material, color, function and technology, as a result, the humidifier industry is facing many new challenges. Increasing brands of humidifiers, how to occupy the sales market of humidifiers? it has become the primary concern of every humidifier producer. A simple and generous humidifier can greatly improve the sales of humidifiers, it is conducive to the brand of humidifier better occupy the market. From the perspective of consumer psychology, how to make the shape of humidifier more visual impact, in order to achieve the ultimate goal of sales.

2. Design Research

Due to the rapid development of the humidifier industry, in order to protect the normal rights and interests of consumers, ensure the stable and orderly development of the market, so the humidifier standard in the humidifier volume, noise, evaporator service life, softener performance and service life, display control and other functions have been stipulated. In addition, this standard is applicable to household and similar humidifiers, it is also suitable for humidifiers used in public places, but it is not suitable for humidifiers used in special environment where corrosive and explosive gases.

2.1. Market Research

Due to the global economic recession, the overall demand of domestic household appliances industry slowed down. In the face of such a grim situation, market growth is still good of humidifier in autumn and winter, presenting a trend of counter-market growth. Each manufacturer has abandoned the characteristics of specializing in product technology in the humidifier, it takes a lot of hard work in the design of appearance, designing different appearances for different age levels and using environments. A battle for market hegemony of smokeless humidifiers is quietly beginning. Therefore, a humidifier that can solve the above problems becomes a necessary solution. So we have the following research.

2.1.1. Market Analysis. At present, humidifier is not a popular product, and it has not formed a huge brand effect, but the market has begun to take shape, It is foreseeable that the market of humidifiers will attract more and more consumers' attention in the near future. Air purifier, some products are equipped with humidifying function, air conditioning fans and other products is an alternative product to humidifier in the market, this makes the design of humidifier more and more important.
Humidifier enterprises can survive only through breakthroughs. For traditional humidifier enterprises, if you want to stay out of the competition, the wisest way is to transform. However, transformation is not easy. At present, 70% of the humidifier market is dominated by YADU and MIDEA brands, other 30% of enterprises are non-mainstream brands. Strong humidifier enterprises should not only do functional design in order to transform, more time should be spent on fashion trends, For example, product shape, humanization, simplicity, space size, use range and market position.

2.1.2. Analysis of Domestic Famous Brands. Advantages of YADU: continuous renewal of indoor environmental protection products for different groups, excellent quality assurance and accumulated brand benefits, expert service.
Advantages of MIDEA: widening market channels, novelty and fashion in appearance design, excellent cost performance.
Integrating the advantages of two brands, humidifier modelling design is mainly based on humanization, simplicity, in line with the trend of the times, in order to win the favor of consumers, on this basis, function design should be done well, to meet the psychological and physiological needs of modern people, continuous improvement of product performance of humidifier.
For the growing market of humidifiers, a high degree of brand concentration may lead to a certain degree of brand monopoly, this monopoly will affect the updating of product technology and the rationalization of product price. In this market pattern, we need more strong brands to compete. Currently, the annual growth rate of household ownership in the household humidifier market in China is only around 3-4%, The United States, Canada, France, Britain and other countries with the same geographical latitude as the north of China, they have higher household ownership of humidifiers, for example, the United States has reached 35 percent.

2.1.3. Price Position. In order to correctly assist market positioning and product positioning, an incomplete statistical survey was conducted on the internet. Through a survey on an e-commerce website, get the product price and profit distribution map of high, middle and low-end market of humidifier in China. From the analysis of the price list, the market sales of low-end products are the largest, high-end products are in the middle of the market, mid-end products have the lowest market sales. Combining the consumption of other commodities, the trend of consumers' bipolar consumption is becoming more and more obvious.
Although the low-cost humidifier products are sold well in the market, but for the next 10 years, China's economy will rise to a higher level, people's pursuit of quality of life will be more refined, people's requirements for product design will also be stricter. The product design of household humidifier will be positioned in the mid-end market, try to design a household humidifier with simple and atmospheric shape.

2.2. User Survey

2.2.1. Investigation Respondent. Mainly white-collar workers aged 25-40, they have certain economic ability and pursue aesthetics.

2.2.2. Investigation Method. Face-to-face interviews and questionnaires. The form of user survey for purchasing humidifier products is mainly downstairs of office buildings, major occasions such as large-scale household shopping malls, and in the form of filling out questionnaires, final statistics. The questionnaire includes the survey of user's living environment, as well as shape research, function research, material research, it is a comprehensive questionnaire.

IOP Conf. Series: Materials Science and Engineering **611** (2019) 012067 doi:10.1088/1757-899X/611/1/012067

2.3. Research Summary
Structural analysis from two aspects of market research and user research shows that, modelling design of humidifier has certain market space, users also need a product that meets all their needs. When designing, we should consider meeting the aesthetic needs of most people, and they feel comfortable using it. The consumer class is defined as urban white-collar workers aged 20-40, gender is predominantly female consumers, they pay more attention to moisturizing and creating a comfortable family environment. The price of humidifier is set at RMB 100-300.

3. Material Analysis
Material is indispensable to human creation, material is one of the basic conditions of human survival. Based on the previous analysis and product positioning, the materials used in different parts of the design of the purified humidifier were determined.

3.1. PP
The appearance of polypropylene plastics is milky white and translucent, it is non-toxic, tasteless and light, excellent bending fatigue resistance, chemical stability and good electrical insulation, forming dimension stability, the thermal expansion is small, mechanical strength, rigidity, transparency and heat resistance are higher than those of polyethylene.

3.2. ABS
ABS plastics synthesize the properties of three components, for example, the rigidity, heat resistance, chemical corrosion resistance and weatherability of acrylonitrile, impact resistance of butadiene. Low temperature resistance, surface gloss, dimensional stability, colourability and machinability of styrenem.

4. Product Orientation and Design

4.1. Product Orientation
Need to meet the user's purchase of styling needs. It can highlight the simplicity and atmosphere of the product, it can also meet the needs of consumers. It can also be a part of family happiness while improving the family environment. Design begins with brainstorming, associate humidifier, think of the words related to it and get the design ideas from them. According to the extracted keywords, the humidifier is designed and the final scheme is obtained. Simple and generous style is the key to this design, Using regular geometric shape, there are not too many free curves, it not only breaks the traditional model of humidifier, but also embodies the simplicity of style. The main color is white, which is integrated with the general home environment. There are other color matching options.

4.2. Design Plan
Simple and generous style is the key to this design, using regular geometric shape, there are not too many free curves, it not only breaks the traditional model of humidifier, but also embodies the simplicity of style. The main color is white, which is integrated with the general home environment. There are other color matching options.

4.3. Product Features
Modelling design of this humidifier, it is modeled by regular geometric bodies with rounded corners, to convey a simple, generous and natural feeling to consumers. It uses white as the main color, bright and comfortable, is shown in Figure 1. The combination of red and black makes it more modern, is shown in Figure 2. Simple appearance gives people a sense of relaxation and leisure. The innovation of this humidifier lies in breaking the conventional humidifier model, it is different from the general humidifier on the market.

Figure 1. Effect picture.

Figure 2. Other color matching options.

Conclusion

Humidification technology determines the principle of humidifier. Through the analysis of humidifier products in the existing market, mastered the advantages and disadvantages of humidifier products at present, preliminary determination of the design task. Through further research on customers' desires and needs, Through brand research, market situation analysis, market segmentation research and the market orientation of products is determined. According to the research history of humidifier products and the trend of other household appliance design, determine the direction of design, that is, the trend of decorative design. Only by persisting in seeking truth from facts, combining theory with practice is the fundamental way to design well.

References

[1] Jin H 2011 *Classification and Principle of Humidifier (Public Power* vol 3*)* pp 43–5 In Chinese
[2] Ye X 2009 *The Humidifier National Standard has been Formally Implemented (Household Electric Appliance* vol 11*)* pp 62–3 In Chinese
[3] Pang C 2011 *China Humidifier Market Research Annual Report 2010-2011 (Household Electric Appliance* vol 2*)* pp 86–7 In Chinese
[4] Zhao L 2006 *The Brand of the Humidifier Becomes More and More Monopolistic and Diversified (Modern Household Electric Appliance* vol 23*)* pp 66–7 In Chinese
[5] Yu S 2004 *Furniture Modelling and Structural Design* (Harbin: Heilongjiang Science and Technology Press) In Chinese

Design of Special Lathe for Taper Turning Slender Shaft

Jiefeng Jiang[1,]* and Xiaodong Huang[2]

[1]Qianjiang College, Hangzhou Normal University, Hangzhou 310036, China
[2]Shanghai United General Technics Co, Ltd, Shanghai 201612, China

*Corresponding author

Abstract. Based on the analysis of the machining process characteristics for special slender shaft work-piece, considering the increase in taper turning efficiency and accuracy, a new clamping scheme for machining both ends of slender shaft is put forward, which also includes using pulling tail-stock instead of traditional pushing, using follow-manipulator instead of follow-rest. Following the clamping scheme, the dual drive device, the tail-stock with pulling device and the manipulators are designed, then the parts are integrated into the special lathe for taper turning slender shaft.
Keywords: slender shaft; special lathe; dual main drive; clamping; manipulator.

1. Introduction

In general, the axis parts with the ratio of length to diameter (L/D) greater than 20 are called slender shaft. The shape of slender shaft is not complicated, but it has the poor rigidity. In the turning, because of lateral cutting force, gravity, cutting heat and other factors, process vibration and bending deformation occur easily, which cause the shape error of taper, drum or saddle[1-3]. Though the technique of error compensation[4,5] can solve the problem, but the error rule must be known, which is difficult to achieve preciously, and it is not often used in application. So turning the slender shaft is a difficult processing problem all along.

The front griping with chuck and rear pushing with lathe center are often used for slender shaft turning. But the clamping scheme especially rear pushing often increases the bending of the work-piece. Studies[6,7] have shown that holding and pulling work-piece could be appropriate to reduce bending deformation in turning, so the pulling plan and device are proposed in the clamping scheme of the special lathe. In order to enhance the rigidity of slender shaft work-piece, the tool follow-rest or center rest is often used[8,9]. Considering taper turning and automatically control requirement, the manipulator is employed in the special lathe, instead of follow-rest, it also play the role of enhancement of processing stiffness.

When both ends of slender shaft need to be machined, the traditional lathe with one powder head usually needs clamping twice and cutting twice, causing low efficiency in mass production. In this paper, a kind of lathe with dual driver for synchronous turning slender shaft is put forward to improve the producing efficiency.

2. Specific Process Requirement

Here special machining is required, which is the original stick (material 55CrSiA) for making the damping spring of truck, the production is massive. The length range of stick is 2000 mm to 3000 mm, the diameter range is 8 mm to 14 mm, and which is the typical slender shaft. A type of slender shafts is shown in figure 1. The total length of the work-piece is 2800 mm, the middle section with 1200 mm and two ends with 50 mm are equal diameter segments, and have no processing need. The variable diameter segments with 750 mm on both sides of the middle section need to be machined, and the maximum diameter is 13.6 mm with tolerance ±0.1 mm, the minimum diameter is 8.3 mm with ±0.1, the taper is about 0.2 degrees. The characteristics is that the machining length is large, the diameter is small. Taper machining cause the cutting volume changing constantly in the process, so the cutting force is constantly changing, which increases the machining difficult and the unstable possibility.

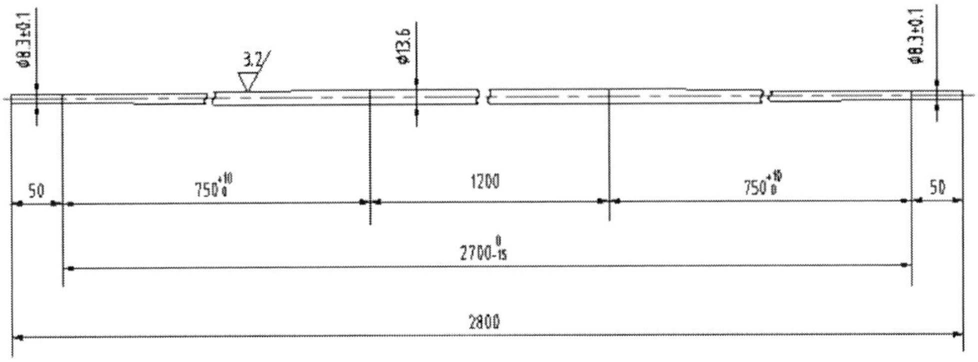

Figure 1. Slender shaft (unit: mm)

3. Clamping Scheme Development

Generally the traditional machining method of the special slender shaft work-piece is turning by universal NC-lathe, as shown in figure 2. The machining process is clamping work-piece, turning one end, disassembling and changing direction, clamping again, turning the other end. Two times of clamping and cutting are needed, so the efficiency is low.

Figure 2. Traditional machining method and machined product

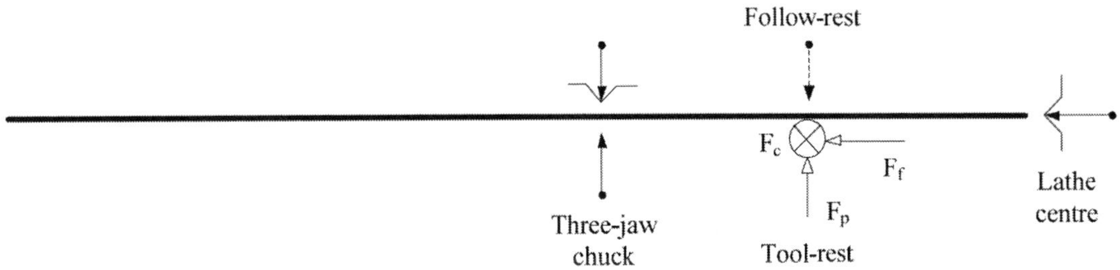

Figure 3. Traditional clamping method

In above machining process, the model of the clamping method is shown in figure 3, the middle section griped with three-jaw chuck, one end pushed with the lathe center and auxiliary support with follow-rest is proposed. According to the stability theory of pressure bar, pushing influence at the lathe centre will increase the instability and bending deformation of the slender shaft. In addition, because the diameter of taper turning segment in work-piece is changing, if the rigid follow-rest is pre-adjusted

IOP Conf. Series: Materials Science and Engineering **611** (2019) 012068 doi:10.1088/1757-899X/611/1/012068

to fit the maximum diameter, the role of the rigidity enhancement is disappeared when tool-rest turns to the smaller diameter position. So the machining accuracy is difficult to guarantee, some unaccepted products occur.

In order to improve the processing efficiency and accuracy, and considering the machining characteristics of the slender shaft, a new processing clamp scheme is put forward, as shown in figure 4. There are two sets of clamping device symmetrically, clamping one time, the work-piece on both ends are machined. So the processing efficiency can be improved greatly. And cutting force both ends of the work-piece is opposite in feeding direction, canceling each other, the stability of machining increases. At the tail-stock, the pulling is adopted instead of pushing to improve the processing rigidity of the slender shaft. Moreover, for the tape turning requirement, follow-manipulator is used as a flexible support instead of rigid follow-rest, and several auxiliary supports using manipulator are proposed to improve the rigidity further. In the machining, the auxiliary support manipulator retracts duly to avoid the feeding tool-rest.

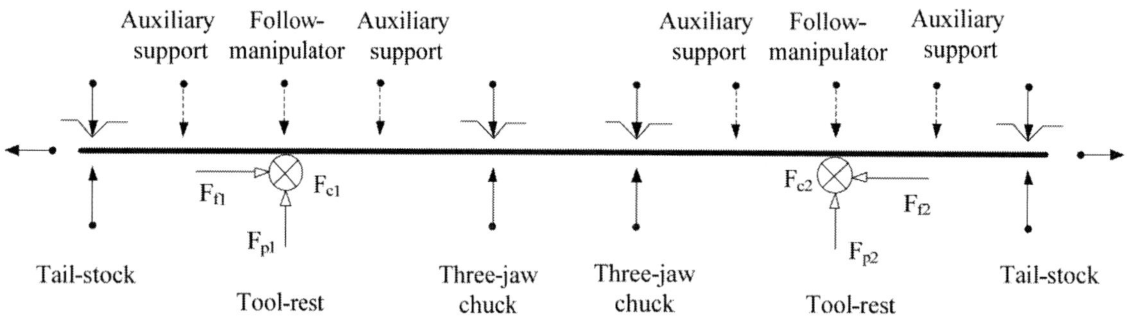

Figure 4. New clamping scheme

4. Structure Design of Special Lathe

According to the special process demand and the new clamping scheme, a special machine tool for synchronizing taper turning on both ends of the slander shaft work-piece is designed. It mainly includes dual main drive device, tail-stock with pulling, auxiliary manipulator, tool holder and discharge device and the base. These parts are integrated with special lathes.

4.1. Dual Main Drive Device

In order to machine both ends of the work-piece synchronously, the dual main drive device is designed, as shown in figure 5. Which is powered by a main motor, and the spindle rotates through the belt transmission. The slender shaft work-piece is clamped by a rotating hydraulic cylinder and a power chuck, so it is easy to realize automatic control. The main drive device is mounted on the slide plate, and can be driven by a linear transmission component (screw-nut) to achieve the relative axial movement, and the relative positions can be changed to adapt to the slender shaft parts of different lengths. when the main drive device moves in axial direction, the spline shaft plays a guiding role.

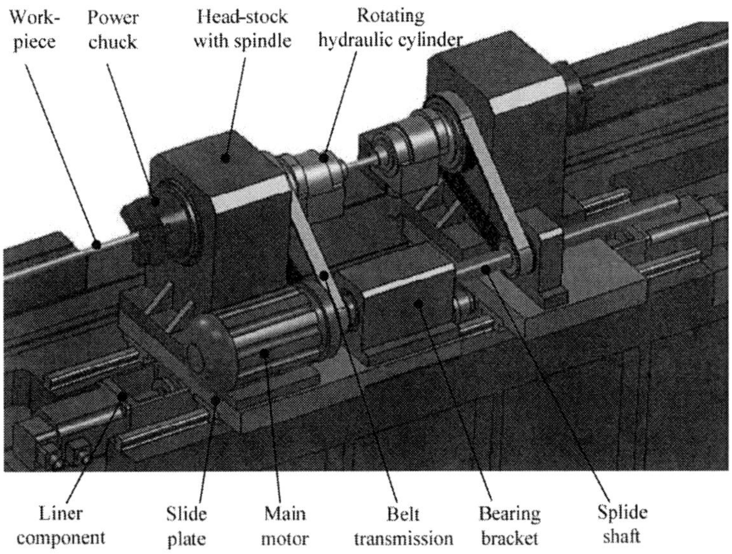

Figure 5. Dual main drive device

4.2. Tail-stock Device

Tail-stock pulling device is designed as shown in figure 6, the slender shaft work-piece is clamped and relax by pneumatic collet chuck. The upper cylinder controls the extending and retracting of the pull rod and pneumatic collet device to fit the different length of slender shaft work-piece, and the distance of extending and retracting is limited by adjustable sleeve. In addition, the upper cylinder also plays the role of pulling work-piece. The lower cylinder is connected the slide plate by the piston rod, which can drive the tail-stock to move in the transverse direction, and the aim is to make the space when the slender shaft work-piece is loaded and unloaded.

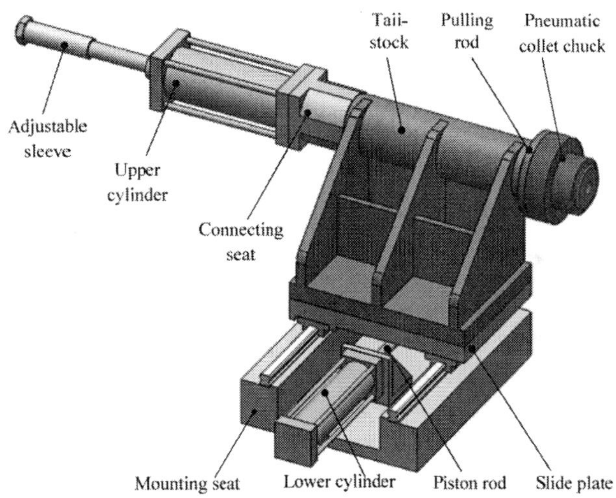

Figure 6. Tail-stock pulling device

4.3. Manipulator

In the special lathe, follow-manipulator is proposed instead of follow-rest, and several same manipulators is used as auxiliary support, improving the rigid of slender shaft in the turning. The structure of manipulator is shown in figure 7, and the manipulator has telescopic function and clamping function. The telescopic action is driven by the pneumatic sliding table installed in the mounting seat or tool-rest. And the clamping function is accomplished by the cylinder, the pushing rod, and two fingers. The finger is jointed with pushing rod and connecting plate, there are rollers in the end of finger to follow the turning of work-piece.

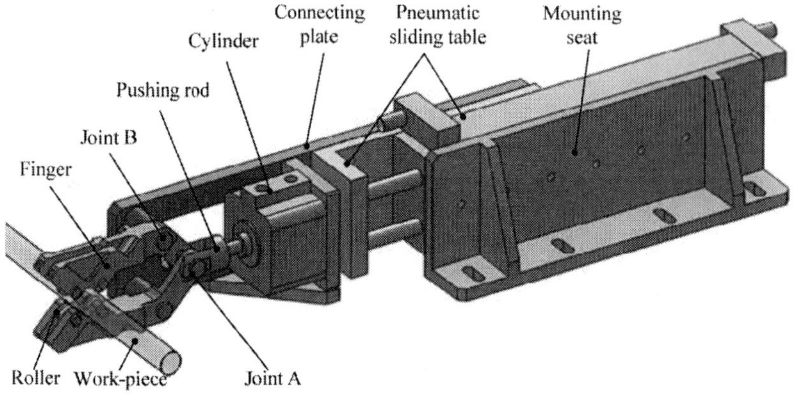

Figure 7. Manipulator

4.4. Special Lathe Integration

Besides the dual main drive device, the tightening tail-stock and the auxiliary support manipulator, the tool carrier and drainage device are designed, and the devices are integrated on the lathe bed as shown in figure 8. In the axial direction of slender shaft, two sets of the same devices is placed symmetrically, and the auxiliary support manipulator is obliquely arranged. The chip room is arranged below the work-piece, and the transportation screw of chip device is installed in the chip room. In the transverse direction, tool feed device are located in the rear of the chip room, so the chip is not easy to fall in the rail of tool feed device. Following the design, special lathe is manufactured and assembled, as shown in figure 9. Next, the special lathe will be tested.

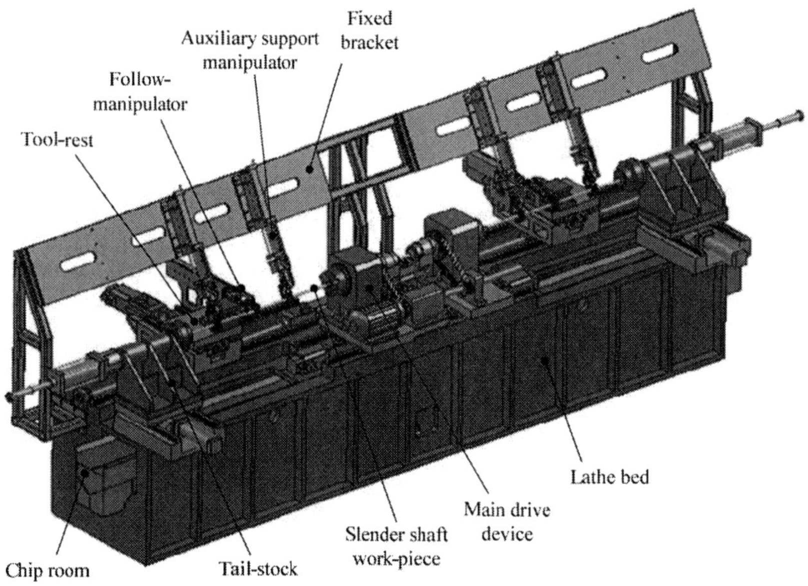

Figure 8. Overall layout of special lathe

Figure 9. Special lathe in assembly stage

5. Conclusions

(1) Considering the processing efficiency of special slender shaft work-piece, a plan of the synchronous turning in both ends is put forward, and then the dual drive device is designed.
(2) From improving the stiffness of work-piece processing, the method of pulling instead of pushing is adopted, then tail-stock tightening device is designed. Follow-manipulator for taper turning replaces follow-rest, and the same manipulator as auxiliary support is used for further improving the stiffness. Several parts are integrated into the special lathe, and now the actual machine is in the stage of assembly. Next time, the test will carry out, and the turning effect of special work-piece will be analyzed in the follow-up study.

Acknowledgments

This study is supported by Basic Public Research Project of Zhejiang Province (LGG18E050018) and General Research Project of Zhejiang Education Department (Y201737162) in China.

References

[1] Lu K, Lian Z, Gu F and Liu H 2018 *Mechanical Systems and Signal Processing* vol 100 pp814-826
[2] Deng Z, Zhang H and Jin L 2014 *Chinese Journal of Construction Machinery* vol 12 pp176-179
[3] Li J and Pang Y 2012 *Modular Machine Tool & Automatic Manufacturing Technique* (2) pp 23-25
[4] Wang T and Wang X 2015 *Manufacturing Technology & Machine Tool* (1) pp 92-95
[5] Zhao H, Yang J and Tong H 2005 *Journal of Shanghai Jiao Tong University* vol 39 pp 117-120
[6] Shen Z 2009 *Machine Building & Automation* vol 38 pp 60-62
[7] Zhu J and Ke L 2015 *Manufacturing Technology & Machine Tool* (1) pp 62-65
[8] Guo J and Han R 2006 *International Journal of Machine Tools and Manufacture* vol 46 pp 1002-1012
[9] Liu X 2013 *Manufacturing Technology & Machine Tool* (6) pp 107-109

IOP Conf. Series: Materials Science and Engineering **611** (2019) 012069 doi:10.1088/1757-899X/611/1/012069

Modeling Design of Instrument Panel

Wei Chen

Jiangsu Open University, Nanjing, China

Abstract. The basic requirements of instrumentation modeling, composition and layout, representation of functional areas, color, text, graphic symbols and signage, materials and coloring technology are discussed. The panel modelling should meet the following basic requirements, that is, the needs of product function, the needs of product uses, quick recognition, adaptability, artistry, and easy installation, adjustment and maintenance; The composition and layout of the panel are mainly panel composition of characteristic rectangular and panel composition of non-characteristic rectangular; The functional area of the panel can be divided into six types: wireframes, brackets, color blocks, spaces, segmentation lines and volume and surface; The color of the panel should consider its main color and matching color; Words, graphic symbols and signs on the panel, as well as the materials and coloring technology used in the panel are also factors that must be considered. The selection of materials is very important.
Keywords: modeling design; instrument panel.

1. Introduction

For instrumentation, panels are used to install control function elements and display function elements. It is a set of keys, knobs, indicator lights, display disk, jack, signage, model and other devices. The panel is the pivot of human-machine information transmission and exchange, is the window of human-machine dialogue, and is the most frequently operated component. Like the face of the product, its panel is the visual center of the entire product, is the core part of product modeling, it will made an important and deep impression on you. Therefore, modeling design of instrument panel greatly affects the using performance and sales of the whole product and we must attach great importance to it [1].

2. Basic Requirements for Panel Modeling

Before proceeding with the modeling design of instrument panel, we must analyze the basic requirements that the panel should meet [2].

1) It should meet the needs of product function

Product function plays a leading and decisive role in the panel structure and modeling. The panel structural, shape and color should meet the functional requirements.

2) It should meet the needs of product uses. The layout of functional components should be reasonable and logical. Their size, shape, position and color should conform to human physiological, psychological habits, to easy to observe, operate, to make people not easy to fatigue, no mistakes, safe and reliable use.

3) It should be quick to recognize. The text, symbols, display, etc. on the panel must be clear, concise, easy to understand, easy to recognize, and fast to read so as to increase efficiency of use.

4) It should be adaptable. Here refers to the adaptation to environmental conditions, it can adapt to mechanical environment (sufficient strength and stiffness, reliable connection, vibration resistance, etc.), climate environment and electromagnetic environment, it can also meet the requirements of corrosion prevention.

5) It should be artistic. The composition, layout, color, and texture of the panel should give people artistic appeal, they have a sense of the times, and should be consistent with the style of the entire machine.
6) It should be easy to install, adjust, and maintain.

3. Composition and Layout of Panel

To design the panel as a decorative painting with reasonable layout, pleasant color and convenient operation, the composition and layout of the panel are the core issues in the modeling design.
Composition and layout should start from the operation of the product itself, and using the certain expression to achieve the purpose of clear instructions and clear at a glance. Composition and layout should guide the people sight line to flow according to the operation procedure, so as to maximize the convenience of users.
Composition refers to the reasonable segmentation of the face plate. Segmentation is based on aesthetic principle and fractal segmentation rule. In addition to meeting the aesthetic requirements, the composition must be closely combined with the functional requirements of the panel, so as to achieve both aesthetic and internal and external coordination, mutual correspondence, reasonable distribution and easy operation. Only in this way can the composition get good results [3].
1) Panel composition of characteristic rectangular
The panel shape is (or is approximately) a square, a medium rectangle, a Φ rectangle and a root-mean-square rectangle, called a characteristic rectangle. Panel composition of characteristic rectangular can be divided according to the "fractal segmentation rule" [1], which can achieve better composition effect. This method is especially suitable for instruments with many operating indicators on the panel.
2) Panel composition of non-characteristic rectangular
The panel that is not a characteristic rectangle is non-characteristic rectangular panel, which is also the most common panel. Its basic principles of composition are:
(1) To avoid monotonous and rigid equivalence segmentation;
(2) The asymmetric segmentation method should be adopted, but the balanced effect of layout should be paid attention to.
This method is the easiest to achieve both harmony and contrast, both change and unity, both natural and vivid visual effects, as shown in Figure 1. It divides the plane into four unequal areas, so that the size of the area has both contrast and overall coordination, to give people a sense of beauty.

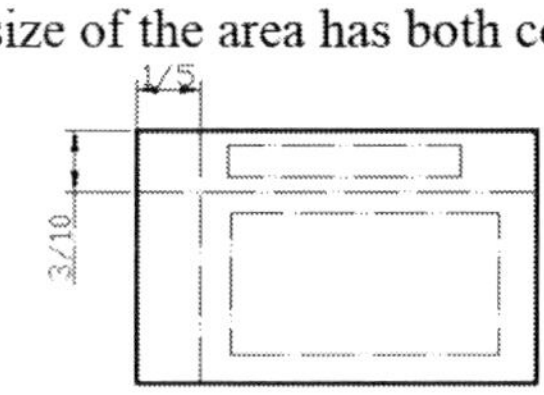

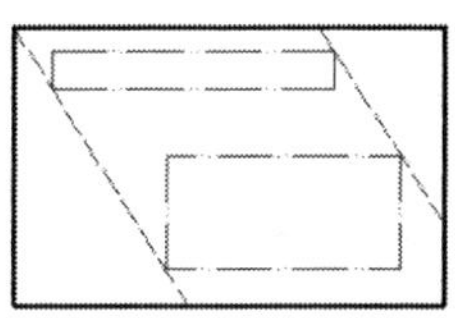
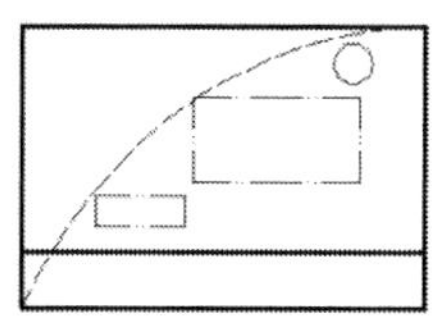

Figure 1. asymmetric segmentation. **Figure 2.** oblique segmentation. **Figure 3.** curve segmentation.
(3) Use of oblique line segment
The oblique line gives people a sense of motion and strength, as shown in Figure 2. This composition can achieve dense contrast, balance, stability and rhythm.
(4) Use of curve segmentation
Figure 3 shows curve segmentation. This kind of segmentation can make the plane composition not only get active dynamic feeling, but also have quiet and stretching visual effect. It looks flexible, exciting and harmonious atmosphere, and produces implicit aesthetic feeling. This segmentation also have flexible changes and exciting, harmonious atmosphere, to produce implicit aesthetic feeling.
In the composition of the specific instrument panel, the reasonable layout should be carried out according to the function of the product and the number of components. It can be a single method or a combination of various methods, but no matter how to layout, we should pay attention to the following issues:
A. The hierarchical relationship of the panel. To highlight the theme is the primary problem to be solved in the design of instrument panel. Subject plays a decisive role in composition and object plays

a foil role. Subject refers to the main functional part, that is, observation center. The key to highlight subject is to deal with the hierarchical relationship of the panel plane, so that the panel presents the effect of three-dimensional space.

Generally speaking, the strong contrast part should be regarded as "foreground" (such as button, knob, name, model, sign, etc.); the medium contrast part, no stimulation part or less stimulation part should be regarded as "middle" (such as text, symbol, etc.); and the soft and implicit part should be regarded as "background" (such as bottom, display screen, etc.). This treatment can make the panel level clear and improve the artistic effect of modeling.

B. The position relationship of the panel plane. The different plane positions of the top and bottom, left and right give people different feeling effect.

C. Blanks in panel layout. The layout of words and symbols on components of the panel should not fill the whole panel with full or even layout, to make people feel overcrowded and breathless. Appropriate blanks such as scattered, sparse, close, real, virtual, should be left, which can eliminate the monotony of the panel and leave room for its functional expansion. The blank is equivalent to the background, which serves as a foil to the panel and enhances the theme.

4. Form of Functional Area of Panel

In addition to conforming to aesthetic principles, the composition and layout of instrument panel should be in line with the operator's human functional characteristics, to make the operator easy to use and obtain the highest operational efficiency and accuracy. Special attention should also be paid to the intrinsic connection of indicators and adjusting control buttons. The various components on the panel should be arranged according to the functional area, that is to say, in accordance with the operator's operational logic. Six forms of expression are generally used in the design of instrument panel to highlight the characteristics of operation function and logic rules. They are:

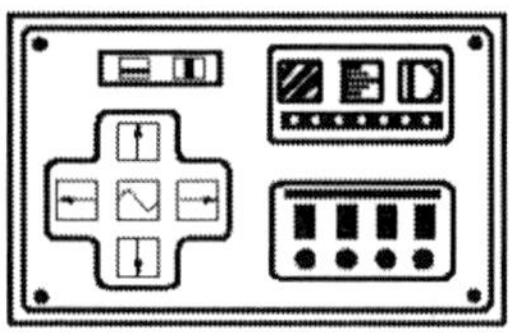

Figure 4. distinguished by wireframes.

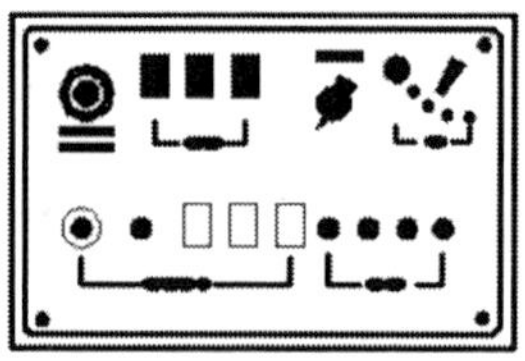

Figure 5. distinguished by brackets.

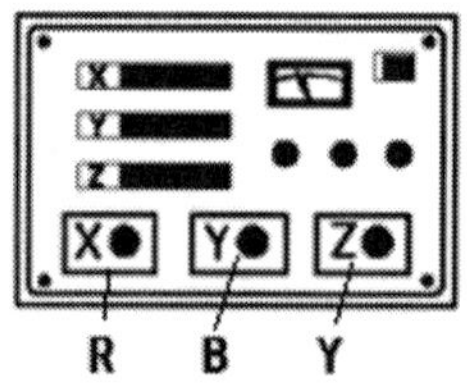

Figure 6. distinguished by color blocks.

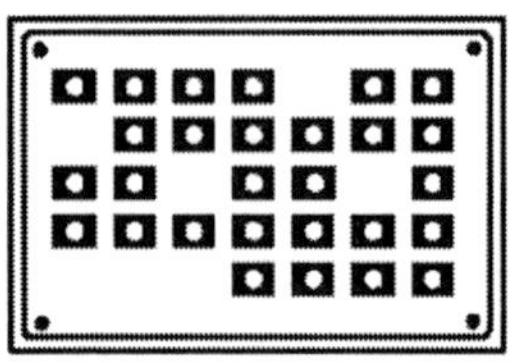

Figure 7. distinguished by blank spaces.

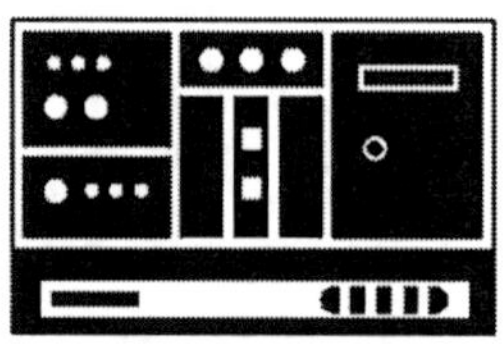

Figure 8. distinguished by segmentation lines.

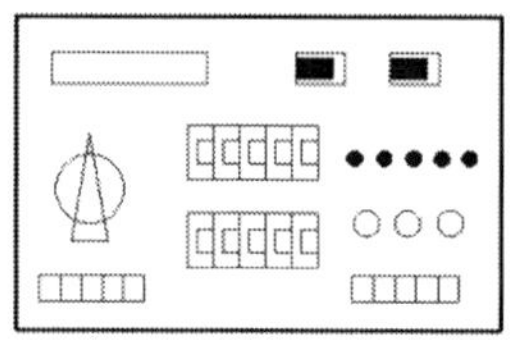

Figure 9. distinguished by volume and surface.

1) Functional areas are distinguished by wireframes (see Fig.4).
2) Functional areas are distinguished by brackets (see Fig.5).
3) Functional areas are distinguished by color blocks (see Fig. 6, where R is red, B is blue, Y - yellow).
4) Functional areas are distinguished by blank spaces (see Fig.7).
5) Functional areas are distinguished by segmentation lines (see Fig. 8).
6) Functional areas are distinguished by volume and surface (see Fig. 9).

The above-mentioned forms of functional areas should be flexibly applied according to the actual situation and the needs of modelling. Wireframe method and bracket method are simple in modeling design, various in technological means and have good appearance effect. Wireframe method and blank space method are suitable for the same or similar size and shape of components. Bracket method and segmentation lines method are suitable for the case of more types of components, larger differences in

size and shape. Color blocks method and blank method are suitable for the case of using touch switch. Volume and surface method is mostly used for engineering plastic panel of injection moulding, to rich in it layers and three-dimensional sense.
Components in each functional area should be fully separated. The more spacing, the less interference, which helps to enhance memory, improve operational efficiency and accuracy.

5. The Color of the Panel

The panel is the "face" of the instrument. How to deal with the color is not only related to the function and use of the product, but also has a great impact on the moulding[4].
1) Main color of panel
Generally, the panel uses mostly cold tone or neutral tone and other monochrome, to give people a concise, elegant and generous feeling.
Monochrome is not monotonous, because there are other colors on the instrument panel, such as indicator lights, buttons, signs and so on, which can make the overall color still rich. Of course, when considering the panel main color, the color of the whole machine always is pay attention to harmony.
2) Color matching of panel
There are many components and character symbols on the instrument panel for users to observe and use. This requires that the color of the panel be cordial and pleasant, lively and generous. Usually, the color with slightly higher purity and moderate value can be chosen.
3) Attention points of panel coloring
(1) Large-side harmony and small-side contrast
For larger instruments and meters, the overall color harmony should be adopted, while the small area contrast color should be used on the panel.
(2) Harmony in comparison.
For middle or smaller instruments and meters, large contrast can also be used to achieve harmony and unity.
(3) The texture of the panel
The texture of the panel should be fine, dull or matte and soft. It is necessary to avoid the appearance of glare-producing electroplated parts on and near the panel, so as to avoid strong visual stimulation.
(4) Corporate image (CI)
It is imperative that modern enterprises must fix a main tone in the product, which is conducive to the sales and display of the product.

6. Words, Graphic Symbols and Signs on the Panel

The design of instrument panel aims at communicating with people and conveying functional information of products. This kind of communication and convey is mediated by words, graphic symbols and signs, so it is also very important for their design[5].
1) Word. The words on the panel should be clear, standard. Other descriptions and figures should be typesetted with standard Chinese bold and Song-style photographs, except that the trademark is a special graphic or text for registration. The English letters should be isoline of different sizes and thicknesses to form a rhythmic composition.
2) Graphic symbol. Various types of graphic symbols are widely used in modern indicative means. It replaces and supplies transmission of text signals with a highly generalized, concise and vivid graphical language. It is intuitive and you know what it means when you see a graphic symbol. It has a strong versatility and is not limited by the country, region, language and cultural level. Many graphic symbols have become the common language of users.
Most graphic symbols have national standards, so national standards should be adopted in design.
3) Sign. Here is the trademark and nameplate of the product. A trademark is a commodity mark and cannot be changed at will. A nameplate is a product model and name. They are well designed and play a good decorative role and a strong role in publicity, so we must also attach great importance to them.

7. Panel Material and Its Coloring Technology

The main materials for making instrument panel are aluminum plate, engineering plastics and plastic-bonded aluminum sheet[6].

1) Aluminum plate. This material is the most widely used in instruments. Different texture panels can be obtained by electrochemical treatment, dyeing, sandblasting, wire drawing, polishing and painting. Aluminum plate is light in weight, long in life. Its treatment methods are simple and varied, suitable for single piece, small batch and large-scale production. The common coloring processes have fine printing and offset printing.

2) Engineering plastics. Engineering plastics of injection moulding can be made into convex and concave three-dimensional panels, which has the characteristics of soft color, harmonious pattern and various colors. It is also a widely used on instrument panel, but it is not suitable for single piece and small batch production, because the cost of manufacturing a set of moulds is very high. Common coloring processes are: PVC transparent plate silk printing, polycarbonate PC colorless transparent plate silk printing and plastic parts direct silk printing.

3) Plastic-bonded aluminum sheet. Plastic-bonded aluminum sheet is formed by combining processed plastics with aluminum sheet through a certain process. It has a variety of colors, soft color, delicate texture, simple and generous, stable and elegant artistic effect. This material is also widely used in instrumentation. The coloring process of plastic-bonded aluminum sheet is similar to that of engineering plastics.

8. Conclusion

The basic requirements of instrument panel modeling are composition and layout, representation of functional area, color, text, graphic symbols and signage, materials and coloring technology. Like the face of the product, its panel is the visual center of the entire product, is the core part of product modeling, it will made an important and deep impression on you. The modeling design of instrument panel has a great influence on the performance modeling and even sales of the whole product, and must be given high attention.

Acknowledgments

This research was supported by mechanical CAD/CAM for project establishment course.

Reference

[1] Li Li 2016*Product Innovative Design* (Beijing: Metallurgical Industry Press)
[2] Wei Chen CCME2018 *Control Panel Modeling Design of Household Electrical Appliances* (USA: DEStech Publication, Inc.) p416-420
[3] Yuanlai-Min 2011*Case modeling and color* (Wuhan: Hubei People's Publishing House)
[4] Weitie-Lai, 2012 *Molding design of electromechanical product* (Shanghai Science and Technology Publishing House)
[5] Niuwang-Zhong 2009 *Design of control panel for household appliances* (New trend of design Vol.6)p13-15
[6] Jing-Yuan Yao 2015 *Product Materials and Design* (Beijing: National Defense Industry Press)

Variable Stiffness Iterative Learning Control for Flexible-Joint Robot

Jiankun Pang*, Wu Wei and Qiuda Yu

Dept. of Automation Science and Engineering South China University of Technology No. 381 Wushan Road Tianhe District, Guangzhou, China

*Corresponding author

Abstract. Flexible-Joint (FJ) robots have attracted researchers' attention in these years for their good compliance and the safety. But as a matter of fact, it is hard to design a suitable controller with stiffness adaptation for FJ robots due to the model uncertainties and the nonlinear systems when using elastic components. In this paper, an iterative learning control method for flexible-joint robot with model uncertainties in both robot dynamics and actuator dynamics is proposed, which involves the use of mechanical elastic elements with adjustable stiffness. Adjusting the stiffness of the joints by using the positions of the roller installed inside of the actuator, the joints generate fixed periodic motions and reduced the tracking error. We show the theory of the variable stiffness and the adaptive law, which involve the iterative learning controller. Simulation results also demonstrate that we achieve a perfect result while generating the desired motions.
Index Terms: iterative learning; stiffness adaptation; flexible-joint robot; motion tracking.

1. Introduction

Over the past three decades, FJ robots have been the focus of numerous researchers because of their good compliance and safety. It is safer to interact with the outside world whether it is the FJ robot itself or the person interacting with it. Even if it is subjected to external shocks, it still proves robust to damage, which makes FJ robots better able at adapting to outside environments and enhances their work safety and reliability.

The FJ robot is a very complex dynamic system whose dynamic equations are characterized by nonlinearity, strong coupling, and consolidation. For the study of the dynamics of flexible arms, the establishment of the model is extremely important. The FJ robot is not only a rigid-flexible nonlinear system, but also a nonlinear system in which the system dynamics and control characteristics are coupled. The parametric uncertainties, nonlinear dynamics, and other disturbances will arise in the control procedure. These inaccuracies may lead to poor performance when controlling the FJ robots, thus any practical controller design should not ignore these influences.

A lot of researchers have developed controllers for robots using different control methods. E. Gurkan [1] and C. W. Park [2] may use fuzzy control to deal with the uncertainties of the FJ robots. C. J. B. Macnab [3], C. Kwan [4] and F.Abdollahi [5] used neural networks (NN) control to solve this problem. As a matter of fact, many controller parameters are considered while using neural network learning and some large networks may require much more training data, which would lead to cumbersome calculations [8][9]. In this case, it is of great need to obtain a highly configured computer or other processors, which limits application scope. And it is difficult to guarantee the fast convergence of the NN controllers.

Moreover, the robotic manipulator with active adjustment capability can select the optimal stiffness according to the environment in actual work. If the space is narrow or there are many obstacles, the robot should actively reduce the stiffness and reduce the damage caused by accidental collision [13][14]. If it is in a relatively open work space, the robot should increase the stiffness to meet the operational accuracy requirements. In the future, the manipulator or biped robot should have self-adjusting ability to work in various environment, so as to meet the needs of different working conditions or scenes. Therefore, we try to meet the various conditions by actively controlling the flexibility of the manipulator.
This paper introduces an iterative learning control method [10][12] for FJ robot with model uncertainties in both robot dynamics and actuator dynamics, which involves the use of mechanical elastic elements with adjustable stiffness. Adjusting the stiffness of the joints by using the positions of the roller installed inside of the actuator, the joints generate fixed periodic motions and reduced the tracking error. We show the theory of the variable stiffness and the adaptive law, which involve the iterative learning controller. Simulation results also demonstrate that we achieve a perfect result while generating the desired motions.

2. Problem Formulation

This section formulates the problem of the FJ robot and introduces our control objective.

2.1. Dynamics Model of Flexible-Joint Robots

The dynamic model of FJ robots, can be given by

$$M(q)\ddot{q} + C(q,\dot{q})\dot{q} + D_q\dot{q} + g(q) + K(q-\theta) = 0, \tag{1}$$

$$B\ddot{\theta} + D_\theta\dot{\theta} + K(\theta - q) = \tau, \tag{2}$$

where $\theta \in \mathcal{R}^n$ is the vector of actuator with elastic components, $M(q) \in \Re^{n\times n}$ is the inertia matrix, $C(q,\dot{q})\dot{q} \in \Re^n$ denotes the Coriolis and centripetal forces, and $g(q) \in \Re^n$ represents the gravity vector of robot. $K \in \Re^{n\times n}$ is a diagonal stiffness matrix of actuator, $B \in \Re^{n\times n}$ is a diagonal inertia matrix of actuator, $D_q\dot{q} \in \Re^n$ and $D_\theta\dot{\theta} \in \Re^n$ represent the viscous frictions for the manipulator and the actuator, and $\tau \in \Re^n$ is a vector of the input torque exerted on the actuator. Equation (1) represents the rigid robot dynamics, and (2) describes the dynamic model of actuator.
We can also conclude some following important properties of the dynamics:
1). the matrices $M(q)$ and B are positive definite;
2). the matrix $\dot{M}(q) - 2C(q,\dot{q})$ is skew-symmetric;
3). $D_q = diag(d_{q1},\ldots,d_{q2})$and $D_\theta = diag(d_{\theta1},\ldots,d_{\theta2})$, represent the friction matrices, are diagonal and positive definite;

2.2. Control Objective

First of all, we define some pre-specified trajectories, and our control objective is to track them with little error and time, i.e.,

$$q(t) \to q_d(t), \Delta q(t) \to 0 \ as \ t \to \infty$$

where the desired trajectory $q_d(t)$ and its time derivatives $\dot{q}_d(t)$ are bounded. We aim to bounded the error between the desired trajectory and the actual trajectory.
Then, during the process, we hope to adjust the stiffness K to its optimal value, which we will discuss in the next chapter.

$$K \to K_{opt}, q(t) \to q_d(t), \Delta q(t) \to 0 \ as \ t \to \infty$$

IOP Conf. Series: Materials Science and Engineering **611** (2019) 012070 doi:10.1088/1757-899X/611/1/012070

3. The Variable Stiffness of the Actuator

3.1. The Theory of Variable Stiffness

Now we discuss the theory of variable stiffness. In Figure 1, one of the spring leaves was fixed to the big motor on the left, and the other was connected with the joint on the right by using the rotator. Between the motor and the joint, there was a roller, which could move from left to right or move back driven by the small motor. The positions of the roller mean different stiffness of the joint. As we can see, the effective length of the spring was designed by the roller position. Obviously, when the roller positions change, different torques are required to achieve the same relative rotation angles, which make sense the different torsional stiffness [11]. Therefore, the torsional rigidity between the input and output can be adjusted by adjusting the position of the roller [6].

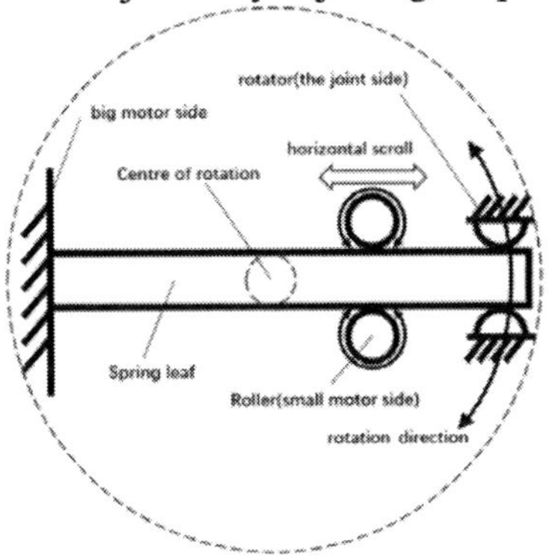

Figure 1. The Internal structure of the actuator.

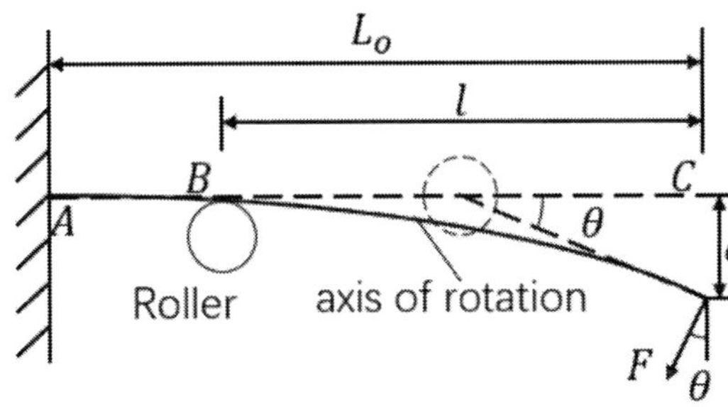

Figure 2. The physical model of the spring.

In Figure 2, L_0 is the whole length of the spring from A to C. The positions of A and C cannot be changed, so we just only change the position of B. The length from B to C means the effective length of the leaf spring, which is signed as l. We assume that the force F is applied at C, which will make a deflection angle θ on the spring leaf.

The joint is a rotating joint that transmits flexible torque. The joint stiffness is defined by the rotational stiffness. The stiffness is calculated as:

$$K = \frac{dT}{d\theta}, \tag{3}$$

where T is the torque at the free end of the spring piece, and θ is the input and output relative torsion angle.

In Figure2, we can get the bending deflection ω, and the relationship between θ and ω is as follows:

$$\theta = sin^{-1}\frac{2\omega}{D}, \tag{4}$$

D is the diameter of the joint.

We can rewrite the equation of ω:

$$\omega = \frac{4 \cdot F\cos\theta \cdot l^2}{Ebh^3} L_0, \tag{5}$$

where E is the Young's modulus of the leaf spring, b is spring leaf width, and h is the thickness of the spring leaf. From (3)(4)(5), we can obtain F as:

$$F = \frac{Ebh^3 D \cdot \tan\theta}{8l^2 L_0}, \tag{6}$$

Then we have,

$$T = F \cdot D/2, \tag{7}$$

So, we can obtain the stiffness K:

$$K = \frac{d\tau}{d\theta} = \frac{Ebh^3 D^2}{16l^2 L_0 \cos^2\theta}, \tag{8}$$

3.2. Stiffness Adaptation

In this section, we discuss the stiffness adaptation. We hope to adjust the stiffness of our actuator during the process without requiring exact parameter values of the robot system. We obtain the adaptive control law $\dot{k}$ for the stiffness K, which means:

$$\dot{k} = \Gamma \cdot \Delta q \cdot y, \tag{9}$$

$$y = \Delta\dot{q} + C\Delta q, \tag{10}$$

where $K = (k_1, \dots, k_n)^T$, $C = \mathrm{diag}(b_1, \dots, b_n) = K_p K_d^{-1}$. $\Gamma \in \Re^{n\times n}$ is a positive definite matrix of adaptive gains, and $\Delta q = (\Delta q_1, \dots, \Delta q_n) = q - q_d$.

4. Iterative Learning Controller Design

We now design the iterative learning controller for the actuator.

The actuator torque is designed by

$$\tau = -K_p \Delta q - K_d \Delta\dot{q} + \Xi_u, \tag{11}$$

where $K_p = diag(k_{p1}, \dots, k_{pn})$ is the angular feedback gain, $K_d = diag(k_{d1}, \dots, k_{dn})$ is the velocity feedback gain, and Ξ_u represents the feedforward torque, which is defined later.

The feedforward torque Ξ_u adopts an iterative learning method, and uses the error correction control information input of the previous or previous (period) operation, so that the repeated task can be better in the next operation. Repeatedly until the output track of the entire time interval tracks the desired trajectory [7].

Therefore, the feedforward Ξ_u of $i \in N$ $(i \geq 2)$ cycle is updated based on the manner of iterative learning control

$$\Xi_u(t) = \Xi_u(t-T) - \beta\{y(t-T) - W_i q_d(t-T)\}, \tag{12}$$

where $\beta \in \Re$ is the learning gain, $W_i = diag(w_{i1}, \dots, w_{in})$, $y = (y_1, \dots, y_n)^T$. The feedforward torque Ξ_u in the first cycle is initialized as $\Xi_u(t) = 0$.

$$w_{ij} = \frac{\int_{iT-T}^{iT} y_i q_{dj} dt}{\int_{iT-T}^{iT} q_{dj}^2 dt} (j = 1, \dots, n), \tag{13}$$

The coefficients $w_{i1}, \dots, w_{in}$ of the $i+1$ cycle $(iT \leq t < iT + T)$ are updated at the beginning of the $i+1$ cycle $(t = iT)$ by using the signals of the i th cycle $(iT - T \leq t < iT)$.

5. Simulation Results

In this section, to illustrate the validity of the controller, we conduct a simulation for the two-link FJ manipulator. The results of the joint trajectory, error, torque and stiffness K are given in the simulation.

5.1. Parameters Setting

The two links of FJ robot have lengths $l_1 = 0.30m$, $l_2 = 0.30m$, distances between their joints and the respective centre of masses, $l_{c1} = 0.15m$, $l_{c2} = 0.15m$, moments of inertia $I_{c1} = I_{c2} = 0.5kg \cdot m^2$, and their masses are $m_1 = 1.5kg$, $m_2 = 1.0kg$. The inertia matrix $B = 0.2I_2 kg \cdot m^2$, where $I_2 \in \Re^{2\times 2}$ is an identity matrix, the friction matrix $D_\theta = I_2 kg \cdot m^2$. The above parameters were the basic elements of the FJ robot, which could be different when using different model.

The desired motion q_d was designed as: $q_{d1} = cos(1.5 * t + 3/\pi)$, $q_{d2} = 1.2 * cos(1.5 * t)$. We define tow different desired motions so as to indicate the adaptive ability of our controller.
The angel and velocity feedback gains for the rigid robot were set as $kp_1 = 25,\ kp_2 = 22.5,\ kd_1 = 15,\ kd_2 = 14.5$. And the learning parameters were set as $\Gamma = [1.5,0; 0,1],\ \beta = 0.005$.

5.2. Results

Simulation results are shown in Figure.3 - Figure.7. Figure.3 and Figure.4 show the trajectories of each joint; Figure.5 shows the tracking errors of each joint; Figure.6 shows the stiffness K. Figure.7 shows the torques that applied in joint 1 and joint 2 using our controller.

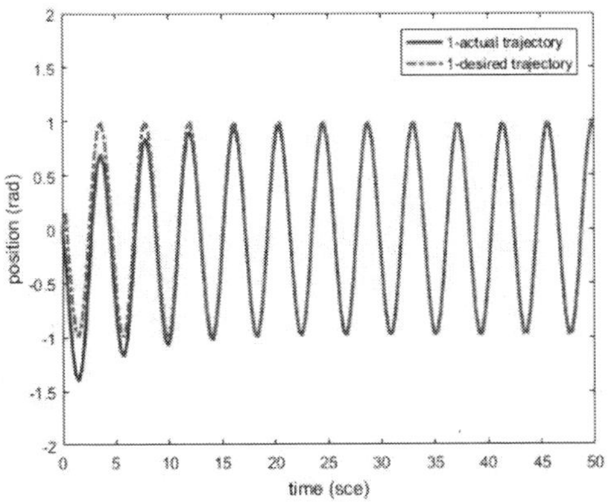

Figure 3. Trajectory for joint 1.

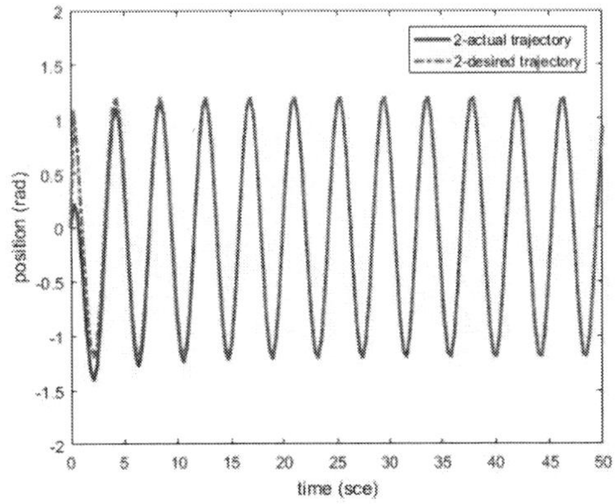

Figure 4. Trajectory for joint 2.

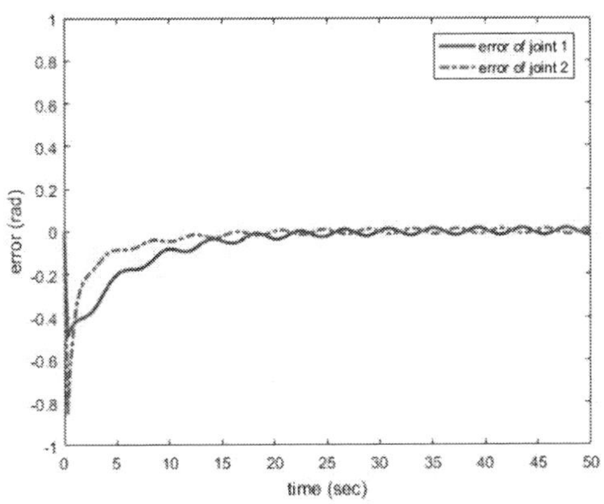

Figure 5. Tracking error for each joint.

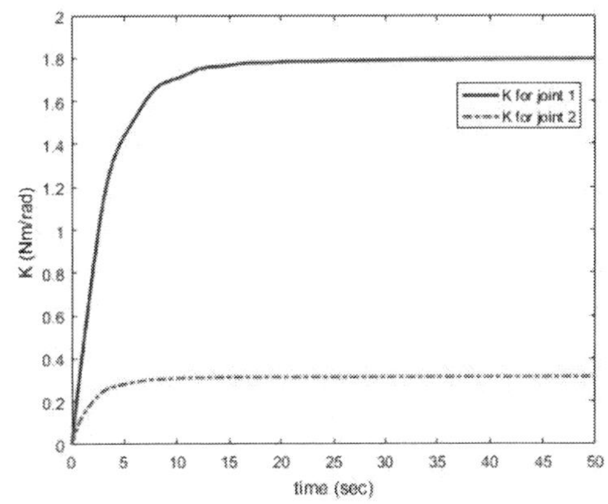

Figure 6. The stiffness K for each joint .

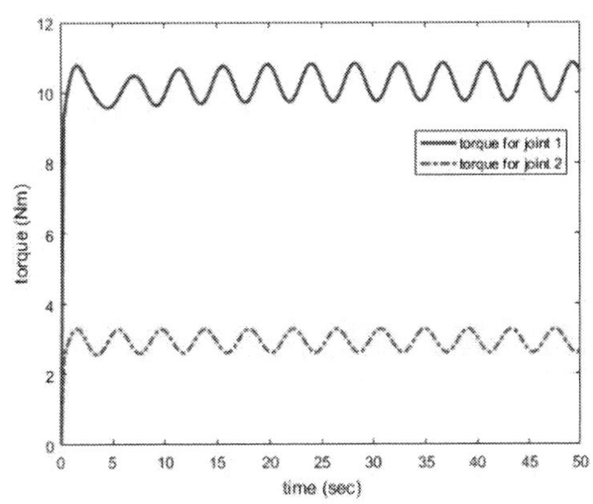

Figure 7. The torque of each joint.

From the simulation results, the average trajectory errors for the iterative learning controller are 0.015 rad and 0.011 rad. The optimal stiffness K of joint 1 is 1.75Nm/rad, and the optimal stiffness K of joint 2 is 0.25 Nm/rad. We set the original stiffness K as 0 Nm/rad because it is easy to discuss the process of the optimal stiffness generation. The torque for joint 1 is 10.5 Nm, and for joint 2 is 2.8 Nm. Obviously, the iterative learning controller increases the tracking accuracy, which represents that the good performances of our control. The iterative learning controller considers the whole dynamic model of the FJ robot, which includes the rigid robot and the actuator. While generating the desired motions, our controller finally find an optimal stiffness for the system without requiring additional parameters of the actuator. Moreover, the iterative learning controller clearly guarantees the convergence of the close-loop system after several period.
The above simulation results showed us the tacking errors became smaller and smaller when the stiffness K reached to its optimal value. Then, we didn't have to adjust many control parameters, which made it much easier to applied.

6. Conclusion

Adjusting the stiffness of the joints by using the positions of the roller installed inside of the actuator, the joints generate fixed periodic motions and reduced the tracking error. We show the theory of the variable stiffness and the adaptive law, which involve the iterative learning controller. Simulation results demonstrate that our controller gives a perfect performance while generating the desired motions. One of our core future works is to apply our controller to control the real-world manipulators with external disturbance.

Acknowledgements

The NSFC (61573148), Science and Technology Planning Project of Guangdong (2015B010919007, 2016A04043012 and 2017B090901043) and the Science and Technology Planning Project of Guangzhou (201604016014 and 201604046015) supported this work.

References

[1] E. Gurkan, S. P. Banks and I. Erkmen. *Stable controller design for the T-S fuzzy model of a FJ robot arm based on Lie algebra*. Proc. IEEE Conf. Decision Control, pp. 4717-4722, 2003

[2] C. W. Park and Y. W. Cho. *Adaptive tracking control of flexible joint manipulator based on fuzzy model reference approach*. Proc. Inst. Elect. Eng. Control Theory Appl., vol. 150, no. 2, pp. 198-204, 2003

[3] C. J. B. Macnab, G. M. T. D'Eleuterio and M. Meng. *CMAC adaptive control of FJ robots using backstepping with tuning functions*. Proc. IEEE Int. Conf. Robot. Autom., pp. 2679-2686, 2004

[4] C. Kwan and F. L. Lewis. *Robust backstepping control of nonlinear systems using neural networks*. IEEE Trans. Syst., Man, Cybern. A, Syst., Humans, vol. 30, no. 6, pp. 753-766, 2000

[5] F.Abdollahi, H.A. Talebi, and R.V. Patel. *A stable neural network based observer with application to flexible-joint manipulators*. IEEE Trans. Neural Netw. vol.17, no. 1, pp. 118-129, Jan.2006

[6] Tao Y, Wang T, Wang Y, et al. *Design and modeling of a new variable stiffness robot joint*[C]2014 International Conference on Multisensor Fusion and Information Integration for Intelligent Systems (MFI). IEEE, 2014.

[7] Arimoto S. *Control Theory of Nonlinear Mechanical Systems*[M]. Oxford University Press, Inc. 1996.

[8] Yoo S J, Park J B, Choi Y H. *Adaptive dynamic surface control of flexible-joint robots using self-recurrent wavelet neural networks*[J]. IEEE Transactions on Systems Man & Cybernetics Part B Cybernetics A Publication of the IEEE Systems Man & Cybernetics Society, 2006, 36(6):1342-55.

[9] Abidi K, Xu J X. *Iterative Learning Control for Sampled-Data Systems: From Theory to Practice*[J]. IEEE Transactions on Industrial Electronics, 2011, 58(7):3002-3015.

[10] Lam H, Leung F, Tam P. *A switching controller for uncertain nonlinear systems*[J]. IEEE Control Systems Magazine, 2002, 22(1):7-14.

[11] *A new variable stiffness robot joint*[J]. Industrial Robot: An International Journal, 2015, 42(4):371-378.
[12] Lijin Fang, Yan Wang. "*Study on the stiffness property of a variable stiffness joint using a leaf spring*", Proceedings of the Institution of Mechanical Engineers, Part C: Journal of Mechanical Engineering Science, 2018.
[13] Yoo, S. J. *Actuator fault detection and adaptive accommodation control of flexible-joint robots*[J]. IET Control Theory & Applications, 2012, 6(10):1497---1507.
[14] Jiang Z H, Shinohara K. *Workspace trajectory tracking control of flexible joint robots based on backstepping method*[C] Region 10 Conference. IEEE, 2017.

IOP Conf. Series: Materials Science and Engineering **611** (2019) 012071 doi:10.1088/1757-899X/611/1/012071

Outgassing Environment of Spacecraft: An Overview

Zilong Jiao[1,2,*], Lixiang Jiang[1,2], Jipeng Sun[1,2], Jianguo Huang[1,2] and Yunfei Zhu[1,2]

[1]Science and Technology on Reliability and Environmental Engineering Laboratory, Beijing, China
[2]Beijing Institute of Spacecraft Environmental Engineering, Beijing, China

*Corresponding author

Abstract. With the advancements of materials science and technology, organic materials found significant applications in aerospace industry. But under the orbital thermal and vacuum environment, organic materials can outgas and the outgassed products and induced environment often degraded performances of spacecraft or its subsystem. In this article, the typical impacts of outgassing environment on spacecraft are explained briefly. Then the outgassing theory, test method, identification of compounds and control measures were reviewed in detail. The need for transient and long-term outgassing model is discussed. The standard test methods of E595 and E1559 are compared. The promising new techniques such as infrared spectrometry and/or mass-spectrometry with thermogravimetric analysis is proposed for the identification of outgassed compounds. The vacuum bakeout and molecular absorber for outgassing control are reviewed and for the last more research work is needed.

1. Introduction

The space industry has made great achievements, and it is inseparable with aerospace materials technology development and breakthroughs. Developments of materials play a strong supporting role for space technology, in turn, the development of space technology has great demand to lead and promote the development of aerospace materials [1]. Deep space exploration, high-resolution Earth observation, manned space program and other projects are put forward even higher requirements for the aerospace material. NASA in 2012 announced technology roadmap, noting in particular high-performance structural and functional materials is one of the areas in the next 10 years need to focus on the study [2].

With the stringent requirements of long life, high reliability, high resolution on future spacecraft, the application of polymeric materials in spacecraft is increasingly demanded. For example, carbon fibre/epoxy composites widely used in structures of spacecraft [3]; a large number of thermal control coatings, thermal insulation materials and others used in the spacecraft's structure and thermal control system [4]; the manufacturing processes of spacecraft's structure also uses adhesives extensively [5].

When in space, the materials used in spacecraft including structural and functional materials must withstand the space environment factors, including high vacuum, rather large temperature variations, ionizing radiation (Earth trapped radiation belts, solar cosmic rays, galactic cosmic rays), solar electromagnetic radiation, micrometeoroids / orbital debris, atomic oxygen, plasma environment and other factors, resulting in chemical, electrical, thermal, optical, mechanical and other performance parameters of materials have been seriously affected, and therefore the need for adaptability of the space environment materials research [6-8].

Polymeric materials can outgas in the space environment, causing mass loss, not only affects the performance of the material itself, the outgassing products and its formed self-atmosphere also have detrimental effects on the spacecraft itself. Therefore, it is important for materials outgassing research. Although research in this area has a lot of achievements, but the lack of systematic summary analysis, and there is still the key issue is not resolved, affect the effective control of the material outgassing. Spacecraft development is mainly concerned with the practice of the total mass loss, compounds of outgassing and control measures in three areas. In this paper, the results of these studies have been carried out systematically summarized in three areas, pointed out the direction of future work.

2. Outgassing Phenomenon and Its Impact

Any materials in a vacuum environment will lose its mass. Material will evaporate; molecules adsorbed on surface will desorb. Also, organic polymer in the production of the polymerization process will contain a large number of low molecular weight additives. The compounds contained within the material will diffuse to the material surface and desorb from the surface. Evaporation of materials is related to its saturated vapour pressure. Polymers used on spacecraft have very low vapour pressure, so in the normal operating temperature, mass loss caused by evaporation are small. Therefore, the main consideration is outgassing caused mass loss.
As can be seen from figure 1, the outgassing results in brown coloured deposits on aluminium panel, which is a clear evidence of exists of outgassing. And there are many lessons learned about outgassing impacts. Outgassing can easily have serious impact for space borne optical system, HV devices, scientific exploration tasks etc.

Figure 1. Contamination from internal outgassing of spacecraft. The brown coloured outgassing deposits on aluminium panel.

2.1. The Impact on Deep Space Exploration Spacecraft

In the first moon orbital flight by Apollo 8, the silicon rubber seals outgassed severely and contaminated large observation window, so had to temporarily move to less contaminated small view window to take picture and video. On Apollo 14 the sealed motor switch failed during flight, the reason is the room temperature vulcanized (RTV) silicone rubber used during operation outgassed silicone compounds with low molecular weight, and outgassed silicone formed carbon particle with petroleum-based lubricant in the brush arc action, increased brush marks lead to motor failure [9].
European comet probe Rosetta mission used a quadrupole mass spectrometer for detection of comets surrounding gas composition. Analysis showed that material outgassing seriously affect the results of the detection of the gas components [10]. In deep space exploration, as the most important feature of the origin of life, water detection is one of the main tasks, and material outgassing of water is the main component of one of them, so its direct impact on mission success.

2.2. The Impact on Spacecraft Optical System

The image quality of charge-coupled device (CCD) detector inside the navigation sensor of NASA's Stardust probe was degraded due to outgassing. Narrow viewing angle camera on Cassini spacecraft also experienced severe flare in image due to outgassing (figure 2), for removing the contamination, had held a long heating degassing [11].

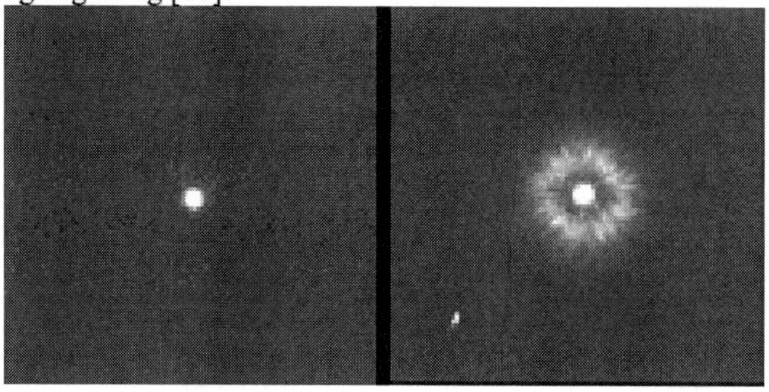

Figure 2. Glow in image of planet due to contamination of spacecraft Cassini: Left: no contamination; Right: with contamination

Space telescope and earth observation satellites are equipped with extremely sophisticated optical systems, materials outgassing product in the low-temperature optical system condenses, causing contamination, so that the optical system transmittance and signal-to-noise (SNR) decrease, seriously affect the performance of the optical system. Hubble Space Telescope and the James Webb Space Telescope have made very stringent contamination control requirements, including control measures of outgassing.

2.3. Impact on the Spacecraft High-voltage Devices

As the spacecraft performance improvement, power increases, the high-voltage devices are used more widely, such as ion propulsion system, a photomultiplier tube (PMT), traveling wave tubes amplifier (TWTA), high voltage solar array and scientific exploration instruments. Vacuum insulation are main insulating method for high-voltage devices, local atmosphere from outgassing easily generated discharge breakdown, seriously affecting the performance and reliability of the satellite [12].

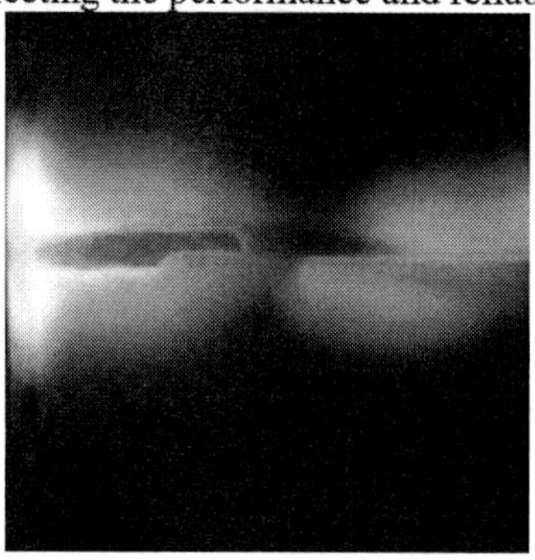

Figure 3. Corona of discharge resulted in high voltage radio frequency system.

A series of satellites in thermal vacuum tests, high-power microwave switch repeatedly appear abnormal temperature rise and along with the output power falling. With dissembling the switch found inside contact ablation. The reason was speculated on outgassing of a large number of non-metallic materials in thermal vacuum environment, formed local atmosphere with higher pressure in the vicinity of high-power microwave switch, in the switch-on process causing the low pressure discharge problems. Besides, outgased contaminants at the same time may condense inside contacts, resulting in a large contact resistance, which may lead to micro-discharge phenomenon, leading to the switch burned [13-14].

3. Materials Outgassing Theory

In order to quantitatively assess the impact of outgassing caused firstly need to determine the amount of outgassing under certain conditions. Bareiss proposed that outgassing rate $\mathrm{d}m/\mathrm{d}t$ is proportional to the remaining material mass m, and the coefficient k related to temperature of materials T, then outgassing rate was expressed as [15]:

$$\frac{\mathrm{d}m}{\mathrm{d}t} = -km \tag{1}$$

$$k = k_0\,\mathrm{e}^{-E_a/RT} \tag{2}$$

$$m = m_0 e^{-kt} \tag{3}$$

Accordingly, the logarithm of total amount of material outgassing $\ln(\Delta m)$ is linear with temperature. Scialdone proved this law by experiments [16]. However, additional experiments found that this equation can only fit best with initial outgassing process, but not for long-term variation of the material outgassing.

Taking into account the long-term changes in the material outgassing controlled by diffusion process, many researchers modelled outgassing process by diffusion equation[18-20].

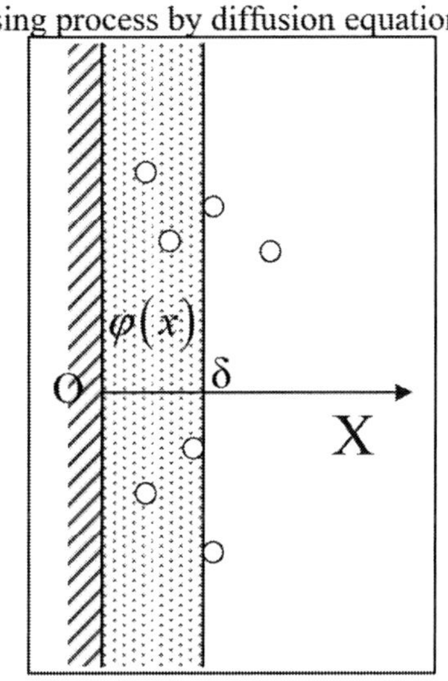

Figure 4. Diffusion model of outgassing

As in figure 4, for one-dimensional diffusion problem, we have:

$$\frac{\partial C}{\partial t} = D\frac{\partial^2 C}{\partial x^2} + f(x,t) \tag{4}$$

There, C is concentration of the volatiles in materials, t is time, x is distance, D is diffusion coefficient, $f(x,t)$ is the volatile source in materials.

Then the outgassing rate can be expressed as follows:

$$q = -D\frac{\partial C}{\partial x}\bigg|_{x=\delta} \tag{5}$$

The outgassing amount at time t is

$$Q(t) = \int_0^t q(\tau)\,d\tau \tag{6}$$

Boundary conditions for the above problem:

$$\begin{cases} C\big|_{x=0} = C\big|_{x=\delta} = 0 \\ C\big|_{t=0} = \varphi(x) \end{cases} \tag{7}$$

The definite solution of the problem is non-homogeneous equations and boundary conditions for the standard homogeneous parabolic equation. Under the given conditions, as long as the initial conditions to ascertain the specific form, it can be solved. Usually suppose that total amount of volatiles isn't change with time, so $f(x,t)$ set to zero. And very often the polymers can outgas not only one kind of volatile compound, but tens of them. But we don't know the initial concentration distribution of them, i.e. the function $\varphi(x)$, and precise value of diffusion coefficient D for every kind of volatiles, that make the equation unresolvable. Then the analysis with diffusion equation can only be qualitative.

In addition to diffusion of volatiles from inner to surface, Roussel et al. suggested that it should be using surface desorption model and internal diffusion model for materials outgassing [21]. The current work is limited to qualitative discussion, it's expected to validate the theoretical analysis with more experimental quantitative results.

In conclusion, theoretical analysis of outgassing have unresolved issues in two aspects. First is the transient material outgassing model. Such as satellite internal pressure changes; the optical system internal outgassing, adsorption, desorption, etc., belong to the transient material outgassing problems. In this aspect focus should on the impact of temperature of the material on outgassing. Second is the long-term outgassing model. Currently satellites can live up to 15 years, but there is no relevant test or measurement data in orbit, make the effective application of the material difficult.

4. Materials Outgassing Test Methods

The American Society for Testing Materials (ASTM) has developed two material outgassing test standards, namely ASTM E-595 "Standard Test Method for Total Mass Loss and Collected Volatile Condensable Materials from Outgassing in a Vacuum Environment" and ASTM E-1559 "Standard Test Method for Contamination Outgassing Characteristics of Spacecraft Materials".

In test method of E-595, the vacuum of test chamber need to be better than 7×10^{-3}Pa, the sample was heated to 125 ℃, test duration is 24h. Test temperature was selected according to spacecraft surface temperatures at extreme conditions, while the degree of vacuum guarantee for molecular flow. The results obtained from the experiments include: material mass losses (total mass loss, TML), the fraction that can deposit of mass loss (collected volatiles of condensable materials, CVCM), the ratio of water to the mass loss (water vapor recovery, WVR). Due to the need for contamination control, generally require the use of TML <1%, CVCM <0.1% of the material.

For test method of E595 there are several deficiencies: 1) TML and CVCM two parameters does not fully reflect the contamination characteristics of outgassing products, i.e. the different degrees of contamination impacts on sensitive surface; 2) In contamination simulation and analysis, it is required dynamic characteristics of outgassing, the relationship between the outgassing, deposition and temperature, while the test method of E595 uses single sample heating temperature, a single collection plate temperature, single test duration, it cannot give dynamic outgassing data; 3) Test temperature and actual using temperature are different; material sample in test are sheet or block, are different with actual state. All these make the test results may not be the same when materials in the orbit. In short, test method of E595 is more an evaluation criterion, not a test standard for outgassing properties.

Recognizing the deficiencies of E595, Air Force Material Laboratory Non-Metallic Material Group contracted to Lockheed Missiles and Space Company to develop a new test method, and provided to the ASTM, became the test method of E1559. After subsequent improvement of this standard method, and now for the 1559-03 revision.

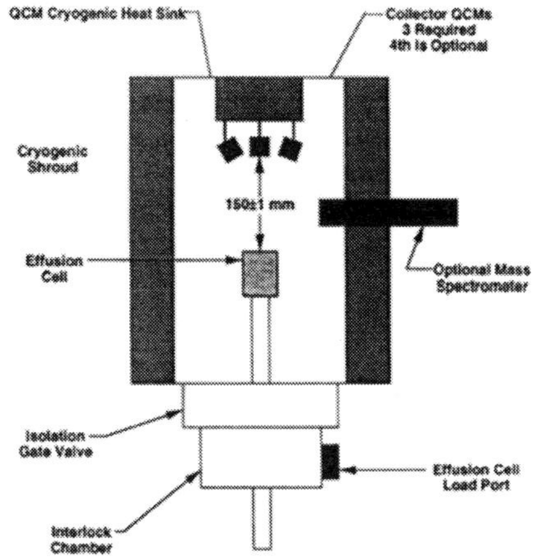

Figure 5. E1559 test configuration diagram.

Test configuration shown in figure 5. Materials placed in temperature-controlled effusion cell. Quartz crystal microbalances (QCM) are temperature-controlled type. A QCM for low temperature, liquid nitrogen cooling, to collect all of its surface impinges outgassing product. TML total mass loss as a function of time is determined by the quality and geometric relationships deposited thereon. The VCM determined by QCM temperature. Inside the test chamber is shroud, i.e. molecular sink. Three QCM were controlled at different temperatures, such that the in-situ material outgassing TML and CVCM at different surface temperatures can be measured. Standard requires temperature control accuracy of QCM should be better than ± 0.5 ℃.

In E-1559 there are two method options: Test Method A and Test Method B. Method A the various test parameters was of the strict rules. 3 QCM operating temperatures are 90K, 160K and 298K. Method B allows the user to select the sample heating temperature, the QCM temperature and other parameters, so as to perform the material outgassing test for specific tasks possible. Two test methods tested samples at three temperatures 398K (125℃), 348K (75℃), 323K (50℃), the duration is 1-5 days.

The test method of E-1559 is characterized in using three different temperatures TQCM, it can provide dynamic data of outgassing in different time and temperature. The use of QCM thermogravimetric analysis (QTGA) also gives composition desorption temperature, the adsorption heat (heat of condensation) and other information about the outgassing component.

5. Identification of Compounds of Materials Outgassing

Materials outgassing products contain complex compounds, and its concentration is extremely low, commonly used analysis methods are mass-spectrometry and infrared spectrometry.

It was found that water is one of the most important product of outgassing, mainly from material processing and storage environment. Relatively loose structure of the polymer material, microscopic pores larger than the diameter of a variety of common gas molecules, therefore polymers can dissolve a variety of gases. Solubility depends on the affinity between the material and the gas molecules. Water is a polar molecule, the affinity is relatively large, so the solubility is high, and this is the reason why water is one of the main products of outgassing.

Chen P T et al. use low temperature cold finger to collect outgassing product in thermal vacuum test, by using mass spectrometry to identify compounds. Through statistical analysis it is found that the molecular weight of more than 100 of outgassing products mainly siloxane and phthalate esters [22]. Yokozawa H. et al., using infrared spectroscopy for outgassing products analysis [23]. The method further improved can be used for real-time analysis of outgassing compounds.

Identification of compounds of material outgassing products should combine with materials outgassing test described in section 4, aimed at obtaining total amounts of different outgassing products. Possible methods include temperature-controlled QCM in combination with mass spectrometry, temperature-controlled QCM in combination with infrared spectroscopy. Vanhove E et

al. presented an experimental method to obtain compound information by coupling QTGA and mass-spectrometry [24]. By using an automated computer program, the outgassing species are separated and their initial masses and characteristic desorption constants are obtained. This technique is called deconvolution by the authors. It is promising, but as the Roussel J-F et al. stated in another paper, there are still some improvements should be made to the experimental configurations, such as temperature control along movement path of molecules etc. [25].
The identification of compounds is expected for further research works.

6. Control Measures of Outgassing

The control measures of materials outgassing includes vacuum bakeout, molecular adsorber, etc. The vacuum bakeout is based on linear relationship between logarithmic of outgassing amount with material temperature described in section 3. It is heating the material to a higher temperature, and artificially remove volatile components in materials, which can effectively reduce the contamination caused by outgassing [26]. Molecular absorber is based on absorption property of the highly porous material which has large specific surface area. Currently more attention was focused on zeolite absorber. NASA GSFC has developed a zeolite absorber coating technology [27-28]. This technology has broad application prospects; it is worth further study.

7. Concluding Remarks

Outgassing is a unique phenomenon of polymer material in space applications. Outgassing can have a negative impact on spacecraft optical systems, high-voltage devices and for scientific exploration missions. Therefore, outgassing and its control measures must be studied in detail. In this paper, the research on outgassing of polymers for space applications in the thermal vacuum environment were reviewed, including theoretical analysis, test method, identification of compounds and control measures.
On the basis of current researches, the next step should be to establish a long-term and transient physical model for outgassing; combine outgassing test with material outgassing compound analysis to create a new test method, aiming at obtaining outgassing characteristics of different products; study control measures of outgassing contamination based the different requirements of materials and spacecraft systems, in particular the application of zeolites and other porous adsorbent material.

References

[1] Li C G, Fu H Z and Yu Q 2002 *Aerospace Materials* (Beijing: Defense Industry) pp 7-8
[2] Wang J Q 2012 *Space Intl.* **7** 35
[3] Chen L M 2005 *Spacecraft Structures and Mechanisms* (Beijing: Science and Technology) pp 63-70
[4] Hou Z Q and Hu J G *Spacecraft Thermal Control – Principle and its Application* (Beijing: Science and Technology) pp 129-140
[5] Chen S H 2004 *Shanghai Aerospace* **3** 39
[6] Miller S K R, Banks B 2010 *MRS Bulletin* **35** 20
[7] Edwards D L, Tighe A P, Eesbeek M V, Kimoto Y and de Groh K K 2010 *MRS Bulletin* **35** 25
[8] Feng W Q 2010 *Spacecraft Envir. Eng.* **27** 139
[9] Malave V, Burkitt B, Riegler B, Johnson R and Thomaier R 2011 *J. of Sp. Roc.* **48** 235
[10] Schläppi B, Altwegg K, Balsiger H, Calmonte U, Hässig M, Hofer L, Jäckel A, Wurz P, Berthelier J J, De Keyser, et al. 2011 Characterization of the gaseous spacecraft environment of Rosetta by ROSINA *Proc. 3rd AIAA Atmospheric Space Environments Conference (Honolulu, US, 27-30 June 2011)*
[11] Haemmerle V R and Gerhard J H 2006 *Cassini Camera Contamination Anomaly: Experiences and Lessons Learned* (Jet Propulsion Laboratory)
[12] Capineri L, Dainelli G, Masterassi M and Dunn B D 2003 *IEEE Trans. Elec. Pack. Manuf.* **26** 294
[13] Liu T X, Luo C, Zhu J T, Yi Z and Yang D S 2014 *Spacecraft Eng.* **23** 47
[14] Zhang H B, Liu T X and Li C J 2011 *Spacecraft Eng.* **20** 125

[15] Bareiss L E 1978 Contamination design analysis approach for spacelab *Proc. 10th Space Simulation Conference (Bethesda, US, 16-18 october 1978)*
[16] Scialdone J J 1976 *An equivalent energy for the outgassing of space materials* (NASA)
[17] Tribble A C, Boyadjjan B, Davis J, Haffner J and McCullough E 1997 *Contamination Control Engineering Design Guidelines for Aerospace Community* (NASA)
[18] Salnikov V A 1994 Vacuum mass loss of materials (mathematical model) *Proc. 6th International Symposium on Materials in a Space Environment (Noordwijk, Netherlands, 19-23 September 1994)* ed T D Guyenne (ESA)
[19] Silver D M 1992 Modeling of spacecraft contamination outgassing as a diffusion-controlled process *Proc. Optical System Contamination: Effects, Measurement, Control III (San Diego, US, 18 December 1992)*, ed A P M Glassford (SPIE) pp 37-45
[20] Girard T J 1993 Algorithm for calculating nonisothermal diffusion-limited outgassing rates *Proc. Damage to Space Optics, and Properties and Characteristics of Optical Glass (San Diego, US, 11 January 1993)* ed J B Breckinridge, A J Marker III (SPIE) pp 114-118
[21] Roussel J-F, Tondu T, Paulmier T, Faye D, Eesbeek M V and Rampini R 2011 *J. of Spacecraft Rockets* **48** 246
[22] Chen P, Hedgeland R and Montoya A 1997 *Statistical Evaluation of Molecular Contamination During Spacecraft Thermal Vacuum Test* (Goddard Space Flight Center)
[23] Yokozawa H, Baba S and Miyazaki E 2012 Evaluation of bakeout effectiveness by optical measurement of a contaminated surface *Proc. Opitcal System Contamination: Effects, Measurements, and Control (San Diego, US, 21 July 2012)*
[24] Vanhove E, Tondu T, Roussel J-F, Faye D and Guigue P 2016 *J. of Spacecraft and Rockets* **53** 1166
[25] Roussel J-F, Vanhove E, Grosjean E, Faye D, Rioland G, Rampini R and Ergincan O 2018 Progress in a physical approach to contamination in Europe *Proc. Systems Contamination: Prediction, Control, and Performance (San Diego, US, 5 November 2018)* ed C E Soares, E M Wooldridge, B A Matheson (SPIE) pp 137-145
[26] Dyer J S, Benson R C, Philips T E, Guregian J J 1992 Outgassing analyses performed during vacuum bakeout of components painted with Chemglaze Z306/9922 *Proc. Optical System Contamination: Effects, Measurement, Control III (San Diego, US, 18 December 1992)* ed A P M Glassford pp 177-181
[27] Abraham N S, Hasegawa M M, Straka S A 2014 Black molecular adsorber coatings for spaceflight applications, *Proc. SPIE Optics and Photonics 2014 (San Diego, US, 19-21 August 2014)*
[28] Abraham N S, Jallice D E 2018 Preliminary testing of NASA's Molecular Adsorber Coating technology for future missions to Mars *Proc. Systems Contamination: Prediction, Control, and Performance (San Diego, US, 5 November 2018)* ed C E Soares, E M Wooldridge, B A Matheson (SPIE) pp 41-46

IOP Conf. Series: Materials Science and Engineering **611** (2019) 012072 doi:10.1088/1757-899X/611/1/012072

Research on Terrain Recognition for Gait Selection of Hexapod Robot

Xiang Zhou[1], Wu Wei[1], Yong Gao[1,*], Kuilin Li[1] and Ruidong Chen[1]

[1]School of Automation Science and Engineering, South China University of Technology, Guangdong, China

*Corresponding author

Abstract. In order to improve the stability of the hexapod robot walking in different terrains, this paper adopts the strategy corresponding to the terrain environment and the motion gait, that is, the robot selects the gait that keeps the robot walk safer and more stable according to the identified terrain. In this paper, using the self-established data set, Terrain6, for terrain classification, the feature extraction of terrain images for hexapod robot is realized firstly based on the Convolutional Neural Networks and the transfer learning. Secondly, according to the stacking fusion method, a terrain recognition model with higher precision is obtained by integrating three terrain classification models include the support vector machine, the naive Bayes and the random forest algorithm. Finally, the experiments show that the hexapod robot selects suitable gait based on the result of terrain recognition to cross complex environment, and the stable and efficient motion of the robot verifies the validity of the research results.

1. Introduction

With its many degrees of freedom, multiple limbs and discrete landing points, the multi-footed robot can achieve stable walking and perform tasks under various complex terrains [1]. As a typical example of a multi-legged walking robot, the hexapod robot has a wide range of motion gait, flexible and variable control scheme, and extremely strong load capacity [2], so it has broad application prospects.
In order to improve the stability and adaptability of the hexapod robot in the unknown complex terrain environment, it is necessary to perceive the terrain environment through the sensors carried by itself [3], and then adopt the corresponding gait control strategy to achieve the desired motion.
During the movement of the robot, the terrain environment picture can be obtained by the camera, and then the color, texture, shape and other characteristic information of the picture are obtained by the classifier model. This kind of terrain classification method based on visual information has been widely used in the field of robotics. In 2011, based on the local texture features of topographic images, Khan et al. selected a random forest classifier with the best overall performance to achieve terrain classification by comparing various classification models [4]. In 2012, Filichkin and Byl of the UCSB Robotics Institute extracted the image features of the terrain environment of the robot based on the Speed Up Robust Feature (SURF) algorithm, generated a relatively compact terrain feature vector through the visual word bag model, and implemented the terrain classification using Support Vector Machines (SVM) classification model. This method has been successfully implanted on the LittleDog robot of Boston Dynamic Robotics [5]. In 2013, for the hexapod robot AMOS II, Steffen Zenker et al. used Scale Invariant Feature Transform (SIFT) or SURF to realize terrain feature detection and implement terrain classification based on SVM [3]. In 2014, for the terrain recognition of outdoor mobile robots, Yuhua Zou et al. experimented with several classifiers of extreme learning machine

(ELM), SVM and nearest neighbor (NN) to evaluate the performance of different image descriptors and classifier combinations [6].

Based on Convolutional Neural Networks, transfer learning and fusion algorithm, this paper classifies and identifies the terrain environment of the hexapod robot, and then selects the corresponding gait mode according to different terrain environments to achieve stable motion in complex environments. The terrain recognition algorithm framework for hexapod robot designed in this paper is shown in Figure 1. The terrain recognition method is divided into offline model training and online terrain recognition. The offline model training is based on the collected terrain dataset to train the recognition model. The online terrain recognition is based on the terrain information collected by the camera carried by the hexapod robot body for real-time terrain classification. Both parts need to use the terrain image of the environment in which the robot is located. Based on the MobileNet Convolutional Neural Networks model in the remote host of the robot control system (a PC with two 1080Ti GPUs), the features of the image are extracted and vectorized [7]. The offline part uses the collected terrain dataset to train three separate models of support vector machine, naive Bayes and random forest, and then use the stacking model fusion method to fuse the three models to obtain the final high-precision terrain recognition model. The online terrain recognition directly utilizes the offline training model, and the environmental terrain image acquired by the camera carried by the hexapod robot body is transmitted to the remote host system through the local area network for online classification and recognition. Finally, according to the results of terrain recognition, the hexapod robot selects the appropriate motion gait and control scheme to complete the specified motion control task.

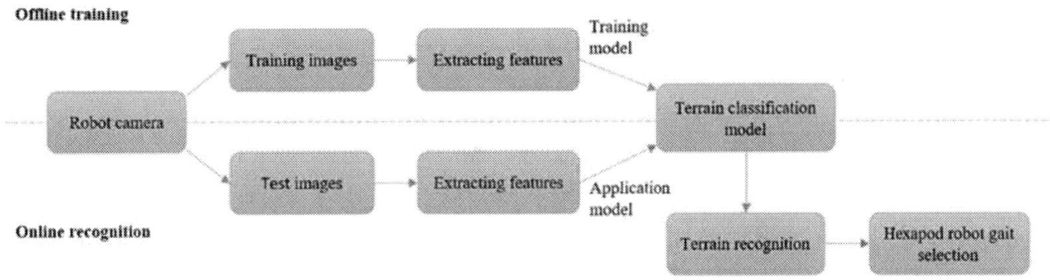

Figure 1. Terrain recognition flow chart.

2. Data Set Establishment

This paper first establishes the data set Terrain6 which is suitable for the terrain classification of the hexapod robot. The data set includes six typical terrains that the subject robot may encounter: tile floor, concrete floor, grassland, sandy land, ground covered with roots, gravel floor, as shown in Figure 2. Each picture is cropped to a pixel value of 256*256, and each type of image contains 250 sheets. Most of the pictures are taken by the camera carried by the robot body under different lighting conditions and weather conditions, and a small part comes from the Internet.

Figure 2. Terrain6 typical terrains.

After collecting the required terrain images, the subject uses image augmentation technology to expand the acquired terrain image dataset by changing the brightness or contrast, random cropping, and rotation of the image to provide more training samples for model training. At the same time, because the augmentation technique randomly changes the sample, the model's dependence on certain attributes can be reduced, thereby improving the generalization ability of the model and preventing the model from overfitting.

3. Feature Extractor Design

Considering the hardware configuration of hexapod robot and the real-time requirements of terrain recognition, this paper uses MobileNet Convolutional Neural Networks to extract terrain image features based on transfer learning method, and then uses the extracted image feature training classifier to realize the recognition of terrain by robot. With classification.

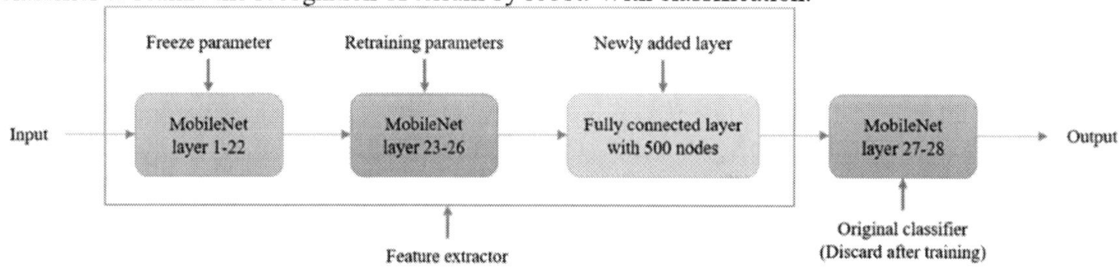

Figure 3. MobileNet neural network based on transfer learning.

The MobileNet Convolutional Neural Networks model based on transfer learning built in this paper is improved by ImageNet large image dataset model. As shown in Figure 3, the model has 28 layers. The 1-26 layers extract features from the image, and the last two layers classify the images based on the extracted features. Since the self-built Terrain6 data set belongs to the image classification problem of small data sets, the transfer learning method is used to freeze the 1-22 layer parameters in the pre-training model [8], and only the 23-26 layer model parameters are retrained. A full connection layer of 500 nodes is added after the 26th layer for extracting picture features. The improved model is trained by using the hexapod robot terrain data set Terrain6, and the network structure parameters after 22 layers are updated to obtain and save the trained model. Finally, the fully connected layer of the original MobileNet neural network classifier part is removed, only retain the feature extraction part of the original model and the improved fully connected layer are added as a new picture feature extractor. The feature of each picture in the data set is represented as a feature vector of length 500, and the picture feature vector set and its corresponding terrain category tag are saved as a training data set of the terrain recognition classifier.

Based on the Keras deep learning framework, the cross entropy loss function commonly used in multi-classification problems is used as the loss function of MobileNet convolutional neural training, and the model is iteratively trained with Adam as the model optimizer. The number of iterations is 30 times. Figure 4 shows the training effect.

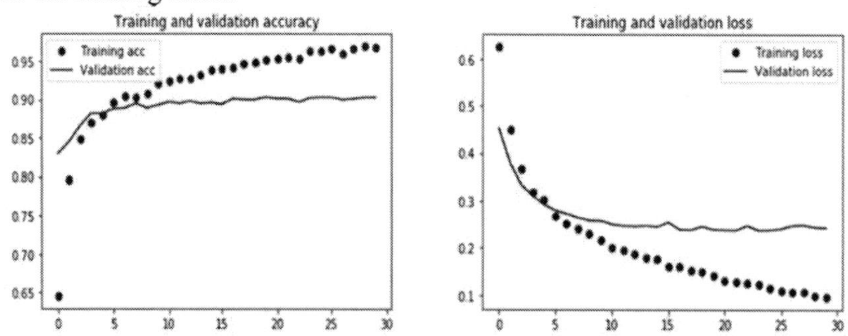

(a) (b)

Figure 4. Training effect of MobileNet neural network: (a) Accuracy, (b) Loss.

It can be seen from Figure 4 that the MobileNet Convolutional Neural Networks based on transfer learning has a good training effect. The accuracy of the model on the training set and the cross-validation set is high (Figure 4 (a)), and the training error is small and tends to converge (Figure 4 (b)).

4. Design and Training of Terrain Classification Model

This section will use the 500-dimensional picture feature vector and picture category label obtained in Section 3 to train the machine learning model for terrain recognition. In this paper, using the accuracy of terrain recognition as an evaluation index, the support vector machine, the naive Bayes and the random forest models are independently trained, and the model parameters are adjusted by 5-folds cross-validation method.

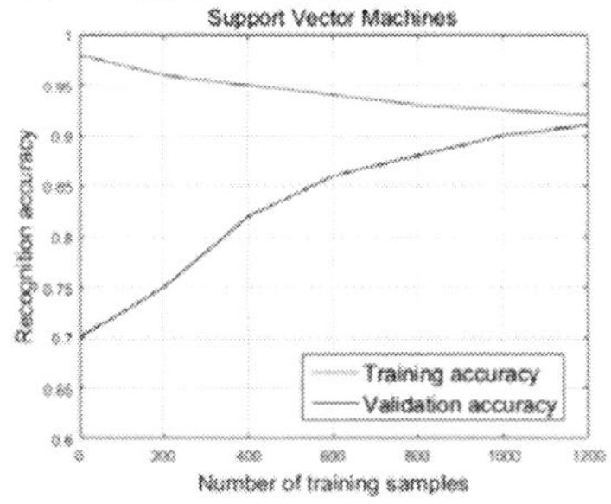

Figure 5. Terrain classifier learning curve based on support vector machine.

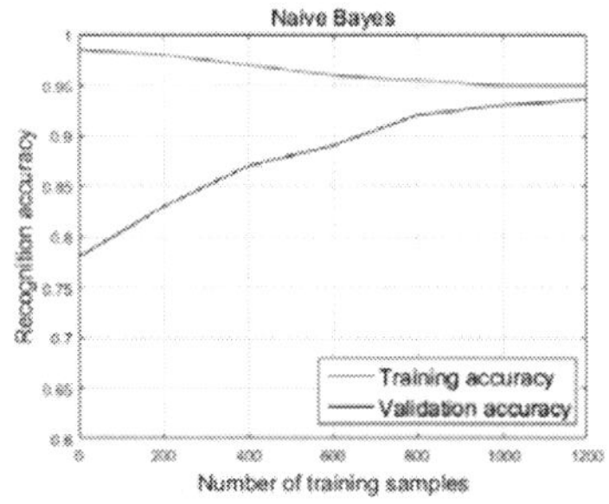

Figure 6. Terrain classifier learning curve based on Naive Bayes classifier.

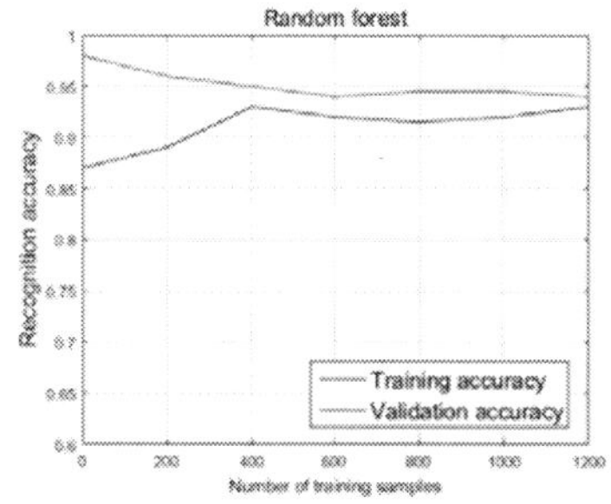

Figure 7. Terrain classifier learning curve based on random forest.

The support vector machine classifier uses a linear kernel as a kernel function. During the training process, the penalty coefficient C of the model is adjusted by cross-validation. When C=8, the model is optimal. The learning curve of the training process is shown in Figure 5. The Naive Bayes classifier is a generation model that is trained using the scikit-learn machine learning library. There are no parameters to be adjusted during the training, and the model training is fast. Its learning curve is shown in Figure 6. The parameters of the random forest model mainly include the number of weak model CART decision trees and the maximum depth of the CART decision tree. Cross-validation shows that when the number of decision trees is 120 and the maximum depth of the decision tree is 15, the model recognition accuracy is the highest, and the learning curve is shown in Figure 7.

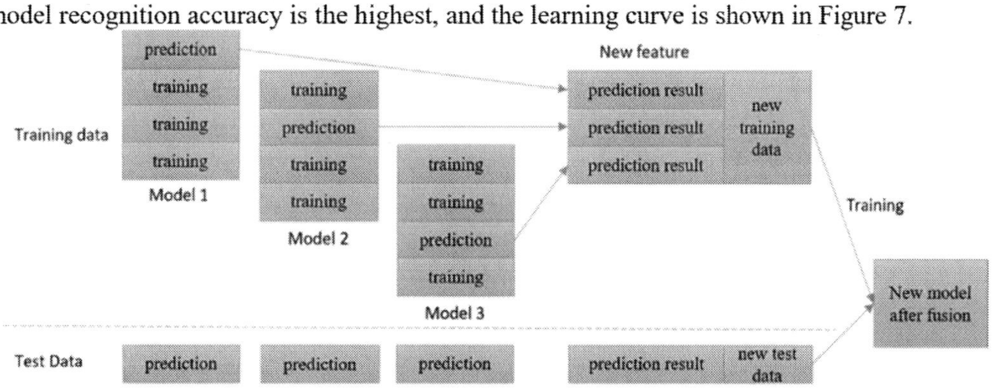

Figure 8. Stacking model fusion architecture diagram.

For the above three independent classification models, the final high-precision model is obtained by using the stacking model fusion strategy [9]. The stacking model fusion algorithm framework is shown in Figure 8. The training process can be broken down into three steps:

1) The terrain data set is disassembled into a training set and a test set according to a 4:1 ratio. At the same time, the training data is randomly and evenly divided into m parts, where m is 4.

2) On the disassembled training set, use the same training set to train the three classifier models described above. In Figure 8, model one is a terrain classifier based on support vector machine, model two is a terrain classifier based on naive Bayesian method, and model three is a terrain classifier based on random forest. The specific training method is to use (m-1) data training models for each individual

model to predict the remaining one. Subsequently, the trained model is predicted using the test set disassembled in step 1.
3) Based on the results obtained in step 2, train a secondary classifier model to obtain the final fusion model. The three training set prediction results are combined into a new training set, and the three test set prediction results are taken as a new test set. Using the new training set and the new test set, a new terrain classification model is trained by logistic regression. This model is a terrain classification model with three independent models.
The accuracy of the identification of each terrain classifier on the test set is shown in Table 1. It can be seen that the terrain classifier model based on stacking fusion proposed in this paper has higher recognition accuracy for terrain.

Table 1. Recognition accuracy of various shape classifiers on test sets.

Classifier type	Recognition accuracy on the test set	Classifier type	Recognition accuracy on the test set
SVM	88.2%	random forest	90.3%
naive Bayes	87.6%	stacking fusion	94.6%

5. Gait Selection of Hexapod Robot and Experiments on Prototype

The experimental platform of the hexapod robot prototype designed in this paper is shown in Figure 9. The robot has a rectangular body structure, six legs are distributed on both sides of the body in an axisymmetric manner, and each leg has three active rotating joints, each of which is driven by a steering gear.

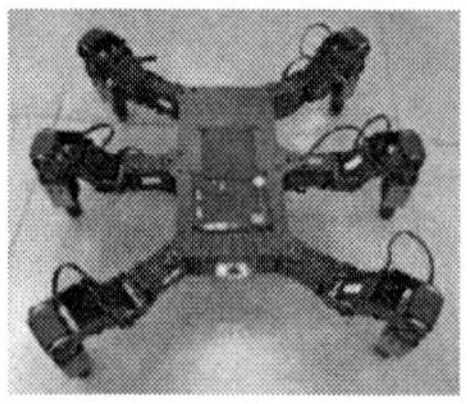

Figure 9. Hexapod robot prototype.

According to the static balance condition of the robotic body, the more legs of the hexapod robot are in contact with the ground, the more stable the robot movement [10], but the slower the walking speed. The optional walking gait of the hexapod robot designed by this subject mainly includes: tripod gait, tetrapod gait and wave gait [11]. The stability of the above three gaits increases sequentially, but the walking speed decreases sequentially.
Considering the terrain recognition results and the motion requirements of the robot: When the local terrain recognition result is flat terrain such as cement floor or tile floor, the robot adopts tripod gait to achieve fast walking; when the local terrain recognition result is grass or sand, the robot should adopt tetrapod gait to walk as fast as possible while keeping the body stable; when the local terrain recognition result is complex terrain such as the ground covered with roots or the gravel floor, in order to ensure the stability of the robot, a wave gait that can be slowly advanced should be adopted.
Figure 10 shows the use of tripod gait when the hexapod recognizes that its location as tile floor. When the hexapod recognizes that its location as the grass, it uses the tetrapod gait, as shown in Figure 11. Figure 12 shows the use of wave gait when the hexapod recognizes its location as the ground covered with roots.

Figure 10. Tripod gait of hexapod robot on tile floor.

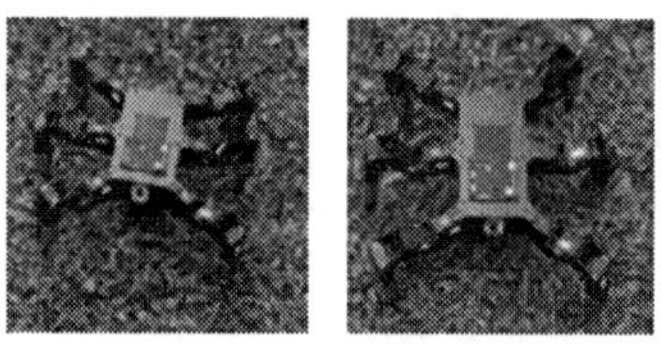

Figure 11. Tetrapod gait of hexapod robot on the grass.

Figure 12. Wave gait of hexapod robot on the ground covered with roots.

6. Conclusion

This paper mainly studies the problem of the terrain recognition and the gait selection of the hexapod robot. Based on the Terrain6, a self-built topographic image dataset, and combined with MobileNet Convolutional Neural Networks and transfer learning, a method of feature extraction of terrain images for hexapod robot was designed. Then, using the stacking fusion method, the three terrain classification models including support vector machine, naive Bayes and random forest are combined to obtain a more accurate terrain recognition model. Finally, the experiment shows that the hexapod robot selects the appropriate motion gait based on the terrain recognition results in three different environments, and verifies the validity of the research results.

In this paper, for the different types of terrain environment identified by the hexapod robot, the fixed rhythm gaits are selected. In the future work, we can consider adding force sensors, integrating foot force analysis and visual information to guide the robot to achieve more stable, safe and efficient adaptive motion under different terrains.

Acknowledgments

This paper supported by the National Natural Science Foundation of China (61573148), the Science and Technology Planning Project of Guangdong Province (2015B010919007, 2016A040403012, 2017B090901043), and the Science and Technology Planning Project of Guangzhou (201604046015, 201604046014, 2017171201).

References

[1] Manhong Li, Minglu Zhang, Jianhua Zhang, et al. 2015 *Review on Key Technology of the Hexapod Robot* (Tianjin: Journal of Machine Design) pp 1-8.
[2] He Zhang 2014 *Research on Force Sensing Based Hexapod Robot and Control for Walking on Uneven Terrain* (Harbin: Harbin Institute of Technology).
[3] Zenker S, Aksoy E E, Goldschmidt D, et al. 2013 *Visual Terrain Classification for Selecting Energy Efficient Gaits of a Hexapod Robot* 2013 IEEE/ASME International Conference on Advanced Intelligent Mechatronics (AIM) (New York: IEEE) pp 577-584.
[4] Khan Y N, Komma P, Bohlmann K, et al. 2011 *Grid-based Visual Terrain Classification for Outdoor Robots using Local Features* IEEE Symposium on Computational Intelligence in Vehicles and Transportation Systems (CIVTS 2011) Proceedings (New York: IEEE) pp 16-22.
[5] Filitchkin P and Byl K 2012 *Feature-based Terrain Classification for Littledog* 2012 IEEE/RSJ International Conference on Intelligent Robots and Systems (IROS) (New York: IEEE) pp 1387-1392.
[6] Zou Y, Chen W, Xie L, et al. 2014 *Comparison of Different Approaches to Visual Terrain Classification for Outdoor Mobile Robots* vol 38 (Amsterdam: Elsevier) pp 54-62.
[7] Krizhevsky A, Sutskever I and Hinton G 2012 *ImageNet Classification with Deep Convolutional Neural Networks* NIPS (New York: Curran Associates Inc) pp 1097-1105.
[8] Gogul I and Kumar V S 2017 *Flower Species Recognition System using Convolution Neural Networks and Transfer Learning* 2017 Fourth International Conference on Signal Processing, Communication and Networking (ICSCN) (New York: IEEE) pp 1-6.
[9] Zicheng Zhou 2018 *The Research and Design of Telecom Customer Credit Model Based on Stacking Model Fusion* (Guangzhou: South China University of Technology).
[10] Ding Kai 2016 *Research on Biomimetic Gait Planning and Locomotion Control System for Hexapod Robots* (Nanjing: Southeast University).
[11] Xiaolin Yin 2013 *Research on Gait Planning and Control Strategy for Biomimetic Hexapod Robot* (Harbin: Harbin Institute of Technology).

Analysis of Harmonic Propagation and Coupling among Wind Farm Groups

Xujuan Miao[1], Jinming Liu[1], Changling Li[1,*], Xinyan Zhang[2], Qidan Xu[1], Jiahui Yue[2] and Ke Sun[2]

[1]State Grid Xinjiang Electric Power Co., Ltd. Economic and Technological Research Institute, Urumqi, China
[2]College of Electrical Engineering, Xinjiang University, Urumqi, China

*Corresponding author

Abstract. Although wind energy itself has the conditions of large-scale development and utilization, power quality has always been the focus and difficulty of wind power generation. Harmonics generated by power electronic equipment in doubly-fed generators may lead to overheating damage and faults of transmission lines and transformer equipment. To improve the safe operation and harmonic control ability of power grid and reduce the loss of electrical equipment, a model of double-fed wind farm group with voltage level of 690V/35kV/110kV was established in Matlab/Simulink. By analyzing the characteristics of harmonic propagation and coupling between wind farms under different conditions, it is found that when the harmonic source was on the fan side, the harmonic is transmitted from the fan side to the grid side. If the harmonic content of wind power station is large, it will have a greater impact on power quality. In addition, due to harmonic coupling, 110 kV bus current has the highest harmonic distortion rate and the highest harmonic content.
Keywords: doubly-fed wind turbine; doubly-fed wind farm group; harmonic propagation; harmonic coupling.

1. Introduction

With the large-scale use of transformers, rotating electrical machines and power electronics in wind power plants, the harmonic hazards caused by non-linear loads are attracting more and more attention from the industry. Harmonics are also increasingly becoming a problem that cannot be ignored in the field of wind power[1]. Regardless of domestic and foreign, harmonic problems have long attracted the attention of practitioners and scholars in the power industry, and have also achieved many achievements.

Foreign research results: The literature [2] through the establishment of a non-traditional doubly-fed wind turbine generator model, simulation analysis, can more intuitively show the impact of harmonic distortion on power quality.Literature [3] outlines the improvement of harmonic evaluation methods based on large-scale wind farms, and also analyzes harmonic phenomena. Literature [4] analyzed the harmonic characteristics of wind farms. Literature [5] discusses whether the harmonics generated by fixed-speed wind turbines are affected by wind speed.

Domestic research results: The literature [6] discusses the use of MATLAB and PSCAD simulation software to simulate the single-number and multi-number DFIG grid-connected cases, and obtain the difference of harmonic currents at the grid-connected points and analyze and distinguish them. Literature [7] discusses the problems existing in the grid connection of single-fan and multi-fan in large-capacity doubly-fed wind power generation system, and proposes a method to solve the problem of grid connection. The literature [8] discusses the relationship between the propagation characteristics of harmonics and the voltage level and the relationship between the harmonic transfer coefficient and the harmonic order. Reference [9] discusses the basic relationship between harmonic voltage propagation and voltage level.

According to the research status at home and abroad, wind farm harmonics research mainly focuses on the localization of internal harmonic sources [10], the influence of electric field voltage level on harmonics and the harmonic propagation characteristics of transmission lines in electric field. From the macro point of view, the study of harmonic propagation and coupling among wind farm groups is relatively rare.

In this paper, based on MATLAB/SIMULINK, wind farms with different capacities are built to form two wind farm groups, and the fast Fourier transform of each wind farm current waveform is performed. By comparing the harmonic amplitude, it can be obtained that when the harmonic source is measured by the fan, the harmonics are transmitted from the fan side to the grid side, which will increase the harmonic current of each power generation field in the power grid and increase the current distortion rate. For wind farms with small installed capacity, the sudden increase of harmonic current content of the power grid has a significant impact on the power quality. The harmonic superposition and cancellation phenomenon between the multi-harmonic sources, harmonic stacking on the 110kV bus bar in this experiment, the current harmonic distortion rate is the largest, and the current with the highest harmonic content is generated.

2. Harmonic Current Permissible Value of Power Grid

In order to ensure the quality of power, countries have promulgated national standards to limit the harmonic content in the power grid to the normal operation of the power system. The allowable harmonic currents of public power grids are shown in Table 1 below.

Table 1. Harmonic current permissible value of public grid.

Standard voltage/kV	Reference Short Circuit Capacity/MVA	Harmonic frequency and allowable value of harmonic current/A							
		2	3	4	5	6	7	8	9
0.38	10	78	62	39	62	26	44	19	21
6	100	43	34	21	34	14	24	11	11
10	100	26	20	13	20	8.5	15	6.4	6.8
35	250	15	12	7.7	12	5.1	8.8	3.8	4.1
66	500	16	13	8.1	13	5.4	9.3	4.1	4.3
110	750	12	9.6	6.0	9.6	4.0	6.8	3.0	3.2

For harmonic analysis, it is necessary to understand the current distortion rate. The formula for calculating the total current harmonic distortion rate is as follows:

$$THD_I = \frac{1}{I_1}\sqrt{\sum_{h=2}^{\infty} I_h^2} = \sqrt{(\frac{I_{rms}}{I_{1_{rms}}})^2 - 1} \tag{1}$$

In the formula, I_1 means the peak value of fundamental current.

3. Wind Farm Group Modeling

Based on the DFIG model that comes with MATLAB/SIMULINK, five wind farms are built to construct two wind farm groups (the wind farm group capacity is 27MW and 22.5MW respectively). The grid voltage is 690V/35kV/110kV and the frequency is 50Hz. The wind farms are connected to the common grid to form a wind farm group. The wind farm and its parameters are shown in Table 2 below:

Table 2. Wind farm installed capacity and wind speed distribution.

Wind farm name	Installed capacity(MW)	Rated wind speed(m/s)
Wind farm A	12	10
Wind farm B	9	5
Wind farm C	6	8
Wind farm D	15	9
Wind farm E	7.5	7

The wind farm A, the wind farm B, and the wind farm C are combined into a wind farm group A; the wind farm D and the wind farm E are combined into a wind farm group B.

4. Prediction Model

Since the wind speed is random in the wind farm, in order to get closer to the actual operating conditions of the wind turbine, a white noise module is added based on the rated wind speed.

4.1. Analysis of Harmonic Characteristics of Doubly-fed Wind Farm Group

Firstly, running the wind farm group model. In order to obtain the harmonic characteristics of the wind farm group, the wind farm output working waveform and the wind farm group grid current waveform are analyzed. The results are shown in the Fig. 1.

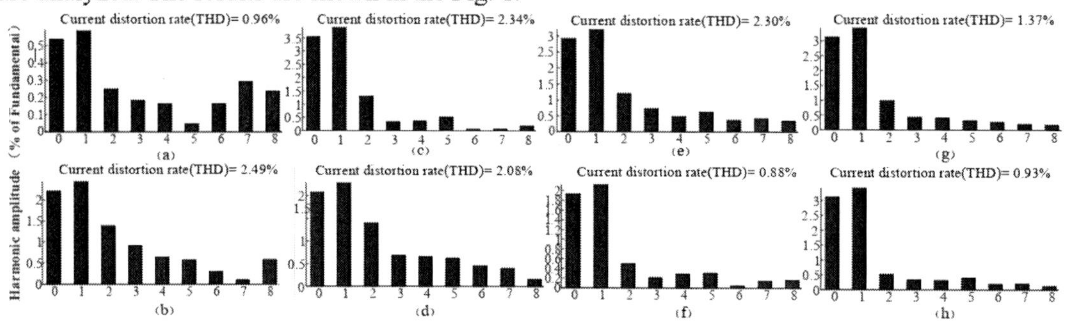

Figure 1. Wind farm current spectrum analysis.

In Fig. 1, a, b, c, d and e respectively represent 690V bus FFT harmonic analysis of wind farm A, wind farm B, wind farm C, wind farm D and wind farm E; f and g respectively represent wind farm groups A and wind farm group B 35kV bus FFT harmonic analysis; h map shows 110kV grid point FFT harmonic analysis. Because it is in normal working state, the harmonic content of the bus bars at all levels is very small, and the distortion rate is also very low. The power quality of each power station and grid side meets the requirements.

The waveforms obtained under normal working conditions are analyzed, and the distortion rate of the bus bars at the same time is statistically tabulated in the table 3:

Table 3. Distortion rate on bus under normal working conditions.

Bus Bar	THD%	Bus Bar	THD%
Wind farm A(690V)	0.96	Wind farm E(690V)	2.30
Wind farm B(690V)	2.49	Wind farm group A (35kV)	0.88
Wind farm C(690V)	2.34	Wind farm group B (35kV)	1.37
Wind farm D(690V)	2.08	Connecting point (110kV)	0.93

IOP Conf. Series: Materials Science and Engineering **611** (2019) 012073 doi:10.1088/1757-899X/611/1/012073

4.2. Analysis of Simulation Results of Harmonic Propagation in Wind Farm Group

In order to analyze the harmonic propagation law of harmonics in the wind farm group, the distribution of the 3rd and 5th harmonics of each bus bar in the wind farm group model is observed after the 3rd and 5th harmonic sources are injected into the wind farm A. The simulation results are as follows the Fig. 2 shows.

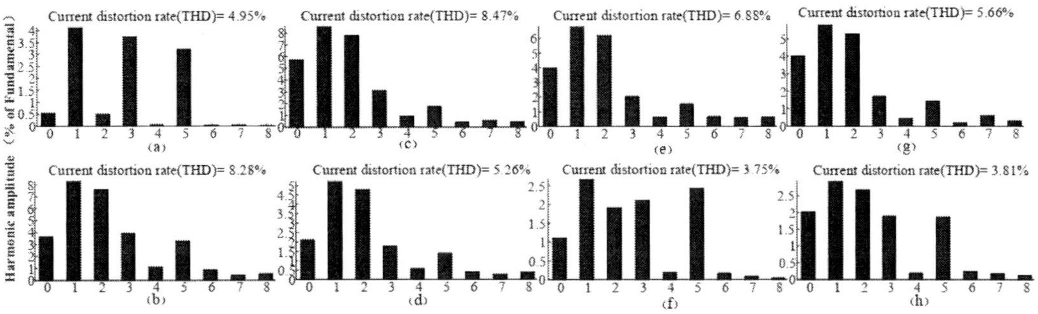

Figure 2. Wind farm current spectrum analysis.

In Fig.1 and Fig.2, each FFT harmonic analysis map still corresponds to the same 690V/35kV/110kV bus. Because the wind farm A generates the 3rd and 5th harmonics, the 3rd and 5th harmonics in the harmonic analysis itself are very high. In the harmonic analysis of the Wind farm group A bus (35kV) and the 110kV grid point of the wind farm group, it can be seen that the content of the 3rd and 5th harmonics is significantly higher than the normal working state.

The distortion rate of the bus bars at the same time is taken for statistical calculation in the table 4:

Table 4. Bus bar distortion rates.

Bus Bar	THD%	Bus Bar	THD%
Wind farm A(690V)	4.95	Wind farm E(690V)	6.88
Wind farm B(690V)	8.28	Wind farm group A (35kV)	3.76
Wind farm C(690V)	8.47	Wind farm group B (35kV)	5.66
Wind farm D(690V)	5.26	Connecting point (110kV)	3.81

Making two sets of data into Fig.3 to compare and analyze:

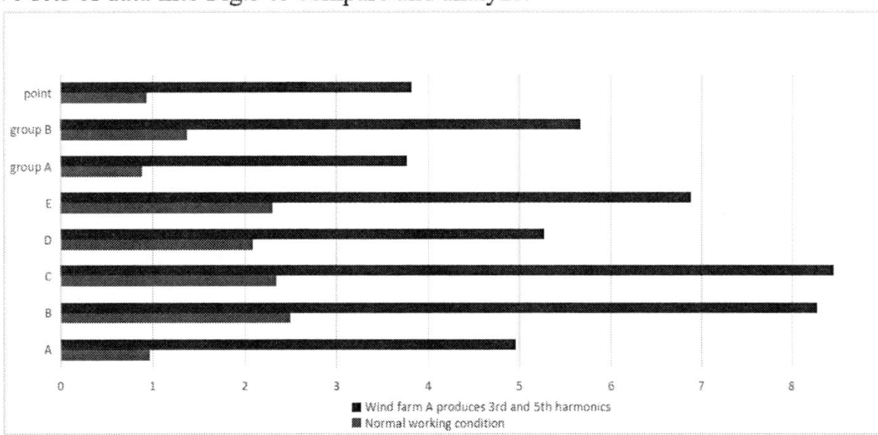

Figure 3. Current distortion rates under different states of wind farm.

By comparing and analyzing the distortion rate, it can be seen that when the 3rd and 5th harmonics are emitted by the wind farm A, the current distortion rates in the wind farm line increase a lot. At the same time, the harmonic current of wind farm A is not only propagated from the fan side along the

690V line to the wind farms B and C, but also through the 35kV bus to another wind farm group B, resulting in increasing THD up to 5%,even to 8%. And the grid connection point has also increased. Obviously, the harmonics of wind farm A also affect the working state of other wind farms to some extent and the power quality is seriously polluted.

In this way, it can be judged that when the harmonic source is on the fan side, the harmonics are transmitted from the fan side to the grid side. When the harmonic content of the fan side is large, the power quality of the entire power grid is greatly affected, and this result agrees with the principle of harmonic propagation.

4.3. Analysis of Harmonic Coupling Simulation Results of Wind Farm Group

In order to analyze the coupling law of harmonics in the wind farm group, after injecting the 3rd and 5th harmonics in the wind farm A and the wind farm D, it can observe the node currents of the 35kV bus and the 110kV bus of the grid, especially the 2nd, 3rd, 4th and 5th harmonics distribution, the simulation results are shown in Fig.4:

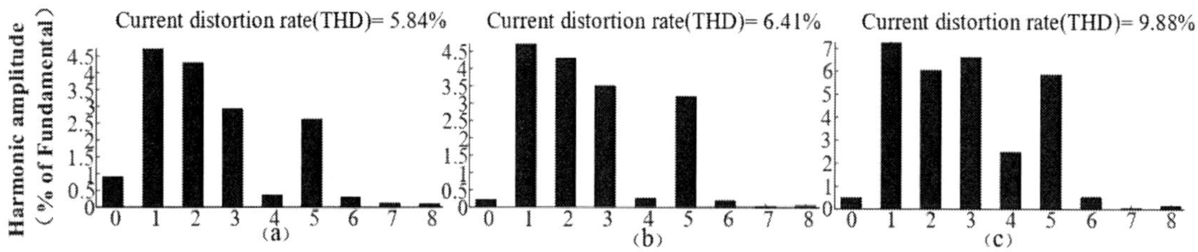

Figure 4. Wind farm group current spectrum analysis diagram.

The diagrams of a and b show 35kV bus FFT harmonic analysis, c shows the 110kV bus FFT harmonic analysis of the grid connection point. It can be seen from Fig.5 that the harmonic content of the 110kV bus bar increases sharply, because of the 3rd and 5th harmonics in each wind farm group. Except for the fundamental wave, the amplitudes of the 2nd, 3rd, 4th and 5th harmonics are very high. Taking the distortion rate of the 35kV bus bar and the 110kV bus bar of the grid point at the same time for statistical calculation.

Table 5. Bus bar distortion rates.

Bus Bar	THD%
Wind farm group A (35kV)	5.84
Wind farm group B (35kV)	6.41
Connecting point (110kV)	9.88

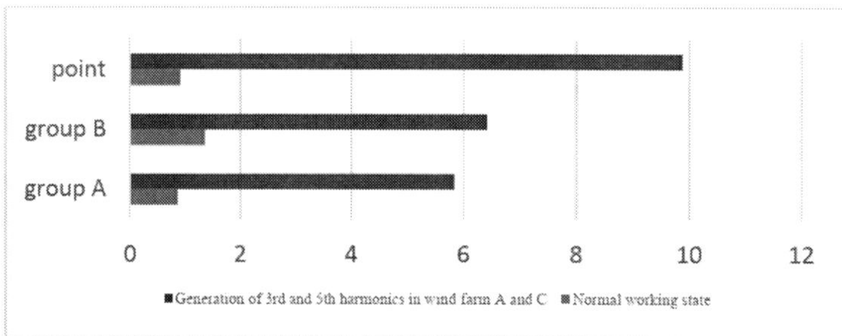

Figure 5. Current distortion rate of wind farm groups under different conditions.

It can be seen that the electric energy of the whole power grid is seriously polluted. The distortion rate of 110 kV current at the grid-connected point is nearly 10%. The normal working state of the grid side is seriously affected.

5. Conclusion

In this paper, based on the principle of harmonic propagation and coupling, the simulation of harmonic propagation and coupling between electric field groups in MATLAB/SIMULINK environment is carried out, and the following conclusions are obtained:

(1) Under normal working conditions, the FFT fast Fourier transform is used to decompose 690V, 35kV and 110kV bus current waveforms in MATLAB/SIMULINK environment. Then the 3rd and 5th harmonic sources are injected into the wind farm A, and the fast Fourier transform is used to harmonize the bus current waveforms of all levels again in MATLAB/SIMULINK environment. By comparing and analyzing the distortion rate THD, it can be found that the harmonic current of wind farm lines propagates from the fan side to the grid side. When the harmonic content of the fan side is large, the power quality of the grid will be greatly affected. With the inflow of common bus, its amplitude and harmonic number show a weakening trend along the line.

(2) By injecting 3rd and 5th harmonics into wind farm A and D and decomposing 35kV and 110kV bus current waveform, it can be found that the current waveform contains higher 2nd, 3rd, 4th and 5th harmonics, resulting in the coupling superposition of harmonics, that is, all harmonic coupling superimposed together, resulting in the current with the highest harmonic content, and generates new harmonic components at the same time. By comparing and analyzing the distortion rate THD, it is found that the harmonic content in the fault line is very large, which will not only affect the current of buses at all levels, but also increase the current harmonic content of adjacent wind farms. The current distortion rate THD of the whole power grid will increase, which will affect its safe and stable operation.

Acknowledgments

This research was financially supported by State Grid Xinjiang Electric Power Co., Ltd. Economic Technological Research Institute (Research and Application of Cost and Cost System for Productive Technological Renovation Projects) and National Natural Science Foundation of China (Grant No.51667018).

References

[1] Han Xiaoqing, Zhang Jian, Zhang Youmin, et al. Harmonic Analysis and Calculation of Wind Farm[J]. Journal of Taiyuan University of Technology, 2009, 40(5): 540-544.

[2] Abniki H, Nateghi S, Ghandehari R, etal. Harmonic Analyzing of Wind Farm Based on Harmonic Modeling of Power System Components [C]. 2012 11th International Conference on Environment and Electrical Engineering (EEEIC), 2012: 667-672.

[3] Hasan K N M, Rauma K, Rodriguez P, etal. An Overview of Harmonic Analysis and Resonances of Large Wind Power Plant [C]. IECON 2011-37th Annual Conference on IEEE Industrial Electronics Society, 2011: 2467-2474.

[4] Kai Yang, Bollen M H J, Wahlberg M. Characteristic and Non-characteristic Harmonics From Wind parks[C]. 21st International Conference on Electricity Distribution (CIRED). Frankfurt, 2011: 251-254.

[5] Pisitpol Chirapongsananurak, SuryaSantoso. Harmonic Analysis for Fixed-Speed Wind Turbines [C]. 2013 IEEE Power&Energy Society General Meeting, 2013: GM2164.

[6] Guo Tao. Harmonic analysis of wind farms with doubly-fed wind turbines[D]. North China Electric Power University (Beijing), 2009.

[7] Zhou Honglin. Research on key issues of grid connection of large capacity doubly-fed wind power generation system [D]. Tsinghua University, 2011.

[8] Xue Junru, Zhao Shichang, Li Xiaowei, et al. Study on the propagation mechanism of harmonics under different voltage levels[J]. Qinghai Electric Power, 2016, 35(3): 38-41.

[9] Zhang Jianjun. Analysis of Harmonic Voltage Transfer Characteristics [J]. High Voltage Apparatus, 2016(10): 130-134.

[10] ZHANG Xiaofeng. Analysis of Current Status of Wind Power at Home and Abroad[J]. Science Technology and Business, 2012(19):168-168.

IOP Conf. Series: Materials Science and Engineering **611** (2019) 012074 doi:10.1088/1757-899X/611/1/012074

Design of Zero-gravity Unloading Process Equipment for Satellite Overlapping Antenna on Ground Test

Yougao Fan[1,2,3,*], Guangtong Liu[1,2], Li Liu[1,2] and Chunsheng Yang[1,2]

[1]Beijing Institute of Spacecraft Environment Engineering, Beijing 100094, China
[2]Beijing Engineering Research Center of the Intelligent Assembly Technology and Equipment for Aerospace Product, Beijing 100094, China
[3]Space Control and Inertial Technology Research Center, Harbin Institute of Technology, Harbin 150001, China

*Corresponding author

Abstract. In this paper, the zero-gravity unloading scheme and the design scheme of the zero-gravity unloading process equipment are proposed for the three-overlapping antennas of a communication satellite. Due to the particularity of the structure, test state and unloading state of the communication satellite antenna, the zero-gravity unloading simulation process equipment commonly used at present cannot meet the test requirements. Based on the analysis of the requirements of zero-gravity deployment test and the process of the three-overlapping antennas zero-gravity unloading, a solution to meet the test requirements is proposed. System performance tests verified the effectiveness of the zero-gravity unloading system.
Keywords: zero-gravity; process equipment; satellite; antenna; ground test.

1. Introduction

Zero-gravity, also known as weightlessness or minus gravity, is one of the most important features of the space environment [1]. Due to the weightlessness of the spacecraft in space, the internal components of the state of stress and the ground is different, so in order to test the performance of the spacecraft and its components on the ground, it must simulate the weightlessness or micro-weight state in space [2].

As a result of the development of space technology, there is an increasing demand for beam-directed manoeuvrability of satellite antennas [3]. Therefore, a large number of satellites are equipped with mechanically movable antennas, and the antennas can be adjusted according to the needs of the mission, and communicate with different satellites or different ground stations[4]. However, due to the influence of ground gravity, the gravity of the mechanical movable antenna may cause damage to the antenna mechanism during the ground test [5]. On the other hand, due to the differences between the zero-gravity environment and the on-orbit, the effectiveness of the test is affected [6]. It may even affect the normal use of the antenna in orbit.

The effects of gravity have three main aspects. Firstly, the influence on the installation accuracy of the antenna; secondly, the influence on the accuracy of the shape of the antenna after unfolding; thirdly, the influence on the driving torque during the unfolding process [7]. As the size of the antenna increases, the impact of the ground gravity environment also increases [8]. Therefore, while developing the antenna, it is necessary to develop a corresponding zero-gravity environment simulation device.

In this paper, taking the three-overlapping antennas of a communication satellite as an example, the design of the antenna unloading scheme and the unloading system is proposed by analysing the working conditions of the ground zero-gravity deployment test of the three-overlapping antennas, which provides a technical basis for the antenna ground test.

2. Requirements for Antenna Deployment Conditions

A communication satellite is parked on a two-axis turntable during the EMC test. The unfolding of the three-overlapping antennas requires separate ground equipment for unloading and unfolding tests. The east and west sides of the satellite face the sides, and the antenna is deployed and locked using ground craft equipment. When the three-overlapping antennas are deployed on the ground, the deployment sequence is as shown in Figure 1.

During the unfolding process, the three-overlapping antennas can be connected to the unloading system through the lifting point respectively. Since they are sequentially deployed in position, there is no interference on the trajectory. During the unfolding of the three-overlapping antennas, the reflector is driven by the antenna pointing mechanism to perform two-dimensional motion, and the position of the unfolding axis moves in a two-dimensional plane, and the envelope of the unfolding axis motion path is an ellipse.

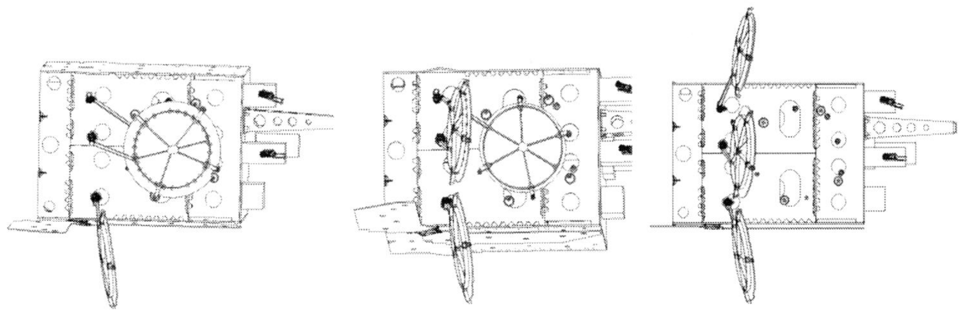

Figure 1. Schematic diagram of three-overlapping antennas

3. Detailed Design of the Antenna Unloading System

3.1. Overall Design

The process equipment used in the zero gravity unfolding test of the three-overlapping antennas includes a two-axis turntable, an antenna gravity unloading system and an antenna fixing device, and the three are not in contact with each other, and the use effect is shown in Figure 2.

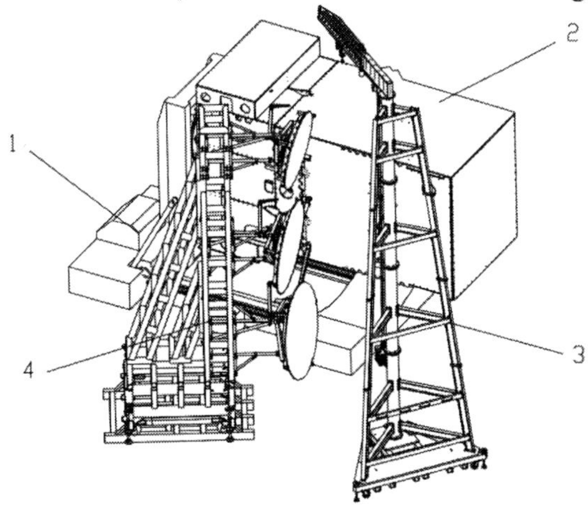

1-Two-axis turntable; 2-satellite; 3-antenna unloading system; 4-antenna fixing device

Figure 2. Schematic diagram of overlapping antenna ground zero-gravity deployment test

IOP Conf. Series: Materials Science and Engineering **611** (2019) 012074 doi:10.1088/1757-899X/611/1/012074

3.2. Structural Form

The main body of the system is fixed by the support of the chassis of the suspension device. The support chassis is installed with a foot and a universal roller to adjust the flatness of the chassis and move the system on the ground respectively. A support column is installed on the support chassis, and the upper end of the column is fixedly connected with an 1 meter extension arm. The cantilever end is equipped with a two-stage deployable arm with a length of 2 meter through the bearings on both ends. The two-stage deployable arm can freely rotate around the first-stage deployable arm with a rotation range of more than ±90°. A linear guide rail is installed at the lower end of the rocker arm, and the travel distance of the guide rail is 1.6 meter. Part of the pulley block is installed on the slide block of the guide rail to realize zero gravity unloading of the unloaded antenna, which can provide a maximum lifting force of 50kg and a lifting height range of 0 to 7 meter. The position of the lifting point can be adjusted automatically with the change of the position of the connecting point.
The antenna unloading system is not in direct contact with the antenna, and only the unloading lifting point is provided, which is basically applicable to the antenna of most satellites. After the antennas deployed, it needs to be locked with the antenna fixture.

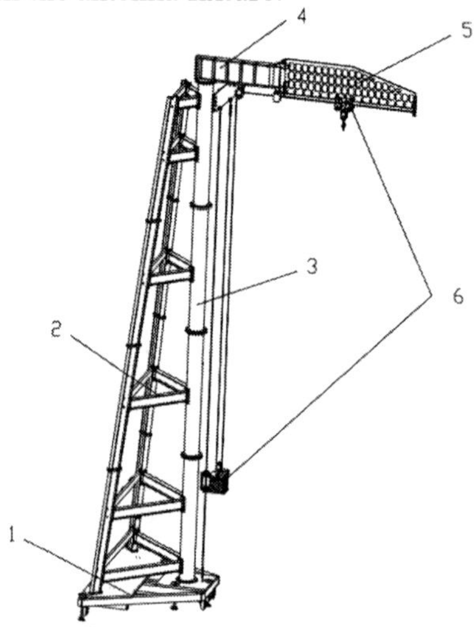

1-the unloading system chassis; 2-the unloading system auxiliary support; 3-the unloading system main body; 4-the first-stage unfolding arm; 5-the second-stage unfolding arm; 6-the hanging system

Figure 3. Schematic diagram of antenna unloading system

3.3. Anti-overturning Design

Since the system is a cantilever structure, the system is stressed at the top during operation. Therefore, the system needs to be anti-overturned. The center of mass of the system is shown in Figure 4. The mechanical model of its anti-overturning design based on the system structure is as follows.

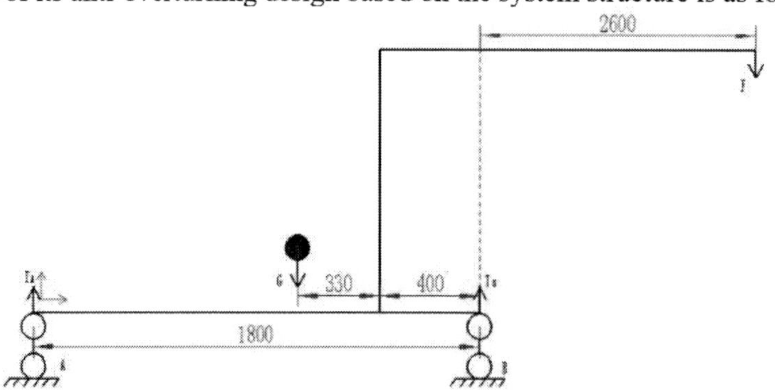

Figure 4. Schematic diagram of system anti-overturning design mechanics

When an external force F is applied to the end of the secondary arm, the critical condition for the whole system to overturn is $T_A = 0$, so $G = 27489N$ .When $T_A = 0$, the force analysis of the system is carried out $\sum M_s = 0$, that is $G \times (330 + 400) - F \times 2600 = 0$, $F = 7718N$. When $F > 7718N$, that is the suspension load greater than 7718N, the whole system will capsize around point B. The designed load of the hanging system is 90kg, so the capacity margin of anti-overturning load of the system is 757%, which can meet the requirements of use.

3.4. Deformation Analysis

Deformation analysis of the suspended part of the antenna unloading bracket system. The total load of the system is 110kg, of which the load is 90kg and the constant force hanging device is 40kg. The hanging point is located at the end of the secondary deployment arm, the middle of the secondary deployment arm and the root of the secondary deployment arm, and the angle between the secondary deployment arm and the primary deployment arm is analyzed at 0 and 90 degrees. The specific results are shown in the Figure 5.

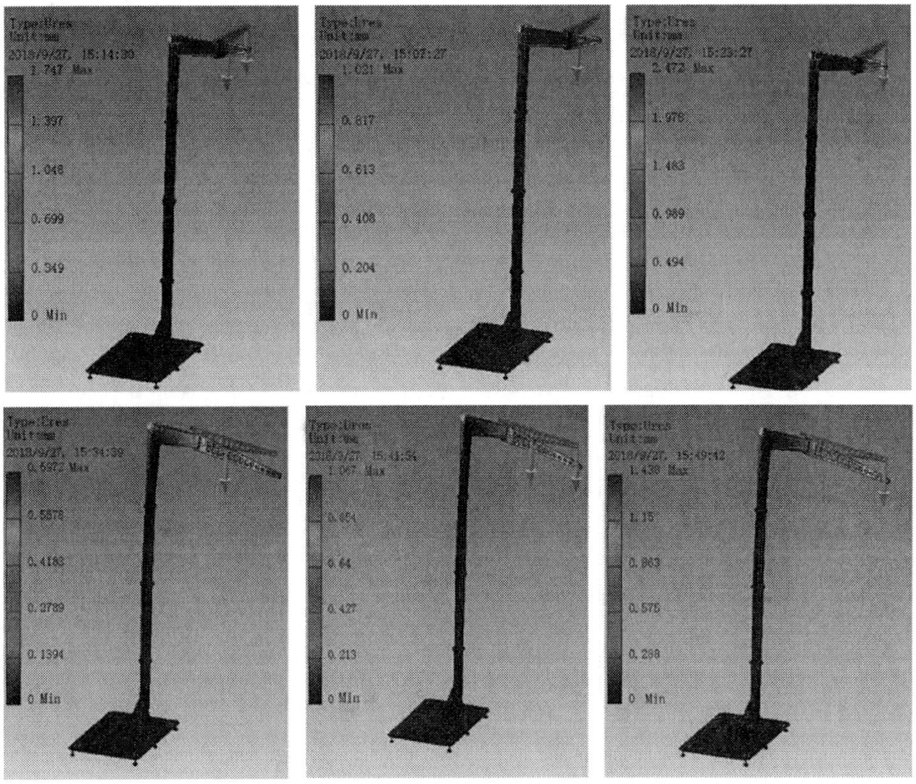

Figure 5. Overall deformation analysis

Table 1. Overall deformation analysis of the hanging device Unit: mm

Working condition	Maximum deformation under rated load		Maximum deformation fluctuation during system oscillation under maximum load
	The angle between the two arms is 0°	The angle between the two arms is 90°	
Stressed on the outermost side	1.438	2.472	1.034
Force midpoint	1.067	1.747	0.68
Stressed on the innermost side	0.697	1.021	0.324

It can be seen from the above table that under the rated load, the maximum deformation is 2.47 mm, and the deformation amount is 1.034 mm during the rotation of the antenna by 90°. This data basically satisfies the requirements of the three-overlapping antennas unloading process.

3.5. Intensity Analysis

The maximum stress of the system occurs at the rocker joint, which is 113.8 MPa, and the rest is between 20 and 50 MPa. According to the material and force of the suspension system, the safety factor of the system is discussed as follows. The minimum safety factor of the system is 1.67, which is located at the maximum stress of the system, that is, the connection of the rocker arm, as shown in Figure 6. In summary, the strength of the suspension system meets the requirements, and the minimum safety factor of the system is 1.67, which can be further optimized to achieve a system safety factor of more than 2.

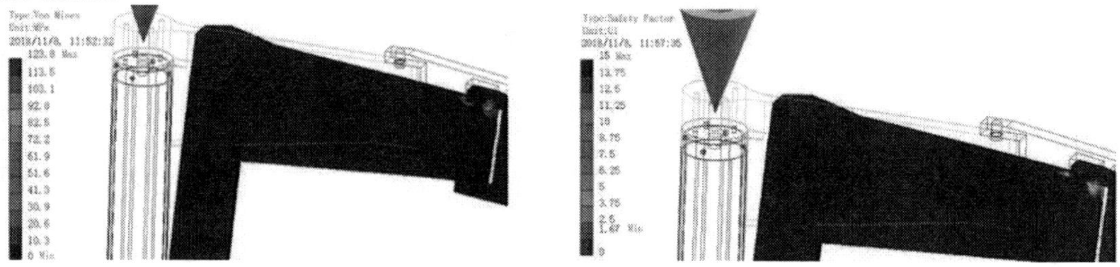

Figure 6. Intensity analysis

4. System Performance Test Verification

4.1. Downwarping Test

The downward deflection test is performed on the deployment arm for the technical requirements of the ground test of the overlapping antenna. The 40kg counterweight is hung at the hanging point, and the measuring points are respectively arranged at the corresponding positions of the second-stage unfolding arm and the first-stage unfolding arm, and the measuring points are measured by the laser tracker and the theodolite.

The test is divided into two states, one of which is that the primary deployment arm and the secondary deployment arm are in a straight line state (denoted as 0° state), and the result is shown in Table 2; the other state is that the primary deployment arm is perpendicular to the secondary deployment arm (recorded as 90 ° state), and the result is shown in Table 2. The measured points on the selected first-stage unfolding arm and second-stage unfolding arm are shown in Figure 7.

Table 2. Deformation test for unloading system

Measuring point	no-load(Z-axis coordinate)		full load(Z-axis coordinate)		Amount of downwarping	
	0° state (mm)	90°state (mm)	0°state (mm)	90°state (mm)	0°state (mm)	90°state (mm)
P1	7202.9600	7202.9801	7201.4369	7200.708	1.5231	2.2721
P2	7203.9147	7203.9247	7202.5573	7201.9971	1.3574	1.9276
P3	7204.5302	7204.5216	7203.3816	7202.7204	1.1486	1.8012
P4	7204.8184	7204.8246	7203.897	7203.4384	0.9214	1.3862
P5	7207.8685	7207.8569	7207.244	7206.7293	0.6245	1.1276
P6	7243.8185	7243.8123	7243.5714	7243.4869	0.2471	0.3254

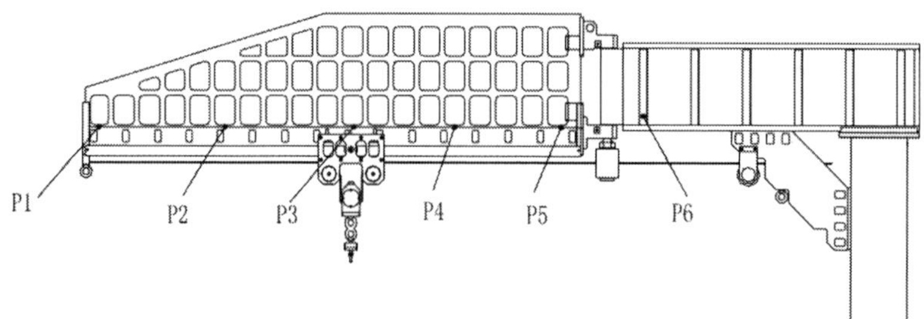

Figure 7. Schematic diagram of unloading system

4.2. Friction Test
For the general technical indicators, the system is tested for translational friction and rotational friction. After hanging 40kg counterweight, turn on the air source, and apply a pulling force parallel to the secondary deployment arm at the end of the weight with a tension meter. When the pneumatic slider follows the movement, the tension value displayed on the dynamometer is the translational friction of the system; The pulling force is applied to the counterweight weight end perpendicular to the second unwinding arm by the dynamometer. When the second unfolding arm starts to rotate along the unwinding axis, the tension value displayed on the dynamometer is the system rotating friction. The test results are shown in Table 3.

Table 3. Friction test for unloading system

Friction type	Friction test site		
	Front of the secondary arm	Middle of the secondary arm	End of the second arm
Translational friction	less than 1N	less than 1N	1N~2N
Rotational friction	1N~3N	1N~3N	1N~3N

5. Conclusion
In this paper, a zero-gravity unloading scheme and a zero-gravity process equipment design scheme for three-overlapping antennas of a communication satellite are proposed. Based on the analysis of the working conditions of ground zero-gravity test of three-overlapping antennas, the theoretical design of zero-gravity unloading process equipment is carried out in detail from the aspects of the overall design, unloading system structure form, anti-overturning design, deformation analysis and intensity analysis. Finally, through the system performance test, the effectiveness of the designed zero-gravity unloading process equipment performance is verified.

References
[1] Attanasio S. D. and Charron J. Development of a sensing system for zero gravity simulation using a robot manipulator to test antenna reflector unfolding in satellites [C]. *MEDER 2018: Mechanism Design for Robotics*, 2018, 66: 41-48.
[2] Silva A. C. and Loureiro G. Quality assurance of complex systems-satellite AIT [C]. *IEEE International Conference on Industrial Engineering and Engineering Management*, 2009, 935-939.
[3] Kartsan I. N., Zelenkov P. V., Tyapkin V. N. et al. Adaptation algorithms for satellite communication systems equipped with hybrid reflector antennas [C]. *IOP Conference Series: Materials Science and Engineering*, 2015, 94: 012010.
[4] Li P., Liu C., Tian Q., Hu H., Song Y. Dynamics of a deployable mesh reflector of satellite antenna: parallel computation and deployment simulation [J]. *Journal of Computational and Nonlinear Dynamics*, 2016, 11(6): 061005-061005-16.
[5] Rao S. K. Advanced Antenna Technologies for Satellite Communications Payloads [J]. *IEEE Transactions on Antennas and Propagation*, 2015, 63(4): 1205-1217.
[6] Brungs, S., Egli, M., Wuest, S. L. et al. Facilities for simulation of microgravity in the ESA ground-based facility programme [J]. *Microgravity Science and Technology*, 2016, 28(3): 191-203.
[7] Guo T., Hu R., Xiao Z. et al. Research to Assembly Scheme for Satellite Deck Based on Robot Flexibility Control Principle [C]. *IOP Conference Series: Materials Science and Engineering*, 2018, 324: 012086.
[8] Dai H., Zhang C., Meng F. et al. Mechanical analysis and verification of the large-scale rack structure about the zero-gravity test of manipulator of the space station [J]. *The Journal of Engineering*, 2019, 2019(13): 186-189.

IOP Conf. Series: Materials Science and Engineering **611** (2019) 012075 doi:10.1088/1757-899X/611/1/012075

Research on Assembly Technology of Complex Flexible Out-cabin Motional Cable for Spacecraft

Chunsheng Yang, Yi Lu, Yu Wang, Feng Xue and Zhenyue Ren

Beijing Institute of Spacecraft Environment Engineering, China

Abstract. Based on the application of flexible cable assembly simulation analysis technology in complex flexible out-cabin motional cable assembly of spacecraft, the working environment and physical characteristics of flexible motional cable are analyzed, reasonable assembly constraints are planned, and cable assembly simulation analysis is described in detail. The research results not only ensure the accuracy and practicability of the design, but also reduce the assembly defects and product failure rate. The actual effect of cable assembly can be accurately judged, the product performance can be improved, the product development cycle can be saved, and the development cost can be significantly reduced.

1. Introduction

Cable assembly technology is widely used in spacecraft assembly integration. There are many factors affecting cable assembly performance. Especially the assembly of large-scale moving cable is basically based on physical prototype test. The simulation assembly test method of prototype is inefficient, with much dependent on manual experience, so it has high cost and long development cycle, which can not meet the development needs of the current aerospace industry. In order to solve these problems, virtual assembly technology of flexible cables has been extensively studied in domestic and foreign institutes to replace the traditional physical model test method. [1]However, in recent years, the research on cable virtual assembly at home and abroad still remains in the aspects of semi-automatic and full-automatic cable, optimization of pipeline layout design algorithm and ergonomics analysis of man-machine interactive wiring in virtual environment, rarely considering cable dynamic assembly process. [2]

Spacecraft cables are complex in variety, with large number and complex laying process. However, cables are always belong to non-linear flexible body with large deformation. [3]Therefore, the assembly technology of flexible moving cables becomes an unpredictable risk point in spacecraft assembly integration design. The assembly effect of cables will directly affect the success or failure of space missions. Relevant research work needs to be carried out. In the stage of process design, detailed process parameters of cable assembly are formed, and the working state of the motional cable on orbit is accurately predicted. Problems such as extrusion, pulling and collision of the cable are avoided, so as to improve the assembly quality and efficiency of flexible moving cable for spacecraft. [4] [5]

Based on the complex flexible cables of a second-generation data relay satellite, this paper studies the simulation analysis technology and application of flexible cables assembly.

2. Analysis of Design Characteristics of Out-cabin Motional Cable

2.1. Structural Characteristics of Extravehicular Motional Cables

The out-cabin cables of a second-generation data relay satellite are fixed as a whole with several cables of different specifications. They are installed outside of the spacecraft. They have the characteristics of long-time and large-angle movement around two directions and large span between fixed points. The cable lies along the outer surface of the spacecraft cabin, passes through the fixing bracket, and then extends into the high-frequency module, as shown in the following figure. A large umbrella antenna is installed on the module. Through the movement of the module with antenna, the user satellite can be positioned accurately and the relay communication between the ground station and the user satellite can be ensured. The two-axis driving mechanism is installed on the side of the high-frequency module, and the other end is installed on the fixing bracket. The dual-axis driving mechanism has two rotating functions (rotating around the azimuth axis of X from -20 degrees to +46 degrees, rotating around the elevation axis of Y from -20 degrees to +20 degrees). By controlling and adjusting these two degrees of freedom, the orientation of the high-frequency module can be determined and continuously changed when it is in orbit. The two-axis driving mechanism rotates around its two orthogonal axes, so that the high-frequency module is separated from the surface of the cabin and rotates in two directions accordingly. Movable cables follow the motion of module to achieve long-term complex follow-up in three-dimensional space.

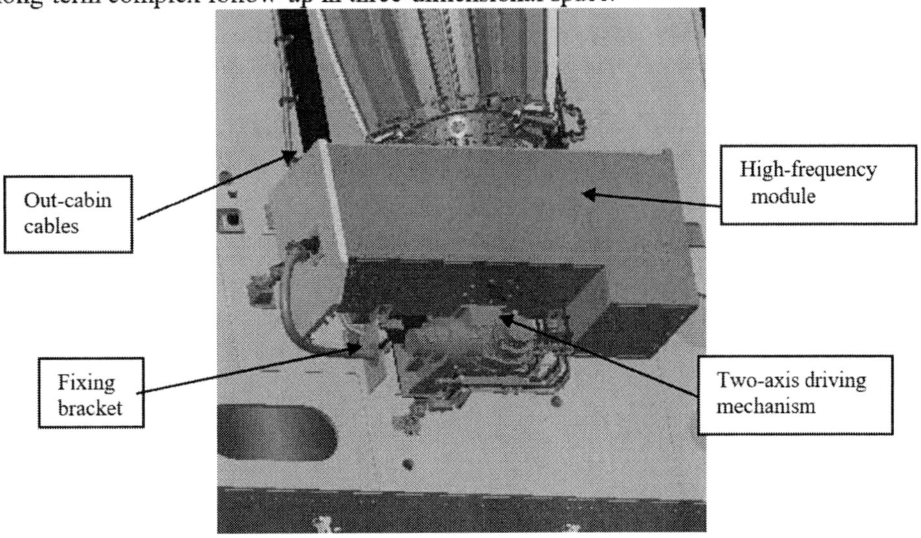

Figure 1. Schematic diagram of position of extravehicular moving cables

Because out-cabin cables are the only transmission channel connecting spacecraft and high-frequency module for data and power signals, so the cables contain a number of low and high frequency cables of different specifications. At the same time, the thermal control and radiation environment requirements for the cabin are arranged outside the cabin. The cables outside the cabin are inlaid with constantan heating wires, and the cables outside are covered with metal shielding tubes and thermal control multilayers. It can be seen that this cable is a complex combination of various materials. The physical parameters related to the motion characteristics need to be obtained by means of experiments.

2.2. Assembling Process Analysis of Out-cabin Motional Cable

Combining the layout trend, installation position, structural characteristics, movement characteristics and fixed point spacing of the cables outside cabin, it can be judged that the assembly of the cables is very important. It is necessary to ensure that the cables follow freely in the course of driving the high-frequency module by the two-axis driving mechanism, without hindering the movement of the modules and the two-axis driving mechanism, without generating additional resistance. Its structural components should not interfere and scrape with others, while the cable itself is not allowed to be abnormally squeezed and pulled, to ensure that the cable can transmit data and power signals stably in the long-term movement process.

Therefore, it is necessary to analyze and study the assembly process of flexible out-cabin motional cables, which mainly includes the following three aspects:

1) Determine the appropriate cable direction, reserved length and fixed mode;

2) To meet the requirement of safe distance, the forces acting on the cables and the clearance between the cables and the rotating structure are analyzed and studied after many times rotation of cables.
3) Under the excitation of harsh vibration environment at launch stage, the force and swing amplitude of cables outside cables under vibration are analyzed.
These characteristics bring great difficulties to process design, and can not meet the above requirements through conventional virtual assembly research. It is necessary to use flexible cable simulation technology to get the instantaneous state of cables under dynamic conditions, and then work out the process parameters that meet the requirements, so as to realize the smooth assembly of cables.

3. Simulation and Analysis Technology of Flexible Cable Assembly

In traditional mode, the assembly analysis of cables is generally based on the judgment of the rationality of cable path in three-dimensional modeling software, and the path planning from the beginning to the end does not take into account of the physical properties of cables, such as elastic modulus, density, Poisson's ratio, etc. As a result, the state of the cable can not be effectively evaluated after the design, such as whether to reach the minimum bending radius, whether to reach the maximum pulling force of the cable and so on. It is impossible to analyze the motion state of the cable when it moves with other parts. As for whether the cable meet the target after assembly, can the conclusion be drawn by assembly and test in practice after the physical prototype has been built. In this way, it not only consumes a lot of time and test costs, but also causes certain losses to the quality of cable products.
The simulation analysis technology of flexible cable assembly is to enrich the cable by acquiring the material property parameters of the cable. It is not a simple curve, but a more realistic spatial motion mechanism. At the same time, it sets the motion constraints of the cable, and the virtual simulation obtains the real work and movement process of the cable. The main approaches are as follows:
1) Using physical experiments to test the real material properties of cable, such as external diameter, tensile strength and so on. For cables with different wiring harnesses but the same material properties, it can be obtained from the data base without physical experiments.
2) Using the special simulation analysis software for flexible cables. After importing the cable model, the real material attributes are assigned to the cable in the model, and the connection form of the cable's beginning and end is pretreated. Flexible simulation engine generates flexible cables in virtual reality environment through cable model transformation. The solver can display the safety of cable tension, bending and torsion in real time, and can obtain safe and reasonable flexible layout and length by reflecting the safety of different color materials, and output the cable as a general three-dimensional data model.

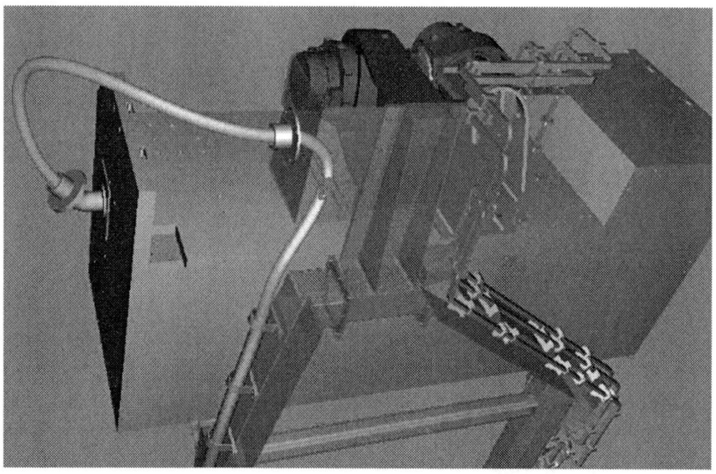

Figure 2. Display effect of flexible cable model

3) Restrict the behavior between flexible cable and motion mechanism. The built-in behavioral engine of moving parts in the software can define the constraints of axial displacement, axial rotation,

composite motion and composite motion of a single component in the model. Behavior constraints can define the range of motion of a single component in translation or rotation, or the behavior of components after reaching the constraint limit, so that components can have more complex behavior, so as to obtain more realistic digital virtual prototype. Through this module, the motion relationship between the components of the spacecraft is established, and the motion of the rigid parts is related to the flexible cables. The motion relationship between the flexible cables and the rigid parts is established, and the state of the cables in the process of following the rigid body motion is investigated.
4) The animation module is used to generate the motion simulation animation of the whole component or component related to the cable. The animation of individual components and assembly will be saved and recorded in the form of key frames, which can support editing, modifying and flipping of key frames, and support multi-level composite motion. In the case of lossless shape accuracy, lightweight publishing of models and scenes can be realized to ensure the fluency of large and complex assembly work.

4. Research on Assembly Process Technology of Out-cabin Motional Cable

In this study, the technology described in the second section of this paper is used to conduct physical tests combined with flexible simulation. First, the physical properties of cables, such as elastic modulus and tensile strength, are calculated by experimental tests. Then the measured physical attribute data are input into the ICIDO simulation software, and the overall layout and motion state of the cable are simulated and analyzed by using the powerful flexible pipeline simulation function of ICIDO software.
Firstly, the initial design state of the flexible cables is analyzed and verified. The design length of the cable in the long-term moving section of the initial design state is 1040mm, the outer diameter of the cable is 20mm, the minimum bending radius is 60mm, and the maximum pulling force is 2144N according to the physical test results.

4.1. Assembling Analysis of Initial Design State

Under the initial state condition, Over bending occurs because the bending radius of the cable is less than the critical value of the minimum bending radius. At the same time, the minimum distance between the cable and the left limit rod is 5.54 mm, and the minimum distance between the cable and the right limit rod is 4.75 mm. So there is no interference between the cable and the two limit devices in the natural state.

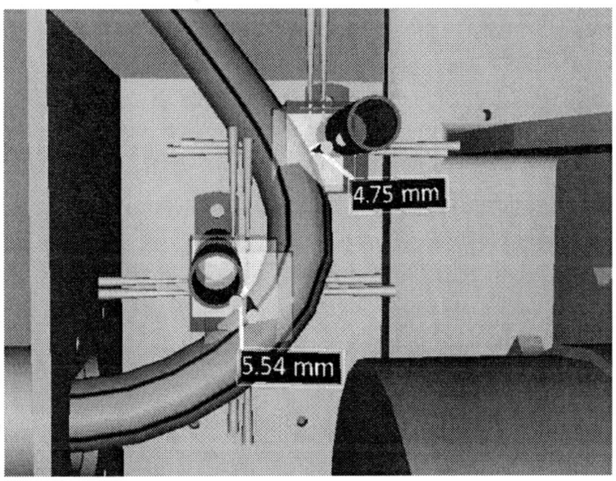

Figure 3. Performance analysis of cables in initial design state after assembly

The cable state of flexible motional cables after rotating around X azimuth axis and Y elevation axis respectively is analyzed.
For example, after rotating around Y elevation axis + 20 degrees, the cable overbends at both upper and lower pipe ports, meanwhile the cable collides with the left limit bar(as shown in figure 4).

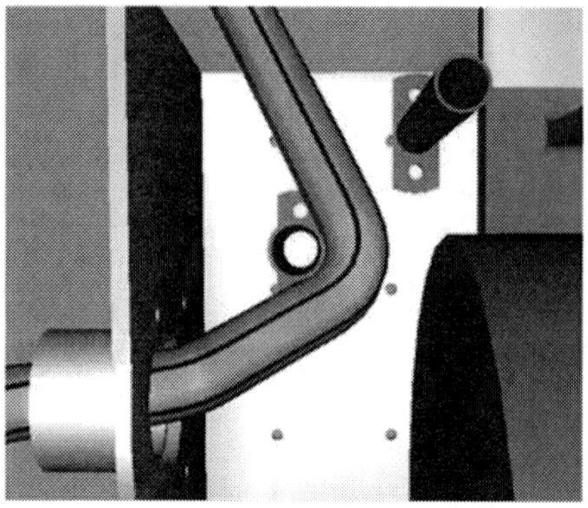

Figure 4. Performance analysis of cables in initial design state after rotating around Y axis

It can be seen that in the initial design state, the assembly of the cables will not only force the cables, but also affect the movement of the high frequency box and the driving mechanism. So the initial design state needs to be optimized.

4.2. Optimizing Process and Analysis of Flexible Cable Assembly Research

According to the analysis results of the previous section, it is necessary to gradually enlarge the length of the moving section of the motional cable, and then carry out multi-state analysis and testing according to the items of the previous section. When the length of the moving section of the cable is gradually adjusted to 1070 mm, according to the analysis of the same items mentioned above, the cable has no over-bending state and has not collided with the two limit poles.

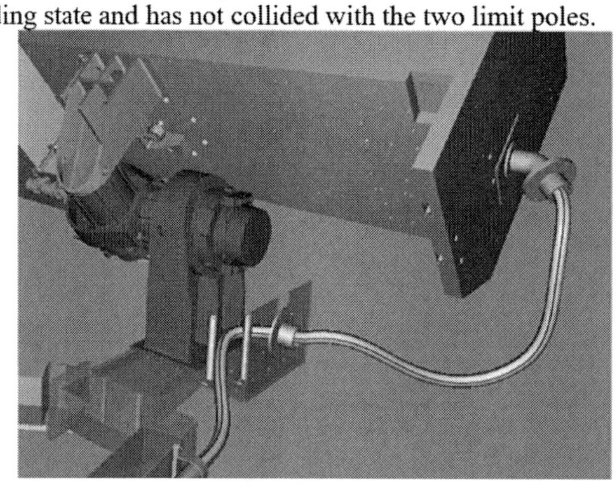

Figure 5. Analysis results after optimizing the moving section of cable with 1070mm length

Table 1. Structural comparison of cables after optimization of design status

No.	Analysis content	Cable movable length 1040mm (initial value before optimization)	1070mm (final value after optimization)
1	Initial assembly, no rotation	The whole cable is in normal condition. The cable has no contact with the two limit poles and no over-bending occurs.	The whole cable is in normal condition. The cable has no contact with the two limit poles and no over-bending occurs.
2	Rotate - 20 degrees around X axis	The current state of the cable is consistent with the initial assembly state.	The current state of the cable is consistent with the initial assembly state.
3	Rotation around X axis + 46 degrees	The cable is partially over-bended and touches the right limit rod.	There is no over-bending phenomenon and cable don't contact with the limit rod.
4	Rotation around Y - 20 degrees	The minimum distance between the cable and the left limit rod is 6.79 mm and the right limit rod is 1.58 mm.	There is no over-bending phenomenon and cable don't contact with the limit rod.
5	Rotation around Y + 20 degrees	The cable collides with the left limit rod.	The cable does not touch the limit rod without over-bending phenomenon.

4.3. Analysis and Research on Vibration Environment of out-cabin Motional Cable

When the cables are launched, the high-frequency module and the two-axis driving mechanism do not rotate. Under the same vibration excitation conditions as the actual situation, the vibration analysis of the cables with the high-frequency module and the two-axle drive mechanism is shown in the following figures.

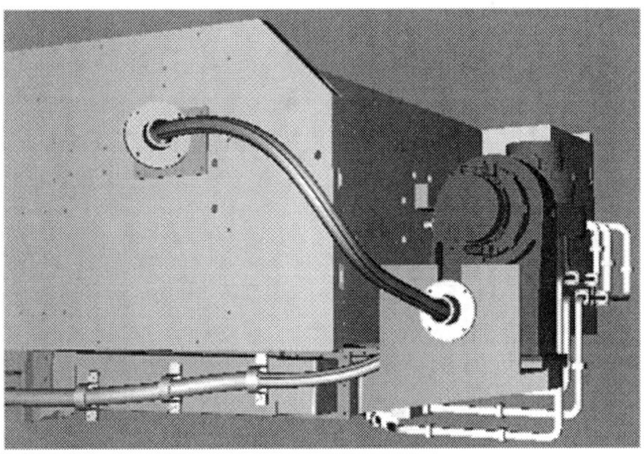

Figure 6. Analysis effect diagram of cables outside cabin under vibration condition

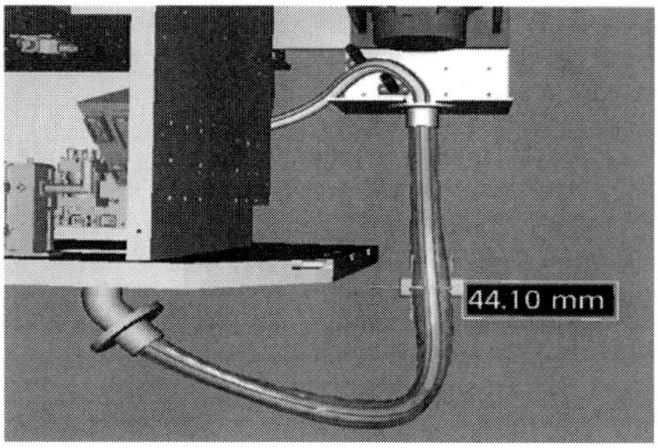

Figure 7. Envelope area formed by vibration of cables outside cabin

The green area in the figure is the envelope area formed by the cable sweeping with the high-frequency module during vibrating. As shown in the figure, the maximum position width of the envelope is 44.10mm, and there is no interference with other structural mechanism components.

4.4. Application Results for Actual Assembly

According to the above research results, through the analysis of cable status of out-cabin motional cables under different working modes, the corresponding measures in the actual assembly process can be obtained.

(1) If the reserved length of the moving section of the cable is not less than 1070 mm, the collision between the moving cable and the right limiting rod occurs when the cable rotates around the X axis + 46 degrees, and the repeated scraping and collision with the left limiting rod when the cable rotates around the Y axis + 20 degrees. Therefore, when transplanting, assembling and binding the fixed cable (moving section), it is necessary to ensure that the length of this section should not be less than 1070 mm.

(2) Considering the possible errors in the assembly process, it is necessary to take corresponding protective measures for the right side of the metal outlet of the moving cable within 126 mm, as shown in the following figure.

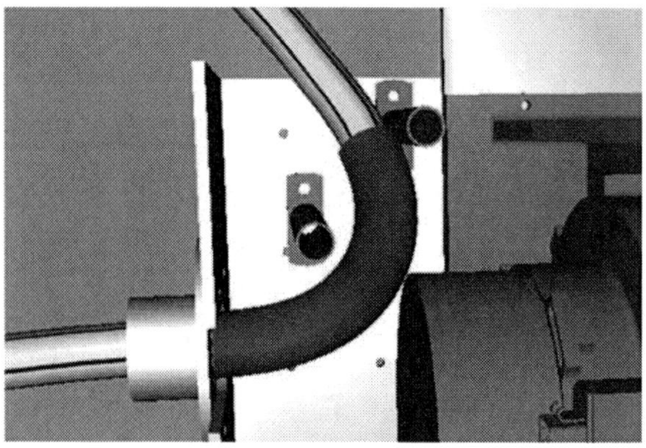

Figure 8. Protection range of cables outside cabins

5. Concluding Remarks

In this paper, the assembly process of complex out-cabin motional cables is analyzed and studied. Obtain the physical properties of cables, design the route of cables, specify constraints, construct the simulation environment for cable assembly and dynamic analysis, and carry out flexible simulation research. In the course of this research, the selection of fixed points, protective measures, the stress condition under each limit state, the permissible bending radius, the range of motion under dynamic conditions, the analysis of the surrounding safety clearance and the oscillation amplitude of the vibration environment of the cable are studied. Then a reasonable and feasible series of technological parameters are formed, and the effect of the cables after assembly can be done. Then we can make accurate analysis and judgment. Through the research of flexible cable assembly technology, it can effectively ensure the accuracy and practicability of design, reduce assembly defects and product failure rate, ensure product assembly quality, significantly save development cycle and reduce development costs. [6] [7]The successful application of this technology in laying complex flexible out-cabin motional cables for second-generation data relay satellites can provide reference and guidance for assembly design of other spacecraft products.

References

[1] Huang Yixin. Research on wiring of spacecraft and virtual prototyping of deformable objects[D]. Harbin Institute of Technology, 2012

[2] Du Hongwang. Physical modeling and configuration simulation for flexible cable[D]. Dalian Maritime University, 2014

[3] Wang Tongsu, Du Baorui. Research on Aircraft Cable Assembly in Virtual Environment. Aeronautical Manufacturing Technology, 2015(19): 52~55

[4] Feng Wei, Zhang Yanlei, etc. The Virtual Simulation Technology in Assembly Process Design of China's Lunar Exploration Project II. Spacecraft Environmental Engineering, 2014 (6): 326-331

[5] Shen Zhong, Zheng Hongtao. Application of digital assembly simulation technology for complicated vehicle based on DELMIA. Missiles and Space Vehicles, 2015(6): 42~45

[6] Ritchie J M, Dewar R G, Simmons J E L. The generation and practical use of plans for manual assembly using immersive virtual reality [J].Proceedings of the Institution of Mechanical Engineers (Part B), 1999, 213: 461-474

[7] Han Bing. Research on digital layout design of cable harness in aerospace products [D]. Harbin Institute of Technology, 2014

Research on Improved Localization and Navigation Algorithm for Automatic Guided Vehicle

Xiaohui Liu[1], Wu Wei[1], Yong Gao[1,*] and Xiaoqin Wei[1]

[1]School of Automation Science and Engineering, South China University of Technology, Guangdong, China

*Corresponding author

Abstract. Aiming at the problem of inaccurate positioning caused by the wheel slippage or "kidnapping" movement of the robot in the positioning navigation, an improved autonomous positioning navigation strategy based on the robot operating system (ROS) is proposed. Firstly, combined with adaptive Monte Carlo localization (AMCL) algorithm and laser-based point-and-line iterative closest/corresponding point (PLICP) pose estimation algorithm, the accuracy and robustness of positioning are effectively improved. Then, based on the path planning strategy combining A* algorithm and dynamic window algorithm (DWA), an improved navigation failure recovery method is proposed and integrated into the ROS navigation framework, which can effectively improve the efficiency of robot positioning navigation and task execution. Finally, the mobile robot model TurtleBot is used as the experimental platform. The simulation experiment and field test demonstrate that the improved algorithm is superior to the original algorithm. The improved algorithm can adapt to the inaccuracy of the odometer and can achieve accurate localization and navigation in long-distance environment.

1. Introduction

The automatic guided vehicle (AGV) is widely used in military, transportation, logistics and other fields due to its good environmental adaptability and anti-interference ability. It is an important branch of mobile robot technology development. With the rapid development of flexible manufacturing, the demand of AGV is also growing in China [1]. Under the premise that the environmental map is known, the positioning and navigation technology is the core issue for the development of AGV, and on which domestic and foreign scholars have carried out extensive research.

AGV navigation means that the AGV controls its speed and angle according to the path offset to ensure that the AGV can reach the target point accurately [2]. AGV navigation can be transformed into path planning and obstacle avoidance problems. In literature [3], a magnetic nail positioning and navigation method based on Kalman filtering and PD control is proposed which has high accuracy but large limitation on complex paths. In literature [4], an improved Markov localization algorithm based on Gaussian kernel function is proposed. Although it can solve the abduction problem of robots, it has a large amount of computation and great restrictions on AGV positioning in non-structural environment. Fox D and Burgard W [5] proposed the dynamic window algorithm, which has better obstacle avoidance effects for obstacles in static environments.

Aiming at the above problems, this paper proposes an improved positioning and navigation algorithm based on ROS framework by using laser, and verifies its effects through simulation experiments and field tests.

2. Improved Positioning Algorithm

Based on the known map, the purpose of the positioning is to determine the current position of the robot in the map, and the positioning accuracy directly affects the accuracy of subsequent navigation. Commonly used positioning methods are mainly divided into relative positioning and absolute positioning. Relative positioning determines the current pose by measuring the distance and direction of the robot relative to the initial position. Time drift may occur, however, because the relative positioning is based on the accumulation of measured values to achieve accurate positioning, which is not suitable for long-distance and long-time positioning. Therefore, this paper adopts the commonly used probability-based absolute positioning method-adaptive Monte Carlo localization method, and proposes corresponding improvement methods.

2.1. Adaptive Monte Carlo Localization Algorithm

Guided by probability statistics theory, Monte Carlo localization (MCL) method is based on particle filtering [6]. The method uses the distribution of particle samples to represent the confidence of pose, and continuously updates the particles and their weights according to the motion and observation model of the robot to realize the particle filter estimation, the algorithm process is shown in Figure 1. However, with the continuous iteration, the conventional MCL method will have more serious particle degradation, particle shortage and other phenomena, so it can't solve the abduction problem of the robot. For these problems, AMCL emerges as required [7]. Many scholars have proposed different improved methods, one of which is to combine the conventional MCL with the Kulbback-Leibler Divergence (KLD) sampling method to adjust the size of the particle set over time to form adaptive Monte Carlo localization algorithm.

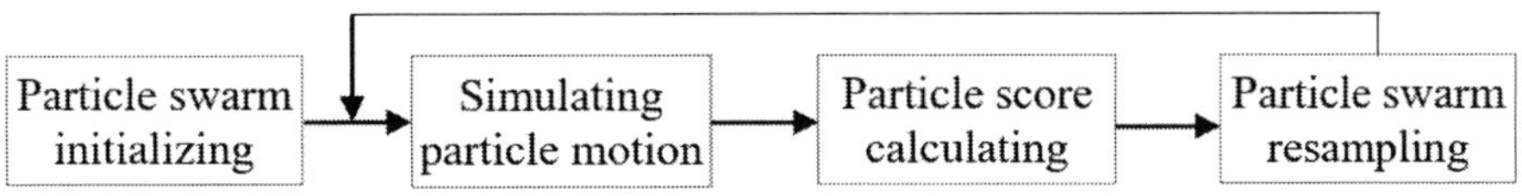

Figure 1. Process of Monte Carlo localization algorithm.

2.2. Improved Adaptive Monte Carlo Localization Algorithm

During the experiment, it was found that when the robot moved, the odometer data error would be large if the ground is uneven or the wheels slips, bringing difficulties to the robot positioning. In order to avoid the above problems and improve the positioning accuracy, a positioning method based on PLICP pose estimation and AMCL method is proposed, as shown in Figure 2.

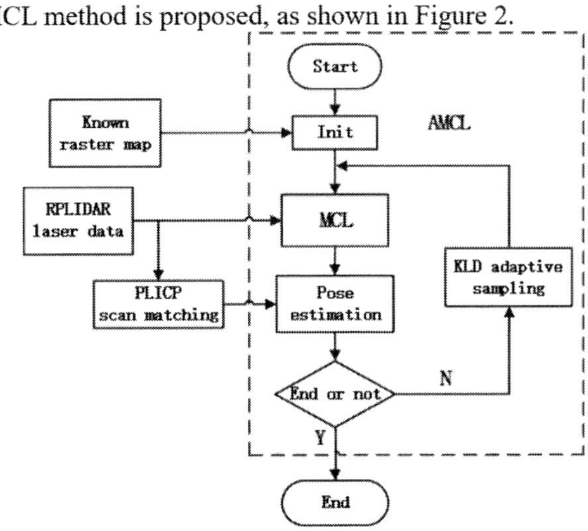

Figure 2. Improved PLICP-AMCL algorithm flow chart.

PLICP is a point-to-line match method, which means minimizing the Euclidean distance from a point to the straight line rather than from a point to a point [8]. The core idea of the improved PLICP-AMCL positioning method is: based on the AMCL method, when the odometer error of the robot exceeds the set threshold, the positioning estimation no longer uses the odometer positioning data, but uses PLICP for relocation. This can improve the robustness of positioning. Since the improved algorithm is a linear fusion of AMCL and PLICP methods, the improved algorithm will not change the convergence of the original algorithm [9, 10].

3. Improved Navigation Algorithm

The navigation of the robot can be transformed into the path planning problem. The path planning is further divided into global path planning based on prior information and local planning problem based on sensor information.

3.1. Improved Navigation Failure Recovery Method

When the robot deviates from the navigation path and the positioning is lost, the original navigation failure recovery method is to let the robot rotate in place, then let the sensor acquire the surrounding environmental information and match the map until the relocation is successful. However, this method may cause the robot to continuously rotate in place, not only wasting time, but also usually making the robot unable to reach the target point from the lost navigation.
Therefore, this paper proposes an improved navigation failure recovery method: when the robot navigation is lost (positioning failure), let the robot rotate a few turns in place to update the local cost map, then move the robot to a small distance in the direction without obstacles. After stopping, let the robot get global positioning and re-plan a path to the target point. The method can effectively improve the efficiency of the robot positioning navigation and performing tasks.

3.2. Integrated Localization and Navigation Frame

This paper builds the navigation framework to realize the laser-radar-based navigation experiment based on the ROS platform which includes the navigation algorithm package configuring global and local path planning. In this paper, A* algorithm, which combines the advantages of Dijkstra algorithm and Breadth-First Search (BFS) algorithm [11], realizes global path planning, and DWA, which is suitable for unstructured static environment, realizes local path planning [12]. The input of the navigation algorithm package includes: target setting, raster map, RPLIDAR laser data, robot positioning, coordinate transformation, odometer information. The improved ROS navigation framework is shown in Figure 3. After the robot's initial pose is determined, the robot's navigation target can be determined on the map via the "2D Nav Goal" button in RVIZ.

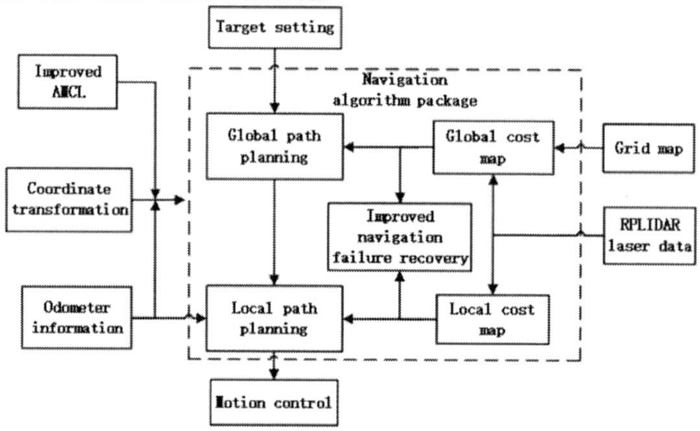

Figure 3. Improved ROS navigation framework.

4. Simulation and Experiment

4.1. Simulation Analysis

This paper uses the Gazebo platform and the RVIZ visual interface to simulate the robot positioning navigation. The open source TurtleBot model, equipped with a virtual laser sensor, is introduced into

Gazebo as a robot experiment platform, which can realize omnidirectional scanning in a 2D plane with a scan radius of 8m, as shown in Figure 4(a). The bottom of the robot is equipped with two left and right driving differential wheels and two front and rear driven universal wheels. The photoelectric encoder is mounted on the differential wheel, which can calculate the displacement of the robot by the number of measured pulses per unit time. The robot's experiment environment built in Gazebo is shown in Figure 4(b), including five obstacles. The grid map created through its own laser sensor is displayed in RVIZ, as shown in Figure 4(c). The small green arrow in the figure indicates the orientation of the 500 particle samples distributed around the robot.

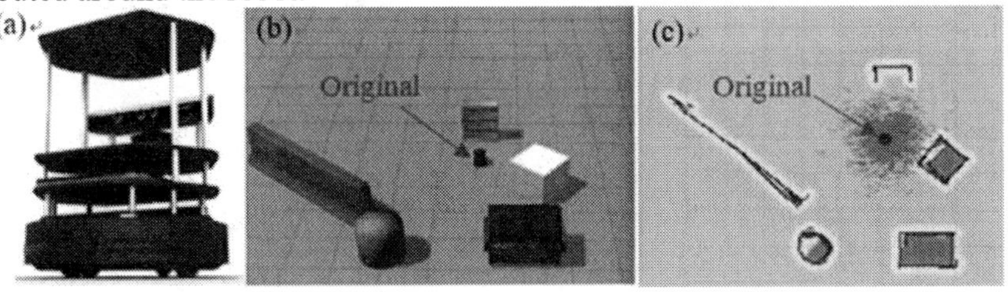

Figure 4. The initial position of the navigation.

After setting a target point for the robot in the grid map (the red five-pointed star logo in Figure 5), the RVIZ automatically plans an excellent navigation path for the robot (the green curve in Figure 6). When the robot moves to the target point, the navigation effect of using the original and improved positioning navigation algorithm is respectively shown in Figure 5 and Figure 6.

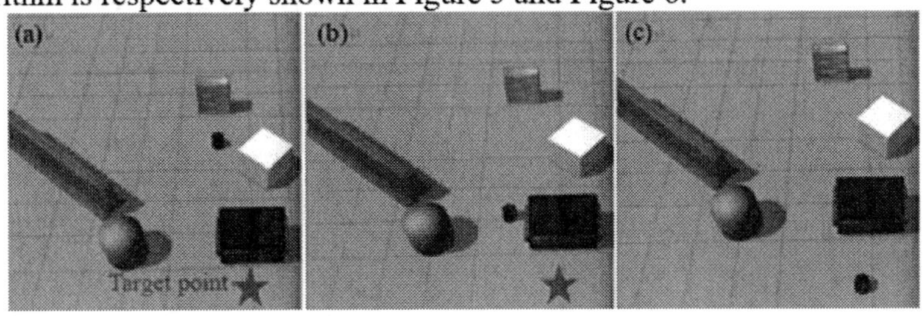

Figure 5. The navigation effect of the robot obtained by the original algorithm at (a) starting point, (b) path point and (c) target point.

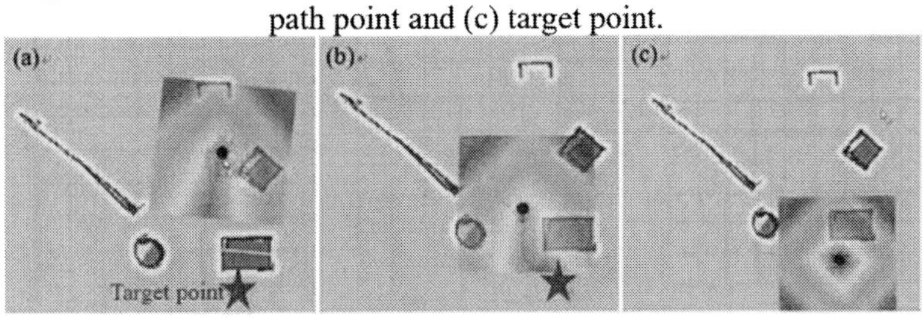

Figure 6. Navigation effect diagram of the robot obtained by the improved algorithm at (a) starting point, (b) path point and (c) target point.

4.2. Field Test

The robot platform used in this paper is the TurtleBot mobile robot developed by YUJIN Company, as shown in Figure 7(a). Its mechanical structure and hardware configuration are similar to the simulation model. The laser-borne radar sensor is the RPLIDAR A2 developed by SLAMTEC, which enables 360-degree rotational scanning in the plane. The environment in which the robot is located consists of a laboratory and the outside corridors, as shown in Figure 7(b)-(h). Since the RPLIDAR laser radar can only scan 2D plane, a 2D raster map is constructed, as shown in Figure 8.

The initial position of the robot is set at the turning aisle (d). Its logo on the grid map is shown by the black dot in Figure 9(a). The small green arrow indicates the particle distribution and the corresponding pose. Three target points (red five-pointed star in Figure 9(b)-(d)) are set in the grid map, and the robot sequentially reaches the specified target point and moves along the path by using the positioning navigation algorithm.

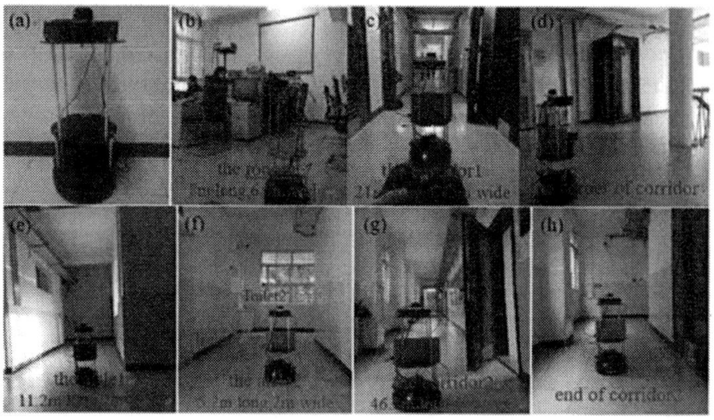

Figure 7. The robot platform used in the field test (a), and the test environment (b)-(h).

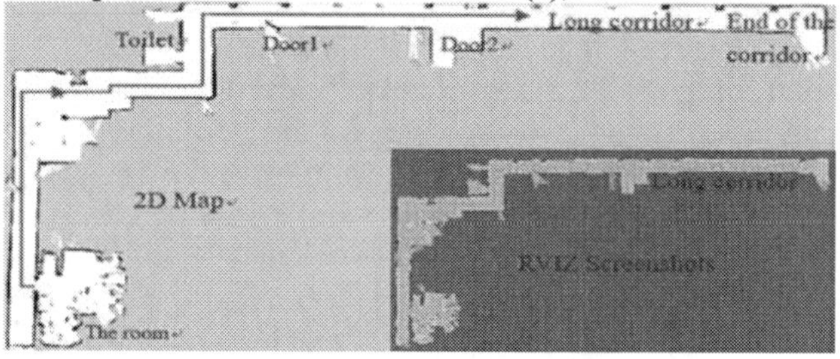

Figure 8. 2D map of test environment.

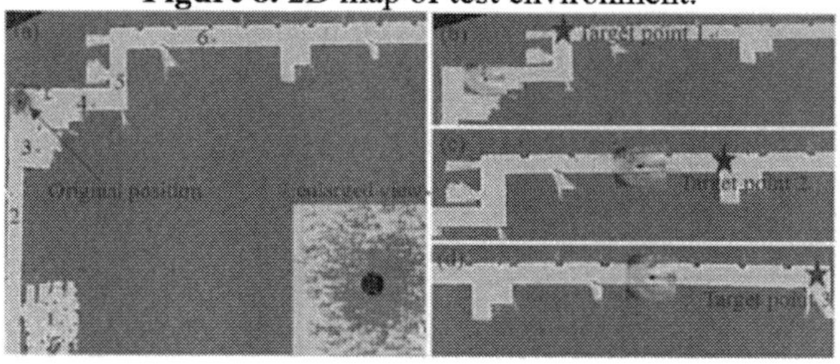

Figure 9. The robot's initial position (a) and its motion process (b) from start point to target point1; (c) from target point1 to target point2; (d) from target point2 to target point3.

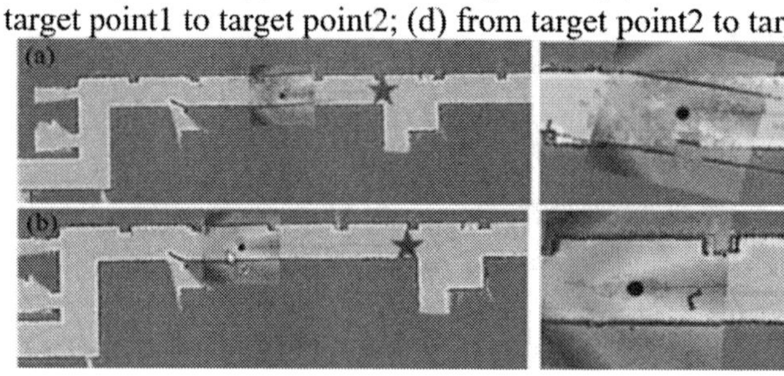

Figure 10. Positioning navigation comparison based on (a) initial algorithm and (b) improved algorithm.

In the navigation process from the target point 1 to 2, the navigation effect of the original algorithm and the improved algorithm of the present invention is compared, as shown in Figure 10. It can be seen that the original algorithm has a high degree of particle dispersion and significant deviation, and the laser scanning does not match the current map, resulting in an error in positioning navigation. The improved algorithm makes the laser scanning and map coincidence degree, the particles are concentrated around the current position of the robot. The positioning is relatively accurate, and the robot can accurately reach the target point.

In the long corridor environment, the robot has been tested many times. It can be found that when the wheels slip or the robot is "kidnapped", the original algorithm will lead to the positioning drift and the positioning loss phenomenon, and the improved algorithm has strong adaptability. This allows the robot to quickly adapt to long-distance navigation and reach the target point accurately.

5. Conclusion

This paper mainly studies the improved localization and navigation algorithm to improve the localization accuracy of mobile robot and avoid navigation failure. Firstly, This paper combines the traditional AMCL method and the PLICP method to solve inaccurate positioning problem of the robot caused by wheel skidding and collision. Secondly, using the ROS framework integrated with PLICP-AMCL algorithm, A* algorithm, DWA algorithm and improved navigation failure recovery method, the robot is successfully controlled along the path of independent planning, effectively improving the efficiency of positioning and navigation. Finally, the TurtleBot robot is used as the experimental platform for comparative simulation and physical experiments. The results show that the improved positioning and navigation algorithm has better accuracy and robustness. In the future, we can consider the fusion of the algorithm in the text with Inertial Measurement Unit (IMU) sensor information and depth camera information to further improve the autonomy of mobile robots in complex environments' positioning and navigation.

Acknowledgments

This work was supported by the National Natural Science Foundation of China (61573148), the Science and Technology Planning Project of Guangdong Province (2015B010919007, 2016A040403012, 2017B090901043), and the Science and Technology Planning Project of Guangzhou (201604046015, 201604046014, 2017171201).

References

[1] Sergio R González and Ivan Mondragón 2017 *Manufacturing Control Architecture for FMS with AGV: A State-of-the-Art* (Advances in Automation and Robotics Research)

[2] Iñigo RM and Alley D 1991 *Algorithms for path planning, navigation and guidance of an AGV* vol 7 (Robotics and Autonomous Systems) pp 309-326

[3] Song Z and Wu X 2016 *A new method of AGV navigation based on Kalman Filter and a magnetic nail localization* Int. Conf. on Robotics and Biomimetics (New York: IEEE) pp 952-957

[4] Hao Li and Wenhua Ye 2018 *Markov localization algorithm based on Gaussian kernel function in feature map* vol 24(Computer Integrated Manufacturing Systems) pp 1081-1088

[5] Fox D Burgard W Thrun S 1997 *The dynamic window approach to collision avoidance* vol 4 (IEEE Robotics & Automation Magazine) pp 23-33

[6] Rohde J and Jatzkowski I 2016 *Vehicle pose estimation in cluttered urban environments using multilayer adaptive Monte Carlo localization* Int. Conf. Information Fusion (New York: IEEE) pp 1774-1779

[7] Fengping C and Qiyao F 2018 *Research on real-time positioning based on adaptive Monte Carlo algorithm* vol 44 (Computer Engineering) pp 28-32

[8] Almeida J and Santos V M 2012 *Real time egomotion of a nonholonomic vehicle using LIDAR measurements* vol 30 (Journal of Field Robotics) pp 129-141

[9] Censi A 2008 *An ICP variant using a point-to-line metric* Proc. Int. Conf. on Robotics and Automatio (Pasadena) pp 19-23

[10] Thrun S Fox D Burgard W 2008 *Robust Monte Carlo localization for mobile robots* (A msterdam: Artificial Intelligence) pp 103-104

[11] Yajie Wu and Jingdong Yang 2017 *Mobile robot path planning based on A* algorithm* vol 30 (Electronic Science and Technology) pp 124-127

[12] Enqiang Zhou 2019 *Research on Positioning and Navigation Technology of Mobile Robot based on SLAM Algorithm* Int. Conf. Machinery, Materials and Computing Technology (Chongqing) pp 500-505

IOP Conf. Series: Materials Science and Engineering **611** (2019) 012077 doi:10.1088/1757-899X/611/1/012077

Universal Industrial-applicable Calibration Method of Optical-electronic Triangulation Systems for 3D Geometry Measurements

Sergey Dvoynishnikov*, Vitaly Rakhmanov and Vladimir Meledin

Kutateladze Institute of thermophysics SB RAS, Novosibirsk, Russia

*Corresponding author

Abstract. Universal industrial-applicable calibration method for optical-electronic triangulation systems using structured illumination for 3D geometry measurements is proposed. An analytical and experimental estimations of the measurement error due to the method of phase steps and system calibration were obtained. The method has a simple implementation when measuring the 3D geometry of large objects. It is shown, that proposed method can be effectively used in measurement systems constructed on a standard low-cost element base.

1. Introduction

The development of methods for measuring 3D geometry of complex-profile objects is very important for the modern industrial production technologies [1-2]. The most promising are optoelectronic methods for measuring geometric parameters. Calibration of optoelectronic systems for measuring 3D geometry consists in determining a set of parameters describing the mapping of 2D coordinates of a spatial modulator of an optical source on spatial 3D coordinates in which the measured object is located (world coordinates); and the mapping of world 3D coordinates, in which the measured object is located, on the 2D coordinates of the image of the detector of optical radiation. We can determine the three-dimensional coordinates of a point in the world coordinate system, having received coordinates of the point of the measured object on the image of the detector and knowing the corresponding coordinates in the plane of the projector. Unlike calibration of robotic arms or automated trolleys, the calibration methods proposed and used in this work are static. Calibration is performed once and after that it will be reliable as long as the relative position of the detector and the projector remains unchanged.

In measurement systems using spatially modulated sources of optical radiation, calibration is an extremely important task. Recovery of complete information about the surface geometry is possible if the optical parameters of the detector and the projector of optical radiation are known in the global coordinate system. As a rule, methods for calibrating active-type vision systems are more complex than methods for calibrating passive-type systems, such as stereovision systems.

Most widely used standard technical solution of calibration [3-5] is associated with the execution of a 3-step procedure. These are the determination of the internal parameters of the detector, the determination of the internal parameters of the projector and the determination of the geometric parameters of the Euclidean transformation associated with the projector and detector of optical radiation.

In most works, when determining the internal parameters of the projector and detector of optical radiation, the authors use approximations of geometric optics, which do not take into account the distortions caused by defects and aberrations of optical elements. When using systems in laboratory conditions and using high-quality low-aperture optics, errors associated with optical aberrations will be negligible, but when implementing measurements in industrial technology, they must be taken into account due to the presence of non-stationary phase-inhomogeneous distortions.

2. Method Description

To solve this scientific problem, a method of calibrating an optoelectronic system for measuring 3D geometry has been proposed. Proposed method involves to use a calibration object, which is a vertical rail on a stand, stable in a vertical position (fig. 1). On the rail there are calibration targets in the form of circles. The rail is moved in the horizontal plane and the dependence of the coordinates of the targets on the images, the magnitude of the phase shift of the probe sinusoid and the Cartesian coordinates of the targets in the global coordinate system are memorized.

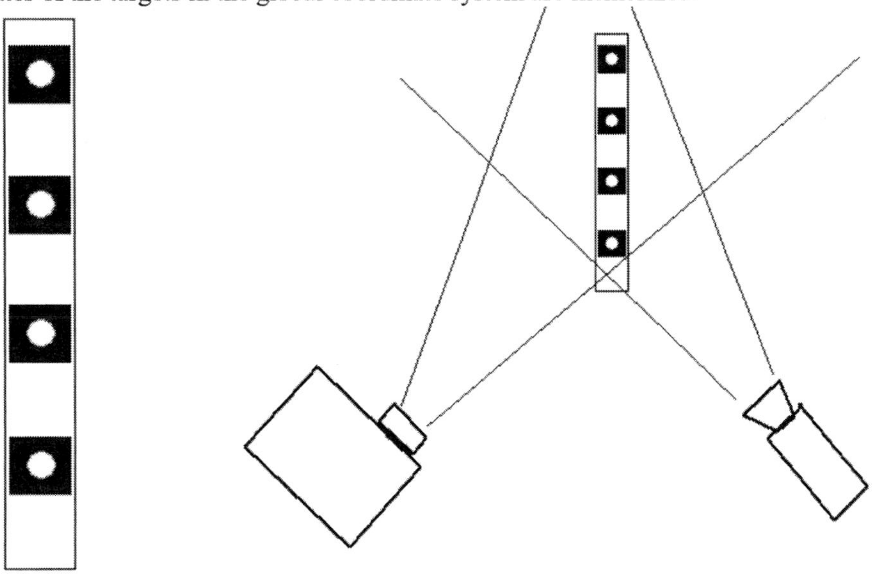

Figure1. The calibration target (a) and the principle of calibration of the system (b).

After measuring and remembering the required number of points in space, based on the data obtained, a regression function is constructed depending on the 3 Cartesian coordinates in the global report system on the magnitude of the phase shift and the coordinates of a point in the image

$$(X,Y,Z) = F(X_c, Y_c, P_p)\,. \tag{1}$$

After that, the target object is measured: the phase shift value is measured for each point on the image of the object and the Cartesian coordinates are determined in the global coordinate system for each image point of the object using the obtained regression calibration function.

The relative measurement error caused by the calibration error using the calibration procedure can be estimated as

$$\Omega = \frac{\delta}{\sqrt{M}}, \tag{2}$$

where δ is the measurement error of the Cartesian coordinates of the calibration object, M is the number of measurements of the calibration object with different Cartesian coordinates, X_c is the resolution of the horizontal component of the optical detector, Y_c is the resolution of the vertical component of the optical detector.

The total relative error of measurement of Cartesian coordinates can be estimated from above as a sum of θ, Ω and an additional error due to the image resolution obtained by using an optical detector:

$$\Omega = \frac{\Delta I}{N \cdot I} + \frac{\delta}{M} + \frac{1}{X_c}. \tag{3}$$

The obtained analytical error estimation allows us to estimate the relative measurement error of the 3D geometry of large-sized objects by triangulation methods using structured lighting. Let the relative error of measuring the brightness of the radiation in the path of projector - detector of optical radiation does not exceed 0.0078 (this error is due to the use of an LCD projector, the color coding depth of which is 8 bits). Let in our implementation of an optoelectronic measuring system the number of radiated highlights having a different initial phase shift, N = 1000. Let the relative error of measurement of the Cartesian coordinates of a point in the calibration process δ= 0.1%. The number of points of the calibration object involved in the calibration M = 200. Let the resolution of the images received by the receiver of optical radiation is 3000x2000 pixels (such resolution is provided by standard digital cameras). Then, the total relative error of measurement of 3D geometry of methods based on optical triangulation using structured radiation can be estimated from above as

$$\Omega = 0.00024 + 0.00007 + 0.0005 = 0.00081 = 0.081\%. \tag{4}$$

3. Experimental Results

An experimental implementation of the proposed universal calibration method was performed. The calibration target is a vertical rail with reference marks located on it. As a result of the calibration, a set of points in space is formed, corresponding to the reference marks that fall into the camera's sector of view and illuminated by a projector.

To simplify the task of determining the 3 Cartesian coordinates of the reference marks, it is assumed that the OZ axis is located vertically, and the OX and OY axes are horizontal and orthogonal to each other.

To solve the problem of determining the Cartesian coordinates of the reference marks in the world coordinate system, the rail is mounted vertically using a vertical plumb located on a calibration target. After that, the distance from the reference targets to the point of intersection of the vertical bar of the calibration target and the plane Z = 0 is measured. This determines Z coordinates in the world coordinate system for the reference marks (since the rail axis is parallel to the OZ axis). After that, it is necessary to calculate the Cartesian coordinates of the vertical bar of the calibration target in the horizontal XY plane. To solve this problem, measurements of distances in the XY plane are carried out.

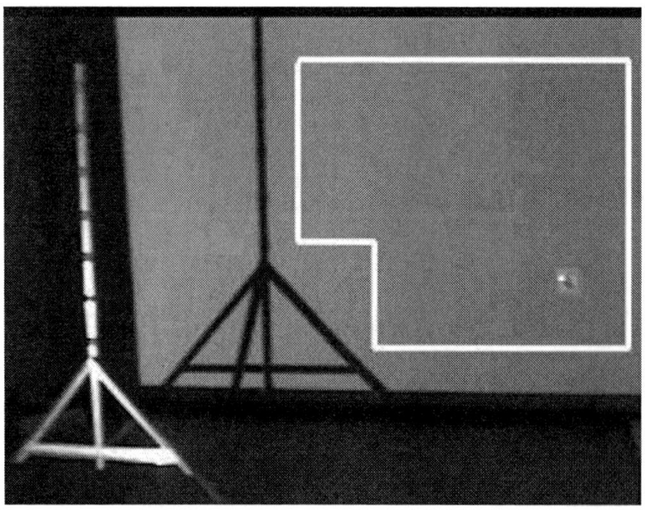

Figure 2. Measured wall fragment (circled in white frame).

IOP Conf. Series: Materials Science and Engineering **611** (2019) 012077 doi:10.1088/1757-899X/611/1/012077

The calibration of the measurement system that implements the optical triangulation method with structured lighting has been performed. According to the results of 45 measurements of the calibration target, it was found that the RMS error in determining coordinates X, Y, Z is about 5 mm. The linear size of the measuring volume was about 1 meter, the relative error of about 0.5%. The error was determined by comparing the coordinates measured manually by the experimenter and measured by the system. This error characterizes the total error in determining the coordinates of the calibration target during calibration.
To assess the accuracy of measurements of three-dimensional geometry, the system conducted an experiment to measure the 3D geometry of a flat surface of a room wall. In fig. 2 in the white frame shows the measured fragment of the wall.
Figure 3 shows the measured 3D wall geometry. The relief charts clearly show the electrical outlet in the lower right corner of the wall fragment.

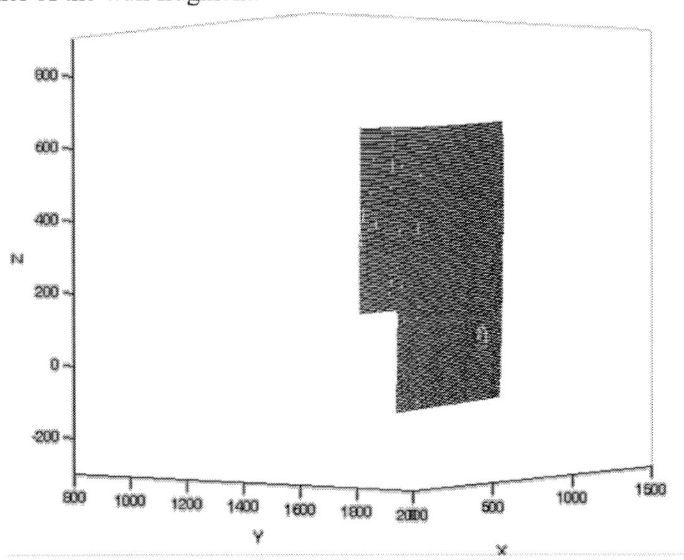

Figure 3. Measured 3D wall fragment geometry.
Figure 4 shows the cutoff deviation along the line Z = 400 mm of the measured 3D geometry of the fragment of the wall from the straight line.

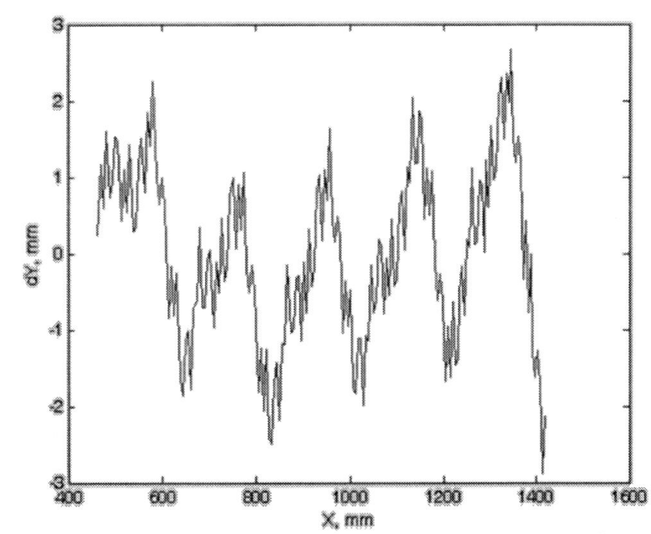

Figure 4. Deviation of the cut Z = 400 mm of the measured 3D geometry of the wall fragment from a straight line.

The standard deviation of the curve shown in Fig. 4 is 1.14 mm. This means that the relative measurement error obviously does not exceed the level of 0.057%.

4. Conclusion

Universal industrial-applicable calibration method for optical-electronic triangulation systems using structured illumination for 3D geometry measurements is proposed. The method has a simple implementation when measuring the 3D geometry of large objects. An analytical and experimental estimations of the measurement error due to the method of phase steps and system calibration were obtained. It was less than 0.1% of the range of measured values when implemented on a standard low-cost element base.

5. Acknowledgement

This research was supported in part by RFBR (project No 18-08-00910) and was carried out under state contract with IT SB RAS.

References

[1] Dvoynishnikov S. V., Rakhmanov V. V., Meledin V. G., Kulikov D. V., Anikin Yu. A., Kabardin I.K. Experimental Assessment of the Applicability of Laser Triangulators for Measurements of the Thickness of Hot Rolled Product // Measurement Techniques. 2015, 57(12). –P. 1378-1385.

[2] Dvoinishnikov S.V., Anikin Y.A., Kabardin I.K., Kulikov D.V., Meledin V.G. An Optoelectronic Method of Contactless Measurement of the Profile of the Surface of Large Complexly Shaped Objects // Measurement Techniques. 2016. 59(1). - P. 21-27.

[3] De Piero F. W., Trivedi M. M. 3-D computer vision using structured light: Design, calibration, and implementation issues // Advances in Computers, 1996, V.43. – P.243–278.

[4] McIvor A.M., Valkenburg R.J. Calibrating a structured light system // Aukland: Industrial Research Limited Report 362, 1995. – 17 p.

[5] Abdel–Aziz, Y., Direct linear transformation into object space coordinates in close–range photogrammetry // Proceedings of the Symposium on Close–Range Photogrammetry, Urbana, Illinois, 1971. - P. 1–18.

Research on Optimum Laying Method for Increasing Underground Power Cable Ampacity

Dan Pang[1], Haipeng Wu[2,*], Bin Dai[1] and Changpeng Zhao[1]

[1]Changchun Power Supply Company, Jilin Electric Power Company, Changchun, China
[2]College of Electrical Engineering, Northeast Electric Power University, Jilin, China

*Corresponding author

Abstract. When the cables are laid with equal spacing, its ampacity is limited due to the mutual heat effects. In order to increase the ampacity of the cable group, an optimum laying method with unequal spacing is proposed in this paper. The constraint function is that the temperature of all cables is lower than the upper limit, and the objective function is that the ampacity of cables is maximum. The constrained problem is transformed to an unconstrained problem by constructing a penalty function. The results show the effectiveness of the proposed method. The ratio of the increased ampacity increases with the increase of cable numbers.

1. Introduction

When power cable is running, the components of core, insulating layer and sheath will generate heat due to loss, which will increase the temperature of each part of the cable. For XLPE insulated cables, the current carrying capacity corresponds to the constant load current at 90℃ of core temperature. Current carrying capacity is an important parameter in cable design and operation. Constant load current above current capacity will shorten cable life and reduce operational reliability. Below the current capacity, cable transmission capacity cannot be fully utilized. [1-4]. The IEC-60287 standard provides a current-carrying calculation formula for a direct-buried cable, which is highly accurate for soils with uniform thermal conductivity. [5]. When the cables are laid in clusters, the mutual heat effect will result in high temperature rise of cables in the central region and low temperature rise of cables in the edge region if the equal spacing is adopted. The difference in temperature rise will result in low utilization rate of edge cables, thereby limiting the overall transmission capacity of cable clusters. In order to solve this problem, the artificial fish swarm algorithm is used to arrange the cables with unequal spacing. The spacing of cables from the centre to the edge decreases from large to small, which reduces the mutual heat effect of the central cables and increases the mutual heat effect of the edge cables, and reduces the temperature difference of the cable cores at different locations, so as to improve the current carrying capacity of the cable cluster. In order to better reflect the advantages of unequal spacing arrangement of cables, this paper uses IEC method to calculate and compare the carrying capacity of multi-loop cables laid with medium spacing and optimized with unequal spacing in uniform soil.

2. Calculating Method of Ampacity of Power Cable Cluster

IEC-60287 gives the formula for calculating the current carrying capacity of directly buried AC cables as follows:

$$I = \sqrt{\frac{(\theta_{\max} - \theta_c) - W_d\left[0.5T_1 + n(T_2 + T_3 + T_4)\right]}{RT_1 + nR(1 + \lambda_1)T_2 + nR(1 + \lambda_1 + \lambda_2)(T_3 + T_4)}} \tag{1}$$

In the formula, I is the current in the cable, A;θ_{max} is the maximum working temperature for the cable,℃; θ_c is the ambient medium temperature,℃; W_d is the dielectric loss of the insulation layer per phase of the cable,W/m; T_1、T_2、T_3、T_4 are thermal resistance of insulating layer, insulation shielding, outer shield and surrounding medium; R is the unit length resistance of each phase of cable core, $\lambda_1$and λ_2 are the ratio of metal sleeve loss and cable armor loss to core loss, respectively.

If N identical cables are laid in the soil and the load current is the same, the mutual heat effect between cables should be considered. The equivalent thermal resistance T_4 of the K-th cable around the cable is obtained by the mirror method [6].

$$T_4^{'} = \frac{\rho_{T4}}{2\pi}\ln\left(\frac{4L}{D_e} \times \frac{d'_{1K}}{d_{1K}} \times \frac{d'_{2K}}{d_{2K}} \ldots\ldots \times \frac{d'_{NK}}{d_{NK}}\right) \tag{2}$$

where ρ_{T4} is the thermal resistance coefficient of the soil, K • m /W; L is the laying depth of the cable, m; D_e is the outer diameter of the outer sheath of the cable, m; d_{1K}, d_{2K} , ⋯, d_{NK} are respectively cable 1, 2,..., the distance from the center of N to the center of the cable, m; d'_{1K} , d'_{2K} , ⋯, d'_{NK} are the distances from the mirror center of the cable 1, 2, ..., N to the center of the cable K, respectively.

3. Cable Group Optimization Based on Artificial Fish Swarm Algorithm

3.1 Mathematical Model

The cable cluster is laid in the way of Fig. 1, and A, B, and C indicate the three-phase and the lower-angle standard circuit number of each circuit. In this paper, the Artificial Fish Swarm Algorithm (AFSA) is used to find the optimal laying position of the cable [7]: The cable cluster current I and the core operating temperature[θ_i]and the cable laying position coordinates [x, y] are variables, and I maximum is the objective function. The constraint conditions are: 1), the operating temperature of each cable core is lower than the maximum allowable value of 90℃; 2), the adjacent two loop cable spacing is not less than the cable outer diameter, that is, the horizontal coordinate of each loop is greater than the previous loop abscissa The sum of the outer diameters of the cables. This problem can be described by equations (3) to (4):

$$\text{Objective function: } Max \quad I = f(x, y) \tag{3}$$

$$Restrictions: \begin{cases} \theta_i \le 90°C & (i = 1, 2, \cdots, N) \\ x_j + D_e < x_{j+1} & (j = 1, 2, \cdots, m) \\ 0 \le x_j \le width \end{cases} \tag{4}$$

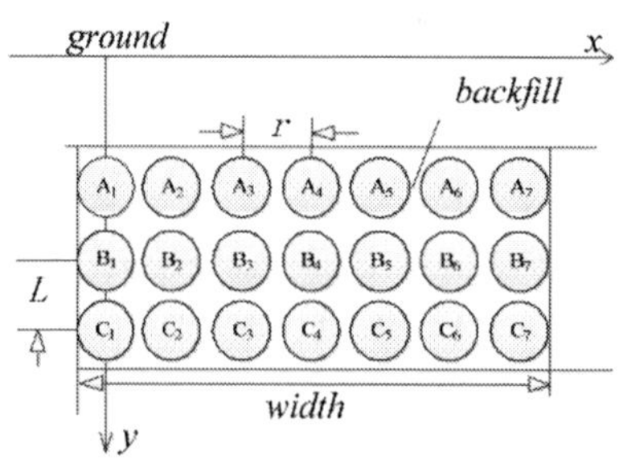

Figure 1. Map of laying for identical and single-core cables

where m is the number of loops in a single-core cable cluster, N is the total number of cable clusters, D_e is the outer diameter of the cable, *width* is the laying width of the cable group. This problem assumes that the four boundaries of the section of the cable cluster laying area are fixed, and the x-coordinates of the three-phase cables per loop are the same, so the main work is to use the artificial fish swarm algorithm to determine the optimal coordinates of the x-direction of each loop.

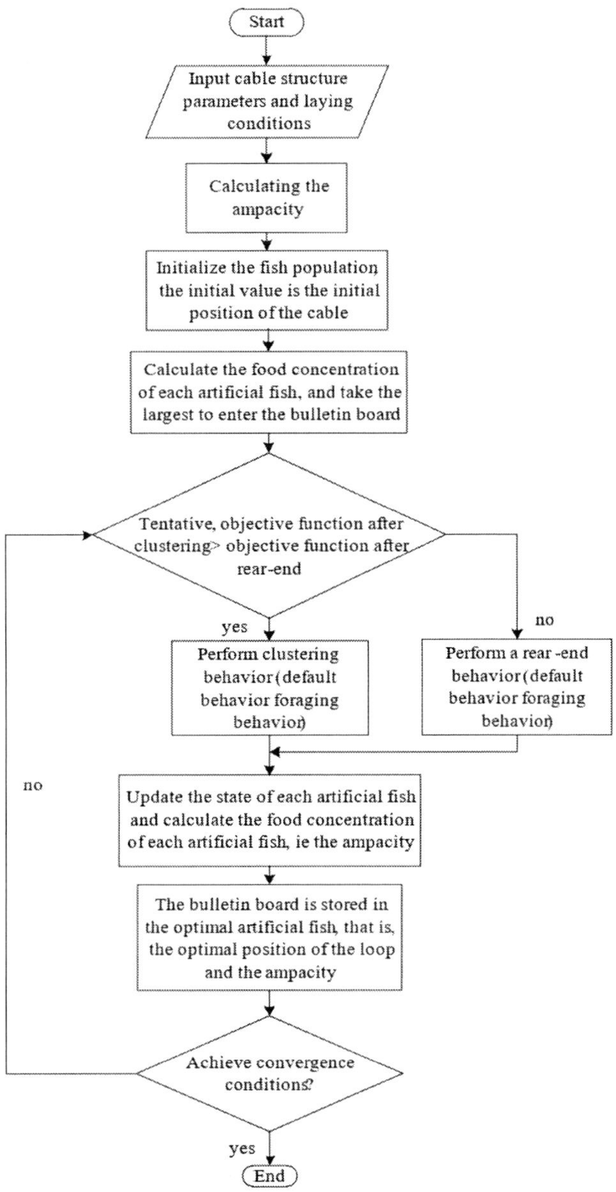

Figure 2. Calculation flowchart of power cables collocation optimization based on AFSA

3.2 Processing of Constraints

In this paper, the penalty function method is used to construct a generalized objective function with penalty terms, and the constrained problem is transformed into an unconstrained problem [8].

Abstract mathematical model:

$$\begin{aligned} Min \quad & -I = -f(x) \\ s.t. \quad & h_1(x) = x_j + D_e - x_{j+1} < 0 \\ & h_2(x) = x_j - width \le 0 \qquad (j = 1, 2, \cdots, m) \\ & h_3(x) = -x_j \le 0 \end{aligned} \tag{5}$$

where $h_1(x)$, $h_2(x)$ and $h_3(x)$ are functions constructed using constraints. The out-of-bounds constraint of the inequality constraint is added to the objective function $f(x)$ in the form of a penalty term, and the fitness function (ie, the penalty function) $F(x)$ of the artificial fish swarm algorithm is constructed:

$$Min \quad F(x) = -f(x) + c_i \sum_{i=1}^{3} H_i(x) \tag{6}$$

where the second term in the right is the penalty term, the penalty coefficient c_i is greater than zero and decreases as the number of iterations increases.

The program flow of the cable group optimization design based on the artificial fish swarm algorithm is shown in Figure 2.

4. Optimization Layout Calculation Results Analysis

The calculation object is YJLW03 64/110 kV power cable. Tab. 1 shows the cable structure parameters. The soil thermal resistance coefficient is 1.0K·m/W, and the original soil temperature in the laying area is 20℃. Tab. 2 shows the cable core temperature, circuit spacing and current carrying capacity after equidistant laying and optimized placement.

Table 1. Structure parameters of the power cable

Cable structure	Parameter	Thermal resistance coefficient/((K·m)·W^{-1})	Cable structure	Parameter	Thermal resistance coefficient/((K·m)·W^{-1}))
Conductor diameter /mm	30.3	0.0025	Appropriate thickness of water blocking layer /mm	2×1.5	6.0
Conductor shielding thickness /mm	1.3	3.5	Wrinkle aluminum sleeve thickness /mm	2.0	0.0042
Insulation thickness /mm	16.5	3.5	Outer layer thickness /mm	5.5	3.5
Insulation shielding thickness /mm	1.0	3.5	Cable outer diameter /mm	99.1	—

It can be seen from Tab. 2 that the cables are equally spaced, the average cable core temperature is 83 °C, the temperature of the middle cable core is obviously higher than that of the edge cable, and the maximum temperature difference can reach 17 °C. After optimal placement, the average core temperature increased to 87 °C and the maximum temperature difference was reduced to 6 °C. For XLPE insulated cables, the difference between the core temperature and 90 °C reflects the current carrying capacity of the cable to be excavated. The closer the average temperature is to 90 C, the higher the overall utilization rate of the cables is. In Fig. 1, the optimal layout of the 7-loop cable is 5.8% higher than that of the equal-pitch arrangement. Fig. 3 shows the improvement of the current carrying capacity of the optimal arrangement than the equidistant laying when the cable group circuit number m is increased according to the method of Fig. 1. When the number of circuits is greater than 3, the transmission capacity of the optimized laying increases with the number of circuits, and the unequal spacing optimization technology makes more sense for multi-loop cable cluster laying.

Table 2. Comparison of Ampacities and core temperatures between equal and unequal spacing

Number of loops			1	2	3	4	5	6	7	Vertical spacing /mm
Equal spacing	Core Temp. /℃	A	73	80	83	84	83	80	73	0
		B	78	85	89	90	89	85	78	193
		C	77	84	87	88	87	84	77	193
	Circuit spacing r/mm		0	233	233	233	233	233	233	—
	Ampacity /A						446			
Unequal spacing	Core Temp. /℃	A	84	87	87	86	87	87	84	0
		B	87	90	90	89	90	90	87	93
		C	84	87	87	88	87	87	84	293
	Circuit spacing /mm		0	100	240	360	360	240	100	—
	Ampacity /A						472			

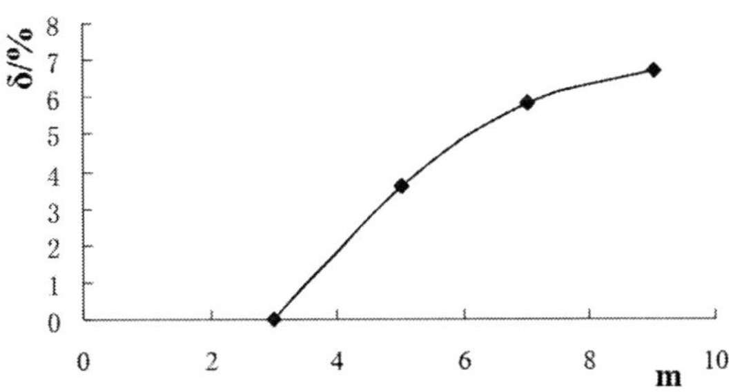

Figure 3. The increased rate of ampacity as a function of the number of loops

5. Conclusion

(1) Adopting equal spacing arrangement, the mutual heat effect between cables makes the central cable of the cable cluster hot, while the edge cable of the cable cluster cold.

(2) Unequal spacing arrangement of cable spacing is gradually reduced from the center to the edge of the cable cluster, which can reduce the temperature difference of the entire cable cluster core.

(3) Unequal spacing arrangement of cable clusters excavates the conveying capacity of edge cables and improves the transmission capacity of cable clusters.

References

[1] Nahman J 2018 *Electrical Power and Energy Systems* **103** 310–316
[2] Yongchun Liang 2013 *Proc. Int. Conf. on PSEC* **1** 1–4
[3] Chepurnenko A S, Yazyev S B and Evtushenko A I 2017 *Proc. Int. Conf. on Industrial Engineering, Applications and Manufacturing* **1** 1–4
[4] Wang Jian, Jin Huan, Ma Xiao and Bai Xu 2018 *Proc. Int. Conf. on 12nd International Conference on Smart Grid and Smart Cities* **1** 1–7
[5] IEC 60287-1-1 *Electric cables-calculation of the current rating-part 1* 2014.
[6] Zheng Zhaoji and Wang Ming 1983 *High voltage cable line* (Beijing: China Water & Power Press)
[7] Li Xiaolei 2003 *PH.D. thesis* (Hangzhou: Zhejiang University)
[8] Zhang Guangcheng 2005 *Computational Metbods for Nonlinear Optimization* (Beijing: Higher Education Press)

Calculation Method of Cable Ampacity Based on Field-circuit Combined Mode

Dan Pang[1], Haipeng Wu[2,*], Bin Dai[1] and Xiaoyan Wang[1]

[1]Changchun Power Supply Company, Jilin Electric Power Company, Changchun, China

[2]College of Electrical Engineering, Northeast Electric Power University, Jilin, China

*Corresponding author

Abstract. Electricity is usually transmitted through power cables in densely populated areas. It is significant to predict the ampacity and temperature field accurately to ensure safe and economical operation of cables. In this paper, a field-circuit combination algorithm is proposed. The equivalent thermal circuit method and the finite difference method are effectively combined to calculate the temperature of cable and soil area respectively. This method is more accurate than the equivalent thermal circuit method and faster than the traditional numerical method. It is proved that the new method in this paper can meet the needs of engineering by comparing with the results of equivalent thermal circuit method in homogeneous soil and coordinate combination method in inhomogeneous soil respective.

1. Introduction

Accurate calculate cable ampacity is great significance to improve the economics of cable applications with the wide application of cables in power grids construction. The equivalent thermal circuit method for calculating cable ampacity is in accordance with the IEC-60287 standard recommended by the International Electrotechnical Commission[1]. When the actual laying environment conditions are different from the standard conditions, the correction factor is usually used for correction. The selection of the correction factor is a very complicated problem due to lack of reliable correction basis. Numerical calculation method has advantages of low cost and ability to simulate complex working conditions, which has high practical value for improving the calculation accuracy of ampacity. The numerical methods commonly used to calculate temperature field include: finite element method [2-4], boundary element method [5], finite difference method [6], and simulated thermal charge method [7]. The equivalent thermal circuit method provided by IEC standard is a rapid method with high accuracy when dealing with cable ampacity in uniform soil [8], however the calculation error in non-uniform soil environment is large. The calculation accuracy of numerical method is not affected by soil environment complexity but it's slow. Combining the advantages of the two, this paper use the finite difference method to calculate the soil temperature field in the cable laying area, and combines the equivalent thermal circuit method to obtain the core temperature quickly and accurately.

2. Equivalent Thermal Circuit Method for Calculating Cable Core Temperature

The core, the insulation layer, the metal sheath, etc. will generate losses in running cables, and heat is generated to form a heat flow field. Each layer structure can be represented by an equivalent thermal resistance when heat flow is conducted outward through the various layers of cable. Take the insulating layer shown in Figure 1 as an example. θ_c and θ_{sh} are the temperature (℃) of the inner and outer surfaces of the insulating layer respectively; D_c and D_{sh} are the inner and outer diameters (m) of the cable insulation layer respectively; *Wc* is the core loss (W); ρ_T is the thermal resistance of the insulating layer coefficient (K•m/W); R_1 is the thermal resistance of the insulation layer (K•m/W). Assume that the cable consists of core, insulating layer, water blocking layer, metal sheath, and outer sheath. Figure 2 shows the overall equivalent thermal path of the cable. W_{in} and $\sigma_1 W_c$ are the insulation layer loss and the metal sheath loss (W) respectively. R_2, R_3 and R_4 respectively represent the thermal resistance of the water blocking layer, the outer sheath and the surrounding medium (the thermal resistance of the metal portion is negligible). If the total number of cables is *M*, the depth of a cable is *L*, and the thermal resistance coefficient of the cable group adjacent to the soil is ρ_{T4}, then the R_4 of the cable is[8]:

$$R_4 = \frac{\rho_{T4}}{2\pi}\ln\left(\frac{4L}{D_{ou}}\times\frac{d'_{1K}}{d_{1K}}\times\frac{d'_{2K}}{d_{2K}}\cdots\cdots\times\frac{d'_{MK}}{d_{MK}}\right) \text{ (Which does not include } \frac{d'_{KK}}{d_{KK}} \text{ items)} \quad (1)$$

d_{1K}, d_{2K}……d_{MK} is the distance from the center of the cable 1, 2...*M* to the center *K* of the cable; d'_{1K}, d'_{2K}……d'_{MK} is the distance from the mirror center of the cable 1, 2...*M* to the center *K* of the cable, and the D_{ou} is the outer diameter of the outer sheath of the cable. (m), the heat balance equation for each cable is[8]:

$$\theta_c = \theta_a + R_1\left(W_c + \frac{1}{2}W_{in}\right) + R_2(W_c + W_{in}) + (R_3 + R_4)(W_c(1+\sigma_1) + W_{in}) \quad (2)$$

θ_a is the temperature of the medium surrounding the cable. Using the equivalent thermal circuit method to calculate the core temperature, the soil in the laying area is required to have a single thermal conductivity. If the thermal conductivity of the laying area soil is not unique, the temperature of the core cannot be accurately determined. In this case, a numerical method is needed to solve the problem.

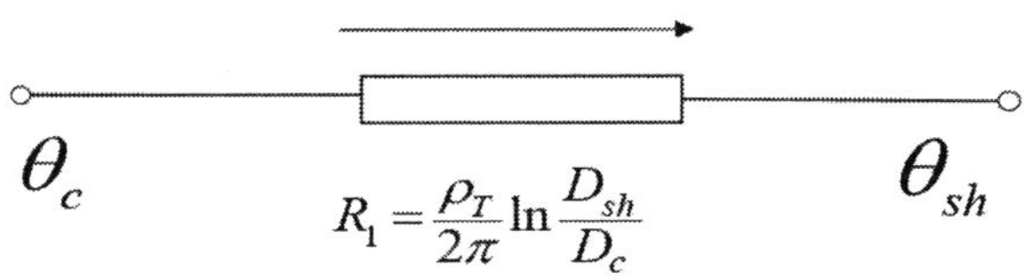

Figure 1. The equivalent thermal circuit of insulating layer

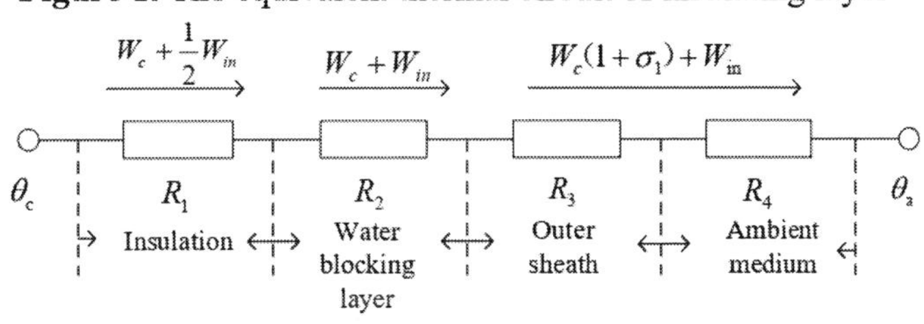

Figure 2. The equivalent thermal circuit of a cable

3. The New Method- field-circuit Combination Algorithm

Calculate temperature field in cable laying area by finite difference method [6]. The calculation area including the cable is divided into right angle grids, the soil and cable layers in the area are divided into different grids. The thermal conductivity of the corresponding materials in the grid is assigned to the right grid, where the core is located as heat source. In order to improve the calculation accuracy and reduce the calculation amount, the cable area adopts an encrypted grid. The farther from the cable, the more sparse the grid is. The temperature error of the cable core obtained by this method is large, but the soil temperature error around the cable is small. At the periphery of the cable, take a few layers of soil concentric with the cable and the same thickness, the total thickness is *d*, as shown in Figure 4, the thermal resistance of the soil layer can be expressed as:

$$R_4 = \frac{\rho_{T4}}{2\pi} \ln \frac{(D_{ou} + 2d)}{D_{ou}} \tag{3}$$

Cable thermal differential equations, initial conditions and boundary conditions can accurately describe the transient temperature field distribution of a particular cable.The heat conduction differential equation of the cable transient temperature field in the Cartesian coordinate system shown in Figure 3 is [9]:

$$\rho C_p \frac{\partial T}{\partial t} = K[\frac{\partial^2 T}{\partial x^2} + \frac{\partial^2 T}{\partial y^2}] + Q \tag{4}$$

Where ρC_P is the product of the medium density and the specific heat at a certain point in the field, T is the temperature, K is the thermal conductivity of the medium, and Q is the heat source.The heat transfer equation can be discrete into:

$$a_P T_P + a_E T_E + a_W T_W + a_N T_N + a_S T_S = b + \rho C_p \frac{T^0_{i,j}}{\Delta t} \tag{5}$$

In the formula: $a_P = \frac{\rho C_P}{\Delta t} + 2K(\frac{1}{\Delta x_1 \Delta x_2} + \frac{1}{\Delta y_1 \Delta y_2})$; $a_E = -\frac{2K}{\Delta x_2(\Delta x_1 + \Delta x_2)}$; $a_W = -\frac{2K}{\Delta x_1(\Delta x_1 + \Delta x_2)}$; $a_N = -\frac{2K}{\Delta y_1(\Delta y_1 + \Delta y_2)}$; $a_S = -\frac{2K}{\Delta y_2(\Delta y_1 + \Delta y_2)}$; $b = Q$.

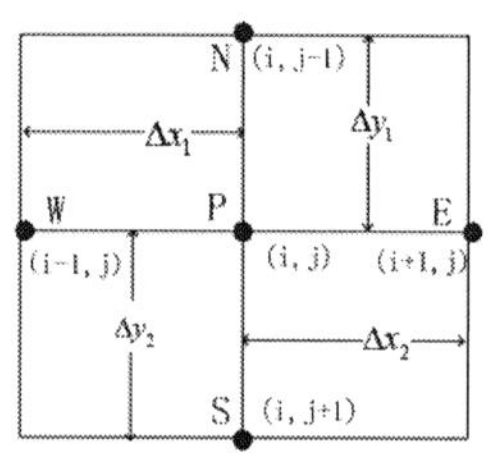

Figure 3. Grid of rectangular coordinates

Based on the temperature field of the right-angle grid, the quadratic interpolation is used to obtain the temperature of the outermost layer of the soil layer. The average temperature of the soil layer is:

$$\theta_a = \frac{1}{12} \sum_{i=1}^{12} \theta_{a_i} \tag{6}$$

Substituting this result into equation (2) in combination, the core temperature can quickly and accurately determine by the equivalent thermal circuit method. This method can not only solve the problem of uneven soil which cannot be solved by the equivalent heat path method, but also the calculation efficiency of the numerical method is significantly improved.

Since the insulation loss and metal sheath loss are much less than the core loss, the former two can be equivalent to the core loss. In addition, in order to improve the calculation speed more effectively under the premise of ensuring the calculation accuracy, the layer structure other than the cable core can be processed by the harmonic averaging method [10]:

The thickness of each layer is equivalent to a medium, and its thermal conductivity λ_T can be expressed as:

$$\lambda_T = \frac{\frac{\ln(r_n / r_i)}{2\pi}}{\sum_i^n \frac{\ln(r_i / r_{i-1})}{2\pi \lambda_i}} \tag{7}$$

λ_T is the equivalent thermal conductivity obtained by the harmonic averaging method; $i = 1$ is the insulating layer, $i = n$ is the outer sheath; λ_i is the thermal conductivity of the layer *i*; r_i is the radius of the layer *i*.

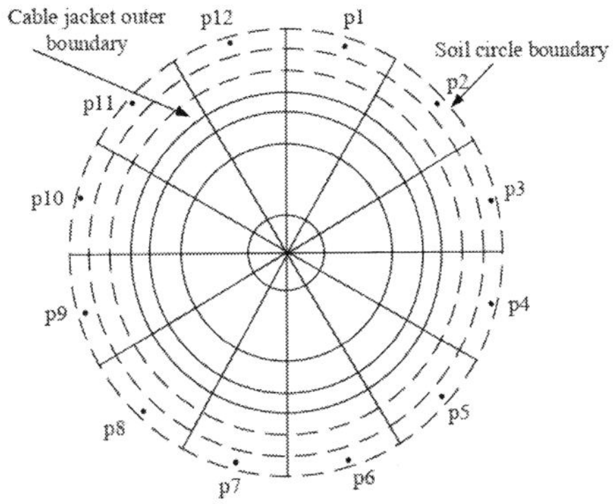

Figure 4. Temperature of the soil around the cable

4. Ampacity Calculation and Method Verification

The load current of each cable is the same. The heat dissipation of each cable is affected by the embedding mode and other cables. The ampacity should be based on the cable with the worst heat dissipation condition. In order to effectively reduce the number of iterations, the following steps can be used to calculate the ampacity:

(1) First set t_m and Δt to the maximum allowable steady-state temperature of the cable core and the allowable error respectively, then obtain the initial value of the load current I_O by the image method[7];

(2) Input I_O into the program to find the steady-state temperature t of the cable core. If $t > t_m$, recalculate I_O-ΔI as a new I_O, otherwise reverse I_O+ΔI as a new I_O;

(3) When iteratively calculates to $t_m - \Delta t < t < t_m$, I_O is the ampacity of the cable group.

In order to verify the correctness of the method, the ampacity of 66kV direct buried cables with single-circuit and double-circuits is calculated, and the calculation results are compared with the IEC standard and coordinate combination method [6]. The calculation object is XLPE insulated cable. The nominal cross section of the copper cable core is 630mm^2. The metal sheath is grounded by three-section cross-connection. The soil temperature in the laying area is 25℃, the thermal conductivity is 0.83 W/m•K, and the thermally conductive coefficient of back-filled sand is 0.3W/m•K. The laying method is shown in Figure 5.

Table 1 shows the calculation results of the three methods, in which I_1 and I_2 are the ampacity of single and double circuits respectively, and t_1 and t_2 are the calculation time of single and double circuits, respectively.

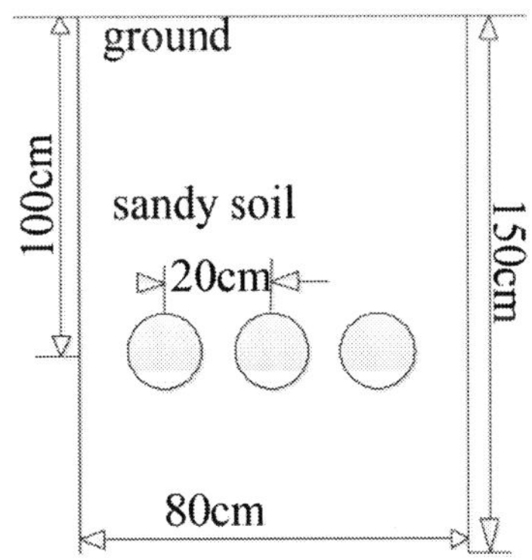

Figure 5. Laying method of 66kV single circuit cable

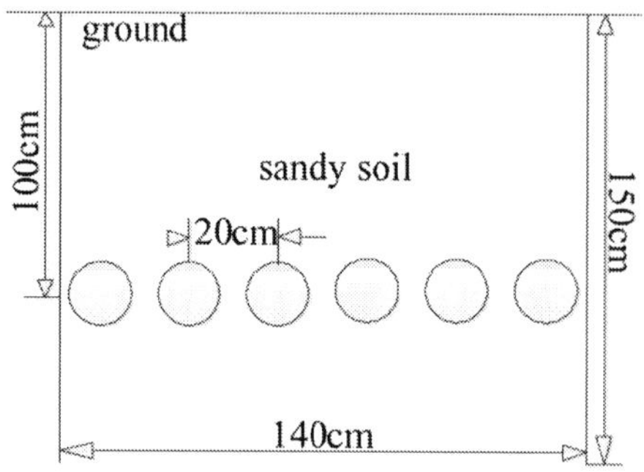

Figure 6. Laying method of 66kV double circuit cable

Table 1. Ampacity comparison of different methods

Calculation method	**Uniform soil** I_1(A)	I_2(A)	t_1(min)	t_2(min)	**Backfill sand** I_1(A)	I_2(A)	t_1(min)	t_2(min)
IEC standard	954	802	—	—	—	—	—	—
Field-circuit combination	963	817	8.8	27.5	715	579	5.8	30.0
Coordinate combination	970	826	36.5	77.0	706	565	51.3	200.0

As shown in Table 1, it can be seen that when the soil of laying area is uniform, the calculation results of the field-circuit combination method and the coordinate combination method are in good agreement with the IEC standard; when the sand is backfilled, the field-circuit combination method and the coordinate combination method are in good agreement. It is worth noting that the calculation time of the field-circuit combination method is significantly lower than the coordinate combination method.

5. Conclusion

Compared with the IEC standard method and the coordinate combination method, the field combination method proposed in this paper has the advantages of high numerical calculation accuracy and high analytical calculation efficiency. It can solve the ampacity of cable group in complex soil condition, and the calculation efficiency is obviously higher than the simple numerical calculation.The calculation of the field-field combination method for the current-carrying cable carrying capacity of clustered soil has important practical significance and high engineering application value.

References

[1] IEC 60287-1-1 *Electric cables-calculation of the current rating-part 1* 2014.
[2] Wang Youyuan, Chen Rengang and Chen Weigen 2008 *High Voltage Engineering* 34 3086-3092
[3] Liang Yongchun, Wang Zhongjie and Liu Jianye 2010 *High Voltage Engineering* 36 763-768
[4] Nahman J 2018 *Electrical Power and Energy Systems* 103 310–316
[5] Gela G and Day J J 1988 IEEE Transactions on Power Delivery 3 1341–1347
[6] Gaggido G, Otero A F and José cidrás 2003 IEEE Transactions on Power Delivery 18 667–677
[7] Liang Yongchun, Li Yanmu and Li Yanming 2008 Proceedings of the CSEE 28 129-134
[8] Zheng Zhaoji and Wang Ming 1983 *High voltage cable line* (Beijing: China Water & Power Press)
[9] YU Jianli, SUI Linlin, TIAN Jiping, et al. 2013 Journal of Northeast Dianli University: Natural Science Edition, 33 43-46.
[10] Liang Yongchun, Li Yanmu and Chai Jinai 2007 *Transactions of China Electrotechnical Society* **22** 185-190

The Research on Key Technologies of Power Quality Comprehensive Control in Complex Distribution Network

Shuanghu Wang[1,*], Hongxu Yin[2] and Shun Liang[3]

[1,2,3]NARI Group Corporation. Nanjing, Jiangsu, China
[1,2,3]NARI Technology Development Co., ltd. Nanjing, Jiangsu, China

*Corresponding author

Abstract. As new types of power sources and loads such as distributed photovoltaic and electric vehicle charging facilities are connected to the distribution network, the distribution network has become more variable and complex, and the simple distribution network has evolved into a complex distribution network. The access of new elements, especially the application of power electronic devices, has a great impact on the power quality of the distribution network. The technical framework of power quality comprehensive management of complex distribution network is proposed, and the corresponding technical route is put forward from three aspects: the influence of power electronic equipment application on power quality of distribution network, power electronic equipment modeling of distribution network, flexible monitoring of power quality and rapid control.
Keywords: complex distribution network; comprehensive control; power quality; power electronic equipment.

1. Introduction

With the rapid development of renewable energy technologies and power electronics technologies, new types of power sources and loads such as distributed photovoltaics and electric vehicle charging piles have been connected to the distribution network. The distribution system has changed from a single power supply radial network to a power supply and load. The multi-power network not only changes the traditional network structure of the distribution network, but also increases the uncertainty of the power flow due to the randomness and intermittentness of the output of the distributed power supply, and the distribution network becomes more varied and complex. Simple distribution network has evolved into a complex distribution network.

The access of distributed power supplies, electric vehicle charging facilities, flexible DC equipment and other equipment is inseparable from the application of power electronic equipment. The existence of a large number of power electronic devices brings harmonic interference and imbalance of phase load to the distribution network. Power quality problems such as voltage amplitude fluctuations. Based on the research on the characteristics of power quality of complex distribution network, this paper grasps the influence mechanism and comprehensive treatment technology of a large number of applications of power electronic equipment on the power quality of distribution network, and guarantees the distribution network to provide users with safe, reliable and high-quality electric energy.

2. Summary of Research Status at Home and Abroad

2.1. The Research on Power Quality Characteristics of Complex Distribution Network

D. Perera et al. of the University of Wollongong in Australia proposed a voltage flicker analysis method for a doubly-fed wind turbine to study the attenuation characteristics of various load types of the distribution network. The effect of the fluctuations, the simulation results show that the constant power load and constant current load will increase the voltage flicker of the load node. The Power Electronics Application Center of the Knoxville Electric Power Research Institute studied the effects of single-phase nonlinear loads on power quality and considered the effects of the North American and European low-voltage power system network topologies on harmonic currents. Muhyaddin JHRawa and David WPThomas of the Department of Electrical Systems and Optics, University of Nottingham, UK, studied the effects of personal computers on power quality by studying the background voltage distortion and the percentage of non-linear load to the total load. Accurately quantify the impact of PC clusters on grid computer quality by the influence of harmonic content in distribution network.

Liu Shuming and Li Qionglin from Henan Electric Power Research Institute conducted detailed analysis, summarization and summary of more than ten typical power quality interference sources such as traction load, electric arc furnace load, intermediate frequency furnace load and electrolytic load in medium and high voltage distribution networks. The power quality data of various nonlinear users is obtained. Pan Shu of Chongqing University studied the influence of distributed power supply access on voltage fluctuation and flicker of distribution network. The paper points out the causes of voltage fluctuation and flicker caused by distributed power supply. The volatility of distributed energy output and distributed power supply. There are two voltage-current harmonics caused by medium power electronic equipment, and the calculation method of voltage flicker attenuation and transmission in distribution network is given. The Electric Power Research Institute of Yunnan Electric Power Research Institute (Group) Co., Ltd. conducted a comprehensive evaluation of the impact of distributed power grid connection, and carefully analyzed the characteristics of power quality of distribution network after distributed energy access. Liu Qing of Guangdong Power Grid Corporation studied the charging mode and the power quality characteristics of the charging station. By measuring the power quality data of a single charging pile, the power quality characteristics of the charging pile were analyzed.

Research at home and abroad focuses on the influence of access on the power quality of the distribution network in terms of distributed power supply or load single dimension, and thus derives the power quality characteristics of the distribution network. The research on the characteristics of power quality under the trend of power supply-grid-load three-dimensional analysis, such as nonlinearization of distribution network, frequent network reconstruction, and capacitive system power supply, is still in its infancy.

2.2. Research on Modeling Technology of New Power Electronic Equipment in Distribution Network

Foroozan Ghassemi and others of the British National Transmission Company proposed a simplified method for harmonic analysis model to simplify the calculation of harmonic analysis. The model of the low-voltage wind farm is simplified according to the two-port network theory. The accuracy of the wave analysis. Danish East Energy has designed a harmonic analysis model for offshore wind farms that considers the effects of high-voltage submarine cables used to connect the coast on harmonic propagation. Walker Geoff of Queensland University of Technology in Australia takes the inverter of photovoltaic power plant as the core, and simplifies the components such as photovoltaic modules, filters and step-up transformers according to the requirements of the inverter topology, and obtains a simplified model of the photovoltaic power plant. For simple harmonic analysis, but does not reflect dynamic harmonic variation processes.

Mi Fuli of Harbin Institute of Technology conducted an in-depth analysis of the harmonic problems of direct-drive wind farms. The components of the wind farm were mainly modeled and analyzed by grid-connected converters. From the working principle of the converters, different frequencies were used. The harmonic characteristics of the downconverter are established, and the high frequency harmonic model and the low frequency harmonic model are respectively established. Zhu Yongqiang

of the State Key Laboratory of New Energy Power System and others considered the impact of multiple PV inverters on power quality after parallel connection, and further considered the influence of collectors, filters and other collector systems on the harmonic transmission of the power grid, and established a large-scale photovoltaic power station. Harmonic transfer network model. Zhao Wei and others of Guangdong Electric Power Company Electric Power Research Institute established a single three-phase uncontrollable rectification charger model. Furthermore, the harmonic characteristics of multiple charging piles are further analyzed, and the harmonic propagation in the distribution network is analyzed by establishing the harmonic current model of the distribution network and the harmonic impedance accurate model of the transmission line. The State Key Laboratory of Power Transmission Equipment and System Security and New Technology of Chongqing University uses the established single charger and charging station simulation models to analyze their impact on power quality, focusing on current harmonic content and current total harmonics. The distortion rate varies with the charging power of electric vehicles and the variation law with the number of charging machines.
The research of domestic and foreign scientific research institutions focuses on the establishment of distributed energy-related power electronic equipment models, and the research methods of new power electronic equipments for complex distribution networks such as electric vehicle charging equipment, charging facilities, and multi-port flexible equipment are at The starting stage.

2.3. Research on Flexible Monitoring and Rapid Control Technology of Power Quality in Complex Distribution Network

Kansas State University uses DSP chips for power system power quality monitoring, highlights the unique features of DSP chips, discusses the issues involved in adopting system architecture, selecting DSPs, and developing application software, making it ideal for real-time applications. select. RPK Lee of the City University of London in the United Kingdom used the power quality database provided by the American Academy of Electric Power as a benchmark to improve the designed power quality detection system. Finally, the Web-based large-grid multi-channel power quality monitoring system was designed to be available to users in the power grid. Together with the grid company to obtain monitored data to jointly address power quality issues. L.Xu et al. of the University of Glasgow in the UK designed a quality management system that focuses on two- and multi-stage pulse width modulation (PWM) controlled voltage source inverters (VSIs) with embedded control software. On the DSP board, flexibility and reliability are improved, and experimental results verify the effectiveness of the system.
Sun Yi, Tang Liangrui and Gong Gangjun of North China Electric Power University followed the trend of network monitoring of power monitoring, and proposed a design scheme of online monitoring system for power quality. The system monitors each monitoring point online through the virtual instrument-based power meter and passes the data. Real-time display, local display and web display give the power department access to real-time detailed data. A number of units such as State Grid Electric Power Research Institute and Naval Engineering University have jointly developed a power quality monitoring system based on ARM+DSP dual system and software based on multi-agent technology. The model of this system is well adapted to the distribution of power system. The development trend of the structure and power market, and can meet the current power grid quality monitoring needs, has a preliminary application in Zhejiang and Hubei power grids. In terms of comprehensive management of power quality, Hunan University, Wei Wei, Tu Chunming, and Cheng Ying proposed a comprehensive power quality control system for harmonics and reactive power in high-voltage distribution networks. The system consists of injection hybrid active power filtering. The device (IHAPF) and the thyristor control reactor (TCR) are composed, and a control method based on the system for comprehensive dynamic control of harmonics and reactive power is proposed.
Researches at home and abroad have focused on the development of power quality parameters. However, there are few studies on the coordinated control methods for multi-power quality control devices in complex distribution networks.

3. Research on Comprehensive Quality Control Technology of Complex Distribution Network Power Quality

3.1. The Overall Framework of Research on Comprehensive Control of Power Quality

The overall framework of the comprehensive control of power quality in complex distribution networks is shown in Figure 1.

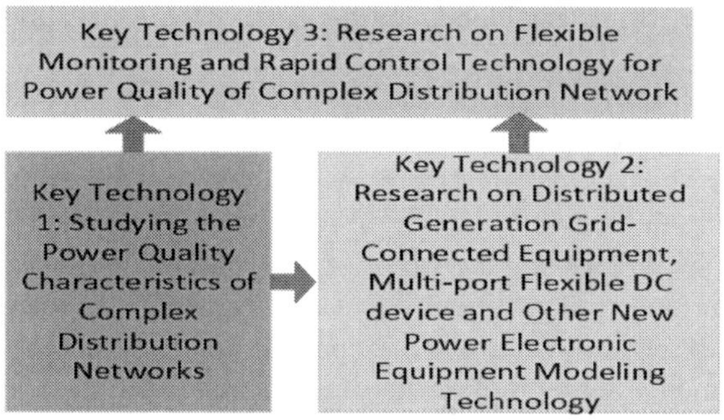

Figure 1. The overall Research framework.

There are three key technologies involved: 1) studying the power quality characteristics of complex distribution networks; 2) studying the modeling techniques of distributed power grid-connected equipment, multi-port flexible lines and other new power electronic equipments in distribution networks; 3) research complex Flexible monitoring and rapid management of power quality in distribution networks. It is the basis of mastering the influence mechanism of distributed energy and diversity load on the power quality of distribution network. It is the key to accurately model new power electronic equipment such as multi-port soft and straight, and master the power quality parameters of complex distribution network. Based on the rapid detection method, a power quality coordinated control technology and a comprehensive management method based on the operational data of the power system are proposed.

3.2. Technical Route for Comprehensive Research on Power Quality

3.2.1. Technical Research Route on Power Quality Characteristics of Complex Distribution Networks. Firstly, combined with the development characteristics of nonlinear distribution of complex distribution network, the power quality characteristics of nonlinear distribution network are analyzed from the three dimensions of power supply-grid-load. Among them, non-linear power supply such as photovoltaic power generation and gas turbine power generation is analyzed from the power supply angle. From the perspective of power grid, the characteristics of power quality of AC/DC hybrid distribution network, especially DC distribution network, are analyzed. From the perspective of load, the access of equipment such as subway and electric vehicle charging load is analyzed. The power quality characteristics of the distribution network; secondly, analyze the power quality characteristics of the distribution network when the distribution network topology changes frequently in the scenarios of distribution network load transfer, transformer economic operation, distribution automation fault handling, etc.; again, from the urban power grid Power supply cable, DC side capacitor, AC side filter capacitor, etc. make the distribution network parameters capacitive analysis of the power quality characteristics of the distribution network; Finally, summarize the power quality characteristics of the complex distribution network.

3.2.2. Technology Route for Modeling New power Electronic Equipment in Distribution Network. The research technology of modeling technology for new power electronic equipment in distribution network is shown in Figure 3.

Firstly, combing the working characteristics of representative power electronic equipment in complex distribution networks, such as distributed photovoltaic on the power supply side, multi-port soft on the grid side, and electric vehicle charging facilities on the load side; secondly, from the power electronic equipment pair Research on the influence mechanism of voltage quality of distribution network, the influence of access location on voltage, and the influence of working mode on voltage quality of distribution network. The impact of power electronic equipment on voltage quality of complex distribution network is studied. The influence of position on the current quality of distribution network, the influence of working mode change on current quality and the influence of power electronic equipment interaction on current quality. Three dimensions analyze the influence of power electronic equipment on the current quality of complex distribution network; The research and analysis of power grid electronic equipment modeling technology, the modeling method of electric power vehicle charging facilities, multi-port flexible equipment and other complex power grid new power electronic equipment is proposed, and the rationality of the model is guaranteed by simulation..

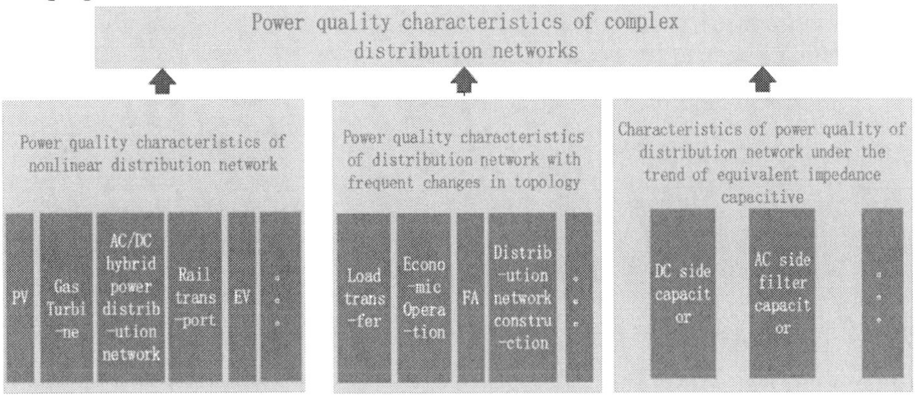

Figure 2. Technical route on power quality characteristics of complex distribution network.

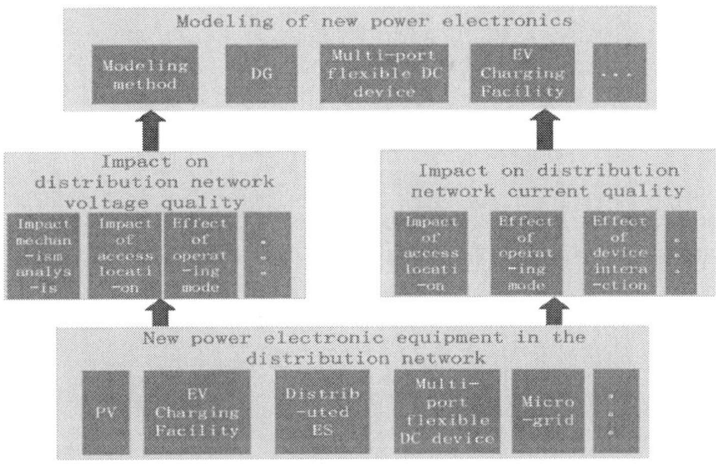

Figure 3. Technical route of modeling technology for new power electronic equipment.

3.2.3. Research Route for Flexible Monitoring and Rapid Control Technology of Power Quality in Complex Distribution Network. The research route of flexible monitoring and rapid management of power quality in complex distribution network is shown in Figure 4.

Firstly, based on the theory of instantaneous harmonic detection and instantaneous reactive power detection, a fast detection method for power quality parameters of complex distribution network is proposed, and the harmonic components of the power grid and the reactive current of three-phase fundamental are analyzed. Then, the power quality is obtained. The optimal configuration method of power quality equipment for regional distribution network is proposed in terms of configuration principle, type, number of devices and location of device access. Thirdly, considering the actual situation of power grid and the equipment configuration of power quality equipment, multi-electric energy of complex distribution network is proposed. The coordinated control method of the quality device; Finally, from the aspects of data acquisition module design, communication module design, decision analysis module design, etc., the intelligent coordination control device based on the regional power distribution network power quality management device is developed.

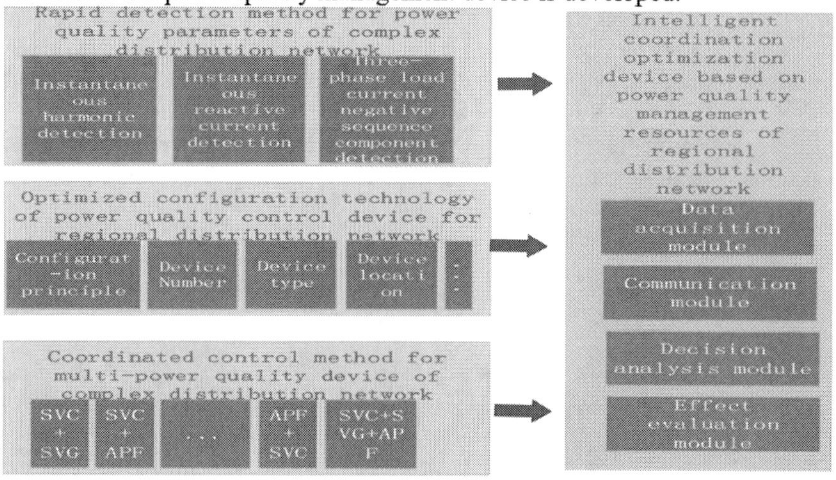

Figure 4. Technical route for flexible monitoring and rapid management of power quality in distribution network.

4. Conclusion

This paper analyzes the solutions proposed by domestic and foreign experts and scholars in the face of the impact of large-scale access of power electronic devices on power quality of distribution network, and proposes a research framework and detailed technical route for comprehensive management of power quality in complex distribution networks. The feasibility, relevant research ideas and technical methods can effectively guide scientific research institutions and colleges and universities to carry out research work on power quality comprehensive management technology of modern distribution network.

Acknowledgments

This work was financially supported by "National Key Research and Development Program of China" (2018YFB0905000).

References

[1] XU Zhiheng, Zhang Fan, Xu Dong. Research on Control Strategy of Comprehensive Treatment of Distribution Network [J]. Journal of State Grid Technology College, 2017, 20(2):5:9
[2] LIN Guiyuan.Discussion on Comprehensive Control of Power Quality in Medium Voltage Distribution System [J]. electric equipment, 2018(6):15-17
[3] WU Jun. Analysis on Power Quality of Smart Distribution Grid. Power Capacitor & Reactive Power Compensation, 2016, 37 (6):35-39
[4] PAN Aiqiang, SHI Shanshan, DONG Ruian, et al. Power quality characteristics and detection concerning in microg-rid [J]. High Voltage Apparatus, 2015 (6):133-138
[5] GAO Ciwei, ZHANG Liang. A survey of Influence of Electrics Vehicle Charging on Power Grid. Power System Technology.2011, 35 (2):127-131
[6] WANG Jinli, SHENG Wanxing, SONG Qipeng, et al. Research on Intelligent Power Quality Monitoring and On-Line Simulation for Distribution Network. Power System Technology, 2014, 38(2):515-519
[7] Jiang Zhonghua, Luo An, Zhao Wei, et al. A simulation system of real-time power quality monitoring and harmonic suppression for distribution network [J]. Power System Technology, 2010, 34(8):18-24

A Analysis of Attack and Defense Mobile Ad Hoc Network Based on OPNET

Jiahao Dai* and Xiaofeng Xu

[1]Science and Technology on Communication Information Security Control Laboratory
[2]China Electronic Group Corporation No.36 Institute, Zhejiang Jiaxing, China

*Corresponding author

Abstract. Due to the broadcast nature of wireless communications, the damage to mobile ad hoc networks is becoming more and more common. We propose a network stability model that can effectively reflect the network performance when the network is damaged in real time. The Ad Hoc network model is built by using the node model supporting the IEEE802.11 standard provided by OPNET software to simulate the network performance of the network under different degrees of damage. This model provides an effective way for the simulation analysis of network performance under the condition of Ad Hoc network damage.
Keywords: OPNET; Ad Hoc; networks stability; network defense.

1. Introduction

In recent years, the rapid development and application of the Internet in our lives has become an indispensable part of people's lives. With the development of countries in the political, economic and cultural fields, the Internet plays an important role in these areas, driving political, economic and cultural development. According to the 42nd "China Internet Development Statistics Report" [1], the most authoritative report on domestic network security, mobile device access traffic consumption in the first half of 2018 reached 26.6 billion GB, an increase of 199.6%. However, while people enjoy the convenient communication services brought by the network, cyber attacks also emerge in an endless stream, seriously jeopardizing the interests of users. For example, the famous ransomware incident in 2017, using the local Internet access rights to upload local information and download the encrypted private key and public key, using the private key and public key to encrypt the file, dealing with the virus developer himself, it is almost impossible for others to decrypt. With the maturity of cyberattacks, a variety of new deficiencies are emerging, and the destructive power is getting more and more amazing. Therefore, it is imperative to analyze network destruction behavior.

The mobile ad hoc network [2] is a self-organizing network with a highly dynamic topology, nodes arbitrarily moving, and no control center. It is independent of a fixed infrastructure, and mobile nodes can communicate freely through distributed management. As a distributed network, a mobile ad hoc network can provide mutual communication between terminals without using or facilitating the use of existing network infrastructure (such as base stations, APs). Since the terminal's transmit power and wireless coverage are limited, if two terminals that are far apart have to communicate, they must use other nodes for packet forwarding, so that a wireless multi-hop network is formed between the nodes. Therefore, for the traditional key-based encryption and decryption technology, due to the broadcast nature of wireless communication, malicious users can conveniently eavesdrop on the transmission of keys, and the destruction against the self-organizing network is more and more common.

Many scholars have carried out research on the damage caused by mobile ad hoc networks. In [3], by analyzing the attack types of mobile ad hoc networks and constructing attack trees initiated from malicious nodes, a FSM-based intrusion detection algorithm is designed by using the idea of finite state machine. The intrusion detection system using this algorithm can pass neighbors. The monitoring of the nodes detects various attacks of the nodes in real time. In [4], the corresponding protection research is proposed for the two attacks faced by the routing protocol AODV, and analyzes the flooded and tamper-attacked of three different network models (chained network, star network, mesh network). Literature [5] analyzes the security problems brought by the exposed topology of mobile ad hoc networks, and proposes a secure multi-path routing protocol with topological concealment. The protocol takes into account the time factor and path length factor to achieve the safest shortest path determination. In [6], a security problem based on the IEEE 802.11 MAC protocol of the mobile ad hoc network is proposed. A DDoS attack method based on attack set is proposed. This method can accurately attack the target node and reduce the detection rate of attack behavior.

Although many literatures have analyzed the vulnerability of mobile ad hoc networks, most of the literature only considers the threats faced by ad hoc networks, and analyzing the performance of ad hoc networks is also an indispensable aspect; in addition, most of the literature The security problem is only considered theoretically, and the reality of the networking situation is more complicated than the theory. Therefore, in this paper, we first propose a network stability model, which can effectively reflect the network performance when the network is damaged in real time. Next, we use OPNET to simulate the network performance of the network under different degrees of damage and give recommendations for maintaining network stability in the event of damage. Finally, we conclude this paper.

2. Networks Stability Model

2.1. Network Security State Model

Remember that the network state in the network at time t is $x(t)$, we can get the Logistic model that $x(t)$ satisfies:

$$\begin{cases} \dot{x}(t) = rx(1-\frac{x}{N}) \\ x(0) = N_0, \end{cases} \tag{1}$$

Where r is the coefficient and N is the optimal state of the network under finite resources. The solution of equation (1) is obtained by the separation variable method.

$$x(t) = \frac{N}{1+e^{-rt}(N-N_0)/N_0} \tag{2}$$

The equation (2) has two equilibrium points, $x_1 = 0, x_2 = N$ where x_1 is unstable and x_2 is globally stable in the positive half-axis.

2.2. Network Destruction State Model

Establish an equation for network state compliance in the event of network failure and analyze the conditions for stable network conditions.

Assume that the number of damaged nodes in the unit time is proportional to the network state $x(t)$. With the proportional coefficient k, k can be further decomposed into $k = qE$. E is called the damage strength and is measured by the controllable parameters such as the number of damages, q is called the coefficient of destruction. In order to facilitate us to take $q = 1$, the number of damaged nodes per unit time is $h(x) = Ex(t)$. $h(x)$ is a constant that represents a particular way of breaking. Thus, the network state satisfies the equation

$$\dot{x}(t) = rx\left(1-\frac{x}{N}\right) - Ex \tag{3}$$

This is a first-order Riccati-type nonlinear equation, also known as the Scheafer model.

We want to obtain the steady state of the network and maintain stable conditions, that is, the trend of the network state after time t. There is $dx/dt = 0$ at the equilibrium point, and equation (3) has two equilibrium points $x1 = 0$, $x2 = N(1 - E/r)$.

- When $E < r$, $x2$ is a positive equilibrium point. Equation (3) can be expressed as

$$x(t) = -x(x - x_2) \tag{4}$$

  It is easy to know that when $0 < x < x2$, $x(t) > 0$, when $x > x2$, $x(t) < 0$, which means the equilibrium solution $x1$ is unstable, and $x2$ is a stable equilibrium solution. That is, in the case of the destruction strength $E < r$, the network state will be stabilized at the level of $x2$, and the destruction ability is stabilized at the extent of $Ex2$.

- When $E > r$, $x(t) < 0$, the network performance continues to drop, eventually leading to network paralysis. How to maintain the stability of the network. In another words, how to minimize network damage? Mathematically, the expected damage effect is minimized under the condition of $x(t) = 0$ or $rx(t)(1-x(t)/N) = Ex(t)$, and the damage consumption can be understood. To destroy the number of nodes as $h(x) = Ex(t)$, the problem can be described as the following optimization problem:

$$Hmax = maxEx(t) \tag{5}$$

  The constraint is $rx(t)(1-x(t)/N)-Ex(t) = 0$. This optimization problem can be attributed to the maximum value of the quadratic function $h(E) = NE(1-E/r)$ of E. By deriving we can get the maximum damage strength $Emax = r/2$, and the number of broken nodes is $hmax = rN/4$. The breaking strength $Emax$ is a strategy to obtain the best network damage effect.

3. Simulation Model

3.1. Network Model

In this paper, we consider the network model of adhoc [7]. As shown in Figure 1, we assume that the simulation area is a plane of 1km*1km, there are four limited moving areas in the plane, and there are 10 mobile nodes in each area. All mobile nodes can move freely in the limited area of 300m*300m. If the distance between two nodes is greater than 100m, the two nodes cannot communicate with each other. The moving speed of each node is any random number in the range of 0-10m/s. During the simulation, the nodes randomly stay, or the nodes move to the edge of the circular area, and the time of each stay is 10s, then the node again Random direction and velocity motion until the end of the simulation, the simulation time is 1h. Nodes located in different areas cannot communicate directly. Data must be exchanged through the corresponding routers in their respective areas. Each router can communicate with each other. That is, if node 1 wants to communicate with node 20, it must first transmit data to node 2 by node 1, then to router 5 by router 2, and finally to node 20 by router 5. The SITL module is implemented by the wireless network card device. The mobile ad hoc network in the OPNET simulation environment is connected to two external computers. The routing protocol uses the AODV routing protocol. The semi-physical simulation environment realizes data communication between different network segments by a semi-physical simulation method.

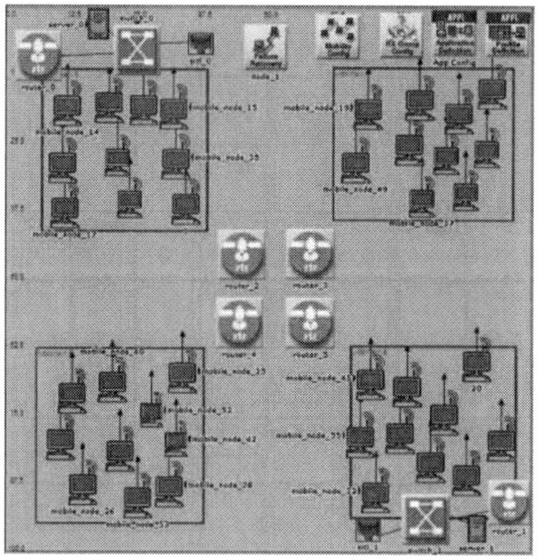

Figure 1. Network model with lots of mobile nodes and several routers.

3.2. Node Model

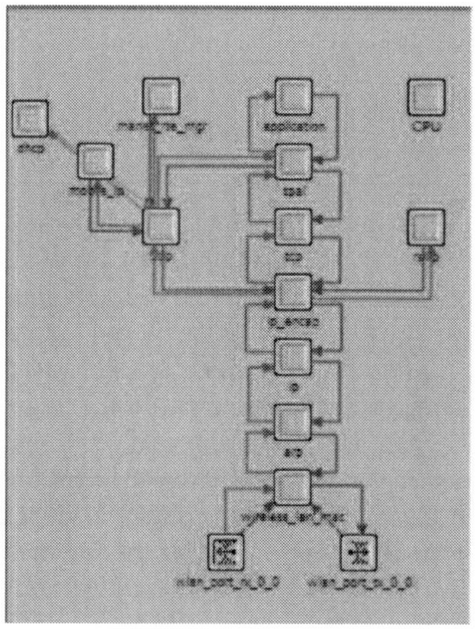

Figure 2. Node model.

The nodes in the network are randomly distributed in a limited area and move randomly, so each mobile node has the same node model, as shown in Figure 2. In order to obtain the network performance parameters in the simulation process in real time, the simulation rewrites the code in the wlan wkstn node model and puts the obtained data into the database. The process module functions in the node model are as follows:

- Wlan port rx (receive) module: adopts a non-directional reception mode to obtain data frames of other nodes and transmit them to the MAC module;
- Wla port tx (send) module: uses a non-directional reception mode to transmit data frames to other mobile nodes;
- Wireless lan mac module: emulation link layer random access channel protocol, channel access protocol is IEEE 802.11 protocol, access mechanism is CSMA/CA, and counts the network performance parameters such as network throughput and delay;
- Tcp/ip module: processing the obtained tcp/ip data packet;

- Application module: set a random destination address for the packet introduced by the upper layer, generate a service request to the routing layer according to the ICI format of the internal communication interface, and the packet is sent to the tcp/ip module together with the ICI.

3.3. Process Model

Each process module in the node model is implemented by a process model, where each process model is a finite state machine implemented by c code. In the mobile ad hoc network, the selection of the MAC protocol needs to consider its service performance for integrated services such as video, voice, and data. The MAC layer emulation protocol of the selected Ad hoc network is IEEE 802.11 MAC with distributed coordination function protocol, the wireless mac module process model for setting the channel access protocol in the node model is shown in Figure 3.

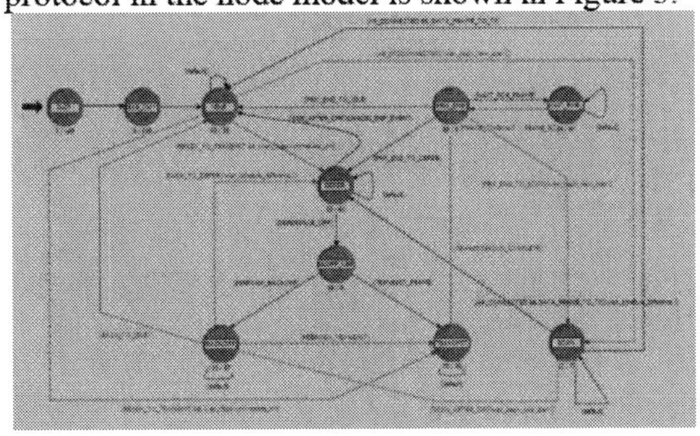

Figure 3. Process model.

Source process: which is used to forward the obtained packets. When the emulation system starts, the sent information is initialized in the INIT state, and then the GENERATE obtains the data packet according to the given function according to the given function, and sends the data packet out. When the network status triggers the STOP and DISABLED events, the transmission is terminated.

Receive process: which is used to collect data generated in the simulation and analyze the simulation. When the simulation system is started, in the initialization INIT state, the simulation amount is counted through the registration of the corresponding function, and the necessary configuration is performed on the receiving module. After initialization, it directly enters the DISCARD state and waits for the next event to arrive. The DISCARD state is used to split the data packet and count the network simulation amount. After waiting for it, it returns to the DISCARD state and waits for the next trigger.

Routing process: which is used to simulate the packet forwarding, path allocation, and arbitration. When the simulation system starts, the processed information is initialized in the INIT state first, and then enters the state machine IDLE_0. If the PKT_ARR or PK_SEND_INTRPT event is triggered at a certain time, the route v1 function will be executed. If the CHANNEL_STAT_UPDATE or PK_SEND_FLAG event is triggered at a certain time, the PK_RESET event is executed to complete the routing of the packet.

4. Simulation Results

Based on the above model and simulation network, the network model was verified by simulation experiments. Simulation results will be given and the simulation results will be analyzed.

As shown in Figure 4, the relationship between network performance (mainly including network traffic, network data loss, network delay and routing request) under different failure conditions is given. Figure 4 (a) shows that as the number of damaged nodes increases, the network traffic remains unchanged first, followed by a rapid decrease and a slow decrease. This is because the node that was damaged at the beginning is a non-critical node, which has little effect on the network structure. Then, as the important node is destroyed, the network traffic drops drastically; when the damaged node increases to a certain extent, the network becomes simpler and simpler. The flow rate is also slowly reduced. Figure 4 (b) shows the relationship between the number of broken nodes and data loss. The loss of data is related to the stability of the network. The more stable the network, the less data is lost.

We can see that as the number of damaged nodes increases, the lost data first remains stable and then decreases rapidly. This is because at the beginning of the network there are too many nodes, the network structure is too complicated, resulting in serious data loss; and after destroying some nodes, the network structure becomes simple and tends to be stable. Interestingly, when the number of damaged nodes is 21, the data loss suddenly increases. This is due to too many damaged nodes, the basic network structure is destroyed, and the route needs to be searched again.

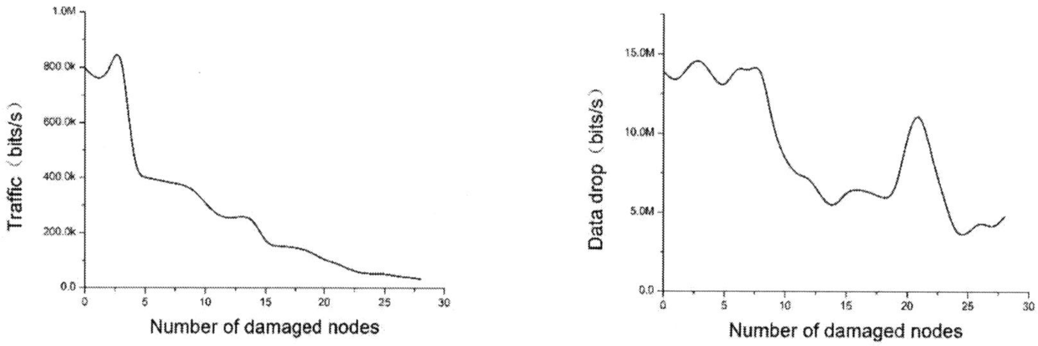

(a) (b)

Figure 4. (a) Impact of the number of damaged node on the traffic of network; (b) Impact of the number of damaged node on the data drop of network.

5. Conclusions

For the destruction of mobile ad hoc networks, this paper uses OPNET to build a network structure to simulate the network performance of the network under different degrees of damage. The simulation results show that the network damage nodes affect network traffic, network data loss, network delay and routing request performance. The network stability model can effectively reflect the network performance under the damaged situation and give suggestions on how to maintain the network stability.

References

[1] C. I. N. I. Center, "The 42nd Statistical Report on Internet Development in China," Internet World, 2018.7.
[2] W. Daoyuan, T. Hui, and Z. Ping, "Research and application of mobile self-organizing network," Telecommunications Science, vol. 24, no. 6, pp. 60–64, 2008.
[3] Y. Xia, L. Junshan, and L. Rong., "Research on classification of mobile ad hoc network attack based on fsm detection," Modern Electronic Technology, vol. 33, no. 21, pp. 89–92, 2010.
[4] Y. Yanfang, "Research on the protection of self-organizing cyber attacks," University of Electronic Science and Technology.
[5] H. Qi, Z. Jiao, and Z. Yujun, "Topologically Hidden MANET Secure Multipath Routing Protocol," Journal of Software, vol. 22, no. 5, 2011.
[6] X. Rui, Z. Zemao, and H. Fei, "MAC layer DDoS attack in mobile ad hoc networks based on attack set," Journal of Computer Applications, vol. 28, no. 3, pp. 1148–1151, 2011.
[7] X. Lei, L. Xiaohui, and F. Hongyu., "Modeling and simulation of adhoc network based on opnet," Computer Engineering, vol. 35, no. 1, pp. 123–125, 2009.

InGaN/GaN/AlGaN RTD with Inherent Extreme Low Leakage Current & High Current Rising Rate

Haipeng Zhang[1,2,*], Xing He[1], Mi Lin[2,*], Zhonghai Zhang[2], Fu Chen[1], Zifa Zhou[1], Weifeng Lü[2] and Ying Wang[2]

[1]School of Physics & Electronic Information/Jiangxi Province Engineering Technology Research Center of Intelligent Electric Vehicle Parts & Components, Shangrao Normal University, Shangrao, 334001, China;
[2]School of Electronics & Information/School of Microelectronics, Hangzhou Dianzi University, Hangzhou, 310018, China.

*Corresponding author

Abstract. In order to explore better diode candidates for gate controlled FET applications at about ±0.1 V DC power supply with extreme low subthreshold swing, current-voltage characteristics of several InGaN/GaN/AlGaN Resonant Tunneling Diodes (RTDs) were investigated by device simulation with Non-Equilibrium Green's Function (NEGF) method on Technology Computer Aided Design (TCAD) platform. Their electric switching characteristic parameters were extracted from the simulation results and contrasted with the corresponding results reported recently. The comparison results indicate that sample 1—with thicknesses of its $In_{0.21}Ga_{0.79}N/GaN/In_{0.14}Ga_{0.86}N/Al_{0.1}GaN$ layers are respectively 4-3-4 times of each crystal constant itself—is the best potential candidate of them with inherent forward (backward) threshold voltage at 50 mV (-20.3 mV), off current at 2.4×10^{-52} μA/μm, on-off current ratio at 3e8 and extreme high inherent forward (backward) current rising rate with its reciprocal at 5.90 mV/dec(-2.39 mV/dec). Sample 5 —with thicknesses of its $In_{0.21}Ga_{0.79}N/GaN/In_{0.14}Ga_{0.86}N/Al_{0.1}GaN$ layers are respectively 5-4-5 times of each crystal constant itself— and sample 6 — with thicknesses of its $In_{0.21}Ga_{0.79}N/GaN/In_{0.14}Ga_{0.86}N/Al_{0.1}GaN$ layers are respectively 5-6-5 times of each crystal constant itself— are better optional candidates with relatively lowered on-off current ratio at 6×10^{4} (1.4×10^{7}) and higher inherent forward (backward) current rising rate with their reciprocals at 10.46 mV/dec(-4.18 mV/dec) and 7.00 mV/dec (-2.80 mV/dec).
Keywords: InGaN/GaN/AlGaN, RTD; leakage current; current; rising rate; on-off current ratio.

1. Introduction

As Si-based MOSFETs continue scaling down for LVLP (low voltage low power) logic applications, the degradation of switching characteristic of conventional Si MOSFETs has been worse and worse. To overcome this issue, several novel gate structure FETs and III-V family based FETs have been brought out [1]-[6]. Y. Moriata et al.[1] reported the experimental realization of complementary p- and n- tunnel FinFETs on a Si CMOS platform. The FinFETs are featured of SS(subthreshold swing) of 58mV/dec(decade) and 56mV/dec and on/off-current ratios of 2×10^{6} and 3×10^{4} (with V_D=|0.2| V) respectively. To constrain SS degradation, G. P. Marco & B. Sylvan[2] proposed a novel method, in which an InAs/GaSb/InAs quantum well was inserted into the source region of the GaSb/InAs HFET(hetero-junction tunnel-FET) along the transport direction. Such a design option was predicted to provide on/off-current ratios larger than 10^{7} even at V_{DD} = 300 mV. S.-J. Chang et al.[4] reported the characteristics of InGaAs-based independent double-gate FinFETs with $Al_2O_3/LaAlO_3$ as gate dielectric. It exhibits better performance in terms of the on/off current ratio, SS, off current, and channel mobility than the single-gate counterpart. Its SS is lowered down to about 250mV/dec. C.-S. Lee et al.[5] proposed a Schotky gate InAlN/AlN/GaN/Si HFET and an Al_2O_3-MOS gate InAlN/AlN/GaN/Si HFET with backside substrate metal-trench structure. The present $In_{0.18}Al_{0.82}N$/AlN/GaN MOS-HFET (Schottky-gate HFET) has demonstrated improved maximum drain–source current density $I_{DS,max}$ of 1.08 (0.86) A/mm @ V_{DS} = 8 V, gate-voltage swing of 4 (2) V, on/off-current ratio I_{on}/I_{off} of 8.9×10^8 (7.4×10^4), SS of 140 (244) mV/dec, turn-on voltage (V_{on}) of 4.2 (1.2) V. J.-T, Lin et al.[6] proposed a RGTFET (recessed-gate tunneling field-effect transistor) to improve the on current of TFETs by increasing the tunnel area with line tunneling. For optimal device structures, their proposed n-TFET reaches 1.44×10^{-6} A/μm of on current and 3.22×10^9 on/off current ratio. A minimum SS SS_{min} = 28.3 mV/dec and an average swing SS_{avg} = 59.8 mV/dec over seven orders of drain current are achieved. In addition, complementary TFET inverters show good noise margins of NM_H = 65 mV (38.5 % V_{DD}) and NM_L = 77 mV (32.5 % V_{DD}) and also a high voltage gain even at V_{DD} = 0.2 V .

In order to improve switching characteristic of transistors, we tried to explore a novel kind of diode candidates for gate controlled transistor applications. Through synthetically using energy engineering, built-in polarization field and applied field quantum control methods to regulate electron transportation, it was successfully discovered that nanometer GaN/InGaN/AlGaN-based RTDs are featured of inherent and ideal switching characteristic although without gate control. Its forward (backward) switching voltage is about 0.05 V (0.021 V), forward (backward) on/off-current ratio (I_{on}/I_{off}) is up to 1×10^{12} (2.3×10^{12}), forward (backward) power dissipation is reduced down to about -430 dBm/μm^2 (-420 dBm/μm^2), forward (backward) leakage current is less than 10^{-58} A/μm^2 and the reciprocal of forward (backward) current rising rate is lowered to about 5 mV/dec (2 mV/dec). Thus, we name it as EHCRR-RTD (RTD with extreme high current rising rate).

2. Structure of the Proposed EHCRR-RTD

The propsed EHCRR-RTD consists of a sandwiched $In_{0.21}Ga_{0.79}N/GaN/In_{0.14}Ga_{0.86}N/Al_{0.1}Ga_{0.9}N/In_{0.07}Ga_{0.93}N$ DBS(Double Barriers) structure on wurtzite InGaN/GaN substrate, of which the top surface is N-original surface as illustrated in Figure 1.

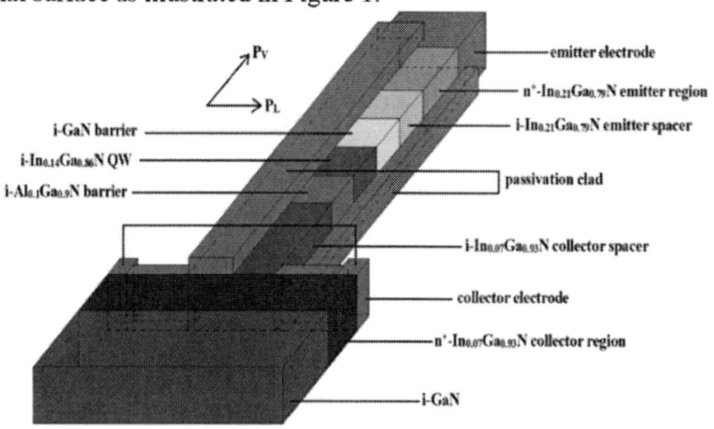

Figure 1. Device structure of GaN/InGaN/AlGaN-based EHCRR-RTD cell

In Figure 1, P_L means lateral position and P_V means vertical position. Thickness of each sandwich layers was set in terms of integer times of its crystal constant itself along $\bar{c}$ ($(0\ \ 0\ \ 0\ \ \bar{1})$) orientation and listed in Tab. I. Width of the device core is set at 1μm along the normal direction of $P_V \times P_L$ plan. The crystal constants of $In_{0.14}Ga_{0.86}N$ and $Al_{0.1}Ga_{0.9}N$ layers were calculated according to Vegard's law [7]-[8] for three-element nitride as Equaiton (1) and listed in Table 1 also.

$$c(A_x B_{1-x} N) = xc(AN) + (1-x)c(BN) \tag{1}$$

Table 1. Thicknesses of Collector Spacer Layer,sandwich Layers and Emitter Spacer Layer of EHCRR RTD Samples

Material layer Name	Emitter spacer layer $In_{0.21}Ga_{0.79}N$	GaN	$In_{0.14}Ga_{0.86}N$	$Al_{0.1}Ga_{0.9}N$	Collector spacer layer $In_{0.07}Ga_{0.93}N$
crystal constant (nm)	0.529	0.519	0.526	0.517	0.522
	Unit(time of crystal constant along $\bar{c}$ orientation)				
Sample 1	38	4	3	4	40
Sample 2	38	2	6	2	40
Sample 3	38	3	2	3	42
Sample 4	38	3	3	3	41
Sample 5	38	5	4	5	42
Sample 6	38	5	6	5	40

The 50 nm thick N^+ doped $In_{0.07}Ga_{0.93}N$ collector region was firstly grown with MBE (Molecular Beam Epitaxy) on semi-insulated GaN substrate. Its doping concentration was set at $1.0\times10^{18}cm^{-3}$ in uniform distribution. Secondly, an intrinsic-$In_{0.07}Ga_{0.93}N$ (i-$In_{0.07}Ga_{0.93}N$) collector spacer layer was grown on it. Thirdly, $Al_{0.1}Ga_{0.9}N$/i-$In_{0.14}Ga_{0.86}N$/GaN sandwich layers were grown in turn. Fourthly, an intrinsic-$In_{0.21}Ga_{0.79}N$ (i-$In_{0.21}Ga_{0.79}N$) emitter spacer layer was grown on the sandwich. Thicknesses of collector spacer layer, sandwich layers and emitter spacer layer were also listed in Table 1. After that, a 50 nm thick N^+ doped $In_{0.21}Ga_{0.79}N$ emitter region was grown on the emitter spacer layer. Its doping concentration was also set at $1.0\times10^{18}cm^{-3}$ in uniform distribution as illustrated in Figure 1, too. Then, a passivation clad of SiO_2 was deposited on the naked surfaces and contact holes were etched. Finally, metal film was deposited and etched to form electrodes as illustrated in Figure 1. The polarization charge sheet densities on i-$In_{0.21}$GaN/GaN,GaN/i-$In_{0.14}$GaN,i-$In_{0.14}$GaN/$Al_{0.1}$GaN and $Al_{0.1}$GaN/i-$In_{0.07}$GaN interfaces were calculated out as $-1.4075\times10^{13}cm^{-2}$,$5.4686\times10^{12}cm^{-2}$,$-1.3102\times10^{13}cm^{-2}$ and 7.4010×10^{12} cm^{-2} respectively.

3. Electric Characteristics of EHCRR-RTD Samples

The conduction band edge of the proposed EHCRR-RTD appears declined and distorted on both sides of its DBS structure in thermal equilibrium state owing to built-in field by polarization effects of the GaN, AlGaN and InGaN epitaxial layers similar to those of HVRTD(high voltage RTD) in [9] with the same material layers and original surface of substrate and MQW(multi-quantum well) RTD in [10-11] with different material layers and original surface of substrate. Only electron transportation is considered for inner-band tunneling by considering influences from both bias voltage and polarization field during simulation. As a result, the polarization effects lead to sheet polarization charges at each heterogeneous junction interface. Along the opposite direction of growth direction of the hetero-structures (z-axis as indicated in Figure 1 by P_V), the relationship between the vertical quantum transport of electron wave function $\psi(z)$ and their corresponding BSELs(bound state energy levels) E_j under built-in polarization field and applied electric field can be described by coupled 1D Schrödinger and Poisson equations. While the latter can be simplified with linear approximation for these cases instead of parabolic approximation for HVRTDs since the power supply voltage is set at about ±0.1 V. Through properly adopting the combinational modulation method of conduction band engineering, built-in polarization field and applied electric field of GaN/InGaN/AlGaN DBS structure, nearly ideal switching characteristics of the EHCRR-RTD can be obtained. Figure 2 (a)-(d) illustrated distribution of conduction band bottom,distribution of energy levels in equilibrium and at ±0.1V bias respectively. Figure 2 (e) illustrates the simulated results of current density-bias voltage characteristics for the six EHCRR-RTD samples listed in Table 1. Simulations were carried out through solving the coupled Schrodinger and Poisson equations in numerical NEGF (non-equilibrium Green's function) method with TCAD software. Their electric switching characteristic parameters were extracted form Figure 2 (e) and outlined in Table 2 together with corresponding results in [1]-[6] by contrast.

As can be seen from Figure 2 and Table 2, sample 3 of the EHCRR-RTDs is not a good candidate for gate controlled FET applications since its on/off current ratio is too small. Among other samples, sample 2 and sample 4 are featured with the highest on/off current ratio up to 4×10^{11}-5×10^{11} and the lowest leakage current down to 2×10^{-57}-1×10^{-59} μA/μm. However, they will not be better candidates for gate controlled FET applications due to their lower turn-on voltage although their reciprocals of current rising rates are lower than SS of conventional MOSFET and results of [1], [3], [6]. Reciprocals of current rising rates of sample 5 and sample 6 are also much lower than them with relative small on/off current ratio at about 10^4-10^7 and with higher turn-on voltage. Therefore, they are better candidates for gate controlled FET applications. While reciprocal of current rising rates of sample 1 is lower than those of sample 5 and sample 6 with relative higher on/off current ratio up to 10^8-10^9 and the same high turn-on voltage. Therefore, sample 1 is the best candidate for ±0.1 V power supply gate controlled FET applications. Moreover, the leakage current of GaN/InGaN/AlGaN-based EHCRR-RTD is about 30+ orders less than the lowest off-current result reported recently in [6], which also makes it possible for extreme low power dissipation applications and electrical isolation applications.

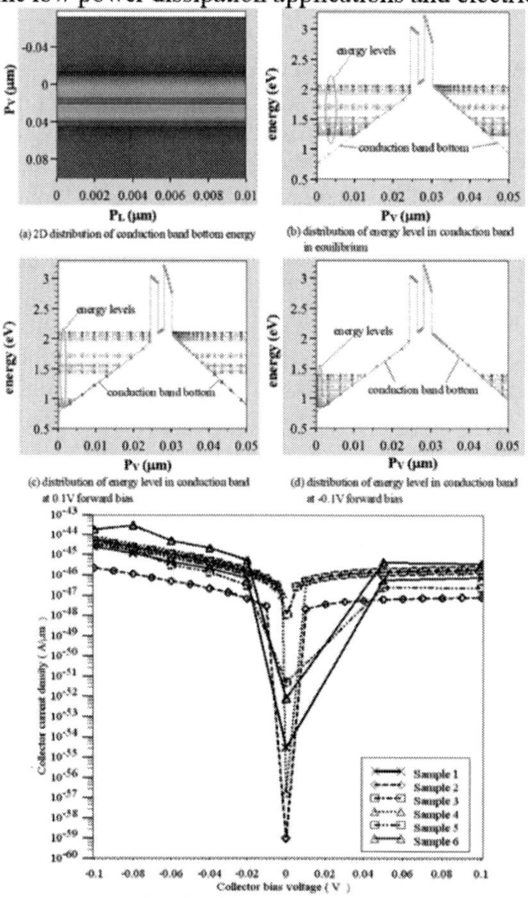

(e) current density-bias voltage characteristics

Figure 2. Distributions of energy levels in conduction band of EHCRR RTD sample 1 and current density-bias voltage characteristics of EHCRR RTD samples.

Table 2. Contrast on switching characteristic parameters of EHCRR RTD samples and FETs in [1]-[6]

Device Name	Power supply	Forward (backward) V_{on} or V_{th}	I_{on}/I_{off}	leakage current or I_{off}	Reciprocal of current rising rate or SS
Unit	V	mV	-	μA/μm	mV/dec
Sample 1	±0.1	50/(-20.3)	3×10^{8}	1×10^{-52}	5.90/(-2.39)
Sample 2	±0.1	10.1/(-10.3)	4×10^{11}	1×10^{-59}	0.87/(-0.89)
Sample 3	±0.1	34.87/(-5.36)	3.4×10	1×10^{-48}	22.77/(-3.50)
Sample 4	±0.1	10.1/(-2.15)	5×10^{11}	2×10^{-57}	0.86/(-0.18)
Sample 5	±0.1	50/(-20)	6×10^{4}	1×10^{-51}	10.46/(-4.18)
Sample 6	±0.1	50/(-20)	1.4×10^{7}	1×10^{-52}	7.00/(-2.80)
Si FinFETs (N/P channel)[1]	1	200	$2\times10^{6}/3\times10^{4}$	$1\times10^{-7}/1\times10^{-6}$	58/56
GaSb/InAs TFETs [2]	0.3	200	1×10^{7}	1×10^{-2}	<60
SDP JLFET [3]	1	600	1×10^{7}	$1\times10^{-6}\sim1\times10^{-4}$	55
InGaAs DG FinFET [4]	1	-70~80	1×10^{3}	2×10^{-2}	33.3
Al_2O_3-dielectric InAlN/AlN/GaN/Si MOS-HFET [5]	10	4200	8.9×10^{8}	1.2×10^{-6}	140
Si RGTFET [6]	0.2	200(-300)	3.22×10^{9}	4.5e-10	28.3~60

4. Conclusion

Through the above investigations, GaN/$In_{0.14}Ga_{0.86}N$/AlGaN quantum barrier diodes with $In_{0.07}Ga_{0.93}N$ collector and $In_{0.21}Ga_{0.79}N$ emitter are featured with inherent extreme low off current (I_{off}), extreme high current rising rate and an acceptable turn-on voltage when their corresponding layer thicknesses are 4-3-4, 5-4-5 and 5-6-5 expressed in term of crystal constant of each layer. Therefore, they will provide better candidates as the channel structures for extreme low SS TFET applications.

References

[1] AY. Morita,T. Mori, K. Fukuda, et al. IEEE IEDM, Dec. 2014, pp.9.7.1-9.7.4.
[2] AG. P. Marco, et al. *Journal of The EDS*, 3(3): 115-121, 2015.
[3] M. P. V. Kumar et al. *IEEE Trans.* ED, 62(11): 3541-3546, 2015.
[4] S.-J. Chang et al. *IEEE EDL*, 38(4): 441-444, 2017.
[5] C.-S. Lee et al. *Journal of the EDS*, 6(1): 68-73, 2018.
[6] J.-T. Lin et al. *IEEE Trans.* ED, 65(2): 769-775, 2018.
[7] V. Fiorentini et al. *Appl. Phys. Lett.*, 80(7):1204-1206, 2002.
[8] S. Sandeep et al. *Applied Physics Express*, 7(9): 095201, 2014.
[9] H. Zhang,L.Geng,M.Lin et al. IEEE IPFA, P7.8:1-5, July 2018.
[10] H. Zhang,Q. Zhang,M.Lin et al. *Journal of Semiconductors*, 39(7): 074004(11pages), July 2018.
[11] Y Liu, B Gao and M Gong. Journal of Physics: Conf. Series 864 (2017) 012022.

Research on Two-stage RGV Scheduling Based on Simulated Annealing and Tabu Search

Jing Guan[1,*], Chao Quan[2], Yiming Ma[1] and Jianxun Tang[1]

[1]Sino-European Institute of Aviation Engineers, Civil Aviation University of China Tianjin, China
[2]China Aviation Industry Tianjin Aeronautical Machinery and Electricity, Tianjin, China

*Corresponding author

Abstract. The scheduling problem of an intelligent machining system, with multiple Computerized Numerical Control Machine(CNCs) and one Rail Automatic Guidance Vehicle(RGV) with manipulator as its core, in a deterministic environment is studied. It is necessary to determine the ratio of the number of CNCs responsible for the first stage to the number of CNCs responsible for the second stage, the location and the moving path of the manipulator. Starting from the scheduling theory of parallel machines, the RGV algorithm based on simulated annealing is used to determine the optimal route of RGV by analyzing the type of machines and the processing time of work pieces. The experimental results show that the scheduling rule can effectively improve the efficiency of intelligent workshop.
Keywords: quasi-simulated annealing algorithm; tabu search; RGV scheduling; Roulette.

1. Introduction

RGV is mostly used in stereoscopic warehouse. Some literatures have studied RGV in warehouse. Li Guo [1] et al. proposed a scheduling rule based on complete traversal path planning (CCPP) for RGV under uncertain environment. Chen Yingge and Xianlong [2] use 0-1 integer programming to study such problems. Cao Hongtao et al. [3] used dynamic programming to solve similar problems. [4] Study the application of logistics system.

Multiple (even) CNCs are divided into two columns and emitted at equal distances. Each CNC can only process one job at a timeInitially, RGV is in the initial position between CNC1# and CNC2#, and all CNCs are idle. Under normal working conditions, the idle CNC sends the feeding demand signal to RGV. RGV chooses a CNC as its feeding according to the demand signal. During the processing time, RGV can complete the feeding work according to the feeding request of other idle CNC. When a CNC machine is finished, RGV arrives at the position, completes the cutting and new raw material feeding, and then the finished material is cleaned in the cleaning tank and placed on the feeding conveyor belt to make it leave the processing system. The RGV returns to its original position until all the jobs that need to be processed leave the system.

Many CNCs share the two following processes[5-9] of processing job. Each job must go through two processes before it can leave the processing system. RGV is dispatched to meet CNC's feeding and cutting requirements, which makes the whole system the most efficient. The intelligent system works continuously for one day (8 hours) per shift to determine the ratio of the number of CNC responsible for the first stage to the number of CNC responsible for the second stage, the location and the moving path of the manipulator.

2. Hypothesis

(1)For the convenience of studying the problem, we make the following assumptions. The transmitter will not fail and will not affect RGV scheduling. When RGV is ready for feeding, the conveyor belt conveys the material to the designated position. When RGV is ready for feeding, the conveyor belt can quickly remove the finished job.(2) RGV only feeds CNC when CNC does not process any workpiece in the initial stage of processing; RGV only feeds CNC when CNC no longer has other tasks in the final stage of processing; RGV always feeds CNC at the same time in the process of processing.(3) For convenience of calculation, it is assumed that CNC is idle when the system starts to run.(4) When calculating the maximum completion time, the conveyor belt transportation time after the last work piece is discharged is neglected.

3. Symbolic Description

n	Number of jobs to be Processed
N_i	Number of CNC signals emitted at point i at the same time
I	· CNC Number Set for the First Procedure in Two-Stage Problem
J	CNC Number Set for Second Procedure in Two-Stage Problem
S	CNC label set
signals	set of the labels of the signaling machine
signals1	··set of the labels of the signaling machine in I
signals2	set of the labels of the signaling machine in J
T_p^1	The Processing time of the first procedure in the two-stage problem
T_p^2	Processing time of the second process in two-stage problem
$T_{c,j}$	Time required for loading and unloading of the jth machine
T_{ws}	The time required to clean a job
T_{Fj}	Uploading or unloading time of machine j
T_w	Time required for RGV movement
$T_w(i,j)$	The time required for RGV to move from i position to j position
t_{ij}	The time when the first job begins to be processed on CNC j
t_{ij}^1	The processing time of job i begins with the first stage machine CNC j
t_{ik}^2	job i processing time starting from the second stage machine CNC k

$$x_{ij}=\begin{cases}1, \text{ i is processed on machine j}\\ 0, \qquad \text{else if}\end{cases} \quad i\in\{1,2,\cdots,n\}, j\in S \tag{1}$$

$$x_{ijk}=\begin{cases}1, \text{job i is processed on j in the first stage and on machine j in the second stage,}\\ \qquad i\in\{1,2,\ldots,n\}, j\in I, k\in J\\ 0, \text{else if}\end{cases} \tag{2}$$

C_{max} Maximum completion time

4. Model Establishment and Solution of Two-stage RGV Model

4.1. Selection of Machine Number and Location Determination

In the two-stage processing process, the number of machines responsible for each stage needs to be determined first. According to the data analysis given in the title, the order of magnitude of processing time is obviously larger than that of loading and unloading time and cleaning time in the whole production process, so the classification basis is only machine processing time.

4.2. Mathematical Model of Two-stage RGV Scheduling Problem

$$\text{Min } C_{\max} = \sum_{j\in I}\sum_{k\in J} x_{njk}(t_{ik}^2 + T_p^2 + T_{Fk} + T_{ws})$$

$$s.t.\begin{cases} \sum_{k\in J}\sum_{j\in I} x_{ijk} = 1 & i=1,2,\ldots n \quad (1) \\ \sum_{i=1}^{n} x_{ijk} \le 1 & j\in I,\ k\in J \quad (2) \\ \sum_{j\in I}\sum_{k\in J} t_{ik}^2 x_{ijk} \ge \sum_{k\in J}\sum_{j\in I}(t_{ij}^1 + T_{p,j}^1 + T_{c,j})x_{ijk}, & i=1,2,\ldots n \quad (3) \end{cases}$$

4.3. Two-stage RGV Scheduling Algorithm

The position efficiency is only related to the position, and is independent of the number of signals at the location. Let the evaluation function of the current position be $T_w(i,j)$, Obtaining the Probability Transfer Function $\exp(-T_w(i,j)/T)$, In this way, the RGV algorithm based on simulated annealing can be used to study RGV scheduling problem.

Constructing an evaluation function for other locations that takes account of both the number of calls and the distance of calls p_i, to evaluate the superiority of the next choice of location i: $p_i = \dfrac{T_w * + N_i T_{ws} * + N_i T_p * + N_i T_{Fj} *}{N_i}$. A transformation probability function similar to the simulated annealing probability function in the simulated annealing algorithm is introduced. P_i : $P_i = \exp(-p_i/T)$

4.3.1. Algorithmic Steps. Step1. Initially, all CNCs are empty, and RGV is between CNC1 and CNC2.

Step2. If the RGV's location sends the feeding signal, the raw material will be discharged in situ.

Otherwise, the evaluation functions of three places except RGV are calculated.

Step3. Calculate the transition probability function, and take the maximum two values as the optimal and suboptimal.

Step4. Generates a random number with a larger value and compares it with that. If the value is larger, the local optimal solution is abandoned and the location of the sub-optimal solution is chosen as the moving target location; otherwise, the local optimal position is chosen as the target location.

Step5. Compare the time consumed at this time with the time limit (8 hours).

If it is greater than or equal to the time limit, it will stop working and end production.

Otherwise, return to Step2.

4.3.2. Outer Tabu Search Algorithm Steps. Step1. Given an initial empty tabu table A, the output solution of the inner "RGV algorithm based on simulated annealing" is selected as the initial solution X.

Step2. If T is more than 8 hours, stop the calculation and output the results.

Step3. Select the non-taboo candidate set N (X) from the neighborhood of X. Select a solution X_next with the lowest evaluation value in N (X) and make X = X_next. Then update A and repeat Step2.

IOP Conf. Series: Materials Science and Engineering **611** (2019) 012083 doi:10.1088/1757-899X/611/1/012083

4.3.3. Two-stage RGV Scheduling Problem Algorithm Steps. Step1. Initially, all CNCs are empty and RGV is between CNC1 and CNC2.
Step2. Start the outer tabu search algorithm.
Step3. Call the inner layer "RGV algorithm based on simulated annealing idea".
Step4. Determine whether to terminate, if so, output the result, and if not, return to step 2.

4.4. Solution and Conclusion of Two-stage RGV Static Scheduling Problem
The following three sets of data are given for different working procedures and different machine processing time (unit second):

Parameters	Group 1	Group 2	Group 3
Time required for RGV to move one unit	20	23	18
Time required for RGV to move two units	33	41	32
Time required for RGV to move three units	46	59	46
The time required to complete a process in CNC processing	560	580	545
The time required for CNC to complete the first of the two processes	400	280	455
The time required for CNC to complete the second of the two processes	378	500	182
loading or unloading of RGV for machine 1357	28	30	27
loading or unloading of RGV for machine 2468	31	35	32
RGV cleaning time	25	30	25

4.4.1. The first group of data. In the first set of data, four machines are arranged in the first stage. In the second group of data, three machines are arranged. For the third group of data, six machines are arranged in the first stage.
The first group of data, RGV access path is: 1 - > 2 - > 5 - > 6 - > 3 - > 4 - > 7 - > 8 cycles.In 8 hours, the number of job processed is 240. The second group of data, RGV access path: 1 - > 6 - > 5 - > 4 - > 3 - > 8 - > 7 - > 2 cycles. In 8 hours, the number of jobs processed is 224. The third group of data, path: 7 - > 8 - > 1 - > 2 - > 5 - > 6 - > 3 - > 4 cycles. In 8 hours, the number of jobs processed is 215.
The following conclusions can be drawn:
(1) The RGV access loops in the choice of three kinds of data in static problems. This is because it is relatively simple in machine environment. Once an optimal scheduling strategy is determined, the optimal solution of the problem can be obtained by a long-term cycle of the scheduling strategy.
(2)According to the different processing time of the two stages, the number of jobs calculated by the algorithm varies, but the change is not significant, which shows that the algorithm has better generality.In the first set of data, the shortest time for each job to complete two-stage processing is 280+500+30+35+30=875. In 8 hours, a pair of machines (two-stage and two-machine) can process up to 32 jobs and up to 32*8=256 jobs. The gap between the solution and the boundary is 8/128=6.25%.In the second set of data, the shortest time for each job to complete two-stage processing is 350+30+280=345, and the maximum processing time is 83.48 in 8 hours, so the upper bound of the whole system processing room is 250. The gap between the solution we obtained and the boundary of the problem is 10.4%.
(3)In the second group of data, the upper bound of the system is 337, and the gap is 35.6%. This may be due to the complexity of the machine environment, the boundaries we choose are too large, or the algorithm may be unstable to the data in the special machine environment.

5. Conclusion

In this problem, the order of magnitude of processing time is obviously larger than the loading and unloading time and cleaning time, so machine processing time is used as the basis for distinguishing machine types. Based on the principle that the average processing time of each type of machine is as equal as possible, the number of each type of machine is determined, and the location of each type of

machine is determined according to the minimum RGV movement in the scheduling process. It is concluded that the first type of machine is 1-3-5-7, the second type is CNC2-4-6-8, the first type is 1-5-7, the second type is CNC2-3-4-6-8, and the second type is CNC2-3-4-6-8. The third type is CNC1-3-4-5-7-8 and the second type is 2-6.
The mathematical model of this problem is given clearly. Based on this, a combined algorithm of simulated annealing and tabu search is designed to solve the two-stage problem. The inner layer RGA scheduling algorithm based on simulated annealing algorithm gives a near optimal RGA scheduling strategy, and the outer tabu search algorithm improves the solution to make it close to the optimal as soon as possible. Through the analysis of calculation results, we found that the first group of data, RGV access path is: CNC1 - > 2 - > 5 - > 6 - > 3 - > 4 - > 7 - > 8 cycle, the number of processed jobs is 240, the gap between the number of processed jobs and the upper bound of the number of processed jobs in the system is 6.25%; the second group of data, RGV access path is: CNC1 - > 6 - > 5 - > 4 - > 3 - > 8 - > 7 - > 2 cycle. For the third group of data, RGV access path is CNC7 - > 8 - > 1 - > 2 - > 5 - > 6 - > 3 - > 4 cycle, and the number of processed jobs is 215.

Acknowledgments
This paper is supported by the Fundamental Research Funds for the Center Universities NO. 3122014C022. Fund for the Doctoral Program of CAUC 08qd02S.

Reference

[1] Li Guomin, Gao liang, and Li Xinyu. Dynamic Scheduling of Track Automatic Guidance Vehicles in Uncertainty Environment. China Mechanical Engineering, 2019 (8).
[2] Chen Yingge, Xianlong. Research on dynamic scheduling strategy of intelligent RGV. Scientific Research, 2019 (1).
[3] Cao Hongtao et al. RGV Scheduling Strategy Based on Dynamic Programming Algorithms in Fault Occurrence, Science and Technology Innovation, 2019 (8).
[4] Xu Jianmei, Lou Dongmei, Wang Jiefang and Yuan Hao. Application of shuttle truck in automatic logistics system [J]. Logistics technology and application. 2009 (07).
[5] Tang Hengyong, Zhao Chuanli, Introduction to Sorting, Science and Technology Publishing House, 2002.
[6] Cheng Zhenmin, Zhang Xijuan, and Li Hongxing. An approximate algorithm for parallel machine scheduling with workpiece preparation time. Journal of Beijing Normal University, 2019 (8).
[7] Zhang Zhicong et al. Parallel machine scheduling with weighted average process time is optimized. Modern Manufacturing Engineering, 2007 (9).
[8] Li Tieke, and Wang Berlin, Research on the Nature of Two-stage Flow Shop Scheduling Problem with Limited Waiting Time, Journal of Management Engineering, 2014 (8).
[9] Shi Qingtao, Research on Batch Scheduling of Single Procedure Parallel Machines Based on Genetic Algorithms, Dalian University of Technology, 2014.

Simulation and Analysis of Conductive Electromagnetic Interference in Near-Field Magnetic Coupling

Chin-Hsiung Lee[1,]* and Qinjun Hu[2]

[1]Sino-Euro Aviation College, Fuzhou Polytechnic, China
[2]State Grid Zhangzhou Power Supply Co., China

*Corresponding author

Abstract. Inductors, common-mode inductors, transformers and other magnetic components are important components of power converters. Their layout and wiring may affect the ability to suppress EMI noise. As a result of miniaturization and higher power density, the composition of power converter becomes more and more compact, which leads to the near-field coupling effect between components. In our study, the mechanism of near-field magnetic coupling between magnetic elements is analyzed, and the accuracy of the parameters of near-field magnetic coupling is further verified by experiments.
Keywords: near-field magnetic coupling; simulation analysis; EMI; conducted EMI.

1. Background

Due to miniaturization and increased power density, the composition of the power converters have become considerably compact, which leads to near field coupling effects among the components [1-2]. Near-field coupling could be divided into magnetic field coupling and electric field coupling. The coupling among the magnetic components of power converters is mainly magnetic field coupling. It is reflected in the common-mode inductance, the parasitic inductance between the common-mode inductance and the capacitance loop, and between the common-mode inductor and other magnetic elements in the main circuit. [3-6]

2. Analysis of the External Magnetic Field of Magnetic Elements

The linkage of the magnetic field leakage from magnetic components is the main source of the near magnetic field coupling among the components. In this paper, a three-dimensional electromagnetic simulation software HFSS is used to analyze the spatial distribution of the magnetic fields from the magnetic components of the common mode inductors and transformers and their influences on other magnetic components.

2.1. Magnetic Field Distribution of Common Mode Inductor in Common Mode

Figure 1(b) is a HFSS three-dimensional simulation model, which is based on the size of a real common mode inductor (as shown in Figure 1(a)). The yellow part is windings, and the grey part is ferrite magnetic ring. The magnetic field distribution around the common mode inductor is obtained by setting the same current excitation on the inductor and conducting simulations. The result, the amplitude of the magnetic field around the common mode inductor, is shown in Figure 2.

Figure. 3 is a magnetic equivalent circuit diagram of common mode inductor and other magnetic components, where Rcm1 and Rcm2 are the equivalent reluctance in the two sides of the common mode inductor magnetic ring, Rg1 and Rg2 are the equivalent reluctance of the air loop, and R is the equivalent

IOP Conf. Series: Materials Science and Engineering **611** (2019) 012084 doi:10.1088/1757-899X/611/1/012084

reluctance of other magnetic elements. Since the common mode inductors are excited by common mode currents in the same amplitude and phase, it is possible that the magnetic potentials produced by these two common mode current excitations would be opposite to the magnetic flux reversal produced by other magnetic components, and would thus be offset. Therefore, the magnetic field produced by the common mode inductor does not affect the magnetic elements around it.

(a) Physical diagram. (b) 3-D simulation model.

Figure 1. Common mode inductor.

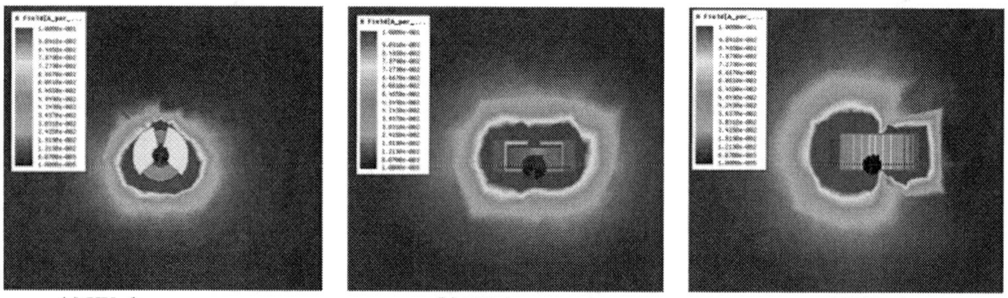

(a) XY plane (b) YZ plane (c) XZ plane

Figure 2. Distribution diagram of magnetic field intensity around common mode inductors.

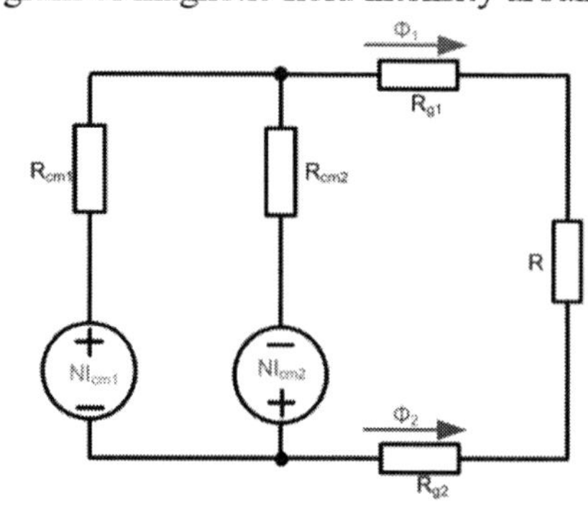

Figure 3. External magnetic equivalent circuit diagram of common mode inductor.

2.2. Magnetic Field Distribution of Common Mode Inductor in Differential Mode

The model in Figure 4 is the same as Figure 1, but the direction of the current excitation is different. In Figure 4, current excitation with opposite direction was set to simulate the magnetic field distribution under differential mode current. The obtained distribution of the differential mode magnetic field around the common mode inductor is shown in Figure 5.

The differential mode magnetic equivalent circuit diagram of the common mode inductor is shown in Figure 6, where Rdm1 and Rdm2 are the equivalent reluctance in the two sides of the common mode inductor magnetic ring, Rg is the equivalent reluctance of the air in the magnetic ring, Rg1 and Rg2 are the equivalent reluctance of the air loop from the magnetic ring to other magnetic components outside the magnetic ring, and R is the equivalent reluctance of other magnetic elements. Common mode inductors are excited by differential mode current in the same amplitude but inverse phase. According to the superposition theorem, the magnetic flux generated by two differential mode currents excited by the common mode inductor is the same as the flux produced by other magnetic components. It's a superimposed relationship. Therefore, the magnetic field generated by the differential mode of the common mode inductor will affect the magnetic components around it.

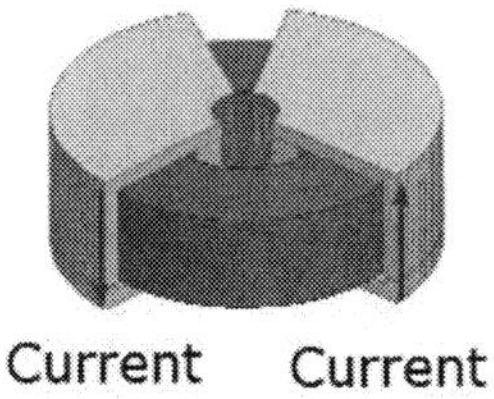

Figure 4. Three dimensional simulation model of common mode inductor.

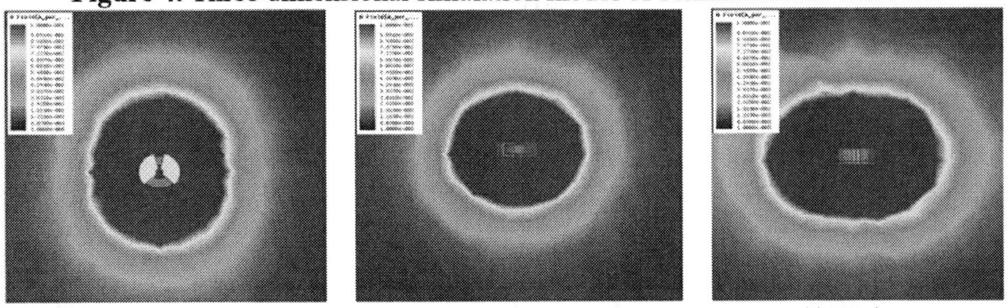

(a) XY plane (b) YZ plane (c) XZ plane

Figure 5. Distribution diagram of magnetic field intensity around common mode inductors.

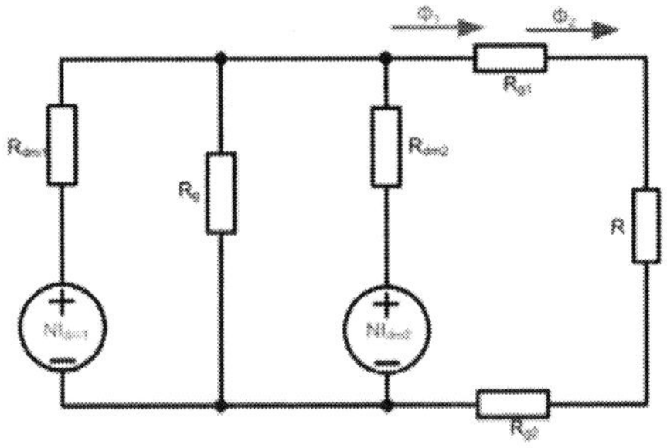

Figure 6. Magnetic equivalent circuit diagram.

3. Near Field Coupling and Decoupling of Magnetic Elements

The magnetic field coupling model can be expressed by the mutual inductance M. Therefore, in the near field coupling model of the inductor and capacitor of the EMI filter, it is a form of controlled voltage source, as shown in Figure 7. The current I in the figure is the current flowing through the interference source.

Because the EMI filter has many magnetic components and capacitor elements, the coupling relationship between them is more complex, which is not conducive to the analysis of their mutual influence. According to the principle of the circuit, we can decouple the components of the coupling relationship, and then analyze the coupling effect of the near field, just as the decoupling method of the basic circuit principle shown in Figure 8 and Figure 9.

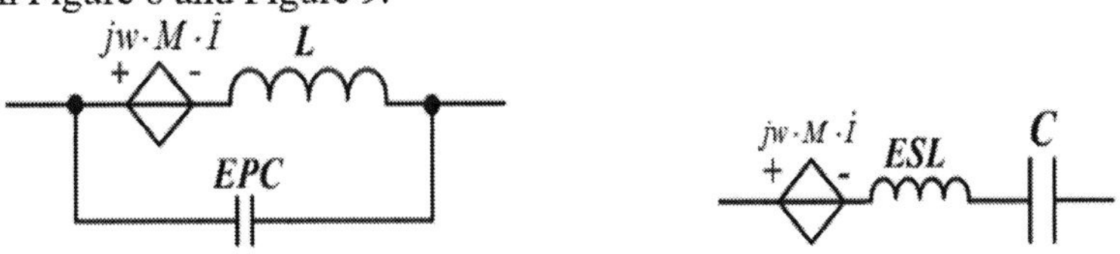

(a) Near field coupling model of inductor. (b) Near field coupling model of capacitor.

Figure 7. Near field coupling model of inductor and capacitor.

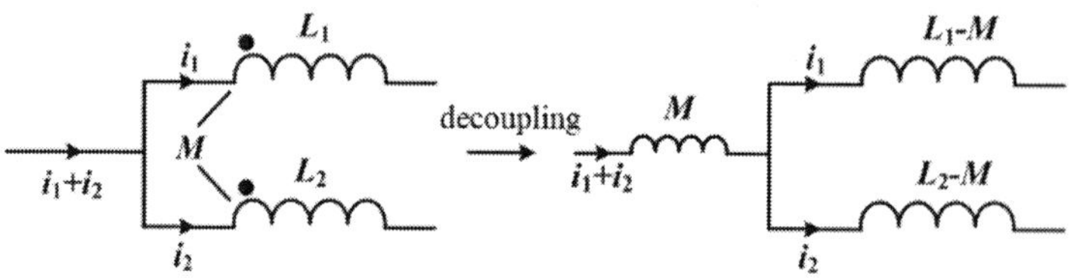

Figure 8. Positive decoupling diagram of inductor.

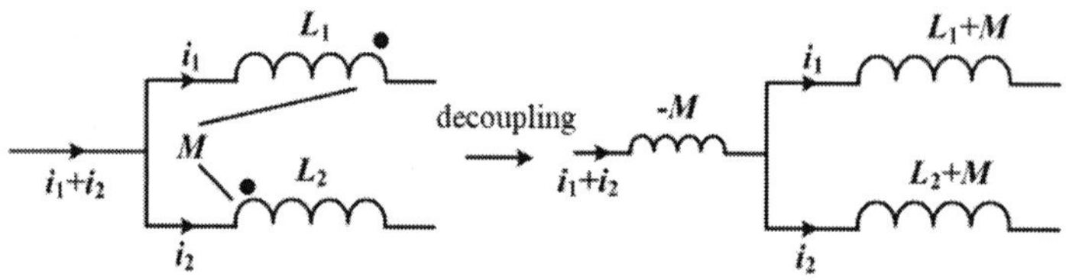

Figure 9. Negative decoupling diagram of inductor.

The two inductors L1 and L2 have mutual inductance M, which are positive coupling and negative coupling, respectively. The decoupling of the circuit can be realized through the equation.

$$\begin{cases} U_{AB} = L_1 \dfrac{di_1}{dt} \pm M \dfrac{di_2}{dt} \\ U_{AC} = L_2 \dfrac{di_2}{dt} \pm M \dfrac{di_1}{dt} \end{cases} \quad (1)$$

$$\begin{cases} U_{AB} = (L_1 \mp M)\dfrac{di_1}{dt} \pm M \dfrac{d(i_1+i_2)}{dt} \\ U_{AC} = (L_2 \mp M)\dfrac{di_2}{dt} \pm M \dfrac{d(i_1+i_2)}{dt} \end{cases} \quad (2)$$

4. Simulation Method for Extracting Near Field Coupling Parameters of Common Mode Inductors

HFSS is a powerful electromagnetic field simulation software, which is widely used in various engineering electromagnetic fields. Based on the Maxwell differential equation and using the finite element method, the calculation of the engineering electromagnetic field is transformed into a matrix solution. In this way, the three-dimensional model of HFSS is used to simulate the near field coupling of two common mode inductors.

The structure of the EMI filter of the prototype is shown in Figure 10(a). According to the placement of EMI filter, the 3D model is built in HFSS software, as shown in Figure 10(b). In order to facilitate the modeling and accurate simulation of the actual inductance, the winding of the common mode inductor model is made of a single turn, and the result is multiplied by the number of turns when calculating the mutual inductance.

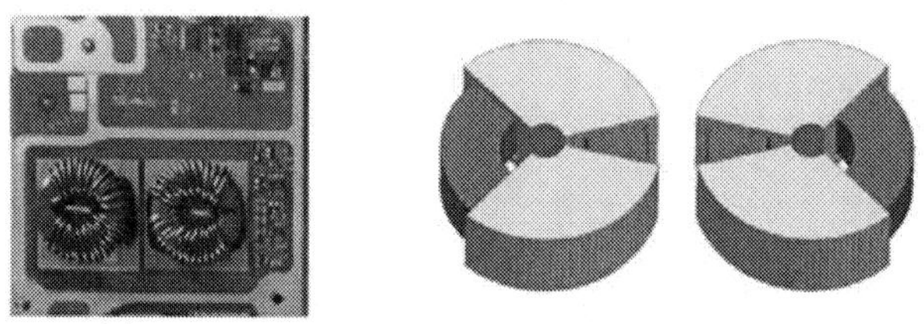

(a) (b)

Figure 10. A schematic diagram of the experimental filter and 3D model.

The magnetic coupling parameters of the simulated EMI filter must be excited by the common mode inductor, and the direction of the corresponding current is set in the HFSS simulation model, as shown in the red arrows of Figure 11.

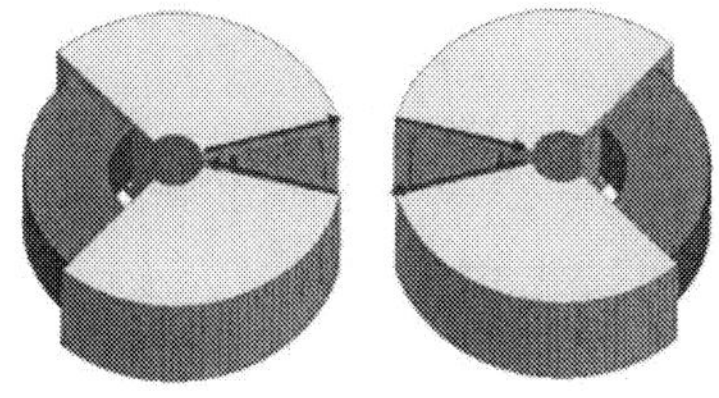

Figure 11. Current direction of common mode inductor model.

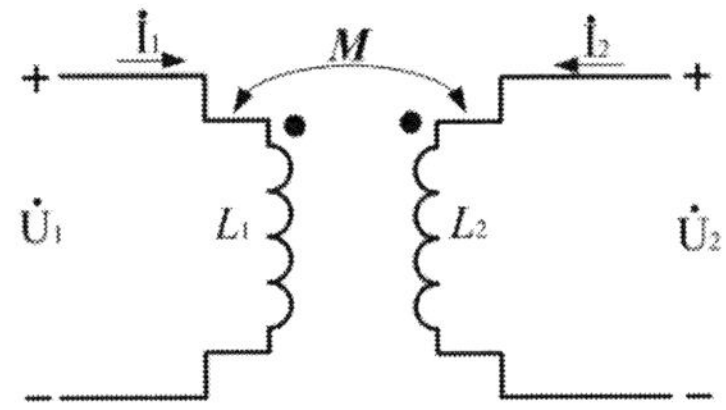

Figure 12. The schematic diagram of the coupled two devices in HFSS simulation.

As the results of simulation with HFSS software are S, Y and Z parameters, the mutual inductance between the two devices can be calculated based on the Z parameters obtained by simulation. This is because the two devices with near field coupling can be regarded as a two port network, as shown in Figure 12. This two port network satisfies the Equations (5) and (6), where

$$\dot{U}_1 = Z_{11} \cdot \dot{I}_1 + Z_{12} \cdot \dot{I}_2 \tag{3}$$

$$\dot{U}_2 = Z_{21} \cdot \dot{I}_1 + Z_{22} \cdot \dot{I}_2 \tag{4}$$

$$Z_{21} = Z_{12} = \left.\frac{\dot{U}_1}{\dot{I}_2}\right|_{\dot{I}_1=0} = j\omega \cdot M \tag{5}$$

Therefore,

$$M = \frac{Z_{12}}{j\omega} \tag{6}$$

Table 1. Simulation parameters of mutual inductance (Unit: nH)

M1	M2	M3	M4
65.57	65.75	65.44	65.99

The simulated Z parameters between the common mode inductor CM1 and the CM2, as shown in Figure 13, are brought into the Equation (6) to get the near field coupling parameters between the two common mode inductors, as shown in Table 1.

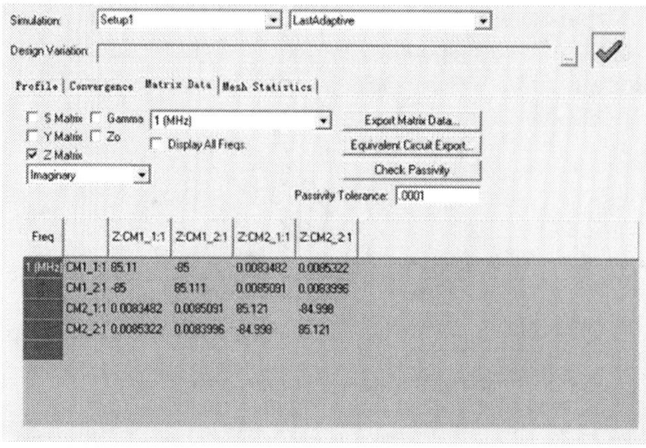

Figure 13. Simulation results of near field coupling of EMI filter.

According to the simulation results in Table 1, we can see that the values of M1, M2, M3 and M4 are almost the same. Based on the decoupling model of T filter near field coupling, the parasitic inductance of the X capacitor in the two common mode inductor is 0.263 μH. The resonant frequency of the X capacitor in the two common mode inductor decreases, and the suppression ability of the EMI filter to the differential mode is reduced. At the same time, the mutual inductance parameter of the common mode inductor is far less than that of the differential mode component, so it can be neglected.

5. Conclusion

Firstly, the leakage of magnetic field and its influence on other magnetic components are analyzed. It can be seen from the analysis that the magnetic field of the differential mode component of the common mode inductor is disturbed by the external magnetic field as the common mode inductor works. Secondly, the near-field magnetic coupling mechanism between magnetic elements is analyzed, and the X capacitor is decoupled through the circuit decoupling principle. After decoupling, a mutual inductance on the X capacitor is superimposed with the ESL of the X capacitor itself. Finally, the magnetic coupling parameters of the common mode inductor are simulated by the three-dimensional electromagnetic field simulation software HFSS, and the accuracy of the simulation is verified by experiments.

Acknowledgments

This work was supported by Scientific Research Startup Foundation of Fuzhou Polytechnic under project RCQD201802.

References

[1] Li Longtao 2012. Modeling and effectiveness evaluation of switching power transmission EMI. Dissertation, Harbin Institute of Technology.

[2] Zhang Yu, Chen Qingbin, Chen Wei 2013. Study on EMI simulation of switched power supply. Proceedings of the Twentieth Annual Conference of China Power Society, pp 1205-1209.

[3] Ma Yinfei 2012. Research on the prediction of switched power transmission EMI. Dissertation, Hebei University of Technology.

[4] Chen Qingbin, Chen Wei 2012. Evaluation method for common mode conducted noise suppression capability of transformer in switching mode power supply. Chinese Journal of Electrical Engineering, 18, pp 73-79.

[5] Xu Ke 2012. Analysis and suppression technology of common mode electromagnetic interference in power electronic devices. Dissertation, Chongqing University.

[6] Chen Qingbin 2012. Research and application of magnetic field conduction EMI characteristics and filter near-field coupling characteristics of switching power supply. Dissertation, Fuzhou University.

Development of Solar-Panel Monitoring Method Using Unmanned Aerial Vehicle and Thermal Infrared Sensor

Dongho Lee[1] and Jonghwa Park[1,*]

[1]Department of Agricultural and Rural Engineering, Chungbuk National University, Rep. of Korea

*Corresponding author

Abstract. Interest in solar energy, which is clean energy, is growing due to continuously high oil prices because of the unstable situation in the Middle East and the explosion of the Fukushima nuclear-power plant in Japan. South Korea has made various efforts to develop renewable energy, and facilities using solar energy are rapidly growing. Periodic inspection of solar panels is very important for the effective management of photovoltaic (PV) modules. In this study, a method to monitor and diagnose faulty solar modules through an Unmanned Aerial Vehicle (UAV)-based thermal infrared camera and GIS spatial analysis is discussed. First, the image of the PV module was taken by installing an RGB camera and thermal infrared camera on the rotor blade of the UAV. The images were generated by using Pix4D SW, and the normal operation of the PV module was diagnosed using the orthoimage. In particular, thermal infrared cameras are very sensitive to reaction speed, unlike optical sensors, so it is very important to find appropriate monitoring methods, such as shooting speed. Therefore, we suggest a way to utilize an optical sensor and thermal infrared sensor together through various attempts. In the images, GIS spatial analysis was used to extract abnormal heat or faulty modules. Using the Virtual Reference Service (VRS) survey method, UAV images could be 3D-mapped with a very precise temperature, within 2 cm in x, y, and z direction errors. The rapid development of PV power-generation facilities and of monitoring technologies is a very important factor for improving energy efficiency. The result of this study is providing a stable solar-panel-monitoring technology by comparing features of ground observation and UAV-based thermal infrared sensor observation. The developed monitoring technology was able to accurately analyze the failure point and temperature of the PV module.
Keywords: solar panel; monitoring; unmanned aerial vehicle; thermal infrared; VRS survey.

1. Introduction

The greatest feature of solar-power generation is the fact that the energy source is endless and clean. There is no danger of generating air pollutants such as CO_2 (carbon dioxide), SOx (sulfur oxide), and NOx (nitrogen oxides) during power generation, as in the case with thermal-power generation that generates electricity by burning petroleum. If the annual power generation of a 1 kW system is 1000 kWh, the CO_2 reduction effect of the crystalline–silicon solar cell is 541.5 kg/year/kW. The amount of crude-oil reduction is 227 *l*/year/kW. The lifetime of a solar-power generation system is comparatively longer, and the service life of the solar cell module used for solar-power generation is now more than 20 years. A solar module is the key component of a photovoltaic (PV) system. The module converts solar energy into electrical energy by connecting several solar cells and semiconductor devices, in series or in parallel. Therefore, the module is the most expensive component of the PV system and determines the lifetime of the entire system. The lifetime of the solar cell module is about 10 to 20 years, depending

IOP Conf. Series: Materials Science and Engineering **611** (2019) 012085 doi:10.1088/1757-899X/611/1/012085

on the manufacturing method, and there is the advantage of the system scale being able to freely be determined from small to large depending on installation location, with little maintenance cost. However, if one of the modules fails, it affects the entire module and causes a reduction in efficiency. Modules are installed in a large area, and require a lot of maintenance and a large workforce. Therefore, finding a way to quickly monitor a module is critical to saving costs and achieving stable efficiency.

The location of a solar-power plant is very important for the efficient operation of solar-power generation. The decision for the location is carried out by analyzing the location of the solar-power plant by applying a hierarchical-analysis or correlation-analysis method based on GIS. PV power plants are affected by various factors, and are damaged or deteriorate in efficiency. At the beginning of the damage, the cell is overloaded, and abnormal heat generation occurs; then, the whole module is short-circuited and overall output is lowered. Therefore, it is important to use a monitoring system together with periodic inspections of the solar-power plant to prevent this. Monitoring and failure of solar-power generation has mainly been done by utilizing the workforce. However, since the solar module of the solar-power plant is generally located at a high altitude, it is necessary to carry out the current test for the solar module and array by ascending to the solar cell panel or, because of the risk, determining the temperature change point by using the thermal camera. It takes much time to monitor the solar module due to the solar-power plant having a large area and a high installation site, and there are various problems in safety. Recently, various monitoring methods have been proposed. One of them is to use Unmanned Aerial Vehicles (UAVs) and thermal infrared sensors. However, this is due to the fact that a thermal infrared sensor is expensive, and the lack of technology to stably photograph the thermal infrared sensor mounted on the UAV.

In this study, an experiment was conducted to automatically diagnose the failure of the module of a solar-power plant by installing an optical sensor based on a UAV rotor blade and a thermal infrared sensor.

The purpose of the study was to develop a technology to analyse and diagnose the temperature difference by GIS spatial analysis for each solar module by superimposing the solar-module layer on the temperature information acquired by the UAV-based thermal infrared sensor.

2. Materials and Methods

2.1. Field Monitoring Area

In order to develop a monitoring method for solar modules using a UAV-based thermal infrared sensor, the target of this study was a PV solar-power station located in Gagok-ri, Ochang-eup, Cheongju-city, Chungbuk, South Korea. The solar-power plant in the target area is composed of 36 pieces of 300 W class modules, and can produce 10.8 kW power in total. Figure 1 shows the location of the solar-power plant, and the observation point is marked with a red circle. The Virtual Reference Service Real-Time Kinematic (VRS-RTK) survey was performed by selecting 6 GCPs for the ground reference point.

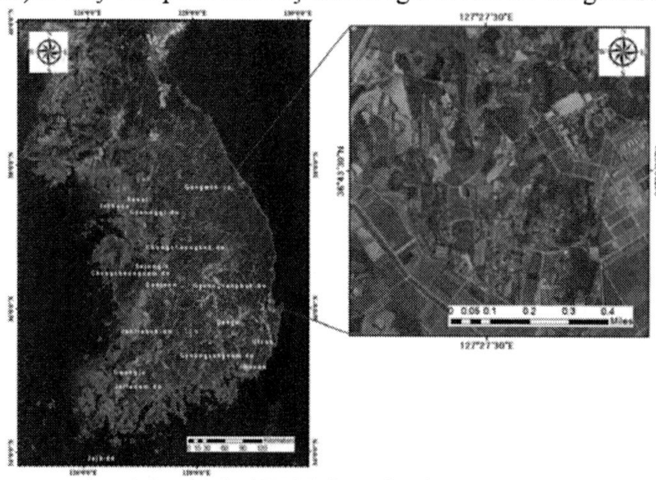

Figure 1. Unmanned aerial vehicle (UAV)-based solar-panel monitoring location map.

2.2. Materials

The aerial observations of the solar- panel monitoring site were observed using DJI Inspire 2 (Table 1). Inspire 2 is equipped with a "vision positioning system" that only lands in a safe place, and a system that detects obstacles not only in the front but also up and down. In addition, the Inspire 2 was designed to allow two cameras to be installed so that the flight time could be extended and the solar panel could be observed in a dual-camera- attached structure. Observation sensors were installed with an RGB sensor and thermal infrared sensor on the UAV. The specifications of the used UAV and the specifications of the thermal infrared sensor are shown in Tables 1 and 2, respectively. The thermal infrared camera used for the field survey is shown in Table 3.

Table. 1 Inspire 2 specifications

Item	Specifications
Weight	3290 g
Diagonal Distance (Propeller excluded)	605mm
Max Flight time	25 min
Max speed	94km/h
Image sensor	Zenmuse X5s (red, green, blue)
Battery type	LiPo 6S/ 4280mAh

Figure 2. Inspire 2.

Table. 2 FLIR Vue Pro R specifications

Item	Specifications
Resolution	640 × 512
Size	63×45×45 mm
Temperature range	–25 to 135℃
Spectral range	7.5 to 13.5 μm
Field of view (FOV)	32°× 26°
Output Type	Radiometric JPEG, TIFF

Figure 3. FLIR Vue Pro R and UAV-based thermal infrared camera image.

Table. 3 FLIR T-420 specifications

Item	Specifications
Weight	880 g
Size	106×201×125 mm
Temperature range	–20℃ to 650℃
Spectral range	7.5 to 13μm
Field of view (FOV)	25°× 19°
Image mode	Thermal/Visual/P-i-P (Resizable and movable)/MSX Thumbnail Gallery
Battery type/ operating time	Li-lon/ >4 hours, display shows battery :

Figure 4. FLIR T-420 Thermal infrared camera using field survey.

2.3. Methodology

In order to diagnose the monitoring and failure of PV modules, module boundaries must be layered for each array. In case of the failure of a solar-module diagnosis using a UAV-based thermal infrared sensor, the resolution of the thermal infrared sensor may be lower than that of the RGB image, resulting in errors in the module boundary. Therefore, it is necessary to use imaging and merging technology to grasp the precise boundary of the PV module. The orthoimage and PV module layers were imaged at the same time with a 2 cm resolution RGB camera. DJI's Inspire 2 model was used for orthoimage and 3D terrain modelling. Since the UAV (Inspire 2) and the thermal image sensor (Vue Pro R) are not compatible with each other, they were modified so that they could be separately mounted and made a gimbal to fly together. UAVs were controlled by the Pix4d Capture application and commanded the shooting of the RGB sensor through the set redundancy during flight. Since the thermal image sensor cannot control the shooting with the automatic flight program, the shooting setting and control were performed separately. After setting the thermal infrared shooting environment, the photographing angle was adjusted through the gimbal setting, and the image was taken at the set time interval according to the automatic flight path by simultaneously commanding the shooting with the UAV take-off. UAV-based aerial-photographing procedures were performed in the order of 1) preplanning and data collection, 2) sensor calibration, 3) route setting and photography, 4) image correction, and 5) image matching.

2.4. Thermal Infrared Orthoimage Temperature Correction Method

Accuracy evaluation of the UAV-based thermal image sensor was performed by regression analysis of the obtained results from the comparative field survey. and the image was corrected with the obtained regression equation.

3. Results and Discussion

3.1. Optimal UAV Flight Condition for Thermal Infrared Orthoimage Production

Unlike ordinary optical sensors, thermal imaging is not easy with the Pix4d mapper software. Therefore, in the process of producing an orthogonal thermal infrared image, there are many cases where image fusion fails along with the error of calibration failure. In order to solve these problems, it is important to find flight conditions that have the best image quality through various conditions, such as image overlap, interval, and flight speed (Table 4).

Optimal flight conditions were investigated through 16 repeated test flights over three days, 23 and 28 November, and 1 December 2018 (Table 4). As a result of the flight-condition change test, it was found that an orthogonal thermal infrared image was possible when the image overlap was 85%, shooting time interval was 3 seconds, and flight speed was 2 m/s.

Table 4. Success and failure of image merges according to flight conditions.

Day	Image overlap (%)	Interval (sec)	Flight speed (m/s)	Image merging	Success and failure
23 November 2018	65	1	4	×	Calibration failure
23 November 2018	70	1	4	×	Calibration failure
23 November 2018	75	1	4	×	Calibration failure
23 November 2018	80	1	4	×	Calibration failure
23 November 2018	85	1	4	×	Calibration failure
28 November 2018	75	2	3	×	Calibration failure
28 November 2018	75	3	3	Δ	Partial success
28 November 2018	75	4	3	×	Calibration failure
28 November 2018	85	2	3	Δ	Partial success
28 November 2018	85	3	3	Δ	Partial success
28 November 2018	85	4	3	×	Calibration failure
1 December 2018	85	3	2	○	Success
1 December 2018	85	3	2	○	Success
1 December 2018	85	3	3	Δ	Partial success
1 December 2018	85	3	3	○	Success
1 December 2018	85	3	4	×	Calibration failure

IOP Conf. Series: Materials Science and Engineering **611** (2019) 012085 doi:10.1088/1757-899X/611/1/012085

3.2. Improved Accuracy through VRS-RTK Survey Method

In this study, GCP targets were placed and surveyed. Error between the RGB image and the thermal image was minimized through performing image correction by inputting the coordinates of the GCP point in the image joining process through the VRS-RTK survey method. As a result of improving the positional accuracy of the two images, the orthoimages of the visible light and the thermal external image could be overlapped and analyzed at the same time (Figure 5). UAV images using the VRS-RTK survey method enabled very precise 3D mapping in the x, y, and z directions within 2 cm when creating the digital surface model (DSM).

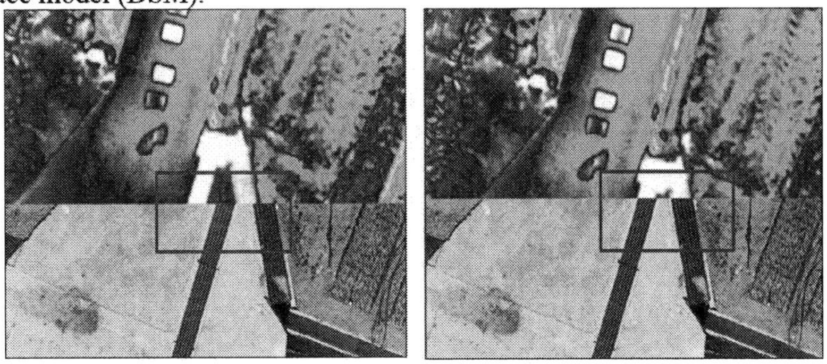

Figure 5. Comparison results before and after position-accuracy correction using the Virtual Reference Service Real-Time Kinematic (VRS-RTK) method.

3.3. Correlation Analysis between UAV-based Thermal Image and Ground Thermal Imaging Sensor

The thermal-infrared-camera images installed on the UAV were acquired as R-JPEG and TIFF images, and the correlation between the temperature information and the DN value, and the temperature information measured by the thermal camera on the ground were investigated. The R-JPEG and TIFF images based on the UAV were highly correlated, as shown in Figure 3, as a result of the ground survey and regression analysis. In particular, TIFF-type files had higher correlation than the R-JPEG files.

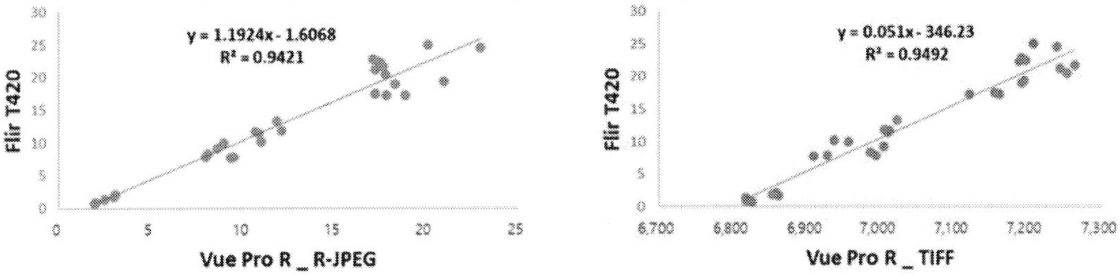

Figure 6. Correlation Analysis between UAV-based and ground thermal imaging sensor.

3.4. Development of Solar-panel Monitoring Application Technology

There have not been many examples of orthoimages that were equipped with thermal cameras in UAVs. In the case of PV solar applications, there is a study on a PV solar-module fault-diagnosis experiment using a fixed-wing UAV. Most of these studies were performed by installing a rubber cover on a solar module to determine whether the solar module failed or not, and calculating the average temperature for each module based on the temperature information obtained from the thermal infrared camera and the solar-module layer. A fixed-wing UAV is advantageous in that it can be taken lightly and for long enough to shoot a large area. However, since it is difficult to cope with unexpected situations, a fixed-wing UAV flies at higher altitudes than rotor-type UAVs. Therefore, the image resolution of the fixed-wing UAV is lower than that of the solar-panel module. In this study, we tried to solve these problems by observing the precise temperature at the low altitude of 30 m using a rotor-type UAV. However, most of the problems are caused by the frequent occurrence of overall-performance degradation due to the failure of one cell, so it is important to solve the problems related to that [3]. Figure 4 shows the result of orthoimage production of the RGB and the thermal infrared images. As shown in Figure 8, it was concluded that it is possible to not only analyze the cell unit, but also the array unit through the production of the orthoimage using the UAV. Cell-by-cell analysis can detect cool spots that are not identified in the RGB image, indicating that the bypass diode of the solar panel is defective. The point where the temperature drops like in the image may not be a problem in the short term, but if the phenomenon continues to periodically occur, we are able to detect the problem and take action.

Table 5. RGB and thermal infrared image acquisition conditions and results.

Sensor	Shooting altitude (m)	Number of shots	Spatial resolution (cm)	Max. Temp. (℃)	Min. Temp. (℃)
RGB	30	189	0.7	-	-
Vue Pro R	30	424	2.8	57.66	-0.92

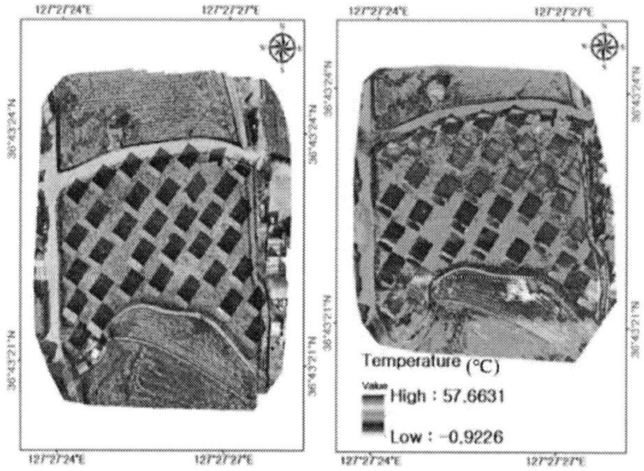

Figure 7. Result of orthoimage production of RGB and thermal infrared images.

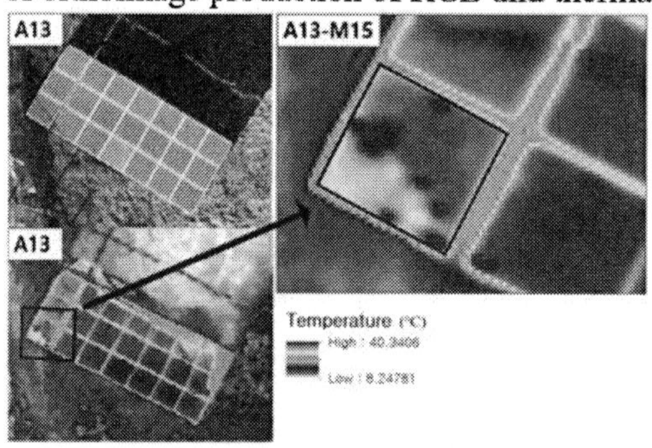

Figure 8. Case of performance-degradation point based on UAV-based thermal-infrared-image analysis.

4. Conclusion

In this study, to solve the compatibility problems between UAVs and thermal image sensors, a mount and gimbals that could be separately mounted, and data-acquisition equipment necessary for orthoimage production were constructed. The thermal infrared image joining method proposed various conditions of automatic flight, and a thermal-image control method and the optimal condition of thermal infrared orthoimages under various conditions of image overlap, image-acquisition interval, and flight speed. UAV photogrammetry enables highly precise 3D temperature mapping in the x, y, and z directions within 2 cm when creating a temperature DSM using the VRS-RTK surveying method by placing the GCP target on the path. We proposed precise temperature-monitoring technology for each cell and array by superimposing the RGB image and a thermal infrared image, and presented application examples in the field.

References

[1] Brewer J, Ames D P, Solan D, Lee R, and Carlisle J 2015 *Using GIS analytics and social preference data to evaluate utility-scale solar power site suitability* Renewable Energy vol.81, pp.825-836.

[2] Park J K and Park J H 2016 *Applicability evaluation of agricultural subsidies inspection using Unmanned Aerial Vehicle*, Journal of the Korean Society of Agricultural Engineers, vol.58, issue 4, pp. 1-9 26.

[3] Zefri Y, Eikettani A, Sebari I, and Lamallam S A 2018 *Thermal Infrared and Visual Inspection of Photovoltaic Installations by UAV Photogrammetry—Application Case: Morocco* Drones vol.2, pp.1-24.

IOP Conf. Series: Materials Science and Engineering **611** (2019) 012086 doi:10.1088/1757-899X/611/1/012086

Analysis of Vegetation Characteristics of Regional Soybean and Paddy Rice Fields Using UAV Images and Farm Maps

Dongho Lee[1] and Jonghwa Park[1,*]

[1] Department of Agricultural & Rural Engineering, Chungbuk National University, Rep. of Korea

*Corresponding author

Abstract. Remote sensing (RS) and GIS together provide an efficient and inexpensive way to assess vegetation characteristics and cultivated areas. Precision vegetation mapping provides important information for understanding crop conditions through the calculation of the vegetation index. In this study, a multispectral sensor was installed on an eBee (Sensefly), which is suitable for wide area photography taken by an unmanned aerial vehicle (UAV). UAV imagery acquisition was used to obtain spatial information on soybean and paddy rice cultivation areas in a wider region. The study area selected was Gimje, which is in the southwest of South Korea. Based on the designation of main areas according to the Ministry of Agriculture, Food, and Rural Affairs (MAFRA), the area was selected considering the cultivation area, production volume, number of farm households, and self-sufficiency ratio. The main producing areas are soybean cultivation areas defined as a city or county with a cultivation area of 30~1500 ha or more, respectively. The identification of the cultivation area of soybeans was made based on images taken in early September, when the growth characteristics of soybeans were clearly distinguished from other crops. The accuracy of the soybean and paddy rice cultivation area identification was improved by using the farm map (FM) prepared by the MAFRA in South Korea. The NDVI was analyzed based on the two periods of high growth activity of soybean and paddy rice. The accuracy of the cultivated area was improved by using the UAV images and the FM prepared by the MAFRA. As a result, this study shows that UAV images can be used more effectively by linking GIS with the FM as a basic map.
Keywords: remote sensing; vegetation; soybean; paddy rice; farm map; unmanned aerial vehicle.

1. Introduction

Soybean is a crop that has three major nutritional components: protein, fat, and carbohydrate. Among them, the percentage of protein is much higher than that in meat (20%) and wheat (10%). Therefore, soybean is known as a high-protein food. The global soybean market was estimated at $305.70/ton as of 7 September 2018. As the awareness of the health benefits of soybeans is growing globally, related markets have been growing rapidly due to the surge in demand for soybean-related products. In addition, the demand for soybeans in South Korea is increasing due to the globalization of Korean culture, known as the Korean Wave. The growth and competitiveness of the soybean market is thus growing in South Korea. Considering this point, research on soybean crops, the diversification of pest management, and quality control technology are urgent points of research in South Korea. Korea has four seasons with seasonal characteristics. There are many rainy and cloudy days during the growth period of soybean, and it is difficult to acquire spatial information of crops in the hot summer. The technology that has been suggested as a solution to this problem is the utilization of UAVs (unmanned aerial vehicles) and sensors. We have been dealing with various problems related to crops by employing UAVs equipped with

IOP Conf. Series: Materials Science and Engineering **611** (2019) 012086 doi:10.1088/1757-899X/611/1/012086

sensors in agricultural applications for 6~7 years [1,2]. Korea's agriculture includes a wide variety of crops grown in a short period of time in a small area, cultivating another crop after the first harvest. In order to understand the growth characteristics of crops, it is very important to understand the characteristics of spatial distribution over a wide area in a certain period via continuous monitoring. Since 2014, the Ministry of Agriculture, Food, and Rural Affairs (MAFRA) of South Korea has been producing and operating a farm map (FM) for the efficient use of spatial information in agriculture. However, it is costly and time-consuming to update and upgrade the FM. To overcome these problems, there we propose a way to improve efficiency by combining the FM and UAV imagery. The purpose of this study is to apply images of agricultural areas photographed by a UAV to the FM for the cultivation and management of crops such as soybean and paddy rice.

2. Materials and Methods

2.1. Field Experimental Area

This study was carried out during 2018 in soybean and paddy rice cultivation area (35°44'~48'N, 126° 47') located in the area of Gimje city, Jeonbuk, South Korea (Figure 1). The study area is located in Honam Plain, a granary zone. Topography formation is alluvial plain, and soils are composed of Jeonbuk Tong, so they are mainly used as paddy fields. This area is the largest rice and soybean producing area in the country, with the Dongjin River running to the west and the Wonpyeong Stream flowing in the center, and irrigation channels related to farming are developed. The study area is 36.8 square kilometers which consists of 48 villages, the number of households is 1,500, and the population is 2,800.

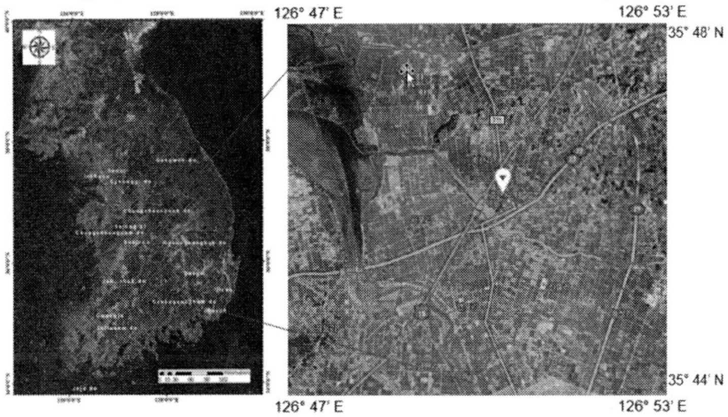

Figure 1. Location map of the study area.

2.2. Materials

The aerial images of the UAVs used in this study were collected using a small, lightweight fixed wing UAV (eBee, Parrot, France) with a body weight of around 700 g. This aircraft can automatically set the flight path and is equipped with an automatic landing function, so it is safe and easy to take images. The vegetation index was measured using a sensor (S110, Cannon, Japan) capable of photographing green, red, and near-infrared (center wavelength: green 525 nm, red 625 nm, NIR 850 nm). The aerial photographs were taken three times (28 May, 28 August, and 16 October 2018), and the spectral characteristics of the sites were investigated using a spectroradiometer (ASD).

The captured images were combined with location information on the elevation, direction, etc. of the aviation body using an image pre-processing program (Emotion, Parrot, Swiss) (Image Tagging). Image synthesis and reflectance were calculated using mosaic image and reflectance image in the Pix4D mapper (Pix4D, Swiss) program. Figure 2 shows the process of the image being taken by the UAV and multispectral sensors [3]. Table 1 shows the specifications of the employed UAV system and sensors. The UAV measurement accuracy was minimized by using the FM and the GPS-RTK method.

IOP Conf. Series: Materials Science and Engineering **611** (2019) 012086 doi:10.1088/1757-899X/611/1/012086

2.3. Methodology

The farm map (FM) is a cropland electronic map that is consistent with satellite or aerial images, filling the need for the production of a map of agricultural land with the identification of specific crop fields. The need for an FM was raised due to the inconsistency between the cadastral map of South Korea and the actual farmland boundaries. The accuracy of the FM image reading was improved by a professional reader who read the image data and divided the boundaries. The image reading was separated into seven land use classifications: rice paddy, crop field, orchard, facility, ginseng planting area, resting area, and abandoned farming area. Figure 3 shows the flow chart of the research conducted using UAV images and the FM.

Figure 2. The process of the image taken by UAV and sensors

Table 1. UAV (unmanned aerial vehicle; eBee) specifications

Item	Specifications
Weight	690 g
Size	960 x 800 x 160 mm
Material	EPP foam
Sensors	GPS, IMU, Pitot
Image sensor	Cannon Powershot S110
	(Red, Green, Blue, NIR)
Max flight time	50 min
Flight altitude	260m
Ground resolution	0.08m

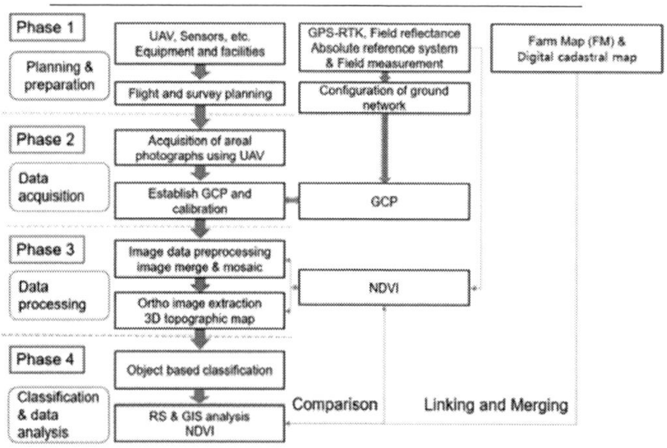

Figure 3. Flow chart of field measurement and analysis

The growth conditions of soybean and paddy rice were investigated by selecting 30 parcels. Field surveys were conducted, investigating nine items such as plant length and height. The NDVI was calculated as the difference between the reflectance in NIR and Red wavelengths, as shown in Eq. (1):

$$\mathrm{NDVI} = \frac{NIR - Red}{NIR + Red} \quad (1)$$

where *NIR* is the near-infrared wavelength band and *Red* is the visible red wavelength band. The NDVI values ranged from −1 to +1.

3. Results and Discussions

3.1. Precision Farmland Mapping

The cadastral map is a plan map made by the state to indicate the location, lot number, ground point, and boundaries of farmland. However, the cadastral map in the agricultural field has the problem of not showing accurate agricultural boundaries. Most of the study area is agricultural area, but it is difficult to grasp the cultivation area and yield because land ownership relation is shown separately from the cadastral map of cropland, simply because land ownership relations are complicated. Therefore, we used the FM, which is a farmland electronic map that divides the boundaries of agricultural fields and agricultural realities through aerial imagery, satellite imagery, and field survey. The FM consists of major items such as the PNU code, parcel coordinate, read code, readjustment attribute of arable land, and parcel area (Figure 4). We used the WGS 1984 UTM Zone 52 as the coordinate system of the UAV for the unification of the coordinate system. Figure 5 is a DSM (digital surface model) composed of synthesized RGB images captured by UAV in the study area. The FM showed very high accuracy because the study area was composed mostly of agricultural areas.

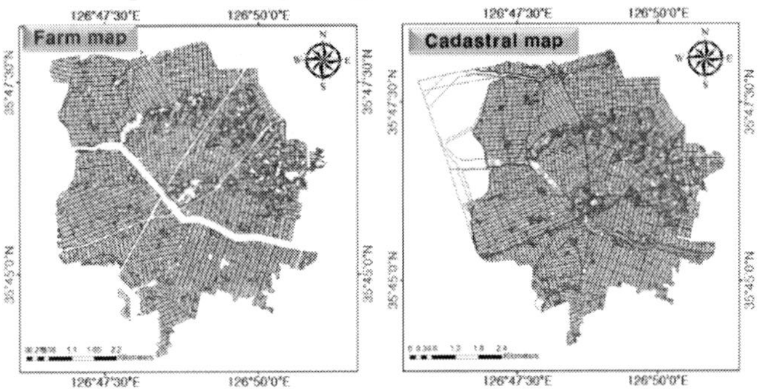

Figure 4. Comparison of the farm map (FM) and cadastral maps boundaries of farmland.

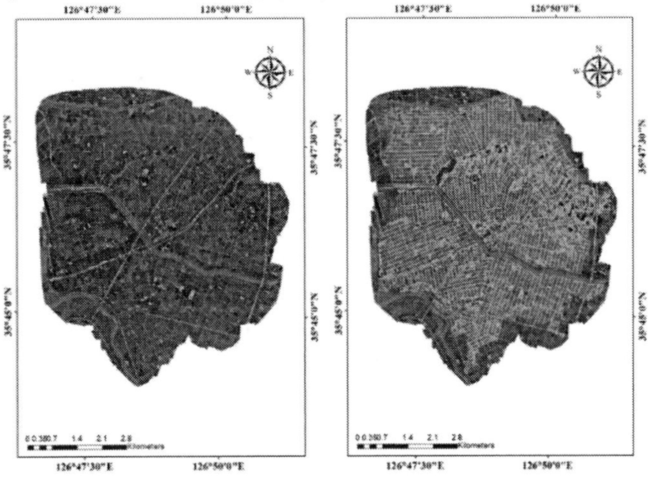

Figure 5. Comparison of the farm map (FM) and digital surface model (DSM) of RGB images acquired by UAV.

3.2. Variation Characteristics of Vegetation Index of Soybean and Paddy Rice

The growth characteristics of soybean vary greatly depending on cultivation area and seeding time (Figure 6). In this study, the vegetation index (NDVI) analysis was determined for two different growth stages (R4 and R7) of vegetation activity. The sowing time was similar for across the area but the difference was about 15 days depending on previously cultivated crops and cultivars. In seeds sown in mid-June showed a higher vegetation index than those sown in late June. The NDVI was close to 0 at the early stage of growth. In addition, 2018 was considered to be largely characterized by the effects of high temperature. The growth characteristics of soybean and paddy rice cultivation areas were analyzed by field survey using a spectroradiometer (ASD) (Figure 7) and the NDVI was determined using images taken by UAV. As a result, there was a slight difference in each parcel, but the field survey value obtained by the spectroradiometer (ASD) tended to be higher than that obtained by UAV. These results indicate that the observation distance is high and the atmospheric effect between the sensor and soybean field affects the accuracy of the information obtained by UAV.

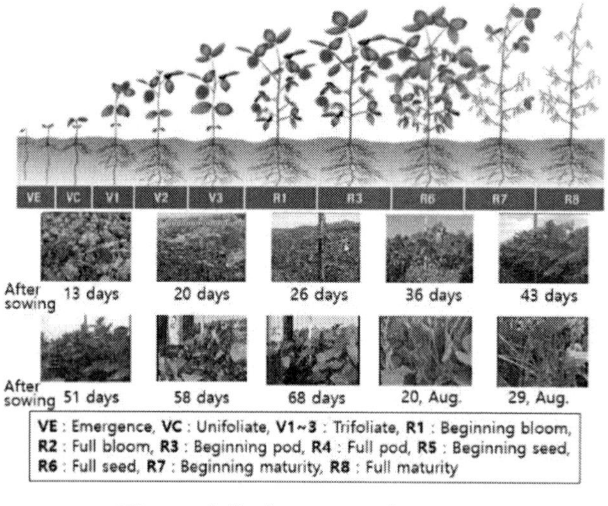

Figure 6. Soybean growth stages.

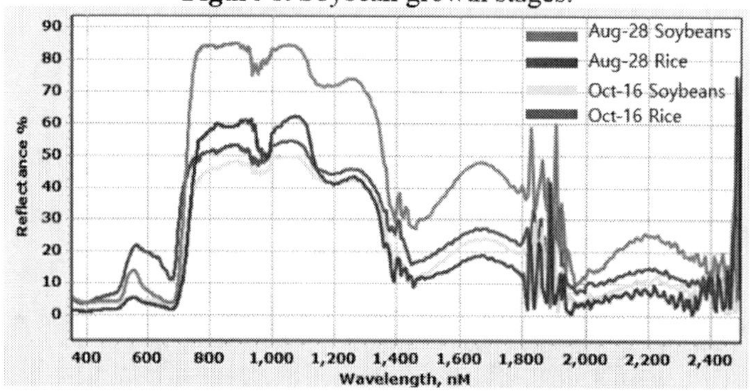

Figure 7. Spectral reflection characteristics of soybean growth stages.

3.3. NDVI of UAV and FM Merged Imagery

R4-NDVI obtained by the merging of UAV and FM images showed very good growth characteristics of soybean (Figure 8). The R4 growth stage is the period in which the soybeans are formed full pods and the activity of soybean plants is the most active. The NDVI of this period showed a very high value overall (Figure 9). The NDVI distribution histogram taken at the R7 stage showed that the soybean plants were mature and the soybean activity decreased, so the NDVI was lower than that in the R4 stage (Figure 10).

IOP Conf. Series: Materials Science and Engineering **611** (2019) 012086 doi:10.1088/1757-899X/611/1/012086

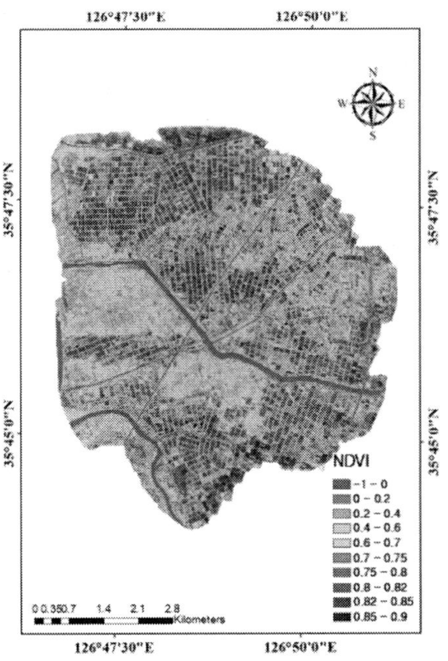

Figure 8. Spatial distribution of NDVI measured by UAV.

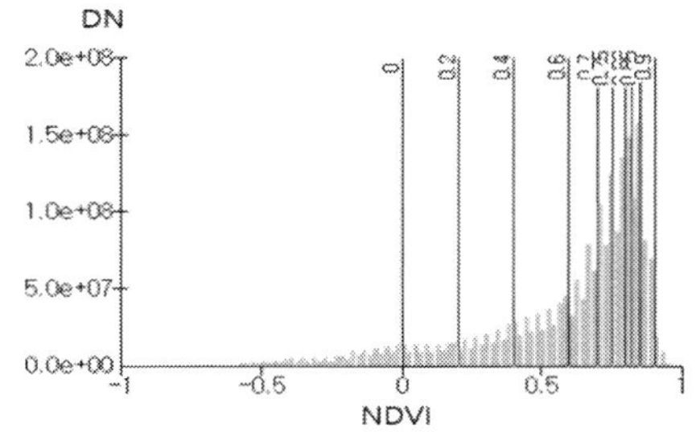

Figure 9. NDVI distribution histogram of the R4 growth stage (28 August 2018).

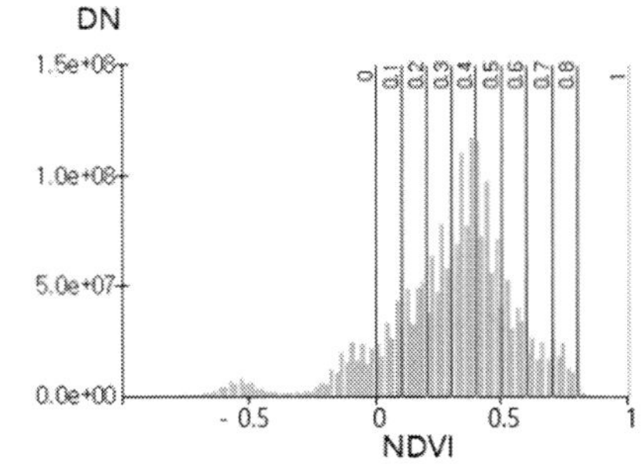

Figure 10. NDVI distribution histogram of the R7 growth stage (16 October 2018)

3.4. Extraction of Cultivation Area Using a Hybrid (Hierarchical and Object-based) Classification Technique

Table 2. Error matrix analysis results.

Classification		UAV image classification: Bean (Yellow)	Bean (Green)	HARVEST	Rice	Row Total	User Accuracy(%)
Field Survey	Bean (Yellow)	548	23	1	70	642	85.36
	Bean (Green)	1	121	-	2	124	97.58
	HARVEST	3	-	81	4	88	92.05
	Rice	67	4	11	242	324	74.69
	Column Total	619	148	93	318	1,178	
	Producer Accuracy(%)	88.53	81.76	87.10	76.10		
Overall accuracy(%)		84.21					
Kappa Index		0.75					

The classification of crops was made by object classification based on the spectral characteristics of Red, Green, Blue, and the NDVI for each parcel. Since Seopo-ri, Juksan-myeon, Gimje-si cultivates only two crops (soybean and paddy rice), the object class was divided into four classifications. An accuracy analysis was performed using an error matrix. This method is mainly used to evaluate the accuracy of land cover classification based on satellite and UAV images. The classification accuracy was correct for 992 parcels out of 1178 parcels. The overall accuracy was 84.21% and the Kappa coefficient was 0.75. The classification accuracy showed Bean (Green) > Harvest > Bean (Yellow) > Rice. The accuracy was 97.58% > 92.05% > 85.36% > 74.69%, respectively. Therefore, the classification accuracy of soybean, which showed normal growth, was the highest.

IOP Conf. Series: Materials Science and Engineering **611** (2019) 012086 doi:10.1088/1757-899X/611/1/012086

4. Conclusion

Vegetation monitoring is very important to evaluate current growth conditions and to identify potential vegetation vitalities. Due to their low cost, high precision, and simplicity, UAVs have been widely used in the evaluation of agricultural fields and the quantification of growth conditions. UAVs are gaining popularity for the purpose of vegetation cover estimates, as they can reduce the effort, time, and additional interpretation of fieldwork in the future. We acquired NDVI values from UAV image data responding to site-based surveys. This study was carried out to investigate characteristics of growth stages of soybeans and to investigate the effects of conventional field survey methods compared to UAV techniques. We found that it is important to know the exact timing of the greatest growth change and to know how to use various combinations of information to determine growth characteristics. This study shows that the area of cultivated paddy rice and soybean can be identified accurately. In order to predict the growth characteristics and yield of soybean and paddy rice, the proposed UAV image and FM integrated method can be very useful.

Acknowledgments

This work was carried out with the support of "Cooperative Research Program for Agriculture Science & Technology Development (Project No. PJ014049022018)" Rural Development Administration, Republic of Korea.

References

[1] Park J K and Park J H 2015 *Crops classification using imagery of unmanned aerial vehicle (UAV)*, Journal of the Korean Society of Agricultural Engineers, vol.57, issue 6, pp. 91-97.

[2] Park J K and Park J H 2016 *Applicability evaluation of agricultural subsidies inspection using Unmanned Aerial Vehicle*, Journal of the Korean Society of Agricultural Engineers, vol.58, issue 4, pp. 1-9.

[3] Francesco N and Fabio R 2014 *UAV for 3D mapping applications: a review*, Applied Geomatics, vol. 6, pp.1–15.

IOP Conf. Series: Materials Science and Engineering **611** (2019) 012087 doi:10.1088/1757-899X/611/1/012087

Design and Implementation of HVDC System Power Automatic Control

Yi Liu[1,*], Guohui Rao[2], Haoran Zhang[3] and Ruibo Shao[4]

[1,2,3,4] XJ Electric Co. Ltd, Xuchang, Henan 461000, China

*Corresponding author

Abstract. With the increasing number of UHVDC transmission projects in China, the design, construction, operation management and equipment manufacturing level of China's HVDC transmission projects are also in the leading international level. But the equipment of converter station in our country is basically in operation state, and the dispatch adjusts power frequently according to various reasons. Manual power modification by operators is not only cumbersome and prone to make mistakes, but also easy to incur bad records for inquiring. In this paper, through an automatic power technology, the standard power curve of dispatching is imported into the operator control system, which runs automatically through the operator control system. The automatic power curve setting process realized by the method can reduce the manual operation, reduce the probability of misoperation, and improve the operation reliability. Operators are very satisfied with this automatic power curve, which greatly improves the efficiency of converter station.

1. Introduction of Automatic Power Curve Technology

The power control of DC transmission system is a link of the DC transmission control system. It is a control function of the pole control level, which keeps the power transmitted by the DC system at a given value[1]. Usually, the control of the DC power transmission power is achieved by changing the current setting value of the DC current regulator[2].

According to the adjustment of power grid operation mode and the change of daily load, it needs to adjust the transmission power under the arrangement of the dispatching center every day for the actual operation of HVDC project.[3]. The main parameters of the adjustment are the power value that needs to be transmitted at a specified time point.

Power control usually has two modes of operation control: One is manual control, the power setting value and power lifting rate are manually input by the operator. The others is Automatic control, the power setting is automatically changed according to the pre-programmed DC transmission power (usually in units of days). When number of the power adjustment points is relatively large, the computer sends the power plan value according to the time point, which can reduce the workload of the operating personnel.

The main principle of the automatic power realization method is that the power curve is saved in the server after setting, and the corresponding power instructions are sent after the arrival of the time point. In addition to the power setting value, the slope value and the advance value of power need to be calculated during the automatic power control[5].

Three conditions for automatic power curve execution: First, the control level of the execution station of the automatic power curve is the system level master station; Second, the operator control position must be currently controllable; Third: Double-click power mode is automatic mode.

The automatic power plan file is a daily power transmission plan issued by the dispatching center corresponding to a HVDC transmission project[6]. The file format is Excel file. The content of the file is power transmission plan arranged by time points. It usually has one power point every 15 minutes and 96 power points per day.

The production planning workstation is connected to the production planning system of the dispatch center and automatically receives the automatic power plan file (Excel format) issued from the dispatch center[7]-[9].

According to the requirements of the power plan of the dispatching center, the automatic power plan setting workstation completes the setting of the execution sequence of the automatic power control in the station. The main parameters that need to be input are the time and reference value of the target power and the power slope value in the process of adjustment[10]-[14].

The reverse isolation device is used to transmit data to devices located in the station LAN under the condition of ensuring physical isolation of the station LAN network (only text files can be transmitted). In the actual project, the computer that receives the automatic power from the dispatch center is the production planning workstation, and the automatic power is set to the server to be executed by the automatic power plan setting workstation. The two workstations are not allowed to connect to a network for security reasons. In the process of automatic power plan setting, there are two ways to input it into the server at the automatic power plan setting workstation after the operator receives the automatic power curve. The first is manual input after printing, and the second is import after recording files to CD-ROM (for security reasons, it is forbidden to use U disk). The second is less workload than the first. The work flow diagram is shown in Figure 1.

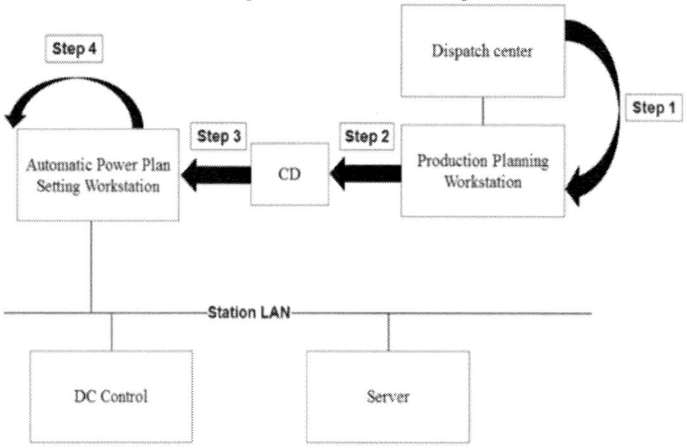

Figure 1. Schematic diagram of the existing automatic power tuning process

At present, the load scheduling is becoming more and more precise. There may be multiple curve adjustments on the same day. It needs to reset the automatic power curve to the server each time, and the process of receiving the curve of the planned workstation is automatic (obtained from dispatching), but there is no hint after receiving, so it needs to be checked and updated manually by the operator, because the power adjustment has timeliness. It needs to check whether there are new curves to be sent out regularly. It also needs to adjust the curves after receiving, which takes up a part of manpower and time.

2. System Design for Automatic Power Curve

This paper provides a method of automatic power curve import. This method detects new power curve on the production planning workstation, transfers the curve conversion format to the automatic power plan setting workstation through isolation device, and imports the curve to the server for execution after confirmation by operators. The logic diagram of automatic power curve import in this paper is shown in figure 2. The implementation method includes the following steps:

1) Monitoring the power plan file receiving directory at the production planning workstation. When a new power plan file is detected, the content of the file is read and converted into a text format file, which is stored in the transmission directory of the reverse isolation device.
2) The text format file is transmitted to the automatic power plan setting workstation by the reverse isolation device.
3) In the automatic power plan setting workstation, the file transfer directory of the reverse isolation device is monitored. When the new file transfer is detected, the operator is notified by event mechanism.
4) After the operator receives the event notification, the newly received files are imported into the server on the automatic power plan setting workstation (the reason for manual import is the regulation requirement, and the command import must be confirmed by the operator on duty). The automatic power control is executed on time.
The purpose of this paper is to simplify the automatic power curve setting process, reduce the workload of operators and improve work efficiency through format conversion and automatic transmission based on the traditional automatic power function.

3. Implementation of the Way

3.1. Design Scheme

In this chapter, we will talk about the technical scheme in detail with the illustration. Automatic power plan automatic import implementation example:
Suppose the dispatch center sends an automatic power curve plan for a certain day, and the file format is an Excel file. This file is automatically transferred to the production planning station of the converter station through the production planning system (step 1 in Figure 3). The steps to perform the automatic power curve in the station are as follows:

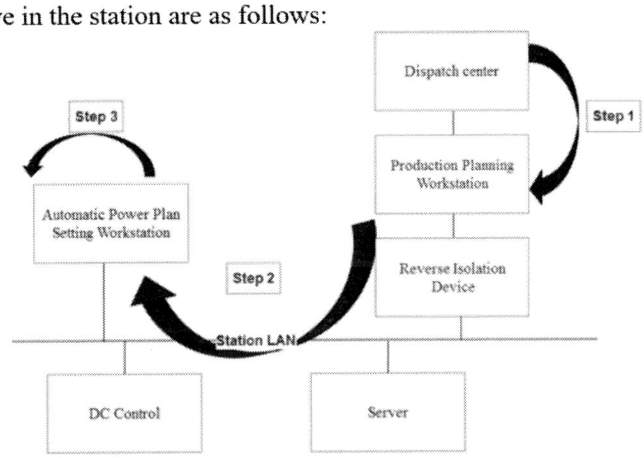

Figure 2. Schematic diagram of the improved automatic power tuning process.

- Perform new file detection on the production planning workstation. After the file in the embodiment is successfully received, start the file format conversion and save the converted file to the reverse isolation device configuration directory;
- Automatic transfer the file to the automatic power plan setting station by the reverse isolation device;
- Perform new file detection on the automatic power plan setting workstation, and send an event notification to the operator when the new file transfer is completed;
- The operator who receives the event notification, through the original function or common means, confirms that the file is correct and then imports the curve into the system, and the system executes according to the time point.

In the above process, except that the last step must be manually confirmed according to the procedure, the other steps are automatically completed, and no manual intervention is required after the configuration is completed.

3.2. Specific Implementation

Example: automatic power table dispatched by dispatch center (daily) is shown in table 1.

time	Power value
750	18:45:00
1000	19:45:00
1524	20:30:00

Table 1. An example of a daily power schedule

When the system auto power has been set, this is a example of the system auto power curve.

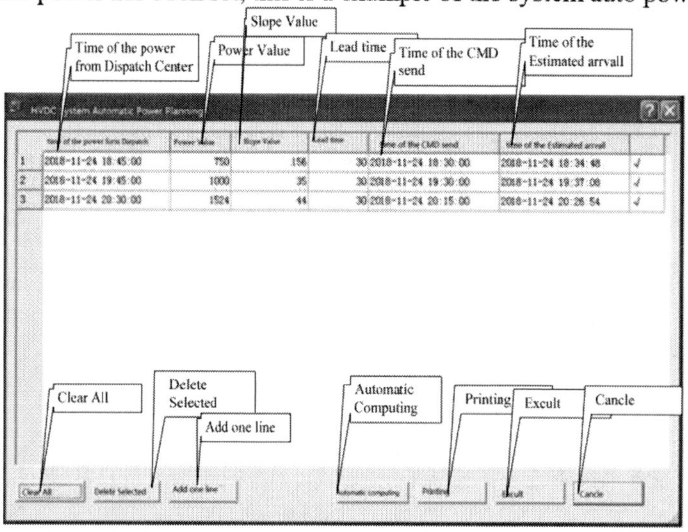

Figure 3. Automatic power schedule generated by the operator control system.

This system provided a function named "copy today's curve to tomorrow", according to that the curve between today and tomorrow should be the same or only a little change. Using this function, little work should be done by the operator and less error should be made by the operator. The principle of this function is only to modify the time of the next command, that time increased by 24 hours.

4. Summary

Based on the vigorous development of UHVDC transmission project, the number of converter stations is increasing, and the number of HVDC operators is reduced. It takes three to five years to train a qualified HVDC operator. Therefore, reducing the burden of operators and making the monitoring system intelligent and automated are the subjects that our converter stations have been studying. According to the system designed in this paper, the automatic power data sent by dispatch center can be converted into text files, and then imported into the monitoring system through the way of import. Then the relevant reference values of the automatic power transmitted by the monitoring system are raised at the corresponding time, and the corresponding information is executed and fed back by the control system.As a new technology, the automatic power curve technology discusses the method and design steps of the automatic power curve in a small space, and has done related experiments.

According to the implementation method described in the present paper, the automatic power curve of the dispatch center is converted into a text format in the implementation method, and automatically transmitted to the automatic power plan setting workstation through the reverse isolation device.

However, with the vigorous development of UHVDC transmission in our country and the increasing investment in the Internet of Things in the power grid, we need to constantly design intelligent control and intelligent products to meet the growing needs of users.

References

[1] Multiterminal HVDC with thyristor power-flow controller. Veilleux, Etienne,Ooi, Boon-Teck. IEEE Transactions on Power Delivery. 2012
[2] Prospects of multilevel VSC technologies for power transmission. Gemmell B,Dorn J,Retzmann D,et al. IEEE PES Transmission and Distribution Conference and Exposition . 2008
[3] Modulares Stromrichterkonzept für Netzkupplungsanwendung bei hohen Spannungen. Marquardt R,Lesnicar A,Hildinger J. ETG-Fachtagung . 2002
[4] New Concept for High Voltage–Modular Multilevel Converter. MARQUARDT R,LESNICAR A. Proceedings of the 35th IEEEAnnual Power Electronics Specialists Conference . 2004
[5] Caprivi Link HVDCInterconnector: Comparison Between Energized System Testing andReal-Time Simulator Testing. MAGG T G,MANCHEN M,KRIGE E,et al. 2012 CIGRE Session . 2012
[5] Caprivi Link HVDCInterconnector: Comparison Between Energized System Testing andReal-Time Simulator Testing. MAGG T G,MANCHEN M,KRIGE E,et al. 2012 CIGRE Session . 2012
[6] Power Flow Control in DC Distribution System. Mohamed A,Mohammed O. Proceedings of theNorth American Power Symposium (NAPS) . 2010
[7] Mluti-terminal HVDC grid with power flow controllability. Veilleux E,Ooi B T. CIGRE Paris . 2012
[8] A current flow controller for use in HVDC grids. Barker C D,Whitehouse R. The 10th International Conference on AC and DC Power Transmission . 2012
[9] Planning theNext Nelson River HVDC Development Phase Considering LCC vs.VSC Technology. JACOBSON D A N,WANG P,KARAWITA C,et al. 2012 CIGRE Session . 2012
[10] Power Flow Control in DC Distribution System. Mohamed A,Mohammed O. Proceedings of theNorth American Power Symposium (NAPS) . 2010
[11] Mluti-terminal HVDC grid with power flow controllability. Veilleux E,Ooi B T. CIGRE Paris . 2012
[12] A current flow controller for use in HVDC grids. Barker C D,Whitehouse R. The 10th International Conference on AC and DC Power Transmission. 2012
[13] Planning theNext Nelson River HVDC Development Phase Considering LCC vs.VSC Technology. JACOBSON D A N,WANG P,KARAWITA C,et al. 2012 CIGRE Session . 2012
[14] Trans Bay Cable—World's First HVDC System Using MultilevelVoltage-Sourced Converter. WESTERWELLER T,FRIEDRICH K,ARMONIES U,et al. 2010 CIGRE Session. 2010

PECVD SiO_2/Si_3N_4 Double-layer Electrets for Application in Micro-devices

Shumin Wang, Yapin Zhang, Jin Liu, Xudong Zou and Jinwen Zhang*

Institute of Microelectronics, Peking University, Beijing 100871, China

*Corresponding author

Abstract. In this paper, the influence of different environmental conditions, substrates, micromachining processes, areas and patternings on the performances of PECVD SiO_2/Si_3N_4 double-layer electrets were studied for its application in micro-devices. Various samples were prepared and then charged by the negatively corona charging method. The charge decays at either 250°C or 95%RH were observed to reveal the chargeability and charge stability. Finally, a micro power generator with patterned PECVD SiO_2/Si_3N_4 double-layer electret was presented briefly as demonstration. The results show that all these conditions that conventionally or even unavoidably happen in the fabrication process of electret micro-devices had more or less impact on the electret performance. It is crucial that the electret should be as large area as possible and kept from the micromachining processes and humid condition to the best. Our micro power generator with 2mm rectangular array of PECVD SiO_2/Si_3N_4 double-layer electret had the 5.9μW output power at 20Hz and 0.7g.

1. Introduction

Due to its stable charge storage and other useful properties, electrets have been widely applied in electric-acoustic transducers [1], electret filters [2], electret dosimeters [3], and etc.. Silicon-based inorganic electrets, such as SiO_2, Si_3N_4 and SiO_2/Si_3N_4 double–layer, have been well studied [4-6] for their high charge stability and compatibility to IC and micromachining technologies. Especially, it was proved that the SiO_2/Si_3N_4 double–layer electrets had better performance than other two single layer ones [7].

SiO_2/Si_3N_4 double layers are prepared traditionally by thermal oxidation or APCVD/LPCVD on silicon substrates. Compared to these high temperature deposition methods, PECVD has the advantages of relatively high deposition speed, low temperature, low residual stress and compatibility with both silicon and non-silicon substrates. Our primary experimental results had proved the good chargeability and stable charge storage of PECVD SiO_2/Si_3N_4 double–layer electrets on glass substrates [8-10].

However, when the electret is utilized in micro-devices its area is much smaller than that in material property research and it even has to be patterned such as rectangle array, which may influence its performance significantly. Additionally, the electret must go through various micromachining processes inevitably for the micro-device fabrication. In this paper, we mainly investigate the influence of different substrates, micromachining processes, areas and patterning on the performance of PECVD SiO_2/Si_3N_4 double-layer electrets. Finally, a micro power generator with patterned this kind of electret was presented briefly as demonstration.

2. Samples Preparation

4-inch Pyrex-7740 glass wafers were used as substrates. Only the sample for comparison of different substrate materials was prepared on silicon substrate. Cr/Au (30nm/100nm) as lower electrode were sputtered at first. Then, 1μm SiO_2 and 50nm Si_3N_4 were deposited by PECVD in turn. For all samples, SiO_2 was deposited at a speed of 120nm/min, and Si_3N_4 at a speed of about 15nm/min at 300°C (substrate temperature). Finally, Al (500nm) as back electrode was sputtered on the back of the substrates to form a good contact with ground.

All these wafers were cut into 3cm×3cm pieces as samples except for those for studying on the effect of different areas and patterning on the electret performance. Samples of different areas are 1cm×1cm, 2cm×2cm, 3cm×3cm and the whole 4inch wafer. The rectangle arrays of electret are made of same width and gap of periodic stripes, including 1mm, 500μm, 200μm, and 50μm, on 1cm×1cm substrate.

Table 1. Different conventional micromachining processing conditions.

Process	Condition
Oxygen plasma stripping	10min
Stripping with acetone	10min
Fuming nitric acid stripping	15min
H_2SO_4 and H_2O_2	100°C, 30min
H_3PO_4, HAc and HNO_3	39°C, 30min
I_2 and KI	33°C, 30min

The micromachining process for the electret patterning can be mainly divided into two groups: (1) photolithography with different photo-resistor (PR) stripping processes, such as oxygen plasma, acetone and fuming nitric acid; (2) conventional cleaning or wet etching processes, including H_2SO_4 and H_2O_2 (common cleaning liquid), corrosive of Au (I2 and KI) and corrosive of Al (H3PO4, HAc and HNO3). The processing conditions are listed in table 1. After wet processing, all samples were flushed by DI water and then dried by nitrogen gas.

Before charging, all samples were ultrasonically cleaned and dehydrated at 150°C for one hour. Then, they were stored in silica gel desiccators at room temperature to be insulated from moisture and contaminants in outside. These samples were negatively corona charged at 80°C for 30min with only the back electrode grounded. The tip voltage is -5kV, and the grid voltage is -400V. The distance between the tip and sample surface is about 15mm, and the grid is 2mm above the sample.

3. Measurement Methods

The Surface potential was measured by Trek 347 voltmeter with TRE-6000B-7C probe. Five points of every sample were measured and the average surface potential was calculated. The absolute value of the surface potential was adopted in all figures and context of this paper for convenience. The average surface potential after 30min heat treatment (V_{30min}) at 250°C was used as initial surface potential. Then, all samples went through 250°C heat-treatment for 300min in the oven and the surface potential was measured every 30min (V_t) except for the long-time-observation samples. The normalized data were obtained by V_t/V_{30min} for comparing the charge stability of different samples. At last, High humidity charge decay was operated at 95%RH, 25°C and surface potential was measured every 30min for all samples except the long-time-observation sample. According to our previous research [8], the performance of PECVD SiO_2/Si_3N_4 double-layer electrets in high humidity condition can be improved by heat treatment significantly. Therefore, all the samples were heat treated at 250°C for 3hr before high humidity charge decaying. And so the normalized data were obtained by V_t/V_{3hr}.

4. Results and Discussions

4.1. Surface Potential during Long Time Observation

The surface potential of samples charged at typical conditions was observed at room temperature for 38 days and measured once several days (figure 1). They were kept in three kinds of circumstances of: (I) silica gel desiccators except measuring time; (II) outside; (III) outside after 250°C 3hr heat treatment.

After 38 days, samples of (I), (II) and (III) still had 89%, 61.5% and 95.5% of their initial surface potential, respectively. The surface potentials of both samples (I) and (III) descended slowly and linearly, but that of sample (II) decreased rapidly at first 5 days and then the trend slowed down and became linear. Sample (II) has the highest drop in initial surface potential up to 40V/day, revealing that ions and water molecules in outside can impair the charge stability effectively. After several days, its dropping velocity decreased and kept about 1.3V/day. Although the surface potential of sample (III) dropped just 0.3V/day much slower than 1V/day of sample (I), sample (III) paid more than 150V surface potential decreasing during 250°C 3hr heat treatment as cost. Sample (II) always shows fastest charge losing than others and even lower surface potential than sample (III) after one month. Therefore, if the PECVD SiO_2/Si_3N_4 electret has to work outside proper heat treatment must be done.

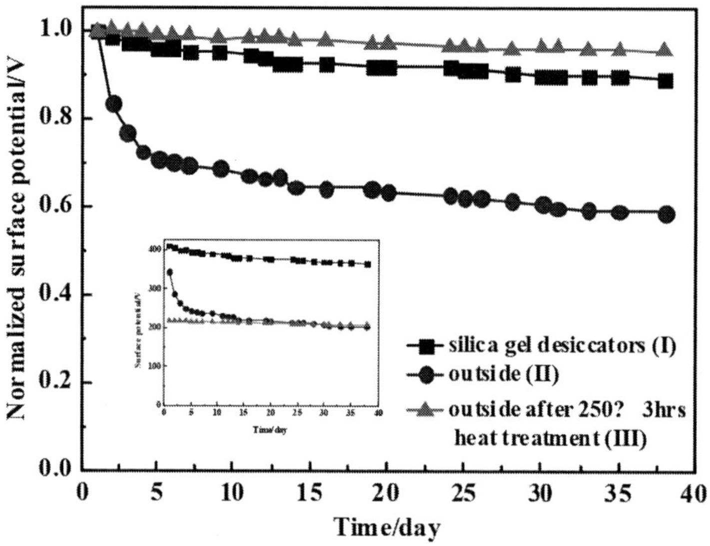

Figure 1. Normalized charge decay of three samples kept in different environmental conditions (inset is charge decay curve).

4.2. Influence of Different Substrates

The silicon and glass are two main kinds of substrates in micro-devices. Figure 2 exhibits that PECVD SiO_2/Si_3N_4 double-layer electret on glass has the similar charge stability at high temperature and humidity with that on silicon and even a little better chargeability (inset of figure 2(a)). It indicates that substrate has no impact on the performance of PECVD SiO_2/Si_3N_4 double-layer electret.

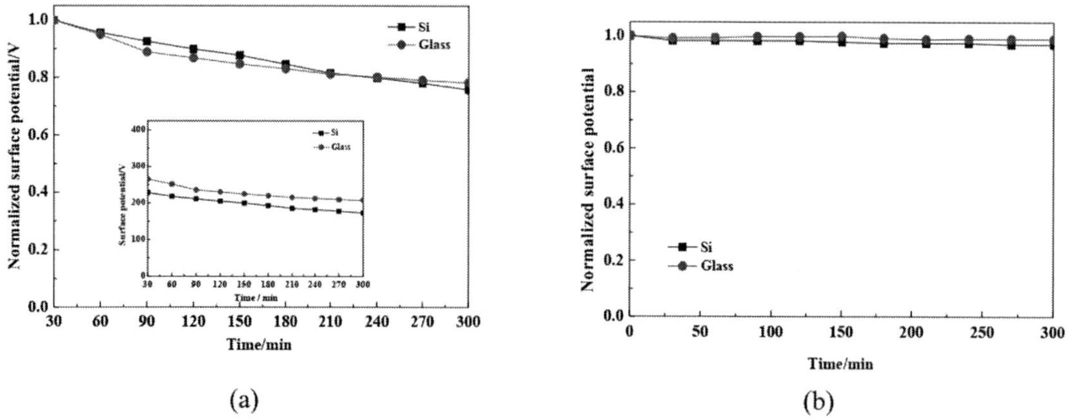

Figure 2. Normalized charge decay of PECVD SiO_2/Si_3N_4 double-layer electret prepared on silicon and glass substrates: (a) at 250°C (inset is charge decay curve); (b) at 95%RH.

IOP Conf. Series: Materials Science and Engineering **611** (2019) 012088 doi:10.1088/1757-899X/611/1/012088

4.3. Influence of Different Micromachining Processes

Figure 3 shows the charge decay curves of samples processed by different PR striping techniques and that of unprocessed one. Initial surface potentials (V30) of processed samples were much lower than unprocessed one (inset of figure 3(a)), indicating that these PR stripping processes decrease the chargeability of SiO_2/Si_3N_4 double-layer electret. Although they have no reaction with SiO_2/Si_3N_4, acetone and fuming nitric acid may impact on the surface state resulting in the trapped charge decreasing. Whether in bulk or on the surface, it is well known that ions can depress the performance of electret seriously. The fuming nitric acid processing made the surface potential decrease much more than the acetone, because it is unavoidable that the water in air or attached on the sample surface was absorbed and ionized somewhat. As can be seen from the figure, samples processed by acetone and fuming nitric acid have similar charge stability at high temperature and humidity with unprocessed samples. Whereas, both the surface potential and the charge stability were decreased significantly by the oxygen plasma stripping processing. The reason may be that the oxygen ion can bombard the sample surface for over-etching and possibly left as recombination defects.

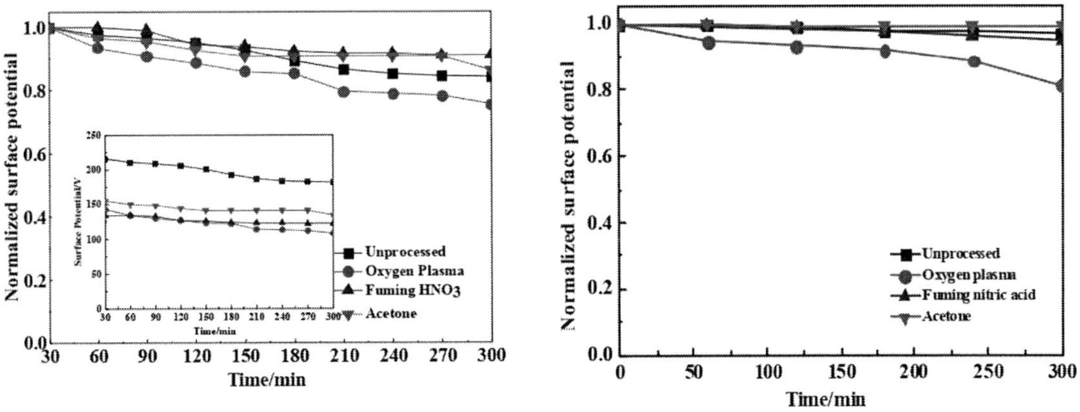

(a) (b)

Figure 3. Normalized charge decay of samples processed by photolithography with different stripping techniques: (a) at 250°C (inset is charge decay curve); (b) at 95%RH.

Figure 4 shows the charge decay curves of samples processed by conventional cleaning or wet etching processes and that of unprocessed one. It can be seen that all of these processes decrease the chargeability of SiO_2/Si_3N_4 double-layer electret (inset of figure 4(a)). And all these samples also have similar charge stability at high temperature and humidity with unprocessed samples. This phenomenon may be explained by above analysis about fuming nitric acid with which these solutions would have similar impact on the SiO_2/Si_3N_4 double-layer electret.

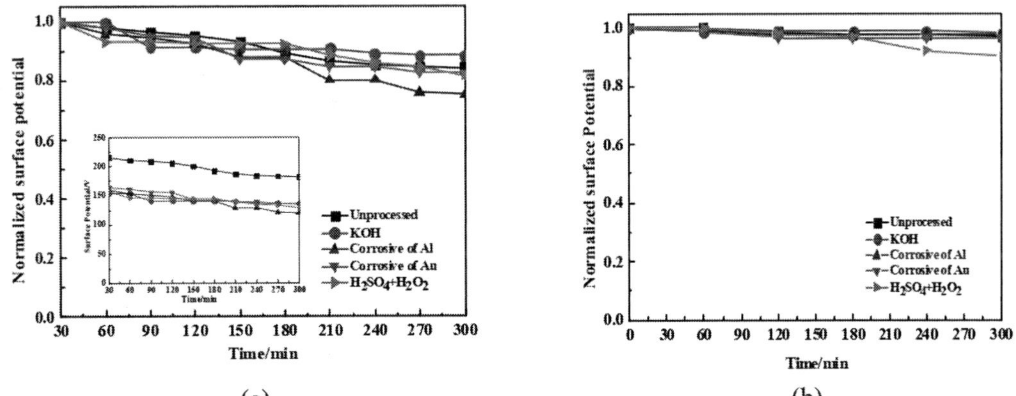

(a) (b)

Figure 4. Normalized charge decay of samples processed by various etching techniques: (a) at 250°C (inset is charge decay curve); (b) at 95%RH.

4.4. Influence of Different Areas and Patterning

4.4.1. Different Areas. It can be seen that the chargeability decreasing with the decreasing area with the order of 4inch wafer (diameter=10cm) the largest of initial surface potential (V_{30}), 3cm×3cm and 2cm×2cm samples the smaller, and 1cm×1cm sample the smallest (inset of figure 5(a)). This phenomenon of may result from the ratio of edge to area. There are three major routes of charge decay: surface, edge and bulk. The closer to the edge of the sample, the easier the charge vanishes. Small area sample has large ratio of edge to area and its charge decays seriously. All four samples exhibit similar normalized charge decay, revealing that there is no floating charge after 30min isothermal charge decaying. And they all kept higher than 95% initial surface potential after 5hr at 95%RH (as shown in figure 5 (b)) indicating that they have high stability in high humid condition. Therefore, decreasing the sample area can depress the chargeability but has no impact on the charge stability.

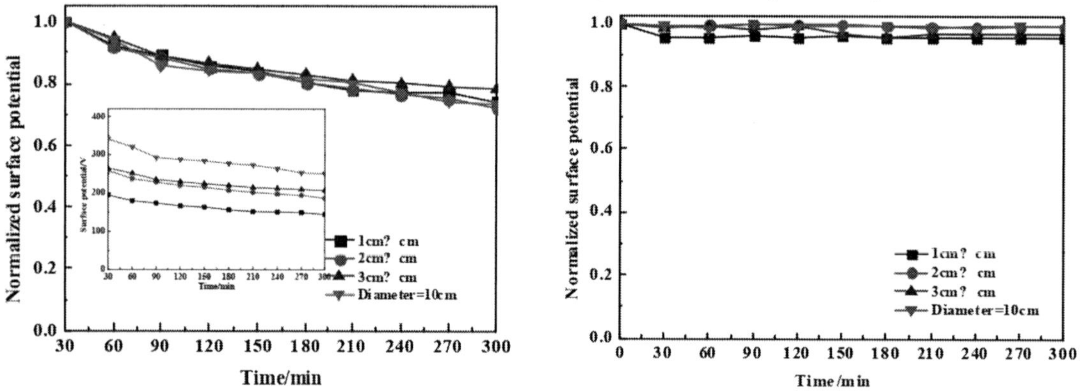

(a) (b)

Figure 5. Normalized isothermal charge decay of samples with different areas: (a) at 250°C (inset is charge decay curve); (b) at 95%RH.

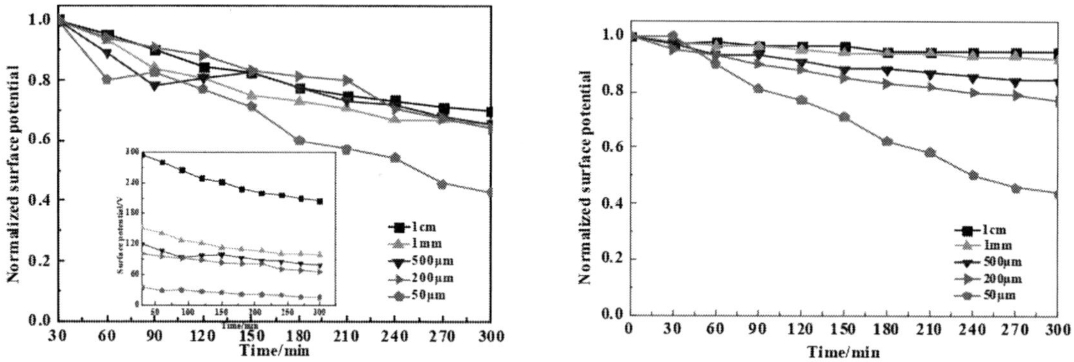

(a) (b)

Figure 6. Normalized isothermal charge decay of samples with different pattern areas: (a) at 250°C (inset is charge decay curve); (b) at 95%RH.

4.4.2. Patterning. The surfaces potentials of the samples with rectangular arrays are much lower than that of without one (i.e. the whole 1cm×1cm electret) (inset of figure 6(a)). The chargeability decrease may result from two aspects. One is that patterning largely increases the ratio of edge to area as discussed above. The other is that the total area of these rectangular array samples is just half of the whole 1cm×1cm one. If the chargeability is same and the charge distributes symmetrically, the total charge and surface potential in any of these samples should be half of that of the whole 1cm×1cm one at best. The sample with 1mm width of rectangular bar did show the surface potential about half of the whole 1cm×1cm one, revealing that they have same chargeability. Whereas, the surface potential of other samples are lower than the half of that of the whole 1cm×1cm one and decrease significantly with the reducing width of rectangular bar. The sample with 50μm width bar only has about 30V surface potential. It can be deduced that the chargeability shows a significant reduction when the width of rectangular bar is smaller than 1mm. Most samples almost have similar charge stability at 250°C except the sample with 50μm width bar which is significantly lower than others (as shown in figure 6(a)). The charge stability at high humidity of the whole 1cm×1cm sample is almost same with that of 1mm width bar one. However, the charge stability at 95%RH decrease obviously with the decreasing rectangular bar width (as shown in figure 6(b)).

Both the chargeability and charge stability were impacted by rectangular bar width although the total area is the same, which is conflicted with the results of figure 6. And the narrower the rectangular bar is the more serious this phenomenon is. This is because that the ratio of edge and area increases with the decreasing of the rectangular bar width. The edge area contains much more defects than the bulk and also play the role like the surface. And it must be considered that the edge of patterned samples had to directly experience RIE and BHF wet-etching techniques to obtaining the rectangular shapes with different widths. These etching processes made the edge area suffered and augmented the defects furthermore.

4.5. Micro Power Generator with Patterned Electret

A micro power generator for vibration energy harvesting from the environment using patterned PECVD SiO_2/Si_3N_4 double-layer as electret was proposed by us [11]. The device was fabricated by simple micromachining technology suitable for batch production. According to above results, the electrets was patterned as strips with a width of 2mm and a length of 27mm. When the oscillation frequency and acceleration are respectively set to 20Hz and 0.7g and the external load resistance is 626kΩ the peak-to-peak charge output of micro power generator is 72nC and the power output is 5.9μW, illustrating that the vibration energy was successfully harvested and transformed to electricity energy by our device.

5. Conclusion

In this paper, we mainly study on the influence of different environmental conditions, substrates, micromachining processes, areas and patternings on the performances of the PECVD SiO_2/Si_3N_4 double-layer electret for its application in micro-devices. Long-term observation implies that hermetic packaging is the best option, and if the electret has to work outside proper heat treatment is necessary. Different substrates have no obvious impact on the performance of electret. Wet processes having no reaction with SiO_2/Si_3N_4 double-layer can make the chargeability decrease significantly but not the charge stability. However, both the chargeability and charge stability can be depressed by oxygen plasma stripping. Decreasing the area of sample can depress the chargeability but has no impact on the charge stability. Both the chargeability and charge stability decreases with the decreasing width of rectangular bar when rectangular bar is narrower than 1mm because of the ratio of edge and area increasing and suffering from the patterning processes of RIE and BHF wet-etching techniques on the edge. A micro power generator with 2mm rectangular array using PECVD SiO_2/Si_3N_4 double-layer electret has the 5.9μW output power at 20Hz and 0.7g.

References

[1] Y. C. Chen, W. C. Ko, H. L. Chen, W. J. Wu, P. Z. Chang, and C. K. Lee. "A Thin Light Flexible Electromechanically Actuated Electret-Based Loudspeaker for Automotive Applications," *IEEE Transactions on Industrial Electronics*, vol. 59, pp. 4140-4147, 2012.

[2] J. Kim, J. P. Hinestroza, W. Jasper, and R. L. Barker. "Application of Electrostatic Force Microscopy on Characterizing an Electret Fiber: Effect of Tip to Specimen Distance on Phase Shift." *Fibers and Polymers*, vol. 12, pp. 89-94, 2011.

[3] T. Maleki, and B. Ziaie. "A Corona-Charged Self-Biased Radiation Dosimeter." *IEEE Electron Device Letters*, vol. 31, pp. 767-769, 2010.

[4] H. Amjadi, "Charge storage in double layers of thermally grown silicon dioxide and APCVD silicon nitride," *Dielectrics and Electrical Insulation, IEEE Transactions on*, vol. 6, pp. 852-857, 1999.

[5] V. Leonov, P. Fiorini, and C. Van Hoof, "Stabilization of positive charge in SiO2/Si3N4 electrets," *IEEE Transactions on Dielectrics and Electrical Insulation*, vol. 13, pp. 1049-1056, 2006.

[6] W. Olthuis and P. Bergveld, "On the charge storage and decay mechanism in silicon dioxide electrets," *Electrical Insulation, IEEE Transactions on*, vol. 27, pp. 691-697, 1992.

[7] H. Amjadi and C. Thielemann, "Silicon-based inorganic electrets for application in micromachined devices," *IEEE Transactions on Dielectrics and Electrical Insulation*, vol. 3, pp. 494-498, 1996.

[8] Z. Chen, Z. Lv, J. Zhang, "PECVD SiO2/Si3N4 double layers electrets on glass substrate," *IEEE Transactions on Dielectrics and Electrical Insulation*, vol. 15, pp. 915-919, 2008.

[9] J. Zhang, X. Zou, and Y. Zhang, "Improvement on the Performance of the PECVD SiO2/ Si3N4 Double-Layer Electrets," *IEEE Transactions on Dielectrics and Electrical Insulation*, vol.18, pp. 456-462, 2011.

[10] Xudong Zou, and Jinwen Zhang, "Study on PECVD SiO2/Si3N4 Double-Layer Electrets with Different Thicknesses," *Science in China, Series E Technological Sciences*, vol.54, pp. 2123-2129, 2011.

[11] Z. Yang, J. Wang, and J. Zhang, "A High-Performance Micro Electret Power Generator Based on Microball Bearing", *Journal of Micromechanics and Microengineering*, vol. 21(6): 065001, 2011.

Precise Inspection Method of Solar Photovoltaic Panel Using Optical and Thermal Infrared Sensor Image Taken by Drones

Jeongsoo Park[1] and Dongho Lee[2,*]

[1]Department of Electronic Engineering, Sogang University, Rep. of Korea
[2]Department of Agricultural & Rural Engineering, Chungbuk National University, Rep. of Korea

*Corresponding author

Abstract. The inspection of the solar panel using the drone has already been put into practical use. However, this method requires an initial investment cost as compared with the conventional method, and it may take a cost such as image processing. In addition, there are challenges such as the ability to carry out the necessary maintenance of the law even if technically possible. Nowadays, the introduction of drones is proceeding in such a way as to check the points of abnormalities through simple flight. The advantage of this method is that safety enhancement and inspection speed are greatly improved compared to the case of performing the inspection by the conventional method. For companies, the use of drones and thermal infrared sensors for inspection can be used to reduce the population and to reduce the problems of recruitment, and to improve the efficiency and reduce the cost by simplifying the inspection procedure with fewer personnel will be. The purpose of this study is to develop a method to accurately detect abnormal points by making thermal infrared images of orthographic images by improving the inspection method through simple flight of drone. The study investigated whether it is possible to precisely check the point where the problem is located under the condition of insufficient solar radiation at the site installed by a small developer. We propose a method that can accurately distinguish between normal and abnormal panel points in images taken using drones and thermal infrared sensors, and how to accurately detect abnormal points by focusing on problem panels. Abnormal points were found to be very accurate and efficient to find and inspect panel more than the temperature range value of each panel.
Keywords: inspection, solar panel, drone, photovoltaic, thermal infrared.

1. Introduction

Global energy demand is growing rapidly with economic growth. Solar energy is growing as a renewable energy source for the realization of a sustainable society that respects the environment. Especially, the solar power generation that is introduced will last more than 20 years after installation, but it is very important to detect and respond to the problem immediately in order to maintain stable generation over the long term. In general, monitoring systems for detecting anomalies are introduced in various forms. So far, the monitoring system is predominantly the threshold value determination. For example, if the power generation falls below a certain value, it is determined that the power generation is abnormal, and the solar power generation manager is notified and checked. However, the amount of power generation depends on various factors such as the season, the time zone, the installation area, and the surrounding environment of the power plant. Therefore, most of the tasks that require analysis and judgment by humans based on the results of the anomaly are lacking in reliability. In this background, we tried to develop a monitoring system that can detect abnormalities by using measured temperature values using drones and thermal infrared sensors and to identify the cause of abnormalities. When the power generation deterioration is detected, the cause (such as a fuse or a breaker, a shadow of a tree and a grass or a deterioration of the solar panel itself) is specified, and a check is made so that the abnormality can be immediately recognized.

Drones are a technique that utilizes the excellent accessibility of drones for surveying, and it solves the problem of workable days, economics and convenience which have been pointed out as problems of conventional aerial photogrammetry. Recognizing the utility of drones, the Korean government published an "Enactment of the Guidelines for Public Survey of Utilization of Unmanned Aerial Vehicles" in February 2018, and instituted institutional procedures to allow drones to be recognized as the outcome of public surveys. The principle of drones surveys is the same as the principle of aerial photogrammetry using the principle of "collinear conditional expression" which has been going on since 1849. However, the drones are characterized by the accuracy of interpretation due to the development of artificial intelligence mapping technique that produces unique economic, universality, accessibility, and interpretation of photographed images.

Photovoltaic (PV) facilities in Asia are often installed on a large scale in homes, factories, or large areas. Especially, 2/3 of Korea's land is composed of mountains, so there are many solar facilities utilizing hilly areas. These areas are disadvantageous in that accessibility such as inspection and repair is disadvantageous, and it is difficult to utilize the measuring equipment and expensive labour costs are incurred. In addition, ground observation is time consuming and efficient and accurate monitoring methods are required to be developed. Image processing methods using drones and thermal infrared sensors are emerging as solutions to these problems. Image processing technology using drones has been developed rapidly in recent years mainly in the field of surveying. These technologies have recently been rapidly applied to agriculture and have begun to be applied to solar power plant inspection [1].

The purpose of this study is to propose a method to detect the problem of solar power generation facility by using difference of thermal characteristics in response to sensor by acquiring high precision orthoimage by attaching optical and thermal infrared sensor to drone.

2. Materials and Methods

2.1. Study Area

In this study, a small scale solar power generation facility was selected as a study area. The study area was a power generation facility applied to a hill located in a rice growing area and consisted of two rows on a slight slope. Figure 1 shows the location of the study area at Ochang-eup, Cheongju, Chungbuk in Republic of Korea.

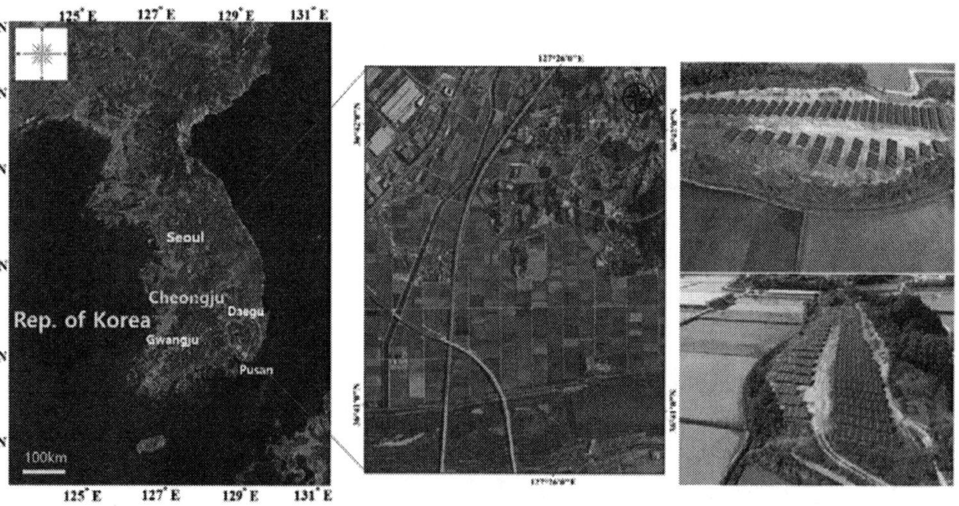

Figure 1. Location of study site

2.2. Materials

The drone used DJI's Inspire 2 (China). The Inspire 2 is a rotor drones that have good stability, long flight times, and payloads that can be equipped with optical and thermal infrared sensors. The optical sensor uses a DJI's Zenmuse X5s (Red, Green, Blue) model with a very high resolution of 5280×2970. The thermal infrared sensor is a product of FLIR's Vue Pro R (USA) model with a resolution of 640 x 512. The two sensors were installed with different GPS devices with different resolution and reaction speed. Image processing was performed using Sensefly's Fix4D Capture and Mapper and the images were processed using R, MATLAB, and GIS tool.

2.3. Causes, Faults and Accidents Occurring in PV Power Generation Facilities

2.3.1 Broken part of panel. In the case of PV generation in the field, wild animals may collide with bases when they are installed in mountainous areas or places with little humanity. Animals such as deer and fox, as well as birds, may collide. There are also cases where the birds are dropped as well as the shit, by dropping them on the panel. There are cases where wild animals eat cables connecting panels and equipment, so you should not install fences or anything.

2.3.2 Discoloration of panel appearance. If there is a scratch that can be seen by the outline, the panel is faintly brown or dark colour, so it is often visible to the naked eye. The reason for discoloration is that the developed heat passes through the part well and the heat accumulates. The portion where this heat is piled up is called "hot spot". Hot spots due to poor connection of electrical circuits and wiring in modules and systems. Hot spot caused by adhesion of bird droppings and fallen leaves. The hotspots are hot at around 90 degrees, and when viewed with an infrared camera, they are reddish, so you can find unexplained failures in the naked eye [2].

2.3.3 Snail trail (Snail-like shape) phenomenon. The Snail trail phenomenon occurs only in silicon solar panels. Small damage (microcracks) that are hard to find with the naked eye. Even if the snail trail occurs, the generation does not go down, so observation is necessary. Micro cracks are known to occur when installing or transporting solar cells and placing heavy objects on the panels.

2.3.4 Solar power plant fire. Solar power generation systems can cause fires due to electrical leakage or shorts due to various factors. The following may be considered as factors that cause a fire. Aging due to age deterioration, initial failure, mounting failure. Fire caused by insulation failure at the end of wiring. Discharge due to wiring touching metal parts such as module mounts. These problems caused by human errors such as "inappropriate area development", "inappropriate design" and "inappropriate construction" are widely confirmed regardless of low or high pressure. In addition, weeds with strong fertility and the effects of plants with a high degree have been reported in many cases, and several cases of fires have been reported among the failure cases.

2.3.5 Necessity of new abnormal phenomenon identification technology. As mentioned above, problems caused by various causes have been grasped by ground observation. If part of the cell connection in the panel is broken or weak, the voltage balance will be broken and the current will be partially concentrated and generate heat. This is called a hotspot. Hot spots are difficult to see visually from the front and are very difficult to see from the back, so we use a thermographic camera to find out where the heat is rising locally.
The difficult part of this inspection is that if the sunshine intensity is weak, the degree of heat generation of the hot spot is low, and it is not only difficult to detect the failure point, but sometimes it cannot be said as defective even if the heat generation point is found. This means that it is necessary to make a decision based on basic knowledge about the operation of the PV system.

2.4. Methodology

The research process proceeded as follows.

IOP Conf. Series: Materials Science and Engineering **611** (2019) 012089 doi:10.1088/1757-899X/611/1/012089

First, select the study area and establish a drone flight plan. Secondly, we monitor the time of inspection through simulation flight to the image shooting point. Thirdly, we use the optical and thermal infrared sensor at the same time to find the altitude and resolution for obtaining the optimal orthographic image. Fourth, by using a drones and a thermal infrared sensor, it is possible to capture the time available for shooting even if the insufficient amount of solar radiation is available. Fifthly, the captured optical and thermal infrared sensor images were analysed by orthoimaging. Sixth, we proposed a method to find the abnormal panel accurately using statistical values through the orthoimage. In the seventh, we proposed a method to accurately locate the problematic points for the abnormal panel found.

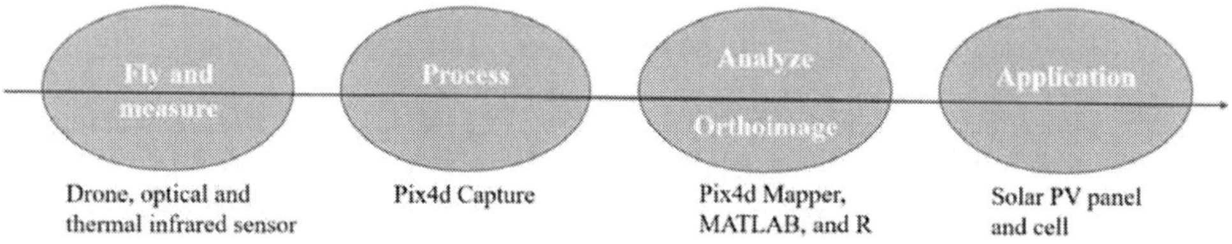

Figure 2. Procedure and application method of research

3. Results and Discussions

3.1. Orthoimages by Optical and Thermal Infrared Sensor

When a PV panel is photographed using an optical and a thermal infrared sensor mounted on a drone, it is possible to construct a topographic map having three-dimensional coordinate information by analysing a plurality of overlapping points. The shooting altitude in the image production process is a factor that greatly affects the resolution of the captured image [3]. Although images with high resolution Ground Sample Distance (GSD) are acquired at low altitude, the length of the flight path and the number of photographic images required for shooting the same area are excessively increased. Therefore, this case greatly affects the photographing operation and the mapping of the images. In this study, the performance of the sensor installed in the drone, the limitation on the current aviation safety law, and the condition of obtaining the optimal image in various flight conditions were investigated at the altitude of 30m. Drones photogrammetry gave vertical and lateral redundancy as high as 85% due to instability of drone posture due to light weight.

Mapping works in the following order. Initial Processing, (2) Point cloud generation, (3) GCP input and correction, (4) Mesh generation process, (5) Orthoimage by optical and thermal infrared sensor, it corrects the geometric distortion such as height difference and tilting of the photographed image, and matches all inputted images, and it is made into an image photographed at right angles to the ground in the air (Figure 3).

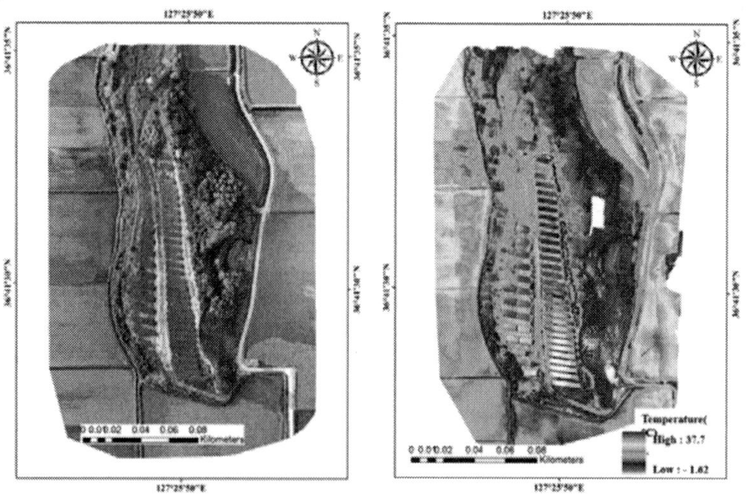

Figure 3. Orthoimages by optical sensor and thermal infrared sensor

IOP Conf. Series: Materials Science and Engineering **611** (2019) 012089 doi:10.1088/1757-899X/611/1/012089

3.2. Examination of Inspection Method of Solar Panel

For the precise inspection and diagnosis of the PV panel, it is necessary to use the photographed images to accurately identify the problematic points and take countermeasures. First, the PV panels were numbered in the same way from the top right to the bottom as shown in Figure 4. Statistical analysis was applied to find the problems in each row of panels. The measured panel data and statistical values of each panel are listed in Table 1, and the problematic panel has a large range value. The range is the value that indicates the spread of the temperature data, minus the minimum value. If the temperature is distributed at a low level, it is confirmed that the temperature at the boundary of the solar cell such as the boundary part is low, or when a problem corresponding to 2.3.1 or 2.3.3 occurs. In these cases, the corresponding arrays are A7 and A17 as shown on the Table 1.

Table 1. The results analyzed by applying the statistical method

	vars	n	mean	sd	median	trimmed	mad	min	max	range
A1	1	9736	20.25	0.62	20.28	20.26	0.48	17.96	25.79	7.82
A2	2	11604	22.19	0.7	22.14	22.21	0.63	19.83	23.84	4
A3	3	13755	21.64	0.51	21.69	21.68	0.53	18.75	23.14	4.39
A4	4	13333	21.63	0.39	21.68	21.66	0.36	15.61	22.68	7.07
A5	5	13527	21.27	0.45	21.35	21.3	0.34	16.76	**29.98**	**13.22**
A6	6	13451	21.09	0.36	21.11	21.1	0.27	11.92	22.8	**10.88**
A7	7	13426	21.13	0.44	21.16	21.14	0.29	14.09	**30.26**	**16.17**
A8	8	12951	21.3	0.31	21.36	21.33	0.23	19.25	22.75	3.51
A9	9	12753	20.8	0.4	20.84	20.81	0.25	19.2	25.36	6.16
A10	10	13364	20.72	0.48	20.67	20.68	0.29	19.63	**31.31**	**11.68**
A11	11	13589	20.45	0.35	20.43	20.43	0.24	18.97	25.09	6.12
A12	12	13642	19.99	0.39	19.95	19.95	0.27	17.26	26.15	8.89
A13	13	13235	19.69	0.36	19.72	19.68	0.34	18.67	24.85	6.18
A14	14	13482	19.47	0.4	19.48	19.44	0.3	18.14	24.49	6.35
A15	15	15789	19.25	0.36	19.23	19.23	0.32	17.18	24.03	6.85
A16	16	15738	19	0.35	18.97	18.98	0.35	17.23	21.31	4.08
A17	17	18176	18.93	0.8	18.85	18.85	0.47	12.56	22.92	**10.37**
A18	18	18395	18.8	0.58	18.72	18.74	0.41	11.69	**25.63**	**13.94**
A19	19	17836	18.49	0.51	18.41	18.44	0.43	15.35	22.64	7.29
A20	20	19936	18.49	0.53	18.43	18.45	0.51	17.39	22.36	4.98

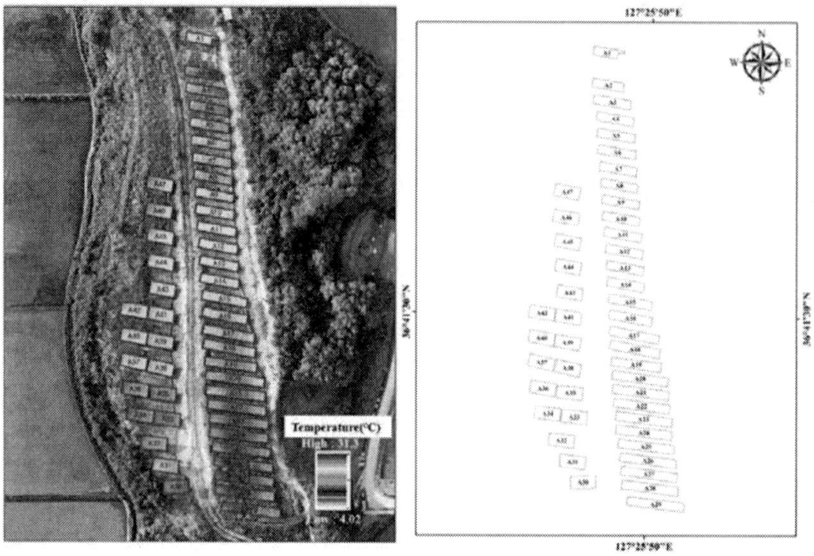

Figure 4. Applied numbering of photovoltaic panels

When the maximum value of the temperature range is very high, it is found that it corresponds to 2.3.2. and 2.3.5. The study site is located at the point where two years have elapsed since the installation, but it is found that A5, A7, A10 among the target panels are in these cases.

3.3. *Precision Review of Abnormality Panel*

Figure 5 shows the comparison between the optical and the thermal infrared image for the abnormal panel obtained through the statistical analysis. Normal operating panels exhibit a homogeneous temperature distribution, but conditions 2.3.1 and 2.3.3 indicate that the panel has a lower temperature distribution than the surrounding panels. The panel corresponding to the hot spot corresponds to the condition of 2.3.2 and 2.3.5, which is relatively higher than that of the normal panel as shown in Figure 5. As described above, the range of the problematic panel, such as abnormal panel, was analysed to indicate 10 or more. Therefore, it is analysed that precise inspection should be performed on panels with a range of 10 or more.

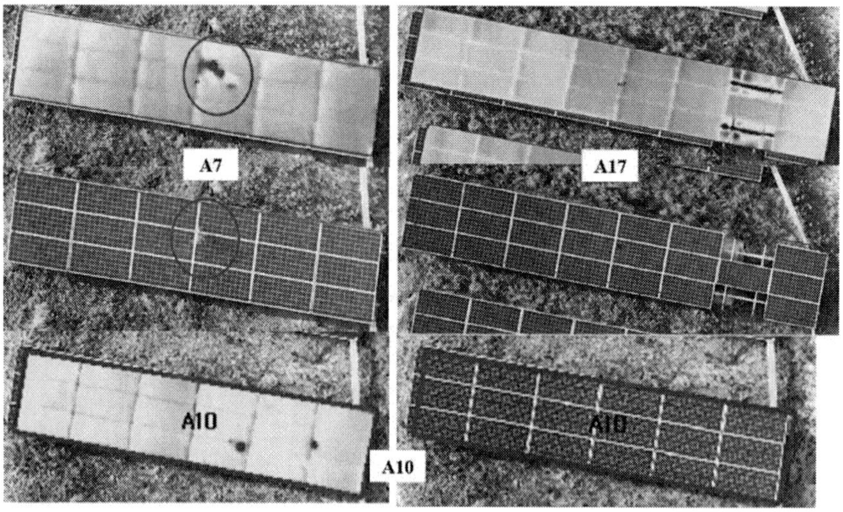

Figure 5. Comparison of optical and thermal infrared images for abnormal panels

4. Conclusion

Precise inspection and diagnosis of solar panels is a very important part of the solar power generation industry, which is in rapid demand and supply. The existing labour-intensive methods are costly, time-consuming, and dangerous, requiring the integration of new technology needs that combine the fourth industrial revolution technology. The purpose of this study was to investigate the problems of solar panels using drones, optical and thermal infrared sensors. As a result of the study, it was analyzed that the thermal image taken by the drone can accurately identify the problematic point and can take corrective action. In particular, the abnormal panel has a very large range of statistical values compared to the normal panel. Therefore, it is concluded that intensive close inspection should be performed on panels with a range of 10 or more.

References

[1] Park, J.K.; Das, A.; Park, J.H. A new approach to estimate the spatial distribution of solar radiation using topographic factor and sunshine duration in South Korea. *Energy Conversation and Management* 2015, 101, 30-39.

[2] Zefri, Y.; Eikettani, A.; Sebari, I.; Lamallam, S.A. Thermal Infrared and Visual Inspection of Photovoltaic Installations by UAV Photogrammetry—Application Case: Morocco. *Drones* 2018, 2, 1-24.

[3] Park, J.K.; Lee, D.H.; Shin, H.S.; Park, J.H. Paddy rice inventory studies using drone imagery on a small town area in South Korea. Proc. SPIE 10777, Land Surface and Cryosphere Remote Sensing IV, 1077718 (22 October 2018).

Integrated Software Development Methodology for Real-time Precise Control of UAV

Jung Kyu Park[1], Eun Young Park[2] and Heung Seok Jeon[3,*]

[1]Dept. of Computer Software Engineering, Changshin Univ., Changwon-si, Korea
[2]Dept. Biomedical Laboratory science, Shinhan University, Uijeongbu-si, South Korea
[3]Dept. of Computer Engineering, Konkuk University, Chungju-si, Korea

*Corresponding author

Abstract. Recently, drones have been widely used in education and industry. In this paper, we propose a navigation and control algorithm to develop unmanned drones quickly. To limit the proposed framework, we analysed the existing drone solution and simulation framework. We also used Hardware-in-the-Loop (HIL) simulation for rapid development. HIL will simulate physical systems that interface with embedded control equipment on real-time hardware. The code generated in this process can be experimented on a hardware device so that the control performance of the drones can be verified without draining the drones in the outdoor environment. Using the proposed method, you save time and money on drone development while providing stability.

1. Introduction

A major issue in the automotive market in recent years was the introduction of autonomous driving technology. From three years ago, self-driving cars have been steadily increasing. Many research institutes are predicting that unmanned vehicles will replace people's vehicles. This technical success can provide many benefits for our lives. The use of this technology also provides safety and efficiency in operations [1].

As mentioned above, the autonomous system can be used in situations where it is difficult for humans to work or in extreme environments where it is impossible to work. Autonomous systems use the latest sensors to process events using collected information, so they can handle more events than people in the same place. For example, in a surveillance system, autonomous systems are more efficient because they can be processed faster by using more information than humans. In addition to the guard system, autonomous systems are being used in a variety of fields such as military operations, crop cultivation, and fire monitoring [2]. Nowadays, there are new problems to be solved as they are used in general fields besides special fields, and studies are being done simultaneously to solve problems.

It can be seen that the representative automobile of autonomous driving is made by the development of new sensor due to the development of IT technology. For example, in the field of animation, you need to be able to bring new advances in the computer and graphics technologies to the market. For this reason, it is recommended that new frameworks be used to exclude existing frameworks in order to develop and validate new research systems. The most useful tool for system development and performance evaluation is a simulation environment for quick feedback [3].

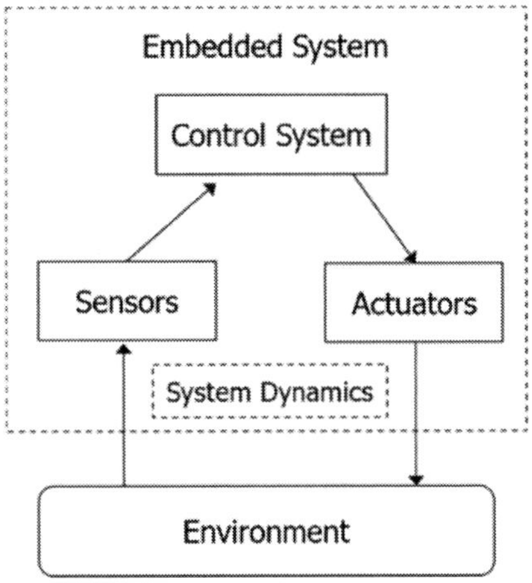

Figure 1. Architecture of embedded system

An autonomous vehicle equipped with a computer for automatic driving can be said to be an embedded system [4, 5]. Unlike general-purpose computer systems, embedded systems are equipped with CPUs and sensors that are specially designed hardware and software that performs control functions and is used for control functions (Figure 1). These embedded systems require special requirements to obtain simulation experiments and verification results. In the general case, when you try to develop a specific algorithm, it is a real requirement to implement a real simulation environment. However, in the embedded field, there are many things that are impossible with the previous implementation. It is important to keep in mind the computer hardware and software capabilities that perform the operation and control of the physical part of the actual system, which may have produced accurate results.

In the simulation environment, the embedded system in Figure 1 can be divided into two areas. At the top, the control system is responsible for control, while the bottom sensor, the actuator, and the environment are responsible for the infrastructure. Simulations can be viewed as data flows in these two areas. In the lower area, the sensor data input from the outside is input and output. In the upper area, the calculation for the control is performed using the given sensor data.

Two simulation methods can be used when the theoretical content is applied to the actual implementation in the control system. The first is the Software-in-the-Loop (SIL) and the next is the Hardware-in-the-Loop [5, 6, 7].

The SIL runs within the computer and consists of programming code. In other words, all simulation elements (input data and output, control) are all simulated within the computer system. To perform this simulation correctly, you need to take the code that runs on the embedded system and run the experiment on a system that can handle this code. This is called SIL because the code responsible for control is part of the software. HIL, on the other hand, uses both the computer that performs the simulation and the hardware device on which the algorithm code is run. The hardware device responsible for the control is called the HIL [8].

The advantage of SIL is that it does not require the expensive embedded board to be tested, and there is no board damage. It also has the advantage of being able to debug easily since it is experimented in the computer. However, HIL has difficulty debugging because it has to conduct a real hardware experiment. The disadvantage is that it is not an experiment on a computer such as a board performance, so it is difficult to predict the performance on an actual board. For this reason, you should know exactly what your application needs when setting up your experiment. SIL can also be useful because of its ease of implementation. HIL can be more difficult than SIL and requires accurate understanding of the hardware, but can be used for accurate performance measurements.

In this paper, HIL simulation was used [9]. This is because the HIL provides an environment for developing new algorithms targeting real hardware and for evaluating performance. While SIL simulation can be performed more easily than HIL, it can be difficult to conduct experiments considering the actual environment. For example, SIL can not catch various bugs, such as jitter bugs or bugs that only appear in certain situations, as they do on hardware.

2. Software System Architecture for UAV

As mentioned earlier, when designing hardware using HIL simulation, various components should be considered. In this environment, the application must be installed on the simulator computer, and the user interface application must be connected to the embedded board. The paper consists of a workstation and a drone board that can perform various experiments. Figure 2 shows the relationship between the internal elements in the proposed system.

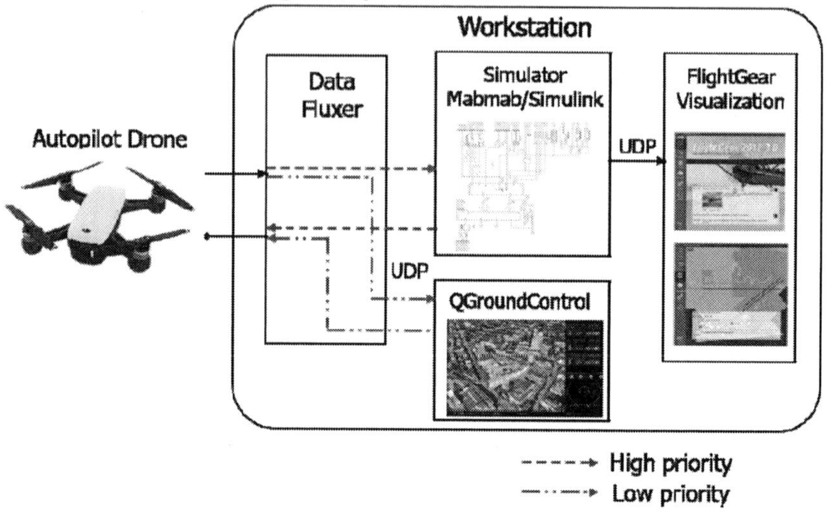

Figure 2. System architecture

The new idea in the proposed method is to split the data flow into two channels using the autopilot board. The data flow associated with the SIL must be of high priority because it simulates "real world" information representing the real-time flow. However, there is little real-time constraint on the information the user communicates.

The two data flows are processed by specialized data flux program. This program is written in C ++, uses *ptask* to perform timing analysis and uses the results to evaluate system performance. We also used the FligthGear Flight simulation program to simulate the virtual environment throughout the system. In the internal design, UDP socket communication was implemented using Matlab / Simulink. Through the implementation of the functions described above, it is possible to visually check the operation of the unmanned drones. The data transfer from the board to the simulation workstation uses the UART interface and the data majors between the programs running inside the workstation used UDP sockets. There are three reasons for using UDP without using TCP on your workstation.

1. No internal connection required
   Typically, TCP uses a three-stage handshake before sending data. This process has the problem of delaying data transmission. UDP is advantageous for internal data transmission because it does not use handshake method.

2. Data transfer overhead is small
   UDP is a transport layer connectionless protocol. Therefore, unlike TCP, there is no connection setup, and congestion control is not used, which is faster than TCP. Also, the segment header is smaller than TCP, which is 8 bytes and 20 bytes.

3. High transfer rate
   UDP has the advantage of faster congestion control than TCP. However, since there is no guarantee of data transmission, packet loss may occur. However, UDP detects a minimal error through the Checksum field in the header.

The basic UPD is a non-connection protocol, but it ensures the reliability of the connection through program implementation. In the following chapters a little more is explained [10].

2.1. Component of Simulation

Simulation components are a key part of the system that must reproduce the actual behavior. The components included in the system are shown in Figure 1. In experiments, system dynamics are represented by physical laws governing the motion of bodies in space. The sensor is mounted on the unmanned drones and the actuator is the four DC motors mounted on the drones.

We used Matlab / Simulink to implement the simulation components as described above. The reason for using Matlab / Simulink in the experiment is because of the problem with the original simulator. This program is used in many fields to model various systems. It also does not limit the time spent using the CPU to process data and other program elements. In the experiment, the FlightGear Flight simulation engine was integrated to meet the requirement to visualize the simulation state. Here are some reasons to use the FlightGear engine: First, this engine is open source and well documented. Second, a variety of API offerings enable you to quickly develop new systems. Third, it supports the integrated development environment, allowing easy debugging of programs. Fourth, it is easy to update APIs that users can use. Finally, the engine is updated periodically and can be reliably used.

3. Base Station

In this paper, QGroundControl program was used to implement Ground Station. QGroundControl is one of the solutions included in the Dronecode project under Linux Foundation [12].

The base station is a key component of the system because it communicates with the unmanned drones and the user interface is installed. This allows you to check whether the unmanned drones are operating or not. The base station allows you to filter the sensor values, determine the value of the Flight Control parameter, and pass specific commands to the unmanned drones. In QgroundContol, the user can specify the movement path of the unmanned drones as a drag on the map of the screen and load the contents into the autopilot. Telemetry data and sensor values that show the status of the drones can also be viewed directly on the screen and saved for replay.

Base stations using QGroundControl use the MAVLink protocol and support a variety of autosavigation boards such as PX4 Pro, ArduPilot, and pxIMU [11, 12]. It can also be run on a variety of operating systems, including Windows, Linux and MacOS.

QGroundControl is a software framework that helps you develop drone-related solutions as quickly as possible. This allows you to develop your program source in a modular fashion and make it easy to modify or add features later. Developers can easily incorporate new equipment into the system or easily modify the graphical interface using newly developed graphical tools. One of the greatest features of this base station is that it uses the MAVLink protocol to communicate with the unmanned drones. The MAVLink protocol is currently being used by many autonomous navigation boards and is open source, so it can be modified for development purposes [14].

4. Framework for Simulation

4.1. Simulation

The simulation component is responsible for two things. It carries out the actuator command created on the autopilot board and also provides the actual sensor data. The sensor data is input to the autopilot board and used for the simulation loop. Simulation components can be roughly divided into three parts.

1. Model of UAV Dynamics
   This part models the dynamics of an unmanned drones. And it calculates the orbit based on the actuator command input to the system.

2. Model of Sensor
   The sensors are mounted on an unmanned drone and provides the current status and external information of the drones. All sensor data is transmitted wirelessly to the base station.
3. Model of Environment

The environmental model describes the external environment in which the drones operate. The external environment includes current temperature, air density, and gravitational acceleration.

4.2. Model of UAV Dynamics

The UAV used in this study assumes a quad-copter with four rotors. In studies involving quadrotors, several methodologies have been proposed that relate to kinetic models for movement. Several studies have studied the model for detailed analysis and modeling [6-8]. As with these studies, advanced control algorithms are required to accurately describe the model. Especially if the UAV to be used is to be used without modification for a long time, a particularly accurate model is required.

This study has no restrictions on specific UAVs with specific settings. We have chosen a very simple geometric model to develop dynamic equations for UAVs. This assumes that the proposed model can be applied to most UAVs. The main frame of the quadrotor can be built with a cross-shaped frame and can consist of four motors and a propeller of the same length. Due to the nature of the quadcopter, which uses four motors as propulsive forces, it is difficult to keep the entire level at all during initial operation and flight, since they affect each other as well as themselves. Therefore, the balance value can be implemented through PID control to achieve stable centered flight.

4.3. Model of Sensors

In order to fly the drone, information such as position, altitude, speed, direction, and obstacles are basically required. This information is essential for controlling unmanned aerial vehicles. It is also important that the drones fly safely and hover.

Most of our drones are equipped with 5 ~ 6 sensors basically. It is almost standard in consumer drones such as 3 axis accelerometer, 3-axisgyroscope, magnetometer, barometer, GPS sensor and distance meter.

The Matlab / Simulink programming environment provides many tools for developing various sensor models. Inside, there are various libraries available to use sensor blocks already developed and used. The sensor data used in the autopilot's algorithm to simulate the function simulator has been simulated.

1) Magnetometer

A magnetometer is a compass sensor. Measure the magnetic north and send the direction information of the drone to the drones' CPU. This sensor is built in to drones with GPS capability. The movement of the drone can be grasped by combining the GPS position information, the magnetometer direction information, and the accelerometer movement information. The use of GPS drone is limited in this latitude above, since measurement of magnetic north is impossible at above 70 degrees north latitude. This sensor is basically a compass. In this reason, it is affected by magnetic objects in the vicinity. Electric power lines that generate electromagnetic waves, steel structures such as electronic devices and automobiles also have an effect. If a magnetometer error occurs, it can be resolved by moving slightly from the current position.

2) GPS Sensor

All aircraft use satellite GPS signals to measure the exact location and altitude of the aircraft. In recent years, as the GPS reception sensor has become cheaper, it has been installed in smart watches as well as hobby drones. The top GPS satellites were used for military purposes in the United States and Russia. Currently, however, it is open to commercial use and is used in various fields such as all aircraft, unmanned aerial vehicles, and automobiles. Of course, our GPS map and navigational map (NAVI) we use every day are also thanks to this GPS signal. In Matlab, the sensor model takes local coordinates and uses Simulink to convert latitude, longitude, and altitude.

5. Experiments

In this chapter, we measured the overall performance of the proposed system framework. In this experiment, we experimented with simulations at workstations, measured the time of simulation related processes, and analyzed the results according to time constraints. By analyzing the system time results, we confirmed that the system was operating correctly, and we identified the problem and identified areas that need to be corrected. Using the experimental data, we evaluated the meantime,

IOP Conf. Series: Materials Science and Engineering **611** (2019) 012090 doi:10.1088/1757-899X/611/1/012090

standard deviation time, best and worst time of a particular job. Through such mathematical analysis, we analyzed the performance of the proposed framework.

5.1. Experiment Setup

In the experiment to derive the results, the autopilot board was connected to the data management program and the workstation on which FlightGear was performed [13]. To ensure workstation performance, the Matlab / Simulink program was run on another computer connected to the local network. We also made additional connections with the board to monitor system status through the shell screen. We built the experimental environment and performed the simulation several times, logging the data for analysis. Statistical analysis was performed on the stored data using Matlab.

5.2. Experiment Results

Figures 3 and 4 show the execution time of each task. Analysis of the values reveals that the internal data flow is being delivered without significant delay during the operation. However, as can be seen from the graph, the standard deviation of about 3700us is large even if the average value is in the allowable range of the delay time. As shown in Figure 5, the internal-arrival time of the simulator's sensor data can well exceed the requested 4000us. Figure 6 shows the delay time over time during the simulation.

We analyzed the simulator response time data and found negative data. The time required to get a response from the simulation is greater than the 5ms sampling time on the autopilot board. The average response time is about 11ms and the standard deviation is 12ms. This problem can be expected to be a difference in the way Simulink implements UDP blocks.

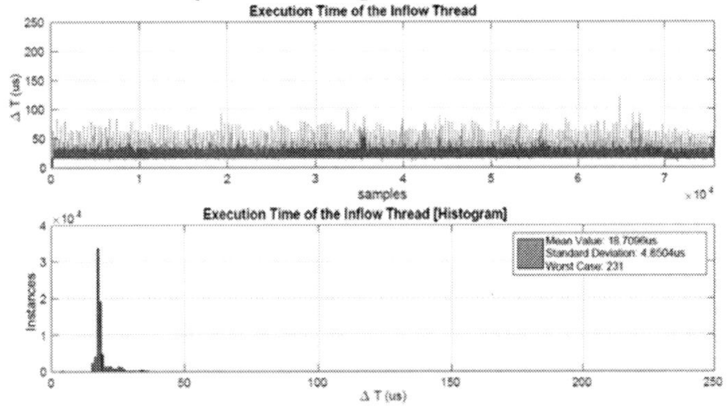

Figure 3. Execution time of the inflow job

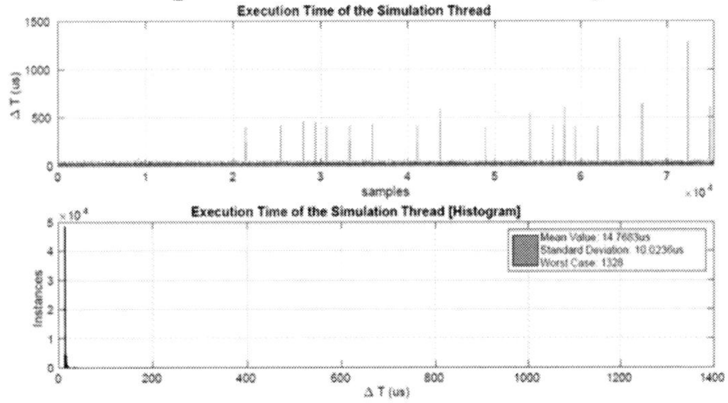

Figure 4. Execution time of the simulation job

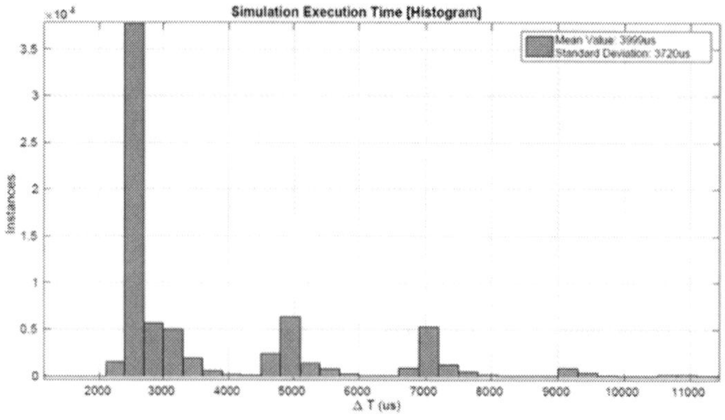

Figure 5. Execution time of the detailed simulation Job

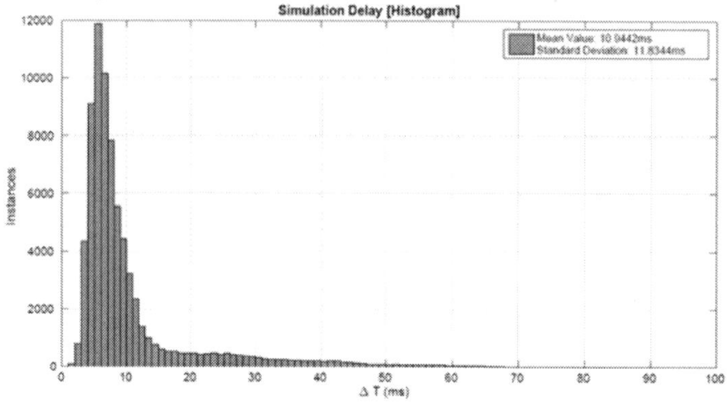

Figure 6. Response Time of the Simulation Component

6. Conclusions
The proposed UAV framework in the paper suggests a way to quickly develop navigation and control algorithms. Developers can run on the control board targeted by the algorithm they develop and see that it works correctly. This is because timing constraints can be met using timing precision. Considering execution time, the loop simulator can check the accuracy of the simulation flow itself to ensure the function of the hardware. Experiments have shown that the simulator generates a constant, non-negligible delay. This problem is related to the UDP implementation issue in Matlab / Simulink.

Acknowledgments
This work was supported by the National Research Foundation of Korea (NRF) grant funded by the Korea government (MSIT) (No. 2018R1C1B5046282).
This work was supported by Changshin University Research Fund of 2019(Changshin-2019-32)

References
[1] Kyungbok S, Kyoungwook M and Jeongdan C. 2018 *Proc. Int. Conf. on Advanced Communication Technology (ICACT) (Gangwon-do)* (IEEE) pp. 300-302
[2] Kwangjae L, Heewook K, Tae Chul H and Jae Young A 2017 *Proc. Int. Conf. on Information and Communication Technology Convergence (ICTC) (Jeju)* (IEEE) pp. 1222-1224
[3] Prabhu Jyot S and Rohan de S 2018 *Proc. Int. Conf. on Information and Communications Technology (ICOIACT) (Yogyakarta)* (IEEE) pp. 168-173
[4] N. Nithya, P. Vinothkumar, Liza M K and S. Riyas Parveen 2017 *Proc. Int. Conf. on Emerging Devices and Smart Systems (ICEDSS) (Tiruchengode)* (IEEE) pp. 68-71
[5] Wang Ting T, Bai B, Yuan Yi F and Zhu Yong W 2018 *Proc. Int. Conf. on Mechatronics and Automation (ICMA) (Changchun)* (IEEE) pp. 1972-1977

[6] Rongxiao W et al. 17 2017 *Proc. Int. Symp. on Distributed Simulation and Real Time Applications (DS-RT) (Rome)* (IEEE) pp. 1-4
[7] Muhammad Faisal S, Ahmad B and Hasnain A 2018 *Proc. Int. Conf. on Applied Sciences and Technology (IBCAST) (Parkistan)* (IEEE) pp. 340-346
[8] Nkemdilim A O and Matthew C T 2016 *IEEE Transactions on Control Systems Technology (IEEE)* vol. 24, no. 6, pp 1980-1992
[9] ZeFang H and Long Z 2018 *Proc. Int. Conf. on Information Technology, Networking, Electronic and Automation Control Conference (ITNEC) (Chengdu)* (IEEE) pp. 1254-1258
[10] Shuhang Z, Hongliang Z, Boya D and Lingyang S. 2019 *IEEE Transactions on Wireless Communications (IEEE)* vol. 19, no. 3, pp 1346-1359
[11] PX4 Autopilot Project, https://pixhawk.org/start
[12] QGroundControl, http://qgroundcontrol.com/
[13] FlightGear Flight Simulator, http://www.ightgear.org/
[14] Simple multirotor simulator with MAVLink protocol support, https://github.com/DrTon/jMAVSim